Toxic Cyanobacteria in Water

Toxic Cyanobacteria in Water

A Guide to Their Public Health Consequences, Monitoring and Management

Second Edition

Edited by

Ingrid Chorus and Martin Welker

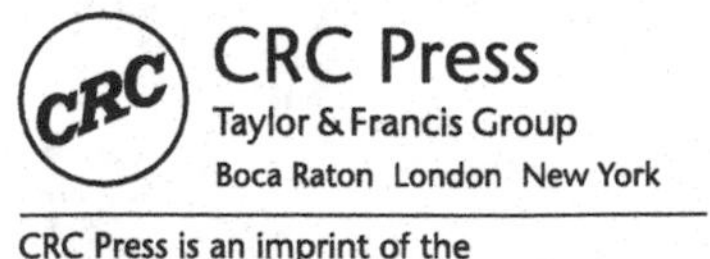

Second edition published 2021
by CRC Press
2 Park Square, Milton Park, Abingdon, Oxon, OX14 4RN

and by CRC Press
6000 Broken Sound Parkway NW, Suite 300, Boca Raton, FL 33487-2742

First issued in paperback 2022

First edition published by CRC Press 1999

Suggested citation of this book:
Chorus, I, Welker M; eds. 2021. Toxic Cyanobacteria in Water, 2nd edition. CRC Press, Boca Raton (FL), on behalf of the World Health Organization, Geneva, CH.

Suggested citation of individual chapters (example):
Humpage AR, Cunliffe DA. 2021. Understanding exposure: Drinking-water. In: Chorus I, Welker M; eds: Toxic Cyanobacteria in Water, 2nd edition. CRC Press, Boca Raton (FL), on behalf of the World Health Organization, Geneva, CH.

CRC Press is an imprint of Informa UK Limited

Publisher's Note
The publisher has gone to great lengths to ensure the quality of this reprint but points out that some imperfections in the original copies may be apparent.

British Library Cataloguing-in-Publication Data
A catalogue record for this book is available from the British Library

Library of Congress Cataloging-in-Publication Data
Names: Chorus, Ingrid, editor. | Welker, Martin, 1963- editor.
Title: Toxic cyanobacteria in water : a guide to their public health
consequences, monitoring and management / edited by Ingrid Chorus and
Martin Welker.
Description: Second edition. | Boca Rataon : CRC Press, an imprint of
Informa, 2021. | Includes bibliographical references and index.
Identifiers: LCCN 2020031428 (print) | LCCN 2020031429 (ebook) |
ISBN 9780367533311 (hardback) | ISBN 9781003081449 (ebook) |
ISBN 9781000262049 (epub) | ISBN 9781000262032 (mobi) | ISBN 9781000262025
(adobe pdf)
Subjects: LCSH: Cyanobacteria. | Drinking water—Microbiology. |
Cyanobacterial blooms. | Cyanobacterial toxins.
Classification: LCC QR99.63 .T67 2021 (print) | LCC QR99.63 (ebook) |
DDC 579.3/9—dc23
LC record available at https://lccn.loc.gov/2020031428
LC ebook record available at https://lccn.loc.gov/2020031429

ISBN: 978-0-367-53331-1 (hbk)
ISBN: 978-0-367-53332-8 (pbk)
ISBN: 978-1-003-08144-9 (ebk)

DOI: 10.1201/9781003081449

Typeset in Sabon
by codeMantra

Visit the Taylor & Francis Web site at
http://www.taylorandfrancis.com

and the CRC Press Web site at
http//www.crcpress.com

Contents

Foreword

The way we manage our waterbodies determines the extent to which cyanobacteria proliferate. While for some waterbodies the past decades have seen some progress in controlling the excessive nutrient loads that result in eutrophication and cyanobacterial blooms, pressures on many others are increasing, through population growth, urbanisation, changes in agricultural land use and climate. In this context, the 2030 Agenda for Sustainable Development includes Sustainable Development Goal 6, which recognises the importance of protecting and restoring water-related ecosystems (target 6.6), improving ambient water quality (target 6.3) and safe (and affordable) drinking-water for all (target 6.1).

Cyanobacterial blooms have been recognised as an environmental concern since they began to occur widely in many countries in the 1960s. An awareness of their public health significance grew during the 1980s as their toxicity became increasingly understood, including as cause of the deaths of exposed livestock, pets and wild animals, and cases of human illness were attributed to exposure to cyanotoxins following recreational activities or drinking-water consumption.

In 1999, the World Health Organization (WHO) developed its first drinking-water guideline value for a widely occurring cyanobacterial toxin, microcystin-LR. The WHO also published the first edition of *Toxic Cyanobacteria in Water*, largely written by the pioneers of cyanotoxin science. Since then, this document has been widely used by regulators for the development of national policies for managing cyanotoxin risks, by local public health services in implementing measures to protect public health and by academia for teaching and planning research. Since 1999, cyanotoxin research has grown exponentially, and the new knowledge generated in these two decades has improved our basis for assessing the health risks caused by toxic cyanobacteria. We now know more about the range of cyanotoxins – from their occurrence to potential health effects – and thus can set priorities more effectively. This enabled the WHO to develop guideline values for further cyanotoxins, including for short-term and for recreational exposure. The approach of developing site-specific Water Safety Plans, initially promoted by WHO in

2003 in the Guidelines for Safe Recreational Environments and 2004 in the Guidelines for Drinking-water Quality, now provides a platform for bringing together the wide range of expertise and stakeholder interests needed to understand the causes of blooms and to develop the most effective and sustainable context-specific strategy for controlling them.

Such a joint effort requires communication between managers, stakeholders and experts from a range of fields and with diverse backgrounds. This second edition of *Toxic Cyanobacteria in Water* brought together a correspondingly wide range of expertise of authors from many countries, with experience from very different types of waterbodies harbouring toxic cyanobacterial blooms. This book resulting from this collective effort strives to facilitate communication between those developing strategies to prevent blooms and human exposure through water by providing the necessary tools, background and guidance. This book includes an introduction to the basics about cyanobacteria and their toxins, an overview of human exposure routes, guidance for assessing risks to human health and preparing for short-term responses to prevent human exposure, as well as guidance on effective management and monitoring from catchment to the end user.

It is hoped that readers find this second edition of *Toxic Cyanobacteria in Water* useful for developing longer-term, sustainable approaches bridging environmental management and public health, as countries strive towards the realisation of their commitments under the 2030 Agenda for Sustainable Development.

Acknowledgements

The World Health Organization (WHO) wishes to express its appreciation to the numerous colleagues who contributed to the preparation and development of this book, including the colleagues named below.

This second edition of *Toxic Cyanobacteria in Water* evolved under the guidance and support of several expert meetings held between 2008 and 2017, beginning with meetings of the WHO Guidelines for Drinking-water Quality (GDWQ) working groups on chemicals as well as on protection and control. Specific consultations to evaluate drafts of this book were held with cyanotoxin experts in July 2016 and March 2017, with the latter meeting also including experts involved in WHO's work on chemical aspects of the GDWQ and Guidelines for Safe Recreational Water Environments.

EDITORS:

Ingrid Chorus, formerly German Environment Agency, Germany[1]
Martin Welker, Independent Consultant, Germany[1]

LEAD AUTHORS:

Sandra M. F. O. Azevedo, Universidade Federal do Rio de Janeiro, Brazil
Andreas Ballot, Norwegian Institute for Water Research, Norway
Luděk Bláha, Masaryk University, Czech Republic[2]
Justin Brookes, University of Adelaide, Australia
Mike Burch, SA Water, Australia
José Capelo Neto, Universidade Federal do Ceará, Brazil
Neil Chernoff, United States Environmental Protection Agency, USA[2]

[1] Also lead author.
[2] Also peer-reviewed specific chapters.

David Cunliffe, Department for Health and Wellbeing South Australia, Australia[2]
Lesley V. D'Anglada, Independent Consultant, USA
Daniel Dietrich, University of Konstanz, Germany[2]
Elisabeth J. Faassen, Wageningen University & Research, the Netherlands
Jutta Fastner, German Environment Agency, Germany
Amanda Foss, Greenwater Laboratories, USA
Gesche Grützmacher, Berliner Wasserbetriebe, Germany
Donna J. Hill, United States Environmental Protection Agency, USA
Lionel Ho, Allwater, Australia
Andrew Humpage, formerly South Australian Water Corporation, currently Adelaide University, Australia
Bastiaan W. Ibelings, Université de Genève, Switzerland[2]
Ralf Junek, German Environment Agency, Germany
Triantafyllos Kaloudis, EYDAP - Athens Water Supply and Sewerage Company, Greece[2]
Sondra Klitzke, German Environment Agency, Germany
Rainer Kurmayer, University of Innsbruck, Austria
Linda A. Lawton, Robert Gordon University, UK[2]
Blahoslav Maršálek, Masarykova Univerzita, Czech Republic
James S. Metcalfe, Institute of Ethnomedicine, USA
Rory Moses McKeown, World Health Organization, Switzerland[2]
Gayle Newcombe, formerly SA Water, Australia
Nicholas J. Osborne, University of Queensland, Australia
Judit Padisák, Pannon Egyetem, Hungary
Heather Raymond, Ohio State University, USA[2]
Blake A. Schaeffer, United States Environmental Protection Agency, USA
Emanuela Testai, Istituto Superiore di Sanità, Italy[2]
Andrea Törökné, National Accreditation Authority, Hungary
Erin Urquhart, National Aeronautics and Space Administration, USA
Leticia Vidal, Consultant, Uruguay[2]
Susanna A. Wood, Cawthron Institute, New Zealand[2]
Bojana Žegura, National Institute of Biology, Slovenia
Matthias Zessner, Technische Universität Wien, Austria

AUTHORS OF TEXTBOXES PROVIDING SPECIFIC ASPECTS OR CASE STUDIES:

Eduardo Andrés, Ministerio de Vivienda, Ordenamiento Territorial y Medio Ambiente, Uruguay
Rafael Bernardi, Ministerio de Vivienda, Ordenamiento Territorial y Medio Ambiente, Uruguay
Beatriz Brena, Universidad de la República de Uruguay, Uruguay
Wayne W. Carmichael, Wright State University, USA[2]

Elisa Dalgalarrondo, Ministerio de Vivienda, Ordenamiento Territorial y Medio Ambiente, Uruguay
Andrea A. Drozd, Universidad Nacional de Avellaneda, Argentina
Cesar García, Ministerio de Vivienda, Ordenamiento Territorial y Medio Ambiente, Uruguay
Natalia Jara, Ministerio de Vivienda, Ordenamiento Territorial y Medio Ambiente, Uruguay
Gary Jones, University of Canberra, Australia
Evanthia Mantzouki, Université de Genève, Switzerland
Carolina Michelena, Ministerio de Vivienda, Ordenamiento Territorial y Medio Ambiente, Uruguay
Juan Pablo Peregalli, Ministerio de Vivienda, Ordenamiento Territorial y Medio Ambiente, Uruguay
Giannina Pinotti, Ministerio de Vivienda, Ordenamiento Territorial y Medio Ambiente, Uruguay
Helena L. Pound, University of Tennessee, USA
Raquel del Valle Bazán, Universidad Nacional de Córdoba, Argentina
Steven W. Wilhelm, University of Tennessee, USA

ADDITIONAL EXPERTS WHO PROVIDED INSIGHTS, WROTE TEXT, PROVIDED PEER REVIEW, AND/OR PARTICIPATED IN MEETINGS:

Yasuhiro Asada, National Institute of Public Health, Japan
Cécile Bernard, Museum National d'Histoire Naturelle, France
Myriam Bormans, Université de Rennes, France
Alexander F. Bouwman, PBL Netherlands Environmental Assessment Agency, the Netherlands
Cristina Brandão, University of Brasilia, Brazil
Teresa Brooks, Health Canada/Government of Canada, Canada
Phil Callan, Consultant, Australia
Richard Carrier, Health Canada/Government of Canada, Canada
Richard Charron, Health Canada/Government of Canada, Canada
Andrea Cherry, Health Canada/Government of Canada, Canada
Geoffrey A. Codd, University of Dundee, UK
Rosario Coelho, Aguas do Algarve, Portugal
Joseph Cotruvo, Joseph Cotruvo & Associates, USA
Jennifer De France, World Health Organization, Switzerland
Elke Dittmann, University of Potsdam, Germany
Alexander Eckhardt, German Environmental Agency, Germany
Ian Falconer, University of Adelaide, Australia
John Fawell, Cranfield University, UK
Valerie Fessard, BioAgroPolis, France

Valerie Fessard, BioAgroPolis, France
Ambrose Furey, Cork Institute of Technology, Ireland
Anastasia Hiskia, NCSR Demokritos, Greece
Jean-Francois Humbert, Sorbonne Université, France
Fernando Antonio Jardin, COPASA, Brazil
Edwin Kardinaal, KWR Watercycle Research Institute, the Netherlands
David Kay, Aberystwyth University, UK
Adam Kovacs, International Commission for the Protection of the Danube River, Austria
Japareng Lalung, Universiti Sains Malaysia, Malaysia
Joanna Mankiewicz-Boczek, UNESCO, Poland
Stephanie McFadyen, Health Canada/Government of Canada, Canada
Kate Medlicott, World Health Organization, Switzerland
Jussi Meriluoto, Åbo Akademi University, Finland
Zakaria A. Mohamed, Sohag University, Egypt
Teofilo Monteiro, formerly World Health Organization Pan-American Health Organization, Peru
Choon Nam Ong, National University of Singapore, Singapore
Catherine Quiblier, Museum National d'Histoire Naturelle, France
Mohd Rafatullah, Universiti Sains Malaysia, Malaysia
Bettina Rickert, German Environmental Agency, Germany
Angella Rinehold, World Health Organization, Switzerland
Piotr Rzymski, Poznan University of Medical Sciences, Poland
Nico Salmaso, Fondazione Mach-Istituto Agrario di S. Michele all'Adige Hydrobiology, Italy
Ciska Schets, National Institute for Public Health, the Netherlands
Oliver Schmoll, World Health Organization Regional Office for Europe, Germany
David Sheehan, Coliban Water, Australia
Ian Stewart, Food and Water Toxicology Consulting, Australia
Assaf Sukenik, Israel Oceanographic and Limnological Research, Israel
Marc Troussellier, Museum National d'Histoire Naturelle, France
Thomas Waters, United States Environmental Protection Agency, USA
Hartmut Willmitzer, Technische Universität Dresden, Germany
Ping Xie, Institute of Hydrobiology - Chinese Academy of Sciences, China
Gordon Yasvinski, Health Canada/Government of Canada, Canada
Arash Zamyadi, University of Montreal, Canada
Kristin Zoschke, Technische Universität Dresden, Germany

The development of this document was coordinated and managed by Jennifer De France and Rory Moses McKeown of WHO. Bruce Gordon (WHO) and, in the earlier phase of developing the document, Jamie Bartram (formerly WHO) provided strategic direction. The support from Public Utilities Board, Singapore, for assistance with coordinating peer review is also acknowledged.

WHO also gratefully acknowledges the financial support provided by the Federal Environment Agency of Germany and the Public Utilities Board, the National Water Agency, a statutory board under the Ministry of Environment and Water Resources, Singapore; Foreign, Commonwealth & Development Office, United Kingdom; and the Environmental Protection Agency, United States of America.

THE FOLLOWING MATERIALS APPEAR WITH PERMISSION FROM:

PHOTOGRAPH CREDITS:

Figure 3.5D-I and III: Ballot, A., Ramm, J., Rundberget, T., Kaplan-Levy R. N., Hadas, O., Sukenik, A., Wiedner, C. (2011): Occurrence of non-cylindrospermopsin-producing Aphanizomenon ovalisporum and Anabaena bergii in Lake Kinneret (Israel). J. Plankt. Res. 33: 1736–1746, by permission of Oxford University Press.

Figure 3.5F-V and VI; Figure 3.5H-I and III: Ballot A., Sandvik M., Rundberget T., Botha C. J., and. Miles C. O. (2014): Diversity of cyanobacteria and cyanotoxins in Hartbeespoort Dam, South Africa. Marine and Freshwater Research 65:175–189. © Copyright CSIRO Australia 2014.

Figures 5.6, 5.7, 5.8: Ingrid Chorus, formerly Federal Environment Agency, Germany.

Figure Box 15.1: Intendencia de Montevideo, Uruguay.

Figure Box 15.2(1): Hernán Ezequiel Castro. Grupo Especial de Rescate y Salvamento (G.E.R.S.) Calamuchita. Dirección de Bomberos de la Policía de la Provincia de Córdoba.

Figure Box 15.2(2): Ministerio de Salud, Argentina.

Figure Box 15.2(3): Ministerio de Salud, Argentina.

Note – additional photograph credits are listed in Chapter 3.

DISCLAIMER

Chapter 11 has been reviewed by the National Exposure Research Laboratory of the United States Environmental Protection Agency and approved for publication. Mention of trade names or commercial products does not constitute endorsement or recommendation for use by the United States Government. The views expressed in Chapter 11 and in section 2.7 are those of the authors and do not necessarily reflect the views or policies of the United States Environmental Protection Agency.

Editors

Ingrid Chorus completed her PhD at the Technical University, Berlin, supporting a collaborative ecosystem study of a highly eutrophic urban lake, and in the 1980s, she studied the restoration of heavily eutrophicated lakes. In 1991, Ingrid became head of the German Federal Environment Agency's unit on drinking-water resources and began her focus on cyanotoxin research and management. From 2007 until 2018, she led the agency's Department for Drinking-Water and Swimming-Pool Hygiene. A key focus of her work was developing and implementing the WHO Water Safety Plan approach, particularly towards catchment and waterbody management.

Martin Welker started his career as a plankton ecologist at the Institute of Freshwater Ecology, Berlin. His PhD thesis focused on cyanobacteria and their toxins, with a particular emphasis on the release and degradation of microcystins by biotic and abiotic processes. As a postdoc at the Technical University of Berlin and Institute Pasteur, Paris, he explored the diversity of cyanobacterial metabolites and their biosynthesis. In 2006, Martin joined AnagnosTec, contributing to the development of microbial identification systems for clinical diagnostics. In 2010, Martin joined bioMérieux as senior scientist where he works on clinical microbiological diagnostics and research.

Chapter 1

Introduction

Ingrid Chorus and Martin Welker

CONTENTS

Safe drinking-water is of paramount importance for human health. Throughout history, access to drinking-water has been a prerequisite for the development of civilisations – and the loss of access often a key factor for their decline. Recognising the vital role of drinking-water for public health, the World Health Organization (WHO) dedicates a significant share of its efforts to promote the safety of water for today and for the future (Onda et al., 2012). With a human population reaching 10 billion by the mid of this century, the pressure on global drinking-water resources will not cease, and ongoing efforts in research, management and governance are needed to recognise, understand and mitigate health risks associated with water use. This includes further uses of water involving human exposure, particularly for recreation, and, depending on specific local or regional circumstances, also for irrigating crops, cooling water or dust suppression, for example.

Health hazards recognised in water today comprise infectious microorganisms (e.g., bacteria, viruses and protozoa causing gastrointestinal diseases), geogenic substances (e.g., arsenic, fluoride, uranium), industrial and agricultural chemicals (e.g., perfluorinated chemicals [PFCs], pesticides) and toxins produced by cyanobacteria – the subject of the first edition of "Toxic Cyanobacteria in Water" (Chorus & Bartram, 1999) and of the present volume.

Among the hazards considered in the Guidelines for Drinking Water Quality (GDWQ; WHO, 2017), infectious microorganisms are the most significant causes of mortality on a global scale, causing a substantial burden of disease via diarrhoeal illnesses such as cholera, cryptosporidiosis or retroviral enteritis (James et al., 2018; Roth et al., 2018; Prüss-Ustün et al., 2019). In contrast, the contribution of toxic chemicals in water to morbidity and mortality is rarely acute, and aside from a few geogenic chemicals, the impacts on health are less visible and less clearly attributable to chemicals. This applies particularly for carcinogenic compounds, the impacts of which accumulate over time. Thus, considering global causes of mortality and morbidity, data to estimate the disease burden through exposure to chemicals in water are typically lacking.

This is also true for cyanobacterial toxins: only a relatively low number of recorded cases of acute human intoxication are clearly attributable to these toxins (Wood, 2016). Nonetheless, like with other toxins potentially found in drinking-water, exposure to low, subacute concentrations is possible because drinking-water is an indispensable part of the human diet, and hence, exposure is difficult to avoid – abstinence or replacement from alternative sources for longer periods is not a feasible option in most settings.

Compared to other agents that may occur in water and that are covered in the GDWQ, the occurrence and behaviour of cyanotoxins is fundamentally different and consequently requires different management approaches. On the one hand, the producing cyanobacteria need to be addressed as microorganisms that can proliferate in surface waters – but which are not in themselves infectious – requiring measures to reduce their occurrence that shift management from microbiology to ecology. On the other hand, their toxins are chemicals and need to be addressed as such, including the derivation of values for maximally tolerable concentrations and the development of technical methods to reduce their concentration through drinking-water treatment.

Other unique characteristics include:

- Cyanotoxins are among the most toxic naturally occurring compounds: lethal doses are in the same range as some toxins from mushrooms (amanitin, phaloidin) or plants (aconitine, strychnine, atropine).
- Cyanotoxins occur worldwide in many lakes, reservoirs and rivers used as sources of drinking-water or for recreational activity.
- Contact with toxic cyanobacteria is difficult to avoid without implementing severe restrictions: most people who enjoy swimming in natural waters most likely have been in contact with toxic cyanobacteria.
- The occurrence of toxic cyanobacterial blooms is often not perceived as a danger by the public in the same way as a spill of an industrial toxin or chemical with the same hazard potential would be, because it may be regarded as "natural" and hence innocuous.
- Cyanotoxins are produced naturally within surface waters and are not, like most chemicals for which guideline values have been set or proposed, directly introduced by human activity. For many of the anthropogenic contaminants, legislation regulating their use and release into the environment has successfully reduced concentrations

in ground or surface waters an approach that is not practicable for cyanobacterial toxins.
- The control of toxigenic cyanobacteria is complex and typically requires efforts on scales beyond the water supply and waterbody with its immediate environment, potentially including the management of entire catchments and requiring longer-term investments (e.g., in sewage management) as well as political decisions with wider impact (e.g., on fertiliser use).

Thus, cyanobacteria and their toxins pose specific challenges, and guidance with respect to their management warrants a dedicated WHO publication.

Cyanobacteria have been present in natural ecosystems since the Precambrian Era, some 2 billion years ago (Wilmotte, 1994), and the production of cyanotoxins is probably an equally ancient characteristic (Rantala et al., 2004). The first scientific report on toxic cyanobacteria dates from the late 19th century (Francis, 1878), but earlier historical records have been interpreted as similar poisoning events (Codd et al., 2015). Studies on cyanobacterial toxins in lake sediments found microcystins (Zastepa et al., 2017) and cylindrospermopsin (Waters, 2016) in layers deposited well before the 20th century. In comparison with more recent sediments, in most cases, the assumed historic concentrations were, however, much lower than those found in today's eutrophic lakes.

In large parts of the world, waterbody eutrophication started accelerating in the middle of the 20th century, in the wake of urbanisation and industrialisation. Since that time, massive cyanobacterial blooms have occurred in many lakes and reservoirs in which this phenomenon was not known before. Therefore, it is not the biosynthesis of toxins itself that created a new health hazard, but the more recent significant proliferation of toxic cyanobacteria in waterbodies as a result of human activities. This health hazard most probably will gain growing importance as cyanobacterial blooms are expected to increase at the scale at which eutrophication is expected to increasingly occur in many regions of the world (Huisman et al., 2018).

Whether or not global warming is likely to increase cyanobacterial proliferation depends on specific conditions in a particular waterbody. In order to support the inclusion of climate change scenarios in risk assessment and management (e.g., water safety planning), this book includes information on how these conditions may influence cyanobacterial growth and bloom formation.

Cyanobacteria can produce a huge diversity of secondary metabolites, the biosynthetic pathways of which are known for a number of individual compounds or compound classes, respectively. Only a small share of the known metabolites shows toxic effects, but these cyanotoxins have caused numerous cases of poisoning of farm or wild animals, which demonstrate their toxic potential (Wood, 2016; Svirčev et al., 2019) and which suggests that animal illnesses and deaths are sentinel events for human health risks (Hilborn & Beasley, 2015). A large body of evidence from experimental studies with laboratory animals has elucidated their mode of action: some cyanotoxins are highly neurotoxic and others can damage the liver, kidney or other organs when ingested.

Epidemiological studies have looked for chronic effects in human populations exposed to toxic cyanobacteria, and indeed, a number of studies since the mid-19th century associate symptoms with cyanotoxin exposure. The key caveat of several of these anterior studies is the lack of data on the dose to which the population might have been exposed and a lack of analytical tools for detecting other hazards at that time, such as molecular techniques for the detection of pathogenic viruses. However, although our current knowledge may question some of the epidemiological evidence frequently quoted to highlight the cyanotoxin hazard, the evidence from animal experiments is clear and sufficient to derive guideline values for a range of cyanotoxins.

In this respect, cyanotoxins are in line with most other substances for which World Health Organization (WHO) has set guideline values: this is not typically done because a substance has been widely shown to cause human illness or result in fatalities through water consumption, but rather because a substance has significant toxic properties and water is recognised as a relevant pathway for exposure. Given the widespread occurrence of cyanobacteria – as compared to the occurrence of many purely anthropogenic contaminants in water – cyanotoxins are likely to occur more widely and more often in concentrations of potential concern than many of the other chemicals considered in the Guidelines for Drinking Water Quality (WHO, 2017).

1.1 DOCUMENT PURPOSE AND SCOPE

The second edition of "Toxic Cyanobacteria in Water" presents the state of knowledge regarding the impact of cyanobacterial toxins on health through the use of water and provides guidance on assessing and managing the risks of cyanobacteria and their toxins in order to protect drinking-water sources and recreational waterbodies. It further provides an overview of exposure through other important sources, including food, use of dietary supplements and through dialysis.

This edition is an update of the first edition of this publication, which was published 20 years ago (Chorus & Bartram, 1999). In addition to updating the state of knowledge specifically related to cyanobacteria and their toxins, this updated edition accounts for developments in and best practices for water supply management, namely, water safety planning, as well as the broader state of knowledge on climate change, eutrophication and others.

Water safety planning (see Box 1.1) is a comprehensive preventive risk assessment and risk management approach, and is a critical component of WHO's Framework for Safe Drinking-Water, to most effectively ensure drinking-water safety. Most importantly, the Water Safety Plan (WSP) approach systematically addresses all steps in a water supply from catchment to consumer (Bartram et al., 2009).

While the concept of WSP development is tailored to drinking-water supplies, many of its elements can be applied to the assessment and management of other potential exposure routes. For food safety – fish and shellfish in the context of this volume – the related concept of HACCP (Hazard Analysis Critical Control Points; from which WSPs were developed) applies and can readily be linked to WSP elements. The WSP approach is currently being developed for application to other areas of water management, that is, as Sanitation Safety Plans and Recreational Water Safety Plans. Among the hazards relevant to water, cyanobacteria are often likely to expose people through multiple pathways, and adopting a WSP approach will provide the most effective approach to protecting their health.

BOX 1.1: DEVELOPING A WATER SAFETY PLAN (WSP)

Drinking-water safety often relies heavily on the verification of compliance to water quality standards. However, by the time laboratory results show non-compliance, the population served will already have consumed the water and become exposed – and in the case of pathogens, many people may thus become ill. Therefore, "end-of-pipe" monitoring alone is insufficient to guide management decisions. The WSP approach shifts the emphasis of drinking-water quality management to a holistic risk-based approach that covers all processes from catchment to consumer which are crucial for maintaining drinking-water quality.

A WSP is specifically developed for the individual water supply. The process of developing it means

1. describing the system to identify and **analyse the hazards** and the **hazardous events** that are likely to cause the hazard to occur, and to **assess the health risks** they may present, as well as the **system's performance** in controlling these hazards and hazardous events;
2. to identify which additional barriers – the **control measures** – could be implemented to control these risks at different levels: the catchment, in the waterbody and at the offtake, in drinking-water treatment, in distribution networks and in households. Further, to **validate**

that **control measures** are appropriate for the intended purpose and achieve their respective contribution to mitigate the risks;

3. to ensure that control measures are working as intended by implementing a **monitoring system** that effectively indicates whether the **system's performance** is within the operational limits set for the respective measure. This requires the definition of **critical limits** for monitoring results, as well as setting up **corrective actions** to take immediately if values were outside these critical limits;
4. to **document** these steps and **revise** the whole system at regular intervals, that is, to assess whether the risk assessment is still adequate, the system's design takes account of the control of all relevant risks, and performance of the whole system is satisfactory;
5. to **verify** the outcome – that is, that drinking-water quality actually meets the targets set – in respect to the topic of this book, by targeted analysis of cyanotoxin concentrations in finished drinking-water.

Going through these steps effectively requires preparation, particularly forming a **team** including the technical expertise needed for the assessments and the stakeholders needed for making decisions. Preparation also includes **describing the supply system** from catchment to consumer, identifying **who will be exposed** to the water (particularly with respect to sensitive subpopulations) and obtaining the full endorsement and support from senior management for developing the WSP.

The WSP concept focuses attention on **risk assessment** and on **process control**. It is an operational system of quality management. This structured, systematic approach to process control is particularly useful for managing cyanotoxin risks, as it provides a platform for including expertise for the management of the catchment and waterbody, as well as interests of stakeholders involved in source water management.

World Health Organization (WHO) describes the WSP concept in Chapter 4 of Guidelines for Drinking Water Quality (GDWQ):

https://www.who.int/publications/i/item/9789241549950

and provides a WSP manual with practical guidance for individual settings:

https://apps.who.int/iris/handle/10665/75141

More WHO resource materials on water safety planning are available at https://www.who.int/teams/environment-climate-change-and-health/water-sanitation-and-health/water-safety-and-quality/water-safety-planning

1.2 TARGET AUDIENCE

"Toxic Cyanobacteria in Water" is intended for use by all those working on toxic cyanobacteria, with a specific focus on public health protection. It intends to empower professionals from different disciplines to communicate and cooperate for sustainable management of toxic cyanobacteria, for example:

- for public health professionals, including those in the fields of water supply and the management of recreational water, by providing detailed information on cyanobacteria and their ecology as well as on the management of catchments, waterbodies and water supplies;
- for ecologists and catchment and waterbody managers, by providing information on the public health impacts of cyanobacteria and their toxins.

This publication may also be useful to academia for a basic understanding of the current state of knowledge – and its gaps – and thus of possible research needs. This volume is not intended to replace textbooks on limnology, taxonomy, bacteriology or physiology that provide much more detail on issues such as eutrophication control, cyanobacterial diversity, toxin biosynthesis or toxicity mechanisms, and for further information, readers are referred to other sources quoted in the respective chapters, as well as to the WHO guidance on "Protecting Surface Water for Health" (Rickert et al., 2016) and "Guidelines for Safe Recreational Water Environments" (WHO, 2003).

1.3 DOCUMENT STRUCTURE AND OVERVIEW

This volume includes five key sections:

- introduction to cyanobacteria and their toxins (Chapters 2–4);
- understanding and assessing potential exposure routes (Chapter 5);
- guidance on control measures for cyanotoxin hazards (Chapters 6–10);
- overview of methods for sampling and analysis (Chapters 11–14);
- guidance on cyanotoxin-specific aspects of public surveillance, incident management and communicating cyanotoxin risks to the public (Chapter 15).

Chapter 2 includes detailed descriptions of the cyanotoxin groups that are relevant to human health. Although the chemically diverse cyanotoxins share the feature of being toxic to mammals, their respective modes of action are quite diverse. Each of the sections on a group of cyanotoxins in Chapter 2 summarises the state of knowledge on chemical structure, toxicity and mode of action, producing cyanobacteria and biosynthesis, occurrence and environmental fate. For those cyanotoxins for which WHO proposes Guideline Values or "Health-Based Reference Values" (i.e., for microcystins, cylindrospermopsin, anatoxin-A and saxitoxin), the respective section summarises the considerations for the derivation of these values, referring to the respective WHO background documents for more detailed information.

Chapter 3 introduces cyanobacteria as organisms which occur naturally in a large variety of habitats, from ultraoligotrophic oceans to deserts – and in a broad diversity of freshwaters. It further briefly describes the limited number of taxa known to contain metabolites of relevance to human health.

Chapter 4 describes the main environmental conditions that may lead to blooms of cyanobacteria, among which elevated nutrients concentrations are a key precondition which is important to understand when developing management strategies to control blooms. It includes further environmental conditions that determine the dominance of specific cyanobacterial taxa, their temporal dynamics and their spatial heterogeneity.

Chapter 5 reviews the available scientific and epidemiological evidence for each relevant route for human exposure to hazardous concentrations of cyanotoxins: drinking-water (section 5.1), recreational and occupational activity (section 5.2), food (section 5.3), renal dialysis (section 5.4) and cyanobacteria as food supplements (section 5.5). For exposure through drinking-water or recreation, it proposes Alert Level Frameworks to guide timely management responses to elevated concentrations of either cyanobacteria or their toxins. These frameworks help focus operational monitoring for two purposes – to minimise the risk of unnoticed exposure, but also to avoid inefficient monitoring efforts where risks are likely to be low.

A prerequisite to choosing the locally most effective approach is to understand and characterise the individual water-use system. Cyanotoxin occurrence and exposure can be reduced and managed by measures that act at different levels. Chapter 6 therefore describes the steps to take for assessing the given conditions and developing the locally most effective management approach, aligned with the water safety planning framework. Cyanotoxins are most effectively and sustainably controlled by avoiding conditions which lead to cyanobacterial proliferation. Figure 1.1 illustrates the conditions that lead to elevated concentrations of cyanotoxins, and the subsequent chapters describe control measures that can be taken to minimise the cyanotoxin hazard at different levels: nutrient loads from the catchment (Chapter 7), nutrient concentrations and other conditions in the waterbody (Chapter 8), the selection of sites for drinking-water offtake or of recreational areas (Chapter 9) and the treatment of raw water to produce drinking-water (Chapter 10).

The guidance given in Chapters 6–10 intends to support developing a locally specific approach to controlling cyanotoxin occurrence. This is

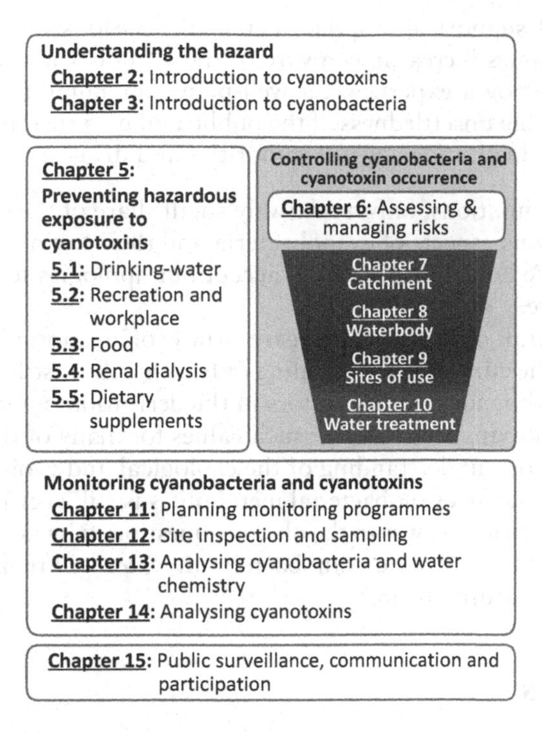

Figure 1.1 Overview of the structure of "Toxic Cyanobacteria in Water" 2nd edition.

supported by monitoring cyanobacteria and cyanotoxin occurrence. Chapter 11 outlines the planning of monitoring programmes. This includes a section on remote sensing, a technology that is becoming increasingly accessible, and, although it will likely not replace field work, offers opportunities to optimise targeted monitoring. Chapter 12 gives guidance for on-site field work. Laboratory analytical methods for cyanobacteria and associated water chemistry are presented in Chapter 13, from microscopic determination of cyanobacterial biomass to nutrient analyses and recently developed molecular methods. The analysis of cyanotoxins themselves is discussed in Chapter 14, from chemical and biochemical methods to a critical assessment of bioassays that have been applied in cyanotoxin research and their role in future cyanotoxin research. These chapters are not intended to replace specialised textbooks and standards by giving details on how to perform specific analytical methods; rather, in order to support the development of efficient monitoring programmes, they provide an overview of widely used methods and techniques, together with their respective laboratory requirements, specific advantages and shortcomings.

Chapter 15 discusses public health surveillance, incident planning and response, as well as public communication and participation to develop awareness and support appropriate personal decisions on water use. As the public becomes increasingly aware of the cyanotoxin risk through the media and their own experience, a well-planned communication strategy can prevent undue unsettledness of the public and help sustain public confidence in health institutions, water authorities and drinking-water supplies.

Due to space limitations, only a relatively small share of the many valuable studies on various aspects of cyanobacteria and their toxins can be cited in this volume. We hope our colleagues accept our apologies for the selection we had to make.

Despite several decades of intensive study of cyanotoxins, questions remain open: the current understanding of health risks is still sketchy – as is in part reflected by uncertainty factors in the derivation of guideline values for some cyanotoxins and a lack of such values for many of their structural variants. Also, our understanding of the ecological and evolutionary value of toxic and bioactive cyanobacterial metabolites is still very limited – there is still no satisfactory answer to the obvious question: why do cyanobacteria produce toxins? Hence, toxic cyanobacteria in water will remain an important subject for future research.

REFERENCES

Bartram J, Correales L, Davison A, Deere D, Drury D, Gordon B et al. (2009). Water safety plan manual: step-by-step risk management for drinking-water suppliers. Geneva: World Health Organization:103 pp. https://apps.who.int/iris/handle/10665/75141

Chorus I, Bartram J (1999). Toxic cyanobacteria in water: a guide to their public health consequences, monitoring and management. London: E & FN Spoon, on behalf of WHO. 400 pp.

Codd GA, Pliński M, Surosz W, Hutson J, Fallowfield HJ (2015). Publication in 1672 of animal deaths at the Tuchomskie Lake, northern Poland and a likely role of cyanobacterial blooms. Toxicon. 108:285.

Francis G (1878). Poisonous Australian lake. Nature. 18:11–12.

Hilborn E, Beasley V (2015). One health and cyanobacteria in freshwater systems: animal illnesses and deaths are sentinel events for human health risks. Toxins. 7:1374–1395.

Huisman J, Codd GA, Paerl HW, Ibelings BW, Verspagen JM, Visser PM (2018). Cyanobacterial blooms. Nat Rev Microbiol. 16:471.

James SL, Abate D, Abate KH, Abay SM, Abbafati C, Abbasi N et al. (2018). Global, regional, and national incidence, prevalence, and years lived with disability for 354 diseases and injuries for 195 countries and territories, 1990–2017: a systematic analysis for the Global Burden of Disease Study 2017. Lancet. 392:1789–1858.

Onda K, LoBuglio J, Bartram J (2012). Global access to safe water: accounting for water quality and the resulting impact on MDG progress. Int J Environ Res Public Health. 9:880–894.

Prüss-Ustün A, Wolf J, Bartram J, Clasen T, Cumming O, Freeman MC et al. (2019). Burden of disease from inadequate water, sanitation and hygiene for selected adverse health outcomes: an updated analysis with a focus on low- and middle-income countries. Int J Hyg Environ Health. 222:765–777.

Rantala A, Fewer D, Hisbergues M, Rouhiainen L, Vaitomaa J, Börner T et al. (2004). Phylogenetic evidence for the early evolution of microcystin synthesis. Proc Natl Acad Sci USA. 101:568–573.

Rickert B, Chorus I, Schmoll O (2016). Protecting surface water for health. Identifying, assessing and managing drinking-water quality risks in surface-water catchments. Geneva: World Health Organization:178 pp. https://apps.who.int/iris/handle/10665/246196

Roth GA, Abate D, Abate KH, Abay SM, Abbafati C, Abbasi N et al. (2018). Global, regional, and national age-sex-specific mortality for 282 causes of death in 195 countries and territories, 1980–2017: a systematic analysis for the Global Burden of Disease Study 2017. Lancet. 392:1736–1788.

Svirčev Z, Lalić D, Savić GB, Tokodi N, Backović DD, Chen L et al. (2019). Global geographical and historical overview of cyanotoxin distribution and cyanobacterial poisonings. Arch Toxicol. 93:2429–2481.

Waters M (2016). A 4700-year history of cyanobacteria toxin production in a shallow subtropical lake. Ecosystems. 19:426–436.

WHO (2003). Guidelines for safe recreational water environments. Vol. 1: Coastal and fresh waters. Geneva: World Health Organization. https://apps.who.int/iris/handle/10665/42591

WHO (2017). Guidelines for drinking-water quality, fourth edition, incorporating the 1st addendum. Geneva: World Health Organization. https://www.who.int/publications/i/item/9789241549950

Wilmotte A (1994). Molecular evolution and taxonomy of the cyanobacteria. In: Bryant DA, editor: The molecular biology of cyanobacteria. Dordrecht: Kluwer Academic Publishers:1–25.

Wood R (2016). Acute animal and human poisonings from cyanotoxin exposure—a review of the literature. Environ Int. 91:276–282.

Zastepa A, Taranu Z, Kimpe L, Blais J, Gregory-Eaves I, Zurawell R et al. (2017). Reconstructing a long-term record of microcystins from the analysis of lake sediments. Sci Tot Environ. 579:893–901.

Chapter 2

Cyanobacterial toxins

CONTENTS

INTRODUCTION AND GENERAL CONSIDERATIONS

The following sections provide an overview of the individual types of cyanobacterial toxins, focusing on toxins that have been confirmed to, or suggested to have implications for human health, namely, microcystins, cylindrospermopsins, anatoxins, saxitoxins, anatoxin-a(S) and dermatotoxins, the latter primarily produced by marine cyanobacteria. Two further cyanobacterial metabolites, lipopolysaccharides (LPS) and β-methylamino-alanine (BMAA), are discussed in respective sections with the conclusion that the available evidence does not show their proposed toxic effects to occur in dose ranges relevant to the concentrations found in cyanobacterial blooms. A further section includes information on taste and odour compounds produced by cyanobacteria because, while actually not toxic, they sometimes indicate the presence of cyanobacteria. Finally, recognising that there are many cyanobacterial metabolites and further toxic effects of cyanobacterial cells that have been observed which cannot be attributed to any of the known cyanotoxins, a section covers "additional toxicity" and bioactive cyanobacterial metabolites.

The sections on the major toxin types review the chemistry, toxicology and mode of action, producing cyanobacteria and biosynthesis, occurrence and environmental fate. Given the document's scope, the individual sections discuss ecotoxicological data only briefly. This, however, does not imply that cyanotoxins do not play an important role in aquatic ecosystems. Further, possible benefits of toxin biosynthesis for the producing cyanobacteria are currently discussed but not yet understood, and this remains an important field of research but is discussed in this volume only briefly.

For microcystins, cylindrospermopsins and saxitoxins guideline values (GVs) have been derived based on the toxicological data available and considering there is credible evidence of their occurrence in water to which people may be exposed. For anatoxin-a, although GVs cannot be derived due to inadequate data, a "bounding value", or health-based reference value, has been derived. For anatoxin-a(S) and the dermatotoxins, the toxicological data for deriving such values are not sufficient, and hence, no such values are proposed.

BOX 2.1: HOW ARE GUIDELINE VALUES DERIVED?

For most chemicals that may occur in water, including for the known cyanotoxins, it is assumed that no adverse effect will occur below a threshold dose. For these chemicals, a tolerable daily intake (**TDI**) can be derived. TDIs represent an estimate of an amount of a substance, expressed on a body weight basis, that can be ingested daily over a lifetime without appreciable health risk. TDIs are usually based on animal studies because, for most chemicals, the available epidemiological data are not sufficiently robust, mainly because the dose to which people were exposed is poorly quantified and because it is scarcely possible to exclude all confounding factors (including simultaneous exposure to other substances) that may have influenced differences between those exposed and the control group. TDIs based on animal studies are based on long-term exposure, preferably spanning a whole life cycle or at least a major part of it, exposing groups of animals (frequently mice or rats) to a series of defined doses applied orally via drinking-water or gavage. The highest dose for which no adverse effects in the exposed animals were detected is the no observed adverse effect level (**NOAEL**), generally expressed in dose per body weight and per day (e.g., 40 μg/kg bw per day). Sometimes no NOAEL is available, while the lowest observed adverse effect level (**LOAEL**) can be considered in establishing the TDI. The LOAEL is defined as the lowest dose in a series of doses causing adverse effects. An alternative approach for the derivation of a TDI is the determination of a benchmark dose (BMD), in particular, the lower confidence limit of the benchmark dose (BMDL; WHO, 2009a). A BMDL can be higher or lower than NOAEL for individual studies (Davis et al., 2011).

A NOAEL (or LOAEL or BMDL) obtained from animal studies cannot be directly applied to determine "safe" levels in humans for several reasons such as differences in susceptibility between species (i.e., humans *vs.* mice or rats), variability between individuals, limited exposure times in the experiments or specific uncertainties in the toxicological data. For example, for the cyanotoxins for which WHO has established GVs, exposure times did not span a whole life cycle because the amount of pure toxin needed for such a long study – a few hundred grams – was simply not available or would be extremely costly to purchase. To account for these uncertainties, a NOAEL is divided by uncertainty factors (**UFs**). The total UF generally comprises two 10-fold factors, one for interspecies differences and one for interindividual variability in humans. Further uncertainty factors may be incorporated to allow for database deficiencies (e.g., less than lifetime exposure of the animals in the assay, use of a LOAEL rather than a NOAEL, or for incomplete assessment of particular endpoints such as lack of data on reproduction) and for the severity or irreversibility of effects (e.g., for uncertainty regarding carcinogenicity or tumour promotion). Where adequate data is available, chemical specific adjustment factors (CSAFs) can be used for interspecies and intraspecies extrapolations, rather than the use of the default UFs.

The TDI is calculated using the following formula:

$$\mathrm{TDI} = \frac{\mathrm{NOAEL\ or\ LOAEL\ or\ BMDL}}{\mathrm{UF_1} \times \mathrm{UF_2} \times \mathrm{UF_N}\ \ \mathrm{or\ CSAFs}}$$

The unit of TDI generally is the amount of toxin per bodyweight (**bw**) per day, for example, 0.1 μg/kg bw per day.

To translate the TDI to a GV, the following formula is generally used:

$$\mathrm{GV} = \frac{\mathrm{TDI} \times \mathrm{bw} \times P}{C}$$

For **drinking-water** GVs, WHO uses a daily water consumption (*C*) of 2 L and a bodyweight of an adult person of 60 kg as default values, while emphasising that this may be adapted to regional or local circumstances. The fraction of exposure assumed to occur through drinking-water (*P*; sometimes termed allocation factor) is applied to account for the share of the TDI allocated to a specific exposure route. The default *P* for drinking-water is 0.2 (20%). Where there is clear evidence that drinking-water is the main source of exposure, like in the case of cyanotoxins, *P* has been adjusted to 0.8, which still allows for some exposure from other sources, including food. Again, this can and should be adapted if local circumstances propose a different factor to be more appropriate. The unit of a GV is a concentration, for example, 0.8 µg/L.

GVs for drinking-water (using a TDI) are generally derived to be safe for lifetime exposure. This means that **briefly exceeding a lifetime GV** doesn't pose an immediate risk or imply that the water is unsafe. This should be communicated accordingly and is particularly relevant where elevated cyanotoxin concentrations occur only during brief seasonal blooms. To clarify this, WHO has derived GVs for short-term exposure for microcystins and cylindrospermopsins. To differentiate these two GVs for cyanotoxins, these have been designated $\mathbf{GV_{chronic}}$ and $\mathbf{GV_{short\text{-}term}}$. In consequence, a concentration in drinking-water that exceeds the $GV_{chronic}$ up to a concentration of $GV_{short\text{-}term}$ does not require the immediate provision of alternative drinking-water – but it does require immediate action to prevent cyanotoxins from further entering the drinking-water supply system and/or to ensure their efficient removal through improving the drinking-water production process. The $GV_{short\text{-}term}$ provides an indication on how much the $GV_{chronic}$ can be exceeded for short periods of about 2 weeks until measures have been implemented to reduce the cyanotoxin concentration. Derivation of $GV_{short\text{-}term}$ follows a similar approach to development of the traditional GVs. The short-term applicability of these values, however, may result in a different study selected for the identification of the NOAEL or LOAEL (particularly if the $GV_{chronic}$ was based on long-term exposure) and the uncertainty factors (UFs) applied, particularly the UF for related database deficiencies.

For **recreational exposure**, the corresponding GV proposed ($\mathbf{GV_{recreation}}$) takes into account the higher total exposure of children due to their increased likelihood of longer playtime in recreational water environments and accidental ingestion. The default bodyweight of a child and the volume of water unintentionally swallowed are 15 kg and 250 mL, respectively (WHO, 2003), and these are used to calculate the $GV_{recreation}$. The same NOAEL or LOAEL and UFs applied for the $GV_{short\text{-}term}$ are used to calculate the $GV_{recreation}$.

All GVs proposed by WHO may be subject to change when new toxicological data become available. By default, GVs with high uncertainty (UF≥1000) are designated as provisional by WHO. GVs with high uncertainty are more likely to be modified as new information becomes available. Also, a high uncertainty factor indicates that new toxicological data are likely to lead to a higher rather than a lower GV, and thus, the provisional GV is likely a conservative one; that is, it presumably errs on the safe side.

Several national and regional GVs or standards deviate from the values proposed by WHO, due to different assumptions on body weight, estimated water intakes or allocation factors in consequence of specific exposure patterns in certain areas or for specific population groups (and sometimes also due to divergent interpretations of toxicological data). WHO gives guidance on adapting WHO GVs to country contexts in the document, "Developing Drinking-water Quality Regulations and Standards" (WHO, 2018). For more information on GV derivation, see the GDWQ (WHO, 2017) and the Policies and Procedures for Updating the WHO GDWQ (WHO, 2009b).

These values describe concentrations in drinking-water and water used for recreation that are not a significant risk to human health. For some of these toxin groups, it was possible to derive values for lifetime exposure and for others only for short-term or acute exposure (see Table 5.1 for a summary of the values established). The corresponding sections in Chapter 2 present the derivation of these values and a short summary of the considerations leading to them; for an extensive discussion, readers are referred to the cyanotoxin background documents on the WHO Water, Sanitation and Health website (WHO, 2020). For a summary on how guideline values are derived, see Box 2.1, and for further information, see also the "Guidelines for Drinking Water Quality" (WHO, 2017).

REFERENCES

Davis JA, Gift JS, Zhao QJ (2011). Introduction to benchmark dose methods and US EPA's benchmark dose software (BMDS) version 2.1. 1. Toxicol Appl Pharmacol. 254:181–191.

WHO (2003). Guidelines for safe recreational water environments. Vol. 1: Coastal and fresh waters. Geneva: World Health Organization. https://apps.who.int/iris/handle/10665/42591

WHO (2009a). Principles and methods for the risk assessment of chemicals in food. Geneva: World Health Organization. Environmental health criteria 240. https://apps.who.int/iris/handle/10665/44065

WHO (2009b). WHO Guidelines for drinking-water: quality policies and procedures used in updating the WHO guidelines for drinking-water quality. Geneva: World Health Organization. https://apps.who.int/iris/handle/10665/70050

WHO (2017). Guidelines for drinking-water quality, fourth edition, incorporating the 1st addendum. Geneva: World Health Organization:631 pp. https://www.who.int/publications/i/item/9789241549950

WHO (2018). Developing drinking-water quality regulations and standards. Geneva: World Health Organization. https://apps.who.int/iris/handle/10665/272969

WHO (2020). Cyanobacterial toxins: Anatoxin-a and analogues; Cylindrospermopsins; Microcystins; Saxitoxins. Background documents for development of WHO Guidelines for Drinking-water Quality and Guidelines for Safe Recreational Water Environments. Geneva: World Health Organization. https://www.who.int/teams/environment-climate-change-and-health/water-sanitation-and-health/water-safety-and-quality/publications

2.1 HEPATOTOXIC CYCLIC PEPTIDES – MICROCYSTINS AND NODULARINS

Jutta Fastner and Andrew Humpage

The cyclic peptides microcystins and nodularins are frequently found in fresh and brackish waters, and the acute and chronic toxicity of some of them is pronounced. WHO has established provisional guideline values for microcystin-LR in drinking-water and water for recreational use (see below) but recommends that these values may be used for the sum of all microcystins in a sample (see WHO, 2020). Microcystin-LR occurs widely and is presumably one of the most toxic variants of this toxin family, though for most of the other congeners no, or only incomplete, toxicological data exist (WHO, 2003a; Buratti et al., 2017).

2.1.1 Chemical structures

The cyclic heptapeptide microcystins were first characterised in the early 1980s and named after the cyanobacterium *Microcystis aeruginosa* from which they were initially isolated (Botes et al., 1984; Botes et al., 1985; Carmichael et al., 1988). Microcystins share a common structure of cyclo-(*D*-Ala^1-X^2-*D*-$MeAsp^3$-Z^4-$Adda^5$-*D*-Glu^6-$Mdha^7$) in which X and Z are variable *L*-amino acids, *D*-MeAsp is *D-erythro*-β-methylaspartic acid, and Mdha is *N*-methyldehydroalanine (Figure 2.1). The amino acid Adda, abbreviated for (2*S*,3*S*,4*E*,6*E*,8*S*,9*S*)-3-amino-9-methoxy-2,6,8-trimethyl-10-phenyldeca-4,6-dienoic acid, is the most characteristic moiety of microcystins and nodularins (including the structurally related motuporins from sponges) as it appears to occur exclusively in these cyanobacterial peptides. Further characteristics are the presence of *D*-amino acids, which are usually not found in ribosomally synthesised peptides and thus gave an early indication of a nonribosomal synthesis of these peptides (see below).

To date more than 250 different variants of microcystins are fully characterised, with molecular weights in the range of 800–1100 Da. While a comprehensive list of variants is given in Spoof and Catherine (2017) and a detailed review on structural variants can be found in Bouaïcha et al. (2019), this volume lists only the apparently most abundant congeners in Table 2.1. Structural modifications exist in all seven amino acids, but the most frequent variations are substitution of *L*-amino acids at positions 2 and 4, substitution of Mdha by Dhb (dehydrobutyrine) or serine in position 7, and a lack of methylation of amino acids at positions 3 and/or 7 (Figure 2.1). The

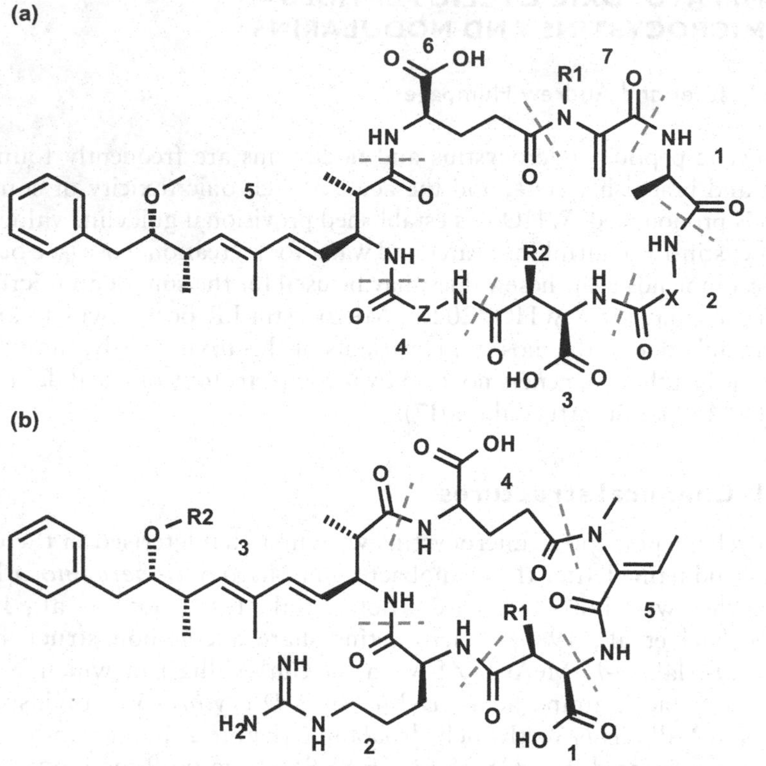

Figure 2.1 Generic structure of microcystins (a, MCs) and nodularins (b, NODs). In MCs, amino acids in positions 2 and 4 given as X and Z for variable amino acids that are generally given by the one-letter code for proteinogenic *L*-amino acids. For example, L=L-leucine. R=L-arginine. A=L-alanine. General structure of MCs is cyclo-(DAla^1-X^2-D-MeAsp^3-Z^4-Adda^5-D-Glu^6-Mdha^7). R1 and R2 is either H (desmethyl-variants) or CH_3. The general structure of NOD is cyclo-(*D*-MeAsp^1-Arg^2-Adda^3-*D*-Glu^4-Mdhb^5). Nodularin: R1. R2=CH_3; D-Asp^{1}Nodularin: R1=H. R2=CH_3; demethyl-Adda^3 Nodularin: R1=CH_3. R2=H. In motuporin, the *L*-arginine in position 2 of nodularin is exchanged by an *L*-valine. For more variants and details on amino acid building blocks, see text. Note that the numbering does not correspond to the biosynthesis pathway that starts with Adda in both MC and NOD, but has been assigned arbitrarily in the first descriptions of the molecule.

principle nomenclature of microcystins is based on the variable amino acids in positions 2 and 4; for example, using the standard one-letter codes for amino acids, microcystin-LR possesses leucine (L) in position 2 and arginine (R) in position 4, respectively (Carmichael et al., 1988). All other modifications vin the molecule are suffixed to the respective variant; for example, [Asp^3] MC-LR lacks the methyl group in position 3.

Nodularins, named after the cyanobacterium *Nodularia spumigena*, are cyclic pentapeptides structurally very similar to microcystins (Rinehart et al., 1988). The chemical structure of nodularin is cyclo-(*D*-MeAsp^1-*L*-arginine^2-Adda^3-*D*-glutamate^4-Mdhb^5), in which the amino acids in positions 1–4 are identical to microcystins and Mdhb is 2-(methylamino)-2-dehydrobutyric acid (Figure 2.1). Structural variability of nodularins is low compared to microcystins: in addition to the unmodified nodularin with the arginine residue, only a few more variants with (de)methylation

Table 2.1 Selection of microcystin structural variants reported in relatively high abundance

Microcystin variant	*Monoisotopic molecular mass (Da)*	*Molecular weight (g/mol)*	LD_{50} *i.p. (oral)*	*Reference*
MC-LA	909.485	910.08	50	Botes et al. (1984) Stoner et al. (1989)
[Asp^3,Dha^7] MC-LR	966.517	967.14	+	Harada et al. (1991a) Namikoshi et al. (1992c)
[Asp^3] MC-LR	980.533	981.16	160–300	Krishnamurthy et al. (1989) Harada et al. (1990)
[Dha^7] MC-LR	980.533	981.16	250	Namikoshi et al. (1992a) Harada et al. (1991b)
MC-LF	985.516	986.18	+	Azevedo et al. (1994) Diehnelt et al. (2006)
MC-LR	994.549	995.19	50 (5000)	Botes et al. (1985) Krishnamurthy et al. (1986) Krishnamurthy et al. (1989) Fawell et al. (1994)
MC-LY	1001.511	1002.18	90	del Campo & Ouahid (2010) Stoner et al. (1989)
[Asp^3,Dha^7] MC-RR	1009.535	1010.16	+	Krishnamurthy et al. (1989) Sivonen et al. (1992a)
[Asp^3] MC-RR	1023.550	1024.19	250	Meriluoto et al. (1989) Namikoshi et al. (1992d)
[Dha^7] MC-RR	1023.550	1024.19	180	Kiviranta et al. (1992)
[Asp^3,(E)-Dhb^7] MC-RR	1023.550	1024.19	250	Sano & Kaya (1995) Sano & Kaya (1998)
MC-LW	1024.527	1025.21	n.r.	Bateman et al. (1995)
[Asp^3,Dha^7] MC-HtyR	1030.512	1031.18	+	Namikoshi et al. (1992b)
[Asp^3] MC-YR	1030.512	1031.18	+	Namikoshi et al. (1992d)
[Dha^7] MC-YR	1030.512	1031.18	+	Sivonen et al. (1992b)

(Continued)

Table 2.1 (Continued) Selection of microcystin structural variants reported in relatively high abundance

Microcystin variant	*Monoisotopic molecular mass (Da)*	*Molecular weight (g/mol)*	*LD_{50} i.p. (oral)*	*Reference*
MC-RR	1037.566	1038.22	600	Namikoshi et al. (1992a) Kusumi et al. (1987)
[Asp³] MC-HtyR	1044.528	1045.21	160–300	Harada et al. (1991a)
[Dha⁷] MC-HtyR	1044.528	1045.21	+	Namikoshi et al. (1992b)
[Asp³,(E)-Dhb⁷] MC-HtyR	1044.528	1045.21	70	Sano & Kaya (1998)
MC-YR	1044.528	1045.21	70	Botes et al. (1985) Namikoshi et al. (1992a)
MC-WR	1067.544	1068.24	150–200	Namikoshi et al. (1992a)

Molecular weight (MW) is given as monoisotopic mass; LD50 in μg/kg body weight intraperitoneal (i.p.) injection in mouse bioassays or by oral dosing (values in parentheses) where data are available.

+: positive toxicity in mouse bioassay; n.r.: not reported.

at the Adda, MeAsp, Mdhb and D-Glu moieties, as well as the non-toxic 6Z-Adda3 stereoisomer equivalent to microcystins, have been identified (Namikoshi et al., 1994; Mazur-Marzec et al., 2006b). Ten nodularins in total have been reported (Spoof & Catherine, 2017).

All microcystins and nodularins are water soluble despite the relatively wide range of hydrophobicity observed especially for microcystins. They are extremely stable and remain potent even after boiling (Harada, 1996).

2.1.2 Toxicity: mode of action

The toxic effects of microcystin, summarised in the following, are described in detail in the WHO Background Document on Microcystins (WHO, 2020; see there for further information and references). In summary, microcystins need a membrane transporter to enter cells – that is, the organic acid transporter polypeptides (OATP) which are expressed particularly in the liver but also in the intestinal tract and in some other tissues. Experiments have shown that when OATP is inhibited or lacking, liver damage is reduced. The essential role of OATP explains why most of the MCs ingested are taken up by the liver. While detoxification occurs in the liver, clearance of MCs seems to take a long time, up to weeks. Once in cells, MCs cause protein phosphatase (PP1, PP2A and PP5) inhibition, resulting in destabilisation of the cytoskeleton followed by cellular apoptosis and necrosis. High acute doses thus cause haemorrhage in the liver due to the damage of sinusoidal capillaries. At low doses (below 20 μg/kg bw) and with repeated long-term exposure, phosphatase inhibition induces cellular proliferation, hepatic hypertrophy and tumour promoting activity.

There is a growing body of evidence indicating harmful microcystin-related neurological and reproductive effects, but the data are not yet robust enough to use as a basis for guideline development.

While some cyanobacterial extracts show genotoxicity, pure microcystins do not, and cellular DNA damage observed after *in vitro* treatments with pure MC may be due to the induction of apoptosis and cytotoxicity rather than direct effects on the DNA. On this basis, IARC has classified microcystins as Group 2B, possibly carcinogenic to humans (IARC, 2010), based on their tumour promoting activity mediated via protein phosphatase inhibition (a threshold effect) rather than genotoxicity.

2.1.3 Derivation of provisional guideline values

The following section is taken directly from the WHO chemicals background document on microcystins which discusses the considerations for the derivation of provisional guideline values for exposure to microcystins in more detail (WHO, 2020). Insufficient data are available to derive a GV for MC variants except MC-LR. The two key oral toxicity studies of the effects of MC-LR on liver toxicity on which human health-based guideline values can be calculated are the following:

- Fawell et al. (1999): Mice of both sexes given MC-LR by gavage at 40 µg/kg bw per day for 13 weeks did not show treatment-related effects in the parameters measured. Only slight hepatic damage was observed at the lowest observed effect level (LOAEL) of 200 µg/kg bw per day in a limited number of treated animals, whereas at the highest dose tested (1 mg/kg bw per day), all the animals showed hepatic lesions, consistent with the known action of MC-LR.
- Heinze (1999): Exposure of male rats (females were not included) to MC-LR in drinking-water for 28 days at doses as low as 50 µg/kg bw per day (identified as the LOAEL) resulted in increased liver weight, liver lesions (with haemorrhages) and increased ALP (alkaline phosphatase) and LDH (lactate dehydrogenase), but no changes were measured in the mean levels of AST (aspartate aminotransferase) and ALT (alanine aminotransferase) which are early markers for hepatotoxicity. Some of the histological effects, including Kupffer cell activation and PAS staining, showed no dose response since all 10 animals at the low and high doses displayed a similar degree of damage.

Although the duration of the Heinze (1999) study was shorter and more applicable to the exposure duration envisaged for application of the short-term guideline value, the advantage of the Fawell et al. (1999) study is that an additional uncertainty factor is not needed for extrapolation from a LOAEL to a NOAEL, which would increase the total uncertainty and reduce the confidence in the derivation of the short-term guideline value. For this reason, the NOAEL derived by Fawell et al. (1999) was selected as the basis for the short-term and recreational guideline values as well as the lifetime guideline value.

The GVs for MC-LR are considered provisional due to inadequacies in the database as reflected in section 6.2 of the background document (WHO, 2020) and the database uncertainty factor (UF) of 1000 for the lifetime GV.

Calculation of provisional tolerable daily intake for microcystin-LR

$$TDI_{chronic} = \frac{NOAEL}{UF} = \frac{40}{1000} \frac{\mu g}{kg} / d = 0.04 \frac{\mu g}{kg} / d$$

where

$TDI_{MC,chronic}$ = tolerable daily intake for chronic exposure
NOAEL = no-observed-adverse-effect level (40 µg/kg bw per day, based on Fawell et al., 1999)
UF = uncertainty factor (1000 = 10 for interspecies variation × 10 for intraspecies variation × 10 for database deficiencies, including use of a subchronic study)

For comparison, if the LOAEL from Heinze (1999) is used as the point of departure and incorporating uncertainty factors of 10 for inter- and intra-species variability and 10 for database uncertainties including the use of a LOAEL (as per WHO policy), then the TDI would be 0.05 µg/kg per day.

Calculation of provisional lifetime drinking-water guideline value for microcystin-LR

$$GV_{chronic} = \frac{NOAEL*bw*P}{UF*C} = \frac{40*60*0.8}{1000*2} \mu g / L = 0.96 \mu g / L \approx 1 \mu g / L$$

where

$GV_{chronic}$ = guideline value for chronic (lifetime) exposure
NOAEL = no-observed-adverse-effect level (40 µg/kg bw per day, based on Fawell et al., 1999)
bw = body weight (default = 60 kg for an adult)
P = fraction of exposure allocated to drinking-water (80%, as other sources of exposure such as air, food and soil are considered minor for lifetime exposure)

UF = uncertainty factor (1000 = 10 for interspecies variation × 10 for intraspecies variation × 10 for database deficiencies, including use of a subchronic study)
C = daily drinking-water consumption (default = 2 L for an adult).

Calculation of provisional short-term drinking-water guideline value for microcystin-LR

$$GV_{short-term} = \frac{NOAEL*bw*P}{UF*C} = \frac{40*60*1.0}{100*2}\,\mu g/L = 12\,\mu g/L$$

where

$GV_{short-term}$ = guideline value for short-term exposure
NOAEL = no-observed-adverse-effect level (40 µg/kg bw per day, based on Fawell et al., 1999)
bw = body weight (default = 60 kg for an adult)
P = fraction of exposure allocated to drinking-water (default for short-term exposure = 100%, as drinking-water is expected to be the most significant source of exposure)
UF = uncertainty factor (100 = 10 for interspecies variation × 10 for intraspecies variation)
C = daily drinking-water consumption (default = 2 L for an adult).

Calculation of provisional recreational water guideline value for microcystin-LR

$$GV_{recreation} = \frac{NOAEL*bw}{UF*C} = \frac{40*15}{100*0.25}\,\mu g/L = 24\,\mu g/L$$

where

$GV_{recreation}$ = guideline value for recreational exposure
NOAEL = no-observed-adverse-effect level (40 µg/kg bw per day, based on Fawell et al., 1999)
bw = body weight (default = 15 kg for a child)
UF = uncertainty factor (100 = 10 for interspecies variation × 10 for intraspecies variation)
C = daily incidental water consumption (default = 250 mL for a child).

The provisional recreational guideline value, aimed to protect from systemic effects, is based on exposure of a child because the lower body weight and higher likely water intake (as a function of body weight) were considered worst case, and on a conservative scenario of a 15 kg child swallowing 250 mL of water (WHO, 2003b).

Considerations in applying the provisional guideline values

The provisional guideline values are based on toxicological data for MC-LR. However, MCs usually occur as mixtures. In the absence of oral toxicity data for other congeners, it is recommended that these values be applied to total MCs as gravimetric or molar equivalents based on the assumption that all MCs have similar toxicity to MC-LR. The kinetic differences among variants mean that further investigation of the oral toxicity of MC variants other than MC-LR is warranted reduce this relevant source of uncertainty.

In some regions, other sources of exposure besides drinking-water can be significant (see section 2.4). This includes food from locations where blooms have a long duration and there is high consumption of locally affected food items (e.g. fish eaten with viscera, or shellfish). In such situations, it may be appropriate to consider reducing the allocation factor for the lifetime and short-term drinking-water GVs based on relative exposure data for the population.

The short-term drinking-water GV is intended to provide guidance on how much the lifetime GV can be exceeded for short periods of about 2 weeks until enhanced water treatment or other measures can be implemented. It is not intended to allow for repeated seasonal exceedances of the lifetime GV.

The short-term drinking-water guideline value is based on exposure of adults. Since infants and children can ingest a significantly larger volume of water per body weight (e.g., up to 5 times more drinking-water/kg bw for bottle-fed infants compared to adults), it is recommended that alternative water sources such as bottled water are provided for bottle-fed infants and small children when MC concentrations are greater than 3 µg/L for short periods, as a precautionary measure.

2.1.4 Production

2.1.4.1 Producing cyanobacteria

Qualitative and quantitative information on microcystin production in particular cyanobacterial species has been gathered through analyses of quasi-monospecific bloom material and, more importantly, large numbers of individual strains isolated from freshwater samples. More recent studies employ sensitive molecular and chemical tools such as PCR, mass spectrometry or ELISA to determine toxins or genes related to their production directly in colonies or filaments picked from water samples. This helps to avoid bias due to eventually selective isolation procedures and allows more detailed studies on the ecology of toxin-producing cyanobacteria. Furthermore, molecular tools are applied to clarify the taxonomic status of toxin-producing cyanobacteria and to complement the identification of toxin producers by verifying the presence of genes encoding toxin biosynthesis (see section 13.6).

Microcystin-producing strains can be found in all higher-level taxa of cyanobacteria, that is, in species belonging to the orders Chroococcales, Oscillatoriales, Nostocales, and Stigonematales; data for the order Pleurocapsales, however, are scarce. Within the orders, the distribution of microcystin occurrence at the level of genera or species is patchy and does not show consistency. Firstly, not all genera of an order produce microcystins; for example in the order Nostocales, microcystins are produced by members of the genera *Dolichospermum* (*Anabaena*) and *Nostoc*, but have never been confirmed for the closely related genus *Aphanizomenon*. Secondly, any particular genus or species may contain both producing (toxigenic) and nonproducing strains. At the time of the publication of this book, microcystin-producing (and nonproducing) strains are known primarily from freshwater species of *Microcystis*, *Planktothrix*, *Dolichospermum*, and *Nostoc* (Sivonen & Jones, 1999; Oksanen et al., 2004; Mowe et al., 2015; Harke et al., 2016; Bernard et al., 2017; Buratti et al., 2017; Table 2.2). Very rarely, microcystins have been reported in single strains from other genera, including *Anabaenopsis, Arthrospira*, *Fischerella*, *Pseudanabaena*, *Phormidium*, *Synechococcus* and *Radiocystis* (Ballot et al., 2005; Carmichael & Li, 2006; Lombardo et al., 2006; Izaguirre et al., 2007; Nguyen et al., 2007; Mohamed & Al Shehri, 2009; Cirés et al., 2014; Table 2.2).

Most of these cyanobacteria are of planktonic nature and some of them, like *Microcystis*, are known for their ability to form surface blooms under favourable conditions (see Chapter 4). Microcystins have also been detected in halophilic *Synechococcus* and *Dolichospermum* (*Anabaena*) from the Baltic Sea (Carmichael & Li, 2006; Halinen et al., 2007).

Microcystin-producing strains of the genera listed above are distributed globally and can be found in tropical, temperate and polar habitats (Hitzfeld et al., 2000; Mowe et al., 2015; Harke et al., 2016) as well as in extreme habitats such as hot springs and hypersaline lakes (Carmichael & Li, 2006; Kotut et al., 2006). Microcystins have also been detected in a symbiotic strain of *Nostoc* in a lichen (Oksanen et al., 2004) and in a soil isolate of *Haphalosiphon hibernicus* (Prinsep et al., 1992).

Nodularins have so far been found largely in strains of the genus *Nodularia*, primarily in *Nodularia spumigena*. Toxigenic strains of *Nodularia spumigena* have been reported from the Baltic Sea, brackish water estuaries and coastal freshwater lakes of Australia, South Africa, New Zealand and Turkey (Bolch et al., 1999; Akçaalan et al., 2009). As with microcystins, both nodularin-producing and nonproducing strains exist in this species (Lehtimäki et al., 1994; Bolch et al., 1999). In addition, single findings of nodularin in *Nodularia sphaerocarpa* from a hot spring, in a symbiotic *Nostoc*, and

Table 2.2 Cyanobacterial taxa potentially producing microcystins and nodularins

Toxin	*Taxon*	*Habitat*
Microcystin	***Microcystis* sp.**	Planktonic
	***Dolichospermum (Anabaena)* sp.**	Planktonic
	Planktothrix agardhii	Planktonic
	Planktothrix rubescens	Planktonic
	Radiocystis sp.	Planktonic
	Arthrospira sp.	Planktonic
	Anabaenopsis sp.	Planktonic
	Calothrix sp.	Planktonic
	Oscillatoria sp.	Planktonic
	Fischerella sp.	Planktonic, benthic
	Annamia toxica	Planktonic
	Synechococcus sp.	Planktonic
	Pseudanabaena sp.	Planktonic
	Phormidium sp.	Planktonic
	Anabaena sp.	Benthic
	Nostoc sp.	Planktonic, benthic, symbiotic (lichen)
	Aphanocapsa sp.	Planktonic
	Plectonema sp.	Benthic
	Leptolyngbya sp.	Symbiotic (coral), periphytic
	Merismopedia sp.	Periphytic
	Haphalosiphon hibernicus	Terrestrial
Nodularin	***Nodularia spumigena***	Planktonic
	Nodularia sp.	Benthic
	Nostoc sp.	Symbiotic (lichen)
	Iningainema pulvinus	Benthic

Only cyanobacteria are listed for which toxin production was verified in cultured strains by NMR, mass spectrometry or by combinations of HPLC-PDA, ELISA, toxicity testing and/or molecular detection of *mcy* genes. References earlier than 1999 are summarised in Sivonen & Jones (1999). In bold are taxa that are known to frequently produce microcystins and that can form blooms.

in the benthic *Iningainema pulvinus* (Nostocales) from Australia have been reported (Beattie et al., 2000; Gehringer et al., 2012; McGregor & Sendall, 2017). Occasionally, nodularin has been detected in pelagic and benthic freshwater ecosystems in which none of the known nodularin producers could be identified, indicating that further species may be identified as nodularin producers in future (Graham et al., 2010; Wood et al., 2012; Beversdorf et al., 2017).

2.1.4.2 Microcystin/nodularin profiles

Although many toxigenic strains simultaneously produce several microcystin variants (Puddick et al., 2014), usually only one to three of them are dominant in any particular strain. It appears that some microcystin variants are more abundant within a certain genus than within others, though this may be biased in some cases by the limited availability of standards as well as the analytical methods used.

Globally, *Microcystis* strains and field samples dominated by *Microcystis* spp. are reported to contain chiefly microcystin-LR, -RR and -YR in varying proportions (Sivonen & Jones, 1999; Vasconcelos, 2001; Gkelis et al., 2005; Kemp & John, 2006; Faassen & Lürling, 2013; Mowe et al., 2015; Beversdorf et al., 2017). Demethylated variants of, for example, [Dha^7]MC-LR and -RR are also observed in *Microcystis* strains, but are less frequently dominant compared to their methylated forms (Vasconcelos, 2001; Gkelis et al., 2005). More hydrophobic microcystins (e.g., MC-LA, MC-LW, MC-LF) can also be found regularly in *Microcystis* strains and field samples; however, high proportions are reported only infrequently (Cuvin-Aralar et al., 2002; Wood et al., 2006; Graham et al., 2010; Faassen & Lürling, 2013; Beversdorf et al., 2017). This picture of a high diversity combined with an abundance of genotypes with a certain microcystin profile has been confirmed *in situ* for some natural *Microcystis* populations from central Europe. Typing of single *Microcystis* colonies from nine European countries by mass spectrometry revealed a high abundance of genotypes producing microcystin-LR, -RR and -YR, while clones with demethylated variants or other microcystins were less abundant (Via-Ordorika et al., 2004). However, exceptions from this overall pattern occur; for example, in *Microcystis* strains and colonies from Finland, demethylated MC-LR and -RR have been seen frequently as dominant variants (Luukkainen et al., 1994; Via-Ordorika et al., 2004), and in one Australian bloom of *Microcystis*, 23 microcystins were detected, none of which was microcystin-LR (Jones et al., 1995).

Planktothrix and some strains of *Dolichospermum* seem to produce only demethylated microcystins (Puddick et al., 2014). In European *Planktothrix agardhii* and *P. rubescens* isolates, cultured strains and field samples primarily produce demethylated variants of microcystin-RR like [D-Asp^3] MC-RR and [D-Asp^3, Dhb^7] MC-RR that have been found as major microcystins (Sivonen et al., 1995; Fastner et al., 1999; Briand et al., 2005; Kurmayer et al., 2005; Cerasino et al., 2016). Various other demethylated microcystins such as [D-Asp^3] MC-LR or [D-Asp^3] MC-HtyR are also found in *Planktothrix* isolates but are rarely the dominant variants (Kosol et al., 2009). As with *Microcystis*, multiple clones with different microcystin profiles exist in natural populations of *Planktothrix* (Welker et al., 2004; Haruštiaková & Welker, 2017).

Only few data exist on microcystin congeners produced by benthic species, and detected variants comprise MC-LR, MC-RR, MC-YR, MC-LA,

[Asp^3] MC-LR as well as unidentified microcystins (Aboal & Puig, 2005; Jungblut et al., 2006; Izaguirre et al., 2007; Fetscher et al., 2015).

Nodularin-R seems the major nodularin present in samples from the Baltic Sea, Turkey and Australia, while other nodularin variants usually seem less abundant (Sivonen et al., 1989; Jones et al., 1994a; Lehtimaki et al., 1997; Mazur-Marzec et al., 2006b; Akçaalan et al., 2009).

2.1.4.3 Biosynthesis

Knowledge of the biosynthesis of microcystins and nodularins has increased since the turn of the millennium. Complete sequences of biosynthesis gene clusters are available for several species, and biochemical pathways are largely understood (Pearson et al., 2016).

Microcystins and nodularins are synthesised by a combined nonribosomal peptide synthetase (NRPS) and polyketide synthase (PKS) pathway, which is well known for the synthesis of peptide antibiotics in bacteria and fungi, including penicillins (Kleinkauf & Döhren, 1996; Dittmann & Börner, 2005). Microcystins are produced by large multienzyme complexes consisting of peptide synthetases, polyketide synthases and tailoring enzymes. These enzymes activate specific amino acids and condense them to peptides. The genes encoding for microcystin synthetases (*mcyA-mcyJ*) have been characterised for *Microcystis*, *Dolichospermum, Fischerella*, *Nostoc* and *Planktothrix* (Tillett et al., 2000; Christiansen et al., 2003; Rouhiainen et al., 2004; Fewer et al., 2013; Shih et al., 2013). The biosynthesis of nodularins is encoded by homologous genes (*ndaA-ndaI)* that have been characterised from *Nodularia* (Moffitt & Neilan, 2004). Both the microcystin and nodularin gene clusters comprise around 50 kb pairs in all investigated species, but differences in the gene order as well as DNA sequence variation in the same modules have been observed. Based on coding nucleotide sequences, Rantala et al. (2004) concluded that microcystin synthetase genes have already been present in an early stage of cyanobacterial evolutionary history.

Microcystin and nodularin production appears to be constitutive in genotypes which have the complete microcystin gene clusters, while it is absent in genotypes lacking the whole or relevant parts of the gene cluster (Christiansen et al., 2008; Tooming-Klunderud et al., 2008). Smaller mutations in single *mcy* genes can lead to genotypes unable to synthesise microcystin (Kurmayer et al., 2004; Christiansen et al., 2006; Fewer et al., 2008).

2.1.4.4 Regulation of biosynthesis

Microcystin contents or cell quota can vary greatly between individual clones within a natural population (e.g., Rohrlack et al., 2001; Akçaalan et al.,

2006). Reported microcystin contents in isolates (cultures) of *Microcystis* and *Planktothrix* range over more than two orders of magnitude, from below 100 µg up to more than 10 mg/g dry weight, from traces up to 20 µg/mm^3 biovolume and from a few to around 1000 fg/cell (Table 2.3).

While qualitative microcystin production is regarded as constitutive, numerous studies on *Microcystis*, *Planktothrix*, *Dolichospermum* and *Nodularia* have investigated to which extent the cell quota may be altered by environmental factors. The environmental factors investigated included temperature, light, pH, macronutrients, trace elements and salinity (reviewed in Sivonen & Jones, 1999; Kardinaal & Visser, 2005; Pearson et al., 2016). Though all the studies showed an effect on microcystin content or cell quota, respectively, they show no consistent pattern in the regulation of the microcystin cell quota. These inconsistencies can partly be explained by large differences between the studies with respect to culture conditions (i.e., batch, semi- and continuous; see Box 4.11), toxin measurement (i.e., ELISA, HPLC), as well as by the biomass proxy to which the toxin content was related (i.e., dry weight, biovolume, cell number, protein or chlorophyll-*a*). Another explanation for the differences in cell quota changes may be a clone-specific binding rate of synthesised microcystins to proteins, which is then not available to conventional analysis, thus leading to underestimation of microcystin contents (Meissner et al., 2013). To add to this complexity, individual clones of the same species can respond differently, even conversely, in microcystin contents to the same cultivation condition (Hesse & Kohl, 2001).

One pattern of changes in the microcystin content which unifies many of the earlier disparate results is the positive correlation between growth and microcystin content under growth conditions limited by nitrogen, phosphorus or light (Kardinaal & Visser, 2005). However, some studies observed this relationship only during exponential growth, while no or a negative relationship existed during the stationary phase (Wiedner et al., 2003; Yepremian et al., 2007), and others could not find any relationship (Jähnichen et al., 2011). Most importantly, however, the results demonstrated that the cell quota varied only within a rather narrow range, that is, by a factor of 2–4 (Table 2.3). In addition to changes in total microcystin cell quota, cultivation factors such as light and nitrogen have also been shown to alter the relative abundance of individual microcystins (Rapala et al., 1997; Tonk et al., 2005; Van de Waal et al., 2009).

More recent work has addressed changes in the microcystin biosynthesis also at the molecular level, but so far without conclusive, comprehensive outcomes. For example, light, iron and nitrogen have been found to either increase or decrease *mcy* transcription in individual strains, though microcystin cell quota do not necessarily reflect change in transcriptional activity (e.g., Kaebernick et al., 2002; Sevilla et al., 2010; Harke & Gobler, 2015).

Table 2.3 Examples of microcystin contents given as mass per dry weight, per biovolume and per cell found in cultured strains

Taxon	*Cond.*	*Origin*	*N*	*Range*	*Reference*
				mg/g DW	
Microcystis	S	TW	6	0.3–10	Lee et al. (1998)
Microcystis	S	GER	10	<0.1–4	Rohrlack et al. (2001)
Microcystis	S	JPN	17	0.6–13	Watanabe et al. (1991)
Planktothrix agardhii	S	FRA	36	0.02–1.86	Yepremian et al. (2007)
P. agardhii	S	Eur, JPN	18	1.2–4.5	Kosol et al. (2009)
P. rubescens	S	Eur, JPN	31	2.9–5.4	Kosol et al. (2009)
Anabaena [a]	S	FRA	2	0.35–1.86	Vezie et al. (1998)
Anabaena [a]	S	FIN	5	1.3–3.9	Halinen et al. (2007)
Anabaena [a]	S	EGY	2	3–3.66	Mohamed et al. (2006)
Anabaena [a]	T.L.N.P	FIN	2	<0.3–7	Rapala et al. (1997)
				$\mu g/mm^3$ BV	
Microcystis	S	KEN	12	0.4–13.8	Sitoki et al. (2012)
P. agardhii	S	Eur, JPN	18	2.3–16.7	Kosol et al. (2009)
P. rubescens	S	Eur, JPN	31	1.1–20.6	Kosol et al. (2009)
Microcystis	L	NLD	1	1.2–2.5	Wiedner et al. (2003)
Microcystis	L.N.P	DEU	3	0.6–5.0	Hesse & Kohl (2001)
P. agardhii	L	FIN	1	~2–3	Tonk et al. (2005)
				fg/cell	
Microcystis	S	KEN	12	17–553	Sitoki et al. (2012)
Microcystis	P	NLD	1	5–20	Ríos et al. (2014)
Microcystis	P	GBR	1	17–97	Ríos et al. (2014)
Microcystis	N	USA	1	70–220	Harke & Gobler (2013)
Microcystis	L	USA	1	47–106	Deblois & Juneau (2010)
Microcystis	N	AUS	1	56–165	Orr & Jones (1998)
P. agardhii	S	GBR	2	75–91	Akçaalan et al. (2006)
P. rubescens	S	GBR	3	104–235	Akçaalan et al. (2006)
P. agardhii	S	Eur, JPN	18	44–343	Kosol et al. (2009)
P. rubescens	S	Eur, JPN	31	27–854	Kosol et al. (2009)
				fg/cell	
Nodularia spumigena	L.N.P	SWE	1	~100–700	Pattanaik et al. (2010)

Cultivation conditions (Cond.) were either one standard (S) or with varying light (L), temperature (T), nitrogen concentration (N) and phosphorus concentration (P). The origin of the analysed strains (N) is given by ISO 3166 country code. Data from some studies have been transformed to units as given here. Note that content in field samples is generally much lower as these consist of a mixture of clones with individual toxin contents ranging from 0 (nonproducers) to 9 (values as reported as maxima in this table; see also section 4.6).

[a] Reported as *Anabaena* but possibly *Dolichospermum* (see Chapter 3).

2.1.5 Occurrence in water environments

Numerous screening programmes conducted during the past 30 years in various parts of the world detected microcystins in 20% to 100% of the samples, with frequencies generally correlated with the trophic state of the sampled water bodies (Bigham et al., 2009). In waterbodies containing potentially toxigenic genera such as *Microcystis* and *Planktothrix*, microcystins were detected in 80–100% of the samples (Fastner et al., 2001; Graham et al., 2010; Gkelis & Zaoutsos, 2014). Thus, any waterbody with these taxa should be assumed to contain microcystins unless analytical results show that this is not the case. In samples dominated by *Dolichospermum*, microcystins were detected less frequently (Chorus, 2001).

Microcystin-producing genotypes can persist in cyanobacterial populations, and thus may be found during the whole growing season, even year-round. This includes not only tropical waterbodies, but also temperate shallow lakes dominated by *Planktothrix agardhii* and stratified, deep lakes harbouring *Planktothrix rubescens* (Pawlik-Skowronska et al., 2008; Mankiewicz-Boczek et al., 2011; Akçaalan et al., 2014; Cerasino et al., 2016).

Early reports on microcystin found in field samples often expressed values as mg or µg per gram dry weight, that is, toxin contents (see box 4.6), most likely due to comparably insensitive methods used at the time and the requirement of large amounts of cell material for toxin analyses. Reported microcystin contents range from a few ng up to – rarely – around 13–15 mg/g dry weight (Sivonen & Jones, 1999). Since the late 1990s, microcystin occurrence has increasingly been reported as concentrations, that is, per volume of water, which is the more relevant unit for cyanotoxin risk assessment.

Average microcystin concentrations in the pelagic water outside scums do not frequently exceed several tens of µg/L (Fastner et al., 2001; Bláha & Maršálek, 2003; Carrasco et al., 2006; Nasri et al., 2007; Graham et al., 2010; Sakai et al., 2013; Gkelis & Zaoutsos, 2014; Chia & Kwaghe, 2015; Mowe et al., 2015; Su et al., 2015; Beversdorf et al., 2017). However, in surface blooms and scums of *Microcystis*, microcystin concentrations can be up to several orders of magnitude higher than in the pelagic water, with the reported maximum values up to 20 and 124 mg/L (Kemp & John, 2006; Wood et al., 2006; Masango et al., 2010; Waajen et al., 2014).

Planktothrix rubescens forms population maxima in the metalimnetic layer with microcystin concentrations usually of only < 1 to 10 µg/L (Jacquet et al., 2005; Akçaalan et al., 2014; Cerasino et al., 2016). Following turnover of the waterbody, the formation of surface blooms has been observed with microcystin concentrations attaining up to 34 mg/L (Naselli-Flores et al., 2007).

Although microcystin concentrations can also reach more than 100 µg/L in blooms of *Planktothrix agardhii*, this species rarely forms surface bloom or scums (Fastner et al., 2001; Wiedner et al., 2002; Catherine et al., 2008; Pawlik-Skowronska et al., 2008; Mankiewicz-Boczek et al., 2011), and hence, microcystin concentrations > 1 mg/L have only rarely been observed.

Nodularins occur frequently in *Nodularia spumigena* populations from both temperate and subtropical environments with recurrent annual toxic blooms, for example, in the Baltic Sea. As with microcystins, concentrations may be several orders of magnitude higher in surface blooms compared to populations entrained homogeneously in the water column. The nodularin content of such blooms ranged from 3.5 to 18 mg/g dw, and concentrations from a few µg in the open water up to 18 mg/L in surface blooms have been reported (Kononen et al., 1993; Heresztyn & Nicholson, 1997; Mazur & Plinski, 2003; McGregor et al., 2012; Sahindokuyucu-Kocasari et al., 2015).

Although the variability of microcystin content within individual clones is limited to a factor of 2–4, the microcystin content of field populations of toxigenic taxa may vary by a few orders of magnitude. This suggests that much, if not most, of the variation in toxin content of monospecific natural blooms is attributable to the waxing and waning of clones of the same species, with clones varying in their toxin contents (see Chapter 4; Briand et al., 2008).

2.1.5.1 Bioaccumulation

Microcystins and nodularins have been detected in common aquatic vertebrates and invertebrates, including fish, mussels, shrimps and zooplankton (Kotak et al., 1996; Freitas de Magalhães et al., 2001; Sipiä et al., 2002; Chen & Xie, 2005; Xie et al., 2005; Ibelings & Havens, 2008). Because of the relevance of these findings for food from aquatic environments, bioaccumulation of microcystins/nodularins in biota and its role in health risk assessment are discussed in the section on food (5.3). The effects and possible bioaccumulation of microcystins on plants are reviewed in Machado et al. (2017).

2.1.6 Environmental fate

2.1.6.1 Partitioning between cells and water

Microcystins and nodularins are primarily found in viable cyanobacterial cells. Experiments with radiolabelled microcystin did not show a substantial export of intracellular toxins from cells under high as well as under low light conditions (Rohrlack & Hyenstrand, 2007). Release to the surrounding water as extracellular (dissolved) toxin is considered to occur mainly during cell senescence, death and lysis.

In laboratory studies, where both intracellular and extracellular microcystins/nodularins have been measured, the general finding was that in healthy cultures, less than 10% of the total toxin pool is extracellular (Lehtimaki et al., 1997; Rapala et al., 1997; Sivonen & Jones, 1999; Wiedner et al., 2003; Jähnichen et al., 2007). Even during log-phase cell growth in culture, a small percentage of cells in the population may be lysing and hence release intracellular microcystins. As cells enter the stationary phase, the increased rate of cell death may lead to an increase in the extracellular fraction.

Accordingly, in growing field populations, no or only little extracellular microcystin has been found. Concentrations of extracellular microcystins measured in such cases mostly range from not detectable to a few µg/L and amount to only a small fraction of the cell-bound toxins in the same samples (Pietsch et al., 2002; Wiedner et al., 2002; Welker et al., 2003; Bláhová et al., 2007; Pawlik-Skowronska et al., 2008). While in ageing or declining blooms large amounts of microcystins are liberated from the cells, the actual concentrations in water depend primarily on dilution and other factors such as adsorption and degradation, rarely reach values exceeding 100 µg/L (Welker et al., 2001). The only report of very high extracellular microcystin concentrations, reaching 1800 µg/L, was following an algicide treatment of a cyanobacterial bloom leading to sudden and complete lysis and thus a massive release of toxins (Kenefick et al., 1993; Jones & Orr, 1994).

For the production of drinking-water, special attention to procedures that potentially could release microcystins from cells is important: oxidation with ozone and chlorine, as well as flocculation and filtration, can lead to a leakage of microcystins from the cells (see Chapter 10; Pietsch et al., 2002; Schmidt et al., 2002; Daly et al., 2007).

2.1.6.2 Chemical breakdown

Microcystins are chemically very stable. They remain potent even after boiling for several hours (Harada, 1996) and may persist for many years when stored dry at room temperature (Metcalf et al., 2012). At near-neutral pH, microcystins are resistant to chemical hydrolysis or oxidation. At 40 °C and at elevated or low pH, slow hydrolysis has been observed, with the times to achieve greater than 90% breakdown being approximately 10 weeks at pH 1 and greater than 12 weeks at pH 9 (Harada, 1996). Rapid chemical hydrolysis occurs only under conditions that are unlikely to be attained outside the laboratory, for example, 6M HCl at high temperature.

Microcystins can be oxidised by ozone and other strong oxidising agents (Rositano et al., 2001; Rodríguez et al., 2007), and degraded by intense ultraviolet (UV) light (Kaya & Sano, 1998). Several studies have investigated the degradation by, for example, photocatalysis, H_2O_2/UV light and the photo-Fenton process (He et al., 2012; de Freitas et al., 2013; Pestana et al., 2015). These processes have relevance for water treatment and are discussed in Chapter 10, but are unlikely to contribute to degradation occurring in the natural environment.

In full sunlight, microcystins undergo only slow photochemical breakdown and isomerisation, with the reaction rate being enhanced by the presence of water-soluble cell pigments, presumably phycobiliproteins (Tsuji et al., 1994). In the presence of such pigments, the photochemical breakdown of microcystin in full sunlight can take as little as 2 weeks for greater than 90% breakdown, or longer than 6 weeks, depending on the concentration of pigment (and presumably toxin, although this has not been tested). A more rapid breakdown under sunlight has been reported in the presence of naturally occurring humic substances which can act as photosensitisers. In an experimental study, approximately 40% of the microcystins was degraded per day under summer conditions of insolation (Welker & Steinberg, 1999). However, since the penetration of active UV radiation is limited in deeper or turbid waters, the breakdown *in situ* is likely to be considerably slower (Welker & Steinberg, 2000). Photosensitised transformation of microcystins has been studied in detail by Song et al. (2007).

The chemical decomposition of nodularin has been studied less intensively; however, their structural similarity suggests similar characteristics as microcystin. Nodularin degradation has also been observed under UV radiation (Mazur-Marzec et al., 2006a).

2.1.6.3 Biodegradation

Microcystins are resistant to eukaryotic and many bacterial peptidases (Botes et al., 1982; Falconer et al., 1986; Harada, 1996); however, some human probiotic bacteria have microcystin-decomposing capability (Nybom et al., 2012).

In contrast, microcystins are susceptible to breakdown by a number of aquatic bacteria. These bacteria appear widespread and have been found in sewage effluent, lake water, lake sediment and river water worldwide (Holst et al., 2003; Edwards et al., 2008). MC-degrading bacteria have also been detected in the mucilage of *Microcystis* colonies (Maruyama et al., 2003). The majority of the microcystin-degrading bacteria isolated to date have been identified as *Sphingomonas* spp. and *Sphingopyxis* spp. belonging to the α-proteobacteria; further microcystin-degrading bacteria are also found among the ß-proteobacteria (e.g., *Pseudomonas aeruginosa*), γ-proteobacteria (e.g., *Paucibacter toxinivorans*), actinobacteria and bacilli (Edwards et al., 2008; Dziga et al., 2013; Li et al., 2017).

Most of these bacteria perform aerobic degradation of microcystins. The degradation pathway and *mir* genes encoding for the involved enzymes have been studied in detail in an isolate of *Sphingomonas* sp. (Bourne et al., 2001). The products of complete bacterial degradation were nontoxic to mice at i.p. doses 10 times higher than the LD_{50} of microcystin-LR. However, other intermediate breakdown products as well as the lack of *mir* genes in some MC-degrading bacteria suggest that multiple aerobic degradation pathways may exist (Amé et al., 2006; Edwards et al., 2008; Dziga et al., 2013). This also applies for possible anaerobic biodegradation, which, however, is far less intensively studied, as only a few bacteria showing anaerobic degradation have been isolated to date (Li et al., 2017).

Degradation of microcystins is often, though not always, characterised by an initial lag phase with little loss of microcystin. This phase was observed in laboratory and field experiments and can last from as little as 2 days to more than several weeks. The duration seems to depend on the previous bloom history of a lake and also on climatic conditions as well as on the concentration of dissolved microcystin (Christoffersen et al., 2002; Hyenstrand et al., 2003; Bourne et al., 2006; Edwards et al., 2008).

Once the biodegradation process commences, the removal of microcystin can be very fast with half-lives of 0.2–5 days for different microcystins, including MC-RR, MC-YR, MC-LR, MC-LW and MC-LF (Lam et al., 1995; Cousins et al., 1996; Park et al., 2001; Christoffersen et al., 2002; Hyenstrand et al., 2003; Ishii et al., 2004; Babica et al., 2005; Amé et al., 2006; Tsuji et al., 2006; Chen et al., 2008; Edwards et al., 2008). Degradation strongly depends on temperature, but is also influenced by the size of the microbial population and initial microcystin concentration (Park et al., 2001; Bourne et al., 2006). Though a more than 90% reduction of microcystin has been observed within a few days, low residual microcystin concentration can occasionally still be observed for weeks especially when initial concentrations were high (Jones et al., 1994b; Bourne et al., 2006).

For nodularin, the degradation by microbial activity was demonstrated in marine and freshwater environments (Heresztyn & Nicholson, 1997; Edwards et al., 2008; Toruńska et al., 2008). The linearisation of nodularin by a *Sphingomonas* strain was demonstrated suggesting a similar degradation pathway as for microcystins (Imanishi et al., 2005; Kato et al., 2007), and *Paucibacter toxinivorans* has also been shown to degrade nodularin (Rapala et al., 1997). Other *Sphingomonas* strains, however, could not degrade nodularin or nodularin-Har, or only in the presence of microcystin-RR (Jones et al., 1994a; Ishii et al., 2004).

REFERENCES

Aboal M, Puig MA (2005). Intracellular and dissolved microcystin in reservoirs of the river Segura basin, Murcia, SE Spain. Toxicon. 45:509–518.

Akçaalan R, Köker L, Gürevin C, Albay M (2014). *Planktothrix rubescens*: a perennial presence and toxicity in Lake Sapanca. Turkish J Botany. 38:782–789.

Akçaalan R, Mazur-Marzec H, Zalewska A, Albay M (2009). Phenotypic and toxicological characterization of toxic *Nodularia spumigena* from a freshwater lake in Turkey. Harmful Algae. 8:273–278.

Akçaalan R, Young FM, Metcalf JS, Morrison LF, Albay M, Codd GA (2006). Microcystin analysis in single filaments of *Planktothrix* spp. in laboratory cultures and environmental samples. Water Res. 40:1583–1590.

Amé MV, Ricardo EJ, Stephan P, Alberto WD (2006). Degradation of microcystin-RR by *Sphingomonas* sp. CBA4 isolated from San Roque reservoir (Córdoba–Argentina). Biodegradation. 17:447–455.

Azevedo SMFO, Evans WR, Carmichael WW, Namikoshi M (1994). First report of microcystins from a Brazilian isolate of the cyanobacterium *Microcystis aeruginosa*. J Appl Phycol. 6:261–265.

Babica P, Bláha L, Maršálek B (2005). Removal of microcystins by phototrophic biofilms. A microcosm study. Environ Sci Pollut Res. 12:369–374.

Ballot A, Krienitz L, Kotut K, Wiegand C, Pflugmacher S (2005). Cyanobacteria and cyanobacterial toxins in the alkaline crater lakes Sonachi and Simbi, Kenya. Harmful Algae. 4:139–150.

Bateman KP, Thibault P, Douglas DJ, White RL (1995). Mass spectral analysis of microcystins from toxic cyanobacteria using on-line chromatographic and electrophoretic separations. J Chromatogr A. 712:253–268.

Beattie KA, Kaya K, Codd GA (2000). The cyanobacterium *Nodularia* in PCC 7804, of freshwater origin, produces [L-Har(2)]nodularin. Phytochemistry. 54:57–61.

Bernard C, Ballot A, Thomazeau S, Maloufi S, Furey A, Mankiewicz-Boczek J et al. (2017). Cyanobacteria associated with the production of cyanotoxins. Appendix 2. In: Meriluoto J, Spoof L, Codd GA, editors: Handbook of cyanobacterial monitoring and cyanotoxin analysis. Chichester: John Wiley & Sons:501–525.

Beversdorf LJ, Weirich CA, Bartlett SL, Miller TR (2017). Variable cyanobacterial toxin and metabolite profiles across six eutrophic lakes of differing physiochemical characteristics. Toxins. 9:62.

Bigham DL, Hoyer MV, Canfield DE (2009). Survey of toxic algal (microcystin) distribution in Florida lakes. Lake Reserv Manage. 25:264–275.

Bláha L, Maršálek B (2003). Contamination of drinking water in the Czech Republic by microcystins. Arch Hydrobiol. 158:421–429.

Bláhová L, Babica P, Maršálková E, Maršálek B, Bláha L (2007). Concentrations and seasonal trends of extracellular microcystins in freshwaters of the Czech Republic–results of the national monitoring program. Soil, Air, Water. 35:348–354.

Bolch CJS, Orr PT, Jones GJ, Blackburn SI (1999). Genetic, morphological, and toxicological variation among globally distributed strains of *Nodularia* (Cyanobacteria). J Phycol. 35:339–355.

Botes D, Viljoen C, Kruger H, Wessels P, Williams D (1982). Structure of toxins of the blue-green-alga *Microcystis aeruginosa*. South African J Sci. 78:378–379.

Botes DP, Tuinman AA, Wessels PL, Viljoen CC, Kruger H (1984). The structure of cyanoginosin-LA, a cyclic heptapeptide toxin from the cyanobacterium *Microcystis aeruginosa*. J Chem Soc Perkin Trans. 1:2311–2318.

Botes DP, Wessels PL, Kruger H, Runnegar MTC, Satikarn S, Smith RJ et al. (1985). Structural studies on cyanoginosins-LR, YR, YA, and YM, peptide toxins from *Microcystis aeruginosa*. J Chem Soc Perkin Trans. 1:2747–2748.

Bouaïcha N, Miles CO, Beach DG, Labidi Z, Djabri A, Benayache NY et al. (2019). Structural diversity, characterization and toxicology of microcystins. Toxins. 11:714.

Bourne DG, Blakeley RL, Riddles P, Jones GJ (2006). Biodegradation of the cyanobacterial toxin microcystin LR in natural water and biologically active slow sand filters. Water Res. 40:1294–1302.

Bourne DG, Riddles P, Jones GJ, Smith W (2001). Characterisation of a gene cluster involved in bacterial degradation of the cyanobacterial toxin microcystin-LR. Environ Toxicol. 16:523–534.

Briand E, Gugger M, Francois JF, Bernard C, Humbert JF, Quiblier C (2008). Temporal variations in the dynamics of potentially microcystin-producing strains in a bloom-forming *Planktothrix agardhii* (cyanobacterium) population. Appl Environ Microbiol. 74:3839–3848.

Briand JF, Jacquet S, Flinois C, Avois-Jacquet C, Maisonnette C, Leberre B et al. (2005). Variations in the microcystin production of *Planktothrix rubescens* (Cyanobacteria) assessed from a four-year survey of Lac du Bourget (France) and from laboratory experiments. Microb Ecol. 50:418–428.

Buratti FM, Manganelli M, Vichi S, Stefanelli M, Scardala S, Testai E et al. (2017). Cyanotoxins: producing organisms, occurrence, toxicity, mechanism of action and human health toxicological risk evaluation. Arch Toxicol. 91:1049–1130.

Carmichael WW, Beasley V, Bunner DL, Eloff JN, Falconer IR, Gorham PR et al. (1988). Naming of cyclic heptapeptide toxins of cyanobacteria (blue-green algae). Toxicon 26:971–973.

Carmichael WW, Li RH (2006). Cyanobacteria toxins in the Salton Sea. Saline Syst. 2:5.

Carrasco D, Moreno E, Sanchis D, Wörmer L, Paniagua T, Del Cueto A et al. (2006). Cyanobacterial abundance and microcystin occurrence in Mediterranean water reservoirs in Central Spain: microcystins in the Madrid area. Europ J Phycol. 41:281–291.

Catherine A, Quiblier C, Yepremian C, Got P, Groleau A, Vincon-Leite B et al. (2008). Collapse of a *Planktothrix agardhii* perennial bloom and microcystin dynamics in response to reduced phosphate concentrations in a temperate lake. FEMS Microbiol Ecol. 65:61–73.

Cerasino L, Shams S, Boscaini A, Salmaso N (2016). Multiannual trend of microcystin production in the toxic cyanobacterium *Planktothrix rubescens* in Lake Garda (Italy). Chem Ecol. 32:492–506.

Chen J, Xie P (2005). Tissue distributions and seasonal dynamics of the hepatotoxic microcystins-LR and -RR in two freshwater shrimps, *Palaemon modestus* and *Macrobrachium nipponensis*, from a large shallow, eutrophic lake of the subtropical China. Toxicon. 45:615–625.

Chen W, Song LR, Peng L, Wan N, Zhang XM, Gan NQ (2008). Reduction in microcystin concentrations in large and shallow lakes: water and sediment-interface contributions. Water Res. 42:763–773.

Chia MA, Kwaghe MJ (2015). Microcystins contamination of surface water supply sources in Zaria-Nigeria. Environ Monit Assess. 187:606.

Chorus I, editors (2001). Cyanotoxins - occurrence, causes, consequences. Berlin: Springer. 357 pp.

Christiansen G, Fastner J, Erhard M, Börner T, Dittmann E (2003). Microcystin biosynthesis in *Planktothrix*: genes, evolution, and manipulation. J Bacteriol. 185:564–572.

Christiansen G, Kurmayer R, Liu Q, Börner T (2006). Transposons inactivate biosynthesis of the nonribosomal peptide microcystin in naturally occurring *Planktothrix* spp. Appl Environ Microbiol. 72:117–123.

Christiansen G, Molitor C, Philmus B, Kurmayer R (2008). Nontoxic strains of cyanobacteria are the result of major gene deletion events induced by a transposable element. Mol Biol Evol. 25:1695–1704.

Christoffersen K, Lyck S, Winding A (2002). Microbial activity and bacterial community structure during degradation of microcystins. Aquat Microb Ecol. 27:125–136.

Cirés S, Alvarez-Roa C, Wood SA, Puddick J, Loza V, Heimann K (2014). First report of microcystin-producing *Fischerella* sp. (Stigonematales, Cyanobacteria) in tropical Australia. Toxicon. 88:62–66.

Cousins IT, Bealing DJ, James HA, Sutton A (1996). Biodegradation of microcystin-LR by indigenous mixed bacterial populations. Water Res. 30:481–485.

Cuvin-Aralar ML, Fastner J, Focken U, Becker K, Aralar EV (2002). Microcystins in natural blooms and laboratory cultured *Microcystis aeruginosa* from Laguna de Bay, Philippines. Syst Appl Microbiol. 25:179–182.

Daly RI, Ho L, Brookes JD (2007). Effect of chlorination on *Microcystis aeruginosa* cell integrity and subsequent microcystin release and degradation. Environ Sci Technol. 41:4447–4453.

de Freitas AM, Sirtori C, Lenz CA, Zamora PGP (2013). Microcystin-LR degradation by solar photo-Fenton, UV-A/photo-Fenton and UV-C/H_2O_2: a comparative study. Photochem Photobiol Sci. 12:696–702.

Deblois CP, Juneau P (2010). Relationship between photosynthetic processes and microcystin in *Microcystis aeruginosa* grown under different photon irradiances. Harmful Algae. 9:18–24.

del Campo FF, Ouahid Y (2010). Identification of microcystins from three collection strains of *Microcystis aeruginosa*. Environ Pollut. 158:2906–2914.

Diehnelt CW, Dugan NR, Peterman SM, Budde WL (2006). Identification of microcystin toxins from a strain of *Microcystis aeruginosa* by liquid chromatography introduction into a hybrid linear ion trap-Fourier transform ion cyclotron resonance mass spectrometer. Anal Chem 78:501–512.

Dittmann E, Börner T (2005). Genetic contributions to the risk assessment of microcystin in the environment. Toxicol Appl Pharmacol. 203:192–200.

Dziga D, Wasylewski M, Wladyka B, Nybom S, Meriluoto J (2013). Microbial degradation of microcystins. Chem Res Toxicol. 26:841–852.

Edwards C, Graham D, Fowler N, Lawton LA (2008). Biodegradation of microcystins and nodularin in freshwaters. Chemosphere. 73:1315–1321.

Faassen EJ, Lürling M (2013). Occurrence of the microcystins MC-LW and MC-LF in Dutch surface waters and their contribution to total microcystin toxicity. Mar Drugs. 11:2643–2654.

Falconer IR, Buckley T, Runnegar MT (1986). Biological half-life, organ distribution and excretion of 125I-labelled toxic peptide from the blue-green alga *Microcystis aeruginosa*. Aust J Biol Sci. 39:17–22.

Fastner J, Erhard M, Carmichael WW, Sun F, Rinehart KL, Rönicke H et al. (1999). Characterization and diversity of microcystins in natural blooms and strains of the genera *Microcystis* and *Planktothrix* from German freshwaters. Arch Hydrobiol. 145:147–163.

Fastner J, Wirsing B, Wiedner C, Heinze R, Neumann U, Chorus I (2001). Cyanotoxin occurrence in Germany: microcystins and hepatocyte toxicity. In: Chorus I, editors: Cyanotoxins: occurrence, causes, consequences. Berlin: Springer:22–37.

Fawell JK, James CP, James HA (1994). Toxins from blue-green algae: toxicological assessment of microcystin-LR and a method for its determination in water. Murlow: Foundation for Water Research:46 pp.

Fawell JK, Mitchell RE, Everett DJ, Hill RE (1999). The toxicity of cyanobacterial toxins in the mouse: I Microcystin-LR. Hum Exp Toxicol. 18:162–167.

Fetscher AE, Howard MDA, Stancheva R, Kudela RM, Stein ED, Sutula MA et al. (2015). Wadeable streams as widespread sources of benthic cyanotoxins in California, USA. Harmful Algae. 49:105–116.

Fewer DP, Tooming-Klunderud A, Jokela J, Wahsten M, Rouhiainen L, Kristensen T et al. (2008). Natural occurrence of microcystin synthetase deletion mutants capable of producing microcystins in strains of the genus *Anabaena* (Cyanobacteria). Microbiology. 154:1007–1014.

Fewer DP, Wahlsten M, Österholm J, Jokela J, Rouhiainen L, Kaasalainen U et al. (2013). The genetic basis for O-acetylation of the microcystin toxin in cyanobacteria. Chem Biol. 20:861–869.

Freitas de Magalhães V, Soares RM, Azevedo SM (2001). Microcystin contamination in fish from the Jacarepaguá Lagoon (Rio de Janeiro, Brazil): ecological implication and human health risk. Toxicon. 39:1077–1085.

Gehringer MM, Adler L, Roberts AA, Moffitt MC, Mihali TK, Mills TJ et al. (2012). Nodularin, a cyanobacterial toxin, is synthesized *in planta* by symbiotic *Nostoc* sp. ISME J. 6:1834.

Gkelis S, Harjunpää V, Lanaras T, Sivonen K (2005). Diversity of hepatotoxic microcystins and bioactive anabaenopeptins in cyanobacterial blooms from Greek freshwaters. Environ Toxicol. 20:249–256.

Gkelis S, Zaoutsos N (2014). Cyanotoxin occurrence and potentially toxin producing cyanobacteria in freshwaters of Greece: a multi-disciplinary approach. Toxicon. 78:1–9.

Graham JL, Loftin KA, Meyer MT, Ziegler AC (2010). Cyanotoxin mixtures and taste-and-odor compounds in cyanobacterial blooms from the midwestern United States. Environ Sci Technol. 44:7361–7368.

Halinen K, Jokela J, Fewer DP, Wahsten M, Sivonen K (2007). Direct evidence for production of microcystins by *Anabaena* strains from the Baltic sea. Appl Environ Microbiol. 73:6543–6550.

Harada K-I (1996). Chemistry and detection of microcystins. In: Watanabe MF, Harada K-I, Carmichael WW et al., editors: Toxic *Microcystis*, first edition. Boca Raton (FL): CRC press:103–148.

Harada K-I, Matsuura K, Suzuki M, Watanabe MF, Oishi S, Dahlem AM et al. (1990). Isolation and characterization of the minor components associated with microcystin LR and RR in the cyanobacterium (blue- green-algae). Toxicon. 28:55–64.

Harada K-I, Ogawa K, Kimura Y, Murata H, Suzuki M, Thorn PM et al. (1991a). Microcystins from *Anabaena flos-aquae* NRC 525-17. Chem Res Toxicol. 4:535–540.

Harada K-I, Ogawa K, Matsuura K, Nagai H, Murata H, Suzuki M et al. (1991b). Isolation of two toxic heptapeptide microcystins from an axenic strain of *Microcystis aeruginosa*, K-139. Toxicon. 29:479–489.

Harke MJ, Gobler CJ (2013). Global transcriptional responses of the toxic cyanobacterium, *Microcystis aeruginosa*, to nitrogen stress, phosphorus stress, and growth on organic matter. PLoS One 8:e69834.

Harke MJ, Gobler CJ (2015). Daily transcriptome changes reveal the role of nitrogen in controlling microcystin synthesis and nutrient transport in the toxic cyanobacterium, *Microcystis aeruginosa*. BMC Genomics. 16:1068.

Harke MJ, Steffen MM, Gobler CJ, Otten TG, Wilhelm SW, Wood SA et al. (2016). A review of the global ecology, genomics, and biogeography of the toxic cyanobacterium, *Microcystis* spp. Harmful Algae. 54:4–20.

Haruštiaková D, Welker M (2017). Chemotype diversity in *Planktothrix rubescens* (cyanobacteria) populations is correlated to lake depth. Environ Microbiol Rep. 9:158–168.

He X, Pelaez M, Westrick JA, O'Shea KE, Hiskia A, Triantis T et al. (2012). Efficient removal of microcystin-LR by UV-C/H_2O_2 in synthetic and natural water samples. Water Res. 46:1501–1510.

Heinze R (1999). Toxicity of the cyanobacterial toxin microcystin-LR to rats after 28 days intake with the drinking water. Environ Toxicol. 14:57–60.

Heresztyn T, Nicholson BC (1997). Nodularin concentrations in Lakes Alexandrina and Albert, South Australia, during a bloom of the cyanobacterium (blue- green alga) *Nodularia spumigena* and degradation of the toxin. Environ Toxicol Wat Qual. 12:273–282.

Hesse K, Kohl J-G (2001). Effects of light and nutrients supply on growth and microcystin content of different strains of *Microcystis aeruginosa*. In: Chorus I, editors: Cyanotoxins: occurrence, causes, consequences. Berlin: Springer:104–115.

Hitzfeld BC, Lampert CS, Spaeth N, Mountfort D, Kaspar H, Dietrich DR (2000). Toxin production in cyanobacterial mats from ponds on the McMurdo Ice Shelf, Antarctica. Toxicon. 38:1731–1748.

Holst T, Jørgensen NO, Jørgensen C, Johansen A (2003). Degradation of microcystin in sediments at oxic and anoxic, denitrifying conditions. Water Res. 37:4748–4760.

Hyenstrand P, Rohrlack T, Beattie KA, Metcalf JS, Codd GA, Christoffersen K (2003). Laboratory studies of dissolved radiolabelled microcystin-LR in lake water. Water Res. 37:3299–3306.

IARC (2010). Ingested nitrate and nitrite, and cyanobacterial peptide toxins. Geneva: International Agency for Research on Cancer.

Ibelings BW, Havens KE (2008). Cyanobacterial toxins: a qualitative meta–analysis of concentrations, dosage and effects in freshwater, estuarine and marine biota. In: Hudnell HK, editors: Cyanobacterial harmful algal blooms: state of the science and research needs. New York: Springer:675–732.

Imanishi S, Kato H, Mizuno M, Tsuji K, Harada K-I (2005). Bacterial degradation of microcystins and nodularin. Chem Res Toxicol. 18:591–598.

Ishii H, Nishijima M, Abe T (2004). Characterization of degradation process of cyanobacterial hepatotoxins by a gram-negative aerobic bacterium. Water Res. 38:2667–2676.

Izaguirre G, Jungblut AD, Neilan BA (2007). Benthic cyanobacteria (Oscillatoriaceae) that produce microcystin-LR, isolated from four reservoirs in southern California. Water Res. 41:492–498.

Jacquet S, Briand J-F, Leboulanger C, Avois-Jacquet C, Oberhaus L, Tassin B et al. (2005). The proliferation of the toxic cyanobacterium *Planktothrix rubescens* following restoration of the largest natural French lake (Lac du Bourget). Harmful Algae. 4:651–672.

Jähnichen S, Ihle T, Petzoldt T, Benndorf J (2007). Impact of inorganic carbon availability on microcystin production by *Microcystis aeruginosa* PCC 7806. Appl Environ Microbiol. 73:6994–7002.

Jähnichen S, Long BM, Petzoldt T (2011). Microcystin production by Microcystis aeruginosa: direct regulation by multiple environmental factors. Harmful Algae. 12:95–104.

Jones GJ, Blackburn SI, Parker NS (1994a). A toxic bloom of *Nodularia spumigena* MERTENS in Orielton Lagoon, Tasmania. Aust J Mar Freshwater Res. 45: 787–800.

Jones GJ, Bourne DG, Blakeley RL, Doelle H (1994b). Degradation of cyanobacterial hepatotoxin microcystin by aquatic bacteria. Nat Toxins. 2:228–235.

Jones GJ, Falconer IR, Wilkins RM (1995). Persistence of cyclic peptide toxins in dried *Microcystis aeruginosa* crusts from Lake Mokoan, Australia. Environ Toxicol Wat Qual. 10:19–24.

Jones GJ, Orr PT (1994). Release and degradation of microcystin following algicide treatment of a *Microcystis aeruginosa* bloom in a recreational lake, as determined by HPLC and protein phosphatase inhibition assay. Water Res. 28:871–876.

Jungblut A-D, Hoeger SJ, Mountfort D, Hitzfeld BC, Dietrich DR, Neilan BA (2006). Characterization of microcystin production in an Antarctic cyanobacterial mat community. Toxicon. 47:271–278.

Kaebernick M, Dittmann E, Börner T, Neilan BA (2002). Multiple alternate transcripts direct the biosynthesis of microcystin, a cyanobacterial nonribosomal peptide. Appl Environ Microbiol. 68:449–455.

Kardinaal WEA, Visser PM (2005). Dynamics of cyanobacterial toxins: sources of variability in microcystin concentrations. In: Huisman J, Matthijs HCP, Visser PM, editors: Harmful cyanobacteria. Berlin: Springer:41–63.

Kato H, Imanishi SY, Tsuji K, Harada K-I (2007). Microbial degradation of cyanobacterial cyclic peptides. Water Res. 41:1754–1762.

Kaya K, Sano T (1998). A photodetoxification mechanism of the cyanobacterial hepatotoxin microcystin-LR by ultraviolet irradiation. Chem Res Toxicol. 11:159–163.

Kemp A, John J (2006). Microcystins associated with *Microcystis* dominated blooms in the southwest wetlands, Western Australia. Environ Toxicol. 21:125–130.

Kenefick SL, Hrudey SE, Peterson HG, Prepas EE (1993). Toxin release from *Microcystis aeruginosa* after chemical treatment. Wat Sci Technol. 27:433–440.

Kiviranta J, Namikoshi M, Sivonen K, Evans WR, Carmichael WW, Rinehart KL (1992). Structure determination and toxicity of a new microcystin from *Microcystis aeruginosa* strain 205. Toxicon. 30:1093–1098.

Kleinkauf H, Döhren H (1996). A nonribosomal system of peptide biosynthesis. FEBS J. 236:335–351.

Kononen K, Sivonen K, Lehtimaki J (1993). Toxicity of phytoplankton blooms in the Gulf of Finland and Gulf of Bothnia, Baltic Sea. In: Smayda TJ, Shimizu Y, editors: Toxic phytoplankton blooms in the sea. Amsterdam: Elsevier Scientific Publications:269–274.

Kosol S, Schmidt J, Kurmayer R (2009). Variation in peptide net production and growth among strains of the toxic cyanobacterium *Planktothrix* spp. Europ J Phycol. 44:49–62.

Kotak BG, Zurawell RW, Prepas EE, Holmes CFB (1996). Microcystin-LR concentration in aquatic food web compartments from lakes of varying trophic status. Can J Fish Aquat Sci. 53:1974–1985.

Kotut K, Ballot A, Krienitz L (2006). Toxic cyanobacteria and their toxins in standing waters of Kenya: implications for water resource use. J Water Health. 4:233–245.

Krishnamurthy T, Carmichael W, Sarver E (1986). Toxic peptides from freshwater cyanobacteria (blue-green algae). I. Isolation, purification and characterization of peptides from *Microcystis aeruginosa* and *Anabaena flos-aquae*. Toxicon 24:865–873.

Krishnamurthy T, Szafraniec L, Hunt DF, Shabanowitz J, Yates JR, Hauer CR et al. (1989). Structural characterization of toxic cyclic peptides from blue-green algae by tandem mass spectrometry. Proc Natl Acad Sci USA. 86:770–774.

Kurmayer R, Christiansen G, Fastner J, Börner T (2004). Abundance of active and inactive microcystin genotypes in populations of the toxic cyanobacterium *Planktothrix* spp. Environ Microbiol. 6:831–841.

Kurmayer R, Christiansen G, Gumpenberger M, Fastner J (2005). Genetic identification of microcystin ecotypes in toxic cyanobacteria of the genus *Planktothrix*. Microbiology. 151:1525–1533.

Kusumi T, Ooi T, Watanabe MM, Takahashi H, Kakisawa H (1987). Cyanoviridin RR, a toxin from the cyanobacterium (blue-green alga) *Microcystis viridis*. Tetrahedron Lett. 28:4695–4698.

Lam AKY, Fedorak PM, Prepas EE (1995). Biotransformation of the cyanobacterial hepatotoxin microcystin- LR, as determined by HPLC and protein phosphatase bioassay. Environ Technol. 29:242–246.

Lee TH, Chen YM, Chou HN (1998). First report of microcystins in Taiwan. Toxicon. 36:247–255.

Lehtimaki J, Moisander P, Sivonen K, Kononen K (1997). Growth, nitrogen fixation, and nodularin production by two Baltic sea cyanobacteria. Appl Environ Microbiol. 63:1647–1656.

Lehtimäki J, Sivonen K, Luukkainen R, Niemelä SI (1994). The effects of incubation time, temperature, light, salinity, and phosphorous on growth and hepatotoxin production by *Nodularia* strains. Arch Hydrobiol. 130:269–282.

Li J, Li R, Li J (2017). Current research scenario for microcystins biodegradation–A review on fundamental knowledge, application prospects and challenges. Sci Tot Environ. 595:615–632.

Lombardo M, Pinto FCR, Vieira JMS, Honda RY, Pimenta AM, Bemquerer MP et al. (2006). Isolation and structural characterization of microcystin-LR and three minor oligopeptides simultaneously produced by *Radiocystis feernandoi* (Chroococcales, Cyanobacteriae): a Brazilian toxic cyanobacterium. Toxicon. 47:560–566.

Luukkainen R, Namikoshi M, Sivonen K, Rinehart KL, Niemelä SI (1994). Isolation and identification of 12 microcystins from four strains and two bloom samples of *Microcystis* spp.: structure of a new hepatotoxin. Toxicon. 32: 133–139.

Machado J, Campos A, Vasconcelos V, Freitas M (2017). Effects of microcystin-LR and cylindrospermopsin on plant-soil systems: a review of their relevance for agricultural plant quality and public health. Environ Res. 153:191–204.

Mankiewicz-Boczek J, Gągała I, Kokociński M, Jurczak T, Stefaniak K (2011). Perennial toxigenic *Planktothrix agardhii* bloom in selected lakes of Western Poland. Environ Toxicol. 26:10–20.

Maruyama T, Kato K, Yokoyama A, Tanaka T, Hiraishi A, Park H-D (2003). Dynamics of microcystin-degrading bacteria in mucilage of *Microcystis*. Microb Ecol. 46:279–288.

Masango MG, Myburgh JG, Labuschagne L, Govender D, Bengis RG, Naicker D (2010). Assessment of *Microcystis* bloom toxicity associated with wildlife mortality in the Kruger National Park, South Africa. J Wildlife Diseases. 46:95–102.

Mazur H, Plinski M (2003). *Nodularia spumigena* blooms and the occurrence of hepatotoxin in the Gulf of Gdańsk. Oceanologia. 45:305–316.

Mazur-Marzec H, Meriluoto J, Pliński M (2006a). The degradation of the cyanobacterial hepatotoxin nodularin (NOD) by UV radiation. Chemosphere. 65:1388–1395.

Mazur-Marzec H, Meriluoto J, Plinski M, Szafranek J (2006b). Characterization of nodularin variants in *Nodularia spumigena* from the Baltic Sea using liquid chromatography/mass spectrometry/mass spectrometry. Rapid Commun Mass Spectrom. 20:2023–2032.

McGregor GB, Sendall BC (2017). *Iningainema pulvinus* gen nov., sp nov.(Cyanobacteria, Scytonemataceae) a new nodularin producer from Edgbaston Reserve, north-eastern Australia. Harmful Algae. 62:10–19.

McGregor GB, Stewart I, Sendall BC, Sadler R, Reardon K, Carter S et al. (2012). First report of a toxic *Nodularia spumigena* (Nostocales/Cyanobacteria) bloom in sub-tropical Australia. I. Phycological and public health investigations. Int J Environ Res Public Health. 9:2396–2411.

Meissner S, Fastner J, Dittmann E (2013). Microcystin production revisited: conjugate formation makes a major contribution. Environ Microbiol. 15:1810–1820.

Meriluoto JAO, Sandström A, Eriksson JE, Remaud G, Grey Craig A, Chattopadhyaya J (1989). Structure and toxicity of a peptide hepatotoxin from the cyanobacterium *Oscillatoria agardhii*. Toxicon. 27:1021–1034.

Metcalf JS, Richer R, Cox PA, Codd GA (2012). Cyanotoxins in desert environments may present a risk to human health. Sci Tot Environ. 421:118–123.

Moffitt MC, Neilan BA (2004). Characterization of the nodularin synthetase gene cluster and proposed theory of the evolution of cyanobacterial hepatotoxins. Appl Environ Microbiol. 70:6353–6362.

Mohamed ZA, Al Shehri AM (2009). Microcystin-producing blooms of *Anabaenopsis arnoldi* in a potable mountain lake in Saudi Arabia. FEMS Microbiol Ecol. 69:98–105.

Mohamed ZA, Al Shehri AM (2010). Microcystin production in epiphytic cyanobacteria on submerged macrophytes. Toxicon. 55:1346–1352.

Mohamed ZA, El Sharouny HM, Ali WSM (2006). Microcystin production in benthic mats of cyanobacteria in the Nile River and irrigation canals, Egypt. Toxicon. 47:584–590.

Mowe MA, Mitrovic SM, Lim RP, Furey A, Yeo DC (2015). Tropical cyanobacterial blooms: a review of prevalence, problem taxa, toxins and influencing environmental factors. J Limnol. 74: 205-224.

Namikoshi M, Choi BW, Sakai R, Sun F, Rinehart KL, Evans WR et al. (1994). New nodularins: a general method for structure assignment. J Org Chem. 59:2349–2357.

Namikoshi M, Rinehart KL, Sakai R, Stotts RR, Dahlem AM, Beasley CR et al. (1992a). Identification of 12 hepatotoxins from Homer lake bloom of the cyanobacterium *Microcystis aeruginosa*, *Microcystis viridis*, and *Microcystis wesenbergii*: nine new microcystins. J Org Chem. 57:866–872.

Namikoshi M, Sivonen K, Evans WR, Carmichael WW, Rouhiainen L, Luukkainen R et al. (1992b). Structures of three new homotyrosine-containing microcystins and a new homophenylalanine variant from *Anabaena* sp. strain 66. Chem Res Toxicol. 5:661–666.

Namikoshi M, Sivonen K, Evans WR, Carmichael WW, Sun F, Rouhiainen L et al. (1992c). Two new L-Serine variants of microcystin-LR and -RR from *Anabaena* sp. strains 202 A1 and 202 A2. Toxicon 30:1457–1464.

Namikoshi M, Sivonen K, Evans WR, Sun F, Carmichael WW, Rinehart KL (1992d). Isolation and structures of microcystins from a cyanobacterial water bloom (Finland). Toxicon. 30:1473–1479.

Naselli-Flores L, Barone R, Chorus I, Kurmayer R (2007). Toxic cyanobacterial blooms in reservoirs under a semiarid mediterranean climate: the magnification of a problem. Environ Toxicol. 22:399–404.

Nasri H, Bouaicha N, Harche MK (2007). A new morphospecies of *Microcystis* sp. forming bloom in the Cheffia dam (Algeria): seasonal variation of microcystin concentrations in raw water and their removal in a full-scale treatment plant. Environ Toxicol. 22:347–356.

Nguyen LTT, Cronberg G, Annadotter H, Larsen J (2007). Planktic cyanobacteria from freshwater localities in ThuaThien-Hue province, Vietnam. II. Algal biomass and microcystin production. Nova Hedwigia. 85:35–49.

Nybom S, Dziga D, Heikkilä J, Kull T, Salminen S, Meriluoto J (2012). Characterization of microcystin-LR removal process in the presence of probiotic bacteria. Toxicon. 59:171–181.

Oksanen I, Jokela J, Fewer D, Wahlsten M, Rikkinen J, Sivonen K (2004). Discovery of rare and highly toxic microcystins from lichen-associated cyanobacterium *Nostoc* sp. strain IO-102-I. Appl Environ Microbiol 70:5756–5763.

Orr PT, Jones GJ (1998). Relationship between microcystin production and cell division rates in nitrogen-limited *Microcystis aeruginosa* cultures. Limnol Oceanogr. 43:1604–1614.

Park HD, Sasaki Y, Maruyama T, Yanagisawa E, Hiraishi A, Kato K (2001). Degradation of the cyanobacterial hepatotoxin microcystin by a new bacterium isolated from a hypertrophic lake. Environ Toxicol. 16:337–343.

Pattanaik B, Wulff A, Roleda MY, Garde K, Mohlin M (2010). Production of the cyanotoxin nodularin - a multifactorial approach. Harmful Algae. 10:30–38.

Pawlik-Skowronska B, Pirszel J, Kornijow R (2008). Spatial and temporal variation in microcystin concentrations during perennial bloom of *Planktothrix agardhii* in a hypertrophic lake. Ann Limnol. 44:145–150.

Pearson LA, Dittmann E, Mazmouz R, Ongley SE, D'Agostino PM, Neilan BA (2016). The genetics, biosynthesis and regulation of toxic specialized metabolites of cyanobacteria. Harmful Algae. 54:98–111.

Pestana CJ, Edwards C, Prabhu R, Robertson PK, Lawton LA (2015). Photocatalytic degradation of eleven microcystin variants and nodularin by TiO_2 coated glass microspheres. J Hazard Mat. 300:347–353.

Pietsch J, Bornmann K, Schmidt W (2002). Relevance of intra-and extracellular cyanotoxins for drinking water treatment. CLEAN–Soil Air Water. 30:7–15.

Prinsep MR, Caplan FR, Moore RE, Patterson GML, Honkanen RE, Boynton AL (1992). Microcystin-LA from a blue-green alga belonging to the Stigonematales. Phytochemistry. 31:4.

Puddick J, Prinsep MR, Wood SA, Kaufononga SA, Cary SC, Hamilton DP (2014). High levels of structural diversity observed in microcystins from *Microcystis*

CAWBG11 and characterization of six new microcystin congeners. Mar Drugs. 12:5372–5395.

Quiblier C, Wood SA, Echenique-Subiabre I, Heath M, Villeneuve A, Humbert JF (2013). A review of current knowledge on toxic benthic freshwater cyanobacteria–ecology, toxin production and risk management. Water Res. 47:5464–5479.

Rantala A, Fewer D, Hisbergues M, Rouhiainen L, Vaitomaa J, Börner T et al. (2004). Phylogenetic evidence for the early evolution of microcystin synthesis. Proc Natl Acad Sci USA. 101:568–573.

Rapala J, Sivonen K, Lyra C, Niemelä SI (1997). Variation of microcystins, cyanobacterial hepatotoxins, in *Anabaena* spp. as a function of growth stimuli. Appl Environ Microbiol. 63:2206–2212.

Rinehart KL, Harada K, Namikoshi M, Chen C, Harvis CA, Munro MH et al. (1988). Nodularin, microcystin, and the configuration of Adda. J Am Chem Soc. 110:8557–8558.

Ríos V, Moreno I, Prieto A, Soria-Díaz ME, Frías J, Cameán A (2014). Comparison of *Microcystis aeruginosa* (PCC7820 and PCC7806) growth and intracellular microcystins content determined by liquid chromatography–mass spectrometry, enzyme-linked immunosorbent assay anti-Adda and phosphatase bioassay. J Water Health. 12:69–80.

Rodríguez E, Onstad GD, Kull TP, Metcalf JS, Acero JL, von Gunten U (2007). Oxidative elimination of cyanotoxins: comparison of ozone, chlorine, chlorine dioxide and permanganate. Water Res. 41:3381–3393.

Rohrlack T, Henning M, Kohl J-G (2001). Isolation and characterization of colony-forming *Microcystis aeruginosa* strains. In: Chorus I, editors: Cyanotoxins: occurrence, causes, consequences. Berlin: Springer:152–158.

Rohrlack T, Hyenstrand P (2007). Fate of intracellular microcystins in the cyanobacterium *Microcystis aeruginosa* (Chroococcales, Cyanophyceae). Phycologia. 46:277–283.

Rositano J, Newcombe G, Nicholson B, Sztajnbok P (2001). Ozonation of NOM and algal toxins in four treated waters. Water Res. 35:23–32.

Rouhiainen L, Vakkilainen T, Siemer BL, Buikema W, Haselkorn R, Sivonen K (2004). Genes coding for hepatotoxic heptapeptides (microcystins) in the cyanobacterium *Anabaena* strain 90. Appl Environ Microbiol. 70:686–692.

Sahindokuyucu-Kocasari F, Gulle I, Kocasari S, Pekkaya S, Mor F (2015). The occurrence and levels of cyanotoxin nodularin from *Nodularia spumigena* in the alkaline and salty Lake Burdur, Turkey. J Limnol. 74: 530-536.

Sakai H, Hao A, Iseri Y, Wang S, Kuba T, Zhang Z et al. (2013). Occurrence and distribution of microcystins in Lake Taihu, China. Sci World J. 2013:7.

Sano T, Kaya K (1995). A 2-amino-2-butenoic acid (Dhb)-containing microcystin isolated from *Oscillatoria agardhii*. Tetrahedron Lett. 36:8603–8606.

Sano T, Kaya K (1998). Two new (E)-2-amino-2-butenoic acid (Dhb)-containing microcystins isolated from *Oscillatoria agardhii*. Tetrahedron. 54:463–470.

Schmidt W, Willmitzer H, Bornmann K, Pietsch J (2002). Production of drinking water from raw water containing cyanobacteria - Pilot plant studies for assessing the risk of microcystin breakthrough. Environ Toxicol. 17:375–385.

Sevilla E, Martin-Luna B, Vela L, Bes MT, Peleato ML, Fillat MF (2010). Microcystin-LR synthesis as response to nitrogen: transcriptional analysis of the *mcy*D gene in *Microcystis aeruginosa* PCC7806. Ecotoxicology. 19:1167–1173.

Shih PM, Wu D, Latifi A, Axen SD, Fewer DP, Talla E et al. (2013). Improving the coverage of the cyanobacterial phylum using diversity-driven genome sequencing. Proc Natl Acad Sci USA. 110:1053–1058.

Sipiä V, Kankaanpää H, Pflugmacher S, Flinkman J, Furey A, James K (2002). Bioaccumulation and detoxication of nodularin in tissues of flounder (*Platichthys flesus*), mussels (*Mytilus edulis, Dreissena polymorpha*), and clams (*Macoma balthica*) from the northern Baltic Sea. Ecotoxicol Environ Safety. 53:305–311.

Sitoki L, Kurmayer R, Rott E (2012). Spatial variation of phytoplankton composition, biovolume, and resulting microcystin concentrations in the Nyanza Gulf (Lake Victoria, Kenya). Hydrobiologia. 691:109–122.

Sivonen K, Jones GJ (1999). Cyanobacterial toxins. In: Chorus I, Bartram J, editors: Toxic cyanobacteria in water. London: E & FN Spoon:41–111.

Sivonen K, Kononen K, Carmichael WW, Dahlem AM, Rinehart KL, Kirivanta J et al. (1989). Occurrence of the hepatotoxic cyanobacterium *Nodularia spumigena* in the Baltic Sea and structure of the toxin. Appl Environ Microbiol. 55:1990–1995.

Sivonen K, Namikoshi M, Evans WR, Carmichael WW, Sun F, Rouhiainen L et al. (1992a). Isolation and characterization of a variety of microcystins from seven strains of the cyanobacterial genus *Anabaena*. Appl Environ Microbiol. 58: 2495–2500.

Sivonen K, Namikoshi M, Evans WR, Gromov BV, Carmichael WW, Rinehart KL (1992b). Isolation and structures of five microcystins from a Russian *Microcystis aeruginosa* strain CALU 972. Toxicon. 30:1481–1485.

Sivonen K, Namikoshi M, Luukkainen R, Fardig M, Rouhiainen L, Evans W et al. (1995). Variation of cyanobacterial hepatotoxins in Finland. In: Munawar M, Luotola M, editors: The contaminants in the Nordic ecosystem, dynamics, processes and fate. Amsterdam: SPB Academic Publishing:163–169.

Song W, Bardowell S, O'Shea KE (2007). Mechanistic study and the influence of oxygen on the photosensitized transformations of microcystins (cyanotoxins). Environ Sci Technol. 41:5336–5341.

Spoof L, Catherine A (2017). Table of microcystins and nodularins. In: Meriluoto J, Spoof L, Codd GA et al., editors: Handbook of cyanobacterial monitoring and cyanotoxin analysis. Chichester: John Wiley & Sons:526–537.

Stoner RD, Adams WH, Slatkin DN, Siegelman HW (1989). The effects of single L-amino acid substitutions on the lethal potencies of the microcystins. Toxicon. 27:825–828.

Su X, Xue Q, Steinman AD, Zhao Y, Xie L (2015). Spatiotemporal dynamics of microcystin variants and relationships with environmental parameters in Lake Taihu, China. Toxins. 7:3224–3244.

Tillett D, Dittmann E, Erhard M, von Döhren H, Börner T, Neilan BA (2000). Structural organization of microcystin biosynthesis in *Microcystis aeruginosa* PCC7806: an integrated peptide-polyketide synthetase system. Chem Biol. 7:753–764.

Tonk L, Visser PM, Christiansen G, Dittmann E, Snelder EOFM, Wiedner C et al. (2005). The microcystin composition of the cyanobacterium *Planktothrix agardhii* changes toward a more toxic variant with increasing light intensity. Appl Environ Microbiol. 71:5177–5181.

Tooming-Klunderud A, Mikalsen B, Kristensen T, Jakobsen KS (2008). The mosaic structure of the *mcyABC* operon in *Microcystis*. Microbiology. 154:1886–1899.

Toruńska A, Bolałek J, Pliński M, Mazur-Marzec H (2008). Biodegradation and sorption of nodularin (NOD) in fine-grained sediments. Chemosphere. 70: 2039–2046.

Tsuji K, Asakawa M, Anzai Y, Sumino T, Harada K-I (2006). Degradation of microcystins using immobilized microorganism isolated in an eutrophic lake. Chemosphere. 65:117–124.

Tsuji K, Naito S, Kondo F, Ishikawa N, Watanabe MF, Suzuki M et al. (1994). Stability of microcystins from cyanobacteria: effect of light on decomposition and isomerization. Environ Sci Technol. 28:173–177.

Van de Waal DB, Verspagen JM, Lürling M, Van Donk E, Visser PM, Huisman J (2009). The ecological stoichiometry of toxins produced by harmful cyanobacteria: an experimental test of the carbon-nutrient balance hypothesis. Ecol Lett. 12:1326–1335.

Vasconcelos V (2001). Freshwater cyanobacteria and their toxins in Portugal. In: Chorus I, editors: Cyanotoxins: occurrence, causes, consequences. Berlin: Springer:62–67.

Vezie C, Brient L, Sivonen K, Bertru G, Lefeuvre JC, SalkinojaSalonen M (1998). Variation of microcystin content of cyanobacterial blooms and isolated strains in Lake Grand-Lieu (France). Microb Ecol. 35:126–135.

Via-Ordorika L, Fastner J, Kurmayer R, Hisbergues M, Dittmann E, Komárek J et al. (2004). Distribution of microcystin-producing and non-microcystin-producing *Microcystis* sp in European freshwater bodies: detection of microcystins and microcystin genes in individual colonies. Syst Appl Microbiol. 27:592–602.

Waajen GW, Faassen EJ, Lürling M (2014). Eutrophic urban ponds suffer from cyanobacterial blooms: Dutch examples. Environ Sci Pollut Res. 21: 9983–9994.

Watanabe MF, Watanabe M, Kato T, Harada K-I, Suzuki M (1991). Composition of cyclic peptide toxins among strains of *Microcystis aeruginosa* (blue-green algae, cyanobacteria). The Botanical Magazine (Shokubutsu Gaku Zasshi). 104:49–57.

Welker M, Christiansen G, von Döhren H (2004). Diversity of coexisting *Planktothrix* (Cyanobacteria) chemotypes deduced by mass spectral analysis of microcystins and other oligopeptides. Arch Microbiol. 182:288–298.

Welker M, Steinberg C (1999). Indirect photolysis of cyanotoxins: one possible mechanism for their low persistence. Water Res. 33:1159–1164.

Welker M, Steinberg C (2000). Rates of humic substance photosensitized degradation of microcystin-LR in natural waters. Environ Sci Technol. 34:3415–3419.

Welker M, Steinberg C, Jones GJ (2001). Release and persistence of microcystins in natural waters. In: Chorus I, editors: Cyanotoxins: occurrence, causes, consequences. Berlin: Springer:85–103.

Welker M, von Döhren H, Täuscher H, Steinberg CEW, Erhard M (2003). Toxic *Microcystis* in shallow lake Müggelsee (Germany) - dynamics, distribution, diversity. Arch Hydrobiol. 157:227–248.

WHO (2003a). Cyanobacterial toxins: Microcystin-LR in Drinking-water. Background document for development of WHO Guidelines for Drinking-water Quality. Geneva: World Health Organization:18 pp. http://www.who.int/water_sanitation_health/dwq/chemicals/cyanobactoxins.pdf?ua=1

WHO (2003b). Guidelines for safe recreational water environments. Vol. 1: Coastal and fresh waters. Geneva: World Health Organization. https://apps.who.int/iris/handle/10665/42591

WHO (2020). Cyanobacterial toxins: Microcystins. Background document for development of WHO Guidelines for Drinking-water Quality and Guidelines for Safe Recreational Water Environments. Geneva: World Health Organization. https://apps.who.int/iris/handle/10665/338066

Wiedner C, Nixdorf B, Heinze R, Wirsing B, Neumann U, Weckesser J (2002). Regulation of cyanobacteria and microcystin dynamics in polymictic shallow lakes. Arch Hydrobiol 155:383–400.

Wiedner C, Visser PM, Fastner J, Metcalf JS, Codd GA, Mur LR (2003). Effects of light on the microcystin content of *Microcystis* strain PCC 7806. Appl Environ Microbiol. 69:1475–1481.

Wood S, Holland P, Stirling D, Briggs L, Sprosen J, Ruck J et al. (2006). Survey of cyanotoxins in New Zealand water bodies between 2001 and 2004. New Zeal J Mar Freshwater Res. 40:585–597.

Wood SA, Kuhajek JM, de Winton M, Phillips NR (2012). Species composition and cyanotoxin production in periphyton mats from three lakes of varying trophic status. FEMS Microbiol Ecol. 79:312–326.

Xie LQ, Xie P, Guo LG, Li L, Miyabara Y, Park HD (2005). Organ distribution and bioaccumulation of microcystins in freshwater fish at different trophic levels from the eutrophic Lake Chaohu, China. Environ Toxicol. 20:293–300.

Yepremian C, Gugger MF, Briand E, Catherine A, Berger C, Quiblier C et al. (2007). Microcystin ecotypes in a perennial *Planktothrix agardhii* bloom. Water Res. 41:4446–4456.

2.2 CYLINDROSPERMOPSINS

Andrew Humpage and Jutta Fastner

The cyanobacterium *Raphidiopsis raciborskii* (the renaming from *Cylindrospermopsis* has been widely accepted; see Chapter 3) first came to notice after the poisoning of 138 children and 10 adults on Palm Island, a tropical island off Townsville in central Queensland, Australia (Byth, 1980). Cultures of the organism were found to produce effects in mice similar to those seen in the human victims (Hawkins et al., 1985). The pure toxin – named cylindrospermopsin – was identified in 1992 (Ohtani et al., 1992).

2.2.1 Chemical structures

Cylindrospermopsins (CYNs, Figure 2.2) are alkaloids comprising a tricyclic guanidino moiety linked via a hydroxylated bridging carbon (C7) to uracil (Ohtani et al., 1992). Four structural variants have been identified (Table 2.4): 7-epi-cylindrospermopsin (7-epi-CYN), 7-deoxy-cylindrospermopsin (7-deoxy-CYN), 7-deoxy-desulpho-cylindrospermopsin and 7-deoxy-desulpho-12-acetylcylindrospermopsin (Norris et al., 1999; Banker et al., 2000; Wimmer et al., 2014). The assignments of the absolute configurations of CYN and 7-epi-CYN have been exchanged, but this has little practical bearing as they are both equally toxic (Banker et al., 2000; White & Hansen, 2005). Pure CYN is a white powder and is very water soluble. It is stable to boiling and a wide range of pH (Chiswell et al., 1999).

2.2.2 Toxicity: mode of action

The toxic effects of cylindrospermopsin, summarised in the following, are described in detail in the WHO Background Document on Cylindrospermopsins (WHO, 2020); see there for further information and references). Based on available studies, the liver, kidneys and erythrocytes may

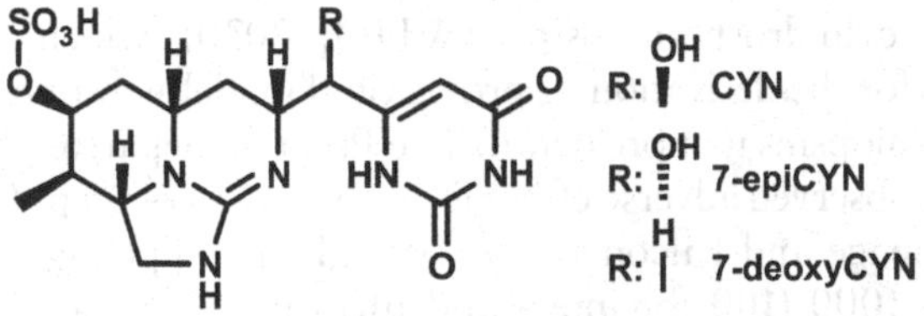

Figure 2.2 Molecular structure of common cylindrospermopsins.

Table 2.4 Congeners of cylindrospermopsin and their molecular masses

Congeners	*Formula*	*Monoisotopic molecular mass (Da)*	*Average molecular weight (g/mol)*
Cylindrospermopsin	$C_{15}H_{21}N_5O_7S$	415.116	415.428
7-Epi-cylindrospermopsin	$C_{15}H_{21}N_5O_7S$	415.116	415.428
7-Deoxy-cylindrospermopsin	$C_{15}H_{21}N_5O_6S$	399.121	399.429
7-Deoxy-desulpho-cylindrospermopsin	$C_{15}H_{21}N_5O_3$	319.164	319.366
7-Deoxy-desulpho-12-acetylcylindrospermopsin	$C_{17}H_{23}N_5O_4$	361.175	361.404

be important targets of CYN toxicity although studies using radiolabelled CYN suggest that it is distributed to all major organs. Skin patch testing produced only mild skin irritation. Since CYNs are hydrophilic molecules, facilitated transport systems mediate their intestinal absorption and uptake into other cell types, including hepatocytes. However, due to the small size of these molecules, a limited passive diffusion through biological membranes is expected. Although not clearly understood, the specific mechanism for toxicity may involve more than one mode of action, depend on the magnitude and frequency of dose, exposure duration, life stage, age or sex of the organism and the duration that an animal is observed post-dosing. At low concentrations, inhibition of protein synthesis (Terao et al., 1994; Froscio et al., 2003) appears to be the primary effect, which is mediated by the parent compound, whereas at higher exposures, CYN toxicity appears to involve metabolites and other mechanisms that are cytochrome P450-dependent. Reactive oxygen species and induction of stress responses may also be involved in the mode of action.

Cylindrospermopsins have been shown to be genotoxic in various mammalian cells and tissues using both *in vitro* and *in vivo* models. The extent and quality of toxicological data on CYN is quite limited, particularly because many studies have used cell extracts rather than pure toxin.

2.2.3 Derivation of provisional guideline values

The following section is taken directly from the WHO chemicals background document on cylindrospermopsins (WHO, 2020) which discusses the considerations for the derivation of provisional guideline values for exposure to cylindrospermopsins in more detail. The Point of Departure has been identified as the no observed adverse effect level (NOAEL) of 30 μg/kg bw per day from the Humpage and Falconer (2003) study. By applying an uncertainty factor (UF) of 1000 (100 for inter- and intraspecies variability and 10 for the lack of chronic toxicity studies and deficiencies in the overall toxicological database), a provisional tolerable daily intake TDI (NOAEL/UF) of 0.03 μg/kg bw per day can be derived. The value is provisional because of deficiencies in the CYN toxicological database, essentially related to the limited availability of studies with purified toxins, lack of in vivo data on reproductive end-points and the unclear role of metabolites, especially related to potential genotoxicity. The Sukenik et al. (2006) 42-week drinking-water study provides supporting qualitative evidence for CYN toxicity, but the experimental design does not allow derivation of a robust reference value (Funari & Testai, 2008). The study by Chernoff et al. (2018) observed many of the same effects as seen previously and demonstrates that the NOAEL is below 75 μg/kg bw per day.

The toxicological database is more limited for CYNs than for microcystin-LR – for example, data on on reproductive effects following oral dosing are lacking. Critically, there is evidence for potential *in vivo* genotoxicity of CYN. However, the lack of chronic dosing studies does not affect derivation of the short-term GV. Therefore, an uncertainty factor of 3 was used to allow for these uncertainties in the derivation of the provisional short-term drinking-water GV and recreational water GV.

For deriving the provisional lifetime drinking-water GV, the fraction of exposure allocated to drinking-water was 80% because drinking-water is expected to be the most likely long-term source of exposure. For deriving the provisional short-term drinking-water GV, the default allocation factor for short-term values of 100% was selected, considering that drinking-water is usually the most likely exposure source.

The provisional recreational water GV, which aims to protect from systemic effects, is based on a conservative scenario of a 15-kg child swallowing 250 mL of water (WHO, 2003).

Calculation of provisional lifetime drinking-water GV for CYN:

$$GV_{chronic} = \frac{NOAEL*bw*P}{UF*C} = \frac{30*60*0.8}{1000*2} \mu g/L = 0.72\ \mu g/L \approx 0.7\ \mu g/L$$

where

$GV_{chronic}$ = GV for chronic (lifetime) exposure

NOAEL = no-observed-adverse-effect level (30 μg/kg bw per day, based on Humpage & Falconer, 2003)

bw = body weight (default = 60 kg for an adult)

P = fraction of exposure allocated to drinking-water (80%, because other sources of exposure, such as air, food and soil, are considered minor)

UF = uncertainty factor (1000 = 10 for interspecies variation × 10 for intraspecies variation × 10 for database deficiencies, including use of a subchronic study)

C = daily drinking-water consumption (default = 2 L for an adult).

Calculation of provisional short-term drinking-water GV for CYN:

To develop a short-term GV, the same logic was applied except that a UF of 3 was used for database limitations:

$$GV_{short\text{-}term} = \frac{NOAEL_{subchronic} * bw * P}{UF * C} = \frac{30 * 60 * 1.0}{300 * 2} \mu g / L = 3 \mu g / L$$

where

$GV_{short\text{-}term}$ = GV for short-term exposure
NOAEL = no-observed-adverse-effect level (30 μg/kg bw per day, based on Humpage & Falconer, 2003)
bw = body weight (default = 60 kg for an adult)
P = fraction of exposure allocated to drinking-water (default for short-term exposure = 100%, as drinking-water is expected to be the most likely source of exposure)
UF = uncertainty factor (300 = 10 for interspecies variation × 10 for intraspecies variation × 3 for database deficiencies)
C = daily drinking-water consumption (default = 2 L for an adult).

Calculation of provisional recreational water GV for CYN:

$$GV_{recreation} = \frac{NOAEL * bw}{UF * C} = \frac{30 * 15}{300 * 0.25} \mu g / L = 6 \mu g / L$$

where

$GV_{recreation}$ = GV for recreational water exposure
NOAEL = no-observed-adverse-effect level (30 μg/kg bw per day, based on Humpage & Falconer, 2003)
bw = body weight (default = 15 kg for a child)
UF = uncertainty factor (300 = 10 for interspecies variation × 10 for intraspecies variation × 3 for database deficiencies)
C = daily incidental water consumption (default = 250 mL for a child).

Considerations in applying the provisional guideline values

The provisional GVs are based on toxicological data for CYN. The limited evidence on the relative potency of other CYN congeners suggests they are probably similar in potency to CYN. Therefore, for assessing risk, as a conservative approach, it is suggested that the sum of of CYNs (on a molar basis), be evaluated against the GV.

In some regions, others sources of exposure besides drinking-water can be significant (see chapter 5). This includes food from locations where blooms have a long duration and there is high consumption of locally affected food items. In such situations, it may be appropriate to consider reducing the allocation factor for the lifetime and short-term drinking-water GVs based on relative exposure data for the population.

The short-term drinking-water GV is based on exposure of adults. Since infants and children can ingest a significantly larger volume of water per body weight (e.g., up to 5 times more drinking-water/kg bw for bottle-fed infants than for adults), it is recommended that alternative water sources such as bottled water are provided for bottle-fed infants and small children when CYN concentrations are greater than 0.7 µg/L even for short periods, as a precautionary measure.

2.2.4 Production

2.2.4.1 Producing cyanobacteria

Cylindrospermopsins (CYNs) have been found in species of Nostocales and Oscillatoriales. Among the Nostocales, *Raphidiopsis* (*Cylindrospermopsis*) *raciborskii, R. curvata, R. mediterranea, Chrysosporum* (*Aphanizomenon*) *ovalisporum, Chrysosporum* (*Anabaena*) *bergii, Aphanizomenon flosaquae, Aphanizomenon gracile* and *Anabaena lapponica,* have been identified as producers (Hawkins et al., 1985; Banker et al., 1997; Li et al., 2001a; Schembri et al., 2001; Preussel et al., 2006; Spoof et al., 2006; McGregor et al., 2011; Kokociński et al., 2013). *Umezakia natans*, a CYN producer from Japan, was originally assigned to the order Stigonematales (Harada et al., 1994), but later genetic analysis suggests that this species belongs to the Nostocales (Niiyama et al., 2011). Cylindrospermopsin producers belonging to the Oscillatoriales are the benthic *Microseira* (*Lyngbya*) *wollei*, benthic *Oscillatoria* (Seifert et al., 2007; Mazmouz et al., 2010), as well as *Hormoscilla pringsheimii* (Bohunická et al., 2015). Producing and nonproducing strains exist within these species.

The CYN-producing species have different regional distribution (de la Cruz et al., 2013). So far only *Raphidiopsis raciborskii* from Australia, New Zealand and Asia have been found to produce CYNs (Saker & Griffiths, 2000; Li et al., 2001b; Wood & Stirling, 2003; Chonudomkul et al., 2004; Nguyen et al., 2017), while none of the *R. raciborskii* strains from North and South America, Africa as well as from Europe have been found to synthesise CYNs (Bernard et al., 2003; Fastner et al., 2003; Saker et al., 2003; Berger et al., 2006; Yilmaz et al., 2008; Fathalli et al., 2011; Hoff-Risseti et al., 2013). CYN-producing *C. ovalisporum* have been reported from strains and/or field samples of Australia, Florida, Turkey, Israel and Spain (Banker et al., 1997;

Quesada et al., 2006; Yilmaz et al., 2008; Akçaalan et al., 2014). In middle and northern Europe, CYN occurrence is largely attributed to the presence of *Aphanizomenon* sp. and *Dolichospermum* spp. (Preussel et al., 2006; Rücker et al., 2007; Bláhová et al., 2009; Brient et al., 2009; Kokociński et al., 2013).

2.2.4.2 Cylindrospermopsin profiles

While earlier studies focused primarily on CYN, data on the presence of 7-deoxy-CYN and 7-epi-CYN are increasingly reported. It appears that strains may contain varying shares of CYN, 7-deoxy-CYN and 7-epi-CYN. In strains and blooms of *R. raciborskii*, as well as strains of *Aphanizomenon* and *Ana. lapponica*, ratios of CYN to 7-deoxy-CYN vary between 0.2 and 5 (Spoof et al., 2006; Orr et al., 2010; Preussel et al., 2014; Willis et al., 2015). In *Microseira* (*Lyngbya*) *wollei* and *Raphidiopsis curvata*, 7-deoxy-CYN has been predominately found (Li et al., 2001a; Seifert et al., 2007). However, growth conditions may alter the ratio of 7-deoxy-CYN to CYN, most probably due to the fact that 7-deoxy-CYN is a precursor of CYN (Mazmouz et al., 2010). 7-Epi-CYN has been detected in *C. ovalisporum* as a minor compound, whereas it was up to threefold more abundant than CYN in some *Oscillatoria* strains (Banker et al., 2000; Mazmouz et al., 2010). No information is available on the distribution and concentration of 7-deoxy-desulpho-cylindrospermopsin and 7-deoxy-desulpho-12-acetylcylindrospermopsin recently found (in addition to CYN) in a Thai strain of *R. raciborskii* (Wimmer et al., 2014).

2.2.4.3 Biosynthesis

The complete gene cluster (*cyr*) for the synthesis of CYN was first sequenced from *R. raciborskii* (Mihali et al., 2008). It spans 43 kb and encodes 15 open reading frames (ORF). The biosynthesis starts with an amidinotransferase and is completed by nonribosomal peptide/polyketide synthetases and tailoring enzymes. Furthermore, the cluster encodes for a putative transporter (*cyrK*) for the export of CYN from the cells (Mihali et al., 2008). A putative NtcA (global nitrogen regulator) binding site has been identified within the *cyr* cluster, suggesting that CYN synthesis is influenced by N metabolism (Mazmouz et al., 2011; Stucken et al., 2014).

Since then, homologous clusters or parts of them have been sequenced from further *R. raciborskii* strains (Stucken et al., 2010; Sinha et al., 2014), *C. ovalisporum* (*aoa*, gene cluster; Shalev-Alon et al. (2002), *Aphanizomenon* sp. (Stüken & Jakobsen, 2010), *Oscillatoria* sp. (Mazmouz et al., 2010) and *Raphidiopsis* sp. (Jiang et al., 2014; Pearson et al., 2016). Differences within the gene cluster between strains comprise the order of the *cyr* genes, flanking genes as well as a sporadic lack of *cyrN* and *cyrO* (Jiang et al., 2014; Pearson et al., 2016).

2.2.4.4 Regulation of biosynthesis

Similar to microcystins, strains differ in the amount of cylindrospermopsins (CYN, 7-epi-CYN and 7-deoxy-CYN) produced, and contents per biomass are in the same range as those of microcystins (Table 2.5). Contents of cylindrospermopsins (CYNs) of some 10 µg up to 9.3 mg/g DW were reported across all producing species (see above) and geographical regions (Saker & Griffiths, 2000; Preussel et al., 2006; Seifert et al., 2007; Yilmaz et al., 2008; Akçaalan et al., 2014; Cirés et al., 2014; McGregor & Sendall, 2015). Cell quota of CYNs for *R. raciborskii* strains range from ~3 to 279 fg/cell (Hawkins et al., 2001; Davis et al., 2014; Willis et al., 2016; Yang et al., 2018), and from ~49 to 190 fg/cell in *C. ovalisporum* (Cirés et al., 2014). Values per unit biovolume are between 0.6 and 3.5 µg CYN/mm³ in

Table 2.5 Examples of cylindrospermopsin contents (not differentiated by congeners, i.e., CYN, 7-epi-CYN and 7-deoxy-CYN) given as mass per dry weight, per biovolume and per cell found in cultured strains

Taxon [a]	*Cond.*	*Origin*	*N*	*Range*	*Reference*
				mg/g DW	
Aphanizomenion sp.	S	DEU	3	2.3–6.6	Preussel et al. (2006)
Chrysosporum ovalisporum	S	USA	1	7.4–9.3	Yilmaz et al. (2008) [b]
C. ovalisporum	S	ESP	6	5.7–9.1	Cirés et al. (2014)
Raphidiopsis raciborskii	T	AUS	4	n.d.–9	Saker & Griffiths (2000) [b]
				µg/mm³ BV	
C. ovalisporum	S	ESP	6	0.9–2.4	Cirés et al. (2014)
R. raciborskii	S	AUS	2	~1–3.5	Saker & Griffiths (2000) [b]
Aphanizomenion sp.	L, T, N	DEU	3	0.3–1.6	Preussel et al. (2009)
				fg/cell	
C. ovalisporum	S	ESP	6	49–190	Cirés et al. (2014)
R. raciborskii	S	AUS	24	91–279	Willis et al. (2016) [b]
R. raciborskii	S	AUS	2	~10–25	Davis et al. (2014) [b]
R. raciborskii	N	CHN	1	45 – 64	Yang et al. (2018)
Oscillatoria sp.[c]	L		1	~3 – 18	Bormans et al. (2014)

Cultivation conditions (Cond.) were either one standard (S), or with varying light (L), temperature (T), nitrogen concentration (N) and phosphorus concentration (P). The origin of the analysed strains (N) is given by ISO 3166 country code. Data from some studies have been transformed to units as given here. Note that content in field samples is generally much lower, as these consist of a mixture of clones with individual toxin contents ranging from 0 (nonproducers) to values as reported as maxima in this table; see also section 4.6.

[a] The taxon given here may deviate from that given in the publication. For changes in taxonomy, see Chapter 3.

[b] Intracellular CYNs only.

[c] Benthic form.

R. raciborskii (Saker & Griffiths, 2000; Hawkins et al., 2001), 0.3 and 1.6 μg CYNs/mm^3 in *Aphanizomenon* sp. (Preussel et al., 2009; Preussel et al., 2014) and 0.9 and 2.4 μg CYN/mm^3 in *C. ovalisporum* (Cirés et al., 2014).

Several studies have investigated the influence of environmental factors on CYN production. Though all of the studies showed an effect on CYN content, no consistent pattern in the regulation of the CYN content emerged. The inconsistencies can partly be explained by differences with respect to culture conditions (i.e., batch and semicontinuous), the biomass proxy to which the toxin content was related (i.e., dry weight, biovolume, cell number, or chlorophyll-*a*) as well as by different reactions of individual strains to the same parameter. A direct linear relationship between total cell quota and growth has been observed in several *R. raciborskii* strains during log phase growth under different light, nutrients and CO_2 conditions with cell quota changing maximally by a factor 2–4 (Hawkins et al., 2001; Davis et al., 2014; Pierangelini et al., 2015; Willis et al., 2015; Yang et al., 2018). This observation in combination with a constant expression of *cyr* genes led to the conclusion that CYN production is constitutive (Davis et al., 2014; Pierangelini et al., 2015; Willis et al., 2015; Yang et al., 2018). However, CYN cell quota decreased substantially (>25-fold) down to nondetectable levels in *R. raciborskii* and to trace levels in *C. ovalisporum* at 35 °C (Saker & Griffiths, 2000; Cirés et al., 2011). This suggests that CYN production may not be constitutive, though this requires further confirmation by following *cyr* transcript levels. Up to 25–30 °C, the influence of temperature on CYN cell quota was moderate (~1.5–2.5-fold) in *Aphanizomenon* sp. and *C. ovalisporum* (Preussel et al., 2009; Cirés et al., 2011). Changes in cell quota between three- and eightfold were found in relation to light in *Oscillatoria* PCC 6506 and *Chr. ovalisporum* and in relation to nutrients in *Aphanizomenon* sp. and *C. ovalisporum* (Bar-Yosef et al., 2010; Cirés et al., 2011; Bormans et al., 2014; Preussel et al., 2014).

Environmental conditions also influence the ratio of 7-deoxy-CYN to CYN. For both *R. raciborskii* and *Aphanizomenon* spp., 7-deoxy-CYN content increased with increasing cell densities under normal growth conditions, while it did not increase or decrease under N-deprived conditions (Davis et al., 2014; Preussel et al., 2014; Stucken et al., 2014).

In contrast to microcystins, a substantial share of cylindrospermopsins is usually and constantly extracellular. In *R. raciborskii* and *Aphanizomenon* sp., up to 20% and in *C. ovalisporum* up to 40% of the total CYNs were extracellular during log-phase growth (Hawkins et al., 2001; Cirés et al., 2014; Davis et al., 2014; Preussel et al., 2014), while in *Oscillatoria* PCC 6505 the extracellular CYNs constantly amounted to more than 50% (Bormans et al., 2014). Furthermore, the extracellular CYNs increased by up to more than twofold in all species during the stationary phase under different treatments (Saker & Griffiths, 2000; Hawkins et al., 2001;

Bormans et al., 2014; Davis et al., 2014). Preussel et al. (2014) found indication of an active release of CYNs under normal growth conditions and showed that the extracellular CYNs did not increase in N-deprived cultures of *Aphanizomenon*.

For water management, it is important that environmental conditions may not only change the cell quota of CYNs, but also change the share of extracellular CYNs (see Box 5.1) as well as the ratio of 7-deoxy-CYN to CYN. Furthermore, it appears that alterations in nutrient concentrations can change strain composition and thus CYN concentrations in the field (Burford et al., 2014).

2.2.5 Occurrence in water environments

Cylindrospermopsins are found globally as a result of the worldwide distribution of producing cyanobacteria, including *Raphidiopsis raciborskii*, *Chrysosporum ovalisporum* and *Aphanizomenon* sp. (Kinnear, 2010; de la Cruz et al., 2013). In Australia, *R. raciborskii* and *C. ovalisporum* are the most abundant CYN producers with a high bloom frequency, though the correlation between CYNs concentration and biovolume is generally weak. Concentrations reported often range between < 1 and 10 μg/L, occasional up to maximally 800 μg/L (Chiswell et al., 1999; Shaw et al., 1999; McGregor & Fabbro, 2000; Shaw et al., 2002). Also in the Mediterranean region and in Florida, CYN occurrence has been often, though not always, associated with *C. ovalisporum*. Concentrations in these regions were from below 10 μg/L up to maximally 202 μg/L (Quesada et al., 2006; Messineo et al., 2010; de la Cruz et al., 2013; Fadel et al., 2014; Moreira et al., 2017). CYN concentrations reported from more temperate regions of Northern America and Europe are often well below 10 μg/L with a maximal concentration of 9–18 μg/L (Rücker et al., 2007; Bláhová et al., 2009; Brient et al., 2009; Graham et al., 2010; Kokociński et al., 2013). The highest CYN concentrations (up to almost 3 mg/L) were reported from Brazil, although these ELISA data need to be confirmed by LC-MS/MS (Bittencourt-Oliveira et al., 2014; Metcalf et al., 2017).

High concentrations in the range of 10–100 mg/L, as observed for microcystins, have not yet been observed for CYNs, most probably due to the fact that CYN-producing species do not accumulate to very high cell densities, in contrast to scums of the abundant microcystin producer *Microcystis* sp., for example. Furthermore, up to 90% of CYN can occur extracellularly in natural waters; outside of scum areas this is then readily diluted by the surrounding water (Rücker et al., 2007).

For water management, it is also important to note that both CYN and 7-deoxy-CYN can be distributed throughout the entire water column with high concentrations also in the hypolimnion (Everson et al., 2011).

2.2.5.1 Bioaccumulation

Bioaccumulation of CYN in (in)vertebrates and plants has been addressed in several studies. The freshwater mussel *Anodonta cygnea* has been shown to accumulate, but partially also depurate, CYN (Saker et al., 2004). Both CYN and deoxy-CYN have been found to bioconcentrate and bioaccumulate up to a factor of 124 in whole aquatic snails; however, the alimentary tract was not separated prior to analysis (White et al., 2006). Saker and Eaglesham (1999) found 4.3 μg/g DW of CYN in the hepatopancreas of crayfish and 1.2 μg/g DW in fish, suggesting a bioaccumulation factor of 2. In contrast, several studies detected no free CYN, probably due to binding of CYN to proteins (Esterhuizen-Londt & Pflugmacher, 2016). As reviewed by Kinnear (2010), biodilution of CYN is likely to occur at higher trophic levels.

Cylindrospermopsin uptake in plants has been found for several crops at environmentally relevant concentrations of ~10–50 μg/L (Kittler et al., 2012; Cordeiro-Araújo et al., 2017; Díez-Quijada et al., 2018; Prieto et al., 2018). While these studies do not indicate substantial bioaccumulation, however, long exposure time with high concentrations of CYN resulted in elevated CYN contents in crops, suggesting their consumption may lead to exceedance of the TDI for CYN (Díez-Quijada et al., 2018). If plants are irrigated with CYN-containing water, assessing potential human exposure through food may require analysing concentrations in crops (see also section 5.3).

More detailed information on bioaccumulation is given in the reviews of Kinnear (2010), de la Cruz et al. (2013) and Machado et al. (2017).

2.2.6 Environmental fate

2.2.6.1 Partitioning between cells and water

The results described above show that cylindrospermopsins may readily leach or be released from intact, viable cells under normal growth conditions and that leakage/release increases greatly when the cells enter the stationary growth phase (Dyble et al., 2006; Bormans et al., 2014; Preussel et al., 2014). CYN leakage/release has also been observed in persistent water blooms of both *R. raciborskii* and *Aph. ovalisporum*, with up to 100% of the total toxin in the water found in the extracellular (dissolved) fraction (Chiswell et al., 1999; Shaw et al., 1999). This is also observed in temperate lakes with *Aphanizomenon* sp. as most probable toxin producer. Where extracellular CYN was detectable, it amounted between 24% and 99% of total CYN (Rücker et al., 2007).

2.2.6.2 Chemical breakdown

Cylindrospermopsin appears to be stable over a wide range of temperatures and pH, whereas only higher temperatures (>50 °C) in combination with alkaline conditions lead to slow degradation (Chiswell et al., 1999; Adamski et al., 2016). It is also relatively stable in the dark and in sunlight, though in sunlight in the presence of cell pigments, breakdown occurs relatively rapidly, being more than 90% complete within 23 days (Chiswell et al., 1999; Wörmer et al., 2010). No data on the stability of 7-epi-CYN and 7-deoxy-CYN apparently exist.

2.2.6.3 Biodegradation

Biodegradation of CYN has been observed for some natural waters (Chiswell et al., 1999; Smith et al., 2008), while for others no biodegradation of CYN was found (Wörmer et al., 2008; Klitzke et al., 2010). This can lead to substantial concentrations of dissolved CYN even weeks or months after the producing organisms have declined (Chiswell et al., 1999; Wiedner et al., 2008). Wörmer et al. (2008) also showed that the previous presence of CYN-producing cyanobacteria may not necessarily lead to CYN biodegradation in a waterbody.

Cylindrospermopsin biodegradation studied with natural bacterial consortia either from lakes or from sediments usually showed a lag phase of 1–3 weeks before biodegradation started (Smith et al., 2008; Klitzke et al., 2010). Repeated dosing of CYN eliminated or substantially shortened the lag phase. Once biodegradation had started, the half-lives reported for mixed consortia from water and sediment were 2–4 days (Smith et al., 2008; Klitzke et al., 2010). Similar half-lives were found for a CYN-degrading *Bacillus* strain which also degraded microcystins (Mohamed & Alamri, 2012), while CYN half-lives in the presence of an *Aeromonas* sp. strain were 6–8 days (Dziga et al., 2016). For both natural bacterial consortia and isolated strains, it has been found that biodegradation is strongest between 20 °C and 35 °C and at pH between 7 and 8 (Smith et al., 2008; Klitzke & Fastner, 2012; Mohamed & Alamri, 2012; Dziga et al., 2016). The biodegradation rate is also strongly influenced by the initial CYN concentration with hardly any CYN biodegradation at concentrations below 1 µg/L (Smith et al., 2008; Mohamed & Alamri, 2012; Dziga et al., 2016). Removal of dissolved CYN from water samples was also observed with probiotic bacteria (*Bifidobacterium longum* 46) with an efficiency of 31% CYN removal within 24 h at 37 °C (Nybom et al., 2008). Degradation of CYN through the activity of manganese-oxidising bacteria – a polyphyletic type of bacteria common in freshwater, for example, *Pseudomonas* sp., *Ideonella* sp. – has been observed (Martínez-Ruiz et al., 2020b). The transformation products showed reduced toxicity to hepatocytes (Martínez-Ruiz *et al.*, 2020a). No studies on the biodegradation of 7-deoxy-CYN and 7-epi-CYN appear to exist.

For water management, it is important to keep in mind that due to the occasionally poor degradation of CYN in surface water, considerable amounts of

CYN may still be present when populations of the producing cyanobacteria have already declined or practically disappeared (see Box 5.1 for an example).

REFERENCES

Adamski M, Żmudzki P, Chrapusta E, Bober B, Kaminski A, Zabaglo K et al. (2016). Effect of pH and temperature on the stability of cylindrospermopsin. Characterization of decomposition products. Algal Res. 15:129–134.

Akçaalan R, Köker L, Oğuz A, Spoof L, Meriluoto J, Albay M (2014). First report of cylindrospermopsin production by two cyanobacteria (*Dolichospermum mendotae* and *Chrysosporum ovalisporum*) in Lake Iznik, Turkey. Toxins. 6:3173–3186.

Banker R, Carmeli S, Hadas O, Teltsch B, Porat R, Sukenik A (1997). Identification of cylindrospermopsin in *Aphanizomenon ovalisporum* (Cyanophyceae) isolated from lake Kinneret. J Phycol. 33:613–616.

Banker R, Teltsch B, Sukenik A, Carmeli S (2000). 7-epicylindrospermopsin, a toxic minor metabolite of the cyanobacterium *Aphanizomenon ovalisporum* from lake Kinneret, Israel. J Nat Prod. 63:387–389.

Bar-Yosef Y, Sukenik A, Hadas O, Viner-Mozzini Y, Kaplan A (2010). Enslavement in the water body by toxic *Aphanizomenon ovalisporum*, inducing alkaline phosphatase in phytoplanktons. Curr Biol. 20:1557–1561.

Berger C, Ba N, Gugger M, Bouvy M, Rusconi F, Coute A et al. (2006). Seasonal dynamics and toxicity of *Cylindrospermopsis raciborskii* in Lake Guiers (Senegal, West Africa). FEMS Microbiol Ecol. 57:355–366.

Bernard C, Harvey M, Briand J, Biré R, Krys S, Fontaine J (2003). Toxicological comparison of diverse *Cylindrospermopsis raciborskii* strains: evidence of liver damage caused by a French C. *raciborskii* strain. Environ Toxicol. 18:176–186.

Bittencourt-Oliveira MdC, Piccin-Santos V, Moura AN, Aaragao-Tavares NKC, Cordeiro-Araújo MK (2014). Cyanobacteria, microcystins and cylindrospermopsin in public drinking supply reservoirs of Brazil. Anais da Academia Brasileira de Ciências. 86:297–310.

Bláhová L, Oravec M, Maršálek B, Šejnohová L, Šimek Z, Bláha L (2009). The first occurrence of the cyanobacterial alkaloid toxin cylindrospermopsin in the Czech Republic as determined by immunochemical and LC/MS methods. Toxicon. 53:519–524.

Bohunická M, Mareš J, Hrouzek P, Urajová P, Lukeš M, Šmarda J et al. (2015). A combined morphological, ultrastructural, molecular, and biochemical study of the peculiar family Gomontiellaceae (Oscillatoriales) reveals a new cylindrospermopsin-producing clade of cyanobacteria. J Phycol. 51:1040–1054.

Bormans M, Lengronne M, Brient L, Duval C (2014). Cylindrospermopsin accumulation and release by the benthic cyanobacterium *Oscillatoria* sp. PCC 6506 under different light conditions and growth phases. Bull Environ Contam Toxicol. 92:243–247.

Brient L, Lengronne M, Bormans M, Fastner J (2009). First Occurrence of Cylindrospermopsin in Freshwater in France. Environ Toxicol. 24:415–420.

Burford MA, Davis TW, Orr PT, Sinha R, Willis A, Neilan BA (2014). Nutrient-related changes in the toxicity of field blooms of the cyanobacterium, *Cylindrospermopsis raciborskii*. FEMS Microbiol Ecol. 89:135–148.

Byth S (1980). Palm Island mystery disease. Med J Aust. 2:40–42.
Chernoff N, Hill D, Chorus I, Diggs D, Huang H, King D et al. (2018). Cylindrospermopsin toxicity in mice following a 90-d oral exposure. J Toxicol Environ Health Part A. 81:549–566.
Chiswell RK, Shaw GR, Eaglesham GK, Smith MJ, Norris RL, Seawright AA et al. (1999). Stability of cylindrospermopsin, the toxin from the cyanobacterium, *Cylindrospermopsis raciborskii*: effect of pH, temperature, and sunlight on decomposition. Environ Toxicol. 14:155–161.
Chonudomkul D, Yongmanitchai W, Theeragool G, Kawachi M, Kasai F, Kaya K et al. (2004). Morphology, genetic diversity, temperature tolerance and toxicity of *Cylindrospermopsis raciborskii* (Nostocales, Cyanobacteria) strains from Thailand and Japan. FEMS Microbiol Ecol. 48:345–355.
Cirés S, Wörmer L, Ballot A, Agha R, Wiedner C, Velazquez D et al. (2014). Phylogeography of cylindrospermopsin and paralytic shellfish toxin-producing nostocales cyanobacteria from mediterranean europe (Spain). Appl Environ Microbiol. 80:1359–1370.
Cirés S, Wörmer L, Timón J, Wiedner C, Quesada A (2011). Cylindrospermopsin production and release by the potentially invasive cyanobacterium *Aphanizomenon ovalisporum* under temperature and light gradients. Harmful Algae. 10:668–675.
Cordeiro-Araújo MK, Chia MA, do Carmo Bittencourt-Oliveira M (2017). Potential human health risk assessment of cylindrospermopsin accumulation and depuration in lettuce and arugula. Harmful Algae. 68:217–223.
Davis TW, Orr PT, Boyer GL, Burford MA (2014). Investigating the production and release of cylindrospermopsin and deoxy-cylindrospermopsin by *Cylindrospermopsis raciborskii* over a natural growth cycle. Harmful Algae. 31:18–25.
de la Cruz AA, Hiskia A, Kaloudis T, Chernoff N, Hill D, Antoniou MG et al. (2013). A review on cylindrospermopsin: the global occurrence, detection, toxicity and degradation of a potent cyanotoxin. Environ Sci Process Impacts. 15:1979–2003.
Díez-Quijada L, Guzmán-Guillén R, Prieto Ortega A, Llana-Ruíz-Cabello M, Campos A, Vasconcelos V et al. (2018). New method for simultaneous determination of microcystins and cylindrospermopsin in vegetable matrices by SPE-UPLC-MS/MS. Toxins. 10:406.
Dyble J, Tester PA, Litaker RW (2006). Effects of light intensity on cylindrospermopsin production in the cyanobacterial HAB species *Cylindrospermopsis raciborskii*. African J Mar Sci. 28:309–312.
Dziga D, Kokocinski M, Maksylewicz A, Czaja-Prokop U, Barylski J (2016). Cylindrospermopsin biodegradation abilities of *Aeromonas* sp. isolated from Rusałka Lake. Toxins. 8:55.
Esterhuizen-Londt M, Pflugmacher S (2016). Inability to detect free cylindrospermopsin in spiked aquatic organism extracts plausibly suggests protein binding. Toxicon. 122:89–93.
Everson S, Fabbro L, Kinnear S, Wright P (2011). Extreme differences in akinete, heterocyte and cylindrospermopsin concentrations with depth in a successive bloom involving *Aphanizomenon ovalisporum* (Forti) and *Cylindrospermopsis raciborskii* (Woloszynska) Seenaya and Subba Raju. Harmful Algae. 10:265–276.
Fadel A, Atoui A, Lemaire BJ, Vinçon-Leite B, Slim K (2014). Dynamics of the toxin cylindrospermopsin and the cyanobacterium *Chrysosporum* (*Aphanizomenon*) *ovalisporum* in a Mediterranean eutrophic reservoir. Toxins. 6:3041–3057.

Fastner J, Heinze R, Humpage AR, Mischke U, Eaglesham GK, Chorus I (2003). Cylindrospermopsin occurrence in two German lakes and preliminary assessment of toxicity and toxin production of *Cylindrospermopsis raciborskii* (Cyanobacteria) isolates. Toxicon. 42:313–321.

Fathalli A, Jenhani AB, Moreira C, Azevedo J, Welker M, Romdhane M et al. (2011). Genetic variability of the invasive cyanobacteria *Cylindrospermopsis raciborskii* from Bir M'cherga reservoir (Tunisia). Arch Microbiol. 193:595–604.

Froscio SM, Humpage AR, Burcham PC, Falconer IR (2003). Cylindrospermopsin-induced protein synthesis inhibition and its dissociation from acute toxicity in mouse hepatocytes. Environ Toxicol. 18:243–251.

Funari E, Testai E (2008). Human health risk assessment related to cyanotoxins exposure. Crit Rev Toxicol. 38:97–125.

Graham JL, Loftin KA, Meyer MT, Ziegler AC (2010). Cyanotoxin mixtures and taste-and-odor compounds in cyanobacterial blooms from the midwestern United States. Environ Sci Technol. 44:7361–7368.

Harada K-I, Ohtani I, Iwamoto K, Suzuki M, Watanabe MF, Watanabe M et al. (1994). Isolation of cylindrospermopsin from a cyanobacterium *Umezakia natans* and its screening method. Toxicon. 32:73–84.

Hawkins PR, Putt E, Falconer IR, Humpage AR (2001). Phenotypical variation in a toxic strain of the phytoplankter, *Cylindrospermopsis raciborskii* (Nostocales, Cyanophyceae) during batch culture. Environ Toxicol. 16:460–467.

Hawkins PR, Runnegar MTC, Jackson ARB, Falconer IR (1985). Severe hepatotoxicity caused by the tropical cyanobacterium (blue-green alga) *Cylindrospermopsis raciborsckii* (Woloszynska) Seenaya and Subba Raju isolated from a domestic water supply reservoir. Appl Environ Microbiol. 50:1292–1295.

Hoff-Risseti C, Dörr FA, Schaker PDC, Pinto E, Werner VR, Fiore MF (2013). Cylindrospermopsin and saxitoxin synthetase genes in *Cylindrospermopsis raciborskii* strains from Brazilian freshwater. PLoS One. 8:e74238.

Humpage AR, Falconer IR (2003). Oral toxicity of the cyanobacterial toxin cylindrospermopsin in male swiss albino mice: determination of No Observed Adverse Effect Level for deriving a drinking water guideline value. Environ Toxicol. 18:94–103.

Jiang Y, Xiao P, Yu G, Shao J, Liu D, Azevedo SM et al. (2014). Sporadic distribution and distinctive variations of cylindrospermopsin genes in cyanobacterial strains and environmental samples from Chinese freshwater bodies. Appl Environ Microbiol. 80:5219–5230.

Kinnear S (2010). Cylindrospermopsin: a decade of progress on bioaccumulation research. Mar Drugs. 8:542–564.

Kittler K, Schreiner M, Krumbein A, Manzei S, Koch M, Rohn S et al. (2012). Uptake of the cyanobacterial toxin cylindrospermopsin in *Brassica* vegetables. Food Chem. 133:875–879.

Klitzke S, Apelt S, Weiler C, Fastner J, Chorus I (2010). Retention and degradation of the cyanobacterial toxin cylindrospermopsin in sediments - the role of sediment preconditioning and DOM composition. Toxicon. 55:999–1007.

Klitzke S, Fastner J (2012). Cylindrospermopsin degradation in sediments–The role of temperature, redox conditions, and dissolved organic carbon. Water Res. 46:1549–1555.

Kokociński M, Mankiewicz-Boczek J, Jurczak T, Spoof L, Meriluoto J, Rejmonczyk E et al. (2013). *Aphanizomenon gracile* (Nostocales), a cylindrospermopsin-producing cyanobacterium in Polish lakes. Environ Sci Pollut Res. 20: 5243–5264.

Li R, Carmichael WW, Brittain S, Eaglesham GK, Shaw GR, Liu YK et al. (2001a). First report of the cyanotoxin cylindrospermopsin and deoxycylindrospermopsin from *Raphidiopsis curvata* (cyanobacteria). J Phycol. 37:1121–1126.

Li R, Carmichael WW, Brittain S, Eaglesham GK, Shaw GR, Mahakahant A et al. (2001b). Isolation and identification of the cyanotoxin cylindrospermopsin and deoxy-cylindrospermopsin from a Thailand strain of *Cylindrospermopsis raciborskii* (Cyanobacteria). Toxicon. 39:973–980.

Machado J, Campos A, Vasconcelos V, Freitas M (2017). Effects of microcystin-LR and cylindrospermopsin on plant-soil systems: a review of their relevance for agricultural plant quality and public health. Environ Res. 153:191–204.

Martínez-Ruiz EB, Cooper M, Al-Zeer MA, Kurreck J, Adrian L, Szewzyk U (2020a). Manganese-oxidizing bacteria form multiple cylindrospermopsin transformation products with reduced human liver cell toxicity. Sci Tot Environ. 729:138924.

Martínez-Ruiz EB, Cooper M, Fastner J, Szewzyk U (2020b). Manganese-oxidizing bacteria isolated from natural and technical systems remove cylindrospermopsin. Chemosphere. 238:124625.

Mazmouz R, Chapuis-Hugon F, Mann S, Pichon V, Méjean A, Ploux O (2010). Biosynthesis of cylindrospermopsin and 7-epicylindrospermopsin in *Oscillatoria* sp strain PCC 6506: identification of the *cyr* gene cluster and toxin analysis. Appl Environ Microbiol. 76:4943–4949.

Mazmouz R, Chapuis-Hugon F, Pichon V, Méjean A, Ploux O (2011). The last step of the biosynthesis of the cyanotoxins cylindrospermopsin and 7-epi-cylindrospermopsin is catalysed by CyrI, a 2-oxoglutarate-dependent iron oxygenase. ChemBioChem. 12:858–862.

McGregor GB, Fabbro LD (2000). Dominance of *Cylindrospermopsis raciborskii* (Nostocales, Cyanoprokaryota) in Queensland tropical and subtropical reservoirs: implications for monitoring and management. Lakes Reserv Res Manage. 5:195–205.

McGregor GB, Sendall BC (2015). Phylogeny and toxicology of *Lyngbya wollei* (Cyanobacteria, Oscillatoriales) from north-eastern Australia, with a description of *Microseira* gen. nov. J Phycol. 51:109–119.

McGregor GB, Sendall BC, Hunt LT, Eaglesham GK (2011). Report of the cyanotoxins cylindrospermopsin and deoxy-cylindrospermopsin from *Raphidiopsis mediterranea* Skuja (Cyanobacteria/Nostocales). Harmful Algae. 10: 402–410.

Messineo V, Melchiorre S, Di Corcia A, Gallo P, Bruno M (2010). Seasonal succession of *Cylindrospermopsis raciborskii* and *Aphanizomenon ovalisporum* blooms with cylindrospermopsin occurrence in the volcanic Lake Albano, central Italy. Environ Toxicol. 25:18–27.

Metcalf JS, Young FM, Codd GA (2017). Performance assessment of a cylindrospermopsin ELISA with purified compounds and cyanobacterial extracts. Environ Forensics. 18:147–152.

Mihali TK, Kellmann R, Muenchhoff J, Barrow KD, Neilan BA (2008). Characterization of the gene cluster responsible for cylindrospermopsin biosynthesis. Appl Environ Microbiol. 74:716–722.

Mohamed ZA, Alamri SA (2012). Biodegradation of cylindrospermopsin toxin by microcystin-degrading bacteria isolated from cyanobacterial blooms. Toxicon. 60:1390–1395.

Moreira C, Mendes R, Azevedo J, Vasconcelos V, Antunes A (2017). First occurrence of cylindrospermopsin in Portugal: a contribution to its continuous global dispersal. Toxicon. 130:87–90.

Nguyen TTL, Hoang TH, Nguyen TK, Duong TT (2017). The occurrence of toxic cyanobacterium *Cylindrospermopsis raciborskii* and its toxin cylindrospermopsin in the Huong River, Thua Thien Hue province, Vietnam. Environ Monit Assess. 189:490.

Niiyama Y, Tuji A, Tsujimura S (2011). *Umezakia natans* M. Watan. does not belong to Stigonemataceae but to Nostocaceae. Fottea. 11:163–169.

Norris RL, Eaglesham GK, Pierens G, Shaw GR, Smith MJ, Chiswell RK et al. (1999). Deoxycylindrospermopsin, an analog of cylindrospermopsin from *Cylindrospermopsis raciborskii*. Environ Toxicol. 14:163–165.

Nybom SM, Salminen SJ, Meriluoto JA (2008). Specific strains of probiotic bacteria are efficient in removal of several different cyanobacterial toxins from solution. Toxicon. 52:214–220.

Ohtani I, Moore RE, Runnegar MTC (1992). Cylindrospermopsin, a potent hepatotoxin from the blue-green alga *Cylindrospermopsis raciborskii*. J Am Chem Soc. 114:7941–7942.

Orr PT, Rasmussen JP, Burford MA, Eaglesham GK, Lennox SM (2010). Evaluation of quantitative real-time PCR to characterise spatial and temporal variations in cyanobacteria, *Cylindrospermopsis raciborskii* (Woloszynska) Seenaya et Subba Raju and cylindrospermopsin concentrations in three subtropical Australian reservoirs. Harmful Algae. 9:243–254.

Pearson LA, Dittmann E, Mazmouz R, Ongley SE, D'Agostino PM, Neilan BA (2016). The genetics, biosynthesis and regulation of toxic specialized metabolites of cyanobacteria. Harmful Algae. 54:98–111.

Pierangelini M, Sinha R, Willis A, Burford MA, Orr PT, Beardall J et al. (2015). Constitutive cylindrospermopsin pool size in *Cylindrospermopsis raciborskii* under different light and CO_2 partial pressure conditions. Appl Environ Microbiol. 81:3069–3076.

Preussel K, Chorus I, Fastner J (2014). Nitrogen limitation promotes accumulation and suppresses release of cylindrospermopsins in cells of *Aphanizomenon* sp. Toxins. 6:2932–2947.

Preussel K, Stüken A, Wiedner C, Chorus I, Fastner J (2006). First report on cylindrospermopsin producing *Aphanizomenon flos-aquae* (Cyanobacteria) isolated from two German lakes. Toxicon. 47:156–162.

Preussel K, Wessel G, Fastner J, Chorus I (2009). Response of cylindrospermopsin production and release in *Aphanizomenon flos-aquae* (Cyanobacteria) to varying light and temperature conditions. Harmful Algae. 8:645–650.

Prieto AI, Guzmán-Guillén R, Díez-Quijada L, Campos A, Vasconcelos V, Jos Á et al. (2018). Validation of a method for cylindrospermopsin determination in vegetables: application to real samples such as lettuce (*Lactuca sativa* L.). Toxins. 10:63.

Quesada A, Moreno E, Carrasco D, Paniagua T, Wörmer L, De Hoyos C et al. (2006). Toxicity of *Aphanizomenon ovalisporum* (Cyanobacteria) in a Spanish water reservoir. Europ J Phycol. 41:39–45.

Rücker J, Stüken A, Nixdorf B, Fastner J, Chorus I, Wiedner C (2007). Concentrations of particulate and dissolved cylindrospermopsin in 21 *Aphanizomenon*-dominated temperate lakes. Toxicon. 50:800–809.

Saker ML, Eaglesham GK (1999). The accumulation of cylindrospermopsin from the cyanobacterium *Cylindrospermopsis raciborskii* in tissues of the Redclaw crayfish *Cherax quandricarinatus*. Toxicon. 37:1065–1077.

Saker ML, Griffiths DJ (2000). The effect of temperature on growth and cylindrospermopsin content of seven isolates of *Cylindrospermopsis raceborskii* (Nostocales, Cyanophyceae) from water bodies in northern Australia. Phycologia. 39:349–354.

Saker ML, Metcalf JS, Codd GA, Vasconcelos VM (2004). Accumulation and depuration of the cyanobacterial toxin cylindrospermopsin in the freshwater mussel *Anodonta cygnea*. Toxicon. 43:185–194.

Saker ML, Nogueira IC, Vasconcelos VM, Neilan BA, Eaglesham GK, Pereira P (2003). First report and toxicological assessment of the cyanobacterium *Cylindrospermopsis raciborskii* from Portuguese freshwaters. Ecotoxicol Environ Safety. 55:243–250.

Schembri MA, Neilan BA, Saint CP (2001). Identification of genes implicated in toxin production in the cyanobacterium *Cylindrospermopsis raciborskii*. Environ Toxicol. 16:413–421.

Seifert M, McGregor G, Eaglesham G, Wickramasinghe W, Shaw G (2007). First evidence for the production of cylindrospermopsin and deoxy-cylindrospermopsin by the freshwater benthic cyanobacterium, *Lyngbya wollei* (Farlow ex Gornont) Speziale and Dyck. Harmful Algae. 6:73–80.

Shalev-Alon G, Sukenik A, Livnah O, Schwarz R, Kaplan A (2002). A novel gene encoding amidinotransferase in the cylindrospermopsin producing cyanobacterium *Aphanizomenon ovalisporum*. FEMS Microbiol Lett. 209:87–91.

Shaw GR, McKenzie RA, Wickramasinghe WA, Seawright AA, Eaglesham GK, Moore MR (2002). Comparative toxicity of the cyanobacterial toxin cylindrospermopsin between mice and cattle: human implications. Proceedings of the 10th International Conference on Harmful Algae, St. Pete Beach, FL. 465–467.

Shaw GR, Sukenik A, Livne A, Chiswell RK, Smith MJ, Seawright AA et al. (1999). Blooms of the cylindrospermopsin containing cyanobacterium, *Aphanizomenon ovalisporum* (Forti), in newly constructed lakes, Queensland, Australia. Environ Toxicol. 14:167–177.

Sinha R, Pearson LA, Davis TW, Muenchhoff J, Pratama R, Jex A et al. (2014). Comparative genomics of Cylindrospermopsis raciborskii strains with differential toxicities. BMC Genomics. 15:83.

Smith MJ, Shaw GR, Eaglesham GK, Ho L, Brookes JD. (2008). Elucidating the factors influencing the biodegradation of cylindrospermopsin in drinking water sources. Environ Toxicol. 23:413–421.

Spoof L, Berg KA, Rapala J, Lahti K, Lepistö L, Metcalf JS et al. (2006). First observation of cylindrospermopsin in *Anabaena lapponica* isolated from the boreal environment (Finland). Environ Toxicol. 21:552–560.

Stucken K, John U, Cembella A, Murillo AA, Soto-Liebe K, Fuentes-Valdés JJ et al. (2010). The smallest known genomes of multicellular and toxic cyanobacteria:

comparison, minimal gene sets for linked traits and the evolutionary implications. PLoS One. 5:e9235.

Stucken K, John U, Cembella A, Soto-Liebe K, Vásquez M (2014). Impact of nitrogen sources on gene expression and toxin production in the diazotroph *Cylindrospermopsis raciborskii* CS-505 and non-diazotroph *Raphidiopsis brookii* D9. Toxins. 6:1896–1915.

Stüken A, Jakobsen KS (2010). The cylindrospermopsin gene cluster of Aphanizomenon sp strain 10E6: organization and recombination. Microbiology. 156: 2438–2451.

Sukenik A, Reisner M, Carmeli S, Werman M (2006). Oral toxicity of the cyanobacterial toxin cylindrospermopsin in mice: long-term exposure to low doses. Environ Toxicol. 21:575–582.

Terao K, Ohmori S, Igarashi K, Ohtani I, Watanabe MF, Harada K-I et al. (1994). Electron microscopic studies on experimental poisoning in mice induced by cylindrospermopsin isolated from the blue-green alga *Umezakia natans*. Toxicon. 32:833–843.

White JD, Hansen JD (2005). Total synthesis of (-)-7-epicylindrospermopsin, a toxic metabolite of the freshwater cyanobacterium *Aphanizomenon ovalisporum*, and assignment of its absolute configuration. J Org Chem. 70:1963–1977.

White SH, Duivenvoorden LJ, Fabbro LD, Eaglesham GK (2006). Influence of intracellular toxin concentrations on cylindrospermopsin bioaccumulation in a freshwater gastropod (*Melanoides tuberculata*). Toxicon. 47:497–509.

WHO (2003). Guidelines for safe recreational water environments. Vol. 1: Coastal and fresh waters. Geneva: World Health Organization. https://apps.who.int/iris/handle/10665/42591

WHO (2020). Cyanobacterial toxins: Cylindrospermopsins. Background document for development of WHO Guidelines for Drinking-water Quality and Guidelines for Safe Recreational Water Environments. Geneva: World Health Organization. https://apps.who.int/iris/handle/10665/338063

Wiedner C, Rücker J, Fastner J, Chorus I, Nixdorf B (2008). Seasonal dynamics of cylindrospermopsin and cyanobacteria in two German lakes. Toxicon. 52:677–686.

Willis A, Adams MP, Chuang AW, Orr PT, O'Brien KR, Burford MA (2015). Constitutive toxin production under various nitrogen and phosphorus regimes of three ecotypes of *Cylindrospermopsis raciborskii* ((Wołoszyńska) Seenayya et Subba Raju). Harmful Algae. 47:27–34.

Willis A, Chuang AW, Woodhouse JN, Neilan BA, Burford MA (2016). Intraspecific variation in growth, morphology and toxin quotas for the cyanobacterium, *Cylindrospermopsis raciborskii*. Toxicon 119:307–310.

Wimmer KM, Strangman WK, Wright JL (2014). 7-Deoxy-desulfo-cylindrospermopsin and 7-deoxy-desulfo-12-acetylcylindrospermopsin: two new cylindrospermopsin analogs isolated from a Thai strain of *Cylindrospermopsis racib*orskii. Harmful Algae. 37:203–206.

Wood S, Stirling D (2003). First identification of the cylindrospermopsin-producing cyanobacterium *Cylindrospermopsis raciborskii* in New Zealand. New Zeal J Mar Freshwater Res. 37:821–828.

Wörmer L, Cirés S, Carrasco D, Quesada A (2008). Cylindrospermopsin is not degraded by co-occurring natural bacterial communities during a 40-day study. Harmful Algae. 7:206–213.

Wörmer L, Huerta-Fontela M, Cires S, Carrasco D, Quesada A (2010). Natural photodegradation of the cyanobacterial toxins microcystin and cylindrospermopsin. Environ Sci Technol. 44:3002–3007.
Yang Y, Chen Y, Cai F, Liu X, Wang Y, Li R (2018). Toxicity-associated changes in the invasive cyanobacterium *Cylindrospermopsis raciborskii* in response to nitrogen fluctuations. Environ Pollut. 237:1041–1049.
Yilmaz M, Phlips EJ, Szabo NJ, Badylak S (2008). A comparative study of Florida strains of *Cylindrospermopsis* and *Aphanizomenon* for cylindrospermopsin production. Toxicon. 52:594–595.

2.3 ANATOXIN-A AND ANALOGUES

Emanuela Testai

Anatoxin-a (ATX) was isolated from strains of *Dolichospermum* (*Anabaena*) *flosaquae* originating from Canada (Carmichael et al., 1975). At the time, several types of toxins (anatoxins a-d) were suspected (Carmichael & Gorham, 1978), of which, however, only one eventually led to the elucidation of the absolute structure (Devlin et al., 1977) for which the suffix "-a" was kept.

Besides ATX, the following also includes information on its variant homoanatoxin-a (HTX), where available. The genetics and biosynthesis of ATX and other neurotoxic substances with a high structural variability produced by some marine cyanobacteria (Aráoz et al., 2010) have been reviewed by Pearson et al. (2016) and Bruno et al. (2017).

2.3.1 Chemical structures

Anatoxins are secondary amine alkaloids (Devlin et al., 1977; Figure 2.3a). The first synthesis of ATX yielded a racemic mixture of stereoisomers with optically positive and negative activity (Campbell et al., 1979). Homoanatoxin-a is a structural variant (differing from ATX by an ethyl-group at the carbonyl-C; Figure 2.3b). It was first synthesised by Wonnacott et al. (1992) just before Skulberg et al. (1992) isolated it from a sample of *Kamptonema* (*Oscillatoria*) *formosum*. Due to its structural similarity to ATX, HTX is most probably produced by the same biosynthetic pathway, with the additional carbon deriving from L-methionine via *S*-adenosyl-methionine (Namikoshi et al., 2004).

Further natural analogues of ATX are dihydroATX (dhATX; Figure 2.3c) and dihydroHTX reduced on C7 and C8, respectively (Smith & Lewis, 1987; Wonnacott et al., 1991).

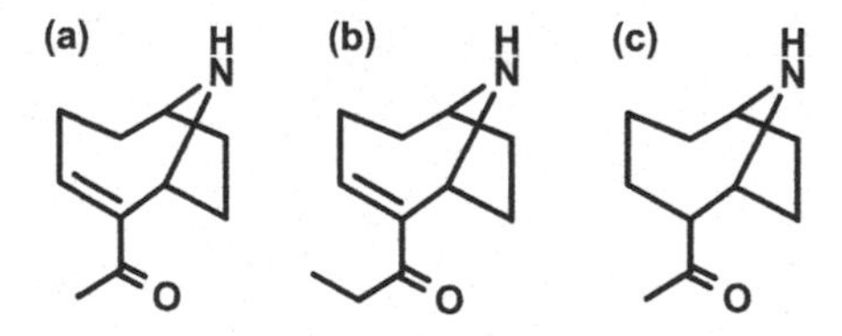

Figure 2.3 Chemical structure of anatoxin-a (a), homoanatoxin (b) and dihydroanatoxin-a (c). Anatoxin-a: molecular mass (monoisotopic): 165.115 Da; molecular weight (average): 165.237 g/mol. Homoanatoxin-a: molecular mass (monoisotopic) 179.131 Da; molecular weight (average): 179.264 g/mol. Dihydroanatoxin-a: molecular mass (monoisotopic) 167.131 Da; molecular weight (average): 167.252 g/mol.

2.3.2 Toxicity: mode of action

The toxic effects of anatoxin-a, summarised in the following, are described in detail in the WHO Background Document on Anatoxin-a (WHO, 2020); see there for further information and references). In summary, ATX is rapidly and passively absorbed after ingestion and widely distributed to different tissues, including the brain. No information about its biotransformation is available but, overall, a low bioaccumulating potential can be anticipated. Anatoxin-a acts as a potent pre- and postsynaptic depolarising agent; it efficiently competes with acetylcholine for nicotinic receptors in neuromuscular junctions and the central nervous system, triggering neurotransmitter release with an increased stimulation of postsynaptic receptors. The cardiovascular system has also been indicated as a target organ. Death through the administration of a lethal ATX dose is due to muscular paralysis and respiratory failure (i.v. LD_{50} = 85 µg/kg bw; i.p. LD_{50} = 260–315 µg/kg bw; oral LD_{50}>5000 µg/kg bw). Acute studies in animals led to deaths within minutes of gavage administration. After the administration of a sublethal single dose, mice readily recovered. Additional effects attributed to ATX in cell cultures include cytotoxic effects, caspase activation, apoptosis, induction of oxidative stress and formation of reactive oxygen species. Diagnosis of ATX and HTX poisoning in dogs and livestock has been reported due to neurotoxic effects after drinking and bathing in waters with ATX-producing cyanobacteria, such as species of the genera *Phormidium*, *Oscillatoria* and *Tychonema*.

On a weight of evidence basis, it can be concluded that ATX has no developmental or teratogenicity potential and is not mutagenic in bacteria. No *in vivo* carcinogenicity studies have been carried out. Regarding effects in humans, neurological symptoms (e.g., headache and confusion/visual disturbance) were reported in 3 of 11 outbreaks associated with cyanobacteria in the USA in 2009–2010 (Hilborn et al., 2014), in which ATX was found in a concentration range of 0.05–15 µg/L, while none of these symptoms were reported in the other 8 outbreaks, where ATX was not detected.

Homoanatoxin-a shows a mode of action and toxicological properties almost identical to its analogue ATX. Dihydro-anatoxin has been suggested as the congener most likely responsible for some dog deaths (Wood et al., 2017). Furthermore, a study indicates that dhATX is about fourfold more toxic than ATX when administered by gavage (Puddick et al., 2021).

2.3.3 Derivation of health- based reference values

The following section is taken directly from the WHO chemicals background document on anatoxins (WHO, 2020) which gives the considerations for the derivation of provisional guideline values for exposure to anatoxin in more detail.

Acute exposure to ATX in animals led to deaths within minutes of gavage administration (Astrachan, Archer & Hilbelink, 1980; Fawell et al., 1999). Since neither of the available repeated toxicity studies identified a nonlethal dose that caused lasting adverse effects, formal guideline values (GVs) (provisional or otherwise) cannot be derived based on the available information. In the 28-day study of Fawell et al. (1999), one of 20 animals in each of two dose groups died without signs that could be attributed to nontreatment effects. If it is conservatively assumed that these animals died from the effects of the toxin, the no observed adverse effect level (NOAEL) would be 98 µg/kg bw per day, but it could be as high as 2.46 mg/kg bw per day if these two animals were excluded (Fawell et al., 1999). Although GVs cannot be derived due to inadequate data, a "bounding value", or provisional health-based reference value, can be derived for short-term exposure using a highly conservative assumption to define the NOAEL at 98 µg/kg. This value is lower than the estimated NOAEL for exposure via drinking-water calculated from data in Astrachan, Archer & Hilbelink (1980) and the i.p. NOAEL for maternal toxicity identified by Rogers et al. (2005).

There is insufficient information to develop a long-term health-based reference value for ATX.

Default assumptions were applied as described in WHO (2009) for deriving the short-term drinking-water value and WHO (2003) for deriving the recreational water value.

Calculation of provisional short-term drinking-water health-based reference value for ATX

$$\text{HBRV}_{\text{short-term}} = \frac{\text{NOAEL}^*\text{bw}^*P}{\text{UF}^*C} = \frac{98\ ^*60\ ^*1.0}{100\ ^*2}\ \mu\text{g}/\text{L} = 29.4\ \mu\text{g}/\text{L} \approx 30\ \mu\text{g}/\text{L}$$

where

$\text{HBRV}_{\text{short-term}}$ = short-term drinking-water health-based reference value
NOAEL = no-observed-adverse-effect level (98 µg/kg bw per day, based on Fawell et al., 1999)
bw = body weight (default = 60 kg for an adult)
P = fraction of exposure allocated to drinking-water (default for short-term exposure = 100%, considering that drinking-water is expected to be the most likely source of exposure)
UF = uncertainty factor (10 for interspecies variation × 10 for intraspecies variation); an uncertainty factor for database deficiencies was not applied since the NOAEL is lower than the i.p. NOAEL for maternal toxicity
C = daily drinking-water consumption (default = 2 L for an adult).

Calculation of provisional recreational water health-based reference value for anatoxin-a

$$HBRV_{recreation} = \frac{NOAEL * bw}{UF * P} = \frac{98 * 15}{100 * 0.25} \mu g/L = 58.8 \mu g/L \approx 60 \mu g/L$$

where

$HBRV_{recreation}$ = recreational water health-based reference value
NOAEL = no-observed-adverse-effect level (98 μg/kg bw per day, based on Fawell et al., 1999)
bw = body weight (default = 15 kg for a child)
UF = uncertainty factor (10 for intraspecies variation × 10 for interspecies variation)
C = daily incidental water consumption (default = 250 mL for a child).

Considerations in applying the provisional health-based reference values

Derivation of the provisional health-based reference values for ATX follows a highly conservative approach. As a result of inadequate data, the provisional health-based reference values derived above do not represent WHO GVs and therefore are not intended for use as scientific points of departure for developing regulations or standards. Nevertheless, a "bounding value" may be useful to guide actions and responses by water suppliers and health authorities. Based on the limited currently available studies of acute and subchronic ATX toxicity, exposure up to the values provided is expected to be safe for adults. Since infants and children can ingest a significantly larger volume of water per body weight (e.g., up to 5 times more drinking-water/kg bw for bottle-fed infants than for an adult), it is recommended that alternative water sources, such as bottled water, are provided for bottle-fed infants and small children when ATX concentrations are greater than 6 μg/L for short periods, as a precautionary measure.

The provisional drinking-water health-based reference value is based on a 28-day repeated dose study and so is applicable for short-term exposure. However, because ATX is acutely toxic, it is recommended that any exposure above this value be avoided.

The provisional health-based reference values are based on toxicological data for ATX. It is recommended that for assessing risk, total ATXs as gravimetric or molar equivalent are evaluated against the health-based reference values, based on a reasonable assumption that HTX has similar toxicity to ATX. There is evidence that dihydro-analogues of ATX and HTX are similarly toxic by the oral route of exposure; hence it would be prudent to include these in determinations of total ATXs, when present.

2.3.4 Production

2.3.4.1 Producing cyanobacteria

Anatoxin was first found in *Dolichospermum* (*Anabaena*) *flosaquae* strains originating from Canada (Carmichael et al., 1975; Devlin et al., 1977) and later in Finland in *Anabaena mendotae* (Rapala et al., 1993), and *D. circinale* and *Anabaena* sp. in Finland and Japan (Sivonen et al., 1989; Park et al., 1993). Since then, many papers have been published reporting its production by several cyanobacteria species in many geographic areas by a variety of cyanobacteria taxa belonging to Nostocales – that is, *Chrysosporum* (*Aphanizomenon*) *ovalisporum, Cuspidothrix, Raphidiopsis* (*Cylindrospermopsis*), *Cylindrospermum, Dolichospermum* (*Anabaena*) *circinale, D. flosaquae* and *D. lemmermannii* – and to Oscillatoriales, that is, *Blennothrix, Kamptonema, Microcoleus, Oscillatoria, Planktothrix, Phormidium* and *Tychonema* (for species names and taxonomic changes, see Chapter 3). Tables 2.6 and 2.7 give examples of ATX contents in strains and concentrations in environmental samples, respectively. For further details, see reviews by Funari and Testai (2008), Pearson et al. (2016), Testai et al. (2016) and Cirés and Ballot (2016).

The production of ATX is species- and strain-specific. It is of interest that the American and European isolates of *D. circinale* investigated so far produce only ATX, while the Australian isolates exclusively produce saxitoxins, even if the two strains are reported to form a phylogenetically coherent group (Beltran & Neilan, 2000).

Homoanatoxin-a was first isolated from a *Kamptonema* (*Oscillatoria*) *formosum* strain in Ireland (Skulberg et al., 1992). Subsequently, it was found to be produced by *Raphidiopsis mediterranea* in Japan and *Oscillatoria* in Norway, isolated from *Microcoleus* (*Phormidium*) *autumnalis* in New Zealand and from species of *Dolichospermum*/*Anabaena* in Ireland (see Testai et al., 2016).

2.3.4.2 Toxin profiles

Anatoxin has been found to be produced alone by *Microcoleus* (*Phormidium*) cf. *autumnalis* (James et al., 1997) as well as coproduced with HTX in *Raphidiopsis mediterranea* (Watanabe et al., 2003), *Oscillatoria* (Araóz et al., 2005), and with microcystins in *Arthrospira fusiformis* (Ballot et al., 2005), *Microcystis* sp. (Park et al., 1993) and *Dolichospermum*/*Anabaena* spp. (Fristachi & Sinclair, 2008). *M. autumnalis* can contain high contents of HTX (together with ATX), showing large differences in toxin contents from week to week, and in some cases also in the same day (Wood et al., 2012). Non-axenic *M. autumnalis* strain CAWBG557 produces ATX, HTX and their dihydrogen derivatives dihydroanatoxin-a (dhATX) and dihydrohomoanatoxin-a (dhHTX; Heath et al., 2014). Dihydro-anatoxin-a

Table 2.6 Neurotoxin contents reported from laboratory cultures of cyanobacteria

Toxin	*Taxon* [a]	*Content in µg/g dw* [b]	*Origin*	*Reference*
ATX	*Oscillatoria sp.*	13 000	FIN	Sivonen et al. (1989)
	Oscillatoria sp.	2713	FIN	Harada et al. (1993)
	Oscillatoria sp.	4000	FIN	Araóz et al. (2005)
	Aphanizomenon sp.	6700	FIN	Sivonen et al. (1989)
	Aphanizomenon sp.	1562	FIN	Harada et al. (1993)
	Cuspidothrix issatschenkoi	(400 fg/cell)	NZL	Wood et al. (2007a)
	C. issatschenkoi	2354 (100 fg/cell)	DEU	Ballot et al. (2010)
	C. issatschenkoi	1683	NZL	Gagnon & Pick (2012)
	Aph. flosaquae	≈6500 [d]	FIN	Rapala et al. (1993)
	Dolichospermum (Anabaena) mendotae	≈9800 [d]		Rapala et al. (1993)
	D. flosaquae	≈8800 [d]		Rapala et al. (1993)
	C. issatschenkoi	(9.4 fg/cell)	NZL	Selwood et al. (2007)
	D. flosaquae (4)	1017 – 13 000	FIN	Sivonen et al. (1989)
	D. flosaquae	13 013	CAN	Harada et al. (1993)
	D circinale	8200	FIN	Gallon et al. (1994)
	D. circinale	4400	FIN	Harada et al. (1993)
	D. circinale (2)	1396 – 3500	FIN	Sivonen et al. (1989)
	Arthrospira fusiformis	0.3	KEN	Ballot et al. (2005)
	Arthrospira fusiformis	10.4	KEN	Kotut et al. (2006)
	Nostoc carneum	156	IRN	Ghassempour et al. (2005)
HTX	*Kamptonema (Oscillatoria) formosum*	n.q.	NOR	Skulberg et al. (1992)
	Microcoleus (Phormidium) autumnalis	(437 fg/cell; ATXeq)	NZL	Heath et al. (2014)
	Raphidiopsis mediterranea	n.q.	JPN	Watanabe et al. (2003)
	Oscillatoria sp. (2)	n.q.		Araóz et al. (2005)
ATX-S	*D. lemmermannii*	29–743	DNK	Henriksen et al. (1997)
	D. flosaquae	n.q.	CAN	Carmichael & Gorham (1978)
	Sphaerospermopsis torques-reginae	n.q.	BRA	Dörr et al. (2010)

(*Continued*)

Table 2.6 (Continued) Neurotoxin contents reported from laboratory cultures of cyanobacteria

Toxin	*Taxon* [a]	*Content in µg/g dw* [b]	*Origin*	*Reference*
STXs	*Aph.* c.f. *flosaquae* [d]	GTX4:≈7 dcGTX2:≈5 neoSTX:≈1 dcSTX:≈0.8 dcGTX3:≈0.5	CHN	Liu et al. (2006b) Liu et al. (2006a)
	Aph. c.f. *flosaquae* [c]	n.q.	USA	Mahmood & Carmichael (1986)
	Aph. gracile	n.q. (ca. 910 STXeq/L)		Pereira et al. (2004)
	Aph. gracile	neoSTX: 500–1600 STX: 550–780 dcSTX: 2.6–5.0 dcNEO: 3.6–6.5	TUR	Yilmaz et al. (2018)
	Aphanizomenon sp.	GTX5+neoSTX: 34.6 fg/cell	PRT	Dias et al. (2002)
	C. issatschenkoi (LMECYA31)	GTX5: 0.80 neoSTX: 0.24 dcSTX: 0.05 STX: 0.05		Pereira et al. (2000) Li et al. (2003)
	D. circinale	1580	AUS	Negri & Jones (1995)
	D. circinale (28)	0.77 fg/cell (STX+deSTX+GTX2/3+deGTX2/3+GTX5+C1/2)	AUS	Pereyra et al. (2017)
	D. circinale	GTX3: 1008 C2: 1545 STXeq: 2553	AUS	Velzeboer et al. (2000)
	D. perturbatum / spiroides	GTX3: 14	AUS	Velzeboer et al. (2000)
	Raphidiopsis raciborskii (2)	STXeq: 0.010	BRA	Lagos et al. (1999)
	R. raciborskii	STX: 0.3	BRA	Castro et al. (2004)
	Planktothrix sp.	n.q. STX	ITA	Pomati et al. (2000)

Numbers following taxa indicate the number of tested strains if more than a single strain was analysed. The taxonomic classification is listed according to the current nomenclature with earlier synonyms given in parentheses (for an overview on recent changes in taxonomy, see Chapter 3).

n.q.: not quantified, only qualitative detection reported.

[a] The taxon given here may deviate from that given in the publication. For changes in taxonomy, see Chapter 3.

[b] If not specified otherwise.

[c] Several strains of Aph. flosaquae have been reclassified as Aphanizomenon sp. or Aph. gracile, respectively.

[d] ≈ Estimated from figure in publication.

Table 2.7 Neurotoxin contents of biomass and concentrations in water reported from environmental samples

Toxin	*Dominant taxa* [a]	*Concentrations/ contents/cell quota*	*Type*	*Origin*	*Reference*
ATX	*Phormidium favosum*	8000 µg/g dw	R	FRA	Gugger et al. (2005)
	Microcoleus. cf. *autumnalis*	444 µg/L 16 µg/g dw	L	IRL	James et al. (1997)
	Dolichospermum sp.	390 µg/L 100 µg/g dw	L	IRL	James et al. (1997)
	Dolichospermum sp. *Aphanizomenon* sp.	13 µg/L intra+extra	L/Res.	DEU	Bumke-Vogt et al. (1999)
	Aphanizomenon sp.	35 µg/g dw	L	RUS	Chernova et al. (2017)
	Cuspidothrix issatschenkoi	1430 µg/L	L	NZL	Wood et al. (2007a)
	Dolichospermum sp. *Aphanizomenon* sp. *Cylindrospermum* sp.	4400 µg/g dw	L	FIN	Sivonen et al. (1989)
	Arthrospira fusiformis	2 µg/g dw	L	KEN	Ballot et al. (2005)
	Anabaena sp. *Art. fusiformis*	223 µg/g dw	L	KEN	Kotut et al. (2006)
	Microcoleus cf. *autumnalis*	0.027 µg/g ww	R	NZL	Wood et al. (2007a)
HTX	*M.* cf. *autumnalis*	0.44 µg/g ww	R	NZL	Wood et al. (2007b)
	Anabaena spp.	34 µg/L	L	IRL	Furey et al. (2003)
dhATX	*M.* cf. *autumnalis*	2118 µg/L	P		Wood et al. (2017)
ATX(S)	*D. lemmermannii*	3300 µg/g dw	L	DNK	Henriksen et al. (1997)
STXs	*D. lemmermannii*	224 µg /g dw STXeq	L	DNK	Kaas & Henriksen (2000)
	D. lemmermannii	930 µg /g dw STXeq 1000 µg /L STXeq	L	FIN	Rapala et al. (2005)
	D. lemmermannii	600 µg/L STX	R	RUS	Grachev et al. (2018)
	D. circinale	4466 µg /g dw STXeq	L/R	AUS	Velzeboer et al. (2000)
	D. circinale	2040 µg STXeq/g dw	L/R	AUS	Humpage et al. (1994)

(*Continued*)

Table 2.7 (Continued) Neurotoxin contents of biomass and concentrations in water reported from environmental samples

Toxin	*Dominant taxa* [a]	*Concentrations/ contents/cell quota*	*Type*	*Origin*	*Reference*
	Planktothrix sp.	181 µg/L STX (intra)	L	ITA	Pomati et al. (2000)
	Aph. flosaquae	4.7 µg/g dw STXeq	Res	PRT	Ferreira et al. (2001)
	Aph. favaloroi	STX: 42 µg/g dw 0.17 fg/cell neoSTX: 17 µg/g dw 0.07 fg/cell	L	GRE	Moustaka-Gouni et al. (2017)
	Aphanizomenon sp.	neoSTX: 2.3 µg/g dw dcSTX: 2.3 µg/g dw dcGTX3: 0.5 µg/g dw	L	CHI	Liu et al. (2006b)
	R. raciborskii	3.14 µg/L STXeq (intra+extra)	Res	BRA	Costa et al. (2006)
	Microseira (Lyngbya) wollei	19–73 µg STXeq/g dw	R	USA	Foss et al. (2012)
	M. wollei	58 µg STXeq/g dw	L/Res	USA	Carmichael et al. (1997)

Contents are given in µg toxin per gram dry weight (dw) or wet weight (ww). For individual studies, maximum values are given. Samples were collected in different types of waterbodies (L: lakes, R: rivers, P: pond, Res: reservoirs) in countries as indicated. For saxitoxins, contents are reported as saxitoxin equivalents (STXeq) in some reports or as individual variants (see text). The taxonomic classification is listed according to the current nomenclature with earlier synonyms given in parentheses (for an overview on recent changes in taxonomy, see Chapter 3).

[a] The taxon given here may deviate from that given in the publication. For changes in taxonomy, see Chapter 3.

has been reported to be produced in amounts much higher than those of ATX by strains of *M. autumnalis* (Wood et al., 2017; Puddick et al., 2021)

The few data available on ATX cell quota range from 90 fg/cell in *Cuspidothrix issatschenkoi* (Selwood et al., 2007) to 500 fg/cell in *M. autumnalis* (Heath et al., 2014). Cell quota detected in *Tychonema bourrellyi* were in a similarly wide range, 10–350 fg/cell (Shams et al., 2015).

The highest contents within the wide variability of ATX contents reported from strains grown as laboratory cultures, in the order of a few mg/g dw, were found in strains of the genera *Oscillatoria, Phormidium, Aphanizomenon, Cuspidothrix* and *Dolichospermum*. The maximum value (13 mg/g dw) was found in *D. flosaquae* and *Oscillatoria* sp., while much lower contents – generally by 1–2 orders of magnitude – of ATX are reported for cyanobacteria of other genera (Testai et al., 2016).

2.3.4.3 Biosynthesis and regulation

Cyanobacteria produce (+)ATX, but no specific studies have addressed the stereoselectivity of the biochemical reaction towards the positive enantiomer.

Anatoxin biosynthesis and regulation have been reviewed in Pearson et al. (2016). Méjean et al. (2009) reported the identification of the first gene cluster coding for the biosynthesis of ATXs (*ana*) within the sequenced genome of *Oscillatoria* sp. PCC 6506, producing mainly HTX. In the following years, five other *ana* clusters were identified within *Dolichospermum/Anabaena* sp. 37, *Oscillatoria* sp. PCC 6407, *Cylindrospermum stagnale* sp. PCC 7417, *Cuspidothrix issatschenkoi* RM-6, *C. issatschenkoi* LBRI48 and *C. issatschenkoi* CHABD3 (Rantala-Ylinen et al., 2011; Shih et al., 2013; Méjean et al., 2014; Jiang et al., 2015).

Each cluster showed general similarities in the protein functions, with a high percentage of identity in nucleotide sequence (with the core genes *anaB-G* being conserved within all strains), but differences in the organisation of genes (Pearson et al., 2016), leading to different toxin profiles between the producing organisms.

The biosynthesis of the ATXs involves a polyketide synthase (PKS) family of multifunctional enzymes with a modular structural organisation as described in Méjean et al. (2014). A detailed biochemical description of the adenylation domain protein AnaC revealed the activation of proline as starter, and not glutamate as previously proposed (Dittmann et al., 2013). The biosynthetic pathway describes AnaB, AnaC and AnaD as acting in the first steps (which have been fully reproduced *in vitro*; Méjean et al., 2009; Méjean et al., 2010; Mann et al., 2011), and AnaE, AnaF, Ana J and AnaG catalysing the following steps, with the latter adding two carbons and methylating the substrate to produce HTX. The release of ATXs may be catalysed by the thioesterase AnaA, although this has not been experimentally verified (Pearson et al., 2016) or a spontaneous decarboxylation step may occur to yield the amine alkaloid ATX (Dittmann et al., 2013).

The molecular regulation of ATX has not been sufficiently studied so far. Under conditions where *anaA*, *anaJ*, *anaF* and *anaG* transcripts were present in *C. issatschenkoi* CHABD3, no ATX was detected (Jiang et al., 2015). This result may indicate that the regulation of ATX occurs at the post-transcriptional level, but interpretation is limited by the lack of investigation of ATX dihydroderivatives production (Pearson et al., 2016).

An influence of light, temperature, phosphorous and nitrogen on cellular ATX content is reported, and it seems that the influence of environmental factors is strain-specific (Harland et al., 2013; Neilan et al., 2013; Boopathi & Ki, 2014; Heath et al., 2014). Overall, the influence of factors, such as light and temperature, reported for the ATX content in *Dolichospermum/Anabaena* and *Aphanizomenon* cultures varies around 2–4-fold, not exceeding a factor of 7 (Rapala & Sivonen, 1998), and a similar range is reported for HTX in relation to phosphorus (Heath et al., 2014). HTX production also seems to be linked to the culture growth phase in *Raph. mediterranea* strain LBRI 48 (Namikoshi et al., 2004). However, the results of most studies were not strongly supported by statistical analyses; furthermore, determining the effect of nutrient limitation requires continuous culture systems or evaluating batch culture data in relation to growth rates, yet in few studies this was done.

2.3.5 Occurrence in water environments

Anatoxin-a has a worldwide distribution that includes temperate, tropical and cold climatic regions (Fristachi & Sinclair, 2008). Although the occurrence of ATX has been less frequently surveyed than that of microcystins, based on the available data, it is evident that a wide variability in ATX contents is reported from environmental freshwater samples (Testai et al., 2016).

In the USA, surveys conducted in Florida in 1999 and 2000 did not detect ATX in most of the samples tested, but the maximum concentration found amounted to 156 μg/L (Fristachi & Sinclair, 2008); in Nebraska, variable ATX concentrations up to 35 μg/L were measured in water samples collected from eight reservoirs between 2009 and 2010 (Al-Sammak et al., 2014), and the highest ATX levels (1170 μg/L) were found in Washington State, where three waterbodies had long-term recurring blooms (Trainer & Hardy, 2015).

In Europe, a monitoring programme on 80 German lakes and reservoirs found ATX in 25% of the surveyed waterbodies and in 22% of water samples with a maximum total concentration of 13.1 μg/L (Bumke-Vogt et al., 1999). In Finland, in a survey of 72 lakes with variable trophic state, nearly half of the blooms dominated by *Dolichospermum* did not contain detectable ATX (Rapala & Sivonen, 1998). Furthermore, in Finland, hepatotoxic blooms have been found to be twice as common as neurotoxic ones (Rapala & Sivonen, 1998). Among 20 Irish lakes investigated, homoanatoxin-a was found in four inland waters dominated by blooms of *Dolichospermum* spp. at concentrations of up to 34 μg/L (Furey et al., 2003).

In Kenya, seven lakes (two freshwater and five alkaline saline waters) and the hot spring mats of Lake Bogoria were investigated for cyanotoxins, and ATX was recorded in almost all of them, at up to 1260 μg/g dw but not as dissolved toxin (Kotut et al., 2006). ATX concentrations up to 2.0 μg/g dw were detected in two alkaline Kenyan crater lakes, dominated by *Arthrospira fusiformis* (Ballot et al., 2005).

A number of publications have addressed the production of ATX by benthic cyanobacteria: the highest toxin concentrations being reported in a river mat sample (8 mg/g dw) in France, formed by benthic *Kamptonema* (*Phormidium*) *formosum* (Gugger et al., 2005). Levels ranging from 1.8 to 15.3 μg ATX/g of lyophilised weight were detected in *Phormidium* biofilms in the Tarn River (France) with high spatiotemporal variability and the highest concentrations being recorded at the end of the summer period (Echenique-Subiabre et al., 2018). The maximum ATX concentration in surface waters reported to date was found in a lake in Ireland (444 µg/L), where no surface blooms were previously observed, and as in the French case, the causative agent was a benthic cyanobacterium (James et al., 1997). Benthic, mat-forming cyanobacteria are common also in New Zealand rivers, frequently populated by *Phormidium*, known to produce ATX and HTX, the latter at contents up to 4400 µg/g dw (Wood et al., 2007b; Wood et al., 2012). In a study motivated by dog deaths, Wood et al. (2017) reported moderate concentrations of ATX (25 µg/L) and high levels of dhATX (2,118 µg/L), indicating that the latter may be present in higher concentrations than estimated so far. These concentrations, however, are associated with benthic grab samples and do not represent concentrations in larger water volumes (see also section 12.8 on benthic sampling). For an example of animal poisoning at a recreational lake and possible implications for human health see also Box 5.6.

Benthic cyanobacterial mats dominated by *Phormidium terebriformis, Microseira* (*Lyngbya*) *wollei, Spirulina subsalsa* and *Synechococcus bigranulatus* in the hot springs at the shore of Lake Bogoria (Kenya) contained MC and ATX (Krienitz et al., 2003). Recently, periphytic and tychoplanktic *Tychonema* have been identified as a producer of ATX and HTX in Italian alpine lakes (Salmaso et al., 2016) and in a German lowland lake (Fastner et al., 2018). However, identification at species level has not always been undertaken for benthic cyanobacteria (Puschner et al., 2008; Faassen et al., 2012), and it seems likely that more HTX-producing *Oscillatoria* or *Phormidium/Microcoleus* populations – and species – will be identified as research continues.

Anatoxin-a occurrence is not limited to freshwater; indeed, it has been found in brackish waters in samples collected off the coast of Poland in the Baltic Sea at the beginning of September (Mazur & Plinski, 2003) and in Chesapeake Bay (USA) at concentrations ranging from 3×10^{-3} to 3 mg/L (Tango & Butler, 2008). Although different planktonic and benthic genera occur and possibly dominate in brackish water (*Nodularia, Aphanizomenon, Microcystis, Dolichospermum, Anabaena* and *Phormidium/Microcoleus*), in these environments ATX seems to be produced exclusively by species formerly assigned to *Phormidium* (Lopes et al., 2014). Moreover, ATX production was found in a benthic marine cyanobacterium (*Hydrocoleum lyngbyaceum*) in New Caledonia (Méjean et al., 2010).

Biocrust-forming cyanobacteria inhabiting the Kaffiøyra Plain (in the Arctic region) are able to synthesise ATX from 0.322 to 0.633 mg/g dw (Chrapusta et al., 2015).

The available data and information have not linked ATX to human poisoning *via* drinking-water (Humpage, 2008). Surveys of cyanotoxins in drinking-water supplies in 1999/2000 across Florida found ATX only in three finished waters with concentrations up to 8.5 µg/L (Burns, 2008). Nevertheless, ATX should not be excluded as a potential human health hazard because some *Oscillatoria* sp. potentially producing ATX can proliferate in facilities and tanks for water storage (Osswald et al., 2007).

2.3.5.1 Bioaccumulation

The issue has been extensively reviewed in Testai et al. (2016). Anatoxin-a has been detected at low concentrations (0.51–43.3 µg/g) in Blue Tilapia fish in Florida (Burns, 2008). However, in Nebraska, this toxin could not be detected in fish from a reservoir although it was present in samples of the water and aquatic plants at the location (Al-Sammak et al., 2014). Concentrations similar to those in Tilapia were found in carp and juvenile trout exposed to high concentrations of ATX in an experimental setting (Osswald et al., 2007; Osswald et al., 2011); when mussels were experimentally exposed to live cells of an *Anabaena* strain (ANA 37), much lower levels were detected in the tissues (Osswald et al., 2008).

A special case of food items potentially containing ATX are "blue-green algal food supplements" (BGAS) that are usually produced from *Spirulina maxima* or *Arthrospira* (*Spirulina*) *platensis* and *Aph. flosaquae*. In *Spirulina*/*Arthrospira*-based BGAS, no direct evidence of the presence of ATX has been reported, but two nontoxic metabolites of this toxin have been found at contents of up to 19 µg/g dw (Draisci et al., 2001). When 39 samples containing the genera *Arthrospira, Spirulina* and *Aphanizomenon* were analysed, three (7.7%) contained ATX at concentrations ranging from 2.5 to 33 µg/g dw (Rellán et al., 2009). See also section 5.4.

2.3.6 Environmental fate

2.3.6.1 Partitioning between cells and water

Anatoxins can be released from producing cells into the surrounding water, but very different results were reported in the ratio between the intra- and extracellular fractions, likely depending on the species and environmental conditions (Testai et al., 2016) as well as on the sensitivity of the analytical method used especially in earlier studies (Wood et al., 2011; Testai et al., 2016). There is currently no evidence that ATXs are released from viable, intact cells to a substantial degree. It may be hence concluded that ATXs are largely confined to viable cyanobacterial cells in the environment and that extracellular release occurs mainly through cell senescence and lysis.

Once released from cells into the surrounding water, ATX can undergo chemical and biological degradation (Rapala & Sivonen, 1998) (see below). This is a challenge for its detection in environmental samples: the presence of ATX degradation products reported in some Finnish lakes at concentrations of 100–710 µg/L for epo-ATX and at 5–150 µg/L for dihydro-ATX (Rapala et al., 2005) indicates that ATX derivatives may serve as indicator of the previous presence of dissolved ATX.

2.3.6.2 Chemical breakdown

In laboratory studies, ATX has been reported to undergo a rapid photochemical degradation in sunlight, under conditions of the light intensity and pH ranges expected to be associated with blooms: Stevens and Krieger (1991) observed the reaction rate to be positively related to both pH and light intensity, with half-lives for photochemical breakdown at pH≥6 of 1.6–11.5 h, whereas at pH of 2, ATX was very stable. Kaminski et al. (2013) showed that ATX was resistant to photosynthetic active radiation with degradation dependent on pH: at low pH (<3), ATX proved stable when stored at room temperature, with minimal (≤3%) losses over a period of 9 weeks, but gradual degradation (≥37% losses) occurred at neutral (pH 7) and high pH (9.5). Anatoxin-a is relatively stable in the dark (Matsunaga et al., 1989), with a half-life of 4–10 days (Stevens & Krieger, 1991), at a pH of 9.

The mouse bioassay results show that regardless of process, photolytic or nonphotolytic, the breakdown products are of reduced toxicity and not antagonistic towards the effects of ATX (Stevens & Krieger, 1991).

In conclusion, once released from cyanobacterial cells and dissolved in water, ATX may degrade faster in water with high pH and further mitigating factors (e.g., microbial activity, elevated temperature), but may generally be more stable than previously assumed.

2.3.6.3 Biodegradation

Biodegradation by bacteria also has an important role: under natural conditions, ATX and HTX are partially or totally degraded and converted to dihydro- and epoxy-derivatives (James et al., 2005). Isolated *Pseudomonas* spp. degraded ATX at a rate of 2–10 µg/mL × day (Kiviranta et al., 1991), organisms in sediments reduced ATX concentrations by 25–48% in 22 days (Rapala et al., 1994), and a laboratory experiment with lake sediments and natural bacteria resulted in a half-life of 5 days (Kormas & Lymperopoulou, 2013).

Dihydroanatoxin-a has been considered the major ATX degradation product, representing from 17% to 90% of the total ATX concentration in the environment (Mann et al., 2011). Its concentrations gradually increased over time, paralleled by a decrease in ATX concentrations (Wood et al., 2011), although the involved enzymatic steps are not fully clarified. However, Heath et al. (2014) found that dhATX can account for 64% of the total intracellular ATX quota, suggesting that it is internally formed and is not only the product of cell lysis and environmental degradation, but is synthesised *de novo* in the cells.

In conclusion, due to the (photo)chemical and biological degradation of ATX and HTX, environmental samples invariably contain large amounts of these derivatives. Similar reactions can be expected to occur within biota, including mammals, although these have so far not been reported. Therefore, both environmental and forensic (e.g., in case of animal poisoning) analyses should also include an investigation of these degradation products.

REFERENCES

Al-Sammak MA, Hoagland KD, Cassada D, Snow DD (2014). Co-occurrence of the cyanotoxins BMAA, DABA and anatoxin-a in Nebraska reservoirs, fish, and aquatic plants. Toxins. 6:488–508.

Aráoz R, Molgó J, De Marsac NT (2010). Neurotoxic cyanobacterial toxins. Toxicon. 56:813–828.

Araóz R, Nghiem HO, Rippka R, Palibroda N, de Marsac NT, Herdman M (2005). Neurotoxins in axenic oscillatorian cyanobacteria: coexistence of anatoxin-alpha and homoanatoxin-alpha determined by ligand-binding assay and GC/MS. Microbiology. 151:1263–1273.

Astrachan N, Archer B, Hilbelink D (1980). Evaluation of the subacute toxicity and teratogenicity of anatoxin-a. Toxicon. 18:684–688.

Ballot A, Fastner J, Lentz M, Wiedner C (2010). First report of anatoxin-a-producing cyanobacterium *Aphanizomenon issatschenkoi* in northeastern Germany. Toxicon. 56:964–971.

Ballot A, Krienitz L, Kotut K, Wiegand C, Pflugmacher S (2005). Cyanobacteria and cyanobacterial toxins in the alkaline crater lakes Sonachi and Simbi, Kenya. Harmful Algae. 4:139–150.

Beltran EC, Neilan BA (2000). Geographical segregation of the neurotoxin-producing cyanobacterium *Anabaena circinalis*. Appl Environ Microbiol. 66: 4468–4474.

Boopathi T, Ki J-S (2014). Impact of environmental factors on the regulation of cyanotoxin production. Toxins. 6:1951–1978.

Bruno M, Ploux O, Metcalf JS, Méjean A, Pawlik-Skowronska B, Furey A (2017). Anatoxin-a, homoanatoxin-a, and natural analogues. In: Meriluoto J, Spoof L, Codd GA et al., editors: Handbook of cyanobacterial monitoring and cyanotoxin analysis. Chichester: John Wiley & Sons:138–147.

Bumke-Vogt C, Mailahn W, Chorus I (1999). Anatoxin-a and neurotoxic cyanobacteria in German lakes and reservoirs. Environ Toxicol. 14:117–125.

Burns J (2008). Toxic cyanobacteria in Florida waters. In: Hudnell HK, editors: Cyanobacterial harmful algal blooms: state of the science and research needs. New York: Springer:127–137.

Campbell HF, Edwards OE, Elder JW, Kolt R (1979). Total synthesis of DL-anatoxin-a and DL-isoanatoxin-a. Polish J Chem. 53:27–37.

Carmichael WW, Biggs DF, Gorham PR (1975). Toxicology and pharmacological action of *Anabaena flos-aquae* toxin. Science. 187:542–544.

Carmichael WW, Evans WR, Yin QQ, Bell P, Moczydlowski E (1997). Evidence for paralytic shellfish poisons in the freshwater cyanobacterium *Lyngbya wollei* (Farlow ex Gomont) comb. nov. Appl Environ Microbiol. 63:3104–3110.

Carmichael WW, Gorham PR (1978). Anatoxins from clones of *Anabaena flos-aquae* isolated from lakes of western Canada. Mitt Int Verein Limnol. 21:285–295.

Castro D, Vera D, Lagos N, García C, Vásquez M (2004). The effect of temperature on growth and production of paralytic shellfish poisoning toxins by the cyanobacterium *Cylindrospermopsis raciborskii* C10. Toxicon. 44:483–489.

Chernova E, Sidelev S, Russkikh I, Voyakina E, Babanazarova O, Romanov R et al. (2017). *Dolichospermum* and *Aphanizomenon* as neurotoxins producers in some Russian freshwaters. Toxicon. 130:47–55.

Chrapusta E, Węgrzyn M, Zabaglo K, Kaminski A, Adamski M, Wietrzyk P et al. (2015). Microcystins and anatoxin-a in Arctic biocrust cyanobacterial communities. Toxicon. 101:35–40.

Cirés S, Ballot A (2016). A review of the phylogeny, ecology and toxin production of bloom-forming *Aphanizomenon* spp. and related species within the Nostocales (cyanobacteria). Harmful Algae. 54:21–43.

Costa I, Azevedo S, Senna P, Bernardo R, Costa S, Chellappa N (2006). Occurrence of toxin-producing cyanobacteria blooms in a Brazilian semiarid reservoir. Braz J Biol. 66:211–219.

Devlin J, Edwards O, Gorham P, Hunter N, Pike R, Stavric B (1977). Anatoxin-a, a toxic alkaloid from *Anabaena flos-aquae* NRC-44h. Can J Chem. 55:1367–1371.

Dias E, Pereira P, Franca S (2002). Production of paralytic shellfish toxins by *Aphanizomenon* sp LMECYA 31 (cyanobacteria). J Phycol. 38:705–712.

Dittmann E, Fewer DP, Neilan BA (2013). Cyanobacterial toxins: biosynthetic routes and evolutionary roots. FEMS Microbiol Rev. 37:23–43.

Dörr FA, Rodríguez V, Molica R, Henriksen P, Krock B, Pinto E (2010). Methods for detection of anatoxin-a (s) by liquid chromatography coupled to electrospray ionization-tandem mass spectrometry. Toxicon. 55:92–99.

Draisci R, Ferretti E, Palleschi L, Marchiafava C (2001). Identification of anatoxins in blue-green algae food supplements using liquid chromatography-tandem mass spectrometry. Food Addit Contam. 18:525–531.

Echenique-Subiabre I, Tenon M, Humbert J-F, Quiblier C (2018). Spatial and temporal variability in the development and potential toxicity of *Phormidium* biofilms in the Tarn River, France. Toxins. 10:418.

Faassen EJ, Harkema L, Begeman L, Lurling M (2012). First report of (homo) anatoxin-a and dog neurotoxicosis after ingestion of benthic cyanobacteria in The Netherlands. Toxicon. 60:378–384.

Fastner J, Beulker C, Geiser B, Hoffmann A, Kröger R, Teske K et al. (2018). Fatal neurotoxicosis in dogs associated with tychoplanktic, anatoxin-a producing *Tychonema* sp. in mesotrophic Lake Tegel, Berlin. Toxins. 10:60.

Fawell JK, Mitchell RE, Hill RE, Everett DJ (1999). The toxicity of cyanobacterial toxins in the mouse: II Anatoxin-a. Human Exp Toxicol. 18:168–173.

Ferreira FM, Soler JMF, Fidalgo ML, Fernández-Vila P (2001). PSP toxins from *Aphanizomenon flos-aquae* (cyanobacteria) collected in the Crestuma-Lever reservoir (Douro river, northern Portugal). Toxicon 39:757–761.

Foss AJ, Phlips EJ, Yilmaz M, Chapman A (2012). Characterization of paralytic shellfish toxins from *Lyngbya wollei* dominated mats collected from two Florida springs. Harmful Algae. 16:98–107.

Fristachi A, Sinclair JL (2008). Occurrence of cyanobacterial harmful algal blooms workgroup report. In: Hudnell HK, editors: Cyanobacterial harmful algal blooms: state of the science and research needs. New York: Springer:45–103.

Funari E, Testai E (2008). Human health risk assessment related to cyanotoxins exposure. Crit Rev Toxicol. 38:97–125.

Furey A, Crowley J, Shuilleabhain AN, Skulberg AM, James KJ (2003). The first identification of the rare cyanobacterial toxin, homoanatoxin-a, in Ireland. Toxicon. 41:297–303.

Gagnon A, Pick FR (2012). Effect of nitrogen on cellular production and release of the neurotoxin anatoxin-a in a nitrogen-fixing cyanobacterium. Front Microbiol. 3:211.

Gallon JR, Kittakoop P, Brown EG (1994). Biosynthesis of anatoxin-a by *Anabaena flos-aquae*: examination of primary enzymic steps. Phytochemistry. 35:1195–1203.

Ghassempour A, Najafi NM, Mehdinia A, Davarani SSH, Fallahi M, Nakhshab M (2005). Analysis of anatoxin-a using polyaniline as a sorbent in solid-phase microextraction coupled to gas chromatography–mass spectrometry. J Chromatogr A. 1078:120–127.

Grachev M, Zubkov I, Tikhonova I, Ivacheva M, Kuzmin A, Sukhanova E et al. (2018). Extensive contamination of water with saxitoxin near the dam of the Irkutsk hydropower station reservoir (East Siberia, Russia). Toxins. 10:402.

Gugger MF, Lenoir S, Berger C, Ledreux A, Druart JC, Humbert JF et al. (2005). First report in a river in France of the benthic cyanobacterium *Phormidium favosum* producing anatoxin-a associated with dog neurotoxicosis. Toxicon. 45:919–928.

Harada K-I, Nagai H, Kimura Y, Suzuki M, Park H-D, Watanabe MF et al. (1993). Liquid chromatography/mass spectrometric detection of anatoxin-a, a neurotoxin from cyanobacteria. Tetrahedron. 49:9251–9260.

Harland FM, Wood SA, Moltchanova E, Williamson WM, Gaw S (2013). *Phormidium autumnale* growth and anatoxin-a production under iron and copper stress. Toxins. 5:2504–2521.

Heath MW, Wood SA, Barbieri RF, Young RG, Ryan KG (2014). Effects of nitrogen and phosphorus on anatoxin-a, homoanatoxin-a, dihydroanatoxin-a and dihydrohomoanatoxin-a production by *Phormidium autumnale*. Toxicon. 92:179–185.

Henriksen P, Carmichael WW, An JS, Moestrup O (1997). Detection of an anatoxin-a(s)-like anticholinesterase in natural blooms and cultures of Cyanobacteria/blue-green algae from Danish lakes and in the stomach contents of poisoned birds. Toxicon. 35:901–913.

Hilborn ED, Roberts VA, Backer L, DeConno E, Egan JS, Hyde JB et al. (2014). Algal bloom-associated disease outbreaks among users of freshwater lakes—United States, 2009–2010. Morb Mortal Wkly Rep. 63:11–15.

Humpage A, Rositano J, Bretag A, Brown R, Baker P, Nicholson B et al. (1994). Paralytic shellfish poisons from Australian cyanobacterial blooms. Mar Freshwat Res. 45:761–771.

Humpage AR (2008). Toxin types, toxicokinetics and toxicodynamics. In: H.K. Hudnell, editor: Proceedings of the Interagency, International Symposium on Cyanobacterial Harmful Algal Blooms (ISOC_HAB): Cyanobacterial harmful algal blooms: State of the science and research needs. New York, USA: Springer:383–415.

James KJ, Crowley J, Hamilton B, Lehane M, Skulberg O, Furey A (2005). Anatoxins and degradation products, determined using hybrid quadrupole time-of-flight and quadrupole ion-trap mass spectrometry: forensic investigations of cyanobacterial neurotoxin poisoning. Rapid Commun Mass Spectrom. 19:1167–1175.

James KJ, Sherlock IR, Stack MA (1997). Anatoxin-a in Irish freshwater and cyanobacteria, determined using a new fluorimetric liquid chromatographic method. Toxicon. 35:963–971.

Jiang Y, Song G, Pan Q, Yang Y, Li R (2015). Identification of genes for anatoxin-a biosynthesis in *Cuspidothrix issatschenkoi*. Harmful Algae. 46:43–48.

Kaas H, Henriksen P (2000). Saxitoxins (PSP toxins) in Danish lakes. Water Res. 34:2089–2097.

Kaminski A, Bober B, Lechowski Z, Bialczyk J (2013). Determination of anatoxin-a stability under certain abiotic factors. Harmful Algae. 28:83–87.

Kiviranta J, Sivonen K, Lahti K, Luukkainen R, Niemelä SI (1991). Production and biodegradation of cyanobacterial toxins - a laboratory study. Arch Hydrobiol. 121:281–294.

Kormas KA, Lymperopoulou DS (2013). Cyanobacterial toxin degrading bacteria: who are they? BioMed Res Int. 2013: 463894.

Kotut K, Ballot A, Krienitz L (2006) Toxic cyanobacteria and their toxins in standing waters of Kenya: implications for water resource use. J Water Health. 4:233–245.

Krienitz L, Ballot A, Kotut K, Wiegand C, Putz S, Metcalf JS et al. (2003). Contribution of hot spring cyanobacteria to the mysterious deaths of Lesser Flamingos at Lake Bogoria, Kenya. FEMS Microbiol Ecol. 43:141–148.

Lagos N, Onodera H, Zagatto PA, Andrinolo D, Azevedo SMFO, Oshima I (1999). The first evidence of paralytic shellfish toxins in the freshwater cyanobacterium *Cylindrospermopsis raciborskii*, isolated from Brazil. Toxicon. 37:1359–1373.

Li R, Carmichael WW, Pereira P (2003). Morphological and 16s rRNA gene evidence for reclassification of the paralytic shellfish toxin producing *Aphanizomenon flos-aquae* LMECYA 31 as *Aphanizomenon issatschenkoi* (Cyanophyceae). J Phycol. 39:814–818.

Liu Y, Chen W, Li D, Shen Y, Li G, Liu Y (2006a). First report of aphantoxins in China – waterblooms of toxigenic *Aphanizomenon flos-aquae* in Lake Dianchi. Ecotoxicol Environ Safety. 65:84–92.

Liu Y, Chen W, Li D, Shen Y, Liu Y, Song L (2006b). Analysis of Paralytic Shelfish Toxins in *Aphanizomenon* DC-1 from Lake Dianchi, China. Environ Toxicol. 21:289–295.

Lopes VM, Baptista M, Repolho T, Rosa R, Costa PR (2014). Uptake, transfer and elimination kinetics of paralytic shellfish toxins in common octopus (*Octopus vulgaris*). Aquat Toxicol. 146:205–211.

Mahmood NA, Carmichael WW (1986). Paralytic shellfish poisons produced by the freshwater cyanobacterium *Aphanizomenon flos-aquae* NH-5. Toxicon. 24:175–186.

Mann Sp, Lombard Brr, Loew D, Méjean A, Ploux O (2011). Insights into the reaction mechanism of the prolyl–acyl carrier protein oxidase involved in anatoxin-a and homoanatoxin-a biosynthesis. Biochemistry. 50:7184–7197.

Matsunaga S, Moore RE, Niemczura WP, Carmichael WW (1989). Anatoxin-a (s), a potent anticholinesterase from *Anabaena flos-aquae*. J Am Chem Soc. 111:8021–8023.

Mazur H, Plinski M (2003). *Nodularia spumigena* blooms and the occurrence of hepatotoxin in the Gulf of Gdańsk. Oceanologia. 45:305–316.

Méjean A, Mann S, Maldiney T, Vassiliadis G, Lequin O, Ploux O (2009). Evidence that biosynthesis of the neurotoxic alkaloids anatoxin-a and homoanatoxin-a in the cyanobacterium *Oscillatoria* PCC 6506 occurs on a modular polyketide synthase initiated by L-proline. J Am Chem Soc. 131:7512–7513.

Méjean A, Paci G, Gautier V, Ploux O (2014). Biosynthesis of anatoxin-a and analogues (anatoxins) in cyanobacteria. Toxicon. 91:15–22.

Méjean A, Peyraud-Thomas C, Kerbrat AS, Golubic S, Pauillac S, Chinain M et al. (2010). First identification of the neurotoxin homoanatoxin-a from mats of *Hydrocoleum lyngbyaceum* (marine cyanobacterium) possibly linked to giant clam poisoning in New Caledonia. Toxicon. 56:829–835.

Moustaka-Gouni M, Hiskia A, Genitsaris S, Katsiapi M, Manolidi K, Zervou S-K et al. (2017). First report of *Aphanizomenon favaloroi* occurrence in Europe associated with saxitoxins and a massive fish kill in Lake Vistonis, Greece. Mar Freshwater Res. 68:793–800.

Namikoshi M, Murakami T, Fujiwara T, Nagai H, Niki T, Harigaya E et al. (2004). Biosynthesis and transformation of homoanatoxin-a in the cyanobacterium *Raphidiopsis mediterranea* Skuja and structures of three new homologues. Chem Res Toxicol. 17:1692–1696.

Negri AP, Jones GJ (1995). Bioaccumulation of paralytic shellfish poisoning (PSP) toxins from the cyanobacterium *Anabaena circinalis* by the freshwater mussel *Alathyria condola*. Toxicon. 33:667–678.

Neilan BA, Pearson LA, Muenchhoff J, Moffitt MC, Dittmann E (2013). Environmental conditions that influence toxin biosynthesis in cyanobacteria. Environ Microbiol. 15:1239–1253.

Osswald J, Azevedo J, Vasconcelos V, Guilhermino L (2011). Experimental determination of the bioconcentration factors for anatoxin-a in juvenile rainbow trout (*Oncorhynchus mykiss*). Proc Int Acad Ecol Environ Sci. 1:77.

Osswald J, Rellán S, Gago A, Vasconcelos V (2007). Toxicology and detection methods of the alkaloid neurotoxin produced by cyanobacteria, anatoxin-a. Environ Int. 33:1070–1089.

Osswald J, Rellan S, Gago A, Vasconcelos V (2008). Uptake and depuration of anatoxin-a by the mussel *Mytilus galloprovincialis* (Lamarck, 1819) under laboratory conditions. Chemosphere. 72:1235–1241.

Park HD, Watanabe MF, Harada K-I, Nagai H, Suzuki M, Watanabe M et al. (1993). Hepatotoxin (Microcystin) and neurotoxin (Anatoxin-a) contained in natural blooms and strains of cyanobacteria from Japanese freshwaters. Nat Toxins. 1:353–360.

Pearson LA, Dittmann E, Mazmouz R, Ongley SE, D'Agostino PM, Neilan BA (2016). The genetics, biosynthesis and regulation of toxic specialized metabolites of cyanobacteria. Harmful Algae. 54:98–111.

Pereira P, Li RH, Carmichael WW, Dias E, Franca S (2004). Taxonomy and production of paralytic shellfish toxins by the freshwater cyanobacterium *Aphanizomenon gracile* LMECYA40. Europ J Phycol. 39:361–368.

Pereira P, Onodera H, Andrinolo D, Franca S, Araujo F, Lagos N et al. (2000). Paralytic shellfish toxins in the freshwater cyanobacterium *Aphanizomenon flos-aquae*, isolated from Montargil reservoir, Portugal. Toxicon. 38:1689–1702.

Pereyra JP, D'Agostino PM, Mazmouz R, Woodhouse JN, Pickford R, Jameson I et al. (2017). Molecular and morphological survey of saxitoxin-producing cyanobacterium *Dolichospermum circinale* (*Anabaena circinalis*) isolated from geographically distinct regions of Australia. Toxicon. 138:68–77.

Pomati F, Sacchi S, Rossetti C, Giovannardi S, Onodera H, Oshima Y et al. (2000). The freshwater cyanobacterium *Planktothrix* sp. FP1: molecular identification and detection of paralytic shellfish poisoning toxins. J Phycol. 36:553–562.

Puddick J, van Ginkel R, Page CD, Murray JS, Greenhough HE, Bowater J et al. (2021) Acute toxicity of dihydroanatoxin-a from *Microcoleus autumnalis* in comparison to anatoxin-a. Chemosphere. 263:127–937.

Puschner B, Hoff B, Tor ER (2008). Diagnosis of anatoxin-a poisoning in dogs from North America. J Vet Diagn Invest. 20:89–92.

Rantala-Ylinen A, Känä S, Wang H, Rouhiainen L, Wahlsten M, Rizzi E et al. (2011). Anatoxin-a synthetase gene cluster of the cyanobacterium *Anabaena* sp. strain 37 and molecular methods to detect potential producers. Appl Environ Microbiol. 77:7271–7278.

Rapala J, Lahti K, Sivonen K, Niemelä SI (1994). Biodegradability and adsorption on lake sediments of cyanobacterial hepatotoxins and anatoxin-a. Lett Appl Microbiol. 19:423–428.

Rapala J, Robertson A, Negri AP, Berg KA, Tuomi P, Lyra C et al. (2005). First report of saxitoxin in Finnish lakes and possible associated effects on human health. Environ Toxicol. 20:331–340.

Rapala J, Sivonen K (1998). Assessment of environmental conditions that favor hepatotoxic and neurotoxic *Anabaena* spp. strains cultured under light limitation at different temperatures. Microb Ecol. 36:181–192.

Rapala J, Sivonen K, Luukkainen R, Niemelä SI (1993). Anatoxin-a concentration in *Anabaena* and *Aphanizomenon* under different environmental conditions and comparison of growth by toxic and non-toxic *Anabaena*-strains - a laboratory study. J Appl Phycol. 5:581–591.

Rellán S, Osswald J, Saker M, Gago-Martinez A, Vasconcelos V (2009). First detection of anatoxin-a in human and animal dietary supplements containing cyanobacteria. Food Chem Toxicol. 47:2189–2195.

Rogers E, Hunter E, Moser V, Phillips P, Herkovits J, Munõz L et al. (2005). Potential developmental toxicity of anatoxin-a, a cyanobacterial toxin. J Appl Toxicol. 25:527–534.

Salmaso N, Cerasino L, Boscaini A, Capelli C (2016). Planktic *Tychonema* (Cyanobacteria) in the large lakes south of the Alps: phylogenetic assessment and toxigenic potential. FEMS Microbiol Ecol. 92:fiw155.

Selwood AI, Holland PT, Wood SA, Smith KF, Mcnabb PS (2007). Production of anatoxin-a and a novel biosynthetic precursor by the cyanobacterium *Aphanizomenon issatschenkoi*. Environ Sci Technol. 41:506–510.

Shams S, Capelli C, Cerasino L, Ballot A, Dietrich DR, Sivonen K et al. (2015). Anatoxin-a producing *Tychonema* (Cyanobacteria) in European waterbodies. Water Res. 69:68–79.

Shih PM, Wu D, Latifi A, Axen SD, Fewer DP, Talla E et al. (2013). Improving the coverage of the cyanobacterial phylum using diversity-driven genome sequencing. Proc Natl Acad Sci USA. 110:1053–1058.

Sivonen K, Himberg K, Luukkainen R, Niemelä S, Poon G, Codd G (1989). Preliminary characterization of neurotoxic cyanobacteria blooms and strains from Finland. Environ Toxicol. 4:339–352.

Skulberg OM, Carmichael WW, Andersen RA, Matsunaga S, Moore RE, Skulberg R (1992). Investigations of a neurotoxic oscillatorialean strain (Cyanophyceae) and its toxin. Isolation and characterization of homoanatoxin-a. Environ Toxicol Chem. 11:321–329.

Smith R, Lewis D (1987). A rapid analysis of water for anatoxin a, the unstable toxic alkaloid from *Anabaena flos-aquae*, the stable non-toxic alkaloids left after bioreduction and a related amine which may be nature's precursor to anatoxin a. Vet Human Toxicol. 29:153–154.

Stevens D, Krieger R (1991). Stability studies on the cyanobacterial nicotinic alkaloid anatoxin-A. Toxicon. 29:167–179.

Tango PJ, Butler W (2008). Cyanotoxins in tidal waters of Chesapeake Bay. Northeastern Nat. 15:403–416.

Testai E, Scardala S, Vichi S, Buratti FM, Funari E (2016). Risk to human health associated with the environmental occurrence of cyanobacterial neurotoxic alkaloids anatoxins and saxitoxins. Crit Rev Toxicol. 46:385–419.

Trainer VL, Hardy FJ (2015). Integrative monitoring of marine and freshwater harmful algae in Washington State for public health protection. Toxins. 7:1206–1234.

Velzeboer RM, Baker PD, Rositano J, Heresztyn T, Codd GA, Raggett SL (2000). Geographical patterns of occurrence and composition of saxitoxins in the cyanobacterial genus *Anabaena* (Nostocales, Cyanophyta) in Australia. Phycologia. 39:395–407.

Watanabe MF, Tsujimura S, Oishi S, Niki T, Namikoshi M (2003). Isolation and identification of homoanatoxin-a from a toxic strain of the cyanobacterium *Raphidiopsis mediterranea* Skuja isolated from Lake Biwa, Japan. Phycologia. 42:364–369.

WHO (2003). Guidelines for safe recreational water environments. Vol. 1: Coastal and fresh waters. Geneva: World Health Organization. https://apps.who.int/iris/handle/10665/42591

WHO (2020). Cyanobacterial toxins: Anatoxin-a and analogues. Background document for development of WHO Guidelines for Drinking-water Quality and Guidelines for Safe Recreational Water Environments. Geneva: World Health Organization. https://apps.who.int/iris/handle/10665/338060

Wonnacott S, Jackman S, Swanson K, Rapoport H, Albuquerque E (1991). Nicotinic pharmacology of anatoxin analogs. II. Side chain structure-activity relationships at neuronal nicotinic ligand binding sites. J Pharmacol Exp Therap. 259:387–391.

Wonnacott S, Swanson K, Albuquerque E, Huby N, Thompson P, Gallagher T (1992). Homoanatoxin: a potent analogue of anatoxin-a. Biochem Pharmacol. 43:419–423.

Wood SA, Holland PT, MacKenzie L (2011). Development of solid phase adsorption toxin tracking (SPATT) for monitoring anatoxin-a and homoanatoxin-a in river water. Chemosphere. 82:888–894.

Wood SA, Puddick J, Fleming RC, Heussner AH (2017). Detection of anatoxin-producing *Phormidium* in a New Zealand farm pond and an associated dog death. New Zeal J Botany. 55:36–46.

Wood SA, Rasmussen JP, Holland PT, Campbell R, Crowe ALM (2007a). First report of the cyanotoxin anatoxin-A from *Aphanizomenon issatschenkoi* (cyanobacteria). J Phycol. 43:356–365.

Wood SA, Selwood AI, Rueckert A, Holland PT, Milne JR, Smith KF et al. (2007b). First report of homoanatoxin-a and associated dog neurotoxicosis in New Zealand. Toxicon. 50:292–301.

Wood SA, Smith FM, Heath MW, Palfroy T, Gaw S, Young RG et al. (2012). Within-mat variability in anatoxin-a and homoanatoxin-a production among benthic *Phormidium* (cyanobacteria) strains. Toxins. 4:900–912.

Yilmaz M, Foss AJ, Selwood AI, Özen M, Boundy M (2018). Paralytic shellfish toxin producing *Aphanizomenon gracile* strains isolated from Lake Iznik, Turkey. Toxicon. 148:132–142.

2.4 SAXITOXINS OR PARALYTIC SHELLFISH POISONS

Emanuela Testai

Saxitoxins (STXs) are natural alkaloids also known as paralytic shellfish poisons (PSP) because they were originally found in molluscs, the consumption of which led to poisonings of humans. The organisms producing this group of toxins are marine microalgae – dinoflagellates of the genera *Alexandrium*, *Gymnodinium* and *Pyrodinium* – as well as freshwater cyanobacteria.

2.4.1 Chemical structures

Saxitoxins, also known as paralytic shellfish poisoning toxins, are a family of 57 analogues (Wiese et al., 2010), consisting of a tetrahydropurine group and two guanidine subunits, representing the tricyclic perhydropurine backbone. Depending on the substitutions in the variable positions, R1-R4, the family can be subdivided into four groups:

1. nonsulphated molecules whose structure is similar to carbamates, including saxitoxins and neosaxitoxin (STX, dcSTX, neoSTX);
2. monosulphated gonyautoxins (GTX 1 to 6 and dcGTX 2 and 3);
3. doubly-sulphated C-toxins (C1-2);
4. variants identified exclusively in strains of *Lyngbya* (*Microseira*) *wollei* (LWTX 1-6) from the USA (Lajeunesse et al., 2012), characterised by the presence of a hydrophobic side chain with an acetate at C13 (LWTX 1-3, 5, 6) and a carbinol at C12 (LWTX 2, 3, 5) instead of a hydrated ketone.

Alternatively, they are grouped on the basis of the R4 substituent into carbamate toxins (STX, neoSTX and GTX1-4), sulphamate toxins (GTX 5-6, C1-4) and decarbamoyl toxins (dcSTX, dcneoSTX and dcGTX1-4) (Figure 2.4, Table 2.8).

Most known STXs are hydrophilic, with the exception of those produced by *L. wollei* in a freshwater environment.

Interconversions among the different STX congeners have been reported, both chemically and enzymatically mediated, and in some cases, this is expected to increase toxicity. Some of the transformations include desulphation (Ben-Gigirey & Villar-González, 2008), oxidation (García et al., 2010), reduction (Oshima, 1995a; Fast et al., 2006), decarbamoylation (Oshima, 1995a; Fast et al., 2006), deacetylation (Foss et al., 2012) and epimerisation (Ben-Gigirey & Villar-González, 2008).

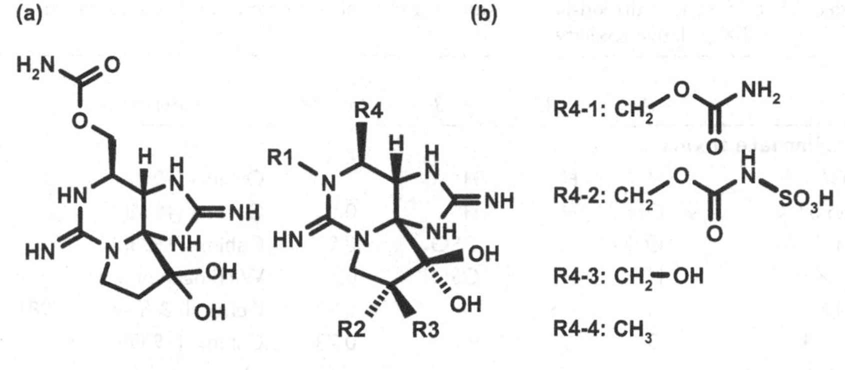

Figure 2.4 Structure of saxitoxin (a) and general structure of saxitoxins (STX) and gonyautoxins (GTX) (b). R4-1: carbamate toxins, including STX and neo-saxitoxin; R4-2: N-sulphocarbamoyl (or sulphamate) toxins, including GTX5 and GTX6; R4-3 decarbamoyl toxins, including dcSTX; R4-4: deoxydecarbamoyl toxins, including deoxydecarbamoyl-STX. For R1, R2 and R3 in particular variants, see Table 2.8.

2.4.2 Toxicity: mode of action

The toxic effects of saxitoxin (STX), summarised in the following, are described in detail in the WHO Background Document on Saxitoxins (WHO, 2020; see there for further information and references). The great majority of reported clinical, epidemiological and toxicological data about STXs have been obtained from cases of poisoning following the consumption of shellfish which accumulate STXs produced by marine dinoflagellates; however, since the chemical structure is the same as that of the STXs produced by cyanobacteria, the toxicological profile is identical. Saxitoxins are readily absorbed by the gastrointestinal tract, rapidly distributed to a range of tissues, including the central nervous system, and undergo rapid excretion mainly in the urine as glucuronides, thus suggesting glucuronidation as a possible detoxication metabolic pathway in animals and humans.

The mechanism of action of STXs is based on Na-channel blocking in neuronal cells and on Ca^{++} and K^{+} channel blocking in cardiac cells. This action prevents the propagation of electrical transmission within the peripheral nerves and skeletal or cardiac muscles. It leads to typical neurologic symptoms such as nervousness, twitching, ataxia, convulsions and muscle and respiratory paralysis, and at a lethal dose, death in animal experiments has been observed within a few minutes; for humans, death through respiratory paralysis has been reported after 2–24 h (FAO, 2004). Depending on the variants, STX toxicity in mice can differ considerably. Carbamate toxins are by far the more toxic and the lack of the carbamoyl group side

Table 2.8 Different saxitoxin-like congeners and their relative toxicity compared to STX (relative toxicity = 1)

	R1	*R2*	*R3*	*Relative toxicity*	*Reference*
Carbamate toxins					
STX	H	H	H	1	Oshima (1995b)
neo STX	OH	H	H	0.93	Oshima (1995b)
GTX1	OH	H	OSO_3^-	0.99	Oshima (1995b)
GTX2	H	H	OSO_3^-	0.41	Wichmann et al. (1981)
GTX3	H	OSO_3^-	H	0.90	Genenah & Shimizu (1981)
GTX4	OH	OSO_3^-	H	0.73	Oshima (1995b)
Sulphamate toxins					
GTX5	H	H	H	0.15	Genenah & Shimizu (1981)
GTX6	OH	H	H	0.07	Oshima et al. (1989)
C1	H	H	OSO_3^-	0.01	Wichmann et al. (1981)
C2	H	OSO_3^-	H	0.17	Oshima et al. (1989)
C3	OH	H	OSO_3^-	0.01	Oshima (1995b)
C4	OH	OSO_3^-	H	0.06	Oshima (1995b)
Decarbamoyl toxins					
dcSTX	H	H	H	0.51	Oshima (1995b)
dcneoSTX	OH	H	H	n.a.	
dcGTX1	OH	H	OSO_3^-	n.a.	
dcGTX2	H	H	OSO_3^-	0.65	Oshima (1995b)
dcGTX3	H	OSO_3^-	H	0.75	Oshima (1995b)
dcGTX4	OH	OSO_3^-	H	0.49	Oshima (1995b)

Where more than one value for i.p. acute toxicity was available for an individual toxin, highest acute toxicity was considered to calculate the relative toxicity. Toxicity of saxitoxins is generally expressed in mouse units (MU), that is, the amount injected toxin which would kill a 20 g mouse in 15 min and is equivalent to 0.18 µg of STX.

n.a.: not available. R1, R2 and R3 refer to the substituent groups as depicted in Figure 2.4.

chain gives rise to a molecule with about 60% of the original toxic activity, whereas C-toxins and LWTXs are characterised by a much lower toxicity.

No robust information on repeated toxicity, genotoxicity, carcinogenicity and reproductive or developmental toxicity is available.

Doses in the range 140–300 µg STXeq/person were reported to induce no or mild symptoms, but variability is pronounced; a case report indicated that ≈300 µg PSP toxin per person may be fatal (FAO, 2004). Mild clinical symptoms (tingling sensation or numbness around lips, gradually spreading to the face and neck) have a quick onset (hours), but may last for days. These symptoms precede prickly sensation in the fingertips and toes, headaches, dizziness, nausea, vomiting and diarrhoea, and distinct muscular weakness. A broad spectrum of effects, from mild to moderate symptoms up to paralysis and death, have been described following ingestions of 460–12 400 μg STXeq/person (FAO, 2004; McLaughlin et al., 2011). This high variability has been attributed to uncertainties in the detection of the actual level of exposure to different STX variants, differences in critical access to rapid health care and differences in individual susceptibility.

2.4.3 Derivation of guideline values

The following section is taken directly from the WHO chemicals background document on saxitoxins (WHO, 2020) which discusses the considerations for the derivation of provisional guideline values for exposure to saxitoxin in more detail. The GV for acute exposure through drinking-water is derived for bottle-fed infants, as the most sensitive subgroup in a population. This is considered appropriate for this cyanotoxin group because the GV is for acute exposure, and there is a relatively small margin of safety, as described below. All other default assumptions were applied as described in WHO (2009, 2017) for deriving the acute drinking-water GV, and in WHO (2003) for deriving the recreational GV.

FAO (2004) identified a LOAEL for mild symptoms of 2.0 μg/kg bw, based on a review of human cases of paralytic shellfish poisoning (PSP). More recently, EFSA (2009) reviewed about 500 cases of human PSP described in case reports that had estimated the consumption of STXs associated with a range of symptoms. This analysis identified a LOAEL for STXeq of 1.5 μg/kg bw by assuming an adult body weight of 60 kg. Because many individuals did not show symptoms at much higher estimated intakes, EFSA (2009) reasoned that the LOAEL must be very near the threshold for effects in sensitive individuals. Therefore an uncertainty factor of 3 was applied to the LOAEL "to estimate a NOAEL", establishing an acute reference dose (ARfD) for STXeq of 0.5 μg/kg bw. An uncertainty factor for intraspecies variation was not applied because documented human cases included a wide spectrum of people (occupation, age, and sex).

The GVs are derived from data from poisoning events caused by mixtures of STXs, with total STXs expressed as STX concentration equivalents (STXeq). The GVs therefore apply to total STXs in a sample, not just the parent compound, STX.

These values are supported by data from animal studies: the use of the lowest acute no observed adverse effect level (NOAEL) for neoSTX of 87 μg/kg bw after gavage administration as a point of departure leads to the derivation of an ARfD for neoSTX of 0.87 μg/kg bw (applying an uncertainty factor of 100). This value is of the same order of magnitude as the reference values obtained with human data (Testai et al., 2016).

Calculation of acute drinking-water guideline value for saxitoxins

$$GV_{acute} = \frac{LOAEL * bw * P}{UF * C} = \frac{1.5 * 5 * 1.0}{3 * 0.75} \mu g/L = 3.3 \mu g/L \approx 3 \mu g/L$$

where

GV_{acute} = guideline value for acute exposure
LOAEL = lowest-observed-adverse-effect level (1.5 µg STXeq/kg, based on the human data on PSP reports)
bw = body weight (default = 5 kg for an infant)
P = fraction of exposure allocated to drinking-water (default for short-term exposure = 100%, considering that drinking-water is expected to be the most likely source of exposure where surface water is used as the source of drinking-water)
UF = uncertainty factor (3, for use of a LOAEL rather than a NOAEL)
C = daily drinking-water consumption (default = 750 mL for an infant).

Calculation of recreational water guideline value for saxitoxin

The calculation is based on a scenario of a child playing in bloom-infested water:

$$GV_{recreation} = \frac{LOAEL * bw}{UF * C} = \frac{1.5 * 15}{3 * 0.25} \mu g/L = 30 \mu g/L$$

where

$GV_{recreation}$ = guideline value for recreational exposure
LOAEL = lowest observed-adverse-effect level (1.5 µg STXeq/kg, based on human poisoning data)
bw = body weight (default = 15 kg for a child)
UF = uncertainty factor (3, for use of a LOAEL rather than a NOAEL)
C = daily incidental water consumption (default = 250 mL for a child).

Considerations in applying the provisional guideline values

As indicated above, for assessing risk, the cumulative detection of both STX and its structural analogues should be evaluated against the GVs.

This is generally expressed as STXeq. STXeq can indicate concentration equivalents – calculated by simple addition of the concentrations of all analogues present, each being quantified against an analytical standard for that analogue. This represents a conservative approach to protect human health in most cases, assuming that all analogues have comparable characteristics and toxicity to STX. An exception is when the more potent neoSTX is the dominant congener present (see below). A more precise, usually less conservative approach is to determine STX toxicity equivalents by multiplying the concentration of each analogue by the respective toxicity equivalence factor (TEF) before addition. Where available, oral toxicities should be used in preference to relative i.p. toxicities. Munday et al. (2013) provides the acute oral toxicities of some analogues while a table of TEFs based on i.p. toxicity in mice has been published by EFSA (2009).

The acute GVs for STXs are based on acute exposure data. A time limit for tolerating concentrations up to 3 µg/L cannot be given because of the lack of data on effects at low doses. Thus, in contrast to other cyanotoxins, short-term and lifetime exposure GVs were not developed, and short-term exceedances of the acute GV should not be permitted. Although there is currently no evidence of health impairments from chronic exposure to low doses of STXs, it is always prudent to implement control measures to reduce the presence of toxic cyanobacterial blooms or their impact on drinking-water supplies as soon as possible (see Chapters 6–10). Limited data show that STX concentrations in drinking-water have almost always been at trace levels (see section 2.4.5), indicating that conventional water treatment is generally effective, provided that cell lysis is avoided (see Chapter 10).

The drinking-water GV for STXs uses an allocation factor of 100% for drinking-water; however, it may be appropriate to consider reducing the allocation factor for drinking-water in locations with increased risk of coincident water and shellfish exposure (marine or freshwater). However, it should be noted that GVs for STX in marine shellfish are comparatively high and, in locations where contamination of shellfish is a concern, drinking-water containing STX would contribute a relatively small additional exposure. Nevertheless, it is recommended that health authorities jointly consider and manage such a scenario, particularly given the relatively steep dose–response relationship for these toxins.

For the drinking-water acute GV, the lower body weight and higher likely water intake of an infant (as a function of body weight) were used because a GV based on adults could allow exposure of infants to a concentration of STXs close to the LOAEL. For a 60 kg adult consuming 2 L of drinking-water per day, a 5-fold higher concentration than the acute GV would be tolerable.

2.4.4 Production

2.4.4.1 Producing cyanobacteria

Saxitoxins are produced by species of marine eukaryotic dinoflagellates within the genera *Alexandrium*, *Gymnodinium* and *Pyrodinium* as well as by cyanobacteria within a range of species and strains belonging to the Nostocales, that is, *Dolichospermum* (*Anabaena*). (Humpage et al., 1994; Velzeboer et al., 2000), *Aphanizomenon* (Ikawa et al., 1982; Sasner et al., 1984; Pereira et al., 2000; Dias et al., 2002) and *Raphidiopsis* (*Cylindrospermopsis*) *raciborskii* mainly in Brazil (Lagos et al., 1999; Molica et al., 2002) and *Scytonema* (Smith et al., 2011) and Oscillatoriales such as *Planktothrix* and *Microseira* (*Lyngbya*) *wollei* (Carmichael et al., 1997; Onodera et al., 1997). From lakes and reservoirs of the southern USA, *Microseira wollei* is known to overwinter in the form of benthic mats and rises to form surface mats during the warmer months (Carmichael et al., 1997).

Cyanobium sp. CENA 142 and *Oxynema* sp. CENA 135 were among 135 strains isolated from cyanobacteria collected from Cardoso Island and Bertioga mangroves for which both molecular analyses and ELISA showed STXs production (Silva et al., 2014). For further details, see the review by Testai et al. (2016) and Cirés and Ballot (2016).

2.4.4.2 Toxin profiles

The production of different STX congeners seems to be strain-specific. Indeed, C1, C2, GTX2 and GTX3 were found as predominant congeners in environmental samples and isolated strains of *Dolichospermum circinale* in Australia, although a hitherto unique toxin composition (exclusively STX and GTX5) was found in a geographically isolated strain from the southwest coast of Australia (Velzeboer et al., 2000). Ferreira et al. (2001) found that two *Aphanizomenon flosaquae* strains and samples of a bloom from a reservoir in Portugal contained a specific STX mixture: GTX4 was the dominant analogue, followed by GTX1 and GTX3. *A. flosaquae* strains in a Chinese lake produced neoSTX, dcSTX and dcGTX3, showing a different toxin profile (Liu et al., 2006a; Liu et al., 2006b), whereas *A. gracile* strains detected in two German lakes produced GTX5, STX, dcSTX and neoSTX (Ballot et al., 2010).

In Brazilian freshwaters, STXs are attributed to *R. raciborskii*. In strains isolated from two reservoirs, the contents of total STXs were similar to those reported in *D. circinale* in Australia (Humpage et al., 1994; Lagos et al., 1999). One of the Brazilian strains showed a toxin profile very similar to that of *A. flosaquae*, while the other produced only STX and GTX2/3. Other toxin profiles were described for the Tabocas Reservoir in Caruaru (NE Brazil) affected by a *R. raciborskii* bloom; several STX analogues

(STX, GTX6, dcSTX, neoSTX and dcneoSTX) were identified but no cylindrospermopsin was detected (Molica et al., 2002). Again in Brazil, Castro et al. (2004) reported a *R. raciborskii* strain isolated from a bloom which contained STX concentrations around 0.3 mg/g DW, which is 4- to 8-fold higher than those of GTX2 and GTX3.

In *Lyn. wollei* strains isolated from a reservoir in southern USA, GTX2 and GTX3 represented the major STX congeners, whereas STX and neoSTX were not detected (Carmichael et al., 1997).

Planktothrix sp. FP1 has been associated with the production of STXs in a lake in Italy, confirmed in the isolated culture; the toxin profile of this strain included STX, GTX2 and GTX3 (Pomati et al., 2000).

Few data have been published on the cellular contents of STXs in different cyanobacteria. Llewellyn et al. (2001) have reported STX cell quota up to slightly more than 450 ng /10^6 cells (i.e., 0.45 pg/cell) in a *D. circinale* strain isolated from an Australian waterbody. Hoeger et al. (2005) estimated the cell quota to be 0.12 pg STXs/cell in *D. circinale.* Higher cell quotas (up to 1300 fg/cell) are reported for a strain of *Scytonema* sp. which, however, has very large cells, and in relation to its biomass, with 119 µg/g dry weight the toxin content of this strain was not exceptionally high (Smith et al., 2011). Cell quota up to 0.034 pg/cell of STXeq. were reported in an *Aphanizomenon* sp. (strain LMECYA 31); in the same culture, very high levels of dissolved STXs were observed in the culture media, especially in the late growth phase, very likely as a consequence of cell lysis and leakage (Dias et al., 2002).

Tables 2.6 and 2.7 give examples of the STX contents of strains and environmental samples, respectively. For further details, see reviews by Funari and Testai (2008), Pearson et al. (2016) and Testai et al. (2016).

2.4.4.3 Biosynthesis and regulation

The saxitoxin biosynthesis gene cluster (*sxt*) was first characterised in *Cyl. raciborskii* T3 by Kellmann et al. (2008); other characterisations followed from other strains, namely, *Dolichospermum circinale* AWQC131C, *Aphanizomenon* sp. NH-5 (Mihali et al., 2009), *Raphidiopisis brookii* D9 (Stüken et al., 2011) and *Lyngbya wollei* (Mihali et al., 2011). All five *sxt* clusters encoded biosynthetic enzymes (*sxtA, sxtG, sxtB, sxtD, sxtS, sxtU, sxtH/T* and *sxtI* which appear to have diverse catalytic functions) plus regulatory genes (*sxtL, sxtN* and *sxtX*) and transporters (Kellmann et al., 2008; Pearson et al., 2010).

Different biosynthetic pathways have been proposed, the most recent by D'Agostino et al. (2014) and reviewed by Pearson et al. (2016), starting with the methylation of acetyl-CoA catalysed by SxtA, followed by a condensation reaction with arginine. Further, the aminotransferase SxtG catalyses the addition of the amidino group from a second arginine residue. The following reactions are cyclisation and desaturation leading to the tricyclic core structure, resulting in decarbamoyl STX (dcSTX). Finally, a carbamoyl group is added to dcSTX by the carbamoyltransferase SxtI, resulting in the finalised STX molecule.

The *N*-sulfotransferase (SxtSUL) can modify STX, GTX2 and GTX3, into GTX5-6, C-1 and C-2, by transferring a sulphate residue from PAPS (3′-phosphoadenosine 5′-phosphosulphate) to the carbamoyl group. SxtDIOX is proposed to catalyse the C11 hydroxylation of STX followed by subsequent *O*-sulphation by SxtSUL for biosynthesis of GTX1-4. A combination of sulphation by SxtSUL and SxtN then leads to biosynthesis of the disulphated C-toxins.

STX congeners are mainly produced during late exponential growth phase in laboratory culture (Neilan et al., 2008). The characterisation of the *sxt* cluster in several genera has enabled the study of molecular mechanisms underlying regulation, based on the identification of the genes *sxtY*, *sxtZ* and ompR putatively involved in regulating the *sxt* cluster, adjacent to the *Raphidiopsis raciborskii* T3 *sxt* cluster (Kellmann et al., 2008). However, so far the direct involvement of the regulatory cluster on STX biosynthesis has not been experimentally demonstrated (Pearson et al., 2016).

Regarding the impact of environmental factors, the analysis of data from Australian field samples suggests that STX production is influenced by environmental factors, particularly alkalinity ($pH > 8.5$), very high ammonia concentration (>1 mg/L) and high conductivity (Neilan et al., 2008). Data from laboratory culture studies further indicate that temperature, culture age, light, pH, salinity and nutrient concentrations affect STX production, although causing a variation of only a 2–4-fold (Sivonen & Jones, 1999; Pearson et al., 2016). However, the impact of a particular environmental modulator strictly depends on strains. As an example, toxin production doubled at higher-than-optimal temperatures with *Aphanizomenon* sp. LMECYA 31 (Dias et al., 2002), but in contrast to this, an increase in toxin content was observed in *Aphanizomenon gracile* UAM 529 (Casero et al., 2014) and *R. raciborskii* C10 (Castro et al., 2004) in response to lower-than-optimal temperature.

2.4.5 Occurrence in water environments

The presence of STX-producing cyanobacterial species has increasingly been published, and they have been found for the first time in many locations, including the Arctic (Kleinteich et al., 2013), New Zealand (Smith et al., 2011), Canada (Lajeunesse et al., 2012) and Europe (Wörmer et al., 2011; Jančula et al., 2014).

Dolichospermum circinale may produce STXs at very high contents (up to 4423 µg STXs/g dw). This species caused one of the world's largest cyanobacterial blooms, involving more than 1000 km of the Murray–Darling River, one of Australia's major river systems, with densities of almost 10^6 cells/mL (Bowling & Baker, 1996). Llewellyn et al. (2001) found that 13 out of 14 *D. circinale* strains isolated from Australian freshwaters (rivers, lakes and dams) produced STXs. *R. raciborskii* is reported to produce similar STX levels in Brazil (Lagos et al., 1999). In Europe, a German survey found STX in 34% of 29 waterbodies tested (Chorus, 2001), while in Danish and Finnish freshwater bodies dominated by *D. lemmermannii*, STX was found in less than 10% of samples (Kaas & Henriksen, 2000; Rapala et al., 2005). Similarly, in 140 lakes in New York State, STXs were detected only in two samples out of nearly 1100 tested, with a maximum concentration of 0.09 μg/L, despite the common occurrence of high biomass blooms of *A. flosaquae* (Boyer, 2008). In Washington State (USA), STXs have been detected in 10 lakes and one pond since 2009, with STX concentrations up to 193 μg/L (Trainer & Hardy, 2015).

Very little has been published about STX in finished drinking-water. Hoeger et al., (2005) found only traces of STX (<0.5 μg/L) in two out of 52 water samples from two water treatment plants in Queensland, Australia, fed with raw waters affected by cyanobacterial blooms of *D. circinale*, containing up to 17.0 μg/L STX.

2.4.5.1 Bioaccumulation

Marine seafood contaminated with STXs is well known to cause foodborne diseases in humans, highlighting that STXs are passed from phytoplankton to higher trophic levels in the aquatic food web. Marine shellfish bioaccumulate STXs by filter-feeding on STX-producing organisms, and many of them exhibit low sensitivity towards these toxins. STXs also accumulate in fish, predatory mammals such as whales and crabs (Negri & Jones, 1995) and other non-filter-feeding seafood such as cephalopods, including the common octopus (*Octopus vulgaris*), the Humboldt squid (*Dosidicus gigas*) and the Australian octopus (*Octopus abdopus*) (Lopes et al., 2013). Accumulation generally occurs in the viscera, but in the common octopus and squid, STX accumulated to the greatest extent (390–2680 mg STXeq/kg) in the digestive gland (Lopes et al., 2014), whereas the arms are the preferential site for bioaccumulation in the Australian octopus (up to 246 mg STXeq/100 g tissue; Robertson et al. (2004)).

In spite of the importance of this issue for possible human health consequences, information on STX occurrence in freshwater organisms is scarce. *Daphnia magna*, a relevant organism for STX transfer along the freshwater food web, is able to accumulate STXs when exposed to *Cuspidothrix issatschenkoi* cells or to lyophilised cyanobacterial material (Nogueira et al., 2004). In the laboratory, the Australian freshwater mussel *Alathyria condola* fed with high densities of neurotoxic *D. circinale* accumulated STX up to 620 µg/100 g of fresh biomass (Negri & Jones, 1995). Another freshwater mussel, *Anodonta cygnea*, exposed to high densities of neurotoxic *C. issatschenkoi* in laboratory experiments accumulated STX to a maximum concentration of 26 µg/100 g fresh biomass (Pereira et al., 2004). Accumulation of STXs has also seen in the freshwater bivalves *Elliptio camoplanatus* and *Corbicula fluminea* after exposure to *A. flosaquae* (Sasner et al., 1984).

It has been reported that due to a slow elimination, the surf clam *Spisula solidissima* can bioaccumulate extremely high quantities of STX (Bricelj et al., 2014). In the clam's marine habitat, the STXs are, however, presumably produced by dinoflagellates rather than by cyanobacteria.

2.4.6 Environmental fate

Data on the release of STXs from viable or senescent cyanobacterial cells are lacking. Only a few studies have investigated the chemical breakdown and biodegradation of dissolved STXs. In the dark at room temperature, STXs undergo a series of slow chemical hydrolysis reactions. The C-toxins lose the *N*-sulphocarbamoyl group to form dc-GTXs, while the dcGTXs, GTXs and STXs slowly degrade to, as yet unidentified, nontoxic products. The half-lives for the breakdown reactions are in the order of 1–10 weeks, with more than 3 months often being required for greater than 90% breakdown. A persistence of 1–2 months has been reported for saxitoxin in surface water (Batoreu et al., 2005). In a laboratory study, several STX toxins in the culture medium were stable for long periods also at around pH 9–10 (Castro et al., 2004).

REFERENCES

Ballot A, Fastner J, Wiedner C (2010). Paralytic shellfish poisoning toxin-producing cyanobacterium *Aphanizomenon gracile* in Northeast Germany. Appl Environ Microbiol. 76:1173–1180.

Batoreu MCC, Dias E, Pereira P, Franca S (2005). Risk of human exposure to paralytic toxins of algal origin. Environ Toxicol Pharmacol. 19:401–406.

Ben-Gigirey B, Villar-González A (2008). Chemical analysis. In: Botana LM, editors: Seafood and freshwater toxins. Boca Raton (FL): CRC Press: 177–196.

Bowling L, Baker P (1996). Major cyanobacterial bloom in the Barwon-Darling River, Australia, in 1991, and underlying limnological conditions. Mar Freshwater Res. 47:643–657.

Boyer GL (2008). Cyanobacterial toxins in New York and the lower Great Lakes ecosystems. In: Hudnell, HK, editor: Cyanobacterial harmful algal blooms: state of the science and research needs. New York: Springer:153–165.

Bricelj VM, Cembella AD, Laby D (2014). Temperature effects on kinetics of paralytic shellfish toxin elimination in Atlantic surfclams, *Spisula solidissima*. Deep Sea Res Part II. Top Stud Oceanogr. 103:308–317.

Carmichael WW, Evans WR, Yin QQ, Bell P, Moczydlowski E (1997). Evidence for paralytic shellfish poisons in the freshwater cyanobacterium *Lyngbya wollei* (Farlow ex Gomont) comb. nov. Appl Environ Microbiol. 63:3104–3110.

Casero MC, Ballot A, Agha R, Quesada A, Cirés S (2014). Characterization of saxitoxin production and release and phylogeny of sxt genes in paralytic shellfish poisoning toxin-producing *Aphanizomenon gracile*. Harmful Algae. 37:28–37.

Castro D, Vera D, Lagos N, García C, Vásquez M (2004). The effect of temperature on growth and production of paralytic shellfish poisoning toxins by the cyanobacterium *Cylindrospermopsis raciborskii* C10. Toxicon. 44:483–489.

Chorus I, editor (2001). Cyanotoxins - occurrence, causes, consequences. Berlin: Springer. 357 pp.

Cirés S, Ballot A (2016). A review of the phylogeny, ecology and toxin production of bloom-forming *Aphanizomenon* spp. and related species within the Nostocales (cyanobacteria). Harmful Algae. 54:21–43.

D'Agostino PM, Song X, Neilan BA, Moffitt MC (2014). Comparative proteomics reveals that a saxitoxin-producing and a nontoxic strain of *Anabaena circinalis* are two different ecotypes. J Proteome Res. 13:1474–1484.

Dias E, Pereira P, Franca S (2002). Production of paralytic shellfish toxins by *Aphanizomenon* sp LMECYA 31 (cyanobacteria). J Phycol. 38:705–712.

EFSA (2009). Scientific Opinion: marine biotoxins in shellfish–saxitoxin group. EFSA J. 1019:1–76.

FAO (2004). Marine biotoxins. Food and Agricultural Organization of the United Nations (FAO), Rome. Food and nutrition paper, 80:278 pp. http://www.fao.org/3/y5486e/y5486e00.htm#Contents.

Fast MD, Cembella AD, Ross NW (2006). In vitro transformation of paralytic shellfish toxins in the clams Mya arenaria and *Protothaca staminea*. Harmful Algae. 5:79–90.

Ferreira FM, Soler JMF, Fidalgo ML, Fernández-Vila P (2001). PSP toxins from *Aphanizomenon flos-aquae* (cyanobacteria) collected in the Crestuma-Lever reservoir (Douro river, northern Portugal). Toxicon. 39:757–761.

Foss AJ, Phlips EJ, Aubel MT, Szabo NJ (2012). Investigation of extraction and analysis techniques for *Lyngbya wollei* derived Paralytic Shellfish Toxins. Toxicon. 60:1148–1158.

Funari E, Testai E (2008). Human health risk assessment related to cyanotoxins exposure. Crit Rev Toxicol. 38:97–125.

García C, Barriga A, Díaz JC, Lagos M, Lagos N (2010). Route of metabolization and detoxication of paralytic shellfish toxins in humans. Toxicon. 55:135–144.

Genenah AA, Shimizu Y (1981). Specific toxicity of paralytic shellfish poisons. J Agric Food Chem. 29:1289–1291.

Hoeger SJ, Hitzfeld BC, Dietrich DR (2005). Occurrence and elimination of cyanobacterial toxins in drinking water treatment plants. Toxicol Appl Pharmacol. 203:231–242.

Humpage A, Rositano J, Bretag A, Brown R, Baker P, Nicholson B et al. (1994). Paralytic shellfish poisons from Australian cyanobacterial blooms. Mar Freshwater Res. 45:761–771.

Ikawa M, Wegener K, Foxall TL, Sasner JJ (1982). Comparison of the toxins of the blue-green alga *Aphanizomenon flos-aquae* with the *Gonyaulax* toxins. Toxicon. 20:747–752.

Jančula D, Straková L, Sadílek J, Maršálek B, Babica P (2014). Survey of cyanobacterial toxins in Czech water reservoirs—the first observation of neurotoxic saxitoxins. Environ Sci Pollut Res. 21:8006–8015.

Kaas H, Henriksen P (2000). Saxitoxins (PSP toxins) in Danish lakes. Water Res. 34:2089–2097.

Kellmann R, Mihali TK, Jeon YJ, Pickford R, Pomati F, Neilan BA (2008). Biosynthetic intermediate analysis and functional homology reveal a saxitoxin gene cluster in cyanobacteria. Appl Environ Microbiol. 74:4044–4053.

Kleinteich J, Wood SA, Puddick J, Schleheck D, Küpper FC, Dietrich D (2013). Potent toxins in Arctic environments–presence of saxitoxins and an unusual microcystin variant in Arctic freshwater ecosystems. Chem-Biol Interact. 206:423–431.

Lagos N, Onodera H, Zagatto PA, Andrinolo D, Azevedo SMFO, Oshima I (1999). The first evidence of paralytic shellfish toxins in the freshwater cyanobacterium *Cylindrospermopsis raciborskii*, isolated from Brazil. Toxicon. 37:1359–1373.

Lajeunesse A, Segura PA, Gélinas M, Hudon C, Thomas K, Quilliam MA et al. (2012). Detection and confirmation of saxitoxin analogues in freshwater benthic *Lyngbya wollei* algae collected in the St. Lawrence River (Canada) by liquid chromatography–tandem mass spectrometry. J Chromatogr A. 1219:93–103.

Liu Y, Chen W, Li D, Shen Y, Li G, Liu Y (2006a). First report of aphantoxins in China – waterblooms of toxigenic *Aphanizomenon flos-aquae* in Lake Dianchi. Ecotoxicol Environ Safety. 65:84–92.

Liu Y, Chen W, Li D, Shen Y, Liu Y, Song L (2006b). Analysis of paralytic shelfish toxins in *Aphanizomenon* DC-1 from Lake Dianchi, China. Environ Toxicol. 21:289–295.

Llewellyn L, Negri A, Doyle J, Baker P, Beltran E, Neilan B (2001). Radioreceptor assays for sensitive detection and quantitation of saxitoxin and its analogues from strains of the freshwater cyanobacterium, *Anabaena circinalis*. Environ Sci Technol. 35:1445–1451.

Lopes VM, Baptista M, Repolho T, Rosa R, Costa PR (2014). Uptake, transfer and elimination kinetics of paralytic shellfish toxins in common octopus (*Octopus vulgaris*). Aquat Toxicol. 146:205–211.

Lopes VM, Lopes AR, Costa P, Rosa R (2013). Cephalopods as vectors of harmful algal bloom toxins in marine food webs. Mar Drugs. 11:3381–3409.

McLaughlin J, Fearey D, Esposito T, Porter K (2011). Paralytic shellfish poisoning: southeast Alaska, May–June 2011. Morbid Mortal Wkly Rep. 60:1554–1556.

Mihali TK, Carmichael WW, Neilan BA (2011). A putative gene cluster from a *Lyngbya wollei* bloom that encodes paralytic shellfish toxin biosynthesis. PLoS One. 6:e14657.

Mihali TK, Kellmann R, Neilan BA (2009). Characterisation of the paralytic shellfish toxin biosynthesis gene clusters in *Anabaena circinalis* AWQC131C and *Aphanizomenon* sp. NH-5. BMC Biochem. 10:8.

Molica R, Onodera H, García C, Rivas M, Andrinolo D, Nascimento S et al. (2002). Toxins in the freshwater cyanobacterium *Cylindrospermopsis raciborskii* (Cyanophyceae) isolated from Tabocas reservoir in Caruaru, Brazil, including demonstration of a new saxitoxin analogue. Phycologia. 41:606–611.

Munday R, Thomas K, Gibbs R, Murphy C, Quilliam MA (2013). Acute toxicities of saxitoxin, neosaxitoxin, decarbamoyl saxitoxin and gonyautoxins 1&4 and 2&3 to mice by various routes of administration. Toxicon 76:77–83.

Negri AP, Jones GJ (1995). Bioaccumulation of paralytic shellfish poisoning (PSP) toxins from the cyanobacterium *Anabaena circinalis* by the freshwater mussel *Alathyria condola*. Toxicon. 33:667–678.

Neilan BA, Pearson LA, Moffitt MC, Mihali K, Kaebernick M, Kellmann R et al. (2008). The genetics and genomics of cyanobacterial toxicity. In: Hudnell, HK, editor: Cyanobacterial harmful algal blooms: state of the science and research needs. New York: Springer:417–452.

Nogueira IC, Pereira P, Dias E, Pflugmacher S, Wiegand C, Franca S et al. (2004). Accumulation of paralytic shellfish toxins (PST) from the cyanobacterium *Aphanizomenon issatschenkoi* by the cladoceran *Daphnia magna*. Toxicon. 44:773–780.

Onodera H, Satake M, Oshima Y, Yasumoto T, Carmichael WW (1997). New saxitoxin analogues from the freshwater filamentous cyanobacterium *Lyngbya wollei*. Nat Toxins. 5:146–151.

Oshima Y (1995a). Chemical and enzymatic transformation of paralytic shellfish toxins in marine organisms. In: Lassus P, Arzul G, Erard E et al., editors: Harmful marine algal blooms. Paris: Lavoisier Intercept:475–480.

Oshima Y (1995b). Postcolumn derivatization liquid chromatographic method for paralytic shellfish toxins. J AOAC Int. 78:528–532.

Oshima Y, Sugino K, Yasumoto T (1989). Latest advances in HPLC analysis of paralytic shellfish toxins. In: Natori S, Hashimoto K, Ueno Y, editors: Mycotoxins and Phycotoxins '88. Amsterdam: Elsevier:319–326.

Pearson L, Mihali T, Moffitt M, Kellmann R, Neilan B (2010). On the chemistry, toxicology and genetics of the cyanobacterial toxins, microcystin, nodularin, saxitoxin and cylindrospermopsin. Mar Drugs. 8:1650–1680.

Pearson LA, Dittmann E, Mazmouz R, Ongley SE, D'Agostino PM, Neilan BA (2016). The genetics, biosynthesis and regulation of toxic specialized metabolites of cyanobacteria. Harmful Algae. 54:98–111.

Pereira P, Dias E, Franca S, Pereira E, Carolino M, Vasconcelos V (2004). Accumulation and depuration of cyanobacterial paralytic shellfish toxins by the freshwater mussel *Anodonta cygnea*. Aquat Toxicol. 68:339–350.

Pereira P, Onodera H, Andrinolo D, Franca S, Araujo F, Lagos N et al. (2000). Paralytic shellfish toxins in the freshwater cyanobacterium *Aphanizomenon flos-aquae*, isolated from Montargil reservoir, Portugal. Toxicon. 38:1689–1702.

Pomati F, Sacchi S, Rossetti C, Giovannardi S, Onodera H, Oshima Y et al. (2000). The freshwater cyanobacterium *Planktothrix* sp. FP1: molecular identification and detection of paralytic shellfish poisoning toxins. J Phycol. 36:553–562.

Rapala J, Robertson A, Negri AP, Berg KA, Tuomi P, Lyra C et al. (2005). First report of saxitoxin in finnish lakes and possible associated effects on human health. Environ Toxicol. 20:331–340.

Robertson A, Stirling D, Robillot C, Llewellyn L, Negri A (2004). First report of saxitoxin in octopi. Toxicon. 44:765–771.

Sasner JJ, Ikawa M, Foxall TL (1984). Studies on *Aphanizomenon* and *Microcystis* toxins. In: Ragelis E, editor: Seafood Toxins. Washington (DC): ACS Publications: 391–406.

Silva CSP, Genuário DB, Vaz MGMV, Fiore MF (2014). Phylogeny of culturable cyanobacteria from Brazilian mangroves. Syst Appl Microbiol. 37:100–112.

Sivonen K, Jones GJ (1999). Cyanobacterial toxins. In: Chorus I, Bartram J, editors: Toxic cyanobacteria in water. London: E & FN Spoon:41–111.

Smith FM, Wood SA, van Ginkel R, Broady PA, Gaw S (2011). First report of saxitoxin production by a species of the freshwater benthic cyanobacterium, *Scytonema* Agardh. Toxicon. 57:566–573.

Stüken A, Orr RJ, Kellmann R, Murray SA, Neilan BA, Jakobsen KS (2011). Discovery of nuclear-encoded genes for the neurotoxin saxitoxin in dinoflagellates. PLoS One. 6:e20096.

Testai E, Scardala S, Vichi S, Buratti FM, Funari E (2016). Risk to human health associated with the environmental occurrence of cyanobacterial neurotoxic alkaloids anatoxins and saxitoxins. Crit Rev Toxicol. 46:385–419.

Trainer VL, Hardy FJ (2015). Integrative monitoring of marine and freshwater harmful algae in Washington State for public health protection. Toxins. 7:1206–1234.

Velzeboer RM, Baker PD, Rositano J, Heresztyn T, Codd GA, Raggett SL (2000). Geographical patterns of occurrence and composition of saxitoxins in the cyanobacterial genus *Anabaena* (Nostocales, Cyanophyta) in Australia. Phycologia. 39:395–407.

WHO (2003). Guidelines for safe recreational water environments. Vol. 1: Coastal and fresh waters. Geneva: World Health Organization. https://apps.who.int/iris/handle/10665/42591

WHO (2020). Cyanobacterial toxins: Saxitoxins. Background document for development of WHO Guidelines for Drinking-water Quality and Guidelines for Safe Recreational Water Environments. Geneva: World Health Organization. https://apps.who.int/iris/handle/10665/338069

Wichmann CF, Niemczura WP, Schnoes HK, Hall S, Reichardt PB, Darling SD (1981). Structures of two novel toxins from *Protogonyaulax*. J Am Chem Soc. 103:6977–6978.

Wiese M, D'agostino PM, Mihali TK, Moffitt MC, Neilan BA (2010). Neurotoxic alkaloids: saxitoxin and its analogs. Mar Drugs. 8:2185–2211.

Wörmer L, Cirés S, Agha R, Verdugo M, de Hoyos C, Quesada A (2011). First detection of cyanobacterial PSP (paralytic shellfish poisoning) toxins in Spanish freshwaters. Toxicon. 57:918–921.

2.5 ANATOXIN-A(S)

Emanuela Testai

Anatoxin-a(S) (ATX(S)) is, despite the similarity of the names, not structurally related to anatoxin-a: while the latter is an alkaloid, ATX(S) is an organophosphate (see below). It received its name during initial studies which isolated multiple toxic fractions from a strain of *Anabaena* sp. to which letters or suffixes were assigned. The "S" in the name denotes a characteristic symptom of exposure in mammals: "salivation". Because of its totally different chemical structure and mechanism of action, Fiore et al. (2020) proposed renaming it to guanitoxin, advocating that the new name should reflect its chemical composition.

2.5.1 Chemical structure

Anatoxin-a(S) is an *N*-hydroxyguanidine methyl phosphate ester with a molecular weight of 252 Da. It is the only known natural organophosphonate besides biomolecules such as DNA, RNA and ATP (Figure 2.5; Mahmood & Carmichael, 1987). No structural variants of ATX(S) have been detected so far.

Anatoxin-a(S) decomposes rapidly in basic solutions but is relatively stable in neutral and acidic conditions (Matsunaga et al., 1989). It is inactivated at temperatures higher than 40 °C (Carmichael, 2001).

2.5.2 Toxicity: mode of action

Anatoxin-a(S) irreversibly inhibits acetylcholinesterase (AChE) in the neuromuscular junctions (but not in the central nervous system) blocking hydrolysis of the neurotransmitter. This results in acetylcholine accumulation, leading to nerve hyperexcitability. The acute neurological effects in mammals are muscle weakness, respiratory distress (dyspnoea) and convulsions preceding death, which occurs due to respiratory arrest (i.p. LD_{50} in mice = 40–228 µg/kg bw,

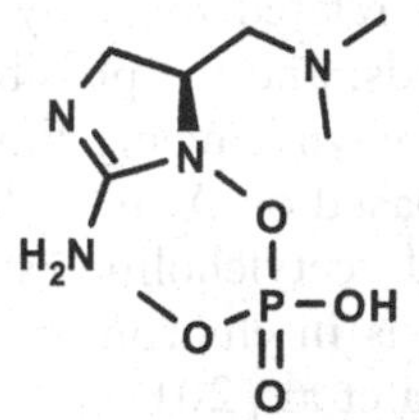

N N N O H₂N O P OH O

Figure 2.5 Chemical structure of anatoxins-a(S). Molecular mass (monoisotopic): 252.099 Da; molecular weight (average): 252.212 g/mol.

lower in rats i.p. LD_{50} = 5.3 mg/kg bw). Viscous mucoid hypersalivation is a typical symptom induced by ATX(S).

Data on oral administration as well as on subchronic and/or chronic toxicity are not available.

2.5.3 Derivation of guideline values for anatoxin-a(S) in water

No toxicological data are available for deriving an acute dose NOAEL or LOAEL as point of departure, and data on subchronic and chronic exposure are also lacking. Therefore, no TDI or guideline value can yet be derived for ATX(S).

New Zealand has established a limit as provisional maximum acceptable value of 1 µg/L for total ATX(S) content in drinking-water (Chorus, 2012).

2.5.4 Production, occurrence and environmental fate

Anatoxin-a(S) has been reported from strains of *Dolichospermum* (*Anabaena*) *flosaquae* from Canada (Carmichael & Gorham, 1978), in both field samples and strains of *D. lemmermannii* from Denmark (Henriksen et al., 1997) and from Portugal (Fristachi & Sinclair, 2008), in *D. flosaquae* from the USA and Scotland (Matsunaga et al., 1989; Codd, 1995), in *D. spiroides* from Brazil (Monserrat et al., 2001), and in *D. crassa* from southern Brazil (Becker et al., 2010).

The available literature on ATX(S) biosynthesis is scant, and the gene cluster responsible for the biosynthesis of ATX(S) has not yet been identified (Pearson et al., 2016). Only the synthesis of the cyclic moiety of ATX(S) has been reported (Matsunaga et al., 1989; Moura & Pinto, 2010).

The precursor for the guanidine group has been proposed to be *L*-arginine, which is hydroxylated at C4, as demonstrated by feeding studies (Moore et al., 1992) with radiolabelled arginine and (4S)-4-hydroxy-arginine, but none of the further steps have been described to date.

The presence of ATX(S) in waterbodies is sparsely documented (Table 2.7); one of the reasons could be related to analytical difficulties such as the absence of analytical standards, and the possible co-occurrence of organophosphate pesticides in the environment, limiting the use of biological tests, including biosensors, based on AChE inhibition (Devic et al., 2002). Indeed, mouse bioassays and acetylcholine esterase inhibition assays may be used to infer ATX(S) levels in environmental samples; however, these tests are not specific (Patocka et al., 2011). This sometimes leads only to a qualitative description of detection, without quantification (Molica et al., 2005). The only chance to use analytical methodologies, overcoming the lack of standards to identify the presence of the toxin, is the LC–MS/MS fragmentation pattern for ATX(S) in cyanobacterial cultures.

Highly variable ATX(S) contents were detected in three Danish lakes dominated by *D. lemmermannii*, reaching maximum contents of 3300 μg/g dw (Henriksen et al., 1997).

The presence of ATX(S) was also suggested by results from acetylcholine esterase inhibition assay in cyanobacterial crusts in Qatar (Metcalf et al., 2012).

Data on chemical breakdown in the natural water environment and biodegradation of this cyanotoxin are not available.

REFERENCES

Becker V, Ihara P, Yunes JS, Huszar VLM (2010). Occurrence of anatoxin-a(S) during a bloom of *Anabaena crassa* in a water-supply reservoir in southern Brazil. J Appl Phycol. 22:235–241.

Carmichael WW (2001). Health effects of toxin-producing cyanobacteria: "The CyanoHABs". Human Ecol Risk Assess. 7:1393–1407.

Carmichael WW, Gorham PR (1978). Anatoxins from clones of *Anabaena flos-aquae* isolated from lakes of western Canada: With 3 figures and 2 tables in the text. Mitt Int Verein Limnol. 21:285–295.

Chorus I (2012). Current approaches to cyanotoxin risk assessment, risk management and regulations in different countries. Dessau: Federal Environment Agency.

Codd G (1995). The toxicity of benthic blue-green algae in Scottish freshwaters. The Scottish Office, Foundation for Water Research, Marlow.

Devic E, Li DH, Dauta A, Henriksen P, Codd GA, Marty JL et al. (2002). Detection of anatoxin-a(s) in environmental samples of cyanobacteria by using a biosensor with engineered acetylcholinesterases. Appl Environ Microbiol. 68:4102–4106.

Fiore MF, de Lima ST, Carmichael WW, McKinnie SM, Chekan JR, Moore BS (2020). Guanitoxin, re-naming a cyanobacterial organophosphate toxin. Harmful Algae. 92:101737.

Fristachi A, Sinclair JL (2008). Occurrence of cyanobacterial harmful algal blooms workgroup report. In: Hudnell HK, editors: Cyanobacterial harmful algal blooms: state of the science and research needs. New York: Springer:45–103.

Henriksen P, Carmichael WW, An JS, Moestrup O (1997). Detection of an anatoxin-a(s)-like anticholinesterase in natural blooms and cultures of Cyanobacteria/blue-green algae from Danish lakes and in the stomach contents of poisoned birds. Toxicon. 35:901–913.

Mahmood NA, Carmichael WW (1987). Anatoxin-a (s), an anticholinesterase from the cyanobacterium *Anabaena flos-aquae* NRC-525-17. Toxicon. 25:1221–1227.

Matsunaga S, Moore RE, Niemczura WP, Carmichael WW (1989). Anatoxin-a (s), a potent anticholinesterase from *Anabaena flos-aquae*. J Am Chem Soc. 111:8021–8023.

Metcalf JS, Richer R, Cox PA, Codd GA (2012). Cyanotoxins in desert environments may present a risk to human health. Sci Tot Environ. 421:118–123.

Molica RJ, Oliveira EJ, Carvalho PV, Costa AN, Cunha MC, Melo GL et al. (2005). Occurrence of saxitoxins and an anatoxin-a (s)-like anticholinesterase in a Brazilian drinking water supply. Harmful Algae. 4:743–753.
Monserrat JM, Yunes JS, Bianchini A (2001). Effects of *Anabaena spiroides* (cyanobacteria) aqueous extracts on the acetylcholinesterase activity of aquatic species. Environ Toxicol Chem. 20:1228–1235.
Moore BS, Ohtani I, Moore RE, Carmichael WW (1992). Biosynthesis of anatoxin-a (s): origin of the carbons. Tetrahedron Lett. 33:6595–6598.
Moura S, Pinto E (2010). Synthesis of cyclic guanidine intermediates of anatoxin-a (s) in both racemic and enantiomerically pure forms. Synlett. 2010:967–969.
Patocka J, Gupta RC, Kuca K (2011). Anatoxin-a (s): natural organophosphorus anticholinesterase agent. Mil Med Sci Lett 80:129–139.
Pearson LA, Dittmann E, Mazmouz R, Ongley SE, D'Agostino PM, Neilan BA (2016). The genetics, biosynthesis and regulation of toxic specialized metabolites of cyanobacteria. Harmful Algae. 54:98–111.

2.6 MARINE DERMATOTOXINS

Nicholas J. Osborne

The dermatotoxic reaction to the marine cyanobacterium "*Lyngbya majuscula*" has been associated with cases of dermatitis in humans, reported since the 1950s (Grauer & Arnold, 1961). This spurred intensive research on natural products revealing hundreds of secondary metabolites supposedly produced by this species (Gerwick et al., 2008). This enormous metabolic diversity was questioned by Engene et al. (2011) who suggest that what has been viewed as a single species or species complex based on morphological criteria in fact represents a multitude of genera and species on the basis of molecular analyses (Engene et al., 2010).

The traditional genus *Lyngbya* consists of several hundred described species (see Chapter 3) of marine and freshwater cyanobacteria with global distribution. It is now proposed that tropical *Lyngbya*-like cyanobacteria are separated from other members of the genus *Lyngbya*, as they have been found to be genetically distinct. The new genus *Moorea*, in particular *M. producens*, largely appears to be synonymous to "*L. majuscula*" (Engene et al., 2012) but other names have been used as synonyms, for example, *Microcoleus lyngbyaceus* (Sims & Zandee van Rilland, 1981). Further genera amended from *Lyngbya* are *Dapis* (Engene et al., 2018) and *Okeania* (Engene et al., 2013b) – and more may follow. However, since it is not possible to retrospectively evaluate the taxonomic assignment of samples in original publications, this chapter gives "*L. majuscula*" in quotation marks whenever the possibility exists that the taxonomic assignment would be different today. Temperate species of "*L. majuscula*" have been recorded, but essentially nothing is known of their toxicity (Hällfors, 2004).

"*L. majuscula*" is a benthic cyanobacterium appearing as clumps of a matted mass of filaments 10–30 cm long, sometimes referred to as "mermaids' hair", that grows to depths of up to 30 m, predominantly in the tropics and subtropics (Izumi & Moore, 1987). Elevated concentrations of iron, nitrogen and phosphorus have been proposed to be drivers of mass development or blooming of this cyanobacterium (Albert et al., 2005).

Among some 200 natural products that have been linked to tropical "*L. majuscula*" (Liu & Rein, 2010; Engene et al., 2013a), some have been found to induce irritant contact dermatitis, that is, to be dermatotoxins: aplysiatoxin (AT; Kato & Scheuer, 1974), (Mitchell et al., 2000), debromoaplysiatoxin (DAT; Mynderse et al., 1977) and lyngbyatoxin A (LTA; Cardellina et al., 1979). Other natural products produced by this cyanobacterium include malyngamides, apratoxins and dolostatins (Todd & Gerwick, 1995; Mitchell et al., 2000; Luesch et al., 2001).

2.6.1 Chemical structures

Lyngbyatoxin A's structure (Figure 2.6) was initially determined in 1979 using samples collected at Kahala Beach, Oahu, Hawaii (Cardellina et al., 1979). An isomer of teleocidin A, first extracted from the actinomycete *Streptomyces medicocidcus*, was found to have an identical structure to LTA (Fujiki et al., 1981), with this organism producing both the 19*R* and 19*S* epimers (i.e., the same chemical formula but different three-dimensional orientations), while in "*L. majuscula*" only the 19*R* epimer was found. Lyngbyatoxin B and lyngbyatoxin C, compounds with similar chemical structure, were extracted from Hawaiian specimens of "*L. majuscula*" (Aimi et al., 1990), as was 12-epi-lyngbyatoxin A and further congeners (Jiang et al., 2014a; Jiang et al., 2014b). Lyngbyatoxin A is more lipophilic than the other lyngbyatoxin, with a mean log *n*-octanol/water partition coefficient of 1.53 (Stafford et al., 1992).

Debromoaplysiatoxin was first isolated in 1977 and the structure derived from extracts of both *Lyngbya gracilis* (reclassified to *Leibleinia gracilis*; see also below) and an inseparable consortium of *Phormidium* (*Oscillatoria*) *nigroviridis* and *Schizothrix calcicola* (Mynderse et al., 1977). The phenolic bis-lactones AT and DAT have similar structures apart from the bromine molecule on the benzene ring (Figure 2.6).

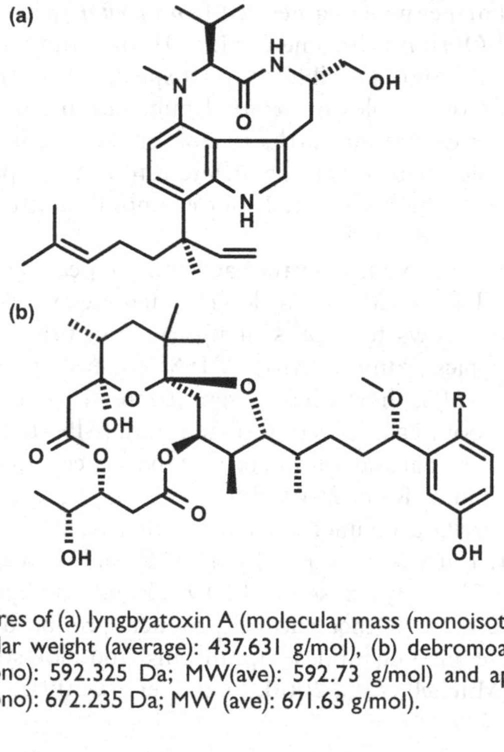

Figure 2.6 Structures of (a) lyngbyatoxin A (molecular mass (monoisotopic): 437.304 Da; molecular weight (average): 437.631 g/mol), (b) debromoaplysiatoxin (R=H; MM(mono): 592.325 Da; MW(ave): 592.73 g/mol) and aplysiatoxin (R=Br; MM(mono): 672.235 Da; MW (ave): 671.63 g/mol).

2.6.2 Toxicity

Although toxicity was first observed in Hawaii in 1912 (Banner, 1959; Osborne et al., 2008), the first confirmed activity by "*L. majuscula*" that caused acute dermatitis was not determined until 1958, via patch testing in humans (Grauer & Arnold, 1961). Banner revealed that the dermatitis was irritant rather than allergenic, and this has been replicated in a later study (Banner, 1959; Osborne et al., 2008): histology of mouse and human skin exposed to either crude extracts of "*L. majuscula*" or its purified toxins showed acute vesicular dermatitis consistent with irritant contact dermatitis after topical application. Microscopic examination described peeling skin and oedema of the epidermis. The dermis was infiltrated with a range of inflammatory cells, including mononuclear cells, neutrophils and eosinophils (Grauer & Arnold, 1961; Osborne et al., 2008). Vesicles contained polymorphonuclear leukocytes and red blood cells, with deep infiltration of the epidermis with polymorphonuclear leukocytes (Grauer & Arnold, 1961). Ito et al. (2002) found LTA to have a minimum lethal dose (LD_{100}) of 0.30 mg/kg (intraperitoneal) in mice.

Applied cutaneously, LTA had an median effective dose ED_{50} (dose causing a biological response in 50% of the sample) with a lower index (reddening) of ~4.8 ng/kg in mice tested via topical application, with DAT and AT showing slightly lower activity (Fujiki et al., 1983). LTA has shown skin penetration rates of 23% and 6.2% for guinea pig and human skin, respectively, within 1 h (Stafford et al., 1992). However, not all of the toxicity in "*L. majuscula*" specimens producing LTA is explained by the concentrations of LTA present, and other factors present thus must be affecting toxicity (Osborne et al., 2008).

Furthermore, animals that feed on toxic "*Lyngbya*" or *Moorea* appear to bioaccumulate toxin: the first report dates back to classical times by Pliny (Plinius, 23–79 AD) who reported toxicity of marine gastropods – the sea hare. Kato and Scheuer (1974) first isolated AT and DAT from their digestive tract. Sea hare (e.g., *Aplysia californica*; Gribble, 1999; *Stylocheilus striatus;* Capper et al., 2006) appear to preferentially feed on cyanobacteria, including "*L. majuscula*" (and possibly other marine cyanobacteria) and to sequester their toxins (Pennings et al., 1996). This bioaccumulation potentially also occurs in other grazers (Capper et al., 2005). Accidental skin contact with chemicals extracted from sea hares led to dermal irritation. While some invertebrate grazers appear to be indifferent to extracts of the cyanobacteria "*L. majuscula*", reef fish are more likely to be deterred (Capper et al., 2006).

For AT, Ito and Nagai (1998) report that dosing of mice at 500, 1000 or 3000 µg/kg intraperitoneally resulted in bleeding in the small intestine, blood loss and pale liver, loss of cells from the stomach, exposure of the lamina propria lumen with small intestine capillaries congested and all villi showing erosion and bleeding. Dosing orally at 0.8 µg/kg, AT induced increased permeability of gastrointestinal vascularisation as well as local inflammation and necrosis. The consequences of the latter were intraperitoneal haemorrhage, hypovolemic liver, small intestinal sloughing and haemorrhage. The activity of AT was proposed to be due to its effect on protein kinase C, not peritonitis (Ito & Nagai, 1998). At an oral or intraperitoneal dose of AT at which 50% of animals were affected, symptoms resembled those of LTA poisoning (Ito et al., 2002).

For DAT, dermal toxicity has also been shown (Solomon & Stoughton, 1978; Osborne et al., 2008). For ear reddening, DAT and AT show highly lower activity than LTA (see above and Fujiki et al., 1983). DAT was originally reported as isolated from *Leibleinia* (*Lyngbya*) *gracilis*, a species classified today in the order Synechococcales. However, a footnote in the publication notes that one taxonomist identified the organism as "*L. majuscula*" (Mynderse et al., 1977). Other authors have suggested that these toxins are also present in seaweed species: in papers reporting chemicals extracted from the red alga *Gracilaria coronopifolia*, the authors suggested the toxicity of the seaweed may be due to epiphytically growing cyanobacteria (Nagai et al., 1996; Nagai et al., 1997). It is still unclear if this is the case, but it is entirely possible as the epiphytic growth of "*L. majuscula*" and other cyanobacteria on seaweeds has been reported worldwide (Moore, 1982; Fletcher, 1995).

Both LTA and DAT have been shown to have tumour-promoting activities via the protein kinase C activation pathway (Nakamura et al., 1989).

In spite of EC_{50} or LD_{50} values given for some of the marine dermatotoxins, no guideline values for their concentration in water used for recreation can be given because, in contrast to the cyanotoxins discussed in sections 2.1–2.5, their exposure pathway is not through ingestion, but through dermal contact, and this is not accessible to quantification for filamentous macroalgae forming mats and rafts.

2.6.3 Incidents of human injury through marine cyanobacterial dermatotoxins

Hawaii 1950–1983

In late 1950s, "*L. majuscula*" was first purported as the agent responsible for an epidemic of acute dermatitis in Hawaii. 125 people were reported suffering dermatitis after swimming at beaches in north-east Oahu, Hawaii, in July and August 1958. After exposure to "*L. majuscula*", swimmers described symptoms similar to a burn, usually appearing underneath swimming costumes in the genital, perianal and perineum areas. Debate continues if the cause of symptoms chiefly at these locations is the thinner epidermis in these areas or extended exposure with cyanobacterium filaments trapped in clothing. Symptoms within a few hours of exposures included erythema and burning followed by deep skin peeling and blistering, which continued for 24–48 h (Grauer & Arnold, 1961).

In 1976, samples of blooming "*L. majuscula*" were found to contain DAT. In 1980 in Oahu, 35 people were affected and developed dermatitis 2–20 h after exposure and with symptoms lasting from 2 to 12 days (Serdula et al., 1982). Both AT and DT were found in samples of "*L. majuscula*" recovered from the ocean (Moore et al., 1984).

In 1983, eye and breathing symptoms were noted in Maui (Anderson et al., 1988). Aerosolised *Lyngbya* fragments were discovered on sampling with high-volume air filters and from waterfront area windows.

Okinawa 1968 and 1973

At Gushikawa Beach, Okinawa, in 1968, 242 of 274 bathers developed a rapid-onset dermatitis. Reported symptoms included rash, itching, burning, blisters and deep peeling of the skin. Sensitive outer areas such the genitals, lips and eyes were usually affected. A later bloom of "*L. majuscula*" was sampled in the same area in September 1973: it caused rashes and blistering in humans and mice (Hashimoto et al., 1976). The compounds extracted and partially characterised had chemical properties similar to those of the uncharacterised toxin found by Moikeha and Chu (1971), and samples collected in the same later were shown to contain DAT and AT (Fujiki et al., 1985).

Queensland, Australia, 1999–2003

A cross-sectional epidemiological survey of residents of Bribie Island, Australia, was undertaken after some evidence of blooms of "*L. majuscula*" in the area. Residents exposed to seawater exhibited symptoms associated with exposure to "*L. majuscula*" (0.6% of the sample population), including redness in the inguinal region, severe itching and blistering (Osborne et al., 2007). The greater surface area of female swimming costumes may explain their increased prevalence of symptoms as compared to men, with an increased entrapment of cyanobacterial strands. Similar epidemiological observations have been reported from nearby Fraser Island (Osborne & Shaw, 2008).

Mortality in humans after the consumption of *Lyngbya* has been reported three times (Sims & Zandee van Rilland, 1981; Marshall & Vogt, 1998; Yasumoto, 1998). "*L. majuscula*" growing epiphytically on the edible endemic Hawaiian alga *Gracilaria coronopifolia* have also been implicated in poisoning from ingestion of the red alga in 1994 (Nagai et al., 1996; Ito & Nagai, 2000). Consumption of "*L. majuscula*" (associated with consuming seaweed) has been associated with an excruciating burning sensation on the patient's lips, anterior part of the oral cavity and the anterior portion of the tongue. Twenty-four hours after consumption, the mucous membranes appeared scalded, swollen and exhibited hyperaemia with several erosive lesions. The patient became free of discomfort after 3 days (Sims & Zandee van Rilland, 1981). It has been postulated that the high incidence of cancers of the digestive system among indigenous Hawaiians may be due to the consumption of seaweed tainted with "*L. majuscula*" (Moore, 1984). Furthermore, an outbreak of respiratory, eye and skin irritations in Mayotte, an island in the Indian Ocean, in 2010 was linked to exposure to cyanobacteria washed on the beach (Lernout et al., 2011).

2.6.4 Biosynthesis and occurrence in the environment

Elements of the biosynthesis of LTA, as well as the genes involved, are reported by Tønder et al. (2004), Read and Walsh (2007), and Edwards and Gerwick (2004). The core of the molecule is synthesised by a nonribosomal peptide synthetase followed by reduction and prenylation steps (Read & Walsh, 2007). Total synthesis of AT and DAT was achieved by Park et al. (1987). Videau et al. (2016) achieved a heterologous expression of LTA in a strain of *Anabaena* sp. (PCC 7120).

"*L. majuscula*" is mainly seen in the tropics and subtropics but has a worldwide distribution (Table 2.9). Different toxicities of samples of this species from around the Hawaii (Grauer & Arnold, 1961) and the Marshall Islands have been noted, where samples taken on the seaward side of the lagoon were more toxic (Mynderse et al., 1977). Similarly, spatial differences in toxins in Moreton Bay, Australia, have been recorded, with DT being produced on the Western side exclusively, and LTA mainly

Table 2.9 Dermatotoxin contents reported for "*L. majuscula*" in µg/g dry weight, collected on various locations around world

Location	*LTA*	*DAT*	*AT*	*Reference*
Ryukyus Islands, Okinawa	240			Hashimoto et al. (1976) [a]
Enewetak Atoll, Marshall Islands		133		Mynderse et al. (1977) [b]
Kahala Beach, Hawaii	200			Cardellina et al. (1979)
Oahu, Hawaii		324	81	Serdula et al. (1982)
Moreton Bay, Australia	n.d.–131	n.d.–43		Osborne (2004)
Maui, Hawaii	10–276	n.d.–0.8		Osborne (2004)
King's Bay, Florida, USA		n.d.–6.31		Harr et al. (2008)
Moreton Bay, Australia	n.d.–39	n.d.–0.3		Arthur et al. (2008)
Big Island, Hawaii	n.d.–168	n.d.–540		Arthur et al. (2008)

n.d.: not detectable. LTA: lyngbyatoxin A; DTA: debromoaplysiatoxin; AT: aplysiatoxin.

[a] probably LTA, not confirmed.

[b] producing organism reported as *Lyngbya* (*Leibleinia*) *gracilis*, but probably was "*L. majuscula*".

being produced on the Eastern ocean side, only 30 km away (Osborne et al., 2002; Osborne, 2004).

Treating DAT and AT with even very mild acid readily leads to dehydration, and the degradation products (anhydrotoxins) do not show the toxicity seen with DAT and AT (Moore, 1984). Hashimoto (1979) reported half of the toxicity of "*L. majuscula*" was lost after 3 h of exposure to ultraviolet radiation, as did Moikeha and Chu (1971). The absence of toxins was noted in seawater surrounding a large bloom of toxic *L. majuscula* in Australia (Osborne, 2004). It appears that the toxins are biodegradable in the environment, but further work is required to explore this.

REFERENCES

Aimi N, Odaka H, Sakai S, Fujiki H, Suganuma M, Moore RE et al. (1990). Lyngbyatoxins B and C, two new irritants from *Lyngbya majuscula*. J Nat Prod. 53:1593–1596.

Albert S, O'Neil JM, Udy JW, Ahern KS, O'Sullivan CM, Dennison WC (2005). Blooms of the cyanobacterium *Lyngbya majuscula* in coastal Queensland, Australia: disparate sites, common factors. Mar Pollut Bull. 51:428–437.

Anderson B, Sims J, Liang A, Minette H (1988). Outbreak of eye and respiratory irritation in Lahaina, Maui, possibly associated with *Microcoleus lyngbyaceus*. J Environ Health. 50:205–209.

Arthur K, Limpus C, Balazs G, Capper A, Udy J, Shaw G et al. (2008). The exposure of green turtles (*Chelonia mydas*) to tumour promoting compounds produced by the cyanobacterium *Lyngbya majuscula* and their potential role in the aetiology of fibropapillomatosis. Harmful Algae. 7:114–125.

Banner AH (1959). A dermatitis-producing algae in Hawaii. Hawaii Med J. 19:35–36.

Capper A, Tibbetts IR, O'Neil JM, Shaw GR (2006). Feeding preference and deterrence in rabbitfish *Siganus fuscescens* for the cyanobacterium *Lyngbya majuscula* in Moreton Bay, south-east Queensland, Australia. J Fish Biol. 68:1589–1609.

Capper A, Tibbetts IR, O'Neil YM, Shaw GR (2005). The fate of *Lyngbya majuscula* toxins in three potential consumers. J Chem Ecol. 31:1595–1606.

Cardellina JHd, Marner FJ, Moore RE (1979). Seaweed dermatitis: structure of lyngbyatoxin A. Science. 204:193–195.

Edwards DJ, Gerwick WH (2004). Lyngbyatoxin biosynthesis: sequence of biosynthetic gene cluster and identification of a novel aromatic prenyltransferase. J Am Chem Soc. 126:11432–11433.

Engene N, Choi H, Esquenazi E, Rottacker EC, Ellisman MH, Dorrestein PC et al. (2011). Underestimated biodiversity as a major explanation for the perceived rich secondary metabolite capacity of the cyanobacterial genus *Lyngbya*. Environ Microbiol. 13:1601–1610.

Engene N, Coates RC, Gerwick WH (2010). 16S rRNA Gene heterogeneity in the filamentous marine cyanobacterial genus *Lyngbya*. J Phycol. 46:591–601.

Engene N, Gunasekera SP, Gerwick WH, Paul VJ (2013a). Phylogenetic Inferences reveal a large extent of novel biodiversity in chemically rich tropical marine cyanobacteria. Appl Environ Microbiol. 79:1882–1888.

Engene N, Paul VJ, Byrum T, Gerwick WH, Thor A, Ellisman MH (2013b). Five chemically rich species of tropical marine cyanobacteria of the genus *Okeania* gen. nov.(Oscillatoriales, Cyanoprokaryota). J Phycol. 49:1095–1106.

Engene N, Rottacker EC, Kaštovský J, Byrum T, Choi H, Ellisman MH et al. (2012). *Moorea producens* gen. nov., sp. nov. and *Moorea bouillonii* comb. nov., tropical marine cyanobacteria rich in bioactive secondary metabolites. Int J Syst Evol Microbiol. 62:1171–1178.

Engene N, Tronholm A, Paul VJ (2018). Uncovering cryptic diversity of *Lyngbya*: the new tropical marine cyanobacterial genus *Dapis* (Oscillatoriales). J Phycol. 54:435–446.

Fletcher R (1995). Epiphytism and fouling in Gracilaria cultivation: an overview. J Appl Phycol. 7:325–333.

Fujiki H, Ikegami K, Hakii H, Suganuma M, Yamaizumi Z, Yamazato K et al. (1985). A blue-green alga from Okinawa contains aplysiatoxins, the third class of tumor promoters. Jpn J Cancer Res. 76:257–259.

Fujiki H, Mori M, Nakayasu M, Terada M, Sugimura T, Moore RE (1981). Indole alkaloids: dihydroteleocidin B, teleocidin, and lyngbyatoxin A as members of a new class of tumor promoters. Proc Natl Acad Sci USA. 78:3872–3876.

Fujiki H, Suganuma M, Tahira T, Yoshioka A, Nakayasu M, Endo Y et al. (1983). Nakahara memorial lecture. New classes of tumor promoters: teleocidin, aplysiatoxin, and palytoxin. Princess Takamatsu Symp. 14:37–45.

Gerwick WH, Coates RC, Engene N, Gerwick L, Grindberg RV, Jones AC et al. (2008). Giant marine cyanobacteria produce exciting potential pharmaceuticals. Microbe. 3:277.

Grauer FH, Arnold HL (1961). Seaweed dermatitis: first report of dermatitis-producing marine algae. Arch Dermatol. 84:720–732.

Gribble GW (1999). The diversity of naturally occurring organobromine compounds. Chem Soc Rev. 28:335–346.

Hällfors G (2004). Checklist of Baltic Sea phytoplankton species. Helsinki: Helsinki Commission Baltic Marine Environment Protection Commission.

Harr KE, Szabo NJ, Cichra M, Phlips EJ (2008). Debromoaplysiatoxin in *Lyngbya*-dominated mats on manatees (*Trichechus manatus latirostris*) in the Florida King's Bay ecosystem. Toxicon. 52:385–388.

Hashimoto Y (1979). Marine toxins and other bioactive marine metabolites. Tokyo: Japan Scientific Societies Press:369 pp.

Hashimoto Y, Kamiya H, Yamazato K, Nozawa K (1976). Occurrence of a toxic blue-green alga inducing skin dermatitis in Okinawa. In: Ohsaka A, Hayashi K, Sawai Y, editors: Animal, plant, and microbial toxins. New York: Plenum:333–338.

Ito E, Nagai H (1998). Morphological observations of diarrhea in mice caused by aplysiatoxin, the causative agent of the red alga *Gracilaria coronopifolia* poisoning in Hawaii. Toxicon. 36:1913–1920.

Ito E, Nagai H (2000). Bleeding from the small intestine caused by aplysiatoxin, the causative agent of the red algae *Gracilaria coronopifolia* poisoning in Hawaii. Toxicon. 38:123–132.

Ito E, Satake M, Yasumoto T (2002). Pathological effects of lyngbyatoxin A upon mice. Toxicon. 40:551–556.

Izumi AK, Moore RE (1987). Seaweed (*Lyngbya majuscula*) dermatitis. Clin Dermatol. 5:92–100.

Jiang W, Tan S, Hanaki Y, Irie K, Uchida H, Watanabe R et al. (2014a). Two new lyngbyatoxin derivatives from the cyanobacterium, *Moorea producens*. Mar Drugs. 12:5788–5800.

Jiang WN, Zhou W, Uchida H, Kikumori M, Irie K, Watanabe R et al. (2014b). A new lyngbyatoxin from the Hawaiian cyanobacterium *Moorea producens*. Mar Drugs. 12:2748–2759.

Kato Y, Scheuer PJ (1974). Aplysiatoxin and debromoaplysiatoxin, constituents of the marine mollusk *Stylocheilus longicauda* (Quoy and Gaimard, 1824). J Am Chem Soc. 96:2245–2246.

Lernout T, Thiria J, Maltaverne E, Salim M, Turquet J, Lajoindre G et al. (2011). Alerte aux cynanobactéries sur la plage de N'Gouja, Mayotte, avril 2010. Bulletin de veille sanitaire. 9:12–14.

Liu L, Rein KS (2010). New peptides isolated from *Lyngbya* species: a review. Marine Drugs. 8:1817–1837.

Luesch H, Yoshida WY, Moore RE, Paul VJ, Corbett TH (2001). Total structure determination of apratoxin A, a potent novel cytotoxin from the marine cyanobacterium *Lyngbya majuscula*. J Am Chem Soc. 123:5418–5423.

Marshall KL, Vogt RL (1998). Illness associated with eating seaweed, Hawaii, 1994. Western J Med. 169:293–295.

Mitchell SS, Faulkner DJ, Rubins K, Bushman F (2000). Dolostatin 3 and two novel cyclic peptides from a Palauan collection of Lyngbya majuscula. J Nat Prod. 63:279–282.

Moikeha S, Chu G (1971). Dermatitis-producing alga *Lyngbya majuscula* Gomont in Hawaii. II. Biological properties of the toxic factor. J Phycol. 7:8–13.

Moore RE (1982). Toxins, anticancer agents, and tumor promoters from marine prokaryotes. Pure Appl Chem. 54:1919–1934.

Moore RE (1984). Public health and toxins from marine Blue- Green Algae. In: Ragelis E, editors: Seafood toxins. Washington (DC): American Chemical Society.

Moore RE, Blackman A, Cheuk C (1984). Absolute stereochemistries of the aplysiatoxins and oscillatoxin A. J Org Chem. 49:2484–2489.

Mynderse JS, Moore RE, Kashiwagi M, Norton TR (1977). Antileukemia activity in the Osillatoriaceae: isolation of Debromoaplysiatoxin from *Lyngbya*. Science. 196:538–540.

Nagai H, Yasumoto T, Hokama Y (1996). Aplysiatoxin and debromoaplysiatoxin as the causative agents of a red alga *Gracilaria coronopifolia* poisoning in Hawaii. Toxicon. 34:753–761.

Nagai H, Yasumoto T, Hokama Y (1997). Manauealides, some of the causative agents of a red alga Gracilaria coronopifolia poisoning in Hawaii. J Nat Prod. 60:925–928.

Nakamura H, Kishi Y, Pajares MA, Rando RR (1989). Structural basis of protein kinase C activation by tumor promoters. Proc Nat Acad Sci USA. 86:9672–9676.

Osborne N (2004). Investigation of the toxicology and public health aspects of the marine cyanobacterium, *Lyngbya majuscula*. Brisbane: Institution. 246 pp.

Osborne N, Webb P, Shaw G (2002). The toxicology and public health aspects of *Lyngbya majuscula* in Queensland, Australia. 10th International conference on harmful algae. WHO: St. Pete Beach (FL), USA: 221.

Osborne NJ, Seawright A, Shaw G (2008). Dermal Toxicology of *Lyngbya majuscula*, from Moreton Bay, Queensland, Australia. Harmful Algae. 7:584–589.

Osborne NJ, Shaw GR (2008). Dermatitis associated with exposure to a marine cyanobacterium during recreational water exposure. BMC Dermatol. 8:5.

Osborne NJT, Shaw GR, Webb PM (2007). Health effects of recreational exposure to Moreton Bay, Australia waters during a *Lyngbya majuscula* bloom. Environ Int. 27:309–314.

Park PU, Broka CA, Johnson BF, Kishi Y (1987). Total synthesis of debromoaplysiatoxin and aplysiatoxin. J Am Chem Soc. 109:6205–6207.

Pennings SC, Weiss AM, Paul VJ (1996). Secondary metabolites of the cyanobacterium *Microcoleus lyngbyaceus* and the sea hare *Stylocheilus longicauda*: Palatability and toxicity. Mar Biol. 126:735–743.

Read JA, Walsh CT (2007) The lyngbyatoxin biosynthetic assembly line: chain release by four-electron reduction of a dipeptidyl thioester to the corresponding alcohol. J Am Chem Soc. 129:15762–15763.

Serdula M, Bartilini G, Moore RE, Gooch J, Wiebenga N (1982). Seaweed itch on windward Oahu. Hawaii Med J. 41:200–201.

Sims JK, Zandee van Rilland RD (1981). Escharotic stomatitis caused by the "stinging seaweed" *Microcoleus lyngbyaceus* (formerly *Lyngbya majuscula*). Case report and literature review. Hawaii Med J. 40:243–248.

Solomon AE, Stoughton RB (1978). Dermatitis from purified sea algae toxin (debromoaplysiatoxin). Arch Dermatol. 114:1333–1335.

Stafford RG, Mehta M, Kemppainen BW (1992). Comparison of the partition coefficient and skin penetration of a marine algal toxin (lyngbyatoxin A). Food Chem Toxicol. 30:795–801.

Todd JS, Gerwick WH (1995). Malyngamide I from the tropical marine cyanobacterium *Lyngbya majuscula* and the probable structure revision of stylocheilamide. Tetrahedron Lett. 36:7837–7840.

Tønder J, Hosseini M, Ahrenst A, Tanner D (2004). Studies of the formation of all-carbon quaternary centres, en route to lyngbyatoxin A. A comparison of phenyl and 7-substituted indole systems. Org Biomol Chem. 2:1447–1455.

Videau P, Wells KN, Singh AJ, Gerwick WH, Philmus B (2016). Assessment of *Anabaena* sp. strain PCC 7120 as a heterologous expression host for cyanobacterial natural products: production of lyngbyatoxin A. ACS Synth Biol. 5:978–988.

Yasumoto T (1998). Fish poisoning due to toxins of microalgal origins in the Pacific. Toxicon. 36:1515–1518.

2.7 β-METHYLAMINO-L-ALANINE (BMAA)

Neil Chernoff, Elisabeth J. Faassen and Donna J. Hill

The nonproteinogenic amino acid, β-methylamino-*L*-alanine (BMAA; Figure 2.7), has been postulated to be a cause of neurodegenerative diseases that affect large numbers of people. However, at the time of publication of this document, this hypothesis is still highly controversial and a number of inconsistencies must be clarified before its role in human disease can be assessed with more certainty. The following section introduces and discusses these.

Interest in BMAA began as a result of a neurological disease known as amyotrophic lateral sclerosis/Parkinsonism dementia complex (ALS/PDC) present in the island of Guam in the Pacific (Arnold et al., 1953; Kurland et al., 1961). ALS/PDC has also been identified in small populations in Irian Jaya (western New Guinea) and Kii Peninsula of Japan. ALS/PDC has a spectrum of symptoms that resemble ALS, Parkinsonism and dementia. Different types of neurological dysfunctions were commonly present in the same individual, and multiple cases were often seen within families. The disease rendered patients incapable of normal movement, produced memory decline, cognitive deficits, and often led to premature death. In Guam, the peak incidence of the disease occurred during the 1950s and has been declining since then (Plato et al., 2002; Plato et al., 2003). The disease seemed limited to the indigenous population or others who had lived in Guam and adopted local customs and diet. ALS/PDC is characterised by hyperphosphorylated tau proteins that may assemble into masses ranging from a few molecules to large amyloid masses that may propagate like prions (Buée et al., 2000; Jucker & Walker, 2013). The altered proteins form neurofibrillary tangles (NFTs), disrupting cell structure associated with loss of function and/or cell death (Walker & LeVine, 2000; Chiti & Dobson, 2006).

2.7.1 Discrepancies introduced by incorrect BMAA analysis

In order to evaluate the possible health risk of β-methylamino-L-alanine (BMAA), one of the crucial elements is an accurate estimation of BMAA levels in environmental and food samples, as well as in tissue of possibly

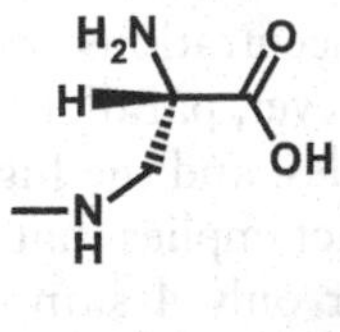

H_2N, O, H, OH, N, H

Figure 2.7 Structure of β-methylamino-L-alanine (BMAA). Molecular mass (monoisotopic): 118.074 Da; molecular weight (average): 118.13 g/mol.

exposed humans. However, one of the major issues impacting the BMAA hypothesis is the use of nonspecific analytical techniques such as liquid chromatography fluorescence detection (LC-FLD) for quantification of BMAA in environmental and human tissue samples. The role of analytical chemistry in the BMAA-human neurodegenerative disease hypothesis is therefore explained first.

As experimentally shown (Faassen et al., 2012), LC-FLD analysis risks misidentification of BMAA. Some cyanobacterial samples tested positive for BMAA when analysed by LC-FLD, while the same samples tested negative when analysed by more reliable mass-specific analytical methods (e.g., liquid chromatography-tandem mass spectrometry, LC-MS/MS). This is in line with the differences found in the literature, in which studies that have used nonspecific analytical techniques for BMAA detection typically report higher percentages of positive samples and/or higher BMAA concentrations than mass spectrometry-based studies. There is now considerable data that indicates shortcomings with many of the analytical approaches used (Cohen, 2012; Faassen et al., 2012; Faassen, 2014; Faassen et al., 2016; Lage et al., 2016; Rosén et al., 2016).

Analytical issues seem to have resulted in a lack of replication in many of the key findings in the BMAA-neurodegenerative disease hypothesis, which will be discussed below. For instance, studies indicating the presence of BMAA in the brains of people who suffered from Alzheimer's disease or ALS (Murch et al., 2004a; Murch et al., 2004b; Pablo et al., 2009) used LC-FLD for quantification, and their results have not been replicated by more recent work using more reliable techniques. Similarly, the suggested universal occurrence of high concentrations of BMAA in cyanobacteria (Cox et al., 2005; Esterhuizen & Downing, 2008; Metcalf et al., 2008) could not be replicated by studies using selective mass spectrometry techniques: these techniques either do not detect BMAA in cyanobacteria or find very low levels (Faassen, 2014; Lance et al., 2018). A key conclusion derived from this body of research is that LC-FLD, along with other optical detection methods that were used in early studies on BMAA in brain tissue, flying fox skin samples (Banack & Cox, 2003a; Banack et al., 2006) and fish (Brand et al., 2010), is not sufficiently selective for BMAA identification and quantification, and should therefore not be used unless positive samples are verified and quantified with a more selective method like LC-MS/MS. An illustrative case in this respect is a study on BMAA concentrations in stranded dolphins (Davis et al., 2019). In this study, BMAA was reported from the brains of 13 of the 14 tested animals, in concentrations ranging from 20 up to 748 μg/g, as quantified by LC-FLD. However, parallel LC-MS/MS analyses were only performed on 4 of the 14 samples, and the highest concentration found was 0.6 μg/g. So although the abstract implies that two orthogonal methods were used throughout the study, for only 4 samples complementary results by LC-MS/MS were available. Moreover, the concentrations found by LC-MS/

MS, which can be found in the supplementary information, are a few orders of magnitude lower than the LC-FLD results reported in the main text, and the only sample that tested negative by LC-FLD tested positive by LC-MS/MS. These discrepancies are not discussed in this chapter, which may leave the reader under the false impression that the high concentrations detected by LC-FLD are valid because they are supposedly confirmed by LC-MS/MS.

2.7.2 The BMAA-human neurodegenerative disease hypothesis

An epidemiological study related the incidence of ALS/PDC to the diets of the Guam population (Reed et al., 1987). Cycad seeds played a large role in the diet of the inhabitants of Guam, the seeds being ground up into flour that was a dietary staple. It was known that ingestion of seeds induced toxicity and they were carefully prepared with repeated washings before use as food but potent toxins like cycasin could be detected in cycad flour (Spencer, 2019). Vega et al. (1968) isolated a nonproteinogenic amino acid, BMAA, from seeds of cycad species utilised as food on Guam and found that it induced neurotoxicity when injected intraperitoneally at high dose levels into chickens or rats. Spencer et al. (1987b) exposed macaque monkeys (*Macaca fascicularis*) to 100–350 mg BMAA·HCl/kg bw × d orally, and observed stooped posture, tremors and weakness in extremities after a month at doses exceeding 200 mg/kg. The amounts of BMAA administered to the monkeys were orders of magnitude greater than the amounts that would have been consumed by people in cycad flour, and a role of BMAA in ALS/PDC was dismissed (Duncan et al., 1990). Other chemicals associated with cycads have been suggested as possible causes of ALS/PDC, including cycasin (methoxymethanol; Spencer et al., 2012) and sterol glucosides (Ly et al., 2007). There was no evidence of cycad consumption in either Irian Jaya or Japan, but it was noted that both areas used cycads for medicinal purposes (Spencer et al., 1987a; Spencer et al., 2005).

Cox and Sacks (2002) postulated that ALS/PDC could be related to the consumption of cycasin and BMAA, produced by cycads, the seeds of which were then eaten by flying foxes (*Pteropus mariannus*) which were subsequently consumed by people (Banack et al., 2006). Cox et al. (2003) reported that symbiotic cyanobacteria (*Nostoc* spp.) in the coralloid roots of cycads produced BMAA and that this was subsequently transported and biomagnified to the outer layer of the seeds, a food item in the diet of flying foxes (Banack & Cox, 2003a). High BMAA contents (mean 3.6 mg/g) were reported from three desiccated skin samples of preserved flying foxes by LC-FLD (Banack & Cox, 2003b), and the authors concluded that people in Guam consumed sufficient numbers of flying foxes to have been exposed to BMAA levels of a similar magnitude as those to which the monkeys in the experiments of Spencer et al. (1987b) were exposed. It was further suggested that the decline in the incidence of ALS/PDC was related to a decline in flying fox populations (Cox & Sacks, 2002). However, the amount of flying foxes consumed by natives is in question since it was a food that appears to have been reserved for special occasions (Lemke, 1992). Borenstein et al. (2007) did not find any positive associations between cycad or flying fox consumption and ALS/PDC in Guam.

Another fundamental issue with this hypothesis is the findings of Foss et al. (2018) who tested skin samples from the identical three preserved flying foxes referred to in the study by Banack and Cox (2003a). LC-MS/MS was used for analysis and failed to identify BMAA in these samples although BMAA was successfully detected in positive controls and spiked samples. These findings support the point raised in section 2.7.1 that BMAA exposure should only be estimated from studies that used selective analytical techniques for identification and quantification.

BMAA in cyanobacteria: The reports of BMAA in the symbiotic cyanobacteria *Nostoc* spp. in the coralloid roots of cycads raised the question of the source of BMAA. Using LC-FLD for quantification, Cox et al. (2005) examined cyanobacteria from different genera and found BMAA in 29 out of 30 the strains. They then postulated that, since BMAA was produced by most cyanobacteria, it should be considered to be a ubiquitous cyanotoxin. Subsequent studies have evaluated the ability of various genera and species of cyanobacteria to produce BMAA and reached different conclusions. Taking only studies into account that use selective, well-documented analytical techniques, reports of the presence of BMAA in cyanobacteria are scarce, and only incidentally, low concentrations are found in cyanobacterial samples (Faassen, 2014; Lance et al., 2018). It was found that BMAA can be produced by diatoms (Réveillon et al., 2016), but more studies are needed to estimate the range of BMAA concentrations in this type of phytoplankton.

Toxicological studies on monkeys and rats: Animal studies of BMAA exposures include primate studies carried out by Spencer et al. (1987b), as discussed above, and Cox et al. (2016) who reported on 32 adult vervet monkeys (*Chlorocebus sabaeus*) exposed orally to 21 mg/kg × d or 210 mg/kg × d of β-methylamino-alanine (BMAA) for 140 days. Although effects are found in the 210 mg/kg × d group, this dose level is unrealistic in terms of any known source of BMAA or suggested route of human exposure.

BMAA concentrations reported from water samples analysed with accurate methods have demonstrated only very low levels when BMAA is detected at all (Lance et al., 2018), and ingestion of cyanobacterial infested waters therefore does not seem to be the most relevant human exposure pathway to BMAA. Considering the reported concentrations in fish and shellfish (Lance et al., 2018), consumption of these foodstuffs seems at present the most likely route of BMAA exposure. Using the data which were selected by Lance et al. (2018) based on their selectivity and well-described quantitation methods, a theoretical weekly human diet consisting of meals of 200 g fish for 6 days and 200 g shellfish for 1 day can be used to estimate the amount of BMAA that would be consumed. Assuming a weekly diet resulting in 6-day exposure of 58 µg BMAA from fish and 540 µg BMAA from a single exposure of shellfish for a total of 598 µg yields an average daily intake of 85.4 µg. For a 60-kg individual, this would be equivalent to 1.42 µg/kg × d. The dose of 21 mg/kg × d (21,000 µg/kg × d) in vervet monkeys after 140 consecutive daily exposures, at which no adverse effects were observed (Cox et al., 2016), was ≈15,000-fold higher. Other issues with the Cox et al.'s (2016) study are that at necropsy, brain homogenates of the vervets were analysed for the presence of BMAA and 14 regions of the brain were analysed for the presence of neurofibrillary tangles (NFT) and β-amyloid deposits. It was concluded that more NFTs were found in high-dose BMAA groups than in the low dose or controls (the data supplied in the paper and supplementary information do not allow estimates of individual variability within groups). In spite of the high dose, the behaviour of animals remained normal and they did not exhibit Parkinsonism or the muscular symptomology observed in the earlier macaque study by Spencer et al. (1987b), although the vervets studied by Cox et al. (2016) were exposed for 140 days, while the macaques studied by Spencer et al. (1987b) exhibited overt toxicity after being exposed to a similar dose only for approximately 45–75 days.

A few rodent studies have been conducted on BMAA. BMAA administered to rats by oral route did not show effects at 500 mg/kg×d for approximately 32 days, or at 1000 mg/kg for approximately 15 days over the course of two months (Perry et al., 1989). BMAA administered to prepubertal rodents by either intraperitoneal ≥500 mg/kg (Seawright et al., 1990; de Munck et al., 2013) or subcutaneous route ≥460 mg/kg (Karlsson et al., 2009) is neurotoxic, but the inappropriate routes of administration and magnitude of the administered levels render these findings difficult to extrapolate to human exposures.

2.7.2.1 ALS/PDC attributed to BMAA versus other manifestations of neurodegenerative disease

An underlying assumption in the BMAA hypothesis of human ALS/PDC effects is that this syndrome encountered in Guam is closely related to other neurodegenerative diseases found globally, but there is evidence contradicting this assumption. Differences between ALS/PDC on Guam and ALS, Parkinsonism and Alzheimer's diseases include the strong familial occurrence (Zhang et al., 1996; Morris et al., 2001) and the common mixed disease syndrome seen in ALS/PDC on Guam (Murakami, 1999), both situations being extremely rare in the other neurological diseases. Additional characteristics indicating that ALS/PDC is distinct from sporadic ALS,

Parkinsonism and Alzheimer's disease include the absence of beta-amyloid plaques that are characteristic of Alzheimer's disease, the absence of ubiquitinated Lewy bodies characteristic of Parkinsonism (Hirano et al., 1961), as well as the absence of the typical ALS/PDC tauopathy in sporadic ALS (Ikemoto, 2000). The individual symptomologies exhibited in ALS/PDC cases have been related to differences in the areas of the central nervous system where the highest densities of the aberrant tau proteins occurred (Hof et al., 1994; Umahara et al., 1994). One other significant difference between ALS/PDC and other neurodegenerative diseases is the presence of a retinal pigment epitheliopathy (RPE) that has only been reported in Guam and Kii Peninsula ALS/PDC cases (Kokubo et al., 2003). The condition manifests itself as linear tracks of retinal depigmentation with intermittent pigment clumping, and the incidence of RPE is significantly higher in ALS/PDC cases than in controls. RPE has not been associated with other diseases elsewhere in the world and is therefore considered part of the ALS/PDC disease postulated to be caused by β-methylamino-alanine (BMAA) (Cox et al., 1989; Steele et al., 2015).

2.7.3 Postulated human exposure and BMAA mechanism of action

BMAA in brain tissue of humans: Reports of BMAA in brain tissue of humans who suffered neurodegenerative diseases are contradictory: three studies of postmortem human brain tissues from people on Guam who had suffered from ALS/PDC, or people in the United States of America and Canada who had either ALS or Alzheimer's disease, reported the presence of BMAA in disease sufferers (39 out of 40) irrespectively of where they had lived, whereas the studies rarely identified BMAA in people (four out of 36) who had not suffered from these neurodegenerative diseases (Murch et al., 2004a; Murch et al., 2004b; Pablo et al., 2009). These studies all utilised LC-FLD to quantify BMAA. In contrast, however, four studies that used mass spectrometry for identification and quantification of BMAA have not found similar incidences and/or levels in brains (Snyder et al., 2009; Combes et al., 2014; Meneely et al., 2016) or cerebrospinal fluids (Berntzon et al., 2015) of people who had suffered from Alzheimer's disease or ALS in the United States of America and Europe. Taking these last four studies together, BMAA was not found in any of 13 ALS/PDC brains/cerebrospinal fluids, and was found in one of 39 brains/cerebrospinal fluids from people who had either ALS or Alzheimer's disease, as well as in three of 20 without disease. When only considering data on BMAA levels in brains or cerebrospinal fluids that have been generated by appropriate analytical techniques, there is little evidence for the hypothesis that BMAA is present in the brains of those suffering from ALS and Alzheimer's disease.

Evidence of human exposure to BMAA has not been well documented. It has been suggested that a cluster of ALS cases in the United States of America was due to proximity to a lake and therefore exposure to BMAA, but this was not based on substantive evidence (Caller et al., 2009). A subsequent study did not show a general linkage between proximity to waterbodies and neurological disease (Caller et al., 2012). Suggestions have been made linking ALS to BMAA inhaled by soldiers in Qatar (Cox et al., 2009), the consumption of blue crabs (Field et al., 2013) and exposure to aerosols from cooling towers (Stommel et al., 2013), but clear evidence supporting these suggestions is not provided.

One of the central questions concerning the BMAA-neurodegenerative disease hypothesis concerns the mechanism by which BMAA would induce these diseases. Protein tangles and deposits are hallmarks of the neurodegenerative diseases discussed in this chapter (Ellisdon & Bottomley, 2004; Jellinger, 2012; Bolshette et al., 2014). These tangles of misfolded proteins include tau proteins in Alzheimer's disease, ubiquinated proteins in ALS and Lewy bodies in Parkinsonism. Dunlop et al. (2013) stated that BMAA is misincorporated into human proteins in place of *L*-serine, but no direct evidence for this is presented. The reported association of BMAA with proteins is not necessarily indicative of incorporation and may simply be due to chemical binding. Glover et al. (2014) examined protein synthesis after co-incubation of BMAA in a cell-free system (PURExpress) in studies where BMAA was substituted for individual essential amino acids. Although the interaction of BMAA and serine is highlighted, the data indicate that BMAA substitution for alanine occurred to a greater extent. BMAA was found to be significantly incorporated into proteins in place of four of the nine additional amino acids for which data are presented. These results may primarily be a reflection of the relaxed fidelity of translation of the PURExpress *in vitro* system, which has been used to facilitate misincorporation of amino acids (Hong et al., 2014; Singh-Blom et al., 2014).

In *in vitro* assays, Beri et al. (2017) and Han et al. (2020) observed that BMAA was not a substrate of human seryl-tRNA synthetase, and therefore, a misincorporation of BMAA instead of serine in proteins as postulated earlier is highly unlikely. Instead, Han et al. (2020) report that BMAA is a substrate for human alanyl-tRNA synthetase, however, with only low rates of product formation despite a 500-fold higher concentration of BMAA compared to alanine. In an *in vivo* assay with *Saccharomyces cerevisiae*, an incorporation of BMAA instead of alanine could not be detected. Notably, the observed rates of mischarging of tRNA with BMAA are within the ranges generally observed for mischarging of aminoacyl-tRNA synthetases with noncognate amino acid – some 10^{-4} errors per codon or tRNA molecule, respectively (Mohler & Ibba, 2017).

Other studies have failed to find indications of misincorporation of BMAA into proteins. van Onselen et al. (2015) compared BMAA and canavanine, a nonproteinogenic amino acid known for its tendency to be misincorporated in proteins. Protein incorporation was evaluated with an *E. coli* expression system using a fragment of a recombinant human protein. In contrast to canavanine, β-methylamino-alanine (BMAA) did not affect cell growth and was not detected in the protein fragment. The authors also showed that the removal of BMAA from bacterial proteins was not accomplished by washing with detergent-containing acid hydrolysis and TCA precipitation, indicating the probability of a strong association with protein surfaces. Similar findings were reported by Okle et al. (2012) who used a human neuroblastoma cell culture and demonstrated BMAA association with proteins after TCA protein precipitation, but not after protein-denaturing SDS gel electrophoresis. Spencer et al. (2016) did not find evidence to support the incorporation of BMAA into proteins in the brains of macaques. Cerebral protein lysates of BMAA-treated animals were analysed after extraction to remove BMAA from denatured proteins, detection was performed with LC–MS/MS, and no incorporation was found.

Rauk (2018) modelled protein folding changes that would have occurred if serine was substituted by BMAA. He concluded that BMAA incorporation instead of serine in proteins would not change conformational characteristics of the β-amyloid peptide and that BMAA was therefore not related to Alzheimer's disease.

2.7.4 Conclusions

The cause(s) of the ALS/PDC in Guam remains a mystery. The existence of the disease in Guam and Rota, but not in other areas where both flying foxes and cycad products are eaten, has not been satisfactorily explained. The possible relationship(s) between the presence of ALS/PDC in Guam, Irian Jaya and the Kii Peninsula remains unknown. Over the course of a decade, the BMAA hypothesis was transformed from one of many concerning the cause of a neurodegenerative disease that occurred on Guam and two other localities, to a global threat purportedly linked not only to ALS/PDC, but also to ALS, Alzheimer's disease and Parkinsonism.

The BMAA-neurodegenerative disease hypothesis is built on four major contentions:

1. BMAA was the primary cause of ALS/PDC due to high levels in food in Guam.
2. The disease is sufficiently similar to ALS, Parkinsonism and Alzheimer's disease to enable BMAA to cause all of these diseases.
3. The environmental/dietary exposure levels outside of Guam are sufficient to cause this disease in humans.
4. BMAA acts through its incorporation into proteins displacing serine.

While this hypothesis may be appealing for its simplicity and universality, these contentions are either disputed by many other studies, or the necessary data to support the hypothesis are not presented. The hypothesis that BMAA caused ALS/PDC was largely based on a primate study that used extremely high dose levels which were postulated to be possible for humans to obtain by the consumption of food with extremely high levels of BMAA. ALS/PDC is a separate neurodegenerative disease that has occurred in several geographically distant and distinct areas. While the sum of its symptoms are similar to other neurodegenerative diseases, the patterns of occurrence are different, the type of aberrant proteins and regions of the brain that are affected are different, and there is no reason for assuming that the same agent acts to induce all of these diseases. Moreover, it seems that human BMAA exposure through food and environment outside Guam is orders of magnitude lower than effective doses administered in animal studies, or postulated to have been consumed by people on Guam. Finally, several well-designed studies have failed to find evidence of BMAA incorporation into proteins.

Research into the cause(s) of ALS/PDC has largely been focused on single factors, but there is little evidence that any of the single factor hypotheses are completely responsible for the disease. There is, however, a possibility that all or most of the different postulated causes, along with the considerable stress on the population of Guam during the World War II occupation, played additive or synergistic roles in the occurrence of ALS/PDC, and a more complex causation should be considered. Mineral imbalance, genetic background, stress-induced physiological alterations and any of several toxins present in cycads may have all played significant roles in the causation of the disease (Chernoff et al., 2017). To solve a problem of this nature is extremely difficult under any circumstances, and this difficulty may increase as the incidence of ALS/PDC lessens in Guam. The evidence for BMAA being the single cause of ALS/PDC in Guam as well as for other unrelated neurodegenerative diseases globally is not convincing.

One can never realistically prove the absence of an effect, but the totality of the evidence for the BMAA-neurodegenerative disease hypothesis at the present time, or better the lack thereof, gives no reasons for immediate concern. The question of mechanisms explaining how one compound can cause four distinctive neurological diseases affecting different regions of the brain and having different proteins associated with the central nervous system changes in different people is a major issue that has yet to be addressed experimentally. BMAA remains an interesting compound, but given the evidence of increasing cyanobacterial and marine algal blooms and various associated toxins in numerous waterbodies globally, there are many other more apparent potential algal toxin health effect issues. Research efforts on BMAA should be balanced with regard to those on the other cyanotoxins.

Although solid exposure data are required for risk assessment, the key question that needs to be answered first is whether the proposed toxic effects of BMAA can be confirmed in health-relevant dose ranges.

REFERENCES

Arnold A, Edgren DC, Palladino VS (1953). Amyothrophic Lateral Sclerosis: fifty cases observed on Guam. J Nerv Ment Dis. 117:135–139.

Banack SA, Cox PA (2003a). Biomagnification of cycad neurotoxins in flying foxes: implications for ALS-PDC in Guam. Neurology. 61:387–389.

Banack SA, Cox PA (2003b). Distribution of the neurotoxic nonprotein amino acid BMAA in *Cycas micronesica*. Bot J Linn Soc. 143:165–168.

Banack SA, Murch SJ, Cox PA (2006). Neurotoxic flying foxes as dietary items for the Chamorro people, Marianas Islands. J Ethnopharmacol. 106:97–104.

Beri J, Nash T, Martin RM, Bereman MS (2017). Exposure to BMAA mirrors molecular processes linked to neurodegenerative disease. Proteomics. 17:1700161.

Berntzon L, Ronnevi L, Bergman B, Eriksson J (2015). Detection of BMAA in the human central nervous system. Neuroscience. 292:137–147.

Bolshette N, Thakur K, Bidkar A, Trandafir C, Kumar P, Gogoi R (2014). Protein folding and misfolding in the neurodegenerative disorders: a review. Revue Neurologique. 170:151–161.

Borenstein A, Mortimer J, Schofield E, Wu Y, Salmon D, Gamst A et al. (2007). Cycad exposure and risk of dementia, MCI, and PDC in the Chamorro population of Guam. Neurology. 68:1764–1771.

Brand LE, Pablo J, Compton A, Hammerschlag N, Mash DC (2010). Cyanobacterial Blooms and the Occurrence of the neurotoxin beta-N-methylamino-L-alanine (BMAA) in South Florida Aquatic Food Webs. Harmful Algae. 9:620–635.

Buée L, Bussière T, Buée-Scherrer V, Delacourte A, Hof PR (2000). Tau protein isoforms, phosphorylation and role in neurodegenerative disorders. Brain Res Rev. 33:95–130.

Caller TA, Doolin JW, Haney JF, Murby AJ, West KG, Farrar HE et al. (2009). A cluster of amyotrophic lateral sclerosis in New Hampshire: a possible role for toxic cyanobacteria blooms. Amyotroph Lateral Scler. 10 (Suppl 2):101–108.

Caller TA, Field NC, Chipman JW, Shi X, Harris BT, Stommel EW (2012). Spatial clustering of amyotrophic lateral sclerosis and the potential role of BMAA. Amyotroph Lateral Scler. 13:25–32.

Chernoff N, Hill D, Diggs D, Faison B, Francis B, Lang J et al. (2017). A critical review of the postulated role of the non-essential amino acid, β-N-methylamino-L-alanine, in neurodegenerative disease in humans. J Toxicol Environ Health Part B. 20:183–229.

Chiti F, Dobson CM (2006). Protein misfolding, functional amyloid, and human disease. Annu Rev Biochem. 75:333–366.

Cohen SA (2012). Analytical techniques for the detection of α-amino-β-methylaminopropionic acid. Analyst. 137:1991–2005.

Combes A, El Abdellaoui S, Vial J, Lagrange E, Pichon V (2014). Development of an analytical procedure for quantifying the underivatized neurotoxin β-N-methylamino-L-alanine in brain tissues. Anal Bioanal Chem. 406:4627–4636.

Cox PA, Banack SA, Murch SJ (2003). Biomagnification of cyanobacterial neurotoxins and neurodegenerative disease among the Chamorro people of Guam. Proc Natl Acad Sci USA. 100:13380–13383.

Cox PA, Banack SA, Murch SJ, Rasmussen U, Tien G, Bidigare RR et al. (2005). Diverse taxa of cyanobacteria produce beta-N-methylamino-L-alanine, a neurotoxic amino acid. Proc Natl Acad Sci USA. 102:5074–5078.

Cox PA, Davis DA, Mash DC, Metcalf JS, Banack SA (2016). Dietary exposure to an environmental toxin triggers neurofibrillary tangles and amyloid deposits in the brain. Proc R Soc B. Biol Sci. 283:2015–2397.

Cox PA, Richer R, Metcalf JS, Banack SA, Codd GA, Bradley WG (2009). Cyanobacteria and BMAA exposure from desert dust: a possible link to sporadic ALS among Gulf War veterans. Amyotroph Lateral Scler. 10:109–117.

Cox PA, Sacks OW (2002). Cycad neurotoxins, consumption of flying foxes, and ALS-PDC disease in Guam. Neurology. 58:956–959.

Cox TA, McDarby JV, Lavine L, Steele JC, Calne DB (1989). A retinopathy on Guam with high prevalence in Lytico-Bodig. Ophthalmology. 96:1731–1735.

Davis DA, Mondo K, Stern E, Annor AK, Murch SJ, Coyne TM et al. (2019). Cyanobacterial neurotoxin BMAA and brain pathology in stranded dolphins. PloS One. 14:e0213346.

de Munck E, Muñoz-Sáez E, Miguel BG, Solas MT, Ojeda I, Martínez A et al. (2013). β-N-methylamino-l-alanine causes neurological and pathological phenotypes mimicking Amyotrophic Lateral Sclerosis (ALS): the first step towards an experimental model for sporadic ALS. Environ Toxicol Pharmacol. 36:243–255.

Duncan MW, Steele JC, Kopin IJ, Markey SP (1990). 2-Amino-3-(methylamino)-propanoic acid (BMAA) in cycad flour An unlikely cause of amyotrophic lateral sclerosis and parkinsonism-dementia of Guam. Neurology. 40:767–767.

Dunlop RA, Cox PA, Banack SA, Rodgers KJ (2013). The non-protein amino acid BMAA is misincorporated into human proteins in place of L-serine causing protein misfolding and aggregation. PLoS One. 8:e75376.

Ellisdon AM, Bottomley SP (2004). The role of protein misfolding in the pathogenesis of human diseases. IUBMB life. 56:119–123.

Esterhuizen M, Downing TG (2008). Beta-N-methylamino-L-alanine (BMAA) in novel South African cyanobacterial isolates. Ecotoxicol Environ Safety. 71:309–313.

Faassen EJ (2014). Presence of the neurotoxin BMAA in aquatic ecosystems: what do we really know? Toxins. 6:1109–1138.

Faassen EJ, Antoniou MG, Beekman-Lukassen W, Blahova L, Chernova E, Christophoridis C et al. (2016). A collaborative evaluation of LC-MS/MS based methods for BMAA analysis: Soluble bound BMAA found to be an important fraction. Mar Drugs. 14:45.

Faassen EJ, Gillissen F, Lürling M (2012). A comparative study on three analytical methods for the determination of the neurotoxin BMAA in cyanobacteria. PLoS One. 7:e36667.

Field NC, Metcalf JS, Caller TA, Banack SA, Cox PA, Stommel EW (2013). Linking β-methylamino-L-alanine exposure to sporadic amyotrophic lateral sclerosis in Annapolis, MD. Toxicon. 70:179–183.

Foss AJ, Chernoff N, Aubel MT (2018). The analysis of underivatized β-Methylamino-L-alanine (BMAA), BAMA, AEG & 2, 4-DAB in *Pteropus mariannus mariannus* specimens using HILIC-LC-MS/MS. Toxicon. 152:150–159.

Glover WB, Mash DC, Murch SJ (2014). The natural non-protein amino acid N-β-methylamino-L-alanine (BMAA) is incorporated into protein during synthesis. Amino Acids. 46:2553–2559.

Han N-C, Bullwinkle TJ, Loeb KF, Faull KF, Mohler K, Rinehart J et al. (2020). The mechanism of β-N-methylamino-l-alanine inhibition of tRNA aminoacylation and its impact on misincorporation. J Biol Chem. 295:1402–1410.

Hirano A, Malamud N, Kurland LT (1961). Parkinsonism-dementia complex, an endemic disease on the island of Guam. II Pathological features. Brain. 84:662–679.

Hof PR, Perl DP, Loerzel AJ, Steele JC, Morrison JH (1994). Amyotrophic lateral sclerosis and parkinsonism-dementia from Guam: differences in neurofibrillary tangle distribution and density in the hippocampal formation and neocortex. Brain Res. 650:107–116.

Hong SH, Kwon Y-C, Jewett MC (2014). Non-standard amino acid incorporation into proteins using Escherichia coli cell-free protein synthesis. Front Chem. 2:34.

Ikemoto AH, Ichiro Akiguchi A (2000). Neuropathology of amyotrophic lateral sclerosis with extra-motor system degeneration: characteristics and differences in the molecular pathology between ALS with dementia and Guamanian ALS. Amyotroph Lateral Scler Other Motor Neuron Disord. 1:97–104.

Jellinger KA (2012). Interaction between pathogenic proteins in neurodegenerative disorders. J Cell Mol Med. 16:1166–1183.

Jucker M, Walker LC (2013). Self-propagation of pathogenic protein aggregates in neurodegenerative diseases. Nature. 501:45.

Karlsson O, Roman E, Brittebo EB (2009). Long-term cognitive impairments in adult rats treated neonatally with beta-N-methylamino-L-alanine. Toxicol Sci. 112:185–195.

Kokubo Y, Ito K, Kuzuhara S (2003). Ophthalmomyiasis-like pigmentary retinopathy in ALS/PDC in the Kii peninsula of Japan. Neurology. 60:1725–1726.

Kurland L, Hirano A, Malamud N, Lessell S (1961). Parkinsonism-dementia complex, en endemic disease on the island of Guam. Clinical, pathological, genetic and epidemiological features. Trans Am Neurol Assoc. 86:115–120.

Lage S, Burian A, Rasmussen U, Costa PR, Annadotter H, Godhe A et al. (2016). BMAA extraction of cyanobacteria samples: which method to choose? Environ Sci Pollut Res. 23:338–350.

Lance E, Arnich N, Maignien T, Biré R (2018). Occurrence of β-N-methylamino-l-alanine (BMAA) and isomers in aquatic environments and aquatic food sources for humans. Toxins. 10:83.

Lemke TO (1992). History of fruit bat use, research, and, protection in the Northern Mariana Islands. Biol Rep, US Fish Wild Serv. 90:135–142.

Ly P, Singh S, Shaw C (2007). Novel environmental toxins: Steryl glycosides as a potential etiological factor for age-related neurodegenerative diseases. J Neurosci Res. 85:231–237.

Meneely JP, Chevallier OP, Graham S, Greer B, Green BD, Elliott CT (2016). β-methylamino-L-alanine (BMAA) is not found in the brains of patients with confirmed Alzheimer's disease. Sci Rep. 6: 36363.

Metcalf JS, Banack SA, Lindsay J, Morrison LF, Cox PA, Codd GA (2008). Co-occurrence of beta-N-methylamino-L-alanine, a neurotoxic amino acid with other cyanobacterial toxins in British waterbodies, 1990–2004. Environ Microbiol. 10:702–708.

Mohler K, Ibba M (2017). Translational fidelity and mistranslation in the cellular response to stress. Nat Microbiol. 2:1–9.

Morris HR, Al-Sarraj S, Schwab C, Gwinn-Hardy K, Perez-Tur J, Wood NW et al. (2001). A clinical and pathological study of motor neurone disease on Guam. Brain. 124:2215–2222.

Murakami N (1999). Parkinsonism-dementia complex on Guam—overview of clinical aspects. J Neurol. 246:II16–18.

Murch SJ, Cox PA, Banack SA (2004a). A mechanism for slow release of biomagnified cyanobacterial neurotoxins and neurodegenerative disease in Guam. Proc Natl Acad Sci USA. 101:12228–12231.

Murch SJ, Cox PA, Banack SA, Steele JC, Sacks OW (2004b). Occurrence of beta-methylamino-l-alanine (BMAA) in ALS/PDC patients from Guam. Acta Neurol Scand. 110:267–269.

Okle O, Stemmer K, Deschl U, Dietrich DR (2012). L-BMAA induced ER stress and enhanced caspase 12 cleavage in human neuroblastoma SH-SY5Y cells at low nonexcitotoxic concentrations. Toxicol Sci. 131:217–224.

Pablo J, Banack SA, Cox PA, Johnson TE, Papapetropoulos S, Bradley WG et al. (2009). Cyanobacterial neurotoxin BMAA in ALS and Alzheimer's disease. Acta Neurologica Scandinavica. 120:216–225.

Perry TL, Bergeron C, Biro AJ, Hansen S (1989). β-N-Methylamino-L-alanine: chronic oral administration is not neurotoxic to mice. J Neurol Sci. 94:173–180.

Plato C, Galasko D, Garruto R, Plato M, Gamst A, Craig U-K et al. (2002). ALS and PDC of Guam Forty-year follow-up. Neurology. 58:765–773.

Plato CC, Garruto RM, Galasko D, Craig UK, Plato M, Gamst A et al. (2003). Amyotrophic lateral sclerosis and parkinsonism-dementia complex of Guam: changing incidence rates during the past 60 years. Am J Epidemiol. 157:149–157.

Rauk A (2018). β-N-Methylamino-l-alanine (BMAA) Not Involved in Alzheimer's Disease. J Phys Chem B. 122:4472–4480.

Reed D, Labarthe D, Chen KM, Stallones R (1987). A cohort study of amyotrophic lateral sclerosis and parkinsonism-dementia on Guam and Rota. Am J Epidemiol. 125:92–100.

Réveillon D, Séchet V, Hess P, Amzil Z (2016). Production of BMAA and DAB by diatoms (*Phaeodactylum tricornutum*, *Chaetoceros* sp., *Chaetoceros calcitrans* and, *Thalassiosira pseudonana*) and bacteria isolated from a diatom culture. Harmful Algae. 58:45–50.

Rosén J, Westerberg E, Schmiedt S, Hellenäs K-E (2016). BMAA detected as neither free nor protein bound amino acid in blue mussels. Toxicon. 109:45–50.

Seawright A, Brown A, Nolan C, Cavanagh J (1990). Selective degeneration of cerebellar cortical neurons caused by cycad neurotoxin, L-β-methylaminoalanine (L-BMAA), in rats. Neuropathol Appl Neurobiol. 16:153–169.

Singh-Blom A, Hughes RA, Ellington AD (2014). An amino acid depleted cell-free protein synthesis system for the incorporation of non-canonical amino acid analogs into proteins. J Biotechnol. 178:12–22.

Snyder L, Cruz-Aguado R, Sadilek M, Galasko D, Shaw C, Montine T (2009). Lack of cerebral BMAA in human cerebral cortex. Neurology. 72:1360–1361.

Spencer P, Fry RC, Kisby GE (2012). Unraveling 50-year-old clues linking neurodegeneration and cancer to cycad toxins: are microRNAs common mediators? Front Genet. 3:192.

Spencer P, Garner C, Palmer V, Kisby G (2016). Vervets and macaques: similarities and differences in their responses to L-BMAA. Neurotoxicology. 56:284–286.

Spencer P, Ohta M, Palmer V (1987a). Cycad use and motor neurone disease in Kii peninsula of Japan. Lancet. 330:1462–1463.

Spencer PS (2019). Hypothesis: etiologic and molecular mechanistic leads for sporadic neurodegenerative diseases based on experience with Western Pacific ALS/PDC. Front Neurol. 10.

Spencer PS, Nunn PB, Hougon J, Ludolph AC, Roy DN, Ross SM et al. (1987b). Guam amyotrophic lateral sclerosis-parkinsonism-dementia linked to a plant excitant neurotoxin. Science. 237:517–523.

Spencer PS, Palmer VS, Ludolph AC (2005). On the decline and etiology of high-incidence motor system disease in West Papua (southwest New Guinea). Mov Disord. 20:S119–S126.

Steele JC, Wresch R, Hanlon SD, Keystone J, Ben-Shlomo Y (2015). A unique retinal epitheliopathy is associated with amyotrophic lateral sclerosis/Parkinsonism-Dementia complex of Guam. Mov Disord. 30:1271–1275.

Stommel EW, Field NC, Caller TA (2013). Aerosolization of cyanobacteria as a risk factor for amyotrophic lateral sclerosis. Med Hypotheses. 80:142–145.

Umahara T, Hirano A, Kato S, Shibata N, Yen S-H (1994). Demonstration of neurofibrillary tangles and neuropil thread-like structures in spinal cord white matter in parkinsonism-dementia complex on Guam and in Guamanian amyotrophic lateral sclerosis. Acta Neuropathol. 88:180–184.

van Onselen R, Cook NA, Phelan RR, Downing TG (2015). Bacteria do not incorporate β-N-methylamino-L-alanine into their proteins. Toxicon. 102:55–61.

Vega A, Bell E, Nunn P (1968). The preparation of L-and D-α-amino-β-methylaminopropionic acids and the identification of the compound isolated from *Cycas circinalis* as the L-isomer. Phytochemistry. 7:1885–1887.

Walker LC, LeVine H (2000) The cerebral proteopathies. Mol Neurobiol. 21:83–95.

Zhang Z, Anderson D, Mantel N, Roman G (1996). Motor neuron disease on Guam: geographic and familial occurrence, 1956–85. Acta Neurologica Scandinavica. 94:51–59.

2.8 CYANOBACTERIAL LIPOPOLYSACCHARIDES (LPS)

Martin Welker

2.8.1 General characteristics of bacterial LPS

Lipopolysaccharides (LPS) are part of the outer membrane of most Gram-negative prokaryotes, including enteric bacteria (Erridge et al., 2002; Raetz & Whitfield, 2002) and also cyanobacteria (Weckesser et al., 1979; Martin et al., 1989). Furthermore, there is evidence that LPS-like compounds can be found in green algae (Armstrong et al., 2002) and chloroplasts of vascular plants (Armstrong et al., 2006). A large body of literature is available on the structure, composition of LPS and their association with adverse health effects, generally focusing on heterotrophic bacteria of clinical relevance (Dauphinee & Karsan, 2006; Bryant et al., 2010; Vatanen et al., 2016).

The structure of all LPS generally follows the scheme given in Figure 2.8. The core structure is highly complex with individual regions showing varying degrees of conservation. In particular, the O-polysaccharide chain is highly variable and is the main characteristic for distinguishing dozens or hundreds of serotypes in some bacterial species, for example, *Escherichia coli* or *Salmonella* sp. (Stenutz et al., 2006). The moiety primarily responsible for the toxicity is lipid A, which is composed of phosphorylated sugar units to which acyl chains of variable length and degree of saturation are linked. Cyanobacterial LPS is different to LPS from Gram-negative heterotrophs as it often lacks heptose and 3-deoxy-*D*-manno-octulosonic acid (or keto-deoxyoctulosonate; KDO), which are commonly present in the core region of the LPS of heterotrophic bacteria. However, since the number of

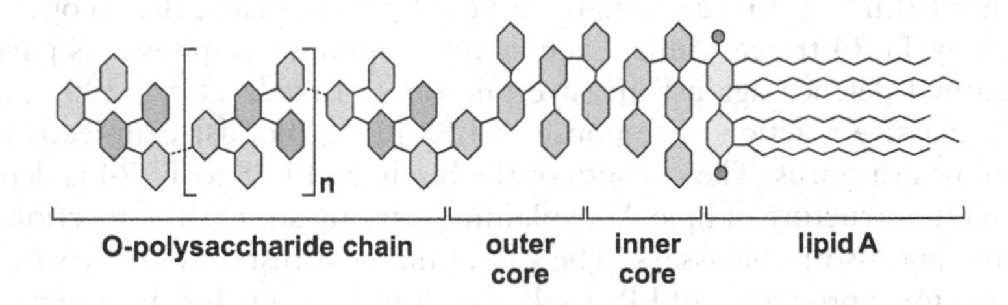

Figure 2.8 Schematic structure of lipopolysaccharides. Lipid A is composed of a highly conserved *D*-glucosamine backbone with variable acyl chains bound to it, the particular structure of which largely determines endotoxicity. The inner core is also highly conserved containing KDO and heptoses, that are, however, less frequent or absent in cyanobacterial LPS. The outer core is moderately variable and contains mostly common sugars such as hexoses or hexaminoses. The O-polysaccharide chain is composed of repetitive units of sugar complexes and is highly variable and responsible for the serological specificity of LPS and is a primary antigen of infective bacteria. (Modified from Erridge et al., 2002.)

well-characterised LPS from cyanobacteria is very limited, species-specific characteristics cannot yet be derived (Durai et al., 2015).

The first cyanobacterial LPS were characterised from two strains of marine *Synechococcus* sp. (Snyder et al., 2009). Structural elucidation of LPS is also available from "*Oscillatoria planktothrix*", confirming the lack of KDO (Carillo et al., 2014). The monosaccharide composition of the O-chain of *Microcystis* sp. LPS was found to be relatively simple (Fujii et al., 2012). Further reports characterise cyanobacterial LPS rather by its bioactivity and less by analyses of chemical structures.

In the literature, the terms "LPS" and "endotoxin" are often used as synonyms, but not always. Occasionally, endotoxin refers to the lipid A part of LPS or refers to an entirely different molecule that is released from cells only upon lysis. The lipid A of different heterotrophic bacterial species is highly variable and expectedly exhibits varying activity in various test systems (Erridge et al., 2002) – as it is presumably the case with cyanobacterial lipid A (Gemma et al., 2016).

The principal clinical relevance of LPS from heterotrophic bacteria is due to its role in sepsis and septic shock, which are potentially life-threatening conditions leading to high numbers of casualties every year worldwide (Hotchkiss et al., 2016). Most cases of endotoxin intoxication occur after systemic infection with Gram-negative bacteria that can lead to sepsis and septic shock when endotoxin is released from cells and enters the blood circuit. There, LPS triggers a signalling cascade in macrophage/endothelial cells that eventually secrete proinflammatory compounds such as cytokines and nitric oxide (Trent et al., 2006).

In particular, LPS is known to bind to one type of so-called toll-like receptors, namely, TLR4 (Bryant et al., 2010), triggering a cascade of cellular reactions that involve the regulation of the expression of a large number of genes (Akira & Takeda, 2004). In healthy individuals, the recognition of LPS by TLR4 triggers innate and adaptive immune responses as part of the normal defence against invasive microbes (Takeda et al., 2003), and only a massive reaction in response to LPS in the bloodstream leads to a critical health status. The strength of the binding of LPS to TLR4 is dependent on the structure of lipid A, explaining varying strength of reactions in patients but also in bioassays. The cascading host response to LPS rather than the toxic properties of LPS itself therefore accounts for the potentially lethal consequences (Opal, 2010). For this reason, LPS (or endotoxin) has been discussed to be classified rather as an (exogenous) hormone than as a toxin in a strict sense (Marshall, 2005). Arguably, LPS is not a secondary metabolite like the known cyanotoxins but a highly variable fraction of a cellular constituent rather than a defined structure.

One precondition of LPS-mediated sepsis is the microbial infection of a body part causing an immune response and, when not controlled, inflammation. The presence of bacteria producing LPS in or on the body is no health risk in itself (Mowat & Agace, 2014). The LPS contained in a healthy gut microbiome is generally no threat and exceeds by far the amount of LPS that triggers a septic shock when circulating in the bloodstream.

Besides their role in acute and life-threatening conditions, LPS play an important role in triggering various signalling pathways in epithelial cells, for example, in the intestine (Cario et al., 2000) or the lung (Guillot et al., 2004). However, the role of LPS (from heterotrophic bacteria) in the regulation of the gut microbiome (d'Hennezel et al., 2017) – the complexity of which has only recently been fully recognised – is beyond the scope of this chapter and the following will focus on cyanobacterial LPS.

2.8.2 What is known about bioactivity of cyanobacterial LPS?

A number of studies on cyanobacterial lipopolysaccharides (LPS) have demonstrated effects in bioassays, for example:

- Mayer et al. (2011) reported several metabolic reactions of rat microglia upon exposure to LPS from *Microcystis*. However, the reactions were considerably less pronounced compared to the reactions observed upon exposure to equimolar concentrations of LPS from *Escherichia coli*.
- Klemm et al. (2018) reported similar reactions of rat microglia *in vitro* after exposure to LPS from *Scytonema*.
- Best et al. (2002) quantified the activities of microsomal and soluble glutathione S-transferases (GST) from zebra fish embryos exposed to LPS from an axenic *Microcystis* strain, cyanobacterial blooms and enteric bacteria. They found a reduction in activity for all types of LPS and concluded that this may reduce the detoxication capacity for microcystins.
- Jaja-Chimedza et al. (2012) also exposed zebra fish embryos to extracts of *Microcystis* strains assumed to contain LPS and found an increase in activity of glutathione-based detoxication enzymes.
- Ohkouchi et al. (2012) exposed a human monocytic cell line to LPS from various heterotrophic and cyanobacteria as well as from microbial consortia to test the inflammatory potential of LPS. The LPSs from an *Acinetobacter lwoffii* culture and from bacterial consortia induced stronger reactions than other LPSs tested, including that of cyanobacteria.
- Macagno et al. (2006) isolated "an LPS like compound" from "*Oscillatoria planktothrix*" that acts as a selective inhibitor of activity induced in dendritic cells through exposure to LPS from *E. coli*. This antagonistic behaviour was found to inhibit LPS-induced toxic shock in mice. In *Limulus* amoebocyte lysate (LAL) assays, activity of the cyanobacterial LPS-like compound was very low at 4 EU/µg compared to 8000 EU/µg

of *Salmonella enterica* serotype abortus equi LPS or 15 000 EU/µg of *Escherichia coli* serotype 055:B5 LPS.

- This cyanobacterial LPS-like compound is a potential inhibitor of *Escherichia coli* LPS-induced inflammatory response in porcine whole blood (Thorgersen et al., 2008).
- Moosova et al. (2019) report a number of proinfammatory effects of LPS extracted from *Microcystis* strains and bloom samples observed in whole-blood in vitro assays, such as induction inflammatory mediators like tumor necrosis factor and interleukins.

These *in vitro* studies demonstrate a wide variety of bioactivities in a number of test systems; yet it is difficult to infer potential *in vivo* bioactivity from these results, especially as to date no study has unambiguously related cyanobacterial LPS to adverse health effects in mammals, including humans, *in vivo*, like this has been demonstrated for microcystin toxicity, for example. Gastrointestinal disorders upon ingestion of cyanobacteria, generally consisting of heterogeneous bloom material, cannot be causally attributed to cyanobacterial LPS as is discussed below.

In most studies that imply an association between observed adverse human health effects and cyanobacterial LPS, this is based more on associative argumentation than on conclusive evidence. Mainly two reports have been influential in advancing the hypothesis that cyanobacterial LPS is a health risk.

Lippy and Erb (1976) reported on an outbreak of gastrointestinal illness that occurred in Sewickley, PA (USA). The epidemiological survey conducted at the time concluded that the outbreak was a water-borne illness and a putative contaminant was thought to have entered the water supply system through an uncovered finished-water reservoir in which cyanobacteria (mainly *Schizothrix*) were present around the time of the outbreak. Although the term "endotoxin" is used only once and only in the context of a general recommendation without making an explicit link to cyanobacteria, in subsequent studies the outbreak is retrospectively linked more or less explicitly to cyanobacterial lipopolysaccharides (Keleti et al., 1979; Sykora et al., 1980). At the time of the outbreak, many aetiological agents were not yet known or were not yet detectable, and hence, the conclusions drawn were possibly influenced by the visible prominence of cyanobacterial blooms compared to, for example, viruses. Also, none of the cyanobacterial toxins with unambiguous adverse health effects treated in this volume were known at the time of this outbreak.

Hindman et al. (1975) reported on an outbreak of pyrogenic reactions in patients being treated at a haemodialysis clinic in Washington, DC. They circumstantially attributed this to cyanobacterial LPS as a cyanobacterial bloom was present in the Potomac River from which the raw water was abstracted. The apparent reason for this connection is that the authors were unable to demonstrate bacterial infections and bacterial contamination of tap water and dialysate was seen only at trace levels. The authors concluded that raw water from the Potomac, affected by a concurrent "algae bloom", was the likely source of LPS or endotoxin. However, no samples were taken to obtain counts of heterotrophs in the raw water (or in the treatment system). In this case also, the connection of cyanobacterial LPS to the adverse health effect is possibly largely based on the prominent visibility of an "algal bloom" – rather than on unequivocal evidence.

Since these early reports, no further studies have unequivocally supported the hypothesis that cyanobacterial LPS poses a risk to human health. Nonetheless, these studies were cited as evidence for this hypothesis until recently. In most of the reports that attribute signs and symptoms to contact with cyanobacterial lipopolysaccharides, as summarised in Stewart et al. (2006), this connection is not well evidenced or given only in general terms such as that "cyanobacterial toxins or LPS can cause adverse health effects".

A study by Lévesque et al. (2016), entitled "Exposure to cyanobacteria: acute health effects associated with endotoxins", suggests a causal relationship between exposure to cyanobacterial LPS (endotoxin) and human illness. The observed health effects consisted of generally mild gastrointestinal symptoms not requiring medical examination. Yet, the statement made in the title is not well supported by the presented data. For example, no information is provided on the taxonomic composition of the cyanobacterial blooms, nor have well-known cyanobacterial toxins been quantified. Further, no attempt was made to analyse water samples for possible heterotrophic pathogens associated with the cyanobacteria – while the authors clearly state that "the hypothesis of a preponderant role of Gram-negative bacteria is attractive" (see also next section) and conclude that "it is possible that the concentration of endotoxins is a proxy of another exposure". In summary, the reported correlation between exposure to cyanobacterial blooms and mild disease does not allow to conclude that specifically cyanobacterial LPS actually played a significant role in this.

2.8.3 Methodological problems of studies on cyanobacterial LPS

Due to their structural complexity, the quantification of LPS in a (cyano) bacterial sample is difficult, and instead of a true molar or gravimetric quantification, a bioassay has been employed. Most studies used the *Limulus* amoebocyte lysate (LAL) assay (Young et al., 1972), with intrinsic uncertainty due to varying activity of LPS from individual strains. The validity of the LAL assay for clinical diagnosis has long been debated due to a supposed lack of specificity but it is still considered the gold standard. Alternatively, pyrogen tests based on human monocytoid cells have been proposed (Hoffmann et al.,

2005). Modern biosensor-based assays are increasingly available (Das et al., 2014; Lim et al., 2015), while modifications of the LAL assay are still in use to detect LPS in clinical samples (e.g., Wong et al., 2016).

The characterisation of LPS from cyanobacteria can only be meaningful if these are extracted from axenic cyanobacterial cultures, that is, cultures free of any contamination with heterotrophic bacteria. Effects of LPS extracted from samples of cyanobacterial blooms cannot be attributed exclusively to cyanobacterial LPS because in field samples, a high diversity of heterotrophic bacteria in high numbers is tightly associated with cyanobacterial cells (Kolmonen et al., 2004; Xie et al., 2016; Yang et al., 2017). Even if the biomass of cyanobacteria in such environmental bloom samples is higher than that of the heterotrophic bacteria, their relative share of LPS is lower because LPS content correlates with cell surface rather than with cell volume. A large number of small heterotrophic bacteria has a higher cell surface than the corresponding biomass of cyanobacteria, so that heterotrophic bacteria are likely to contain more LPS per biomass than cyanobacteria. In consequence, a relevant fraction of the activity in LAL bioassays of field samples is likely partly due to LPS from heterotrophic bacteria (Bláhová et al., 2013).

Considering the highly variable activity of lipopolysaccharides (LPS) of different microbiological origin, a mixture of LPS from an unquantified consortium of (cyano)bacteria does not allow a meaningful toxicological evaluation of one particular and unquantified LPS in this mixture. Bláhová et al. (2013) analysed LPS extracted from cultured *Microcystis* strains and from blooms dominated by *Microcystis*. In the latter, the authors reported higher activity (in LAL assays) and concluded that this is due to the contribution of noncyanobacterial LPS. Rapala et al. (2006) also reported LPS fractions from axenic strains to show a much lower endotoxic activity than LPS fractions from bloom samples dominated by the same species. The same reservation may, although to a lesser extent, also apply to clonal but nonaxenic cultures of cyanobacteria in which the diversity of heterotrophs may be low but their cell numbers can be high and hence also the share of noncyanobacterial LPS in extracts. Interestingly, Moosova et al. (2019) report the opposite, that is, higher activity of LPS extracted from axenic strains compared to LPS from bloom samples. Therefore, unless respective studies explicitly state that cyanobacterial LPS was extracted from an axenic strain, a contamination with heterotrophic LPS needs to be considered when interpreting results.

Heterotrophic bacteria associated with cyanobacterial blooms may not only contribute to the combined amount of LPS but prove to be more important as a direct cause for adverse health effects than the cyanobacteria biomass itself (Berg et al., 2011). For example, *Vibrio cholerae* (Chaturvedi et al., 2015) or *Legionella* spp. (Taylor et al., 2009), the very presence of which may constitute a health risk, have been found associated with cyanobacterial blooms.

LPS in samples is generally reported in endotoxin units per volume (e.g., EU/mL) or per mass of LPS (e.g., EU/mg) with endotoxin units not directly correlated with the gravimetric amount of LPS due to the high variability of less conserved parts of the molecule. Some LPS may consist of a high share of Lipid A, while for others this share may be lower due to a higher share of polysaccharide moieties. Hence, it is very important to understand what the terms "endotoxin", "LPS" or "endotoxic activity" refer to in particular publications. "LPS" is generally reported in gravimetric units, while "endotoxin" is reported either as activity, for example, in *Limulus* amoebocyte lysate (LAL) assays, or in gravimetric units when used synonymous to "LPS".

In most bioassay studies, cyanobacterial LPS has been extracted from cells or samples with organic solvents, generally phenol, and the residue is considered to consist of LPS but often without a further characterisation in terms of purity testing or (partial) structure elucidation. This means that in most studies a fraction of not fully characterised composition is used rather than a defined compound. This is further complicated by the fact that LPS from an individual strain may be a mixture of structural variants. Among Gram-negative bacteria, variations of the polysaccharide chain (Michael et al., 2005) as well as of lipid A (Darveau et al., 2004) have been reported for individual strains.

The extraction procedure to obtain purified LPS needs to be optimised for the particular species under study and may even vary in efficiency when comparing multiple strains of a single species, as Papageorgiou et al. (2004) showed for multiple *Microcystis* strains. For toxigenic cyanobacterial strains, the extraction of LPS is further complicated by the possibility of coextraction of toxins such as microcystins that may bias bioassays when not properly quantified (Lindsay et al., 2009).

2.8.4 Possible exposure routes to cyanobacterial LPS

As discussed above, lipopolysaccharides (LPS) from bacterial heterotrophs becomes a critical health issue when it is released from tissue infections or inflammations and enters the bloodstream. Routes for exposure to cyanobacterial LPS are quite different from such clinical cases: systemic infections with cyanobacteria are very unlikely and have never been reported (in contrast to infections with green algae such as nonautotroph *Prototheca* spp.); thus, an increase of LPS from multiplication of cyanobacteria in the human body can be excluded.

Exposure to cyanobacterial LPS only appears to be possible via the intestinal tract after oral uptake (cells and free LPS), during dialysis (free LPS) or via the respiratory mucosa after inhalation (cells and free LPS). In a review on possible exposure of humans to LPS through drinking-water, Anderson et al. (2002) concluded that two major exposure routes to LPS through drinking-water exist, namely, haemodialysis and inhalation, while other exposure routes such as oral ingestion or skin contact are considered as not being relevant: "occurrences linked to ingestion or through dermal abrasions could not be located" – for LPS neither from heterotrophic bacteria nor from cyanobacterial LPS. Also, the uptake of cyanobacterial LPS through the consumption of cyanobacteria contained in blue green algal dietary supplements or food items (*Aphanizomenon*, *Arthrospira*, *Nostoc*; see section 5.5) so far has not been considered as a health risk. Therefore, possible exposure routes to cyanobacterial LPS can be restricted to haemodialysis and inhalation.

For haemodialysis, it is evident that any exposure to cyanobacterial metabolites and other compounds must be avoided (see section 5.4) and the question whether cyanobacterial LPS pose a threat or not is likely outweighed by the proven direct adverse effect of cyanobacterial toxins such as microcystins. Compared to cyanobacterial toxins, cyanobacterial LPS is presumably of lesser relevance for dialysis-associated health risks, and any measure to avoid exposure to cyanotoxins will inherently also address cyanobacterial LPS.

Inhalation of cyanobacterial LPS remains as a possible exposure route to be considered. The exposure to LPS through inhalation is generally considered to act through free LPS (Anderson et al., 2002). For drinking-water, Gram-negative heterotrophs are generally considered the main source of LPS as these bacteria can proliferate within the treatment system, for example, as biofilms on filters or in distribution pipes from which LPS can be released after cell death. In contrast, cyanobacterial cells are generally removed from raw water at the initial stages of water treatment (see Chapter 10) and cannot or only extremely rarely proliferate in the distribution system. Therefore, respiratory contact to cyanobacteria occurs primarily through intact cells or cell fragments inhaled accidentally during recreational or occupational activity. Inhalation of intact cyanobacterial cells or cell debris may have various effects such as mechanical irritation, tissue damage due to toxins or secondary infections with associated microorganisms (see also section 5.2). Clear evidence of a contribution of cyanobacterial LPS to adverse health effects upon inhalation of cyanobacteria has not been published so far.

2.8.5 Conclusions

There is no doubt that LPS from cyanobacteria affects cell lines or subcellular systems such as the *Limulus* amoebocyte lysate assay in a similar way to LPS from heterotrophic bacteria. However, from this bioactivity *in vitro*, it cannot be concluded that LPS in waterbodies is a human health risk *per se*. LPS contained in aquatic ecosystems, with or without cyanobacteria, are a source of the natural and constant exposure to LPS – as it is the case with

LPS contained in the human gut and skin microbiomes, neither of which pose a direct health risk.

During effective drinking-water treatment, cyanobacterial LPS concentration is very unlikely to increase in the process. Therefore, the exposure to cyanobacterial LPS via consumption of drinking-water as potential health risk can be negated with a fair amount of certainty. Exposure to cyanobacterial LPS via inhalation may equally be irrelevant when considering spray formation, for example, while showering.

Inhalation of spray of water containing cyanobacteria in substantial amounts could be an exposure scenario but most adverse health effects such as inflammation or tissue lesions likely are a consequence of other constituents rather than an effect of cyanobacterial LPS itself. For haemodialysis, water needs to be free not only of any LPS but also of any cyanobacterial toxins to avoid critical exposure and health risks, as described in section 5.4, and ensuring that will inherently include LPS.

In summary, based on the current knowledge, cumulated in several decades of research, cyanobacterial LPS are not likely to pose health risks to an extent known from toxins like microcystins or cylindrospermopsins, in particluar, when considering plausible exposure pathways.

REFERENCES

Akira S, Takeda K (2004). Toll-like receptor signalling. Nat Rev Immunol. 4:499.

Anderson WB, Slawson RM, Mayfield CI (2002). A review of drinking-water-associated endotoxin, including potential routes of human exposure. Can J Microbiol. 48:567–587.

Armstrong MT, Theg SM, Braun N, Wainwright N, Pardy R, Armstrong PB et al. (2006). Histochemical evidence for lipid A (endotoxin) in eukaryote chloroplasts. FASEB J. 20:2145–2146.

Armstrong PB, Armstrong MT, Pardy R, Child A, Wainwright N (2002). Immunohistochemical demonstration of a lipopolysaccharide in the cell wall of a eukaryote, the green alga, *Chlorella*. Biol Bull. 203:203–204.

Berg KA, Lyra C, Niemi RM, Heens B, Hoppu K, Erkomaa K et al. (2011). Virulence genes of *Aeromonas* isolates, bacterial endotoxins and cyanobacterial toxins from recreational water samples associated with human health symptoms. J Water Health. 9:670–679.

Best JH, Pflugmacher S, Wiegand C, Eddy FB, Metcalf JS, Codd GA (2002). Effects of enteric bacterial and cyanobacterial lipopolysaccharides, and of microcystin-LR, on glutathione S-transferase activities in zebra fish (*Danio rerio*). Aquat Toxicol. 60:223–231.

Bláhová L, Adamovský O, Kubala L, Šindlerová LŠ, Zounková R, Bláha L (2013). The isolation and characterization of lipopolysaccharides from *Microcystis aeruginosa*, a prominent toxic water bloom forming cyanobacteria. Toxicon. 76:187–196.

Bryant CE, Spring DR, Gangloff M, Gay NJ (2010). The molecular basis of the host response to lipopolysaccharide. Nat Rev Microbiol. 8:8.

Carillo S, Pieretti G, Bedini E, Parrilli M, Lanzetta R, Corsaro MM (2014). Structural investigation of the antagonist LPS from the cyanobacterium *Oscillatoria planktothrix* FP1. Carbohydr Res. 388:73–80.

Cario E, Rosenberg IM, Brandwein SL, Beck PL, Reinecker H-C, Podolsky DK (2000). Lipopolysaccharide activates distinct signaling pathways in intestinal epithelial cell lines expressing Toll-like receptors. J Immunol. 164:966–972.

Chaturvedi P, Agrawal MK, Bagchi SN (2015). Microcystin-producing and non-producing cyanobacterial blooms collected from the Central India harbor potentially pathogenic *Vibrio cholerae*. Ecotoxicol Environ Saf. 115:67–74.

d'Hennezel E, Abubucker S, Murphy LO, Cullen TW (2017). Total lipopolysaccharide from the human gut microbiome silences toll-like receptor signaling. MSystems. 2:e00046–e00017.

Darveau RP, Pham T-TT, Lemley K, Reife RA, Bainbridge BW, Coats SR et al. (2004). *Porphyromonas gingivalis* lipopolysaccharide contains multiple lipid A species that functionally interact with both toll-like receptors 2 and 4. Infect Immun. 72:5041–5051.

Das A, Kumar P, Swain S (2014). Recent advances in biosensor based endotoxin detection. Biosens Bioelectron. 51:62–75.

Dauphinee SM, Karsan A (2006). Lipopolysaccharide signaling in endothelial cells. Lab Invest. 86:9.

Durai P, Batool M, Choi S (2015). Structure and effects of cyanobacterial lipopolysaccharides. Mar Drugs. 13:4217–4230.

Erridge C, Bennett-Guerrero E, Poxton IR (2002). Structure and function of lipopolysaccharides. Microbes Infect. 4:837–851.

Fujii M, Sato Y, Ito H, Masago Y, Omura T (2012). Monosaccharide composition of the outer membrane lipopolysaccharide and O-chain from the freshwater cyanobacterium *Microcystis aeruginosa* NIES-87. J Appl Microbiol. 113:896–903.

Gemma S, Molteni M, Rossetti C (2016). Lipopolysaccharides in Cyanobacteria: A brief overview. Adv Microbiol. 6:391.

Guillot L, Medjane S, Le-Barillec K, Balloy V, Danel C, Chignard M et al. (2004). Response of human pulmonary epithelial cells to lipopolysaccharide involves Toll-like Receptor 4 (TLR4)-dependent signaling pathways. J Biol Chem. 279:2712–2718.

Hindman SH, Favero MS, Carson LA, Petersen NJ, Schonberger LB, Solano JT (1975). Pyrogenic reactions during haemodialysis caused by extramural endotoxin. Lancet. 2:732–734.

Hoffmann S, Peterbauer A, Schindler S, Fennrich S, Poole S, Mistry Y et al. (2005). International validation of novel pyrogen tests based on human monocytoid cells. J Immunol Meth. 298:161–173.

Hotchkiss RS, Moldawer LL, Opal SM, Reinhart K, Turnbull IR, Vincent J-L (2016). Sepsis and septic shock. Nat Rev Dis Primers. 2:16045.

Jaja-Chimedza A, Gantar M, Mayer GD, Gibbs PD, Berry JP (2012). Effects of cyanobacterial lipopolysaccharides from *Microcystis* on glutathione-based detoxification pathways in the zebrafish (*Danio rerio*) embryo. Toxins. 4:390–404.

Keleti G, Sykora J, Lippy E, Shapiro M (1979). Composition and biological properties of lipopolysaccharides isolated from *Schizothrix calcicola* (Ag.) Gomont (Cyanobacteria). Appl Environ Microbiol. 38:471–477.

Klemm LC, Czerwonka E, Hall ML, Williams PG, Mayer A (2018). Cyanobacteria *Scytonema javanicum* and *Scytonema ocellatum* lipopolysaccharides elicit release of superoxide anion, matrix-metalloproteinase-9, cytokines and chemokines by rat microglia *in vitro*. Toxins. 10:130.

Kolmonen E, Sivonen K, Rapala J, Haukka K (2004). Diversity of cyanobacteria and heterotrophic bacteria in cyanobacterial blooms in Lake Joutikas, Finland. Aquat Microb Ecol. 36:201–211.

Lévesque B, Gervais M-C, Chevalier P, Gauvin D, Anassour-Laouan-Sidi E, Gingras S et al. (2016). Exposure to cyanobacteria: acute health effects associated with endotoxins. Public Health. 134:98–101.

Lim SK, Chen P, Lee FL, Moochhala S, Liedberg B (2015). Peptide-assembled graphene oxide as a fluorescent turn-on sensor for lipopolysaccharide (endotoxin) detection. Anal Chem. 87:9408–9412.

Lindsay J, Metcalf J, Codd G (2009). Comparison of four methods for the extraction of lipopolysaccharide from cyanobacteria. Toxicol Environ Chem. 91:1253–1262.

Lippy EC, Erb J (1976). Gastrointestinal illness at Sewickley, PA. J Am Water Works Assoc. 68:606–610.

Macagno A, Molteni M, Rinaldi A, Bertoni F, Lanzavecchia A, Rossetti C et al. (2006). A cyanobacterial LPS antagonist prevents endotoxin shock and blocks sustained TLR4 stimulation required for cytokine expression. J Exp Med. 203:1481–1492.

Marshall JC (2005). Lipopolysaccharide: an endotoxin or an exogenous hormone? Clin Infect Dis. 41:S470–S480.

Martin C, Codd GA, Siegelman HW, Weckesser J (1989). Lipopolysaccharides and polysaccharides of the cell envelope of toxic *Microcystis aeruginosa* strains. Arch Microbiol. 152:90–94.

Mayer AM, Clifford JA, Aldulescu M, Frenkel JA, Holland MA, Hall ML et al. (2011). Cyanobacterial *Microcystis aeruginosa* lipopolysaccharide elicits release of superoxide anion, thromboxane B2, cytokines, chemokines, and matrix metalloproteinase-9 by rat microglia. Toxicol Sci. 121:63–72.

Michael FS, Li J, Vinogradov E, Larocque S, Harper M, Cox AD (2005). Structural analysis of the lipopolysaccharide of *Pasteurella multocida* strain VP161: identification of both Kdo-P and Kdo–Kdo species in the lipopolysaccharide. Carbohydr Res. 340:59–68.

Mowat AM, Agace WW (2014). Regional specialization within the intestinal immune system. Nat Rev Immunol. 14:667.

Moosová Z, Šindlerová L, Ambrůzová B, Ambrožová G, Vašíček O, Velki M et al. (2019) Lipopolysaccharides from *Microcystis* cyanobacteria-dominated water bloom and from laboratory cultures trigger human immune innate response. Toxins 11:218.

Ohkouchi Y, Tajima S, Nomura M, Itoh S (2012). Comparison of inflammatory responses in human cells caused by lipopolysaccharides from *Escherichia coli* and from indigenous bacteria in aquatic environment. J Environ Sci Health Part A. 47:1966–1974.

Opal SM (2010). Endotoxins and other sepsis triggers. In: Ronco C, Piccinni P, Rosner MH, editors: Endotoxemia and endotoxin shock: disease, diagnosis and therapy. Basel: Karger Publishers:14–24.

Papageorgiou J, Linke TA, Kapralos C, Nicholson BC, Steffensen DA (2004). Extraction of cyanobacterial endotoxin. Environ Toxicol. 19:82–87.

Raetz CRH, Whitfield C (2002). Lipopolysaccharide endotoxins. Annu Rev Biochem. 71:635–700.

Rapala J, Niemelä M, Berg K, Lepistö L, Lahti K (2006). Removal of cyanobacteria, cyanotoxins, heterotrophic bacteria and endotoxins at an operating surface water treatment plant. Wat Sci Technol. 54:23–28.

Snyder DS, Brahamsha B, Azadi P, Palenik B (2009). Structure of compositionally simple lipopolysaccharide from marine *Synechococcus*. J Bacteriol. 191:5499–5509.

Stenutz R, Weintraub A, Widmalm G (2006). The structures of *Escherichia coli* O-polysaccharide antigens. FEMS Microbiol Rev. 30:382–403.

Stewart I, Schluter PJ, Shaw GR (2006). Cyanobacterial lipopolysaccharides and human health - a review. BMC Environ Health. 5:1–23.

Sykora JL, Keleti G, Roche R, Volk DR, Kay GP, Burgess RA et al. (1980). Endotoxins, algae and *Limulus* amoebocyte lysate test in drinking water. Water Res. 14:829–839.

Takeda K, Kaisho T, Akira S (2003). Toll-like receptors. Annu Rev Immunol. 21: 335–376.

Taylor M, Ross K, Bentham R (2009). *Legionella*, protozoa, and biofilms: interactions within complex microbial systems. Microb Ecol. 58:538–547.

Thorgersen EB, Macagno A, Rossetti C, Mollnes TE (2008). Cyanobacterial LPS antagonist (CyP)—A novel and efficient inhibitor of *Escherichia coli* LPS-induced cytokine response in the pig. Mol Immunol. 45:3553–3557.

Trent MS, Stead CM, Tran AX, Hankins JV (2006). Invited review: diversity of endotoxin and its impact on pathogenesis. J Endotoxin Res. 12:205–223.

Vatanen T, Kostic AD, d'Hennezel E, Siljander H, Franzosa EA, Yassour M et al. (2016). Variation in microbiome LPS immunogenicity contributes to autoimmunity in humans. Cell. 165:842–853.

Weckesser J, Drews G, Mayer H (1979). Lipopolysaccharides of photosynthetic prokaryotes. Annu Rev Microbiol. 33:215–239.

Wong J, Davies N, Jeraj H, Vilar E, Viljoen A, Farrington K (2016). A comparative study of blood endotoxin detection in haemodialysis patients. J Inflammation. 13:24.

Xie M, Ren M, Yang C, Yi H, Li Z, Li T et al. (2016). Metagenomic analysis reveals symbiotic relationship among bacteria in *Microcystis*-dominated community. Front Microbiol. 7:56.

Yang C, Wang Q, Simon PN, Liu J, Liu L, Dai X et al. (2017). Distinct network interactions in particle-associated and free-living bacterial communities during a *Microcystis aeruginosa* bloom in a Plateau lake. Front Microbiol. 8:1202.

Young NS, Levin J, Prendergast RA (1972). An invertebrate coagulation system activated by endotoxin: evidence for enzymatic mediation. J Clin Invest. 51:1790–1797.

2.9 CYANOBACTERIAL TASTE AND ODOUR COMPOUNDS IN WATER

Triantafyllos Kaloudis

Cyanobacteria can produce a large number of odorous compounds in water that are usually referred to as water "taste and odour" (T&O) compounds, some of which are specific to cyanobacteria, while others, however, are also produced by other organisms. Odorous metabolites have negative effects especially for drinking-water systems, as they make water unacceptable for consumers but also for tourism, recreational uses and aquaculture enterprises.

Cyanobacterial T&O compounds belong to various chemical groups, including terpenoids, ionones, aldehydes, ketones, sulphurous compounds, amines and others (Watson, 2004). Among these, the most frequently occurring compounds that are associated with seriously unpleasant episodes of off-odours in source waters and water supplies are geosmin and 2-methylisoborneol (MIB).

Geosmin (from the Greek "geo": earth and "osme": odour) is a bicyclic sesquiterpenic compound with an extremely intense muddy/earthy smell that has an odour threshold concentration (OTC, i.e., minimum concentration detected by human nose) of about 4 ng/L in water, while MIB is a bicyclic terpenoid with a strong musty odour and an OTC of 6 ng/L (Young et al., 1996). Geosmin and MIB are responsible for many T&O incidents affecting water supplies, recreational waters and tourism, and they can also accumulate in the lipid tissue of aquatic organisms, for example, of fish, resulting in economic losses to fishery and aquaculture enterprises (Smith et al., 2008).

Many other odorous compounds can be produced by cyanobacteria, such as β-cyclocitral, α- and β-ionones and alkyl sulphides (Jüttner, 1984). This section focuses on the most commonly occurring, that is, geosmin and MIB.

2.9.1 Chemistry and toxicity

Figure 2.9 shows the chemical structures of common cyanobacterial T&O compounds and their characteristic odour. Geosmin and MIB are both tertiary alcohols belonging to the class of terpenoids. Only the (–) enantiomers occur in natural systems, and these are more odorous than the (+) enantiomers (Jüttner & Watson, 2007).

Geosmin and MIB are not considered as health hazards for humans, as it has been shown that environmentally relevant concentrations of both compounds (e.g., ng-μg/L) present no cytotoxicity or genotoxicity (Dionigi et al., 1993; Bláha et al., 2004; Burgos et al., 2014). Furthermore, as these compounds can be sensed by the human nose at extremely low concentrations (low ng/L, see above; Table 2.10), their presence even in low concentrations

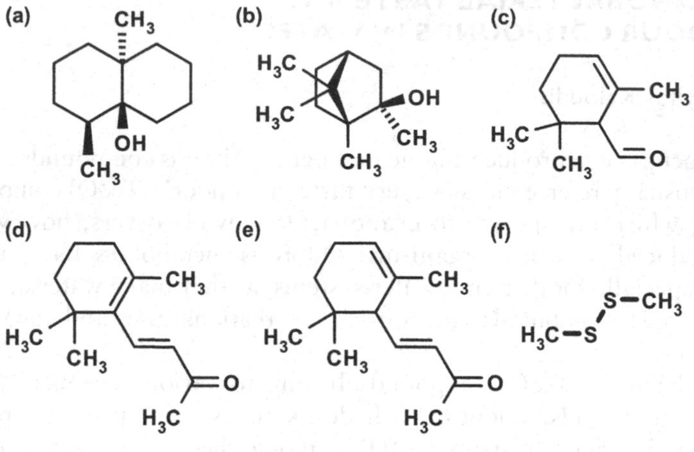

Figure 2.9 Chemical structures of geosmin (a), methyl-isoborneol (b), β-cyclocitral (c), β-ionone (d), α-ionone (e), and dimethyl-disulphide (f). For molecular weights and smell characteristics, see Table 2.10.

makes water unacceptable for consumption, which is the main problem they cause for water supplies. Indirectly, the presence of T&O may affect health if it leads consumers to turn to another, possibly less safe water supply. Other compounds such as β-cyclocitral and α, β-ionones, which are carotenoid breakdown products, also do not pose health safety concerns; in fact, they are used as additives in food or in cosmetic products.

2.9.2 Analysis

Sensory evaluation followed by chemical analysis is generally used for the assessment of T&O compounds produced by cyanobacteria in water (Suffet et al., 2004). Flavour profile analysis (FPA) is a useful sensory evaluation technique, in which a panel of trained assessors describes the character and intensity of the unusual odour (Rice et al., 2017). Panellists can use the water

Table 2.10 Smell characteristics and molecular weights of common cyanobacterial taste and odour substances

	Smell characteristics	*Monoisotopic mass (Da)*	*Molecular weight (g/mol)*
Geosmin	Earthy-muddy	182.17	182.31
2-Methylisoborneol	Musty-mouldy	168.15	168.28
β-Cyclocitral	Tobacco/wood	152.12	152.24
β-Ionone	Violets	192.15	192.30
α-Ionone	Violets	192.15	192.30
Dimethyl disulphide	Septic	93.99	94.20

"Taste and Odour Wheel" (TOW) to associate odour descriptions with groups of chemical compounds that are included in sectors of the TOW. Several sectors contain compounds that are possibly produced by cyanobacteria; for example, earthy/mouldy/musty odours are associated with the sector that contains geosmin and 2-methylisoborneol (MIB) as possible causative agents. The results of FPA-TOW provide guidance for the chemical analysis laboratory regarding which groups of compounds should be specifically targeted.

Detection, identification and quantitation of geosmin, MIB and other T&O compounds is carried out by gas chromatography combined with mass spectrometry (GC-MS). GC-MS techniques, after sample extraction and preconcentration, allow detection and quantitation of T&O at very low concentrations (low ng/L). Confirmation of the identity of odorous compounds is based on mass spectral analysis, retention indices and comparisons with commercially available standards. GC-olfactometry (GC-O) is a supplementary technique, especially for nontargeted analysis, that can provide additional information to identify the compound(s) responsible for the unusual odour (Hochereau & Bruchet, 2004). Efficient extraction of geosmin, MIB and other T&O compounds from water prior to GC-MS can be achieved by techniques such as purge and trap (P&T), solid-phase extraction (SPE), head-space solid-phase microextraction (HS-SPME), stir-bar sorptive extraction (SBSE) and closed-loop stripping analysis (CLSA). These techniques can be optimised so that detection at concentrations below or equal to OTCs can be achieved (Kaloudis et al., 2017).

Molecular methods targeting geosmin and MIB biosynthetic genes of cyanobacteria have been developed and can be applied as additional monitoring tools for the early detection of geosmin and MIB producers in aquatic environments (Giglio et al., 2010; Su et al., 2013; Suurnäkki et al., 2015).

2.9.3 Producing organisms

In aquatic environments, cyanobacteria are considered as the major sources of geosmin and MIB, although these compounds are also produced by actinomycetes that are nonphotosynthetic and largely terrestrial organisms associated with soils (Watson, 2004). Odour compounds from actinomycetes can be washed into surface waterbodies but this process seems to be less relevant in water T&O episodes (Zaitlin & Watson, 2006).

Compilations of cyanobacterial species that produce geosmin and MIB show a variety of primarily filamentous planktonic and benthic producers (Jüttner & Watson, 2007; Krishnani et al., 2008; Smith et al., 2008; Graham et al., 2010). Geosmin- and MIB-producing species belong to the genera *Dolichospermum* (*Anabaena*), *Oscillatoria*, *Phormidium*, *Lyngbya*, *Leptolyngbya*, *Microcoleus*, *Nostoc*, *Planktothrix*, *Pseudanabaena*, *Hyella* and *Synechococcus*. Most of the cyanobacterial species of these genera produce either geosmin or MIB, but there are also species capable of producing both compounds. Similarly to cyanotoxins, production of T&O compounds by cyanobacteria is strain-dependent; therefore, strain isolation and culture or detection of specific gene clusters are required to conclusively identify the T&O producers.

2.9.4 Biosynthesis

Geosmin and MIB are synthesised by terpene synthases. Geosmin is synthesised through cyclisation of farnesyl diphosphate by geosmin synthase (Jiang et al., 2008). MIB is synthesised through methylation of geranyl diphosphate by a methyltransferase, followed by cyclisation to MIB by MIB synthase (Komatsu et al., 2008). The genes associated with biosynthesis of geosmin and MIB from cyanobacteria have been discovered, and this has led to the development of PCR and qPCR methods for the detection of cyanobacteria producers (Suurnäkki et al., 2015). The functions of cyanobacterial T&O compounds for the cells are still largely unknown. It is hypothesised that they may have a role as signalling compounds, in allelopathic interactions or as defensive agents (Zuo, 2019).

2.9.5 Geosmin and MIB concentrations in aquatic environments

The concentrations of geosmin and MIB found in aquatic environments can vary widely, but they are usually below 1 µg/L in surface waters and considerably lower in treated drinking-water. Similarly to cyanotoxins, geosmin, MIB and other cyanobacterial T&O compounds can be cell-bound or dissolved in water, and there is no general consensus in the literature regarding the methods used to discriminate between these fractions or regarding the expression of results (Jüttner & Watson, 2007).

The production and occurrence of cyanobacterial T&O compounds are known to be influenced by various environmental factors, including phytoplankton composition, light intensity, nutrient concentrations, water temperature, pH and dissolved oxygen. Study of these factors is useful for the development of predictive models for T&O incidents (Qi et al., 2012). Geosmin and MIB persist in water and are both only slowly degraded by chemicals or microorganisms, which largely explains their persistence in conventional water treatment processes.

2.9.6 Removal of geosmin and MIB by water treatment processes

Removal of geosmin, MIB and other T&O compounds from water is a great challenge for water utilities, due to the extremely low odour threshold concentrations (OTCs) of some compounds. Conventional water treatment methods (coagulation, flocculation, sedimentation and filtration) are generally ineffective in removing geosmin and MIB from drinking-water. Adsorption with activated carbon (AC), in granular (GAC) or powdered (PAC) forms, is widely used to remove T&O compounds. PAC especially provides the flexibility of application for seasonal, short-term or unexpected T&O episodes. Several factors, including the adsorbent properties such as pore size distribution and surface characteristics and the presence of natural organic matter (NOM) in water, can reduce the effectiveness of removal due to competitive adsorption (Newcombe et al., 2002). Furthermore, biodegradability of geosmin and MIB by several microorganisms has been studied and has been used for biological filtration on sand filters or GAC (Ho et al., 2007).

Common disinfectants and oxidants (e.g., Cl_2, ClO_2, $KMnO_4$) may not completely remove T&O compounds or they may form other odorous by-products, while ozone and a combination of ozone/hydrogen peroxide are shown to be more efficient (Bruchet et al., 2004; Peter & Von Gunten, 2007). A number of studies have focused on the degradation of geosmin and MIB using advanced oxidation processes such as UV/H_2O_2, O_3/H_2O_2, heterogeneous photocatalysis and sonolysis (Antonopoulou et al., 2014; Fotiou et al., 2015). Degradation mechanisms in advanced oxidation processes commonly proceed via the oxidation of T&O compounds by highly reactive oxygen species such as the hydroxyl radical; thus, they are generally more effective than conventional oxidation. However, their application is rather limited mainly due to operational costs.

Such special treatment requires additional investment and operational costs. As cyanobacterial T&O episodes are often seasonal or occasional, modelling the temporal and spatial dynamics of cyanobacteria in water reservoirs in order to prevent and control the growth of producer organisms is therefore considered the most efficient practice for water supplies.

2.9.7 Co-occurrence of T&O compounds and cyanotoxins

Not all cyanobacteria produce toxins and T&O compounds, but, as shown in section 2.9.4, several genera contain one or more cyanotoxin and/or T&O strain producers. Some strains of *Microcystis* also produce microcystins together with β-cyclocitral and alkyl sulphides (Jüttner, 1984). However, cyanobacterial T&O compounds do not inevitably indicate the occurrence of cyanotoxins, since attempts to use T&O parameters as potential indicators of the presence of the toxins have been inconclusive (Khiari, 2017). Nevertheless, when T&O incidents occur in water supplies that use surface water reservoirs, both operators and authorities should be aware that cyanobacteria are a possible cause. As T&O compounds can be sensed at very low concentrations, they can serve as an early warning for further investigations regarding the presence of cyanobacteria and among them, possible cyanotoxin producers.

REFERENCES

Antonopoulou M, Evgenidou E, Lambropoulou D, Konstantinou I (2014). A review on advanced oxidation processes for the removal of taste and odor compounds from aqueous media. Water Res. 53:215–234.

Bláha L, Sabater S, Babica P, Vilalta E, Maršálek B (2004). Geosmin occurrence in riverine cyanobacetrial mats: is it causing a significant health hazard. Wat Sci Technol. 49:307–312.

Bruchet A, Duguet J, Suffe IM (2004). Role of oxidants and disinfectants on the removal, masking and generation of tastes and odours. Rev Environ Sci Biotechnol. 3:33–41.

Burgos L, Lehmann M, Simon D, de Andrade HHR, de Abreu BRR, Nabinger DD et al. (2014). Agents of earthy-musty taste and odor in water: evaluation of cytotoxicity, genotoxicity and toxicogenomics. Sci Tot Environ. 490:679–685.

Dionigi CP, Lawlor TE, McFarland JE, Johnsen PB (1993). Evaluation of geosmin and 2-methylisoborneol on the histidine dependence of TA98 and TA100 *Salmonella* Typhimurium tester strains. Water Res. 27:1615–1618.

Fotiou T, Triantis T, Kaloudis T, Hiskia A (2015). Evaluation of the photocatalytic activity of TiO 2 based catalysts for the degradation and mineralization of cyanobacterial toxins and water off-odor compounds under UV-A, solar and visible light. Chem Eng J. 261:17–26.

Giglio S, Chou W, Ikeda H, Cane D, Monis P (2010). Biosynthesis of 2-methylisoborneol in cyanobacteria. Environ Sci Technol. 45:992–998.

Graham JL, Loftin KA, Meyer MT, Ziegler AC (2010). Cyanotoxin mixtures and taste-and-odor compounds in cyanobacterial blooms from the midwestern United States. Environ Sci Technol. 44:7361–7368.

Ho L, Hoefel D, Bock F, Saint CP, Newcombe G (2007). Biodegradation rates of 2-methylisoborneol (MIB) and geosmin through sand filters and in bioreactors. Chemosphere. 66:2210–2218.

Hochereau C, Bruchet A (2004). Design and application of a GC-SNIFF/MS system for solving taste and odour episodes in drinking water. Wat Sci Technol. 49:81–87.

Jiang J, Saint CP, Cane DE, Monis PT (2008) Isolation and characterization of the gene associated with geosmin production in cyanobacteria. Environ Sci Technol. 42:8027–8032.

Jüttner F (1984). Characterization of *Microcystis* strains by alkyl sulfides and β-cyclocitral. Zeitschrift für Naturforschung C 39:867–871.

Jüttner F, Watson SB (2007). Biochemical and ecological control of geosmin and 2-methylisoborneol in source waters. Appl Environ Microbiol. 73:4395–4406.

Kaloudis T, Triantis TM, Hiskia A (2017). Taste and odour compounds produced by cyanobacteria. In: Meriluoto J, Spoof L, Codd GA et al., editors: Handbook of Cyanobacterial Monitoring and Cyanotoxin Analysis. Chichester: John Wiley & Sons:196–201.

Khiari D (2017). Managing cyanotoxins. Denver (CO): Water Research Foundation. 8 pp.

Komatsu M, Tsuda M, Ōmura S, Oikawa H, Ikeda H (2008). Identification and functional analysis of genes controlling biosynthesis of 2-methylisoborneol. Proc Natl Acad Sci USA. 105:7422–7427.

Krishnani KK, Ravichandran P, Ayyappan S (2008). Microbially derived off-flavor from geosmin and 2-methylisoborneol: sources and remediation. Rev Environ Contam Toxicol. 194:1–27.

Newcombe G, Morrison J, Hepplewhite C, Knappe D (2002). Simultaneous adsorption of MIB and NOM onto activated carbon: II. Competitive effects. Carbon. 40:2147–2156.

Peter A, Von Gunten U (2007). Oxidation kinetics of selected taste and odor compounds during ozonation of drinking water. Environ Sci Technol. 41:626–631.

Qi M, Chen J, Sun X, Deng X, Niu Y, Xie P (2012). Development of models for predicting the predominant taste and odor compounds in Taihu Lake, China. PLoS One. 7:e51976.

Rice EW, Baird RB, Eaton AD, editors (2017). Standard methods for the examination of water and wastewater. Washington (DC): 23nd. American Public Health Association.

Smith JL, Boyer GL, Zimba PV (2008). A review of cyanobacterial odorous and bioactive metabolites: impacts and management alternatives in aquaculture. Aquaculture. 280:5–20.

Su M, Gaget V, Giglio S, Burch M, An W, Yang M (2013). Establishment of quantitative PCR methods for the quantification of geosmin-producing potential and *Anabaena* sp. in freshwater systems. Water Res. 47:3444–3454.

Suffet IM, Schweitze L, Khiari D (2004). Olfactory and chemical analysis of taste and odor episodes in drinking water supplies. Rev Environ Sci Biotechnol. 3:3–13.

Suurnäkki S, Gomez-Saez GV, Rantala-Ylinen A, Jokela J, Fewer DP, Sivonen K (2015). Identification of geosmin and 2-methylisoborneol in cyanobacteria and molecular detection methods for the producers of these compounds. Water Res. 68:56–66.

Watson SB (2004). Aquatic taste and odor: a primary signal of drinking-water integrity. J Toxicol Environ Health Part A. 67:1779–1795.

Young W, Horth H, Crane R, Ogden T, Arnott M (1996). Taste and odour threshold concentrations of potential potable water contaminants. Water Res. 30:331–340.

Zaitlin B, Watson SB (2006). Actinomycetes in relation to taste and odour in drinking water: Myths, tenets and truths. Water Res. 40:1741–1753.

Zuo Z (2019). Why algae release volatile organic compounds – the emission and roles. Front Microbiol. 10:491.

2.10 UNSPECIFIED TOXICITY AND OTHER CYANOBACTERIAL METABOLITES

Andrew Humpage and Martin Welker

Early studies on toxic cyanobacteria largely reported effects of extracts of cyanobacteria, isolated strains or bloom material collected in the field on test systems such as animals and plants. With the purification of individual compounds that cause toxic effects and the elucidation of their structure, whole organisms were often replaced as test systems with cell lines, tissues or enzyme/substrate systems. The mode of action of a number of toxins could be revealed by these tests and eventually lead to a good understanding of the human health risks associated with these toxins.

However, in a number of studies, toxic effects on whole animals or *in vitro* test systems were found that could not be explained by the activity of known and quantifiable cyanobacterial toxins. It is therefore likely that cyanobacteria produce metabolites toxic to humans – as well as animals and plants in general – other than the ones described in sections 2.1–2.6.

This section therefore addresses two aspects of cyanobacterial toxicity beyond the known toxins: compounds produced by cyanobacteria that have shown bioactivity in various test systems and toxic effects of cyanobacterial extracts that cannot be attributed to the well-known compounds. Both aspects are tightly linked and may lead to the identification of further cyanotoxins in future.

2.10.1 Bioactive metabolites produced by cyanobacteria

The cyanotoxins described in sections 2.1–2.6 are only a tiny part of the total diversity of secondary metabolites produced by cyanobacteria. Many of these compounds show bioactivity in organismic or *in vitro* test systems, making cyanobacteria a potentially interesting source of pharmacologically active substances (Burja et al., 2001; Chlipala et al., 2011; Welker et al., 2012; Vijayakumar & Menakha, 2015). It is beyond the scope of this book to review the diversity of cyanobacterial metabolites and their biosynthesis (as far as it is known) and the reader is referred to available reviews (Welker & von Döhren, 2006; Dittmann et al., 2015; Huang & Zimba, 2019). In this context, it is worth to mention that heterologous expression of peptide or polyketide metabolites in cyanobacterial strains has become feasible (Videau et al., 2019; Vijay et al., 2019), potentially offering new opportunities for pharmacological research (Cassier-Chauvat et al., 2017; Stensjö et al., 2018).

Most known metabolites, including the known cyanotoxins, are synthesised by three biosynthetic pathways or hybrids thereof: nonribosomal peptide synthetases (NRPS), polyketide synthases (PKS) or ribosomal synthesis of peptides that are modified post-translationally (Ziemert et al., 2008; Dittmann et al., 2015). These pathways allow the synthesis of virtually hundreds of structural variants of a single basic structure by variations in amino acid composition, modifications such as methylation or dehydration, and others, as has been well documented for microcystins (Catherine et al., 2017). Similar variability is known for several classes of nonribosomally synthesised peptides – for example, cyanopeptolins, aeruginosins and anabaenopeptins (Rounge et al., 2007; Ishida et al., 2009) – and ribosomally synthesised peptides such as microviridins (Philmus et al., 2008). The chemistry and biosynthesis of these peptides and that of the well-known cyanotoxins are very similar, suggesting that their role in contributing to the fitness of the producer organisms is also similar and the high toxicity of some molecules to humans (or mammals in general) is a mere coincidence. The latter point is also supported by the evidence that nonribosomal peptide synthesis by cyanobacteria and in particular, microcystin biosynthesis, is a very ancient trait, dating back to times long before mammals thrived on earth (Christiansen et al., 2001; Rantala et al., 2004). The comparison of gene clusters for biosynthetic pathways for peptide or polyketide synthesis, respectively, revealed a pattern of alternating regions with high conservation of variability between species (Cadel-Six et al., 2008; Ishida et al., 2009; Dittmann et al., 2015). This may indicate that some of the metabolite variability arises from horizontal gene transfer and recombination events (Sogge et al., 2013).

The production of particular metabolites is highly clone-specific, and clones within a population can be described as chemotypes. A high chemotype diversity has been reported for species of *Microcystis*, *Planktothrix*, *Dolichospermum* (*Anabaena*) and *Lyngbya*, for example (Welker et al., 2007; Rohrlack et al., 2008; Leikoski et al., 2010; Engene et al., 2011; Haruštiaková & Welker, 2017; Le Manach et al., 2019; Tiam et al., 2019). Since individual cyanobacterial clones can produce multiple variants of multiple classes of metabolites, a multiclonal bloom of cyanobacteria can contain hundreds of bioactive metabolites (Welker et al., 2006; Rounge et al., 2010; Agha & Quesada, 2014). This diversity makes it difficult to relate an observed toxic effect that cannot be explained by the activity of known (and quantifiable) cyanotoxins to a particular compound in a specific sample. Hence, the key challenges for a comprehensive risk assessment of cyanopeptides are their structural diversity, the lack of analytical standards and complex requirements for their identification and quantification (Janssen, 2019).

For a number of individual cyanobacterial metabolites or groups of metabolites, bioactivity data are available. Toxicity to zooplankton (*Daphnia*) has, for example, been observed for microviridin J (Rohrlack et al., 2004), but no data are available for other organisms or other structural variants. Other frequently occurring peptides, such as cyanopeptolins or anabaenopeptins, have been shown to inhibit proteases of herbivorous zooplankton (Agrawal et al., 2005; Rohrlack et al., 2005; Czarnecki et al., 2006; Schwarzenberger et al., 2010). This indicates that synthesis of these peptides by cyanobacteria may confer a grazing protection for cyanobacterial populations (Savic et al., 2020). However, other compounds isolated from cyanobacteria have been variously described as cytotoxic, immune suppressant or cardioactive, or been shown to inhibit key mammalian enzymes such as acetylcholine esterase, chymotrypsin and trypsin (Humpage, 2008; Nagarajan et al., 2013). Thus, "offtarget" effects also appear to be quite common.

Another hypothesis links the production of diverse (peptide) metabolites to the defence of cyanobacteria against bacteria, phages and parasitic fungi (Gerphagnon et al., 2015). In particular for the latter, evidence has been presented that particular peptides can protect strains of *Planktothrix* from being infected by Chytridomycota (Sønstebø & Rohrlack, 2011). The protection is apparently specific for the *Planktothrix* chemotype as well as for the infectious fungal strains (Rohrlack et al., 2013). This could explain the chemotype diversity and their wax and wane in populations of planktonic cyanobacteria with peptide diversity protecting populations from massive parasitic prevalence in a "Red Queen race" (Kyle et al., 2015). Protection from parasite infection may not be the only selective pressure triggering the high metabolic diversity of cyanobacteria, but surely is an interesting field, last but not the least, for the potential discovery of compounds of pharmacological interest, such as antifungal agents (Chlipala et al., 2011; Welker et al., 2012; Vijayakumar & Menakha, 2015).

Although the structure of hundreds of cyanobacterial metabolites is known, the number of compounds not yet known may be equally high or even higher. This could explain the toxic effects of cyanobacterial extracts that are discussed in the following section.

2.10.2 Toxicity of cyanobacteria beyond known cyanotoxins

A number of researchers have reported toxic effects of cyanobacterial extracts that could not be explained by the compounds verifiably present in the extract. In addition, it has been noted that toxic effects of cyanobacteria that have been attributed to known cyanotoxins may actually have been caused by other toxic compounds (reviewed in Humpage (2008), with later examples included in Humpage (2008), Bernard et al. (2011), Froscio et al. (2011), and Humpage et al. (2012). Such unexplained effects include higher-than-expected acute toxicity in animal bioassays, effects on particular tissues or cell lines that are not observed using known cyanotoxins, and toxic effects which are not in agreement with established mechanisms attributed to known cyanotoxins (Falconer, 2007). For details of toxicity testing and possible pitfalls, see section 14.3.

To further complicate matters, many harmful effects described in human exposure events such as pneumonia and gastrointestinal symptoms, are not easily or solely explainable based on the described effects of cyanotoxins (Stewart et al., 2006). A cyanobacterial bloom provides an ideal habitat for concomitant growth of dependant bacteria, some of which may be pathogenic to humans (Chaturvedi et al., 2015).

From the observations on animals exposed to blooms in waterbodies or cyanobacterial culture material in laboratories, at the time of the publication of this book it appears likely that with the microcystins, cylindrospermopsins, neuro- and dermatotoxins described in sections 2.1–2.6, the most potent and most frequently occurring cyanotoxins have been identified and their principle modes of action characterised. If these are absent or their concentrations are below their respective guideline values, major risks to human health from exposure to cyanobacteria therefore seem unlikely. However, the evidence discussed above also implies that any cyanobacterial bloom may contain further, yet unknown substances or microorganisms that may be hazardous to exposed water users. This is a further reason to avoid exposure to high concentrations of cyanobacterial biomass, regardless of its content of known cyanotoxins.

REFERENCES

Agha R, Quesada A (2014). Oligopeptides as biomarkers of cyanobacterial subpopulations. Toward an understanding of their biological role. Toxins. 6:1929–1950.

Agrawal MK, Zitt A, Bagchi D, Weckesser J, Bagchi SN, Von Elert E (2005). Characterization of proteases in guts of *Daphnia magna* and their inhibition by *Microcystis aeruginosa* PCC 7806. Environ Toxicol. 20:314–322.

Bernard C, Froscio S, Campbell R, Monis P, Humpage A, Fabbro L (2011). Novel toxic effects associated with a tropical *Limnothrix/Geitlerinema*-like cyanobacterium. Environ Toxicol. 26:260–270.

Burja AM, Banaigs B, Abou-Mansour E, Burgess JG, Wright PC (2001). Marine cyanobacteria-a prolific source of natural products. Tetrahedron. 57:9347–9377.

Cadel-Six S, Dauga C, Castets AM, Rippka R, Tandeau de Marsac N, Welker M (2008) Halogenase genes in two non-ribosomal peptide synthetase gene clusters of *Microcystis* (Cyanobacteria): sporadic distribution and evolution. Mol Biol Evol. 25:2031–2041.

Cassier-Chauvat C, Dive V, Chauvat F (2017). Cyanobacteria: photosynthetic factories combining biodiversity, radiation resistance, and genetics to facilitate drug discovery. Appl Microbiol Biotechnol. 101:1359–1364.

Catherine A, Bernard C, Spoof L, Bruno M (2017). Microcystins and Nodularins. In: Meriluoto J, Spoof L, Codd GA et al., editors: Handbook of cyanobacterial monitoring and cyanotoxin analysis. Chichester: John Wiley & Sons:109–126.
Chaturvedi P, Agrawal MK, Bagchi SN (2015). Microcystin-producing and non-producing cyanobacterial blooms collected from the Central India harbor potentially pathogenic *Vibrio cholerae*. Ecotoxicol Environ Saf. 115:67–74.
Chlipala GE, Mo S, Orjala J (2011). Chemodiversity in freshwater and terrestrial cyanobacteria - a source for drug discovery. Curr Drug Targets. 12:1654–1673.
Christiansen G, Dittmann E, Ordorika LV, Rippka R, Herdman M, Börner T (2001). Nonribosomal peptide synthetase genes occur in most cyanobacterial genera as evidenced by their distribution in axenic strains of the PCC. Arch Microbiol. 178:452–458.
Czarnecki O, Lippert I, Henning M, Welker M (2006). Identification of peptide metabolites of *Microcystis* (Cyanobacteria) that inhibit trypsin-like activity in planktonic herbivorous *Daphnia* (Cladocera). Environ Microbiol. 8:77–87.
Dittmann E, Gugger M, Sivonen K, Fewer DP (2015). Natural product biosynthetic diversity and comparative genomics of the cyanobacteria. Trends Microbiol. 23:642–652.
Engene N, Choi H, Esquenazi E, Rottacker EC, Ellisman MH, Dorrestein PC et al. (2011). Underestimated biodiversity as a major explanation for the perceived rich secondary metabolite capacity of the cyanobacterial genus *Lyngbya*. Environ Microbiol. 13:1601–1610.
Falconer IR (2007). Cyanobacterial toxins present in *Microcystis aeruginosa* - More than microcystins! Toxicon. 50:585–588.
Froscio S, Sieburn K, Lau HM, Humpage A (2011). Novel cytotoxicity associated with *Anabaena circinalis* 131C. Toxicon. 58:689–692.
Gerphagnon M, Macarthur DJ, Latour D, Gachon CM, Van Ogtrop F, Gleason FH et al. (2015). Microbial players involved in the decline of filamentous and colonial cyanobacterial blooms with a focus on fungal parasitism. Environ Microbiol. 17:2573–2587.
Haruštiaková D, Welker M (2017). Chemotype diversity in *Planktothrix rubescens* (cyanobacteria) populations is correlated to lake depth. Environ Microbiol Rep. 9:158–168.
Huang I-S, Zimba PV (2019). Cyanobacterial bioactive metabolites – a review of their chemistry and biology. Harmful Algae. 68:139–209.
Humpage A, Falconer I, Bernard C, Froscio S, Fabbro L (2012). Toxicity of the cyanobacterium *Limnothrix* AC0243 to male Balb/c mice. Water Res. 46:1576–1583.
Humpage AR (2008). Toxin types, toxicokinetics and toxicodynamics. In: H.K. H, editors: Proceedings of the Interagency, International Symposium on Cyanobacterial Harmful Algal Blooms (ISOC_HAB): Cyanobacterial harmful algal blooms: State of the science and research needs. New York, USA: Springer:383–415.
Ishida K, Welker M, Christiansen G, Cadel-Six S, Bouchier C, Dittmann E et al. (2009). Plasticity and evolution of aeruginosin biosynthesis in cyanobacteria. Appl Environ Microbiol. 75:2017–2026.
Janssen EML (2019) Cyanobacterial peptides beyond microcystins–A review on co-occurrence, toxicity, and challenges for risk assessment. Water Res 151:488–499.
Kyle M, Haande S, Ostermaier V, Rohrlack T (2015). The red queen race between parasitic chytrids and their host, *Planktothrix*: A test using a time series reconstructed from sediment DNA. PLoS One. 10:e0118738.

Le Manach S, Duval C, Marie A, Djediat C, Catherine A, Edery M et al. (2019). Global metabolomic characterizations of *Microcystis* spp. highlights clonal diversity in natural bloom-forming populations and expands metabolite structural diversity. Front Microbiol. 10:791.

Leikoski N, Fewer DP, Jokela J, Wahlsten M, Rouhiainen L, Sivonen K (2010). Highly diverse cyanobactins in strains of the genus *Anabaena*. Appl Environ Microbiol. 76:701–709.

Nagarajan M, Maruthanayagam V, Sundararaman M (2013). SAR analysis and bioactive potentials of freshwater and terrestrial cyanobacterial compounds: a review. J Appl Toxicol. 33:313–349.

Philmus B, Christiansen G, Yoshida WY, Hemscheidt TK (2008). Post-translational modification in microviridin biosynthesis. Chembiochem. 9:3066–3073.

Rantala A, Fewer D, Hisbergues M, Rouhiainen L, Vaitomaa J, Börner T et al. (2004). Phylogenetic evidence for the early evolution of microcystin synthesis. Proc Natl Acad Sci USA. 101:568–573.

Rohrlack T, Christiansen G, Kurmayer R (2013). Putative antiparasite defensive system involving ribosomal and nonribosomal oligopeptides in cyanobacteria of the genus *Planktothrix*. Appl Environ Microbiol. 79:2642–2647.

Rohrlack T, Christoffersen K, Friberg-Jensen U (2005). Frequency of inhibitors of daphnid trypsin in the widely distributed cyanobacterial genus *Planktothrix*. Environ Microbiol. 7:1667–1669.

Rohrlack T, Christoffersen K, Kaebernick M, Neilan BA (2004). Cyanobacterial protease inhibitor microviridin J causes a lethal molting disruption in *Daphnia pulicaria*. Appl Environ Microbiol. 70:5047–5050.

Rohrlack T, Edvardsen B, Skulberg R, Halstvedt CB, Utkilen HC, Ptacnik R et al. (2008). Oligopeptide chemotypes of the toxic freshwater cyanobacterium *Planktothrix* can form subpopulations with dissimilar ecological traits. Limnol Oceanogr. 53:1279–1293.

Rounge TB, Rohrlack T, Decenciere B, Edvardsen B, Kristensen T, Jakobsen KS (2010). Subpopulation differentiation associated with nonribosomal peptide synthetase gene cluster dynamics in the cyanobacterium *Planktothrix* spp. J Phycol. 46:645–652.

Rounge TB, Rohrlack T, Tooming-Klunderud A, Kristensen T, Jakobsen KS (2007). Comparison of cyanopeptolin genes in *Planktothrix*, *Microcystis*, and *Anabaena* strains: evidence for independent evolution within each genus. Appl Environ Microbiol. 73:7322–7330.

Savic GB, Bormans M, Edwards C, Lawton L, Briand E, Wiegand C (2020). Cross talk: two way allelopathic interactions between toxic *Microcystis* and *Daphnia*. Harmful Algae. 94:101803.

Schwarzenberger A, Zitt A, Kroth P, Mueller S, Von EE (2010). Gene expression and activity of digestive proteases in *Daphnia*: effects of cyanobacterial protease inhibitors. BMC Physiol. 10:6.

Sogge H, Rohrlack T, Rounge TB, Sonstebo JH, Tooming-Klunderud A, Kristensen T et al. (2013). Gene flow, recombination, and selection in cyanobacteria: population structure of geographically related *Planktothrix* freshwater strains. Appl Environ Microbiol. 79:508–515.

Sønstebø JH, Rohrlack T (2011). Possible implications of chytrid parasitism for population subdivision in freshwater cyanobacteria of the genus *Planktothrix*. Appl Environ Microbiol. 77:1344–1351.

Stensjö K, Vavitsas K, Tyystjärvi T (2018). Harnessing transcription for bioproduction in cyanobacteria. Physiol Plant. 162:148–155.

Stewart I, Webb PM, Schluter PJ, Shaw GR (2006). Recreational and occupational field exposure to freshwater cyanobacteria–a review of anecdotal and case reports, epidemiological studies and the challenges for epidemiologic assessment. Environ Health. 5:6.

Tiam SK, Gugger M, Demay J, Le Manach S, Duval C, Bernard C et al. (2019). Insights into the diversity of secondary metabolites of *Planktothrix* using a biphasic approach combining global genomics and metabolomics. Toxins. 11:498.

Videau P, Wells KN, Singh AJ, Eiting J, Proteau PJ, Philmus B (2019). Expanding the natural products heterologous expression repertoire in the model cyanobacterium *Anabaena* sp. strain PCC 7120: production of pendolmycin and teleocidin B-4. ACS Synth Biol.

Vijay D, Akhtar MK, Hess WR (2019). Genetic and metabolic advances in the engineering of cyanobacteria. Curr Opin Biotechnol. 59:150–156.

Vijayakumar S, Menakha M (2015). Pharmaceutical applications of cyanobacteria – A review. J Acute Med. 5:15–23.

Welker M, Dittmann E, von Döhren H (2012). Cyanobacteria as a source of natural products. In: Hopwood DA, editor: Natural product biosynthesis by microorganisms and plants, Part C. Methods in Enzymology. Amsterdam: Elsevier 517:23–46.

Welker M, Maršálek B, Šejnohová L, von Döhren H (2006). Detection and identification of oligopeptides in *Microcystis* (cyanobacteria) colonies: toward an understanding of metabolic diversity. Peptides. 27:2090–2103.

Welker M, Šejnohová L, von Döhren H, Nemethova D, Jarkovsky J, Maršálek B (2007) Seasonal shifts in chemotype composition of *Microcystis* sp. communities in the pelagial and the sediment of a shallow reservoir. Limnol Oceanogr. 52:609–619.

Welker M, von Döhren H (2006). Cyanobacterial peptides - Nature's own combinatorial biosynthesis. FEMS Microbiol Rev. 30:530–563.

Ziemert N, Ishida K, Liaimer A, Hertweck C, Dittmann E (2008). Ribosomal synthesis of tricyclic depsipeptides in bloom-forming cyanobacteria. Ang Chem Int Ed. 47:7756–7759.

Chapter 3

Introduction to cyanobacteria

Leticia Vidal, Andreas Ballot, Sandra M. F. O. Azevedo, Judit Padisák and Martin Welker

CONTENTS

INTRODUCTION

Cyanobacteria are a very diverse group of prokaryotic organisms that thrive in almost every ecosystem on earth. In contrast to other prokaryotes (bacteria and archaea), they perform oxygenic photosynthesis and possess chlorophyll-*a*. Their closest relatives are purple bacteria (Woese et al., 1990; Cavalier-Smith, 2002) – and chloroplasts in higher plants (Moore et al., 2019). Photosynthetic activity of cyanobacteria is assumed to have changed the earth's atmosphere in the Proterozoic Era some 2.4 billion years ago during the so-called **Great Oxygenation Event** (Hamilton et al., 2016; Garcia-Pichel et al., 2019).

Historically, cyanobacteria were considered as plants or plant-like organisms and were termed "Schizophyceae", "Cyanophyta", "Cyanophyceae" or "blue-green algae". Since their prokaryotic nature has unambiguously been proven, the term "cyanobacteria" (or occasionally "cyanoprokaryotes") has been adopted in the scientific literature. A metagenomic study by Soo et al. (2017) revealed that cyanobacteria also comprise groups of nonphotosynthetic bacteria and the taxon Oxyphotobacteria is proposed for cyanobacteria in a strict sense. However, in this volume, the term "cyanobacteria" will be used for photosynthetic, oxygenic bacteria.

3.1 CELL TYPES AND CELL CHARACTERISTICS

As prokaryotes, cyanobacteria lack a cell nucleus and other cell organelles, allowing their microscopic distinction from most other microalgae. In particular, cyanobacteria lack chloroplasts, and instead, the chlorophyll for the photosynthesis is contained in simple thylakoids, the site of the light-dependent reactions of photosynthesis (exception: *Gloeobacter* spp. not possessing thylakoids). Cyanobacteria occur as unicellular, colonial or multicellular filamentous forms. Diverse forms populate all possible environments where light and at least some water and nutrients are available – even if only in very low quantities. Examples for extreme environments in which cyanobacteria can be encountered are caves or deserts (Whitton & Potts, 2000). This volume primarily considers cyanobacteria in the aquatic environments where they may grow suspended in water (i.e., as "plankton"), attached to hard surfaces ("benthos" or "benthic", respectively), or to macrophytes or any other submerged surfaces ("periphytic" or "metaphytic").

Sexual reproduction has not been observed for cyanobacteria; therefore, their only means of reproduction is asexual, through division of vegetative cells.

The morphology of cyanobacterial cells shows a number of characteristics that can be used for microscopic examination and identification: primarily, the shape and size of cells, subcellular structures and specialised cells (Figure 3.1–3.3). Cyanobacterial cells can be spherical, ellipsoid, barrel-shaped, cylindrical, conical or disc-shaped. Some taxa include cells of different shapes. Cyanobacteria do not possess flagella, as are found in many other bacterial or phytoplankton taxa. Nevertheless, many cyanobacteria, in particular filamentous forms, show gliding motility, the mechanism of which is not yet fully understood (Hoiczyk, 2000; Read et al., 2007).

The size of cyanobacteria varies considerably between taxa: more or less spherical cells of unicellular cyanobacteria range in diameter from about 0.2 µm to over 40 µm. In consequence, cell volume may vary by a factor of at least 300 000, making simple cell counts an unreliable parameter for the determination of biomass, especially when reported without differentiation between individual taxa (see Chapter 13). Some filamentous forms have been observed to have cell diameters of up to 100 µm, but as these coin-shaped cells are generally very short, their cell volume is not necessarily much larger than that of other species (Figure 3.2; Whitton & Potts, 2000). The length of filaments (or trichomes; see below) can reach a few millimetres in certain benthic forms. Very small cells of cyanobacteria (in the size range 0.2–2 µm) have been recognised as a significant fraction of

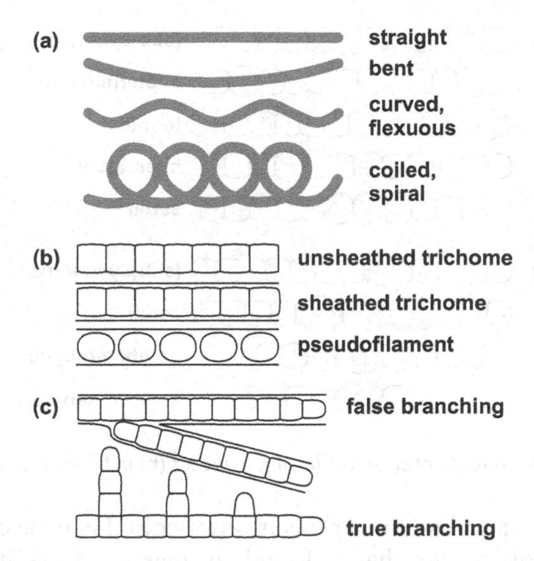

Figure 3.1 Characteristics of cyanobacteria filaments. (a) General shapes; (b) presence of sheaths; (c) branching types.

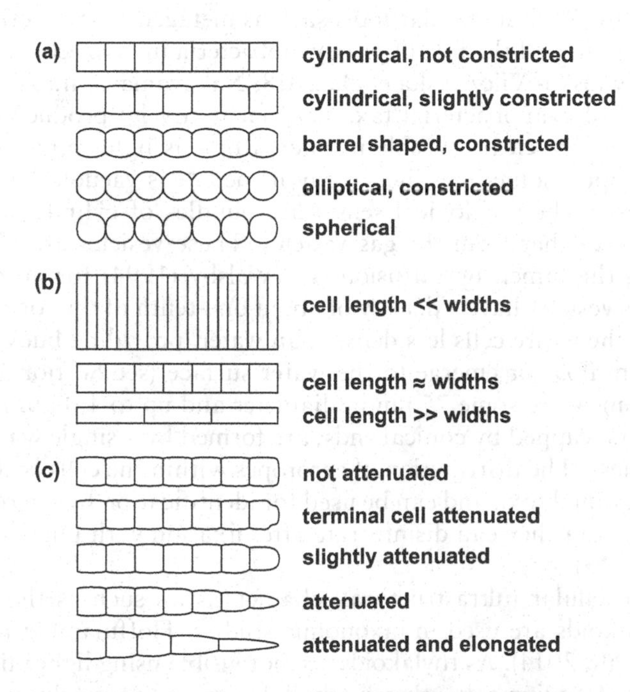

Figure 3.2 Characteristics of cyanobacteria filaments. (a) Cell shapes and arrangement in filaments; (b) cell length-to-width ratios; (c) filament terminal region.

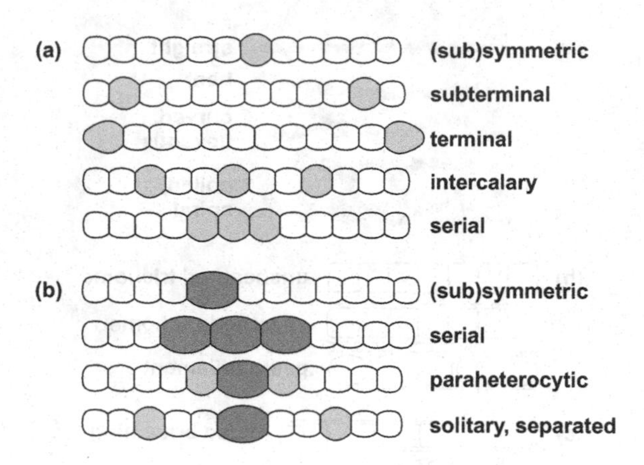

Figure 3.3 Arrangement of heterocytes (a) and akinetes (b) in filamentous cyanobacteria.

the so-called **picoplankton** in various freshwater and marine environments, such as *Prochlorococcus* that is found in huge numbers in the world's oceans (Flombaum et al., 2013). The occurrence of picocyanobacteria in freshwaters is well established (Postius & Ernst, 1999; Stomp et al., 2007) but possibly is underestimated, especially when biomass estimates are based on microscopy. With molecular tools such as metagenomics (section 13.4), our understanding of the role of picocyanobacteria in lake ecosystems may increase (Śliwińska-Wilczewska et al., 2018; Nakayama et al., 2019).

A number of cyanobacterial taxa can (facultatively) produce so-called **aerotopes** that are clearly visible in microscopy as light-refracting structures. Aerotopes (sometimes incorrectly named "gas vacuoles" – they are not vacuoles in the cytological sense) are bundles of cylindrical protein microstructures that form the **gas vesicles**. These vesicles are filled with air entering the lumen by diffusion (see Walsby (1994) for an extensive review). Gas vesicles have a density of about one-tenth of that of water and thus render the entire cells less dense than water, providing buoyancy and making them float or emerge to the water surface (see Section 3.2). The gas vesicles measure some 75 nm in diameter and up to 1.0 µm in length. The cylinders, capped by conical ends, are formed by a single wall layer of 2 nm thickness. The distribution of aerotopes within the cells is characteristic for individual taxa and can be used for identification by microscopical examination, but they can disintegrate after fixation with Lugol's solution (see Chapter 13).

Other subcellular (ultrastructural) characteristics such as the distribution of thylakoids are used in taxonomic studies (Hoffmann et al., 2005; Komárek et al., 2014). As thylakoids are not visible using light microscopy with standard equipment, other methodologies are generally applied for their examination, such as transmission electron microscopy.

In some groups of cyanobacteria (see Table 3.1), specialised cells occur, which are morphologically different from vegetative cells and which can be

Table 3.1 Major groups of cyanobacteria in the taxonomic schemes proposed by Castenholz et al. (2001) and Cavalier-Smith (2002)

Group	*Morphological characteristics*	*Genera (selection)*
Subsection 1 "Chroococcales"	• Unicellular • Colonies with regular or irregular cell arrangement • Embedded in extracellular mucilage	*Aphanocapsa, Gomphospheria, Merismopedia, Microcystis, Synechococcus, Synechocystis, Woronichinia*
Subsection 2 "Pleurocapsales"	• Colonial or filamentous • Reproduction through baeocytes	*Pleurocapsa, Chroococcidiopsis, Cyanocystis*
Subsection 3 "Oscillatoriales"	• Multiplication by hormogonia • Unbranched, linear filaments • No heterocytes or akinetes • Cells typically shorter than broad	*Leptolyngbya, Lyngbya, Microcoleus, Oscillatoria, Phormidium, Planktothrix, Pseudanabaena, Tychonema*
Subsection 4 "Nostocales"	• Multiplication by hormogonia • Nonbranching or false branching • Heterocytes (can be absent in individual filaments) • Akinetes	*Anabaena, Aphanizomenon, Raphidiopsis (Cylindrospermopsis), Cuspidothrix, Chrysosporum, Dolichospermum, Nostoc, Sphaerospermopsis*
Subsection 5 "Stigonematales"	• Multiplication by hormogonia • True branching • Heterocytes (can be absent in individual filaments) • Akinetes	*Chlorogloeopsis, Fischerella, Stigonema*

The morphological characteristics are based on microscopic observation. Exemplary genera are given for subsections.

generally easily recognised by light microscopy (see examples below), that is, heterocytes and akinetes.

Heterocytes are specialised cells that allow the fixation of atmospheric nitrogen, a process also called **diazotrophy** that involves nitrogenases, enzymes capable to reduce nitrogen to ammonium (Berman-Frank et al., 2003). Note that "heterocyte" is the more appropriate term than the traditionally used term "heterocyst" because a "cyst" has another, clearly defined meaning in cytology. Both terms may be seen as synonyms, while in this volume the term "heterocyte" is preferred.

Heterocytes lack the complete photosynthetic apparatus, thus avoiding the production of oxygen which would irreversibly damage nitrogenases (Bothe et al., 2010). Further, they possess a thickened cell wall, which further supports the anoxic intracellular milieu needed for diazotrophy.

Heterocytes often differ in size and shape from vegetative cells. In the microscope, they are generally easily recognised due to their different size and light refraction properties. Their number and the location of heterocytes in filaments can be used for taxonomic determination (Figure 3.3), although heterocyte formation depends on environmental and physiological conditions and may hence vary. They may be completely absent under conditions of ample availability of inorganic nitrogen. For example, *Aphanizomenon* spp. without heterocytes may be confused with *Planktothrix agardhii* if the terminal cells of the filaments are not examined carefully. Some authors suggested that *Raphidiopsis* spp. could be a nonheterocytous stage or type of *Cylindrospermopsis* spp. as recent studies showed both taxa to be phylogenetically very close (Moustaka-Gouni et al., 2009) and should hence be combined (Aguilera et al., 2018).

Akinetes are resting stages that can be found in the same taxa that form heterocytes. They are characterised by a generally (much) larger size compared to vegetative cells and different light refraction in microscopic view. Their cell wall is multilayered, and they often contain granules of glycogen and cyanophycin but generally no polyphosphate granules. Akinete formation and germination is triggered by environmental conditions (Adams & Duggan, 1999).

The position, number and distribution of the heterocytes and akinetes are important morphological characteristics of species and genera. Heterocytes can be in an intercalary position between vegetative cells, that is, in the middle of a trichome, or terminal or subterminal. Akinetes are in an intercalary or subterminal position but generally not terminal. Because the formation of heterocytes and akinetes is triggered by environmental conditions, individual species can appear variable in natural samples or strain cultures. The distribution of these specialised cells also determines the symmetry of the trichome.

3.2 MORPHOLOGY OF MULTICELLULAR FORMS

Most cyanobacterial taxa form multicellular aggregates, and the size and shape of which can be used for the identification of cyanobacteria in freshly collected field samples. In conserved samples, however, these aggregates may disintegrate, rendering identification more difficult (see Chapter 13).

One important characteristic for identification is the type of cell division and the separation of cells following division – or the lack thereof.

In unicellular forms (e.g., *Synechococcus* sp.), dividing cells separate completely and do not form (true) filaments. Some "unicellular" species, however, can form microbial mats or colonies by embedding single cells in a mucous matrix (mucilage). In cultures, species forming colonies in natural environments often grow as singular cells or form aggregates with a morphology that differs clearly from that of naturally occurring colonies. Experimental studies with *Microcystis* sp. indicate that the presence of heterotrophic bacteria triggers the production of extracellular polysaccharide (EPS), a prerequisite for colony formation, while axenic strains generally grow as single cells (Shen et al., 2011; Wang et al., 2016).

In natural populations, colony morphology may change during the seasonal cycle (Reynolds et al., 1981). The arrangement of the cells in a colony can be completely irregular as a result of multiple cell division panes

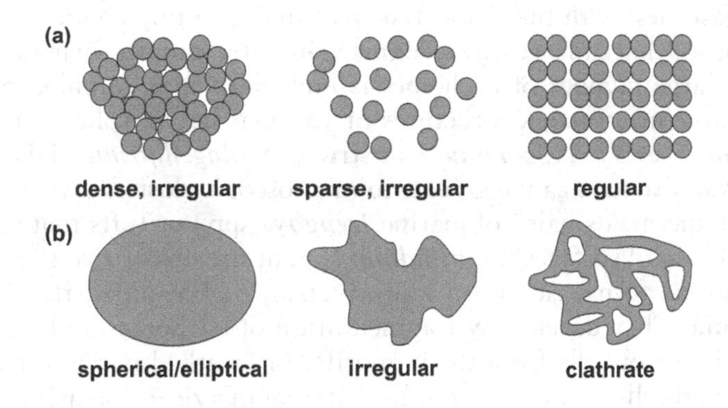

Figure 3.4 Arrangement of cells in colonies (a) and overall shapes of colonies (b).

(e.g., most *Microcystis* sp.) or through series of cell divisions in only two panes, very regular, forming two-dimensional sheets (e.g., *Merismopedia*, Figure 3.4). Colonies can be composed of hundreds or even thousands of cells and reach sizes of some hundred micrometres to a few millimetres, visible with the naked eye. A particular colony form is given when cell division occurs in a single pane, leading to so-called **pseudo-filaments**, linear colonies of singular cells in a mucilaginous sheath, for example, in *Cyanodictyon* spp.

In filamentous forms, cells remain adhered to each other after division, forming chains of connected cells termed "trichomes". Trichomes can be enveloped in a **mucous sheath** in some taxa and are then called **filaments** (Figure 3.1). The general term hence is "trichome", but the term "filament" is often used interchangeably in the literature, as in this volume. Individual cells forming a trichome can be cylindrical, barrel-shaped or nearly spherical, and the corresponding trichomes appear either like a smooth thread or like a pearl necklace, respectively. Trichomes can be completely smooth with cell walls not visible or more or less deeply constricted (Figure 3.2). The cell shape is characteristic for different taxonomic groups, which can also represent different genera (Komárek, 2013). The length-to-width ratio of vegetative cells and the type of connections between them within trichomes are often characteristic for genera. In most filamentous taxa, cell division occurs only in a single division plane, resulting in filament growth by extension in only one dimension. From a single sheath, trichomes can protrude in multiple directions appearing as branched filaments. Such a **"false" branching** combines one-dimensional filaments in a multidimensional manner (e.g., *Scytonema* sp.). Some filamentous taxa perform cell division in more than one plane leading to a **true branching** (e.g., *Fischerella* sp.). Terminal vegetative cells in trichomes often show a shape and size distinctly different from that of cells within the filament (Figure 3.2). This includes elongation, tapered or pointed ends, swelling, or the covering with a **calyptra**, a mucilaginous, cap-like structure.

Multiple filaments can aggregate, forming macroscopically visible clusters. For example, under field conditions, *Aphanizomenon flosaquae* forms clusters (**fascicles**) with the shape and size similar to tiny conifer needles or blades of grass that can be easily recognised in water samples. In many *Nostoc* species, a large number of trichomes is embedded in a common mucilage forming large macroscopic structures of varying shapes (spheres in *Nostoc pruniforme*, sheets in *N. commune* or strings *N. flagelliforme*). Filaments of benthic cyanobacteria can aggregate to macroscopic clusters several decimetres long ("mermaids-hair" of marine *Lyngbya* spp.) or tufts that can completely cover hard surfaces (*Phormidium* spp. in streams or rivers).

Filamentous forms can form reproductive, motile units, the so-called **hormogonia**. They develop by fragmentation of trichomes and the release of short chains of cells from the immotile, unsheathed parental trichome. Hormogonial cells may or may not be different in size and shape from vegetative cells. They show active gliding motion when liberated from filaments and gradually develop into new filaments.

Baeocytes (also spelled beocytes) are small, spherical cells that arise from multiple fissions of a parental cell ("vegetative" cell) and are released after rupture of its fibrous outer wall layer. Baeocytes still contained in the parental cell wall appear as small colonies, and upon release, these reproductive cells develop into vegetative cells.

3.3 CYANOBACTERIAL PIGMENTS AND COLOURS

A key characteristic distinguishing cyanobacteria from other bacteria is that they possess **chlorophyll-*a*** like plant chloroplasts, their major photosynthetic pigment and a variety of carotenoids, the latter acting primarily as photoprotectants to reduce oxidative damage to chlorophyll-*a*. In addition, cyanobacteria possess specific accessory pigments, the **phycobilins** (Tandeau de Marsac, 2003). These pigments are bound to water-soluble proteins, the phycobiliproteins, and occur in variants with different optical properties. **Phycocyanin** is common for all cyanobacteria and appears blue, giving a blue-green colour to many cyanobacteria, hence the classic name "blue-green algae". The blueish colour is especially prominent when cyanobacterial cells have lysed and dissolved phycocyanin stains the water blue. **Phycoerythrin** appears red and is responsible for the reddish or brownish colour of many cyanobacteria, such as *Planktothrix rubescens*. Within the cells, phycocyanins and phycoerythrins absorb certain wavelengths of photosynthetically active radiation (PAR) and transfer the light energy to chlorophyll-*a* in photosystem II, thus extending the wavelength range of light available for photosynthesis (Berman-Frank et al., 2003). In some species, phycobilins are also present in the antenna of photosystem I where they are thought to serve the energy demand of nitrogen fixation (Watanabe et al., 2014).

Cyanobacterial cell colours vary from chartreuse to blue-green to violet-red, depending on the ratio between phycocyanin, phycoerythrin, carotenoids and chlorophyll. By adapting this ratio, cyanobacteria may optimise their efficiency of exploitation of light energy (MacIntyre et al., 2002; Kehoe, 2010). Pigment ratios are often characteristic for a particular species but can also vary between clones or genotypes. They also vary in response to the light spectrum in which the cells (filaments, colonies) are growing, for example, with a higher share of red pigment (phycoerythrin) under conditions of low light intensity (Tandeau de Marsac, 1977; Acinas et al., 2009).

Cyanobacterial water blooms can have a wide variety of colours beyond the typical green or blue-green colour due to varying ratios of chlorophylls, phycocyanin, phycoerythrin and carotenoids. The latter are orange or red in colour, and they can be used to quantify phytoplankton groups, for example, by HPLC (high performance liquid chromatography) analysis of echinenone or canthaxanthin, which are specific for cyanobacteria (Frigaard et al., 1996; Takaichi, 2011). Surface blooms can appear orange, brownish, purple, and light green, among other colours, and have been occasionally reported as suspected "contamination with paint" due to their unexpected colour.

In addition to photosynthetic pigments, cyanobacteria can produce pigments that supposedly protect the cells from intense irradiation, in particular in the UV-wavelength range. These pigments can mask the colour of chlorophyll and phycocyanin, for example, scytonemin, a black pigment produced by *Scytonema* spp. (Dillon & Castenholz, 1999).

3.4 SECONDARY METABOLITES AND CYANOTOXINS

Cyanobacteria can produce a large diversity of secondary metabolites. These are compounds produced by the cells that are not required for the basic cell metabolism, including the compounds considered as cyanobacterial toxins in this volume. These metabolites include polyketides, oligopeptides, lipids, alkaloids and other types of molecules. Many of them show bioactivity in various test systems, but their function for the cyanobacterial cells is not well understood (De Philippis & Vincenzini, 1998; Burja et al., 2001; Gerwick et al., 2001; Welker & von Döhren, 2006; Pereira et al., 2009). It is possible that among the multitude of yet poorly studied metabolites (see section 2.10), further compounds with adverse effects on higher plants and animals – or with therapeutic potential – will be identified in future (Vijayakumar & Menakha, 2015).

A number of cyanobacterial metabolites are considered to serve as UV protection. Scytonemin, an aromatic indole alkaloid (Proteau et al., 1993), has been found primarily in filamentous, sheath-forming types exposed to high UV doses, such as the (semi)terrestrial *Scytonema* sp. (Dillon & Castenholz, 1999). Mycosporine-like amino acids are a diverse family of compounds produced by fungi and several groups of eukaryotic phytoplankton (Oren & Gunde-Cimerman, 2007) as well as by cyanobacteria, including *Microcystis* sp. (Liu et al., 2004; Pathak et al., 2019).

3.5 TAXONOMY OF CYANOBACTERIA

As for all organisms, the key criteria for classifying cyanobacteria in a taxonomic system are phylogenetic relationships that should reflect the grouping of organisms in hierarchical taxa. Taxa (singular: taxon) are thus groups such as orders, families, genera, species, subspecies in which organisms are grouped so that they ideally share a common evolutionary ancestor. Historically, the taxonomic classification was inferred from morphological characteristics of cells and colonies studied by microscopy. In the last few decades, biochemical and molecular methods have been increasingly used in microbial taxonomy. Based on molecular data, a number of classical cyanobacterial taxa have been revised and renamed – and there will be more revisions in future (see Table 3.2).

Table 3.2 Cyanobacterial species names that underwent revision in recent years

New name	*Old name (basionym)*	*Reference*
Chrysosporum bergii	*Anabaena bergii*	Zapomělová et al. (2012)
Chrysosporum ovalisporum	*Aphanizomenon ovalisporum*	Zapomělová et al. (2012)
Cuspidothrix issatchenkoi	*Aphanizomenon issatchenkoi*	Rajaniemi et al. (2005b)
Raphidiopsis raciborskii	*Cylindrospermopsis raciborskii*	Aguilera et al. (2018)
Dolichospermum flosaquae	*Anabaena flos-aquae*	Wacklin et al. (2009)
Dolichospermum circinale	*Anabaena circinalis*	Wacklin et al. (2009)
Dolichospermum crassum	*Anabaena crassa*	Wacklin et al. (2009)
Dolichospermum lemmermannii	*Anabaena lemmermannii*	Wacklin et al. (2009)
Dolichospermum planctonicum	*Anabaena planctonica*	Wacklin et al. (2009)
Dolichospermum smithii	*Anabaena smithii*	Wacklin et al. (2009)
Dolichospermum solitarium	*Anabaena solitaria*	Wacklin et al. (2009)
Dolichospermum spiroides	*Anabaena spiroides*	Wacklin et al. (2009)
Dolichospermum viguieri	*Anabaena viguieri*	Wacklin et al. (2009)
Kamptonema formosum	*Oscillatoria formosum*	Strunecký et al. (2014)
Moorea producens	*Lyngbya majuscula*	Engene et al. (2012)
Microcoleus autumnalis	*Phormidium autumnale*	Strunecký et al. (2013)
Microseira wollei	*Lyngbya wollei*	McGregor & Sendall (2015)
Planktothricoides raciborskii	*Planktothrix raciborskii*	Suda et al. (2002)
Sphaerospermopsis aphanizomenoides	*Aphanizomenon aphanizomenoides*	Zapomělová et al. (2011)
Sphaerospermopsis reniformis	*Anabaena reniformis*	Zapomělová et al. (2011)
Wilmottia murrayi	*Phormidium murrayi*	Strunecký et al. (2011)

The list is not complete but comprises primarily those species that are potentially toxigenic or are closely related to toxigenic species. Note that a new name not necessarily comprises all forms previously published under the old name, and therefore, a retrospective renaming may be critical.

A major challenge for cyanobacterial taxonomy is the lack of a clear concept or definition of a species. Commonly accepted species definitions generally used in bacteriology such as genomic DNA/DNA hybridisation or average nucleotide identity (ANI) are not easily applied to cyanobacteria because these methods require axenic cultures (i.e., pure, clonal cultures free of any other bacteria). For cyanobacteria, taxonomy is further complicated by the fact that two basically different systems of nomenclature have become established, the International Code of Nomenclature for algae, fungi, and plants (ICN) and the International Code of Nomenclature of Bacteria (ICNB) (see Box 3.1). As a consequence, the number of recognised species in a given sample can vary greatly, depending on the scientific background of the person identifying the species and counting them in the sample (Whitton & Potts, 2000; Nabout et al., 2013).

BOX 3.1: NOMENCLATURE OF CYANOBACTERIA: UNSOLVED ISSUES

Historically, cyanobacteria were considered as algae. When their prokaryotic nature was revealed in the mid-20th century (Stanier & Van Niel, 1941), many of the established genera had been already described following the International Code of Nomenclature for algae, fungi, and plants ("botanical code", ICN) – as it was historically the case with heterotrophic bacteria, too. While the nomenclature of heterotrophs and Archaea follows the International Code of Nomenclature of Bacteria ("bacteriological code", ICNB) from 1980 onwards, cyanobacterial nomenclature is treated by both the botanical code and the bacteriological code. Because this is a constant source of confusion, several solutions have been proposed but none found unanimous acceptance (Stanier et al., 1978). In 1985, the "Subcommittee on the Taxonomy of Phototrophic Bacteria" proposed to consider species validly published under the botanical code as valid species in the sense of the bacteriological code, but this proposal was never accepted and the debate is ongoing (Oren & Ventura, 2017). The latest proposals take extreme positions for cyanobacteria, either exclusively following the botanical code (Oren & Garrity, 2014) or exclusively following the bacteriological code (Pinevich, 2015). While taxonomic committees continue to search for a solution, the existing dual nomenclature has consequences in practice when studying toxic cyanobacteria (Komárek, 2006; Komárek, 2011; Gaget et al., 2015a; Gaget et al., 2015b; Dvořák et al., 2018).

Numbers of species: Following the bacterial code, only a very low number of cyanobacterial species are considered as valid bacterial species, while the number of species described following the botanical code is continuously increasing. However, the botanical code is difficult to follow because (botanically) valid descriptions are published in a large variety of scientific journals and not recorded in a central registry. As a result, no comprehensive list of globally accepted species is available.

Example: For the genus *Microcystis*, Algaebase (http://www.algaebase.org/) lists 51 "taxonomically accepted" species (and additionally 62 synonyms and species of unclear status), Cyano database (http://www.cyanodb.cz/) lists 2 species (*M. aeruginosa* and *M. minutissima*), the National Center for Biotechnology Information (NCBI) taxonomy browser (https://www.ncbi.nlm.nih.gov/taxonomy) lists 19 species for which sequences are deposited – while the "list of prokaryotic names with standing in nomenclature" (LPSN, http://www.bacterio.net/index.html) lists 12 species (published under the ICN) but considers *Microcystis aeruginosa* as an illegitimate species name "in need of a replacement" (all accessed April 2020).

Type strains: Formal type strains are not required for species described following the rules of ICN, and in consequence, there are no reference genomic sequences available. This is critical especially for molecular studies that generally rely on designated type strains. Therefore, the taxonomic classification of a deposited sequence depends largely on the depositor's judgement – or misjudgement.

Example: The majority of nucleotide sequences deposited for the genus *Anabaena* in the NCBI GenBank is not assigned to a species, and among those assigned to a species, a large share is classified "cf." (from Latin "*confer*" referring to an unconfirmed classification), for example, *Anabaena* cf. *circinalis* (now *Dolichospermum* cf. *circinale*). Based on these database entries, a reliable molecular identification is not possible.

Global taxonomy: An unambiguous taxonomic scheme for ranks above genera is lacking. Higher ranks are variably labelled as order, sections or subgroups. None of the schemes is formally accepted by the International Committee on Systematics of Prokaryotes (ICSP).

Example: The order Oscillatoriales *sensu* Cavalier-Smith (2002) corresponds largely to Subsection III *sensu* Castenholz et al. (2001) and Section III *sensu* Rippka et al. (1979), each comprising a similar but not identical list of genera. It is not congruent with the order Oscillatoriales *sensu* (Komárek et al., 2014) that includes unicellular taxa like *Cyanothece* but excludes filamentous ones such as *Leptolyngbya* (see also Table 3.3).

Table 3.3 Overview on taxonomic classification systems of cyanobacteria, following either the International Code of Nomenclature for algae, fungi and plants (ICN) or the International Code of Nomenclature of Bacteria (ICNB)

Reference	*Criteria*	*Classification and morphology*	*Observations*
Komárek et al. (2014) (ICN)	*Polyphasic*, including ultrastructure, molecular, genomic	Order Gloeobacterales (U) Order Synechococcales (U/Fi,C) Order Spirulinales (coiled Fi) Order Chroococcales (U/pFi,C) Order Pleurocapsales (U/pFi,Bc) Order Oscillatoriales (U/Fi,fB) Order Chroococcidiopsidales (U/Bc) Order Nostocales (Fi,fB/tB,Ho,Hc,Ak)	New orders: Spirulinales, Chroococcidiopsidales
Hoffmann et al. (2005) (ICN)	*Polyphasic*, including morphology, ultrastructure, molecular	Order Gleobacterales (U) Order Synechococcales (U/Fi,C) Order Pseudoanabaenales (Fi) Order Chroococcales (U,C) Order Oscillatoriales (Fi,fB,Ho) Order Nostocales (Fi,fB/tB,Ho,Hc,Ak)	New orders: Pseudanabaenales, Synechococcales; Some orders with both filamentous and colonial forms; No partitioning of true and false branching in forms with heterocytes; Order Stigonematales is dispersed and former members moved to other orders
Cavalier-Smith (2002) (ICNB, not valid)	Morphology	Order Gloeobacterales (U) Order Chroococcales (U,C) Order Pleurocapsales (U,C,Bc) Order Oscillatoriales (Fi,fB,Ho) Order Nostocales (Fi,fB,Ho,Hc,Ak) Order Stigonematales (Fi,tB,Ho,Hc,Ak)	Classification of cyanobacteria in global bacterial taxonomy; Gloeobacterales (with genus *Gloeobacter*) as sister group of all other cyanobacteria; No details on classification of genera

(*Continued*)

Table 3.3 (Continued) Overview on taxonomic classification systems of cyanobacteria, following either the International Code of Nomenclature for algae, fungi and plants (ICN) or the International Code of Nomenclature of Bacteria (ICNB)

Reference	*Criteria*	*Classification and morphology*	*Observations*
Castenholz et al. (2001) (ICNB, not valid)	Morphology	Subsection 1 (U,C) Subsection 2 (U,C,Bc) Subsection 3 (Fi,fB,Ho) Subsection 4 (Fi,fB,Ho,Hc,Ak) Subsection 5 (Fi,tB,Ho,Hc,Ak)	Largely based on morphological characteristics; Genera known at the time as "form genus"; Largely in agreement with Rippka et al. (1979)
Anagnostidis & Komárek (1985) (ICN)	Morphology	Order Chroococcales (U,C,Be) Order Oscillatoriales (Fi,fB,Ho) Order Nostocales (Fi,fB,Ho,Hc,Ak) Order Stigonematales (Fi,tB,Ho,Hc,Ak)	All nonfilamentous taxa are unified in Chroococcales
Rippka et al. (1979) (ICN)	Morphology	Section I (U,C) Section II (U,C,Bc) Section III (Fi,fB,Ho) Section IV (Fi,fB,Ho,Hc,Ak) Section V (Fi,tB,Ho,Hc,Ak)	Based on morphology, cell organisation and types of cell division
Geitler (1932) (ICN)	Morphology	Order Chroococcales (U,C) Order Chamaesiphonales (U,C,Bc) Order Hormogonales (Fi,fB/tB,Ho,Hc,Ak)	

For each system, the main groups are given and a brief description of applied criteria and a summary of observations.

Morphological characteristics: U: unicellular; Fi: filaments; C: colonies; pFi: pseudofilaments; Bc: baeocytes; fB: false branching; tB: true branching; Ho: hormogonia; Hc: heterocytes; Ak: akinetes. "Not valid" refers to the fact that despite a description following the rules of ICNB, a formal recognition was not granted by taxonomic committees.

The multiple taxonomic systems are a constant source of confusion in academic discussions, albeit of less relevance for practitioners. Nevertheless, an essential understanding of the issue may help to appraise deviating views on taxonomic ratings and to understand why a particular organism is named variably in the literature.

Molecular approaches to cyanobacterial taxonomy are most promising for inferring true phylogenetic relationships. The methods applied involve sequencing of marker genes (16S rDNA, phycocyanin operon), DNA–DNA hybridisation, genome sequencing and biochemical characteristics (fatty acid profiles) or immunological procedures (Wilmotte, 1994; Whitton & Potts, 2000). Preferably, molecular results are combined with other characteristics as the basis for a so-called **polyphasic taxonomy approach** (Vandamme et al., 1996; Komárek, 2016a; Wilmotte et al., 2017).

On the genus level, evolutionary trees based on 16S rRNA gene (Tomitani et al., 2006) sequences are largely in agreement with classifications based on morphological characteristics, in particular if these were re-evaluated and include ultrastructural characteristics (Hoffmann et al., 2005; Komárek, 2006), in particular, the structure and distribution of thylakoids (Mareš et al., 2019).

Most of the species descriptions in the currently available manuals and reference books are based on morphological traits that can be recognised by optical microscopy. Section 13.2 lists taxonomic reviews and keys for the determination of cyanobacteria, focusing on potentially toxigenic taxa. Although some classification systems based on morphological features were published before biochemical and molecular characteristics became important classification criteria, they are still being used (Table 3.2) because new criteria have not sufficiently been consolidated and, particularly, because the identification by microscopy has been the most accessible method for routine analyses. However, for identifying cyanobacteria, it should be considered that their morphological appearance can vary in response to actual growth conditions (phenotypic plasticity).

Today, most species of cyanobacteria have been described following the botanical code of nomenclature based on morphological criteria. Many of the older species descriptions are based on drawings and other pictures that hence cannot be used as fully objective criteria, especially since the botanical code does not require the deposition of a type strain. Possibly the description of cyanobacterial taxa by the bacteriological code would be biologically more appropriate (see Box 3.1) as there is no doubt that cyanobacteria are a monophyletic branch in the global bacterial phylogenetic tree (Woese, 1987; Pace, 1997).

The ambiguity of the definition of cyanobacterial species and the lack of accessible reference material for many species often hamper the unambiguous assignment of cyanobacteria in field samples to a species, especially when molecular methods are applied (section 13.4). For this reason, throughout this volume, taxonomic assignment to the genus level is given preference (e.g., *Microcystis* sp.). In some cases, a more precise identification of a dominant organism to the species level may be useful for a more accurate prediction of toxin occurrence. For example, *Planktothrix agardhii* and *P. rubescens* have both been shown to potentially contain microcystins, but may contain different analogues with different toxicity, typically occur in different types of waterbodies and usually can readily be distinguished by both their colour and cell dimensions.

Practitioners in health authorities with some experience in microscopy can easily learn to recognise the dominant cyanobacterial genera (and in some cases also species), which occur in the region they are monitoring. For a number of taxa, recent revisions have led to a renaming of common taxa, and in a few cases, changes in genus names require close attention for a certain period while old and new names may be used in parallel. Also, some taxa have been reorganised beyond simple renaming of a taxonomic entity. For example, while the organisms described as *Moorea producens* (Engene et al., 2012) formerly were named *Lyngbya majuscula*, we cannot be sure that all organisms referred to as *L. majuscula* in publications prior to 2010 would indeed be classified as *M. producens* today. In this book, we generally refer to the most recent names of species or genera as of 2019 (see Table 3.3 and Salmaso et al. (2016a)), but when referring to older literature, for which allocation of a taxon to the new name risks being wrong, we quote the former name.

In field samples, most cyanobacteria can be readily distinguished from other phytoplankton and particles under the microscope at a magnification of 100× to 400×. The following section describes and depicts the most frequently occurring taxa known to produce toxins.

3.6 MAJOR CYANOBACTERIAL GROUPS

As outlined in Box 3.1, several taxonomic systems exist to group cyanobacteria in a taxonomic scheme as reviewed in more detail in Komárek et al. (2014). Table 3.3 summarises taxonomic schemes for cyanobacteria, starting from the early scheme proposed by Geitler (1932) to the most recent one proposed by Komárek et al. (2014). Several systems avoid the use of nomenclatural categories and instead use groups such as "sections" (Rippka et al., 1979) or "subsections" (Castenholz et al., 2001) instead of orders, discernible by the suffix "*-ales*" in order to reflect the understanding that at least some of these groups do not represent monophyletic units but are defined based on shared morphological characteristics (Ishida et al., 2001; Gugger & Hoffmann, 2004). Based on rapidly increasing genomic sequence information from axenic strains, metagenomic studies and ultrastructural analyses, the taxonomy of cyanobacteria will most probably converge to a truly taxonomic system based on phylogenetic relationships in the near future (Shih et al., 2013).

The classification of taxa proposed by Komárek et al. (2014) is largely based on whole-genome sequences and on ultrastructural characteristics, such as the distribution of thylakoids in the cells. These characteristics are not observable with light microscopy and hence not helpful for the routine analysis of field samples. Further, in this system, classical and morphological characteristics easily observed by microscopy, such as formation of filaments or the presence of sheaths, are less important. For example, the filamentous genus *Pseudanabaena* is grouped together with the unicellular *Synechococcus* in a new order Synechococcales.

For practical purposes, such as the examination of high numbers of samples, earlier taxonomic systems appear more suitable. Therefore, in the following, the taxonomic scheme as proposed by Castenholz et al. (2001) is considered because characteristics like the arrangement of thylakoids or sequences of housekeeping genes are generally not available for monitoring purposes. It is also primarily based on morphological features observed by light microscopy and largely corresponds to the earlier scheme by Rippka et al. (1979) while including more genera. This scheme does not use the nomenclatural definition of categories such as orders and families but rather replaces these with "subsections", "families" and "form genera" that do not reflect monophyletic taxa (and thus is considered invalid in a system of nomenclature based on phylogenetic relationships). However, it provides a temporary system that has the advantage to be a practical, convenient and stable method for the microscopical identification of cyanobacterial strains and samples.

Nonetheless, for molecular methods (see section 13.4), a genome-based taxonomic scheme reflecting phylogenetic relationships may eventually prove to be more suitable, especially once designated type strains or sequences, respectively, are accessible.

3.7 DESCRIPTION OF COMMON TOXIGENIC AND BLOOM-FORMING CYANOBACTERIAL TAXA

The following brief descriptions of common toxigenic and bloom-forming cyanobacteria give an introduction which certainly cannot replace taxonomic keys for their identification (see Chapter 13). Also, the global diversity of cyanobacteria is higher by orders of magnitudes beyond the selection of taxa presented in this chapter. Also, the following section does not include a number of genera and species known to produce toxins but not to form blooms or benthic mats, for example, *Umezakia* sp. and *Fischerella* sp. For the illustration of morphological characteristics, see Figures 3.1–3.4. A regularly updated list of cyanobacterial taxa (and other algae) can be found in AlgaeBase (Guiry & Guiry, 2019) and

CyanoDB (Hauer & Komárek, 2019). The following descriptions consider only morphological characteristics that can be observed by standard light microscopy. Figure 3.5A–U gives microscopic images for most of the taxa. For a brief description of further genera and a short summary of recent taxonomy of cyanobacteria, see also Dvořák et al. (2017).

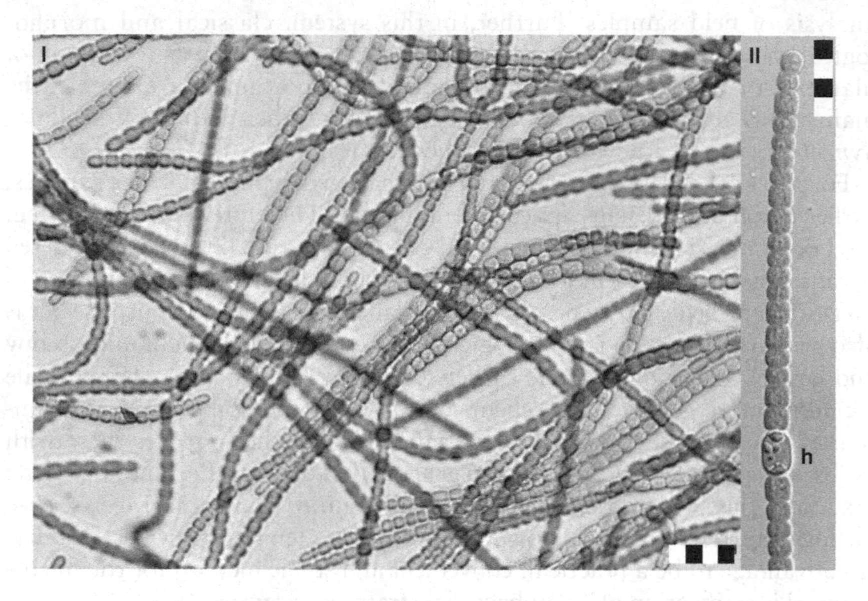

Figure 3.5A *Anabaena sensu stricto* sp. h: heterocyte. Units in scale bars correspond to 5 µm. For origin of individual photographs, see the end of this chapter.

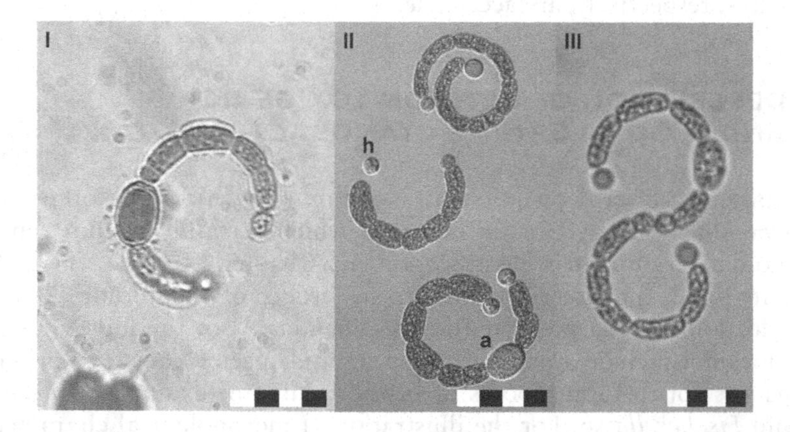

Figure 3.5B *Anabaenopsis* sp. Short, coiled trichomes with terminal heterocytes (h) and symmetric to subsymmetric akinetes (a). Units in scale bars correspond to 5 µm. For origin of individual photographs, see the end of this chapter.

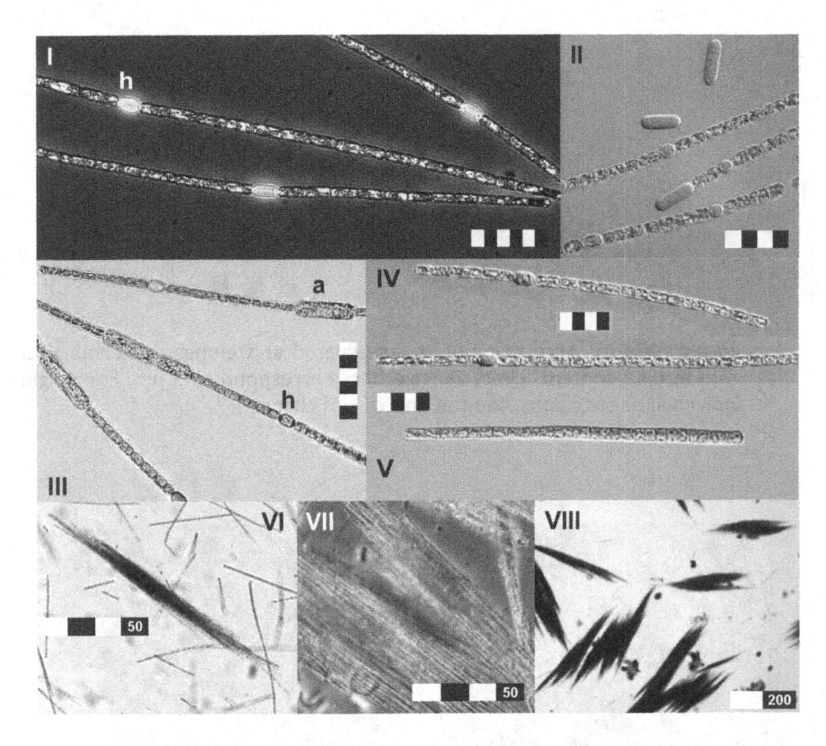

Figure 3.5C *Aphanizomenon* sp. In phase contrast, heterocytes appear highly refractory (I); aerotopes are homogeneously distributed in vegetative cells. Akinetes are much larger than vegetative cells and heterocytes (II–III) and can occur as single cells in cultures (II). Trichomes without heterocytes resemble *Planktothrix* sp. (V). Multiple trichomes of *A. flosaquae* aggregate to macroscopic fascicles (VI–VIII). Units in scale bars correspond to 5 µm if not indicated otherwise. For origin of individual photographs, see the end of this chapter.

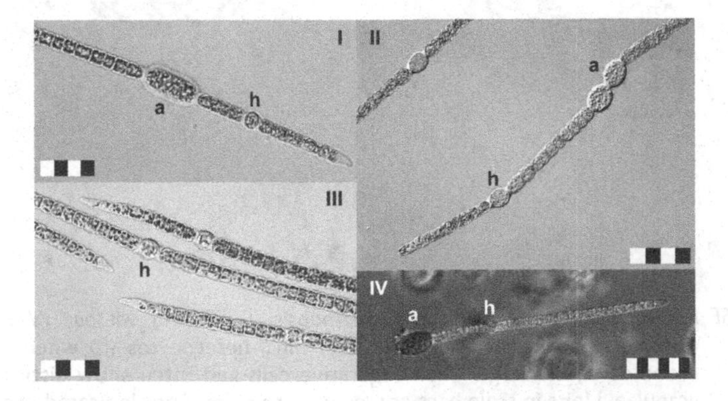

Figure 3.5D *Chrysosporum* sp. Terminal cells are pointed and appear hyaline. Akinetes (a)withdistinctgranulae;h:heterocytes.Unitsinscalebarscorrespondto5µm.For origin of individual photographs, see the end of this chapter.

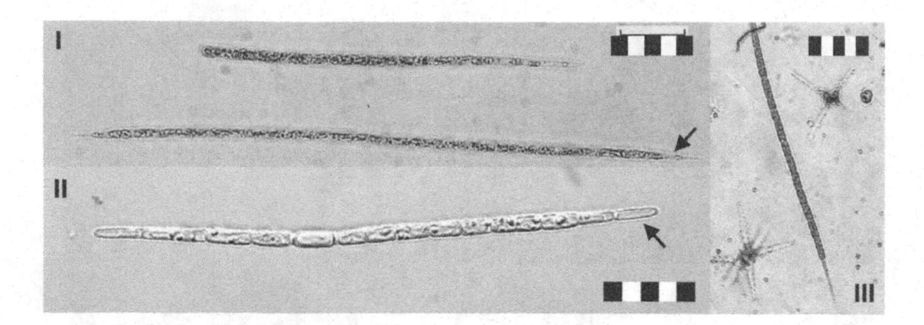

Figure 3.5E *Cuspidothrix* sp. Arrows point to attenuated and elongated terminal cells with hyaline content. Units in scale bars correspond to 5 µm. For origin of individual photographs, see the end of this chapter.

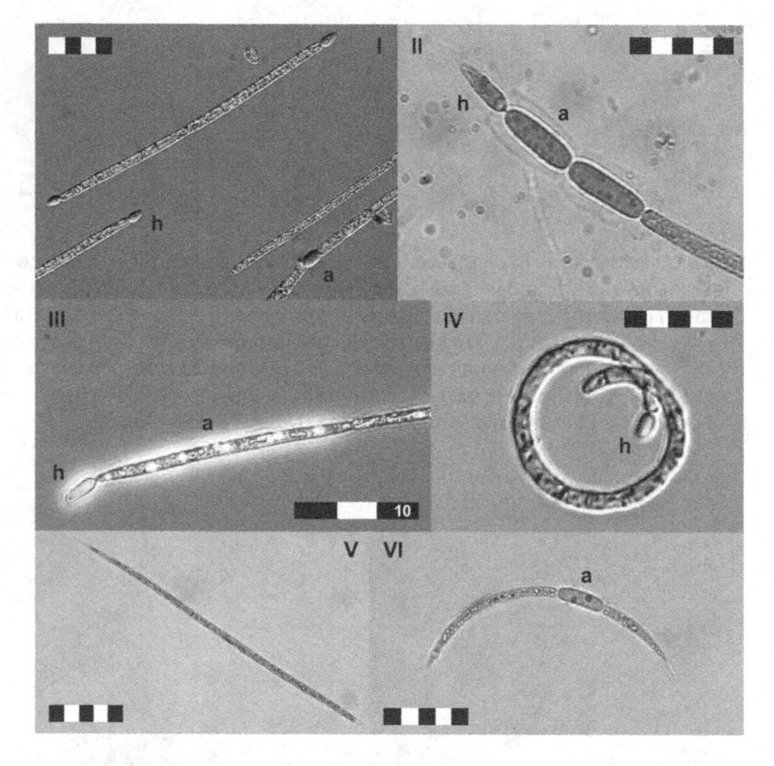

Figure 3.5F *Raphidiopsis* sp. with (*Cylindrospermopsis*; I–IV) and *without* (V–VI) heterocytes. *Raphidiopsis* has typical terminal heterocytes (h) when present. Akinetes (a) are larger than vegetative cells and often show distinct, large granulae. Units in scale bars correspond to 5 µm if not indicated otherwise. For origin of individual photographs, see the end of this chapter.

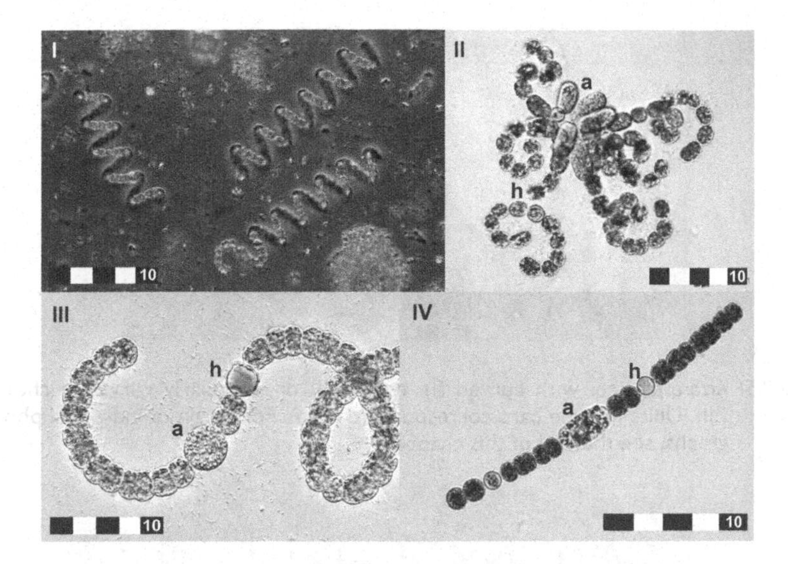

Figure 3.5G *Dolichospermum* sp. *D. crassum* (I), *D. lemmermannii* with aggregated akinetes (a) and short trichomes (II), *D. mucosum* (III) and *D. planctonicum* (IV). Heterocytes (h) with similar size than vegetative cells. Units in scale bars correspond to 10 µm. For origin of individual photographs, see the end of this chapter.

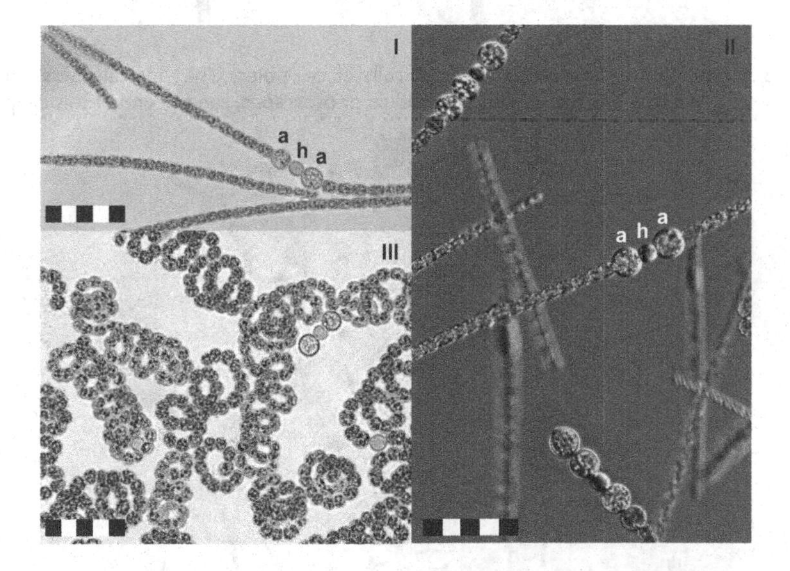

Figure 3.5H *Sphaerospermopsis aphanizomenoides* (I, II) and *S. reniformis* (III). Akinetes (a) typically next to heterocytes (h). Units in scale bars correspond to 5 µm. For origin of individual photographs, see the end of this chapter.

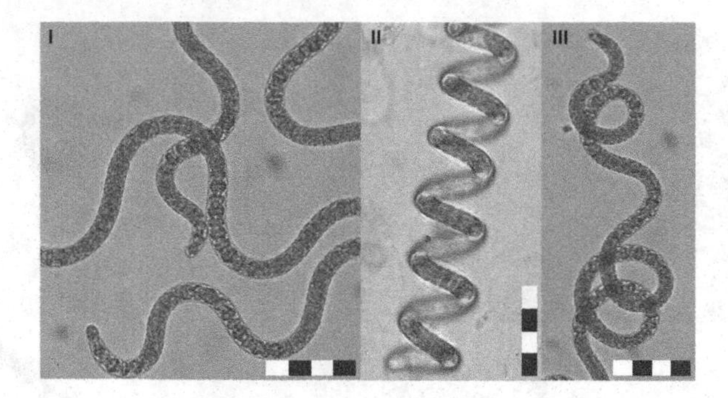

Figure 3.5I *Arthrospira* sp. with curved (I), coiled (II) or irregularly curved trichomes (III). Units in scale bars correspond to 5 µm. For origin of individual photographs, see the end of this chapter.

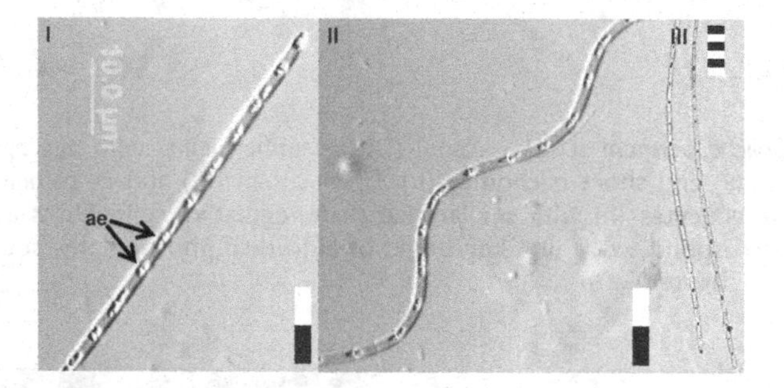

Figure 3.5J *Limnothrix* sp. Aerotopes (ae) typically at cell poles. Units in scale bars correspond to 5 µm. For origin of individual photographs, see the end of this chapter.

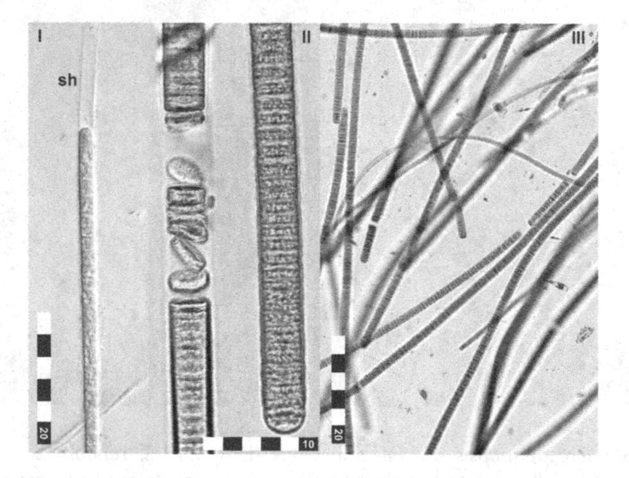

Figure 3.5K *Lyngbya* sp. Cells are enveloped in a sheath (sh) that can be partly empty. Unitsinscalebarscorrespondto5µmifnotindicatedotherwise.Fororiginofindividual photographs, see the end of this chapter.

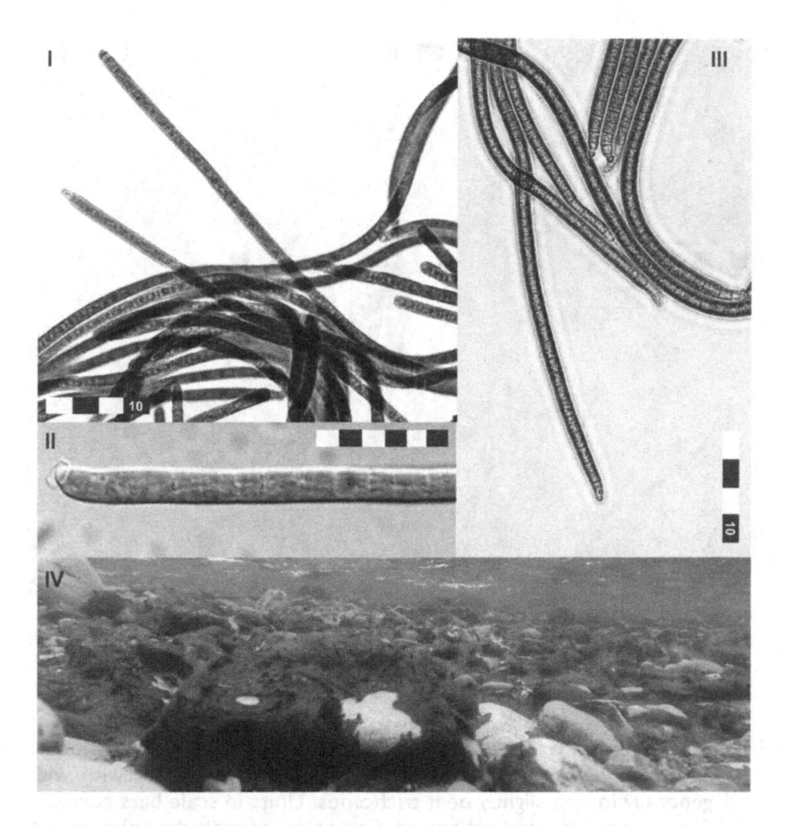

Figure 3.5L *Microcoleus/Phormidium* sp. Trichomes without sheath and typically pointed terminal cells (I–III). In natural habitats, *Microcolcus* (*Phormidium*) *autumnalis* can form dense mats on hard substrates covering large parts of stream beds (IV). Units in scale bars correspond to 5 µm if not indicated otherwise. For origin of individual photographs, see the end of this chapter.

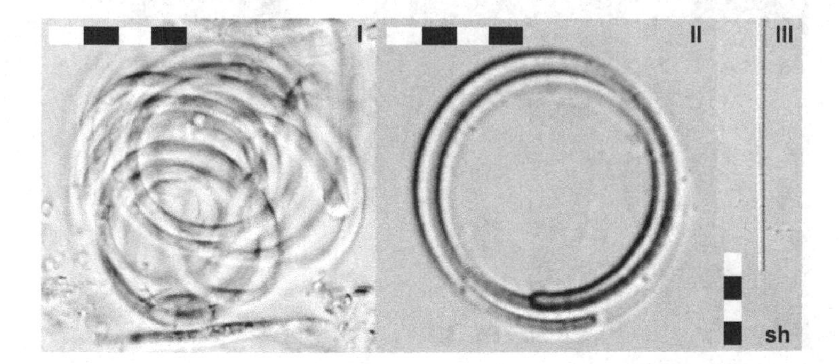

Figure 3.5M *Planktolyngbya* sp. Narrow trichomes in which individual cells can often not be distinguished, surrounded by a fine sheath (sh). Units in scale bars correspond to 5 µm. For origin of individual photographs, see the end of this chapter.

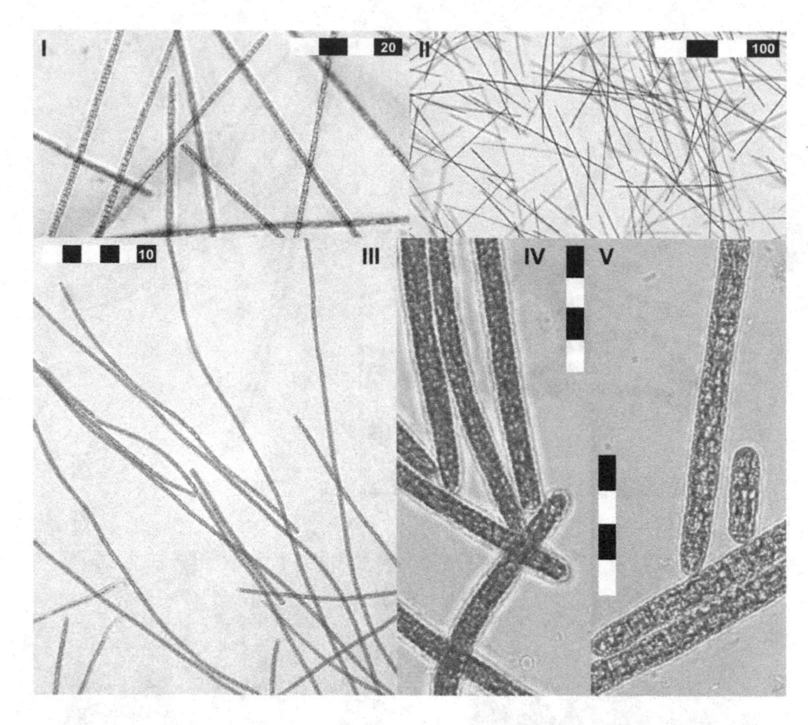

Figure 3.5N *Planktothrix* sp. Green pigmented *P. agardhii* (I, II, IV) with straight trichomes and many aerotopes. Red-pigmented *P. rubescens* (III, V) with wider and generally longer, slightly bent trichomes. Units in scale bars correspond to 5 µm if not indicated otherwise. For origin of individual photographs, see the end of this chapter.

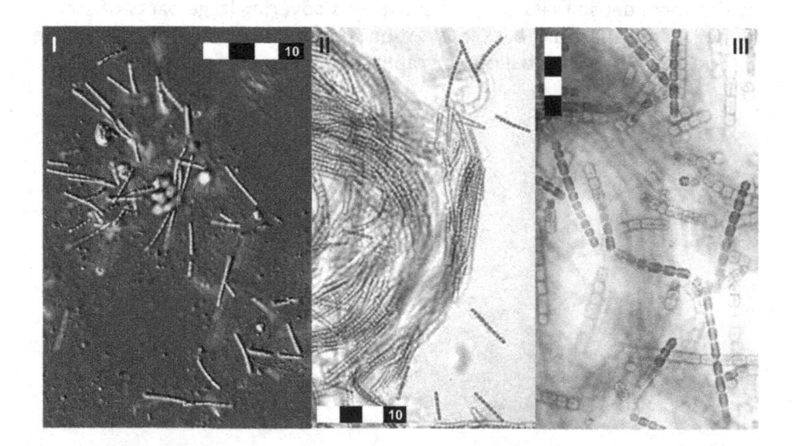

Figure 3.5O *Pseudanabaena* sp. Short trichomes associated with other phytoplankton (I). Red (II)- and green (III)-pigmented strains in culture, forming mats of long trichomes. Units in scale bars correspond to 5 µm if not indicated otherwise. For origin of individual photographs, see the end of this chapter.

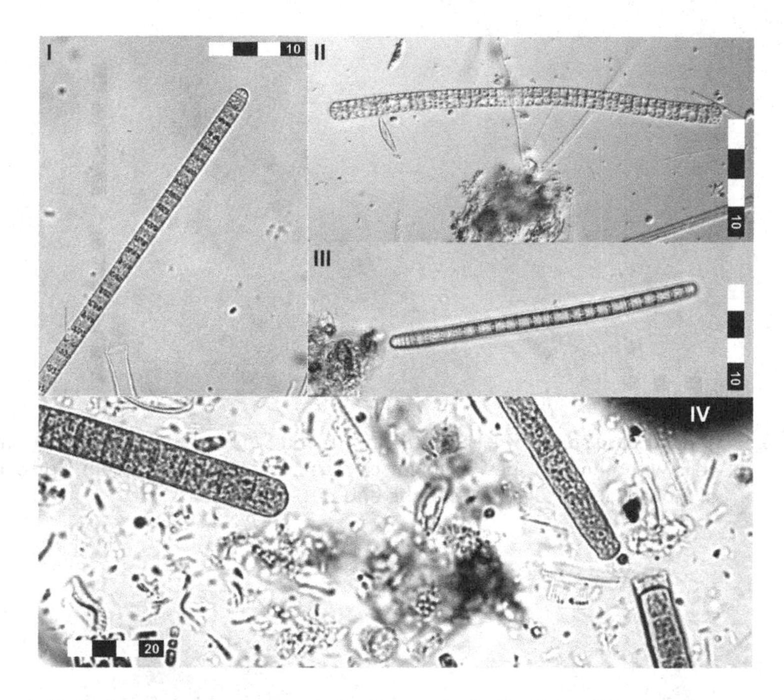

Figure 3.5P *Tychonema* sp. with partly hyaline cells (I–III). Typically, granulae accumulate at the cell walls. IV: Fragments of trichomes in a periphytic sample. Units in scale bars correspond to 10 or 20 µm as indicated. For origin of individual photographs, see the end of this chapter.

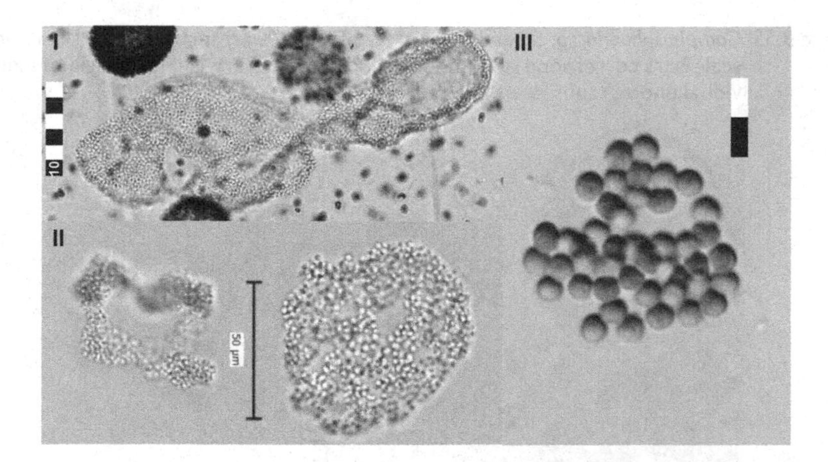

Figure 3.5Q *Aphanocapsa* sp. with small cells without aerotopes. In (I), cells and colonies of *Microcystis* sp. are shown for comparison. Units in scale bars correspond to 5 µm if not indicated otherwise. For origin of individual photographs, see the end of this chapter.

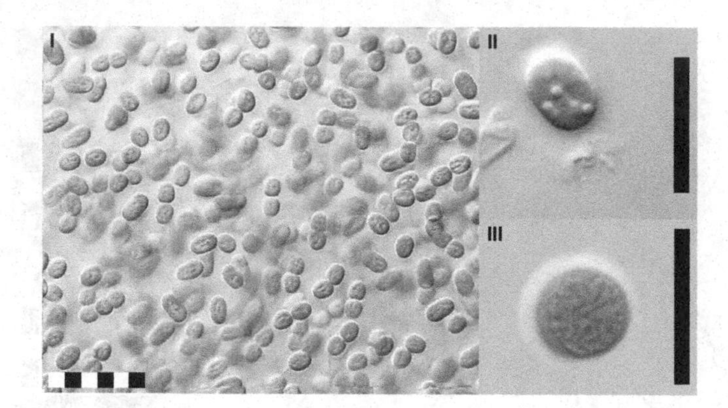

Figure 3.5R *Aphanocapsa* sp. (I), *Synechococcus* sp. (II) and *Synechocystis* sp. (III) with small cells without aerotopes. Units in scale bars correspond to 5 µm. For origin of individual photographs, see the end of this chapter.

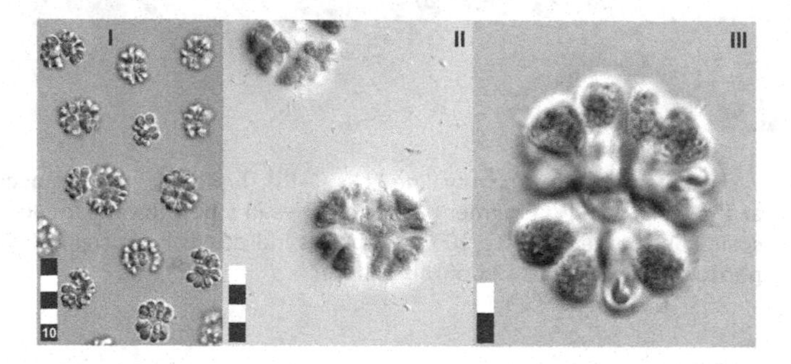

Figure 3.5S *Gomphosphaeria* sp. Small colonies of a few cells arranged radially. Units in scale bars correspond to 5 µm if not indicated otherwise. For origin of individual photographs, see the end of this chapter.

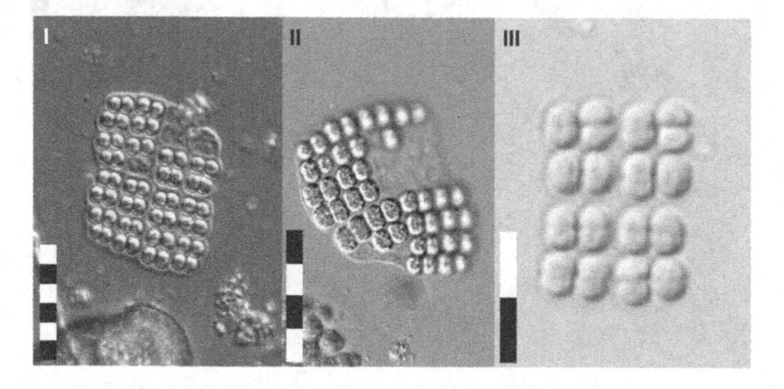

Figure 3.5T *Merismopedia* sp. with highly regular arrangement of cells in flat colonies. Unitsinscalebarscorrespondto5µm.Fororiginofindividualphotographs,seethe end of this chapter.

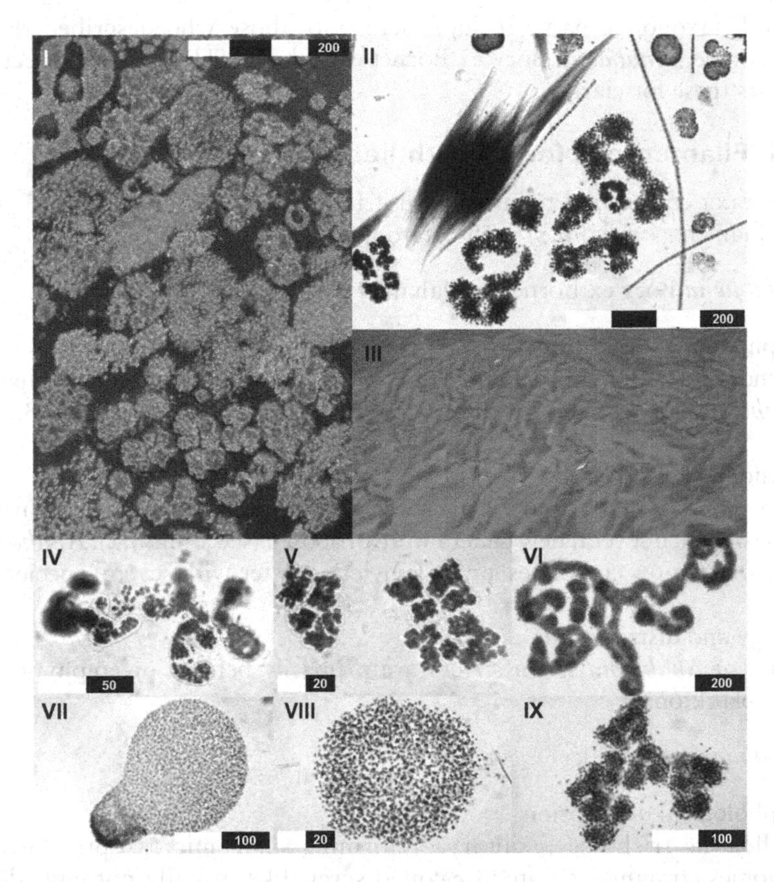

Figure 3.5U *Microcystis* sp. (I) natural sample comprising multiple *Microcystis* form species; (II) typical plankton community consisting of *Aphanizomenon flosaquae* fascicles, *Microcystis* colonies and *Dolichospermum* sp. trichomes; (III) *Microcystis* sp. surface bloom; IV–IX colonies of form species *M. wesenbergii* (IV), *M. viridis* (V), *M. aeruginosa* (VI), *M. flosaquae* (VII), *M. ichtyoblabe* (VIII) and *M. novacekii* (IX); note that the latter five form species have been proposed to be unified as *M. aeruginosa* based on molecular data (Otsuka et al., 2001). For ambiguities of *Microcystis* taxonomy, see also Box 3.1. Units in scale bars as indicated. For origin of individual photographs, see the end of this chapter.

In the following, abbreviations of genera differ occasionally from the general convention to use the initial letter to abbreviate a genus. This is done to avoid confusion, in particular if species epithets are identical. For example, *Aphanizomenon flosaquae* and *Anabaena flosaquae* are abbreviated as *Aph. flosaquae* and *Ana. flosaquae*, respectively (with the latter referred to as *Dolichospermum flosaquae* in the more recent literature). Note that the spelling of species epithets such as 'flos-aquae' has been chenged to non-hyphenated spelling 'flosaquae' in accordance with the International Code of Nomenclature for algae, fungi, and plants (Artcile 60.11; Turland et al., 2018).

The full taxonomic name includes names of those who described them, for example, '*Anabaena* Bory ex Bornet et Flahault'. The following section includes these for clarity.

3.7.1 Filamentous forms with heterocytes

These taxa correspond to subsection 4 in Table 3.2 or "Nostocales" and form filaments (trichomes) with heterocytes and akinetes.

Anabaena Bory ex Bornet et Flahault

Morphological description
The morphology of *Anabaena sensu stricto* largely corresponds to that of *Dolichospermum* except for the consistent lack of aerotopes (Figure 3.5A).

Taxonomic background
The genus *Anabaena* underwent several revisions in recent years in the course of which several new genera were proposed (see Table 3.2). *Anabaena sensu stricto* now comprises a monophyletic cluster with several species.

Ecology and distribution
Species of *Anabaena sensu stricto* are primarily benthic or epiphytic but rarely planktonic.

Anabaenopsis Miller

Morphological description
Free-floating trichomes, solitary or forming small microscopic clusters. Trichomes straight, arcuated or coiled screw-like, usually not embedded in sheaths, but sometimes with a fine sheath. Vegetative cells cylindrical or barrel-shaped, shorter than wide or up to several times longer than wide, pale blue-green, with obligatory or facultative aerotopes, without (rarely) or with constrictions at the cross-walls. Heterocytes develop intercalary in pairs with certain distances from each other. As trichomes often disintegrate after heterocyte maturation between adjacent heterocytes, this typically results in short trichomes with terminal heterocytes. Oval, cylindrical or spherical akinetes develop intercalary, normally distant from heterocytes, but exceptionally adjacent to them. Akinetes generally develop solitary or in pairs, rarely arranged in series with up to five in a row (Figure 3.5B).

Taxonomic background
The genus is clearly defined phenotypically and has been confirmed by molecular analyses. Some species have been transferred to the genera (*Cylindrospermopsis* (now *Raphidiopsis*) and *Cylindrospermum* as they

share a number of characteristics. More details are found in Komárek & Anagnostidis (1989).

Ecology and distribution

All species are planktonic and primarily found in mesotrophic to eutrophic waters, either inland or coastal, with alkaline or slightly saline conditions. Distributed mainly in tropical and subtropical regions but also during summer in temperate zones.

Aphanizomenon Morren ex Bornet et Flahault

Morphological description

Trichomes straight or slightly bent and only slightly narrowing towards the end, generally without sheaths, but sometimes with a very fine sheath. In some species, trichomes tend to form fascicules, that is, macroscopically visible aggregates of multiple trichomes. Vegetative cells cylindrical or barrel-shaped (from 2 to 5 µm in diameter) with variable length/width, often slightly constricted at the cross-wall. Cells contain aerotopes that are distributed evenly in the cells. Terminal cells larger than cells in the trichome, cylindrical with rounded ends or flattened, sometimes with hyaline (transparent) content. Generally, one heterocyte is placed intercalary (i.e., surrounded by vegetative cells) in individual trichomes, rarely 2–4 heterocytes per trichome. Heterocytes cylindrical, spherical or ellipsoidal. The akinetes develop adjacently from heterocyte (paraheterocytic, sometimes distant) forming a subsymmetric trichome. Akinetes often much larger than vegetative cells and are cylindrical, intercalary and solitary (rarely in pairs). Single trichomes of *Aphanizomenon flosaquae* without heterocytes are morphologically very similar to trichomes of *Planktothrix agardhii* (Figure 3.5C).

Taxonomic background

The traditional genus *Aphanizomenon* was recently restricted to a cluster of nine morphospecies (*A. flosaquae*, *A. gracile*, *A. klebahnii*, *A. yezoense*, *A. paraflexuosum*, *A. flexuosum*, *A. slovenicum*, *A. platense* and *A. hungaricum*) based on polyphasic analyses, while other morphotypes formerly placed in the genus were placed in other genera (*Cuspidothrix*, *Sphaerospermopsis* and *Chrysosporum*; (Rajaniemi et al., 2005a; Rajaniemi et al., 2005b; Komárek, 2013)). The main feature that separates these genera from *Aphanizomenon sensu stricto* is that *Aphanizomenon* spp. are able to form parallel fascicules that can reach macroscopic size of several millimetres visible as green, needle-like particles (Komárek, 2013). Further, the subsymmetric filaments are cylindrical, elongated with (almost) hyaline (i.e., transparent) terminal cells. Terminal cells rounded but without distinctly narrowed ends; that are typical for the proposed new genera. For a review of *Aphanizomenon* and related genera and their toxigenicity, see Cirés & Ballot (2016).

Ecology and distribution

Aphanizomenon generally dominate eutrophic, stagnant waters, with low available nitrogen and thermal stratification. *A. gracile* is typically found in shallow lakes and reservoirs (Cirés et al., 2017). *A. flosaquae*, the most common species of the genus, occurs mainly in temperate zones. Other species occur only in isolated areas, but no tropical *Aphanizomenon* species are registered.

Chrysosporum Zapomělová et al., 2012

Morphological description

Solitary, straight or slightly bent trichomes with clear constrictions at the cell walls. Vegetative cells vary from nearly cylindrical to barrel-shaped or ellipsoidal. Terminal cells rounded and slightly pointed and partially hyaline (transparent). Solitary and cylindrical heterocytes formed intercalary (i.e., surrounded by vegetative cells). Akinetes characteristically oval with distinct granular contents, and distant from heterocytes, often situated nearly equidistant between two heterocytes (Figure 3.5D).

Taxonomic background

Only two species to date: *Chrysosporum ovalisporum* that has been renamed from *Aphanizomenon ovalisporum* and *Chrysosporum bergii* from *Anabaena bergii* (Zapomělová et al., 2012).

Ecology and distribution

Only limited data on the distribution of both species is available, indicating that these species may occur primarily in temperate to subtropical climates (Cirés & Ballot, 2016).

Cuspidothrix Rajaniemi et al.

Morphological description

Solitary and free-floating trichomes forming heterocytes. Trichomes straight or coiled and characteristically clearly narrowed towards the ends. Cells slightly constricted or nonconstricted at the cross-walls, up to 6 µm wide. Vegetative cells cylindrical with facultative aerotopes, generally much longer than wide. Typical apical (terminal) cells elongated up to several tens of µm, attenuated and acuminate, and mainly hyaline (i.e., transparent). Heterocytes appear only intercalary (i.e., surrounded by vegetative cells), always solitary, cylindrical or elliptical. Akinetes elongated, more or less cylindrical intercalary, solitary or rarely in pairs, close or at a short distance to heterocytes. Trichomes subsymmetric with paraheterocytic akinetes situated on both sides or slightly distant from heterocytes (Figure 3.5E).

Taxonomic background

Cuspidothrix issatschenkoi was renamed from *A. issatschenkoi* based on polyphasic analyses (Rajaniemi et al., 2005a). Other species in the genus have formerly been assigned to *Aphanizomenon*, for example, *Cuspidothrix capricornii* and *Cuspidothrix elenkinii*.

Ecology and distribution

These species are planktonic in mesotrophic to eutrophic stagnant waters. They are rarely found in running waters. They are also present from freshwater to oligohaline and brackish waters.

Raphidiopsis (Fritsch et Rich) Aguilera et al., including *Cylindrospermopsis* Seenayya et Subba Raju

Morphological description

Aguilera et al. (2018) proposed to unify the genera *Cylindrospermopsis* and *Raphidiopsis* and to give preference to the latter respecting the principle of priority. Because the scientific community is increasingly accepting this revision, in this volume the genus *Raphidiopsis* refers to the combination of the former genera *Cylindrospermopsis* and *Raphidiopsis*. However, since the study of Aguilera et al. (2018) is restricted to only a limited number of species, the taxonomic opinion may change again in the future.

The genus *Raphidiopsis* comprises solitary, free-floating trichomes forming heterocytes, except for species of *Rhaphidiopsis sensu* Fritsch et Rich, for example *R. mediterranea*. Trichomes straight, bent or screw-like coiled, but in several species narrowed towards ends and without sheaths. The main characteristic is the terminal position of heterocytes, one or two at each end of the trichome (when not absent). Heterocytes ovoid, conical or drop-like. Trichomes subsymmetric, isopolar but heteropolar when only one heterocyte is present, generally without constrictions at cross-walls. Vegetative cells cylindrical or barrel-shaped, usually distinctly longer than wide, pale blue-green, yellowish or olive-green, facultatively with aerotopes. Terminal cells conical, blunt or sharply pointed. Ellipsoidal or cylindrical akinetes (2–4× longer and about 2× wider than the vegetative cells) develop usually distant from heterocytes, rarely adjacent to apical heterocytes (Figure 3.5F).

Taxonomic background

At present, 18 (morpho)species have been described of which, *Raphidiopsis* (*Cylindrospermopsis*) *raciborskii* appears to be the most important one with respect to cyanotoxin production.

Ecology and distribution

In eutrophic, turbid, warm and polymictic waters. In tropics, the appearance is often connected with nitrogen limitation of phytoplankton. All species are planktonic in lakes of the pantropical region except *Raphidiopsis raciborskii* that dispersed into the temperate region during the last 100–150 years (Padisák et al., 2016) where it can form dense, suspended water blooms.

Dolichospermum (Ralfs ex Bornet & Flahault) Wacklin, Hoffman et Komárek

Morphological description

Free-floating trichomes forming heterocytes, straight, slightly curved or flexuous, irregularly or more or less screw-like coiled. Solitary trichomes, rarely joined in irregular clusters (very rarely in fascicules). Trichomes not attenuated towards the ends, without sheaths, sometimes with fine diffluent mucilaginous envelope. Vegetative cells usually clearly constricted at the cross-walls and with many aerotopes distributed throughout the cells. Cylindrical trichomes are isopolar, metameric with respect to heterocytes. Apical cells morphologically similar to other vegetative cells in the filament. Heterocytes form intercalarly, solitary or exceptionally in pairs. They develop from vegetative cells in metameric position. Akinetes are elongated and wider than vegetative cells; they develop paraheterocytically, that is, connected with heterocytes, rarely aside heterocytes from both sides or (more commonly) separated from them by several cells, solitary or up to six in a row (Figure 3.5G).

Taxonomic background

The species of the genus *Dolichospermum* have been recently separated from the genus *Anabaena*. The main criterion for separating these genera is the absence of aerotopes in *Anabaena s. str.*, which is consistent with analyses of genomic sequences (Wacklin et al., 2009). Besides the known toxigenic species such as *D. flosaquae* or *D. circinale*, more than 30 species are given in Wacklin et al. (2009), however, without an individual taxonomic evaluation.

Ecology and distribution

All species planktonic in vegetative state, rarely associated with macrophytes (metaphytic). *Dolichospermum* is found in mesotrophic to eutrophic, both stratified and shallow lakes, generally with low nitrogen concentrations, where it can form blooms and surface scums. Several species are considered to be tropical.

Sphaerospermopsis Zapomělová et al.

Morphological description
Solitary trichomes straight, slightly bent or coiled with constrictions at cell walls. Vegetative cells cylindrical- to barrel-shaped with slightly elongated but not pointed terminal cells, often resembling cells within the trichome. Cylindrical to ellipsoidal heterocytes formed solitary and intercalary (i.e., surrounded by vegetative cells). Akinetes characteristically nearly spherical or ellipsoidal and often occur in groups of two or three, but also singularly. Akinetes are frequently formed adjacently to heterocytes, sometimes on both sides. Fragmentation of trichomes at the akinetes yields trichomes with terminal akinetes (Figure 3.5H).

Taxonomic background
Sphaerospermopsis includes species formerly belonging to *Anabaena* as well as *Aphanizomenon* (Zapomělová et al., 2011; Table 3.2).

Ecology and distribution
Sphaerospermopsis occurs primarily in temperate, subtropical and tropical shallow lakes.

3.7.2 Filamentous forms without heterocytes and akinetes

These taxa correspond to subsection 3 in Table 3.1 or "Oscillatoriales" and form filaments (trichomes) without heterocytes and akinetes. Following more recent taxonomic schemes, the genera *Planktolyngbya* and *Pseudanabaena* are placed in the order Synechococcales (Komárek, 2016b).

Arthrospira Stitzenberger ex Gomont

Morphological description
Solitary trichomes always without heterocytes, free floating or in mats covering hard substrate (microscopic or macroscopic). Trichomes more or less regularly coiled screw-like along their entire length. Generally without sheaths; however, if a sheath is present, it is colourless, tube-like with open ends enclosing single trichomes. Trichomes isopolar, 3–10 µm wide. Cells cylindrical, more or less isodiametric or shorter than wide, pale or bright blue-green or olive-green, in planktonic forms with aerotopes and in benthic forms without aerotopes. Not or only slightly constricted at the visible cross-walls. Trichomes not attenuated or only slightly attenuated towards the ends, with motility due to a rotational movement. Terminal cells widely rounded, usually with thickened outer cell walls or with calyptra (Figure 3.5I).

Taxonomic background

Arthrospira may be confused with *Spirulina*; the main difference is that *Arthrospira* has clearly delimited and visible cells, while in *Spirulina* cross-walls are not clearly visible. Further, *Arthrospira* has wider cells, and the trichomes are coiled in wider spirals compared to *Spirulina* trichomes that are tightly coiled. Eight species of *Arthrospira* are described, of which *A. platensis* seems to be the most frequently occurring one. However, in many reports, a species assignment is not made. Nowicka-Krawczyk et al. (2019) proposed to rename *A. fusiformis* to be renamed to *Limnospira fusiformis*.

Ecology and distribution

Arthrospira is generally found in shallow, turbid environments primarily in tropical and subtropical climates, in brackish or saline (alkaline) waters but occasionally also in freshwater.

Limnothrix Meffert

Morphological description

Trichomes solitary, always without heterocytes, free floating, isopolar. Straight or slightly curved or coiled irregularly screw-like, isopolar, without sheath or with an only very fine, colourless sheath. Trichomes cylindrical and from 1 to 6 µm wide, without or with reduced motility. Cells isodiametrical or longer than wide, unconstricted or slightly constricted at the cross-walls. Colour can range from pale blue-green to brown and orange. Aerotopes characteristically located close to the cross-walls. Apical cells are usually cylindrical, but sometimes conical (Figure 3.5J).

Taxonomic background

Most species of *Limnothrix* were originally placed in the genus *Oscillatoria*. The genus was amended by Meffert (1988) and confirmed by more recent studies (Suda et al., 2002; Komárek et al., 2014). The genus *Limnothrix* includes strains closely related to *Pseudanabaena* (Nishizawa et al., 2010) and is classified in the order Synechococcales by Komárek et al. (2014).

Ecology and distribution

Planktonic or tychoplanktonic, in fresh, mesotrophic to eutrophic, turbid and mixed waterbodies. *Limnothrix redekei* is distributed widely in the temperate zones but does not frequently form blooms.

Lyngbya Agardh ex Gomont, ***Moorea*** Engene et al., ***Microseira*** McGregor and Sendall and **related taxa**

Morphological description

Unbranched trichomes not constricted at cross-walls, enclosed in a firm sheath that is often protruding from trichomes. Cells short, cylindrical or, more often, coin-like (cell width $\gg$ cell length). Often strongly pigmented with a brown-green or blue-green colour, making cell walls difficult to recognise. *Lyngbya/Moorea* forms benthic mats on hard substrates or occurs epiphytic, forming macroscopic structures sometimes described as "mermaid's hair" (Figure 3.5K).

Taxonomic background

Two of the most studied species with respect to cyanotoxins are *Lyngbya majuscula* and *L. wollei*. Toxigenic strains of both species have been studied taxonomically and are proposed to be renamed: one cluster of *L. majuscula* is proposed to be renamed to *Moorea producens* (Engene et al., 2012) and a cluster of *L. wollei* to *Microseira wollei* (McGregor & Sendall, 2015). Further new genera separated from the *Lyngbya* species complex are *Okeania* (Engene et al., 2013) and *Dapis* (Engene et al., 2018). More than 500 species of *Lyngbya* are listed in AlgaeBase, with descriptions of a large share dating from before 1950, that is, without support from molecular data. Expectedly, this group of mainly marine filamentous cyanobacteria forming macroscopic aggregates will be subject to taxonomic revision once molecular and polyphasic analyses are conducted systematically (Engene et al., 2010).

Ecology and distribution

L. majuscula (*Moorea producens*) occurs primarily in brackish or marine habitats in tropical and subtropical zones. *L.* (*Microseira*) *wollei* is found in rivers and streams in temperate to subtropical zones where it forms. Other species of *Lyngbya sensu stricto* are found mostly in freshwaters.

***Phormidium* Kützing ex Gomont, *Microcoleus* Desmazières ex Gomont and related taxa**

Morphological description

Unbranched trichomes generally form fine or thick mats (microscopic to macroscopic) and are rarely solitary. Trichomes isopolar, straight, coiled or waved, usually <10–12 µm wide, facultatively with tube-like, firm, colourless sheaths with open ends. Vegetative cells cylindrical to slightly barrel-shaped, more or less isodiametrical or slightly shorter or longer than wide, constricted or unconstricted at the cross-walls, generally without aerotopes but with refractive granules. Trichomes not attenuated at the ends, sometimes bent or twisted screw-like towards the ends, motile within and outside of sheaths (Figure 3.5L).

Taxonomic background

This genus comprises a large number of species (>400; e.g., *P. nigrum*, *P. autumnale*, *P. fragile*), and the taxonomic status of many has been challenged (Palinska et al., 2011). As a consequence, a number of *Phormidium* species have been assigned to new genera, for example, *Wilmottia* (Strunecký et al., 2011), *Oxynema* (Chatchawan et al., 2012; Strunecký et al., 2014) or to existing genera such as *Microcoleus* (*Phormidium*) *autumnalis* (Strunecký et al., 2013) based on molecular analyses. *Microcoleus anatoxicus* has been reported to produce primarily dihydroanatoxin a (Conklin et al. 2020). For this reason, specimens of this taxon are often reported as *Phormidium* sp. and species assignment in elder publication may be no longer valid from a today's point of view. *Phormidium* may be confused with *Geitlerinema*, *Lyngbya* (*Moorea, Microseira*) and others.

Ecology and distribution
Epiphytic or epilithic in shallow rivers or streams but also in shallow areas in eutrophic standing waters. Due to the uncertain taxonomy and the resulting difficulties for unambiguous species determination, the knowledge on geographic distribution of individual *Phormidium* species is incomplete (Marquardt & Palinska, 2007). Specimens of the genus were found in a variety of latitudes, including extreme cold environments (Strunecký et al., 2012).

Planktolyngbya Anagnostidis et Komárek

Morphological description
Trichomes without heterocytes, solitary, with thin, simple, colourless, but firm sheaths. Isopolar, cylindrical trichomes, narrow, up to 3 µm wide, straight, waved or coiled, generally not narrowed to the ends. Slightly constricted or unconstricted at the cross-walls, and always immotile. Cylindrical cells usually longer than wide (rarely shorter than wide), without aerotopes, pale grey-blue, blue-green, yellowish or olive-green. Terminal cells rounded or narrowed-rounded without a calyptra (Figure 3.5M).

Taxonomic background
The genus was separated from *Lyngbya* by Anagnostidis and Komárek (1988) and confirmed by polyphasic analysis (Komárek et al., 2014) who placed the genus to the order Synechococcales (family Leptolyngbyaceae).

Ecology and distribution
Planktonic species are typical in large, mesotrophic reservoirs. Some species are limited to tropical and warm areas of temperate zones, while several species are presumably nordic.

Planktothrix Anagnostidis et Komárek

Morphological description
Trichomes always solitary, free floating, more or less straight or slightly irregularly waved or curved. In culture, trichomes may form irregular

clusters. Sheaths generally absent; if present, they are fine, colourless and diffluent. Trichomes isopolar, cylindrical, not constricted or slightly constricted at cross-walls. Length of trichomes up to 4 mm, width 3–12 µm. Immotile or sometimes slightly motile (trembling, gliding), slightly attenuated or not attenuated towards the ends, sometimes capitated or with terminal calyptra. Most species with prominent aerotopes take a large share of the cells' volume. Vegetative cells cylindrical or (rarely) slightly barrel-shaped, shorter than wide, up to ±isodiametric or rarely little longer than wide (Figure 3.5N).

Taxonomic background

Planktothrix was originally placed in the genus *Oscillatoria*, from which it was separated due to ecological traits and the formation of large numbers of aerotopes (Anagnostidis & Komárek, 1988).

Several species are described, of which *Planktothrix agardhii* and *P. rubescens* are the most relevant with respect to cyanotoxins. Other species (e.g., *P. mougeotii*, *P. pseudagardhii* and *P. spiroides*) are morphologically similar but have not been reported to form blooms. Some species produce only few aerotopes (*P. paucivesiculata*, *P. serta*; Gaget et al., 2015a).

Ecology and distribution

Generally, planktonic and evenly distributed in the water column in non-stratified, shallow lakes (*P. agardhii*) or cumulated at the thermocline of deep, stratified lakes (*P. rubescens* or, more rarely, *P. mougeotii*), occasionally forming blooms. In the case of blooms of *P. rubescens*, these may accumulate at the metalimnion and not be visible at the surface (see Chapter 4). Both species tolerate low light intensities. The genus is widely distributed in temperate climates, but individual species may show more restricted distribution patterns.

Pseudanabaena Lauterborn

Morphological description

Trichomes without branching and without firm sheaths, sometimes with fine, colourless, diffluent envelopes. Trichomes solitary or agglomerated in very fine mucilaginous mats. Individual trichomes straight, slightly waved or bent, usually not very long, 0.8-3 µm wide, not attenuated at the ends, usually with slight constrictions at the distinct cross-walls. Cells cylindrical, usually longer than wide (sometimes barrel shaped or nearly spherical). The apical cell is cylindrical and rounded at the end or more or less conical up to bluntly or sharply pointed. Generally without aerotopes, but sometimes with aerotopes at the ends of cells. Trichomes may have motility (trembling). Pigmentation often reddish (Figure 3.5O).

Taxonomic background

Pseudanabaena is closely related to *Limnothrix* (Acinas et al., 2009), and some authors refer to this as "*Pseudanabaena/Limnothrix*" group (e.g., Zwart et al., 2005). Both genera have been assigned to the order Synechococcales *sensu* (Komárek et al., 2014); these genera are close to single-cell forms such as *Synechococcus* sp.

Ecology and distribution

Mostly planktonic species, tychoplanktonic or benthic in oligotrophic, mesotrophic up to slightly eutrophic water reservoirs and turbid mixed waters. Short trichomes of *Pseudanabaena endophytica* can often be found attached to colonies of *Microcystis*.

Tychonema Anagnostidis et Komárek

Morphological description

Unbranched, cylindrical trichomes lack a visible sheath of 7–12 µm width not constricted at cross-walls. Cells generally slightly shorter than wide or isodiametrical and appear almost empty except for granulae at the cross-walls or the cell periphery. Trichomes mostly solitarily and lacking motility as observed in *Geitlerinema* or *Phormidium* (Figure 3.5P).

Taxonomic background

Tychonema is a currently recognised distinct genus within Oscillatoriales (Anagnostidis & Komárek, 1988) and has been confirmed by polyphasic analyses (Suda et al., 2002). Four species are described, the distinction of which may be difficult (*Tychonema bornetii*, *T. bourrellyi*, *T. decoloratum* and *T. tenue*).

Ecology and distribution

As the name already indicates, *Tychonema* typically is tychoplanktonic: the trichomes are loosely attached to macrophytes or hard substrate but can be detached due to water movement and become planktonic.

Tychonema is primarily found in mesotrophic lakes of temperate zones where macrophyte stands exist, for example, in assemblages of water moss (Fastner et al., 2018). *T. bourrellyi* is considered as truly planktonic (Salmaso et al., 2016b).

3.7.3 Colonial forms

These taxa correspond to subsection 1 in Table 3.1 or "Chroococcales" and are unicellular with cells embedded in a common mucilage. Following more recent taxonomic schemes, the genera *Aphanocapsa* and *Merismopedia* are placed in the order Synechococcales (Komárek, 2016b).

Aphanocapsa Nägeli

Morphological description
Cells form microscopic (sometimes macroscopic) more or less spherical or irregular colonies with irregularly, loosely or densely distributed cells. Mucilage fine and diffluent, generally colourless but macroscopically colonies appear as firm sheaths. Cells spherical from 1.5 to 6 μm of diameter (hemispherical after division), without own mucilaginous envelopes, generally without aerotopes, sometimes with granular content. Cell division always in two perpendicular planes in successive generations. Some species are morphologically similar to *Microcystis*, except for the lack of aerotopes in *Aphanocapsa* (Figure 3.5Q).

Taxonomic background
It is suggested that planktonic species need revision as the relationship to other form genera like *Microcystis* is not clear.

Ecology and distribution
Periphytic, benthic or metaphytic in stagnant and running freshwater systems, usually with clear water, common in lakes. Often found in late summer in the epilimnion of oligotrophic, deep lakes. Registered worldwide, but several species are ecologically sharply limited and occur in geographically limited areas.

Similar small coccoid taxa are the colony-forming *Aphanothece* and single-celled *Synechococcus* and *Synechocystis* (Figure 3.5R).

Gomphosphaeria Kützing

Morphological description
Cells embedded in a mucilage forming spherical or irregularly oval colonies, sometimes composed of multiple subcolonies. Mucilage with gelatinous stalks that radiate from the centre of the colony to the periphery. The stalks are widened at the ends and envelope individual cells with a thin mucilage layer. Cells elongate (6–12×2–8 μm), radially oriented at the end of stalks. Pigmentation pale or bright blue-green, olive-green or red (Figure 3.5S).

Taxonomic background
At first sight, the genus can be confused with *Snowella* or *Coelosphaerium*; see taxonomic update in Komárek and Anagnostidis (1999).

Ecology and distribution

Generally, found in eutrophic to hypertrophic, small- to medium-sized lakes.

Merismopedia Meyen

Morphological description

Free-floating microscopic colonies, square or rectangular with one layer of cells densely or loosely arranged in a single plane. Larger colonies may be contorted or composed of several subcolonies. Colonies of a few to several cells that divide in two alternating planes, forming groups of 4 or 16 cells that collectively form distinctive, flat colonies with hyaline (transparent), fine envelopes, some species with envelopes surrounding each cell. Cells spherical or elliptical (hemispherical after division), generally pale or bright blue-green content; in a few species in central parts of cells with refractive granulae or aerotopes (Figure 3.5T).

Taxonomic background

Some 20 morphospecies of *Merismopedia* are described, for example, *M. glauca*, *M. punctata* or *M. elegans*. In most ecological studies, only *Merismopedia* sp. is reported.

Ecology and distribution

Planktonic or metaphytic, usually in biotopes with submerged macrophyte vegetation. Temperate habitats: deep and shallow, oligotrophic to eutrophic, medium to large lakes. Common in the epilimnion of mesotrophic lakes in summer.

Cosmopolitan distribution, but several species have clearly ecologically and geographically limited areas of distribution.

Microcystis Kützing ex Lemmermann

Morphological description

Free-floating microscopic or macroscopic colonies, spherical, oval or elongated, in several species clathrate. A large number of species have been described based mainly on cell size and colony morphology. The latter is, however, not available in cultured strains that generally grow as single cells or atypical colonies (Otsuka et al., 2000). The same is true for samples fixed with Lugol's solution in which colonies disintegrate to small clusters or single cells, allowing a differentiation only by cell size. In some species, colonies are composed of subcolonies or multiple more or less separated clusters of cells. All cells densely or sparsely arranged in a common mucilage with the density of cells highly variable in particular species. Mucilage fine, colourless, diffluent or distinct and delimited (e.g., *Microcystis wesenbergii*). Gelatinous envelopes around individual cells are never present. Cells spherical or hemispherical shortly after division, ranging from 2 to 7 μm in diameter or slightly elongated, with many, irregularly arranged aerotopes. Differentiation of morphospecies in samples of natural populations is often uncertain as many colonies show characteristics of more than one morphospecies (Figure 3.5U).

Taxonomic background

The number of described morphospecies varies depending on the reference source chosen (see Box 3.1). Due to phenotypic variability, the status of individual morphospecies and their relationships is largely unclear. The genus *Microcystis* is one of the few cyanobacterial genera that underwent systematic taxonomic revision based on molecular data. As a result, based on genomic DNA homologies, Otsuka et al. (2001) proposed to unify the species *M. aeruginosa*, *M. ichthyoblabe*, *M. novacekii*, *M. viridis* and *M. wesenbergii* in a single species. This is also supported by Harke et al. (2016), but nonetheless, this proposal has not been validated, primarily for formal reasons (Oren & Ventura, 2017).

A particular morphotype occurring in tropical waters has been described as *Radiocystis*. This genus is characterised by cells more or less arranged in radial series protruding from the centre embedded in a mucilage also showing radial structures (Komárek & Komárková-Legnerová, 1993). Genomic sequences such as 16S rRNA or phycocyanin operon are, however, identical to those of *Microcystis* (Vieira et al., 2003). Similarly, *Sphaerocarvum* sp. has been split from the genus *Microcystis* based on a particular colony morphology *in situ* (Azevedo & Sant'Anna, 2003), but has genomic sequences identical to *Microcystis* (Rigonato et al., 2018).

Ecology and distribution

Planktonic, in mesotrophic to eutrophic standing waters that are at least temporally stratified, preferably in shallow or medium depth lakes. Mostly absent or restricted to the shallow basins or to the littoral region in deep, stratified lakes where they can, however, form large blooms like in the North American great lakes. *Microcystis* frequently forms blooms in eutrophic systems, and under conditions of stable thermal stratification, it can form surface blooms and massive scums. Colonies can sink to the bottom and overwinter in the sediment.

Many species with a cosmopolitan distribution, except in subpolar regions but several taxa are restricted geographically due to ecological preferences (van Gremberghe et al., 2011; Harke et al., 2016).

PICTURE CREDITS

Microscopic photographs for Figure 3.5A–U were kindly provided as listed here. All contributions are highly appreciated. Andreas Ballot: 3.5C-II, 3.5D-I, 3.5D-II, 3.5D-III, 3.5F-V, 3.5F-VI, 3.5G-II, 3.5G-IV, 3.5H-I, 3.5H-III, 3.5I-I, 3.5I-II, 3.5I-III. Barry H. Rosen, Florida Gulf Coast University; Fort Myers (Fl), USA (brosen@fgcu.edu): 3.5A-II, 3.5C-V, 3.5E-II, 3.5F-IV, 3.5G-III, 3.5J-I, 3.5J-II, 3.5K-I, 3.5L-II, 3.5M-I, 3.5M-II, 3.5M-III, 3.5P-II, 3.5Q-III, 3.5R-I, 3.5R-II, 3.5R-III, 3.5S-II, 3.5S-III, 3.5T-III. Anja Hoffmann,

Landeslabor Berlin-Brandenburg, Inst. für Lebensmittel, Arzneimittel, Tierseuchen und Umwelt -, Berlin, Germany (Anja.Hoffmann@Landeslabor-BBB.de): 3.5P-I, 3.5P-III, 3.5P-IV. Hydrobiology Area, Central Laboratory, Obras Sanitarias del Estado (OSE), Montevideo, Uruguay: 3.5B-I, 3.5B-III, 3.5E-III, 3.5F-II. Martin Welker: 3.5C-I, 3.5F-III, 3.5L-I, 3.5L-III, 3.5N-III, 3.5N-IV, 3.5N-V, 3.5O-II, 3.5O-III, 3.5U-III, 3.5U-IV, 3.5U-V, 3.5U-VI, 3.5U-VII, 3.5U-VIII, 3.5U-IX. Petr Znachor, Institute of Hydrobiology, Biology Centre CAS, České Budějovice, Czech Republic (znachy@gmail.com; www.fytoplankton.cz): 3.5C-VII, 3.5D-IV, 3.5F-I, 3.5G-I, 3.5H-II, 3.5O-I, 3.5S-I, 3.5T-I, 3.5T-II, 3.5U-I. Radovan Kopp, Mendelova univerzita v Brně, Brno, Czech Republic (Fcela@seznam.cz): 3.5A-I, 3.5B-II, 3.5C-III, 3.5C-IV, 3.5C-VI, 3.5C-VIII, 3.5E-I, 3.5J-III, 3.5K-II, 3.5K-III, 3.5N-I, 3.5N-II, 3.5Q-I, 3.5Q-II, 3.5U-II. Susanna A. Wood, Cawthron Institute, Nelson, New Zealand (Susie.Wood@cawthron.org.nz): 3.5L-IV.

REFERENCES

Acinas SG, Haverkamp THA, Huisman J, Stal LJ (2009). Phenotypic and genetic diversification of *Pseudanabaena* spp. (cyanobacteria). ISME J. 3:31–46.

Adams DG, Duggan PS (1999). Heterocyst and akinete differentiation in cyanobacteria. New Phytol. 144:3–33.

Aguilera A, Gómez EB, Kaštovský J, Echenique RO, Salerno GL (2018). The polyphasic analysis of two native *Raphidiopsis* isolates supports the unification of the genera *Raphidiopsis* and *Cylindrospermopsis* (Nostocales, Cyanobacteria). Phycologia. 57:130–146.

Anagnostidis K, Komárek J (1985). Modern approach to the classification system of Cyanophytes. 1-Introduction. Arch Hydrobiol Algol Stud. 38–39:291–302.

Anagnostidis K, Komárek J (1988). Modern approach to the classification system of Cyanophytes 3- Oscillatoriales. Arch Hydrobiol Algol Stud. 80 (50–53):327–472.

Azevedo TMP, Sant'Anna CL (2003). *Sphaerocavum*, a new genus of planktic Cyanobacteria from continental water bodies in Brazil. Algol Stud. 109:79–92.

Berman-Frank I, Lundgren P, Falkowski P (2003). Nitrogen fixation and photosynthetic oxygen evolution in cyanobacteria. Res Microbiol. 154:157–164.

Bothe H, Schmitz O, Yates MG, Newton WE (2010). Nitrogen fixation and hydrogen metabolism in cyanobacteria. Microbiol Mol Biol Rev. 74:529–551.

Burja AM, Banaigs B, Abou-Mansour E, Burgess JG, Wright PC (2001). Marine cyanobacteria-a prolific source of natural products. Tetrahedron. 57:9347–9377.

Castenholz RW, Wilmotte A, Herdman M, Rippka R, Waterbury JB, Iteman I et al. (2001). Phylum BX. Cyanobacteria. In: Garrity G, Boone DR, Castenholz DW, editors: Bergey's Manual® of systematic bacteriology. New York: Springer:473–599.

Cavalier-Smith T (2002). The neomuran origin of archaebacteria, the negibecterial root of the universal tree and bacterial megaclassification. Int J Syst Evol Microbiol. 52:7–76.

Chatchawan T, Komárek J, Strunecký O, Šmarda J, Peerapornpisal Y (2012). *Oxynema*, a new genus separated from the genus *Phormidium* (Cyanophyta). Cryptogam Algol. 33:41–60.

Cirés S, Ballot A (2016). A review of the phylogeny, ecology and toxin production of bloom-forming *Aphanizomenon* spp. and related species within the Nostocales (cyanobacteria). Harmful Algae. 54:21–43.

Cirés S, Delgado A, González-Pleiter M, Quesada A (2017). Temperature influences the production and transport of saxitoxin and the expression of *sxt* genes in the cyanobacterium *Aphanizomenon gracile*. Toxins. 9:322.

Conklin KY, Stancheva R, Otten TG, Fadness R, Boyer GL, Read B et al. (2020) Molecular and morphological characterization of a novel dihydroanatoxin-a producing *Microcoleus* species (cyanobacteria) from the Russian River, California, USA. Harmful Algae 93:101767.

De Philippis R, Vincenzini M (1998). Exocellular polysaccharides from cyanobacteria and their possible applications. FEMS Microbiol Rev. 22:151–175.

Dillon JG, Castenholz RW (1999). Scytonemin, a cyanobacterial sheath pigment, protects against UVC radiation: implications for early photosynthetic life. J Phycol. 35:673–681.

Dvořák P, Casamatta DA, Hašler P, Jahodářová E, Norwich AR, Poulíčková A (2017). Diversity of the cyanobacteria. In: Hallenbeck P, editor: Modern topics in the phototrophic prokaryotes. Cham: Springer.

Dvořák P, Jahodářová E, Casamatta DA, Hašler P, Poulíčková A (2018). Difference without distinction? Gaps in cyanobacterial systematics; when more is just too much. Fottea. 18:130–136.

Engene N, Coates RC, Gerwick WH (2010). 16S rRNA Gene heterogeneity in the filamentous marine cyanobacterial genus *Lyngbya*. J Phycol. 46:591–601.

Engene N, Paul VJ, Byrum T, Gerwick WH, Thor A, Ellisman MH (2013). Five chemically rich species of tropical marine cyanobacteria of the genus *Okeania* gen. nov.(Oscillatoriales, Cyanoprokaryota). J Phycol. 49:1095–1106.

Engene N, Rottacker EC, Kaštovský J, Byrum T, Choi H, Ellisman MH et al. (2012). *Moorea producens* gen. nov., sp. nov. and *Moorea bouillonii* comb. nov., tropical marine cyanobacteria rich in bioactive secondary metabolites. Int J Syst Evol Microbiol. 62:1171–1178.

Engene N, Tronholm A, Paul VJ (2018). Uncovering cryptic diversity of *Lyngbya*: the new tropical marine cyanobacterial genus *Dapis* (Oscillatoriales). J Phycol. 54:435–446.

Fastner J, Beulker C, Geiser B, Hoffmann A, Kröger R, Teske K et al. (2018). Fatal neurotoxicosis in dogs associated with tychoplanktic, anatoxin-a producing *Tychonema* sp. in mesotrophic Lake Tegel, Berlin. Toxins. 10:60.

Flombaum P, Gallegos JL, Gordillo RA, Rincón J, Zabala LL, Jiao N et al. (2013). Present and future global distributions of the marine Cyanobacteria *Prochlorococcus* and *Synechococcus*. Proc Natl Acad Sci USA. 110:9824–9829.

Frigaard N-U, Larsen KL, Cox RP (1996). Spectrochromatography of photosynthetic pigments as a fingerprinting technique for microbial phototrophs. FEMS Microbiol Lett. 20:69–77.

Gaget V, Welker M, Rippka R, de Marsac NT (2015a). A polyphasic approach leading to the revision of the genus *Planktothrix* (Cyanobacteria) and its type species, *P. agardhii*, and proposal for integrating the emended valid botanical taxa, as well as three new species, *Planktothrix paucivesiculata* sp. nov., *Planktothrix tepida* sp. nov., and *Planktothrix serta* sp. nov., as genus and species names with nomenclatural standing under the ICNP. Syst Appl Microbiol. 38:141–158.

Gaget V, Welker M, Rippka R, Tandeau de Marsac N (2015b). Response to: Comments on: "A polyphasic approach leading to the revision of the genus *Planktothrix* (Cyanobacteria) and its type species, *P. agardhii*, and proposavl for integrating the emended valid botanical taxa, as well as three new species, *Planktothrix paucivesiculata* sp. nov.(ICNP), *Planktothrix tepida* sp. nov.(ICNP), and *Planktothrix serta* sp. nov.(ICNP), as genus and species names with nomenclature standing under the ICNP". Syst Appl Microbiol. 38:368–370.

Garcia-Pichel F, Lombard J, Soule T, Dunaj S, Wu SH, Wojciechowski MF (2019). Timing the evolutionary advent of cyanobacteria and the later Great Oxidation Event using gene phylogenies of a sunscreen. mBio. 10:e00561–00519.

Geitler L (1932). Cyanophyceae. In: Rabenhorst L, editor: Kryptogamenflora von Deutschland. Leipzig: Akad. Verlagsges:673-1196.

Gerwick WH, Tan LT, Sitachitta N (2001). Nitrogen-containing metabolites from marine cyanobacteria. Alkaloids Chem Biol. 57:75–184.

Gugger MF, Hoffmann L (2004). Polyphyly of true branching cyanobacteria (stigonematales). Int J Syst Evol Microbiol. 54:349–357.

Guiry MD, Guiry GM (2019). AlgaeBase. Galway: National University of Ireland.

Hamilton TL, Bryant DA, Macalady JL (2016). The role of biology in planetary evolution: cyanobacterial primary production in low-oxygen Proterozoic oceans. Environ Microbiol. 18:325–340.

Harke MJ, Steffen MM, Gobler CJ, Otten TG, Wilhelm SW, Wood SA et al. (2016). A review of the global ecology, genomics, and biogeography of the toxic cyanobacterium, *Microcystis* spp. Harmful Algae. 54:4–20.

Hauer T, Komárek J (2019). CyanoDB.cz 2.0- On-line database of cyanobacterial genera. České Budějovice: University of South Bohemia & Academy of Sciences of the Czech Republic.

Hoffmann L, Komárek J, Kaštovský J (2005). System of cyanoprokaryotes (cyanobacteria)–state in 2004. Algol Stud. 117:95–115.

Hoiczyk E (2000). Gliding motility in cyanobacteria: observations and possible explanations. Arch Microbiol. 174:11–17.

Ishida T, Watanabe MM, Sugiyama J, Yokota A (2001). Evidence for polyphyletic origin of the members of the orders of Oscillatoriales and Pleurocapsales as determined by 16S rDNA analysis. FEMS Microbiol Lett. 201:79–82.

Kehoe DM (2010). Chromatic adaptation and the evolution of light color sensing in cyanobacteria. Proc Natl Acad Sci USA. 107:9029–9030.

Komárek J (2006). Cyanobacterial taxonomy: current problems and prospects for the integration of traditional and molecular approaches. Algae. 21:349–375.

Komárek J (2011). Some current problems of modern cyanobacterial taxonomy. Fottea. 11:1–7.

Komárek J (2013). Cyanoprokaryota Part 3: Heterocystous genera. In: Büdel B, Gärtner G, Krienitz L et al., editors: Süsswasserflora von Mitteleuropa. Heidelberg: Springer Spektrum:1131.

Komárek J (2016a). A polyphasic approach for the taxonomy of cyanobacteria: principles and applications. Eur J Phycol. 51:346–353.

Komárek J (2016b). Review of the cyanobacterial genera implying planktic species after recent taxonomic revisions according to polyphasic methods: state as of 2014. Hydrobiologia. 764:259–270.

Komárek J, Anagnostidis K (1989). Modern approach to the classification system of Cyanophytes 4-Nostocales. Arch Hydrobiol Suppl. 82:247–345.

Komárek J, Anagnostidis K (1999). Cyanoprokaryota 1. Chroococcales. In: Ettl H, Gerloff I, Heynig H et al., editors: Süßwasserflora von Mitteleuropa. Jena: Gustav Fischer.

Komárek J, Kaštovský J, Mareš J, Johansen JR (2014). Taxonomic classification of cyanoprokaryotes (cyanobacterial genera) 2014, using a polyphasic approach. Preslia. 86:295–335.

Komárek J, Komárková-Legnerová J (1993). *Radiocystis fernandoi*, a new planktic cyanoprokaryotic species from tropical freshwater reservoirs. Preslia. 65:355–357.

Liu Z, Häder DP, Sommaruga R (2004). Occurrence of mycosporine-like amino acids (MAAs) in the bloom-forming cyanobacterium *Microcystis aeruginosa*. J Plankton Res. 26:963–966.

MacIntyre HL, Kana TM, Anning T, Geider RJ (2002). Photoacclimation of photosynthesis irradiance response curves and photosynthetic pigments in microalgae and cyanobacteria. J Phycol. 38:17–38.

Mareš J, Strunecký O, Bučinská L, Wiedermannová J (2019). Evolutionary patterns of thylakoid architecture in cyanobacteria. Frontiers Microbiol. 10.

Marquardt J, Palinska KA (2007). Genotypic and phenotypic diversity of cyanobacteria assigned to the genus *Phormidium* (Oscillatoriales) from different habitats and geographical sites. Arch Microbiol 187:397–413.

McGregor GB, Sendall BC (2015). Phylogeny and toxicology of *Lyngbya wollei* (Cyanobacteria, Oscillatoriales) from north-eastern Australia, with a description of *Microseira* gen. nov. J Phycol. 51:109–119.

Meffert M-E (1988). *Limnothrix* Meffert nov. gen. The unsheathed planktic cyanophycean filaments with polar and central gas vacuoles. Algol Stud. 50–53:269–276.

Moore KR, Magnabosco C, Momper L, Gold DA, Bosak T, Fournier GP (2019). An expanded ribosomal phylogeny of cyanobacteria supports a deep placement of plastids. Front Microbiol. 10.

Moustaka-Gouni M, Kormas KA, Vardaka E, Katsiapi M, Gkelis S (2009). *Raphidiopsis mediterranea* Skuja represents non-heterocytous life-cycle stages of *Cylindrospermopsis raciborskii* (Woloszynska) Seenayya et Subba Raju in Lake Kastoria (Greece), its type locality: evidence by morphological and phylogenetic analysis. Harmful Algae. 8:864–872.

Nabout JC, Rocha BdS, Carneiro FM, Sant'Anna CL (2013). How many species of Cyanobacteria are there? Using a discovery curve to predict the species number. Biodivers Conserv. 22:2907.

Nakayama T, Nomura M, Takano Y, Tanifuji G, Shiba K, Inaba K et al. (2019). Single-cell genomics unveiled a cryptic cyanobacterial lineage with a worldwide distribution hidden by a dinoflagellate host. Proc Nat Acad Sci USA. 116:15973–15978.

Nishizawa T, Hanami T, Hirano E, Miura T, Watanabe Y, Takanezawa A et al. (2010). Isolation and molecular characterization of a multicellular cyanobacterium, *Limnothrix/Pseudanabaena* sp. strain ABRG5-3. Biosci Biotechnol Biochem. 74:1827–1835.

Nowicka-Krawczyk P, Mühlsteinová R, Hauer T (2019). Detailed characterization of the *Arthrospira* type species separating commercially grown taxa into the new genus *Limnospira* (Cyanobacteria). Sci Rep. 9:694.

Oren A, Garrity GM (2014). Proposal to change General Consideration 5 and Principle 2 of the International Code of Nomenclature of Prokaryotes. Int J Syst Evol Microbiol. 64:309–310.

Oren A, Gunde-Cimerman N (2007). Mycosporines and mycosporine-like amino acids: UV protectants or multipurpose secondary metabolites? FEMS Microbiol Lett. 269:1–10.

Oren A, Ventura S (2017). The current status of cyanobacterial nomenclature under the “prokaryotic” and the “botanical” code. Antonie van Leeuwenhoek:1–13.

Otsuka S, Suda S, Li R, Matsumoto S, Watanabe MM (2000). Morphological variability of colonies of *Microcystis* morphospecies in culture. J Gen Appl Microbiol 46:39–50.

Otsuka S, Suda S, Shibata S, Oyaizu H, Matsumoto S, Watanabe MM (2001). A proposal for the unification of five species of the cyanobacterial genus *Microcystis* Kützing ex Lemmermann 1907 under the Rules of the Bacteriological Code. Int J Syst Evol Microbiol. 51:873–879.

Pace NL (1997). A molecular view of microbial diversity and the biosphere. Science. 276:734–740.

Padisák J, Vasas G, Borics G (2016). Phycogeography of freshwater phytoplankton: traditional knowledge and new molecular tools. Hydrobiologia. 764:3–27.

Palinska KA, Deventer B, Hariri K, Lotocka M (2011). A taxonomic study on *Phormidium*–group (cyanobacteria) based on morphology, pigments, RAPD molecular markers and RFLP analysis of the 16S rRNA gene fragment. Fottea. 11:41–55.

Pathak J, Ahmed H, Singh SP, Häder D-P, Sinha RP (2019). Genetic regulation of scytonemin and mycosporine-like amino acids (MAAs) biosynthesis in cyanobacteria. Plant Gene. 17:100172.

Pereira S, Zille A, Micheletti E, Moradas-Ferreira P, De Philippis R, Tamagnini P (2009). Complexity of cyanobacterial exopolysaccharides: composition, structures, inducing factors and putative genes involved in their biosynthesis and assembly. FEMS Microbiol Rev. 33:917–941.

Pinevich AV (2015). Proposal to consistently apply the International Code of Nomenclature of Prokaryotes (ICNP) to names of the oxygenic photosynthetic bacteria (cyanobacteria), including those validly published under the International Code of Botanical Nomenclature (ICBN)/International Code of Nomenclature for algae, fungi and plants (ICN), and proposal to change Principle 2 of the ICNP. Int J Syst Evol Microbiol. 65:1070–1074.

Postius C, Ernst A (1999). Mechanisms of dominance: coexistence of picocyanobacterial genotypes in a freshwater ecosystem. Arch Microbiol. 172:69–75.

Proteau P, Gerwick W, Garcia-Pichel F, Castenholz R (1993). The structure of scytonemin, an ultraviolet sunscreen pigment from the sheaths of cyanobacteria. Experientia. 49:825–829.

Rajaniemi P, Hrouzek P, Kastovska K, Willame R, Rantala A, Hoffmann L et al. (2005a). Phylogenetic and morphological evaluation of the genera *Anabaena*, *Aphanizomenon*, *Trichormus* and *Nostoc* (Nostocales, Cyanobacteria). Int J Syst Evol Microbiol. 55:11–26.

Rajaniemi P, Komárek J, Willame R, Hrouzek P, Kaštovská K, Hoffmann L et al. (2005b). Taxonomic consequences from the combined molecular and phenotype evaluation of selected *Anabaena* and *Aphanizomenon* strains. Algol Stud. 117:371–391.

Read N, Connell S, Adams DG (2007). Nanoscale visualization of a fibrillar array in the cell wall of filamentous cyanobacteria and its implications for gliding motility. J Bacteriol. 189:7361–7366.

Reynolds CS, Jaworsky GHM, Cmiech HA, Leedale GF (1981). On the annual cycle of *Microcystis aeruginosa* Kütz emend. Elenkin. Philos Trans R Soc Lond B. 293:419–477.

Rigonato J, Sant'Anna CL, Giani A, Azevedo MTP, Gama WA, Viana VF et al. (2018). *Sphaerocavum*: a coccoid morphogenus identical to *Microcystis* in terms of 16S rDNA and ITS sequence phylogenies. Hydrobiologia. 811:35–48.

Rippka R, Deruelles J, Waterbury JB, Herdman M, Stanier RY (1979). Generic assignments, strain histories and properties of pure cultures of cyanobacteria. Microbiology. 111:1–61.

Salmaso N, Akçaalan R, Bernard C, Elersek T, Krstić S, Pilkaityte R et al. (2016a). Cyanobacterial species and recent synonyms. In: Meriluoto J, Spoof L, Codd GA et al., editors: Handbook on cyanobacterial monitoring and cyanotoxin analysis. Chichester: John Wiley & Sons:489–500.

Salmaso N, Cerasino L, Boscaini A, Capelli C (2016b). Planktic *Tychonema* (Cyanobacteria) in the large lakes south of the Alps: phylogenetic assessment and toxigenic potential. FEMS Microbiol Ecol. 92:fiw155.

Shen H, Niu Y, Xie P, Tao M, Yang X (2011). Morphological and physiological changes in *Microcystis aeruginosa* as a result of interactions with heterotrophic bacteria. Freshwater Biol. 56:1065–1080.

Shih PM, Wu D, Latifi A, Axen SD, Fewer DP, Talla E et al. (2013). Improving the coverage of the cyanobacterial phylum using diversity-driven genome sequencing. Proc Natl Acad Sci USA. 110:1053–1058.

Śliwińska-Wilczewska S, Maculewicz J, Barreiro Felpeto A, Latała A (2018). Allelopathic and bloom-forming picocyanobacteria in a changing world. Toxins. 10:48.

Soo RM, Hemp J, Parks DH, Fischer WW, Hugenholtz P (2017). On the origins of oxygenic photosynthesis and aerobic respiration in Cyanobacteria. Science. 355:1436–1440.

Stanier RY, Sistrom WR, Hansen TA, Whitton BA, Castenholz RW, Pfennig N et al. (1978). Proposal to place the nomenclature of the cyanobacteria (blue-green algae) under the rules of the international code of nomenclature of bacteria. Int J Syst Bacteriol. 28:335–336.

Stanier RY, Van Niel CB (1941). The main outlines of bacterial classification. J Bacteriol. 42:437–466.

Stomp M, Huisman J, Voros L, Pick FR, Laamanen MJ, Haverkamp THA et al. (2007). Colourful coexistence of red and green picocyanobacteria in lakes and seas. Ecol Lett. 10:290–298.

Strunecký O, Elster J, Komárek J (2011). Taxonomic revision of the freshwater cyanobacterium "*Phormidium*" *murrayi*= *Wilmottia murrayi*. Fottea. 11:57–71.

Strunecký O, Komárek J, Elster J (2012). Biogeography of *Phormidium autumnale* (Oscillatoriales, Cyanobacteria) in western and central Spitsbergen. Polish Polar Res. 33:369–382.

Strunecký O, Komárek J, Johansen J, Lukešová A, Elster J (2013). Molecular and morphological criteria for revision of the genus *Microcoleus* (Oscillatoriales, Cyanobacteria). J Phycol. 49:1167–1180.

Strunecký O, Komárek J, Smarda J (2014). *Kamptonema* (Microcoleaceae, Cyanobacteria), a new genus derived from the polyphyletic *Phormidium* on the basis of combined molecular and cytomorphological markers. Preslia. 86:193–208.

Suda S, Watanabe MM, Otsuka S, Mahakahant A, Yongmanitchai W, Nopartnaraporn N et al. (2002). Taxonomic revision of water-bloom-forming species of oscillatoroid cyanobacteria. Int J Syst Evol Microbiol. 52:1577–1595.

Takaichi S (2011). Carotenoids in algae: distributions, biosyntheses and functions. Mar Drugs. 9:1101–1118.

Tandeau de Marsac N (1977). Occurrence and nature of chromatic adaption in cyanobacteria. J Bacteriol. 130:82–91.

Tandeau de Marsac N (2003). Phycobiliproteins and phycobilisomes: the early observations. Photosynth Res 76:197–205.

Tomitani A, Knoll AH, Cavanaugh CM, Ohno T (2006). The evolutionary diversification of cyanobacteria: molecular-phylogenetic and paleontological perspectives. Proc Natl Acad Sci USA. 103:5442–5447.

Turland NJ, Wiersema JH, Barrie FR, Greuter W, Hawksworth D, Herendeen PS *et al.*, editors (2018) International Code of Nomenclature for algae, fungi, and plants (Shenzhen Code) adopted by the Nineteenth International Botanical Congress Shenzhen, China, July 2017. Koeltz Botanical Books, Gashütten. https://www.iapt-taxon.org/nomen/main.php

van Gremberghe I, Leliaert F, Mergeay J, Vanormelingen P, Van der Gucht K, Debeer AE et al. (2011). Lack of phylogeographic structure in the freshwater cyanobacterium *Microcystis aeruginosa* suggests global dispersal. PLoS One. 6:e19561.

Vandamme P, Pot B, Gillis M, De Vos P, Kersters K, Swings J (1996). Polyphasic taxonomy, a consensus approach to bacterial systematics. Microbiol Rev. 60:407–438.

Vieira JM, Azevedo MT, Azevedo SMFO, Honda RY, Correa B (2003). Microcystin production by *Radiocystis fernandoi* (Chroococcales, Cyanobacteria) isolated from a drinking water reservoir in the city of Belem, PA,) Brazilian Amazonia region. Toxicon. 42:709–713.

Vijayakumar S, Menakha M (2015). Pharmaceutical applications of cyanobacteria – a review. J Acute Med. 5:15–23.

Wacklin P, Hoffmann L, Komárek J (2009). Nomenclatural validation of the genetically revised cyanobacterial genus *Dolichospermum* (Ralfs ex Bornet et Flahault) comb. nova. Fottea. 9:59–64.
Walsby AE (1994). Gas vesicles. Microbiol Rev. 58:94–144.
Wang W, Shen H, Shi P, Chen J, Ni L, Xie P (2016). Experimental evidence for the role of heterotrophic bacteria in the formation of *Microcystis* colonies. J Appl Phycol. 28:1111–1123.
Watanabe M, Semchonok DA, Webber-Birungi MT, Ehira S, Kondo K, Narikawa R et al. (2014). Attachment of phycobilisomes in an antenna–photosystem I supercomplex of cyanobacteria. Proc Natl Acad Sci USA. 111:2512–2517.
Welker M, von Döhren H (2006). Cyanobacterial peptides - Nature's own combinatorial biosynthesis. FEMS Microbiol Rev. 30:530–563.
Whitton BA, Potts M (2000). The ecology of cyanobacteria: their diversity in time and space. Heidelberg: Springer.
Wilmotte A (1994). Molecular evolution and taxonomy of the cyanobacteria. In: Bryant DA, editors: The molecular biology of cyanobacteria. Dordrecht: Kluwer Academic Publishers:1–25.
Wilmotte A, Laughinghouse IV DH, Capelli C, Rippka R, Salmaso N (2017). Taxonomic identification of cyanobacteria by a polyphasic approach. In: Kurmayer R, Sivonen K, Wilmotte A et al., editors: Molecular tools for the detection and quantification of toxigenic cyanobacteria. Chichester: John Wiley & Sons:79–134.
Woese C (1987). Bacterial evolution. Microbiol Rev. 51:221–271.
Woese C, Kandler O, Wheelis ML (1990). Towards a natural system of organisms: proposals for the domains Archaeea, Bacteria, and Eucarya. Proc Natl Acad Sci USA. 87:4576–4579.
Zapomělová E, Hrouzek P, Řezanka T, Jezberová J, Řeháková K, Hisem D et al. (2011). Polyphasic characterization of *Dolichospermum* spp. and *Sphaerospermopsis* spp. (Nostocales, Cyanobacteria): morphology, 16S rRNA gene sequences and fatty acid and secondary metabolite profiles. J Phycol 47:1152–1163.
Zapomělová E, Skácelová O, Pumann P, Kopp R, Janeček E (2012). Biogeographically interesting planktonic Nostocales (Cyanobacteria) in the Czech Republic and their polyphasic evaluation resulting in taxonomic revisions of *Anabaena bergii* Ostenfeld 1908 (*Chrysosporum* gen. nov.) and *A. tenericaulis* Nygaard 1949 (*Dolichospermum tenericaule* comb. nova). Hydrobiologia. 698:353–365.
Zwart G, Kamst-van Agterveld MP, Werff-Staverman I, Hagen F, Hoogveld HL, Gons HJ (2005). Molecular characterization of cyanobacterial diversity in a shallow eutrophic lake. Environ Microbiol. 7:365–377.

Chapter 4

Understanding the occurrence of cyanobacteria and cyanotoxins

Bastiaan W. Ibelings, Rainer Kurmayer, Sandra M. F. O. Azevedo, Susanna A. Wood, Ingrid Chorus, and Martin Welker

CONTENTS

INTRODUCTION

Cyanobacteria are found in almost every aquatic ecosystem. Yet their presence does not automatically pose a risk to public health. Whether or not toxins reach health-relevant concentrations depends on the taxonomic (and genotypic or clonal) composition of the phytoplankton as well as on the cyanobacterial biomass. Clearly, the risks of elevated biomass, loosely named blooms, are higher under eutrophic conditions. However, certain cyanobacteria, for example, *Planktothrix rubescens*, tend to decrease under eutrophic conditions, while others like species of *Microcystis* may still reach hazardous levels under much reduced nutrient concentrations if a relatively low biomass accumulates as scums near the lake shore. Hence, to assume a very close link between cyanobacterial proliferation and eutrophication is too restrictive for a good understanding of toxin risks. Therefore, for the assessment and management of cyanobacterial hazards to human health, a basic understanding of their properties, their behaviour in natural ecosystems and the environmental conditions that support their excessive growth is important. This chapter provides a general introduction to key traits of cyanobacteria that support their proliferation in aquatic ecosystems. It focuses on cyanobacteria thriving suspended in the water of lakes and reservoirs, that is, those with a planktonic way of life. It also gives information on conditions potentially promoting the growth of benthic cyanobacteria, that is, those growing on surfaces such as rocks, sediments or submersed vegetation. How to use this information for controlling and managing cyanobacterial proliferation is discussed in Chapter 8.

4.1 WHAT MAKES MANY PLANKTONIC CYANOBACTERIA DOMINANT?

To proliferate, cyanobacteria must be able to satisfy their demands for light energy and nutrients, particularly phosphorous (P) and nitrogen (N). They compete for these resources with other phytoplankton and with each other. Cyanobacteria have a number of specific traits that favour their dominance over algal phytoplankton as well as bloom formation in many lakes, rivers and oceans. Some traits are only found in a restricted subset of cyanobacteria, such as the capability to fix atmospheric nitrogen or the regulation of buoyancy.

4.1.1 Nutrient storage and nitrogen fixation

Cyanobacteria can sustain growth during temporal and spatial shortage of dissolved nutrients by storing them – particularly phosphorous (P) and to a lower extent also nitrogen (N) in amounts exceeding their current demands for cell division and growth (Li & Dittrich, 2019). They store surplus P as polyphosphate in the cell, sometimes enough for several cell divisions even when external supplies are depleted below the analytical detection levels.

As a consequence, one cell can multiply into 8 or even 16 cells without any further phosphorus uptake (Reynolds, 2006).

In addition, cyanobacteria can acquire nitrogen in different forms, such as nitrate, nitrite, and ammonium or urea. While under oxidised conditions, N occurs as nitrate and is used as such by cyanobacteria, they can rapidly take up reduced forms of N (ammonium and urea), for example, from sewage inflows or released from sediments or by mussels (Gobler et al., 2016). Furthermore, some cyanobacteria, for instance, the genera *Aphanizomenon* or *Dolichospermum*, can utilise atmospheric N_2, a capacity termed "diazotrophy". This is a distinct advantage for cyanobacteria compared to eukaryotic microalgae at times when other sources of nitrogen are in short supply (Oliver & Ganf, 2000). The widely occurring genera *Microcystis* and *Planktothrix* cannot fix atmospheric nitrogen, but they can benefit from a smaller biomass of co-occurring N-fixing taxa whose degrading cells may leak N into the surrounding water (Salk et al., 2018).

Nitrogen fixation (diazotrophy) in most taxa is located in specialised cells, the heterocytes (see Chapter 3). These cells do not perform photosynthesis since the nitrogenase enzyme complex that carries out nitrogen fixation is inactivated by oxygen released through photosynthesis. Heterocytes are generally easily recognised microscopically due to their different shape and pronounced cell wall, but may appear similar to vegetative cells in some species (Gallon, 2004). Cyanobacterial N_2 fixation can cause considerable N input into aquatic ecosystems and affect global geochemical cycles, as shown, for instance, for the marine cyanobacterium *Trichodesmium* in the Atlantic Ocean (Capone et al., 2005). Diazotrophy, however, requires high amounts of energy for the heterocytes to maintain anoxic conditions inside the cell while being in an oxygenated environment, and therefore, N_2 fixation is limited in turbid water.

Cyanobacteria are able to store excess N mainly in the form of cyanophycin (a copolymer of aspartate and arginine; Simon, 1971; Ziegler et al., 1998). Even species that cannot fix atmospheric N_2 may thus have competitive advantage against microalgae under conditions of low N concentrations (Li et al., 2001a).

4.1.2 Buoyancy, vertical migration, surface scums and metalimnetic layers

The formation of surface blooms or scums (see Box 4.1) is one of the most prominent visual expressions of cyanobacterial development. The cyanobacteria best known for the formation of surface blooms are colony- or filament-forming species of genera such as *Dolichospermum*, *Cuspidothrix*, *Aphanizomenon*, *Microcystis* and *Nodularia*, and under specific circumstances also species of *Planktothrix* and *Raphidiopsis (Cylindrospermopsis)*. Surface blooms may cover an area reaching from a few square metres up to hundreds of square kilometres, such as blooms of *Nodularia spumigena* that can cover large areas of the brackish Baltic Sea (Bianchi et al., 2000).

In order to grow, cyanobacteria, being photoautotrophic organisms, need to spend sufficient time where there is sufficient light. Many species of cyanobacteria can be buoyant through their gas vesicles (see Chapter 3), thus avoiding sedimentation into deeper layers with low or no light availability.

BOX 4.1: BLOOMS, SCUMS, MASS DEVELOPMENTS – SOME DEFINITIONS

In the literature on cyanobacteria, the use of terms describing high cell density varies widely. This book proposes and uses the following definitions (see also Fig. 4.1 and 4.2):

Bloom: high average phytoplankton (i.e., algae and/or cyanobacteria) cell density in a waterbody (in this book, usually referring to blooms of cyanobacteria), also referred to as "mass development" or "proliferation". Blooms can be visually recognised by low transparency and water colour.

Surface bloom: buoyant cyanobacteria accumulating near or at the surface forming visible streaks, sometimes discernible in remote sensing images. Surface bloom formation may occur even where average cyanobacterial cell density is low if vertical and horizontal concentrating mechanisms lead to pronounced cell accumulation at the surface. Surface blooms can occur at large scales covering tens of square kilometres in large lakes or oceans.

Scum or surface scum: massive accumulation of buoyant cyanobacteria at the water surface forming a cohesive layer, often the result of secondary horizontal concentration at the shore, which can reach up to several centimetres thickness. In very dense scums, the surface of the scum can become nearly dry, and this may lead to massive cell lysis, colouring the water blue through the release of phycocyanin. Scums can also be formed by other phytoplankton, like Euglenophytes, but these are generally only a few millimetres thick.

Under eutrophic conditions, water is turbid, rendering light gradients in the water column quite steep and photosynthesis restricted to the uppermost water layer of a few metres or even less than 1 m. The water layer in which light for photosynthesis is available is termed "euphotic zone". If waterbody mixing is deeper than the euphotic zone, phytoplankton cells spend only a limited part of the day in layers with sufficient light for photosynthesis, and light may then limit growth rates. During calm periods, vertical mixing ceases and thermal (micro)stratification develops in waterbodies (Box 4.2). Under these conditions, nonbuoyant plankton sinks out of the euphotic zone, but buoyant cyanobacteria float up, into the near-surface layer, where the cells spend all or most of the daytime in the light. This is typical for colonial cyanobacteria, because due to their large size, they develop high flotation velocities, as can be observed when a water sample is placed in a glass cylinder (Humphries & Lyne, 1988).

BOX 4.2: THERMAL STRATIFICATION

Many lakes develop temperature gradients over depth. Thermal stratification develops when two water layers of differing average temperature are stably separated. This happens because the specific density of water is the highest at 4 °C and decreases at higher as well as at lower temperatures. In spring, water temperature increases at the surface, and above a certain temperature difference onwards, the warmer, less dense water in the upper part of the water column no longer mixes with colder, denser water in deeper parts. Further heating of the upper layer of water eventually leads to a sharp separation of warm and cold water. The layer where the temperature difference over depth is most pronounced is called the **thermocline**. For practical purposes, stratification is defined as a temperature difference of more than one degree centigrade over one metre of water depth. The upper, warm layer is called **epilimnion**, the deeper, cooler layer the **hypolimnion** and the layer around the thermocline **metalimnion**. The more pronounced the temperature gradient, the more stable the stratification, and more energy, for example, wind, is needed to overcome it by deep mixing. Thus, once stable stratification is established, cooling of surface water at night or periods of cold weather will cause some mixing of the surface layer, but not the total circulation of the waterbody. While thermal stratification distinctly separates these water layers, some exchange of water between the epilimnion and the hypolimnion may yet occur (e.g., through eddy diffusion), the extent of which depends on the stability of stratification. This in turn is affected by the ratio of the waterbody's surface area to its depth, wind exposition and frequency of storms, for example.

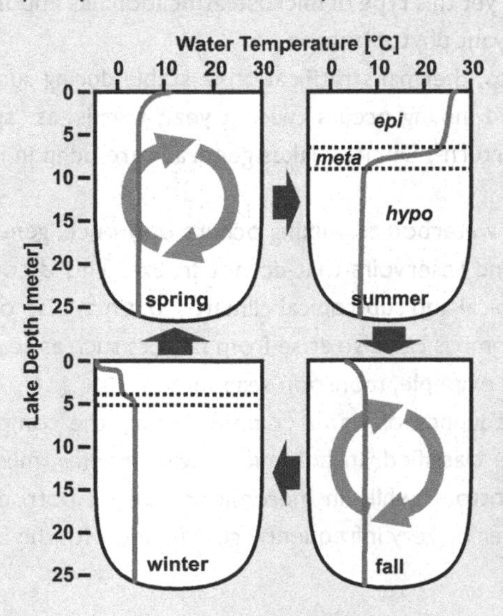

At the end of the summer, the epilimnion successively becomes cooler and once temperatures – and in consequence water densities – of the epi- and hypolimnion have converged, complete mixing of the water column occurs. This process is called mixis or overturn. Further cooling at the surface can lead to an inverse stratification during winter, in particular when an ice-cover is present, because water density decreases at temperatures below 4 °C.

Stratification not only affects water temperature, but also many biological, chemical and physical processes. Particles such as planktonic organisms (dead or alive) and excreta of fish or zooplankton can settle and reach the hypolimnion and the sediment where they decay. This process consumes oxygen and releases minerals, including phosphorus and nitrogen, from the decaying material settling or already deposited on the sediment surface. As there is little exchange between epi- and hypolimnion in a stratified waterbody, substances such as phosphorus and nitrogen accumulate in the hypolimnion. Therefore, thermal stratification also causes chemical and biological stratification, frequently leading to an anoxic hypolimnion that is rich in dissolved inorganic nutrients.

Depending on the stability of thermal stratification and frequency of mixing, several types of lakes are distinguished.

In shallow polymictic waterbodies, stratification occurs irregularly but frequently while lasting only for hours, days or at most for weeks, depending on meteorological conditions such as air temperature, wind speed and direction. Temperature gradients generally are not pronounced, and hence, stratification is not stable, yet this type of microstratification has important effects on, for instance, buoyant phytoplankton.

In dimictic lakes, thermal stratification is stable during summer as well as during winter, and mixing occurs twice a year, that is, as "spring overturn" and "autumn overturn". Dimictic lakes generally are deep in relation to their surface area.

In monomictic waterbodies, mixing occurs only once, generally during the winter, in lakes and reservoirs that do not freeze. This is common for large reservoirs in tropical and subtropical climates, but may also occur in temperate regions. Monomixis can also arise from factors such as seasonal precipitation patterns, for example, monsoon seasons.

Besides the frequency of water column mixing, the completeness of the mixing process is classified. In holomictic waterbodies, mixing is from the surface to the bottom, while in meromictic ones, a bottom layer remains excluded and never or very infrequently gets mixed with the overlaying water layer. This occurs in lakes with a particular morphometry like deep crater lakes that may be classified as dimictic and meromictic.

One of the ongoing and forecasted effects of climate warming is shifts between mixing regimes, for example, from poly- or dimictic to monomictic or, in some situations, less stable stratification if the frequency of storms increases. Given the importance of mixing regimes for lake ecosystem functioning, the consequences can be far reaching.

Buoyancy regulation – that is, a variation in the buoyancy state – depends on the availability of resources such as carbon, nitrogen and phosphorous (e.g., Klemer et al., 1982; Konopka et al., 1987), but most importantly, buoyancy is regulated by light. Cyanobacteria produce more gas vesicles under low irradiance, thus increasing their buoyancy. Under high irradiance, gas vesicles may be diluted by growth, or – after extended light exposure – collapse under high turgor pressure (Kinsman et al., 1991). In addition, cyanobacteria exposed to light produce carbohydrates, the excess of which may be stored as glycogen that acts as ballast, so that buoyancy can decrease to a point where cells start to sink. In darker, deeper layers, the cells consume their glycogen storage as energy source, and thus, they regain buoyancy. This may result in diel changes in buoyancy and vertical migration, with populations being maximally buoyant during the night and a percentage of the cells losing buoyancy during the day (Ibelings et al., 1991; Rabouille & Salençon, 2005; Medrano et al., 2013).

Since the vertical migration of *Microcystis* can span large amplitudes, *Microcystis* can also proliferate in deep lakes. Typically, however, *Microcystis* becomes dominant in eutrophic lakes of intermediate depth, which are sufficiently stratified to allow them to control buoyancy and thus to position themselves vertically. In these ecosystems, *Microcystis* colonies are among the best-adapted phytoplankton to overcome mixing forces (provided these are relatively weak) – allowing the colonies to disentrain from turbulence – and gain a competitive advantage by floating upwards during periods of increased stability (Reynolds, 1994; Dokulil & Teubner, 2000).

In waterbodies with thermal stratification, buoyancy regulation can lead to different accumulation patterns of cyanobacterial biomass and consequently of cyanotoxins, depending on the taxa involved (Figure 4.1, Figure 4.6):

- formation of surface blooms or scums, typical for blooms of *Microcystis*, *Dolichospermum* and *Aphanizomenon*. Formation of a cohesive scum is generally preceded by the occurrence of visible streaks at the surface. Once congregated at the surface, the cells can be further accumulated by wind-drift at leeward sites;

- largely homogeneous distribution of cells in the water column with scum formation occurring only exceptionally under the conditions of extreme water column stability. A typical example is *Planktothrix agardhii*, which only rarely creates surface blooms or scums, but may lead to pronounced greenish turbidity of homogeneously distributed filaments. Further examples include *Limnothrix* and *Pseudanabaena* as well as other taxa that have not been shown to produce known toxins;
- the accumulation of cells in a deep layer (deep chlorophyll maximum), either between upper warm and deeper cold water (i.e., the "thermocline" or "metalimnion"; see Box 4.2) or at low light intensity. Typical examples include the red-coloured *Planktothrix rubescens* in deep, stratified lakes and reservoirs. However, green-pigmented cyanobacteria (*Aphanizomenon flosaquae*, *Dolichospermum*) have also been reported to form metalimnetic layers that may last for several months (Konopka, 1989). In stratified subtropical reservoirs, *Raphidiopsis (Cylindrospermopsis) raciborskii* has been found in deep layers with low light intensity (see section 4.5). Occasionally, such cyanobacteria may leave their position in the metalimnion and accumulate at the surface, in the case of *P. rubescens*, leading to purple surface blooms. Seasonal or storm-induced mixing may also carry them into the upper layers.

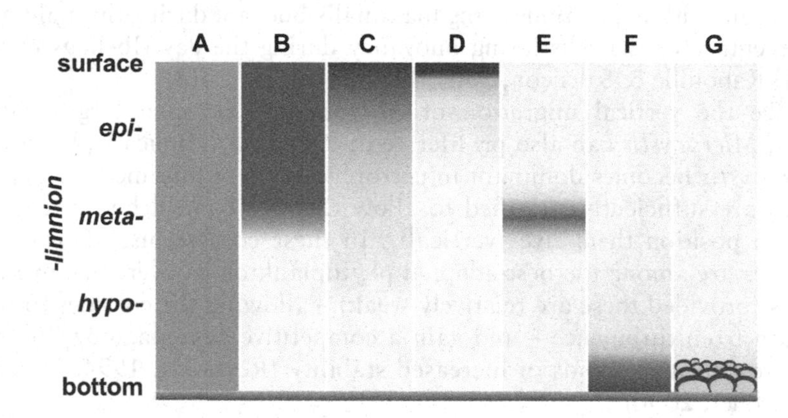

Figure 4.1 Distribution patterns of cyanobacteria (cyanotoxins) in the water column. A: homogeneous distribution, B: homogeneous distribution in the epilimnion of thermally stratified waterbodies, C: gradual concentration of buoyant cyanobacteria towards the surface in a stable water column, D: surface bloom of buoyant cyanobacteria, E: accumulation in a metalimnetic layer in stratified waterbodies, F: accumulation in bottom layers through sedimentation, G: benthic cyanobacteria growing on hard substrate. Transitions between all individual patterns are possible.

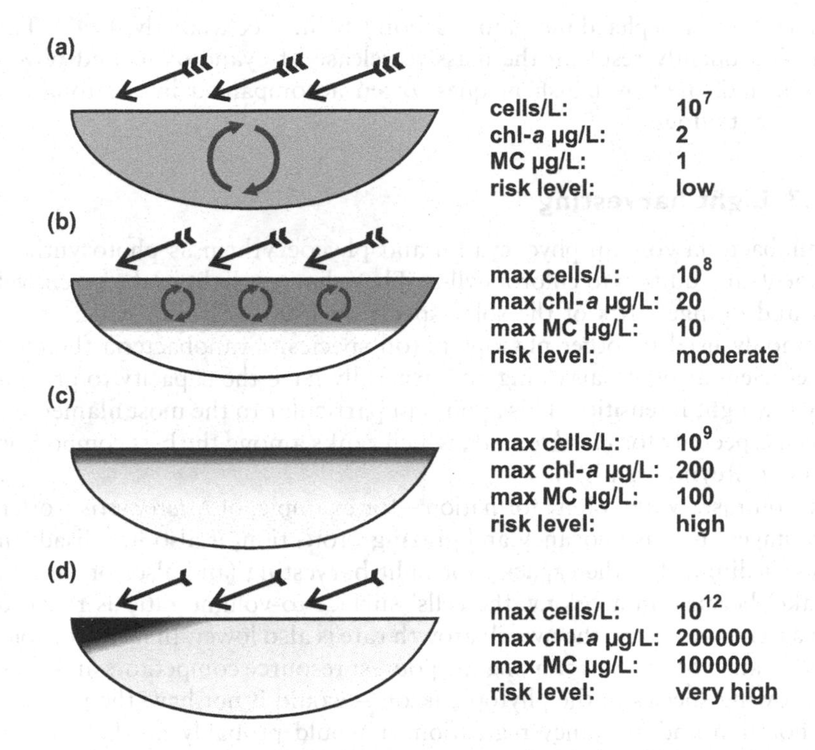

Figure 4.2 Schematic illustration of the formation of a surface bloom and scum of buoyant planktonic cyanobacteria like *Microcystis* spp. The maximum toxin concentrations under given conditions are estimated assuming average cell quota of 100 fg microcystin (MC) per cell (see Box 4.8). (a): strong winds lead to a mixing of the entire waterbody. (b): moderate wind speed allows for (temporary) thermal stratification and buoyant cyanobacteria to accumulate in the epilimnion. (c): Without wind, buoyant cyanobacteria float up and form surface blooms. (d): low wind speed causes the displacement of surface blooms and formation of scums at downwind sites. Note that cell numbers, chlorophyll-*a* (chl-*a*) and MC concentrations are only crude estimates to illustrate the order of magnitude of possible spatial heterogeneities and must not be taken as reference values.

At times, surface blooms can develop into scums when cyanobacteria accumulated at the water surface are further concentrated by horizontal drift, generally at downwind sites (Figure 4.2). Scum formation is typically restricted to parts of a waterbody only, but strongly affects water quality in terms of odour, appearance and sometimes extremely high concentrations of cyanotoxins. When scums are not dissolved after a short time, for example, by wind or wave action, massive cell lysis may set in as a result of unadapted cells being exposed to full sunlight and additional stress factors such as high temperature or depleted inorganic carbon (Ibelings & Maberly, 1998). This can consequently result in the massive release of cyanotoxins and phycocyanin, indicated by bluish plaques, often accompanied by a strong and unpleasant odour.

4.1.3 Light harvesting

Cyanobacteria contain phycocyanin and phycoerythrin as photosynthetic pigments in addition to chlorophyll-*a*. These harvest light in the green, yellow and orange parts of the solar spectrum (500–650 nm), which is not commonly used by other phytoplankton species. Cyanobacteria therefore are efficient at light harvesting and generally have the capacity to grow at very low light intensities. This applies in particular to the most filamentous forms, especially for *Planktothrix*, which ranks among the best competitors for light (Reynolds, 1997).

In contrast, while colony formation – for example, of *Microcystis* – offers advantages such as buoyancy and grazing protection, it also has disadvantages: it diminishes the capacity for light harvesting (and also for nutrient uptake) because in a colony, the cells' surface-to-volume ratio is reduced, and as a consequence, the overall growth rate is also lower. In fact, Reynolds (1997) ranks *Microcystis* among the poorest resource competitors and slowest growing species in the phytoplankton – would it not have the potential for flotation and buoyancy regulation, it would probably not be a strong competitor in the phytoplankton community (Ibelings et al., 1994).

4.1.4 Carbon concentrating mechanisms

Cyanobacteria have a carbon concentrating mechanism (CCM), which allows the CO_2-fixing enzyme RUBISCO to operate efficiently. The first part of the CCM consists of various inorganic carbon uptake systems, two for CO_2 and three for bicarbonate. The carbon, assimilated carbon in the form of bicarbonate, is transported to specific cellular compartments containing the enzyme carbonic anhydrase, known as carboxysomes. Carbonic anhydrase transforms bicarbonate (back) to CO_2, raising it to levels where RUBISCO can perform optimally. Cyanobacterial taxa vary in the combination of carbon uptake systems present in the cell, and this allows them to adapt to environments that differ in the availability of inorganic carbon (Price et al., 2007; Sandrini et al., 2014).

4.1.5 Resistance to grazing and other losses

Cyanobacteria seem well equipped to minimise two loss factors that play a major role in general phytoplankton dynamics: sedimentation and grazing by zooplankton. Resistance to grazing is a widely studied and discussed subject, in part because it may form an important obstacle for the success of biomanipulation (see section 8.7.1).

Toxicity, size of filaments or colonies, and poor nutritional quality may prohibit an efficient grazing on cyanobacteria by zooplankton (Kurmayer & Jüttner, 1999), but how well cyanobacteria are protected against grazing by zooplankton is still under considerable scientific debate, with grazing experiments and field studies showing different outcomes (Sarnelle, 1993; Gragnani et al., 1999). For instance, it was shown that *Daphnia carinata* fed on toxic cyanobacteria, including *Microcystis* colonies (Matveev et al., 1994), while in the microcosm experiments of Mohamed (2001), daphnids preferred green algae and diatoms over toxic *Microcystis*. Some of the initial studies demonstrating cyanobacterial toxicity against grazers seem to have been based upon a restricted selection of zooplankton taxa and cyanobacterial strains (Wilson et al., 2006a), and grazers that originate from lakes with a history of cyanobacteria generally show stronger tolerance against their toxins (Ger et al., 2016). Lemaire et al. (2012) studied how the co-evolutionary history of grazers and cyanobacteria results in specific grazer genotype and cyanobacterium genotype interactions that provide a further explanation for the variable outcome of research on grazing of cyanobacteria. Furthermore, grazing by zooplankton other than cladocerans may lead to very different observations: some rotifers, for example, are able to feed on filamentous cyanobacteria and are adapted to the perceived low food quality of cyanobacteria (Burian et al., 2014).

Likewise, the role of mussels in decimating cyanobacterial populations is still under debate: (toxic) cyanobacterial blooms have reappeared in Lakes Erie and Huron (USA), and this is partially interpreted as a result of zebra mussel (*Dreissena polymorpha*) invasion (Vanderploeg et al., 2001) and their selective rejection of *Microcystis* colonies as food. Dionisio Pires et al. (2005) and Baker et al. (1998) showed a positive selection for cyanobacteria (and other phytoplankton), indicating that grazing effects of mussels on cyanobacteria are not yet well understood (Raikow et al., 2004). Phytoplanktivorous fish are largely absent from temperate lakes, but in (sub)tropical regions, fish such as silver carp and bighead carp have been shown to filter feed on cyanobacteria (Zhang et al., 2008), thus potentially controlling *Microcystis* blooms (reviewed in Triest et al., 2016).

Besides grazing, losses through infection by fungal parasites, phages and heterotrophic bacteria have been recognised as potentially important drivers of cyanobacterial population dynamics (Box 4.3; Steenhauer et al., 2016; Yoshida-Takashima et al., 2012; Rohrlack et al., 2013; Van Wichelen et al., 2010). Respective losses may not affect the entire populations but only selected genotypes, thus contributing to the clonal dynamics in cyanobacterial blooms (Van Wichelen et al., 2016; Agha et al., 2018). Recognition, and in particular, quantification of losses through microbial interactions in waterbodies, is still only in the beginnings.

BOX 4.3: EMERGING INSIGHTS INTO THE ROLE OF CYANOPHAGES

Helena L. Pound, Steven W. Wilhelm

While viruses are well documented in marine systems and virus-induced cell lysis contributes to the conversion of microbial biomass into dissolved organic matter, less is known about their role in freshwaters, particularly for cyanobacteria. Yoshida et al. (2006) isolated and sequenced the first phage capable of lysing *Microcystis aeruginosa*, proposing that it might act as a controlling agent in seasonal blooms. While the effect of phage on growth, proliferation, and biomass of *Microcystis* populations has not been fully elucidated, community-sequencing studies have provided further insight. A metatranscriptomic analysis of a 2014 *Microcystis* bloom in Lake Erie showed an increased frequency of *Microcystis* phage transcripts at the termination of the bloom, with approximately one virus marker transcript for every host marker (Steffen et al., 2017), implying a massive population-level infection. This lytic infection was associated with a spike in detected microcystin concentrations in the raw water of the city of Toledo's drinking-water supply, reaching up to >5 μg/L. The working hypothesis arising from the observation is that virus-infected cells lysed and released cell-associated toxin into the water.

In contrast, a similar analysis of a *Microcystis* bloom in Taihu (Lake Tai, China) in 2014 showed a different effect associated with the *Microcystis* phage. That study observed the formation of lysogens with the association of the phage with the host genome (Stough et al., 2017). In this manner, *Microcystis* hosts may have resisted subsequent infections by other lytic phage, allowing for longer bloom maintenance. Indeed, Kuno et al. (2012) found that repeated exposure to a high diversity of phages leaves traces in the genomes of *Microcystis* strains, including molecular defence mechanisms. These studies provide evidence that viral infections can be relevant both in the dynamics of *Microcystis* blooms and in the release of otherwise cell-bound toxins.

4.1.6 Overwintering strategies

In (sub)tropical regions, cyanobacteria can produce perennial blooms, whereas in temperate ones, they are often subject to pronounced seasonal changes in environmental conditions. Phytoplankton populations – including cyanobacteria – usually decline in winter because a decrease in daily insolation and increased mixing depth minimise light availability for photosynthesis. In the next growing season, the size of the inoculum of overwintering cells can be decisive for the outcome of competition between phytoplankton species for light and nutrients, and overwintering strategies are therefore an important trait for attaining high levels of biomass and/or dominance in the phytoplankton.

Microcystis typically blooms in (late) summer, and its biomass decreases in autumn when growth conditions become less favourable. Part of the colonies can sink to the lake bottom (Reynolds et al., 1981; Takamura et al., 1984). Verspagen et al. (2004) demonstrated that viable colonies, capable of photosynthesis, were present in large numbers in and on the sediment not only in winter but throughout the year. Several studies (Takamura et al., 1984; Boström et al., 1989) found that the total amount of *Microcystis* in the sediment can be much higher than in the water column, even during *Microcystis* blooms. Preston et al. (1980) observed that recruitment of *Microcystis* from the sediment to the water takes place before the establishment of the summer bloom, so that this recruitment may be a crucial factor that initiates the bloom (Ihle et al., 2005).

Filamentous cyanobacteria like *Planktothrix agardhii* seem to follow a different strategy. Even in temperate climates, they may prevail or even bloom throughout the year in eutrophic systems. In some situations, they prevent competing phytoplankton from growing by maintaining the system turbid and light availability too low for microalgae to compete efficiently (Scheffer et al., 1997; Ibelings et al., 2007).

The annual cycle of *Planktothrix rubescens* in Lake Zürich studied by Walsby & Schanz (2002) shows a specific time pattern for this species: deep winter mixing of the lake entrained the metalimnetic population of *P. rubescens*. About half of the filaments survived this winter period, whereas others lost their gas vesicles because of the hydrostatic pressure resulting from deep winter mixing. After annual stratification in spring, the surviving filaments resumed their position in the metalimnion.

A number of cyanobacteria produce specific resting cells called **akinetes** that sink to the sediment where they can survive for extended periods of time (see Chapter 3). These include, in particular, the N_2-fixing taxa *Anabaena*, *Dolichospermum*, *Aphanizomenon* and *Raphidiopsis*. Recruitment and germination of akinetes from the sediment contributes to bloom onset, and prolonged benthic recruitment may strongly promote the presence of respective species (Karlsson-Elfgren et al., 2003). Several factors, especially light and temperature, determine their germination (Karlsson-Elfgren et al., 2004; Wiedner et al., 2007). Cirés et al. (2013) show how blooms of *Dolichospermum circinale* and *D. flosaquae* are both initiated and maintained in the Murray River (Australia) by the germination of akinetes that are present in the sediment, although vegetative cells also overwinter in the water.

4.2 WHERE DO WHICH CYANOBACTERIA OCCUR?

Cyanobacteria causing toxin concentrations of concern for human health are usually planktonic, that is, suspended in the water. However, as mentioned above, some forms grow on surfaces. As discussed in Chapter 3, "benthic" or "epiphytic" cyanobacteria do so continuously until they are physically detached and dislocated to the surface or beaches. "Tychoplanktonic" cyanobacteria inhabit surfaces, including those of submersed aquatic plants, and may at times also be suspended in the water.

4.2.1 Planktonic cyanobacteria

As a taxonomic group, cyanobacteria are highly successful in colonising a wide variety of habitats – with numerous special adaptations to extreme environmental conditions, for example, resistance to high concentrations of copper and zinc, tolerance of low oxygen conditions and of free sulphide, high tolerance of UV-B and UV-C as well as of high temperature (see Whitton, 2012).

Due to their small size and the extremely high abundance, it is frequently argued that microbial species are globally ubiquitous, growing and multiplying where they find suitable conditions ("Baas Becking hypothesis – everything is everywhere but the environment selects"; Finlay, 2002). Accordingly, geographic barriers would not restrict the global distribution of particular taxa, and indeed, cyanobacterial species are cosmopolitan in general and not restricted to geographic regions like, for example, many higher plants (Pridmore & Etheredge, 1987). Later observations, however, suggest that geographic variation in microbial diversity is apparently more common than originally expected (Martiny et al., 2006), which could well be because the environment – which clearly varies between regions – selects indeed, as postulated in the second part of the Baas Becking hypothesis. A (potentially) global distribution of a particular species does, however, not imply genetic homogeneity across the globe. Regional and local differences in genotype occurrence can cause differences in the distribution of (toxic) secondary metabolites. For example, as strains of *Raphidiopsis raciborskii* differ genetically between geographic regions (see below), cylindrospermopsin (CYN) production was found only for Australian strains, while not for European or South American ones (Neilan et al., 2003; Piccini et al., 2011).

While it is impossible to prove that a specific taxon is absent from a given location, documented observations of occurrence provide some indication of which species to expect or not to expect under which conditions. Furthermore, in analogy to plant-sociological groupings for terrestrial plants, phytoplankton, including cyanobacteria, can be grouped in associations of species that typically occur under specific conditions (Reynolds et al., 2002, updated by Padisák et al., 2009). Box 4.4 shows how a grouping on the basis of key functional traits of cyanobacteria helps to reduce – taxonomic – complexity and to understand why certain cyanobacteria occur under certain conditions. Moreover, the close association between cyanobacterial traits and preferred environmental conditions can provide guidance in selecting the most promising measures for successful control of cyanobacterial blooms (Mantzouki et al., 2016).

BOX 4.4: CYANOBACTERIAL FUNCTIONAL GROUPS BASED ON THE CLASSIFICATION BY REYNOLDS (2006)

Evanthia Mantzouki

A functional group is defined by containing species with similar morphological and physiological traits, and ecological functioning (see Padisák et al., 2009 and Salmaso et al., 2015). Reynolds (2006) clustered species on the basis of the typical habitats in which they occur, morphological traits such as shape and dimension or specialised structures such as gas vesicles and heterocytes, as well as functional traits that are linked to resource acquisition, optimum growth temperatures and toxin production. From the extensive list given by Reynolds (2006), the table below (modified after Mantzouki et al., 2016) selects the functional groups with cyanobacteria, listing typical representative taxa and highlighting environmental tolerances and key sensitivities. This may aid in finding suitable management actions to control cyanobacteria. The classification is simplified compared to Reynolds (2006) and Padisák et al. (2009) by combining groups that are found in similar habitats and, more importantly, require similar management actions, that is, the S1/S2, H1/H2 and LO/LM groups. Notable differences in the various preferred habitats of the combined groups are as follows: S2 species prefer warmer mixed and alkaline waters compared to those in group S1; H2 taxa are found in mesotrophic lakes with high light conditions, and although H1 taxa are found in more eutrophic lakes, their nitrogen content is relatively low; LO taxa are found in medium to large lakes, while LM species occur in small to medium-sized lakes. Note that the most frequently occurring, typical taxa are given in bold.

			Traits						
Group	*Habitat*	*Typical Taxa*	*Size/Shape/ Coloniality*	*Resource Acquisition*	*Optimum Temp.*	*Reference*	*Toxins*	*Tolerances*	*Sensitivities – Management Actions*
S_1/S_2	Shallow Turbid mixed layers	***Planktothrix agardhii*** *Limnothrix redekei* *Pseudanabaena limnetica* *Planktolyngbya contorta* *Arthrospira* *Raphidiopsis*	3–>1000 µm/ solitary filaments	Efficient light harvesting	20–27.5 °C	Foy et al. (1976) Lürling et al. (2013)	MCs, CYNs	Light deficiency	Short residence time
S_N	Warm mixed layers	***Raphidiopsis*** *Anabaena minutissima*	5–500 µm/ solitary filaments	Nitrogen fixation Akinetes	29–35 °C	Briand et al. (2004) Chonudomkul et al. (2004) Saker & Griffiths (2000) Mehnert et al. (2010)	MCs, CYNs, ATXs, STXs	Light and N deficiency	Short residence time
K	Shallow lakes or epilimnion of deep lakes Eutrophic	***Aphanocapsa*** *Aphanothece* *Cyanodictyon*	0.2–2 µm/ spherical/ colonies	Colonial	No data		MCs	High pH	Deep mixing

(Continued)

			Traits						
Group	*Habitat*	*Typical Taxa*	*Size/Shape/ Coloniality*	*Resource Acquisition*	*Optimum Temp.*	*Reference*	*Toxins*	*Tolerances*	*Sensitivities – Management Actions*
H_1/H_2	Oligo- to eutrophic Shallow to deep	***Dolichospermum flos-aquae***, *D. spiroides*, *D. circinale* *A. flosaquae* *D. lemmermanni* *Gloeotrichia echinulata*	Solitary filaments	N fixation Buoyancy regulation P storage Akinetes	27–39 °C	Uehlinger (1981) Novak & Brune (1985)	MCs, CYNs, ATXs, STXs	N and C deficiency	Light deficiency – mixing P deficiency – nutrient decrease
L_O/L_M	Oligo- to eutrophic Shallow to deep Small to large	***Microcystis*** *Woronichinia* *Merismopedia*	2–>200 μm/ spherical/ colonies	Buoyancy regulation	28–34 °C	Nalewajko & Murphy (2001) Imai et al. (2009) Thomas & Litchman (2016)	MCs	C deficiency stratification	Light deficiency – mixing
M	Small Eutrophic Intermittent mixing	***Microcystis wesenbergii*** *Sphaerocavum brasiliensis*	2–>200 μm/ spherical/ colonies	Buoyancy regulation	20–29 °C	Imai et al. (2009) Otsuka et al. (1999)	MCs	High insolation	Short residence time – flushing Light deficiency – mixing
R	Mesotrophic Stratified metalimnion	***Planktothrix rubescens*** *P. mougeotii*, *P. limosa* *Planktolyngbya*	3–>1000 μm/ solitary filaments	Buoyancy regulation Efficient light harvesting	15–23 °C	Oberhaus et al. (2007) Bright & Walsby (2000) Zimmermann (1969)	MCs	Light deficiency Strong segregation	Physical instability – mixing

Prochlorococcus is probably the most abundant cyanobacterium, numerically dominating the oligotrophic areas of tropical and subtropical oceans, followed by marine *Synechococcus* and *Trichodesmium*. Planktonic freshwater cyanobacteria contribute only a relatively small percentage of the global cyanobacterial biomass as surface freshwaters are only a small fraction of the global surface water (Garcia-Pichel et al., 2003).

Microcystis species have a cosmopolitan distribution and have frequently been reported from lakes worldwide, from temperate to tropical zones of both hemispheres. They occur in a range of waterbodies, typically (though not exclusively) proliferating at temperatures above 15 °C and under mesotrophic to eutrophic conditions. While thermal stratification may give them a competitive advantage, *Microcystis* occurs both in thermally stratified lakes (generally deeper than 6 m) and in relatively shallow, polymictic waterbodies which show recurrent periods of re-stratification and mixing, allowing *Microcystis* to benefit from its buoyancy (Ibelings et al., 1991).

The widespread filamentous species *Planktothrix agardhii* is a typical and frequently reported species for shallow temperate lakes, particularly in the Northern Hemisphere (Suda et al., 2002). It tolerates a wide range of temperatures, and in some regions (e.g., lowland areas of the Netherlands and northern Germany), it perennially dominates (or dominated) shallow eutrophic and hypertrophic waterbodies for many years (Van Liere & Mur, 1980; Mur, 1983; Rücker et al., 1997). However, it is also reported from subtropical climatic regions in South America (Kruk et al., 2002), Australia (Baker & Humpage, 1994) and from the temperate climatic region of New Zealand (Pridmore & Etheredge, 1987). Because this species is a superior competitor under conditions of low light availability, *Planktothrix agardhii* – sometimes together with *Limnothrix* – generates a positive feedback loop, creating a "shaded" environment in which it can hardly be outcompeted by other phytoplankton (Scheffer et al., 1997).

The nitrogen-fixing genera, *Dolichospermum* and *Aphanizomenon*, typically occur in larger mesotrophic lakes (Dokulil & Teubner, 2000; Reynolds et al., 2002) promoted by surface warming, high light and phosphorus levels together with low nitrogen concentrations (see section 4.2.3). Species of *Dolichospermum* and *Anabaenopsis* have been reported from all five continents, occurring in temperate regions of the Northern and the Southern Hemisphere as well as in subtropical and tropical climatic regions. As the chemical reduction of atmospheric nitrogen is energy demanding, these taxa typically proliferate in environments with high light availability, that is, fairly clear water, rather than in turbid, highly eutrophic systems. Species of *Aphanizomenon* have been widely reported from temperate regions of the Northern Hemisphere (Konopka, 1989; Barker et al., 2000; Porat et al., 2001) as well as from subtropical regions (Porat et al., 2001), from Australia (Baker & Humpage, 1994) and from South

America (Kruk et al., 2002). Frequently, *Aphanizomenon flosaquae* occurs codominant together with *Microcystis* (Teubner et al., 1999).

Raphidiopsis raciborskii has originally been considered a tropical species, first described from Java (Gugger et al., 2005a). *R. raciborskii* was first reported in Europe during the 1930s (Padisák, 1997) and appeared to invade temperate regions of Europe during the 1990s (e.g., Dokulil & Mayer, 1996; Briand et al., 2004). However, newer genetic and ecophysiological analysis demonstrated that *R. raciborskii* shows positive net growth over a wide range of temperatures (20–35 °C) (Briand et al., 2004), and rather than spreading from warm refuge areas across the American and European continents, it shows distinct and geographically separate ecotypes (Gugger et al., 2005b; Haande et al., 2008), which also differ with respect to producing CYNs (Neilan et al., 2003; Piccini et al., 2011; Antunes et al., 2015).

Other potentially toxin-producing cyanobacteria show a more restricted pattern of geographic occurrence. Mass developments of the red-pigmented, filamentous *Planktothrix rubescens* have been reported frequently over several decades but largely from thermally stratified lakes and reservoirs across Europe (e.g., Skulberg & Skulberg, 1985; Feuillade, 1994; Barco et al., 2004; Gallina et al., 2017), North America (e.g., Edmondson & Litt, 1982; Nürnberg et al., 2003) and New Zealand (Pridmore & Etheredge, 1987). In these waterbodies, they typically inhabit the thin layer between warm surface water and the cold deep layer, called the metalimnion. Waterbody mixing can distribute the filaments throughout the whole water column, thus causing a distinctly visible reddish discoloration (Nürnberg et al., 2003). Their preferred habitat in the metalimnion may be the reason why *P. rubescens* populations are sensitive to eutrophication, since having enough light in these depths for growth depends on water being relatively clear in the epilimnion. In peri-alpine lakes such as Lake Hallwil (Switzerland) or Lake Bourget (France), *P. rubescens* (re)appeared as nutrient levels declined and water became clearer and disappeared again only after total phosphorus (TP) had further declined to very low concentrations (e.g., below 10 µg/L in Lake Bourget; Jacquet et al., 2014). In Lake Constance, at similarly low phosphorous levels, metalimnetic blooms of *P. rubescens* have reappeared since 2014, and in other lakes like Lake Zurich (Switzerland), it shows little response to decreases in TP concentrations (Schanz, 1994; Gammeter & Zimmermann, 2001; Posch et al., 2012).

Nodularia spumigena has been reported from marine and brackish waters – typically estuaries and coastal lagoons of Southern Australia (Francis, 1878; Jones et al., 1994) and less frequently in alkaline, brackish lakes (Bolch et al., 1999). In addition, it has been reported from brackish coastal lakes in New Zealand (Carmichael et al., 1988), South Africa and the North Sea (Nehring, 1993). *Nodularia spumigena* regularly forms blooms in the Baltic Sea (Bianchi et al., 2000). Notably, a freshwater strain of *Nodularia* originally isolated from a benthic mat in a thermal source also produced the nodularin (Beattie et al., 2000).

4.2.2 Benthic and tychoplanktonic cyanobacteria

Benthic cyanobacteria generally need a substrate to attach to, like sediment ("epipsammic"), pebbles or stones ("epilithic") or macrophytes ("epiphytic"), while "tychoplanktonic" includes phases of planktonic occurrence. Benthic cyanobacteria occur in a variety of freshwater and marine habitats, including wetlands, lakes, coral reefs and estuaries, hypersaline and geothermal ponds, streams and rivers (Scott & Marcarelli, 2012). Water transparency allowing light to penetrate to the bottom is a prerequisite for the growth. While planktonic cyanobacteria are well known to thrive in eutrophic conditions, benthic cyanobacterial proliferation is common in oligotrophic environments such as alpine lakes or headwater streams with sufficient substrate stability to enable cyanobacterial growth (Scott & Marcarelli, 2012). Where lakes and reservoirs become clearer in consequence of successful reduction of nutrient loads, benthic, epiphytic and/or tychoplanktonic planktonic cyanobacteria may replace planktonic ones, thus causing new exposure scenarios: while the health risk from cyanotoxins in the bulk water declines, a more localised risk may then arise in the vicinity of detached mats of benthic cyanobacteria or of macrophyte tufts containing cyanobacteria – either floating or beached. It is therefore important to observe the possible expansion of such taxa in the course of restoration efforts (Shams et al., 2015).

Some benthic species form mats which can grow on a wide range of substrates such as fine sediment, rocks, artificial substrates and aquatic plants. These cyanobacterial-dominated mats usually range from a few millimetres to several centimetres in thickness, although when environmental conditions are favourable, they can continue to grow and become much thicker. As benthic, cyanobacteria-dominated mats are increasingly being investigated for cyanotoxins worldwide, these are increasingly being found, particularly in wadeable streams, littoral zones of lakes and coastal lagoons (Fetscher et al., 2015; Cantoral Uriza et al., 2017). For both, freshwater and marine benthic taxa, occurrence is often highly variable on small spatial scales of metres or less. In shallow marine environments in tropical and subtropical zones, *Moorea* (formerly *Lyngbya*) and other benthic taxa potentially producing toxins can occur (see also section 2.6).

In freshwater, the dominant cyanobacterial taxa in the mats are usually filamentous Oscillatoriales, including *Oscillatoria*, *Microcoleus*, *Phormidium*, *Microseira*, *Moorea* (formerly *Lyngbya*), *Leptolyngbya*, *Tychonema*, *Calothrix* and *Schizothrix*. The Chroococcales *Aphanothece* and *Synechococcus* are also common components of the mats, with some reports of mats dominated by these genera, for example, in hot spring environments. Among the nitrogen-fixing Nostocales, the most frequently reported genera are *Anabaena*, *Scytonema* and *Nostoc* (Quiblier et al., 2013, Wood et al., 2020). Although dominated by cyanobacteria, the mats usually contain many other organisms (e.g., heterotrophic bacteria and eukaryotic algae) and inorganic matter (e.g., sediment) bound together by extracellular polymeric substances (McAllister et al., 2016). Over 20 species of benthic cyanobacteria are known to produce a range of cyanotoxins, similarly wide as that of planktonic species, that is, MCs, nodularins, CYNs, saxitoxins and anatoxins. Toxicity that could not be attributed to any of the known cyanotoxins has also been identified (Quiblier et al., 2013). As in planktonic cyanobacteria, both toxic and nontoxic genotypes of a given species usually coexist within mats (Cadel-Six et al., 2007) and cyanotoxin content can be highly variable spatially and temporally. Little is known about factors which may upregulate toxin production or promote the dominance of toxic over nontoxic strains. However, again as for the planktonic cyanobacteria, the abundance of toxic genotypes appears to be the key contributor to toxin content variability among benthic cyanobacterial-dominated mats (Wood & Puddick, 2017).

In contrast to planktonic cyanobacteria, research exploring factors that regulate the growth and expansion of benthic cyanobacterial-dominated mats has been relatively limited, and detailed studies are restricted to a few taxa from specific environments. In general, the variables that have been identified as most important in regulating the proliferation of benthic cyanobacterial mats are physical disturbance (wet–dry cycles, wave action, shear stress, abrasion), light, temperature, nutrients and grazing. Once mats are established, they contain diverse microbial communities and these taxa appear to play a vital role in cycling nutrients within the mat (Bouma-Gregson et al., 2019). This is in concert with internal biogeochemical conditions, which have been shown to facilitate the release of phosphorus from sediment trapped in the mats, leading to nutrient conditions within mats that are very different to those of the overlying water. For example, Wood et al. (2015) showed that dissolved reactive phosphorus concentrations in water extracted from benthic cyanobacterial mats was over 300 times higher than in the stream water. These nutrient-rich within-mat conditions probably partly explain how cyanobacterial-dominated mats can form very high biomass in nutrient-replete environments.

A synthesis of research results on the anatoxin-producing mat-forming species *Microcoleus autumnalis* (formerly *Phormidium autumnale*) in wadeable streams in New Zealand shows that *Microcoleus* proliferations are most likely to occur when there is slightly elevated dissolved inorganic nitrogen during the colonisation phase, but when water column dissolved reactive phosphorus is less than 0.01 mg/L. Once established, *Microcoleus*-dominated mats trap sediment, and biochemical conditions within the mats can mobilise sediment-bound phosphorus which is then available for growth. These streams are highly dynamic systems and the mats are primarily removed through shear stress and substrate disturbance, although there is also evidence for autogenic (natural) detachment. As cyanobacterial-dominated mats mature and biomass increases, oxygen bubbles, produced through photosynthesis, become trapped among the cyanobacteria and the extracellular polymeric substances they produce, causing the mat to become buoyant and detach from the substrate.

Tychonema species have repeatedly been recorded in lakes of the temperate zones (Salmaso et al., 2016). *Tychonema* spp. can produce neurotoxins, primarily anatoxins. These filamentous cyanobacteria primarily grows in stands of macrophytes but can occasionally be found in plankton samples and has been identified as the cause of dog deaths where these ingest *Tychonema* together with decaying macrophyte material accumulated at the shoreline (Fastner et al., 2018).

The health risk that benthic cyanobacteria proliferations pose to humans is still relatively unknown. There have been numerous cases of domestic and wildlife poisoning following the ingestion of cyanobacterial mats (Quiblier et al., 2013; McAllister et al., 2016). Anecdotal reports of human illness after recreating in streams containing cyanobacterial proliferations are documented, but conclusive evidence is lacking. As long as the mats are attached to the substrate, the risks of human ingestion are probably limited. However, detached mats often accumulate at the banks of rivers, streams and lakes, where animals are much more likely to consume them (Quiblier et al., 2013; McAllister et al., 2016, Wood et al., 2020). Dogs may be attracted to them by the smell of the decaying material, and numerous cases of dog deaths have been documented, sometimes with cyanobacterial cells and cyanotoxins found in their stomachs (Wood et al., 2007; Fastner et al., 2018). For some species, "free" toxin, that is, dissolved in water, can be detected in lake and stream water, although the concentrations are usually well below drinking-water guideline values (Wood et al., 2018). Assessing risks for human health is challenging in situations where deaths of pets and wildlife have been observed, while the water appears clear and toxin concentrations in the water are low or nondetectable. In such situations, it is best to inform users about the situation, to show what the mats look like and to advise avoiding contact with floating or beached benthic material (see Chapter 15).

4.3 WHICH ENVIRONMENTAL FACTORS FAVOUR CYANOBACTERIAL DOMINANCE?

Phytoplankton communities generally consist of species belonging to diverse taxonomic groups of photosynthetically active microorganisms. Yet the formation of massive, long-lasting blooms to the extent known from cyanobacteria is rarely reported for eukaryotic microalgae such as desmids, chrysophytes or chlorophytes, for example. Specific environmental conditions strongly favour the dominance of certain cyanobacterial taxa over other taxa of both cyanobacteria and microalgae. Understanding these conditions is important for management, allowing us to estimate under which conditions to expect cyanobacterial blooms. Key abiotic conditions determining phytoplankton growth rates and the outcome of competition between species are the availability of resources (nutrients and light), temperature, mixing regime, flushing rate and pH. In addition, biological interactions, including resource and

Table 4.1 Trophic state categories and their definition as given by Vollenweider & Kerekes (1982)

Trophic state	*TP mean* µg/L	*Chl-a mean* µg/L	*Chl-a maximum* µg/L	*Transparency mean* m	*Transparency maximum* m
Ultraoligotrophic	≤ 4	≤ 1	≤ 2.5	≥ 6	≥ 12
Oligotrophic	≤ 10	≤ 2.5	≤ 8	≥ 3	≥ 6
Mesotrophic	10 – 35	2.5 – 8	8 – 25	3 – 1.5	6 – 3
Eutrophic	35 – 100	8 – 25	25 – 75	1.5 – 0.7	3 – 1.5
Hypertrophic	≥ 100	≥ 25	≥ 75	≤ 0.7	≤ 1.5

TP: total phosphorus concentration; Chl-*a*: chlorophyll-*a* concentration; Transparency measured as Secchi depth readings; see Chapter 13.

competition, grazing by zooplankton, phage infection and lysis (Stough et al., 2017; Šulčius et al., 2018; Box 4.3) as well as parasitism (Frenken et al., 2017; Agha et al., 2018), can play a decisive role, and these are included in the discussion below.

4.3.1 The concept of "trophic state"

This concept integrates a classification of levels of nutrient concentrations with their outcome in terms of phytoplankton biomass and turbidity. Initially, Vollenweider & Kerekes (1982) proposed a classification of five trophic states based on data from lakes in temperate climates (Table 4.1). This classification uses the annual mean concentration of total phosphorus (TP) as the chief limiting nutrient in these waterbodies (see section 4.3.2 and Box 4.5), annual mean and maximum concentrations of chlorophyll-*a* as measure of phytoplankton biomass (see Box 4.6), as well as water transparency, measured as Secchi disc reading (see Chapter 13).

BOX 4.5: N LIMITATION *VERSUS* P LIMITATION: SCIENTIFIC CONTROVERSY OR CONSENSUS?

Vollenweider's (1968) keystone report on the scientific background of eutrophication of inland waters starts with the sentence "Nitrogen and phosphorus appear to be the most important among the nutrients responsible for eutrophication". Since then, the debate whether N or P or both are responsible for eutrophication has not ceased and with it, the discussion on approaches how to revert eutrophication is ongoing (Schindler, 2006). During the last decades of the 20th century, eutrophication control of freshwaters largely focused on P with success in many cases. One reason is that P does not, unlike N, have an unlimited atmospheric reservoir: N can be imported into the surface water by N fixation; also in shallow waterbodies, it can be exported by denitrification and thus potentially be lost more quickly than P, which may be recycled between the sediment and water if water exchange rates are low (Shatwell & Köhler, 2019). As phases of N limitation are common in many shallow waterbodies, some authors have concluded that therefore N needs to be reduced together with P (Conley et al., 2009; Paerl et al., 2016).

The controversy of P *versus* N limitation can be exemplified with a series of publications on Lake 227. This small experimental lake has been fertilised for 37 years with constant annual inputs of P and decreasing inputs of N. Even after ceasing N fertilisation, the phytoplankton abundance remained on an "eutrophic" level, leading Schindler et al. (2008) to the conclusion that attempts to reduce eutrophication by reducing N loads are compensated by cyanobacterial N fixation and hence reducing N loading as a management tool may be futile. This conclusion was challenged by re-analysing the same data set by Scott & McCarthy (2010) who found a decline of phytoplankton biomass after N fertilisation stopped with a rate of some 6% per year. With four more years of data, this trend, however, could not be verified, thus supporting the original conclusion (Paterson et al., 2011). In turn, Scott & McCarthy (2011) argued that N limitation may nonetheless be present but will affect phytoplankton biomass only after a time delay. Yet a few years later, Higgins et al. (2018) suggest that N limitation is still not significantly affecting this phytoplankton community.

This example shows how difficult it is to reach unambiguous conclusions since even a small lake is a very complex ecosystem. On a larger scale, Elser et al. (2007) showed that both increased N and P lead to increased biomass in freshwater as well as in marine and terrestrial ecosystems, with most pronounced responses for combined N and P fertilisation. It could therefore be argued that both N and P inputs need to be reduced to effectively revert eutrophication. Arguments held against this proposal (see review in Chorus & Spijkerman, 2020) are that

1. Experiments adding both N and P show that if all other resources, particularly light, are available in surplus, adding both nutrients can lead to higher biomass compared to adding of only N or only P. Such results, however, do not allow the inverse conclusion, that is, that both must be reduced to allow an efficient reduction of phytoplankton biomass. Reducing either P or N alone likely would likely suffice to reduce biomass, and the choice may be driven by practical considerations.
2. Even if N is currently limiting biomass, reducing P sufficiently will shift a waterbody to P limitation.

3. In many situations, P reduction is easier to achieve and results can be predicted with less uncertainty (Chapters 7 and 8).
4. N limitation may shift species composition between cyanobacteria and algal phytoplankton (Harris et al., 2014) and between cyanobacterial taxa (Teubner et al., 1999; Dolman et al., 2012), and such shifts are, as of yet, scarcely predictable. The concern that N limitation may promote N-fixing cyanobacteria (Schindler et al., 2008) has rarely proven relevant in practice.
5. So far, successful cases of lake or reservoir restoration with substantially reduced cyanobacterial biomass have been achieved by reducing P (Fastner et al., 2016; Schindler et al., 2016); success due to reducing N has scarcely been demonstrated (see Shatwell & Köhler (2019) for a successful example).

Jeppesen et al. (2007) contributed to resolving this controversy with data from 56 lakes showing that "if P can be reduced to low concentrations, N is not likely to be of major importance".

Reduction of P loads to many lakes and reservoirs has been successfully achieved with techniques such as P elimination or P precipitation in sewage treatment, appropriate use of mineral fertilisers, reduction of phosphates in laundry detergents or a combination of any of these measures. Therefore, from a practical point of view, P load reduction is often considered a first choice for an efficient management tool to revert eutrophication of lakes and reservoirs and thus to control cyanobacterial blooms.

However, ecological considerations, particularly for downstream coastal areas, may also necessitate controlling N, and when planning measures to control cyanobacteria, it may be effective to take this wider context into account (Chapter 7). Particularly for shallow lakes with several months of N limitation during summer, it is important to avoid any new N loads, and an option to control cyanobacteria may be to seasonally target measures that reduce prevalent N loads (see section 7.1).

This classification of trophic state may be adapted to local or regional circumstances, and systems have been proposed with further "fine-tuning" of the trophic state definitions: regional re-assessments of trophic state boundaries are given, for example, by Nürnberg (1996) for North American Lakes, including the impact of conditions such as lake morphometry, water hardness and fulvic/humid acid concentrations, and by Salas & Martino (1991) for warm-water lakes in Latin America and the Caribbean. Carlson (1977) developed a widely used numerical trophic state index (TSI) ranging from 0 to 100 and including Secchi depth, TP and chlorophyll-*a* concentrations. In spite of controversy over details of the definition of each category and its boundaries, the trophic state concept is widely used in practice. It is valuable for cyanobacterial management, because a few general statements apply to most cases:

- In *oligotrophic* waterbodies, planktonic cyanobacteria are unlikely to attain cell densities that cause hazardous cyanotoxin concentrations. However, benthic or epiphytic cyanobacteria may be present and cover littoral sediments.
- In *mesotrophic* waterbodies, cyanobacterial blooms occur rarely; exceptions include metalimnetic accumulations of *Planktothrix rubescens* (which may be at depths of drinking-water offtakes or come to the surface, increasing risk of exposure), detached mats of benthic cyanobacteria (see above) and – in large waterbodies – recruitment of cyanobacteria (particularly of *Dolichospermum*) from low cell density but very large water volumes and surface areas to visible scums along a downwind shoreline (section 4.5).
- In *eutrophic* and particularly *hypertrophic* waterbodies, cyanobacteria occur frequently and abundantly, often constituting a major share of the total phytoplankton biomass for extended parts of the year.

BOX 4.6: MONITORING TOTAL PHOSPHORUS AS OPPOSED TO SOLUBLE PHOSPHATE FRACTIONS

Considerable confusion prevails in the use of the term "phosphate". Historically, **soluble reactive phosphate (SRP)**, also termed "**dissolved inorganic phosphate**" **(DIP)** or "**orthophosphate**", has been measured and addressed when dealing with phytoplankton growth, because this is the fraction of TP which is directly available for uptake by cyanobacteria and algae. However, recycling of phosphate molecules within the plankton communities is often extremely rapid (within minutes), and phosphate liberated by the degradation of organic material will be rapidly taken up by bacteria and algae. Furthermore, cyanobacteria and algae can store enough phosphate for up to four cell divisions, even if no SRP can be measured. If SRP is found above the detection limits, this means that it is available in excess to the requirements of cyanobacteria and other phytoplankton. The only informational value of such a finding is that P is being "left over" and growth is limited by some factors other than P.

The upper limit of the biomass of cyanobacteria and/or algae that can develop in a given waterbody is, therefore, often largely determined by the amount of P bound within the cells, and **total phosphate phosphorus** is the variable that should be studied for biomass management. Strictly, this variable is not equivalent to **total phosphorus (TP)**, which includes the mineral form (like apatite) unavailable for biological uptake. However, mineral forms are of quantitative importance only in particular waterbodies (e.g., with high silt loading), and for the sake of simplification, TP has become widely used to represent total phosphate phosphorus.

It is of importance to note that concentrations of TP refer to the molecular weight of the phosphorus atom (30.97 g/mol) and not to that of phosphate molecule (PO_4, 94.97 g/mol). Where this is not clearly stated, it has caused considerable confusion in the literature.

"Eutrophication" is the process of nutrient enrichment with the consequence of a massive increase in the biomass of phytoplankton – often in particular of cyanobacteria – and reduced transparency. Eutrophication has become a widespread problem in many regions of the world in the 1960s in the wake of intensified application of mineral fertilisers, intensified animal husbandry as well as urbanisation and increased sewage discharges causing intensified nutrient loading to waterbodies. "Re-oligotrophication" is the reverse of eutrophication, and measures to achieve this started being implemented already in the late 1970s. While in some regions they are increasingly showing success, with the trophic state of waterbodies decreasing, in many regions eutrophication is still increasing, particularly in response to clearing of land for settlements or for agricultural use.

4.3.2 Nutrients

Phytoplankton biomass – that is, microalgae and cyanobacteria – chiefly consists of carbon (C), oxygen (O), hydrogen (H), nitrogen (N) and phosphorus (P). Under the conditions of rapid growth not limited by shortage of nutrients, it contains these elements in the following relative proportions (known as Redfield ratio; Redfield (1934)), expressed either in molar ratios or in mass units:

	C	*N*	*P*
Molar ratios	106	16	1
Mass ratios	42	7	1

Higher ratios of C:P have been suggested by Anderson & Sarmiento (1994) and Sterner et al. (2008). There is no lack of O and H in water, and concentrations of C become limiting only in very acidic water – or sometimes in extremely dense cyanobacterial blooms (Ibelings & Maberly,

1998). Thus, nitrogen (N) and phosphorus (P) are the key nutrients that promote – or limit – the growth of cyanobacteria and other phytoplankton in freshwaters. Although in consequence of the Redfield ratio, biomass production requires only about 1 mole of P per 16 moles of N, in many types of freshwater environments phosphorus is the resource which most frequently limits phytoplankton growth, while N limitation occurs frequently but often not for longer periods of time.

Nitrogen (N) may enter waterbodies leaching from soils particularly after spreading of manure or fertiliser, as run-off from animal feedlots and sewage (see Chapter 7). It is also recycled within a waterbody through the degradation of organic matter and excretion by animals. Cyanobacteria – like microalgae – can take up inorganic dissolved nitrogen in the form of nitrate, nitrite and ammonium. While utilisation of atmospheric nitrogen through diazotrophy has been considered unlikely to fully supply the N demands of an aquatic ecosystem because this process requires a high amount of energy (Conley et al., 2009; Moss et al., 2013; Shatwell & Köhler, 2019), there are, however, case examples even of quite turbid waterbodies like Sandusky Bay (Lake Erie, USA) in which N fixation periodically contributes up to 85% of total N uptake of the phytoplankton (Salk et al., 2018 and literature therein). Aquatic systems loose dissolved inorganic nitrogen through denitrification: under anoxic conditions, microorganisms use nitrate as source for oxidation processes, thus reducing nitrate to atmospheric nitrogen (N_2) which diffuses out of the water into the atmosphere.

Phases of N limitation are common in many waterbodies (reviewed by Sterner, 2008 and Søndergaard et al., 2017), in some pristine mountain lakes (Diaz et al., 2007) and particularly in eutrophic waterbodies and during warmer months in shallow zones with high rates of denitrification at the sediment surface. N limitation is more frequent in some regions than in others (Conley et al., 2009): for example, phytoplankton biomass is N-limited in many lakes in South America (Soto, 2002) or in New Zealand (Abell et al., 2010) – possibly due to lower rates of input from the atmosphere and/or less use of nitrogen fertilisers (Schindler, 2006).

Excessive N concentrations have detrimental effects on aquatic macrophytes that are important competitors of phytoplankton, including cyanobacteria, in shallow lakes (Moss et al., 2013).

When assessing which nutrient is limiting phytoplankton biomass at which time of the year, it is important to realise that N limitation is unlikely if concentrations of dissolved inorganic N are above 30–100 µg/L (Reynolds, 2006; Kolzau et al., 2014).

Phosphorus (P) enters waterbodies from sewage, animal feedlots and soils, particularly if these were fertilised with minerals or manure (see Chapter 7). As it binds to soil particles more effectively than nitrate, the main entry route into waterbodies from land is as surface run-off, often combined with erosion. Carvalho et al. (2013) evaluated a data set from more than 800 lakes of medium or high alkalinity in Europe and show that cyanobacterial biomass generally increases in relation to the concentrations of total phosphorus (TP). Using large data sets from several hundred waterbodies or the response of individual lakes to reduced TP concentrations, several authors (discussed below in section 4.4) have found that cyanobacterial biomass scarcely reaches health-relevant levels at TP concentrations below 10 µg/L. In the concentration range of 20–100 µg/L, TP strongly determines cyanobacterial biomass levels, while at the concentrations of TP exceeding 100 µg/L, curves level off and a further increase in cyanobacterial biomass is rarely observed (in many cases because light then becomes limiting).

P limitation is unlikely if concentrations of soluble reactive phosphorus (SRP) are above 3–10 µg/L of P (Reynolds, 2006; see also discussion in Kolzau et al., 2014).

N-, P- or light-limiting biomass carrying capacity: At any one point in time, one resource will determine the maximum possible amount of phytoplankton biomass – a concept termed "carrying capacity" (Box 4.7), while other resources may be available in excess. In turbid water or during winter in temperate climates, light is usually the limiting resource, while the available N and P would allow a higher level of biomass. The limiting resource may change seasonally, for example, light availability changes – at higher latitudes – in relation to the angular height of the sun and day length, or – as often the case in tropical climates – in relation to turbidity changes caused by pronounced seasonality of the flow regime. Seasonal patterns of light, N and P are likely to be quite specific for a given waterbody. Figure 4.3 shows a conceptual model for a shallow lake in a temperate climate: as light intensity increases in spring, phytoplankton begins to grow and incorporates available N and P into new biomass, depleting P to where it becomes limiting so

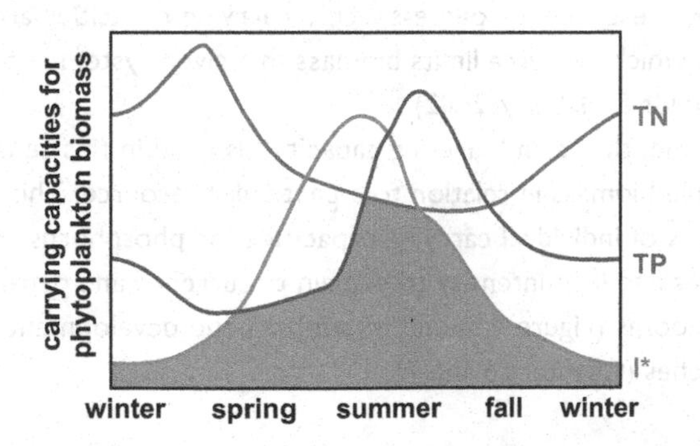

Figure 4.3 Schematic model of the seasonal course of the carrying capacity for phytoplankton biomass in the temperate zone of the Northern Hemisphere. Solid lines indicate carrying capacities determined by light intensity (I*), total phosphorus (TP) and total nitrogen (TN) as the most important limiting factors (Reynolds & Maberly, 2002). The green area shows the resulting composite carrying capacity for phytoplankton biomass. Adapted from Reynolds (1997).

that further increase of biomass is no longer possible – that is, biomass has reached the carrying capacity of the system. In this model case, N is available in excess during most of the year, but by mid-summer, P availability increases again through release from the sediment and N briefly becomes the limiting resource. In autumn, it is once again light that limits carrying capacity. At lower latitudes, the seasonal pattern for the light-limited carrying capacity is less pronounced – or determined by patterns of turbidity caused by monsoon events.

BOX 4.7: THE ECOLOGICAL CONCEPT OF CARRYING CAPACITY

The concept of an ecosystem carrying capacity has a long history in theoretical ecology. Although intuitively easily accessible, the concept involves some complexities due to differences in use in different fields, including ones that are not related to ecology (Sayre, 2008) or ambiguities associated with mathematical modelling (Gabriel et al., 2005; Mallet, 2012; Chapman & Byron, 2018).

In the definition given by Odum (1953), carrying capacity is the number of individuals of a given species that a given habitat can support without being permanently damaged. In this sense, the concept is largely applied to animals such as mammals, bivalves or humans, and considers a long-term equilibrium of the population with the entire environment (Begon et al., 1996).

In plankton ecology, the concept of carrying capacity has been adapted with a somewhat different meaning: for one, it rather refers to the maximal population or biomass density that can be reached at a given point in time – for example, during late summer in temperate lakes – than to a persisting equilibrium population (Reynolds, 1984). Further, carrying capacity is differentiated for each major resource: resource-specific carrying capacities are estimated to determine which resource limits biomass in a given system at a given point in time (Reynolds & Maberly, 2002).

In this volume, the term "carrying capacity" is used in this sense, as maximally attainable biomass in relation to a particular resource. This means analysing dynamics of individual carrying capacities for phosphorus and nitrogen concentrations and light intensity to explain occurrence and dynamics of cyanobacterial blooms (Figure 4.3) and to support the development of management approaches (Chapters 6 and 7).

If this model lake had extremely high concentrations of P and N, its carrying capacity for biomass would remain limited by light year-round: phytoplankton biomass would increase to such high turbidity that cells shade each to the point that no further increase is possible. In shallow lakes, higher biomass concentrations are possible than in waterbodies with deeper mixing because deep mixing entrains cells into darker layers. Using the concentration of chlorophyll-*a* as a measure of phytoplankton biomass, Figure 4.4 illustrates this with a selection of lakes: the deep lakes in the lower panel scarcely touch the curve for the 1:1 relationship of chlorophyll-*a* to TP, whereas some of the shallow ones summer mean values even occasionally exceed this curve. Moreover, at higher phosphorus concentrations, the curves for Lake Constance (Bodensee), Lake Tegeler See and

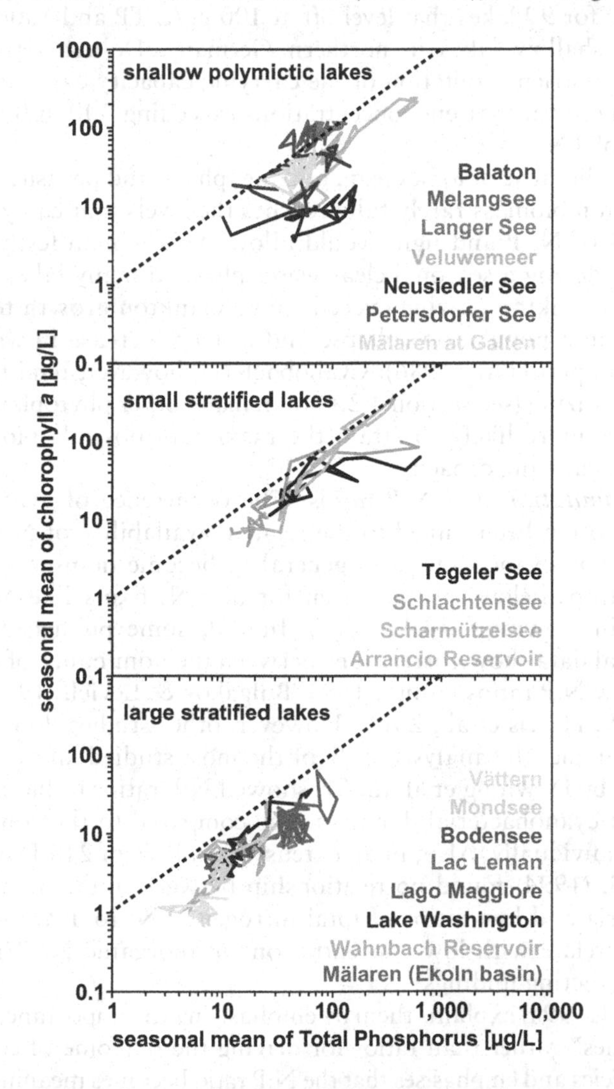

Figure 4.4 Phytoplankton biomass (as chlorophyll-*a*) and total phosphorus (TP) seasonal means (May – October or September) of epilimnion or surface layer in µg/L, connected chronologically for each lake for consecutive years. The dotted line shows the 1:1 relationship of chlorophyll-*a* with TP. See online version for colour code of waterbodies. (Modified from Chorus et al., 2011. Electronic version in colour)

Lake Schlachtensee level off asymptotically at 20–50 µg/L TP rather than showing further increase. Balaton and Neusiedler See show lower means for chlorophyll-*a* because of the strong impact that suspended sediment can have in the large, wind-exposed shallow lakes: in these lakes, suspended sediment particles shade the phytoplankton, which therefore cannot develop as high a biomass per unit phosphorus as it can in other, smaller and less wind-exposed lakes (Fastner et al., 2016). Downing et al. (2001) show curves for 99 lakes that level off at 100 µg/L TP and 1000 µg/L TN. For mostly shallow lakes in northern Germany, Dolman et al. (2012) assume that nutrient limitation of the carrying capacity can fairly reliably be excluded only at nutrient concentrations exceeding 300 µg/L of TP and 4000 µg/L of TN.

Moreover, because of losses (e.g., grazing, phage and parasite infection), phytoplankton biomass rarely fully reaches the levels that carrying capacity in terms of N, P and light would allow. This is manifestly observed particularly during a seasonal clear-water phase in many lakes when loss through zooplankton grazing exceeds phytoplankton growth to a degree that results in a population collapse and a steep increase in water transparency (Lampert et al., 1986). Cyanobacteria, however, tend to be more resistant to grazing (see section 4.2.4) than many other phytoplankton taxa and thus are more likely to attain the maximum possible biomass level given by the carrying capacity.

N or P limitation and N:P ratios: The occurrence of cyanobacterial blooms has often been linked to the relative availability of N in relation to P, proposing cyanobacteria in general to become dominant when the N:P ratio drops below 29 to 1, even for non N_2 fixers like *Microcystis* (discussed in Downing et al., 2001). Indeed, some published field and experimental data show correlations between the dominance of cyanobacteria and low N:P ratios (Smith, 1983; Bulgakov & Levich, 1999; Elser & Urabe, 1999; Harris et al., 2014). However, other studies do not confirm this: for example, the analysis of 99 of the most studied lakes in the temperate zone by Downing et al. (2001) showed N:P ratios to be the weakest predictor of cyanobacterial dominance as compared to the concentration of N or P individually. Also, in an extensive analysis of 210 Danish lakes, Jensen et al. (1994) found no relationship between amounts of N-fixing cyanobacteria and low ratios of total nitrogen (TN) to TP, nor did their biomass correlate with low concentrations of inorganic N. What causes these contradicting findings?

Reynolds (1999a) explains them by emphasising the importance of "absolute quantities" rather than ratios for driving the outcome of competition between species and emphasises that the N:P ratio becomes meaningless when concentrations of either N or P or both are too high to be limiting phytoplankton growth (Reynolds, 1999b). Limitation is highly unlikely if concentrations of TP and TN support a high capacity for phytoplankton biomass (see above in this section), and in many waterbodies, particularly in deep ones, thresholds for the limitation of biomass by TP and TN will be lower. Moreover, if dissolved nutrients are present in concentrations above 5–10 μg/L P or 100 μg/L N, this indicates that these are not limiting the phytoplankton's uptake rates, that is, that phytoplankton cells are saturated (note that in practice for dissolved P, the limit of quantification for most analytical methods corresponds to the threshold for limitation, i.e., 10 μg/L; see Chapter 13). Unfortunately, N:P ratios have been used widely and erroneously without excluding data from such clearly nonlimiting situations. Whether P or N is limiting phytoplankton growth and at which time during the season can be of relevance for water managers who need to prevent further increase in cyanobacterial blooms or find the locally most effective measures to reduce cyanobacterial abundance in a specific waterbody (see section 7.1 and Box 4.5). However, even where N is the limiting resource at times, sufficiently reducing P will turn P into the limiting resource and lastingly limit phytoplankton biomass.

4.3.3 Light availability

Light is exponentially attenuated with increasing depth (following the Lambert–Beer equation), and how steeply it diminishes over depth depends on the water`s constituents. There may be complete darkness already at a few metres depth (Kirk, 1994). The zone in which photosynthesis can occur is termed the "euphotic zone" (Z_{eu}). By definition, the euphotic zone extends from the surface to the depth at which the light intensity amounts to 1% of that at the surface. It is frequently estimated by measuring transparency with a Secchi disc (see Chapter 13) and multiplying the Secchi depth reading by a factor of 2.3–2.5.

Phytoplankton species – including cyanobacteria – differ in the dependency of their growth rate on light. Light availability therefore is an important factor determining the outcome of competition between species and thus whether the phytoplankton biomass consists chiefly of certain microalgae or certain species of cyanobacteria. At low light intensity, some cyanobacterial species (e.g., *Planktothrix agardhii* and *P. rubescens*) can maintain a higher growth rates than other phytoplankton organisms. This is why *P. rubescens* can thrive in the metalimnion of thermally stratified lakes and reservoirs. Also, in waters with high turbidity, *P. agardhii* can outcompete other species, reaching very high biomass, which in turn decreases transparency, thus improving growth conditions for itself in the sense of a positive feedback loop (Mantzouki et al., 2016).

Other cyanobacterial taxa need higher light intensities than *Planktothrix* in order to develop substantial populations, for example, many taxa of the order Nostocales. For *R. raciborskii* in lakes of northern Germany, Wiedner et al. (2007) demonstrated that light intensity is the most important factor determining population growth. *Dolichospermum* species are also known to occur more frequently in fairly clear water, for example, early in the growing season before turbidity increases due to increased phytoplankton density, or in less eutrophic lakes.

Using field data from Lake Victoria, Loiselle et al. (2007) show how the availability of solar irradiation, strongly influenced by mixing depth, determines the carrying capacity in waterbodies with excessive nutrient concentrations and how maximum phytoplankton biomass levels can be estimated from light availability.

4.3.4 Temperature, thermal stratification and mixing

Temperature is a further strong determinant of species composition, acting in two ways: (i) temperature in relation to the depth of a waterbody determines whether it develops thermal stratification or whether the water is mixed down to the sediment and (ii) all species grow faster at higher temperatures, up to a limit, but species differ in the temperature dependency of their growth rate, and for some cyanobacteria, the increase in growth rates as a function of temperature is steeper than for most of their eukaryotic competitors (Visser et al., 2016b).

In continental-scale surveys, cyanobacteria have been shown to dominate at locations with higher temperatures (Kosten et al., 2009; Kosten et al., 2012), and links between blooms and heatwaves (Jöhnk et al., 2008; Huber et al., 2012) have been established, albeit not always in a straightforward manner and partly only based on theoretical considerations (see below). Field data on species occurrence in relation to temperature scarcely allow a distinction between the indirect impact of temperature (i.e., determining the extent of mixing *vs.* stratification) and its direct effect on growth rates, and temperature alone is not a good predictor of blooms.

Often the indirect impacts of temperature are more important than the direct effects (Carey et al., 2012). In temperate climates, many waterbodies develop thermal stratification in spring, as energy input into the surface layer increases with increasing solar insolation and temperature (Box 4.2). Together with irradiation, wind and convection, thermal energy input determines whether waterbodies are well mixed or develop thermal stratification. As discussed above, thermal stratification is of paramount importance for growth and proliferation of all phytoplankton – cyanobacteria as well as microalgae. The strength of stratification, for example, of a deep alpine lake like Lake Zurich (Switzerland) has increased by more than 20% over the last few decades (Livingstone, 2003), compared to a less than 1 °C warming of the surface water of the lake over the same period, thus illustrating the relevance of indirect *versus* direct effects of climate warming. Wagner & Adrian (2009) propose more stable stratification to promote cyanobacterial blooms in eutrophic waterbodies.

Thermal stratification determines

- to which depth an average phytoplankton cell is entrained in the mixed upper layer and thus how much light it is exposed to while being moved across the underwater light gradient;
- the temperature and nutrient gradients that cells and colonies experience while entrained in mixing;
- whether cyanobacterial surface blooms and scums can accumulate at the surface of a waterbody or mixing energy is sufficient to maintain cyanobacteria in suspension;
- whether phytoplankton cells are likely to sink to the bottom through sedimentation (thus getting – at least temporarily – lost from the growing population when light does not penetrate to the sediment), or whether they remain entrained in the mixed layer;
- whether nutrients liberated from decaying biomass are redistributed in the water or whether they accumulate near the sediment or in the hypolimnion.

Huisman et al. (2004) modelled competition between buoyant cyanobacteria and other nonbuoyant phytoplankton as a function of access to light and turbulent mixing, and showed that under weak turbulence buoyant cyanobacteria on average may be positioned closer to the surface than their competitors. Hence, access to light, in particular for bloom-forming cyanobacteria, must always be considered in relation to the mixing processes in the lake. Under sustained, full mixing, *Microcystis* blooms are unlikely to form, an observation strengthened by successful control of these species through artificial lake mixing (Visser et al., 2016a).

Many cyanobacteria have higher temperature optima than many microalgae, in particular *Microcystis* spp. (Reynolds, 1997); thus, despite their relatively slow growth at lower temperatures, *Microcystis* may catch up with competitors in warmer water. Furthermore, elevated temperatures at the sediment surface may promote the recruitment of cyanobacteria from the sediments, as proposed for *Microcystis* (Shapiro, 1997) and *R. raciborskii* (Padisák, 1997; Wiedner et al., 2007). The filamentous cyanobacterium *R. raciborskii* thrives in waters with high temperature and moderately high nutrient levels, particularly in phases of elevated transparency, and was originally considered to be a typical tropical and subtropical species (Dyble et al., 2002). Although some discussions have assigned its spread to temperate climate zones to global warming, an alternative explanation is that the genotypes found on different continents (see above) may also differ in their growth rates in relation to temperature (Gugger et al., 2005b). True extremophiles can be found in hot spring microbial mat communities.

Elevated temperature also speeds up the degradation of organic matter, releasing nutrients for growth.

Most of the studies on cyanobacterial occurrence in relation to temperature focus primarily on population growth while temperature-dependent losses are studied to a lesser extent. Yet, losses may also correlate positively with temperature and outbalance increased growth rates at higher temperatures. For example, Rohrlack (2018) found that *Planktothrix* can proliferate in a low temperature niche that minimises losses through microbial antagonists to a degree that allows bloom formation despite relatively low growth rates.

4.3.5 Hydrodynamics and waterbody morphometry

Waterbody size, depth, shape, wind exposure and water exchange rate are critical conditions determining how water stratifies (layering of water due to density differences caused by heating of surface water or salinity) or moves and how it entrains plankton. In large, shallow lakes, turbulence is often pronounced, particularly if they are exposed to wind, and even highly buoyant cyanobacteria will have little chance for vertical positioning and are mixed throughout the water column. These conditions favour species like *P. agardhii* which do not show pronounced buoyancy regulation but are good competitors for light. Deep lakes generally are thermally stratified with turbulence restricted to the epilimnion. If these lakes are small and wind-sheltered, turbulence may be minimal (with mixing only slight, due to convection as the lake cools nocturnally). However, large stratified lakes and reservoirs may develop internal seiches – with the metalimnion oscillating at regular intervals in consequence of wind moving the surface layer towards one end of the waterbody (Cuypers et al., 2011). This may move species that form distinct layers at the metalimnion, particularly *Planktothrix rubescens*, up and down in the regular rhythm of the seiches, by up to several metres.

The shape of a waterbody (i.e., its morphometry) is an important factor to consider, in particular for planning sites for drinking-water offtake or bathing beaches, and respective monitoring (see Chapter 11). Lake morphometry can be fairly regular, and thus, water quality (including plankton composition) may be fairly homogeneous. Other waterbodies, particularly reservoirs, have highly irregular, fractal shorelines with incised bays where tributaries enter or with more or less separated basins in which water quality varies distinctly between individual sites. In such situations, cyanobacterial blooms can develop or accumulate in one section, not in others.

A further important hydrological characteristic of a waterbody is the water residence time, that is, the theoretical average time a volume of water remains in the waterbody. It is calculated from the in- and outflow discharge, and the lake volume. Water residence times may vary between seasons and years depending on precipitation and other events in the catchment – and for reservoirs also on the abstraction regime. Low water residence times – that is, in the range of days – may dilute cyanobacteria faster than they can multiply. Also, depending on the variability of the quality of inflowing water, low residence times may alter other growth conditions like nutrient concentrations, with short timescales (Romo et al., 2013). Reynolds et al. (2012) show that diluting a lake by 10% per day would cause an exponential loss rate of –0.16 per day, and most phytoplankton species can scarcely reach faster net growth rates.

Both morphological and hydrological conditions also influence nutrient concentrations (Carpenter et al., 1999). Long water residence times – that is, in the range of years – may be an obstacle for reducing lake P concentrations because even at substantial external load reduction, P can be recycled within the waterbody and P concentrations in the sediment will decrease only slowly. For example, Jeppesen et al. (1991) showed that among 27 shallow lakes under restoration in Denmark, cyanobacterial dominance disappeared substantially faster in those with a water retention time of less than half a year.

Stable thermal stratification in deeper lakes can prevent most of the phosphorus released from the sediment from reaching the euphotic zone during the stratified period, whereas in shallow, turbulent lakes, sediment can be resuspended and mixed throughout the water column on windy days, severely reducing water transparency and reducing phytoplankton access to light. For nitrogen, in contrast, mixing down to the sediment particularly in shallow lakes and reservoirs can cause it to become limiting for phytoplankton biomass because higher temperatures at the water–sediment interface enhance the decay of organic substances and in consequence also denitrification rates and hence loss of nitrogen (N_2) to the atmosphere (see section 4.3.2).

4.3.6 pH: acidity and alkalinity

Cyanobacterial blooms are often associated with alkaline conditions. However, this is usually a consequence rather than a cause of these blooms: the uptake of hydrogen carbonate through intense photosynthesis shifts the equilibrium between carbonate and hydrogen carbonate to render the water alkaline (Wetzel, 2001; Lampert & Sommer, 2007). While this can be due to a high biomass of any phytoplankton with a high rate of photosynthesis, it is typically cyanobacterial blooms that reach particularly high biomass concentrations and thus are the cause of high pH. *Vice versa*, cyanobacterial blooms may be expected in waterbodies with pH ranges above 7. Under acidic conditions (pH <6), cyanobacteria are rarely found in sufficiently high cell density to cause detectable levels of cyanotoxins (Chorus & Niesel, 2011). However, given the array of carbon uptake systems available to cyanobacteria, they can be found to dominate at both low and high dissolved CO_2 concentrations (Huisman et al., 2018).

4.4 CAN CYANOBACTERIAL BLOOMS BE PREDICTED BY MODELS?

Two different types of approaches may serve to estimate the likelihood of high cyanobacterial biomass from data on conditions in the waterbody: functional approaches that strive to model the true interactions within the ecosystem and statistical approaches that investigate correlations between cyanobacterial biomass and waterbody conditions. While the advantage of functional modelling is that it depicts causalities, in practice it is limited by the complexity of ecosystem interactions which are challenging to depict in any model. In contrast, while some of the statistical approaches tend to be comparatively simple and straightforward, they only depict correlations, not causalities. Taking correlation for causality is, however, widespread, and when using regressions, it is important to keep in mind that if two factors correlate, it remains unknown whether this is because one is a cause for the other or whether there is a third, perhaps unknown factor causing both factors to increase or decrease in parallel. Correlations are indeed useful to derive hypotheses on the factors determining cyanobacterial biomass; however, before basing management decisions on their results, it is important to investigate and understand the mechanism(s) likely to be causing the correlation. Furthermore, statistical models capture the outcome of all processes determining species dominance and biomass without disentangling their relative importance. Thus, they can be useful for the waterbodies for which they were derived or for those which are similar to the ones used in the model (e.g., within the same ecoregion and similar hydrophysical characteristics), but their predictions are uncertain when applying them to other types of waterbodies.

Functional models strive to overcome this limitation and to be more generally valid by capturing key processes and relationships, including, for example, the dependency of phytoplankton growth rates on environmental conditions, loss processes and competition between some keystone species or phytoplankton groups. A range of such models has been developed that include conditions favouring cyanobacteria. Mooij et al. (2010) give an overview of models developed for the real-time prediction of cyanobacterial scums and blooms from current or recent waterbody and meteorological data, for example, CAEDYM, BLOOM and for shallow, nonstratifying lakes PCLake, PROTECH and SALMO. However, these may fail if processes other than those depicted in the model are important. Predicting cyanobacterial dominance and biomass using generic functional models is not only challenging because of the complexity of waterbody conditions determining the outcome of competition between species, but also because excessively complex models do not necessarily perform better. Reynolds & Maberly (2002) propose a simple model to estimate the maximum possible biomass yield (carrying capacity) in terms of concentrations of chlorophyll-*a* from the concentrations of nitrogen and phosphorus as well as the photon flux density. While this neither includes loss processes nor a differentiation between cyanobacteria and other phytoplankton, it is useful as approach to estimating the maximum conceivable biomass to be expected (and if this is high, rendering the water turbid, a high share of cyanobacteria is likely; see section 4.3). Carvalho et al. (2013) emphasise the value of using nutrient data – particularly those for total phosphorus (TP) – to estimate the likelihood of cyanobacterial blooms because these are much more widely available than phytoplankton data.

Among the statistical approaches, a number of studies addressed the relationship between phytoplankton biomass (often quantified as chlorophyll-*a*) and the concentration of TP, sometimes including total nitrogen (TN). Most of these studies were conducted in temperate climates, and one of the first to be widely used in management is Vollenweider (1976). Since then, a number of studies have been published giving regressions from a varying number of waterbodies relating chlorophyll-*a* to both P and N (see an overview in Yuan & Jones (2020). Phillips et al. (2008) give an overview of 15 earlier studies and show the slopes of a total of 15 regressions to vary between 0.8 and 1.5 for log TP (16 lakes) and 1.0 to 1.4 for log TN (7 lakes); in consequence, the concentration of chlorophyll-α that these regressions predict for 35 µg/L of TP (at 875 µg/L TN) varies by almost a factor of 4, that is, between 5.4 and 19.5 µg/L. Moreover, these authors give new regression equations that they developed from evaluating a data set from >1000 European lakes. These show less scatter of the data for chlorophyll-*a* *versus* TP as compared to chlorophyll-*a* *versus* TN, reflecting the limitation of biomass chiefly by TP in the larger fraction of the lakes included in the regressions.

In particular, Phillips et al. (2008) differentiated between lake types and geographic regions within Europe. While results show no influence of region, the most pronounced result is that the lowest biomass (in terms of chlorophyll-*a*) per unit nutrient is found in deep lakes and the highest biomass in shallow lakes (which were also those with high alkalinity). This is to be expected because, as discussed above, deeper mixing causes more pronounced limitation by light, particularly when a high density of phytoplankton cells causes high turbidity. Also reflecting increasing light limitation at high phytoplankton density, the correlation of chlorophyll-*a* with TP levelled off at concentrations of 100 µg/L TP and 1000 µg/L TN with yet higher levels of both chlorophyll-*a* and nutrient concentrations, found only in very shallow waterbodies. For using such a regression model to estimate target levels for nutrients, Phillips et al. (2008) emphasise that "*managers should use an equation that is derived using data from a lake type that matches as closely as possible the lake they are concerned with*".

For tropical climates, Huszar et al. (2006) evaluated data from 192 tropical lakes and found the relationship between chlorophyll-*a* and TP concentrations to correlate less tightly than for the data from temperate climates, and chlorophyll-*a* concentrations in relation to TP to be slightly

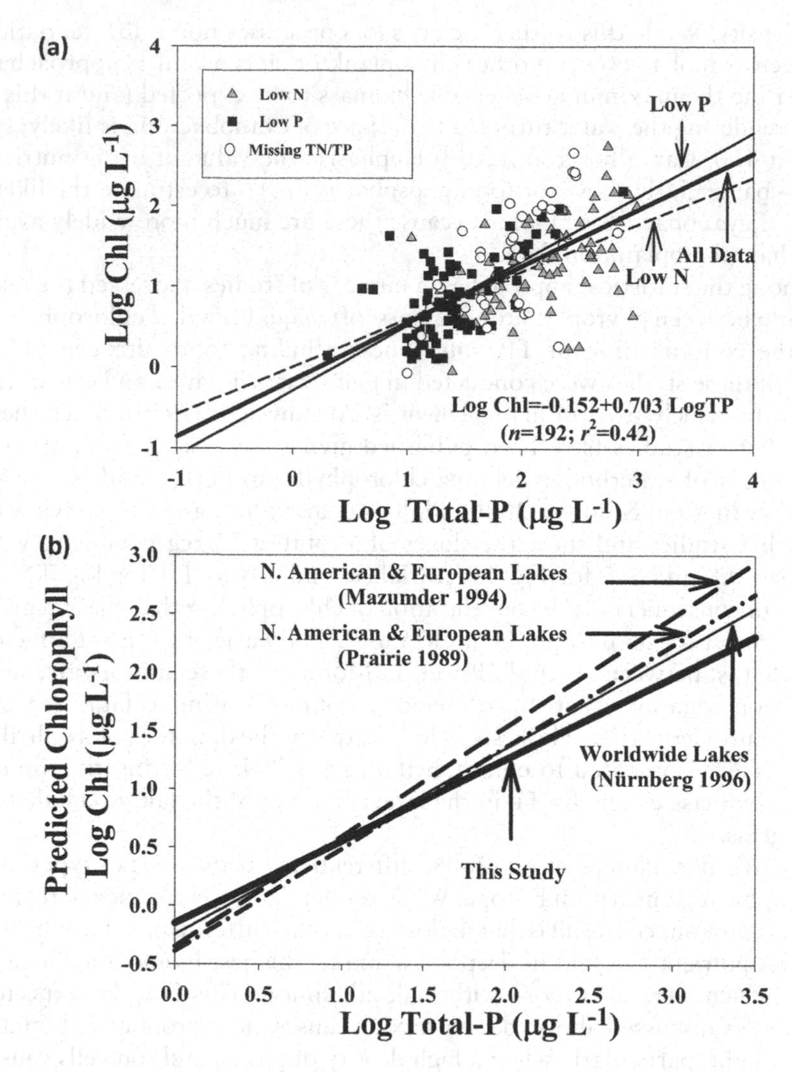

Figure 4.5 (a): annual logarithmic means of chlorophyll *versus* total phosphorus (TP) from 192 tropical and subtropical lakes from Africa, Asia and America with definitions for "low P" being TN:TP by weight >17 and for "low N" <17. (b): comparison of this relationship with regressions published for temperate lakes. (Reprinted with permission from Springer Nature, in: Nitrogen Cycling in the Americas: Natural and Anthropogenic Influences and Controls. Nutrient-chlorophyll relationships in tropical-subtropical lakes: do temperate models fit? Huszar V, Caraco N, Roland F, Cole J. Copyright 2006 Springer Nature. www.springernature.com/gp.)

lower (Figure 4.5). As explanation for these phenomena, Jeppesen et al. (2005) propose a stronger impact of further factors such as light limitation in waterbodies with elevated turbidity and more complex food web interactions. In particular, loss processes may lead to phytoplankton biomass considerably lower than that expected from nutrient concentrations. Mamun & An (2017) show the influence of seasonal Asian monsoon on regressions of chlorophyll-*a* and TP with data from 182 agricultural reservoirs and demonstrate a much stronger correlation of chlorophyll-*a* to TP as compared to TN for these waterbodies.

For practical purposes of predicting biomass levels, it is important to realise that the data points for the individual waterbodies behind these regressions scatter considerably: in the statistical evaluation of Huszar et al. (2006), log(TP) explains only 42% of the variance of log chlorophyll (and a multiple regression approach, including TN, resulted in only moderate improvement to 47% explanatory power). In consequence of this scatter, in any given waterbody observed annual mean chlorophyll concentrations can be several-fold higher or lower than those predicted from total phosphorus (TP)concentrations and the regressions shown in Figure 4.5.

However, this uncertainty in the range of a factor of 3–5 relates to data spanning three orders of magnitude, and thus, the published regressions remain useful for two purposes: (i) for first estimates of the TP target concentrations that are likely to control biomass and (ii) for investigating causes for particularly high or low mean chlorophyll concentrations in relation to those of TP.

Predicting *specifically cyanobacterial biomass* requires an understanding of the conditions in the waterbody which favour their dominance. Trophic state and the concentrations of TP are key: where nutrient concentrations enable high levels of total phytoplankton biomass, this has a high likelihood to consist chiefly of cyanobacteria – in temperate climates particularly during summer and early autumn: Downing et al. (2001) evaluated data from 99 lakes of the temperate zone and found that the risk of cyanobacterial dominance was related to the TP concentration as follows: 0–30 µg/L TP: 0–10% risk of cyanobacterial dominance; 30–70 µg/L TP: ~40% risk; 100 µg/L TP: ~80% risk. The authors further highlight that turbidity – caused by high phytoplankton cell density – increases the risk of cyanobacterial dominance (at concentrations of chlorophyll-*a* above 10 µg/L, transparency tends to decline to less than 1 m).

Data from shallow lakes in north-eastern Germany dominated by filamentous cyanobacteria confirm the pattern described by Downing et al. (2001): total phytoplankton biomass proved to be a good predictor of cyanobacterial biomass, particularly at high total phytoplankton biomass (Figure 4.6). Once biovolumes were higher than 13 mm^3/L, cyanobacteria almost always constituted at least half of this biomass, often more. The causal explanation for this is that these lakes are dominated by filamentous cyanobacteria, chiefly *P. agardhii*, which can outcompete other phytoplankton particularly under the conditions of low light availability (see section 4.2.2). A similar result is reported by Ptacnik et al. (2008) for lakes in northern Europe: at chlorophyll-*a* concentrations above 10 µg/L, cyanobacteria often dominated

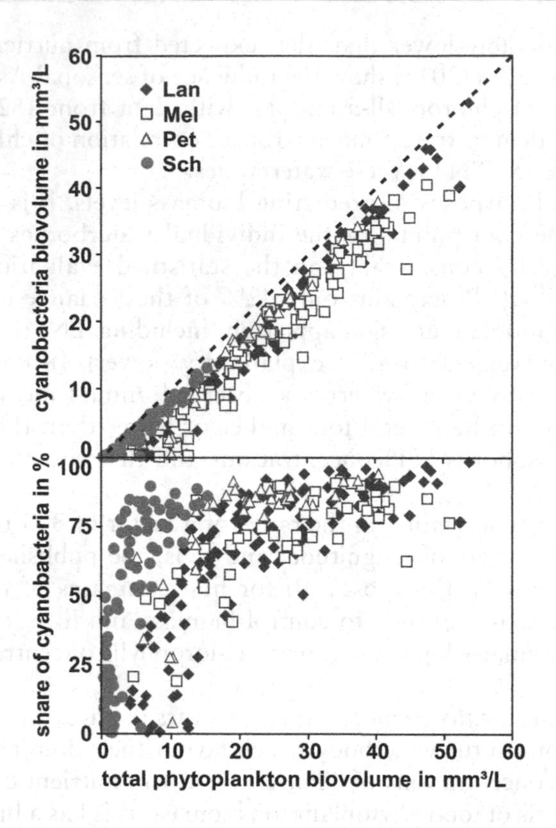

Figure 4.6 Cyanobacterial biovolume in relation to total phytoplankton biovolume for four lakes (Langer See [Lan], Melangsee [Mel], Petersdorfer See [Pet], Scharmützelsee [Sch]) in the same catchment. Upper panel: absolute values of biovolumes; lower panel: relative fraction of cyanobacteria in total biovolume. (Data kindly provided by Brigitte Nixdorf, Brandenburgische Technische Universität Cottbus.)

in samples taken in late summer, and as discussed above, Carvalho et al. (2013) evaluated a data set from more than 800 lakes of medium or high alkalinity in Europe and show a sharp increase in cyanobacterial biomass relative to TP beginning at 10 µg/L, increasing sharply in the range of 20–100 µg/L and levelling off at 100 µg/L.

A further confirmation of the significantly enhanced likelihood of cyanobacteria to be the organisms causing high phytoplankton biomass is given by Chorus et al. (2011) who evaluated a large database with almost 2000 samples from 210 lakes covering six European countries and three ecoregions (Central Plains, Central Highlands and Sicily) for the likelihood of the six most common cyanobacterial taxa to occur, depending on environmental conditions, that is, nutrient concentrations, seasonality, mixing and waterbody depth, turbidity, pH and temperature (Table 4.2). For five

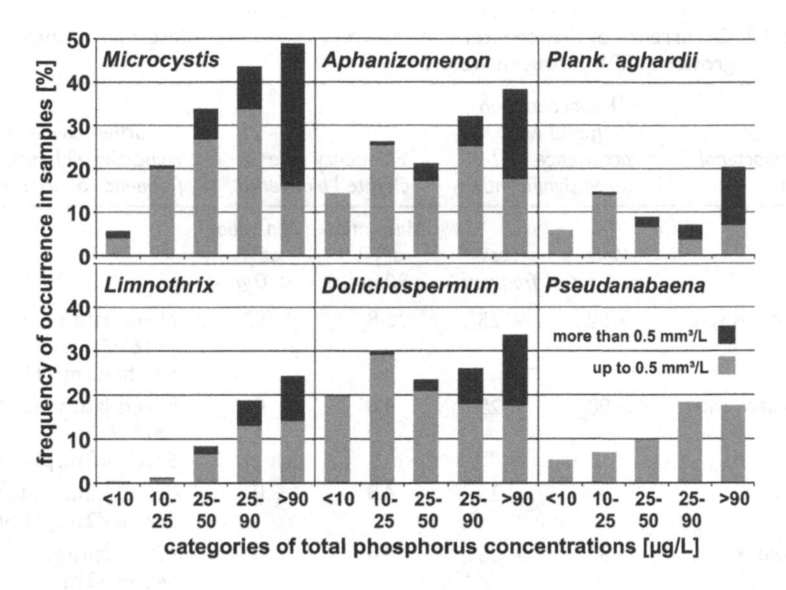

Figure 4.7 Frequency of occurrence of the six most common cyanobacterial taxa in 1928 samples from 210 waterbodies in two biomass categories (biovolumes up to 0.5 mm³/L and >0.5 mm³/L) and in five categories of total phosphorus (TP). Number of samples per category from lowest to highest TP concentrations: 501, 623, 302, 246 and 256. (Modified from Chorus & Niesel, 2011.)

categories of TP concentrations, the frequency of individual taxa (relative to all samples in the database) in samples is shown in Figure 4.7. While these taxa were found in some of the samples even at low TP concentrations, that is, <10 µg/L, with the exception of *Aphanizomenon*, this was only in less than 10% of the samples, and some taxa (*Microcystis* spp. and *Limnothrix* spp.) proved more likely to occur at higher TP concentrations. Importantly, however, cyanobacterial biovolumes above 0.5 mm³/L occurred almost only at TP concentrations above 25–50 µg/L TP and almost exclusively in waterbodies classified as "eutrophic" or "hypertrophic" by the data providers. These cyanobacteria also occurred significantly more frequently at >16 °C in late summer, probably due to the time it takes for them to build up larger populations as the season progresses.

Recknagel et al. (2017) demonstrate that lake-specific models developed from historic data and cyanobacterial cell counts can – using only *in situ* physical–chemical sensor monitoring data – predict blooms quite accurately. Their models (based on hybrid evolutionary algorithms) for Lake Wivenhoe in Australia and the Vaal Reservoir in South Africa forecast blooms exceeding 1 µg/L total MCs 10–30 days ahead of their occurrence. This demonstrates the value of multiseasonal time series to refine more general models for a particular waterbody, as these can improve forecasts for parameters of high relevance for water management.

Table 4.2 Occurrence of cyanobacteria in relation to conditions determining their growth in 201 European waterbodies

Cyanobacterial taxon	*TP concentration [µg/L] with occurrence of the taxon significantly*		*Frequency ratio[a] at elevated biovolume[b]*		*Further conditions coinciding with elevated frequency of occurrence*
	More frequent	*Less frequent*	*Max. ratio at TP >90 µg/L*	*Min. ratio at TP <10 g/L*	
Microcystis spp.	> 50	< 25	3.8	0.2	Mixed, late summer, >16 °C, Secchi <3 m, pH >7
Aphanizomenon spp.	> 90	< 25	4.6	0	Mixed, late summer, >16 °C, Secchi <2 m, pH >6
P. agardhii	> 90	< 25	4.8	0	Late summer, >16 °C, Secchi <2 m, pH >6–7
Limnothrix spp.	> 50	< 25	4.0	0	Mixed, spring, Secchi >3 m
Dolichospermum spp.	> 50	< 25	4.2	0.1	Mixed, late summer, >16 °C, Secchi <3 m, pH >7–8

Source: Adapted from Chorus & Niesel (2011).

[a] "Frequency ratio" is the number of samples with the taxon to the total number of samples in the database (see Figure 4.7 for details); note that these data apply only to this specific database.
[b] "Elevated biovolume" is defined in relation to cell size of the taxon and ranges from >0.1 to >1 mm^3/L biovolume.

4.4.1 Models to predict surface blooms and scums

As discussed in section 4.1.1, the mechanisms of cell accumulation differ between the cyanobacterial genera that form surface blooms (particularly *Microcystis*, *Anabaena*, *Aphanizomenon*) and those that form dense populations in the metalimnion, particularly *Planktothrix rubescens*, but sometimes also *Aphanizomenon*, *Dolichospermum* and *Raphidiopsis*. While the horizontal drift of scums can be roughly inferred from prevalent wind directions (Hutchinson & Webster, 1994; Welker et al., 2003), predicting when cells will rise to the surface is more challenging. In many settings with regularly recurring seasonal populations of surface bloom-forming cyanobacterial species, experience provides good indication of the time of year and type of weather rendering scum formation to be likely.

Beyond experience valid for individual waterbodies, models to predict scum formation from waterbody characteristics and weather forecasts can be useful. These are being developed as vertical migration velocities and buoyancy regulation mechanisms are increasingly becoming understood for frequently occurring cyanobacterial species and genera. For *Microcystis* colonies, models describe the vertical migration of cyanobacteria (e.g., Kromkamp & Walsby, 1990; Visser et al., 1997; Wallace & Hamilton, 2000). The fuzzy-logic-based early warning model developed by Ibelings et al. (2003; see Box 4.8) includes weather forecast (wind speed and direction), cyanobacterial biomass, buoyancy state of the cells and stratification stability to predict the timing and location of cyanobacterial scums several days in advance, giving water managers time to act. A crucial problem with all predictions of scum formation is the trade-off between false negatives (missing blooms in the prediction) and false positives (predicting blooms that then do not occur). Whereas the first may lead to public health risks, the second may lead to economic damage for lake-based operations such as restaurants, campsites and beaches when lakes are closed for recreational purposes. For *P. rubescens*, Walsby (2005) developed a conceptual model to explain the occurrence of surface blooms near shallow, leeward shores arising from populations floating up in the metalimnion.

BOX 4.8: EARLY WARNING FOR THE FORMATION OF *MICROCYSTIS* SCUMS

In the Netherlands, water managers use a scheme in which scums are classified as 1 (light scums/surface blooms) to 4 (severe scums) on the basis of an information sheet showing pictures and descriptions of these categories. For a number of Dutch lakes, observers took notes on the occurrence of scums on a daily basis and used these data to validate an early warning model (Ibelings et al., 2003) that predicts scum formation (time and location) from several factors, including the mid-term weather forecast.

This model estimates cyanobacterial biomass increase using a traditional water quality model, combining this with a fuzzy-logic model to estimate the buoyancy of the cells and to relate this to waterbody mixing or stability of stratification. Fuzzy-logic model allowed qualitative data (expert knowledge) about scum formation to be made available in a quantitative way. The model was first developed for the open water of the large Lake IJsselmeer and then expanded to be applicable to smaller lakes, taking also sheltered areas (harbours, etc.) into account, that is, places where scums are more persistent than in the open water of lakes. It provides lake managers with a weekly bulletin in which the risk of scum formation for the coming week is indicated. The model is continuously being updated and improved (work by Deltares; https://www.deltares.nl).

All models are intended to (i) identify physical conditions of the waterbody that allow for surface bloom formation in a specific lake and (ii) predict the occurrence of surface blooms and scums. Generally, they work satisfactorily for the waterbodies for which they were developed. For the application of scum-forecasting models to a specific waterbody, it is important to consult an expert for limnological modelling.

4.5 WILL CLIMATE CHANGE AFFECT CYANOBACTERIAL BLOOMS?

Evidence for an increase in cyanobacterial blooms in the wake of climate change is increasingly published (e.g., Mooij et al., 2005; Paerl & Huisman, 2009), and there is widespread concern that cyanobacterial blooms will increase in warming waterbodies where these are sufficiently eutrophic to sustain blooms. However, the impact of climate change on growth conditions for cyanobacteria is not straightforward (De Senerpont Domis et al., 2013), and it is important not to confuse the effects of local – and possibly unusual – weather on plankton dynamics in a given lake with climate effects.

Climate change affects water temperatures directly, but also indirectly waterbodies through changes in the strength and duration of thermal stratification (see section 4.3.4), including the impact of (possibly more frequent) storm events, and warming can lead to earlier appearance of zooplankton, changing grazing pressure on phytoplankton (Winder & Sommer, 2012). Elevated atmospheric CO_2 concentrations may stimulate access to CO_2 for surface blooms, which may be limited by inorganic carbon (see Huisman et al., 2018) – but elevated water temperatures simultaneously reduce the solubility of CO_2 in water and thus lead to lower concentrations. Climate change may also impact on the patterns of rainfall and snowmelt and thus on inflow, turbidity and nutrient loading, and it may increase water residence times where periods of drought increase – or reduce water residence time where precipitation increases. Thus, the probable effects of a climate change include mechanisms that may be favourable for cyanobacterial growth as well as mechanisms which are not, and the balance between them will vary strongly between waterbodies and have different effects on different bloom-forming species.

For *Microcystis*, a direct mechanism proposed to cause an increase in its abundance is the more pronounced dependency of growth rates on elevated temperature as compared to most of their competitors among other phytoplankton, which would increase the chances of *Microcystis* to dominate in warmer water. Additionally, increased stability of thermal stratification resulting from enhanced energy input into the surface layer of waterbodies could provide a competitive advantage for species with a mode of active vertical positioning, like buoyant cyanobacteria. In contrast, however, increases in cloud cover and wind speed (as, e.g., predicted for North West Europe) or more frequent storms would weaken thermal stratification of waterbodies and thus reduce this competitive advantage.

Analysis of surface blooms in IJsselmeer showed that scums are absent at wind speeds greater than 2–3 m/s (Ibelings et al., 2003), and in this lake, they may therefore become less frequent in the predicted slightly more windy future (Cheng et al., 2002).

P. agardhii may also be affected by a warmer climate, especially in those hypertrophic lakes where this species forms more or less permanent (year-round) blooms. In Lake Veluwe, the Netherlands, a 15-year bloom of *Planktothrix* was broken by a combination of reduced P loading and flushing, but a series of cold winters with extended ice-cover was also important in reducing the overwintering biomass of this species and giving other species, particularly of planktonic algae, a chance to dominate in spring (Reeders et al., 1998). Jeppesen et al. (2003) suggest that climate warming may increase the probability of lakes to become locked in a turbid state. Heatwave summers have been shown to reduce, rather than expand, the proliferation of *Planktothrix rubescens* in deep lakes. Here, milder winters rather than warm summers promoted cyanobacterial development, an example of ecological memory carrying over environmental effects from one season to the next (Anneville et al., 2015).

The spreading of *R. raciborskii* – previously regarded as subtropical – into temperate regions has been tentatively linked to climate change. Wiedner et al. (2007) showed that this is also not attributable to overall higher summer temperatures but rather to the timing of higher temperatures: resting stages (akinetes) of *R. raciborskii* germinate at 15 °C, and in northern Germany, lake sediments reach this temperature earlier by 2 weeks, when less spring phytoplankton has developed and water is still clearer. The earlier *R. raciborskii* hatches from its resting stages, the more light is available for its rapid growth and the establishment of a large population.

A further aspect is that higher temperatures and higher stratification stability can trigger an increased release of phosphorus from sediments, thus fertilising cyanobacterial blooms. *Vice versa*, however, increased stratification stability can reduce the transport of released phosphorus into the euphotic zone, thus reducing its availability to cyanobacteria possibly resulting in a "climate warming-induced oligotrophication" (Salmaso et al., 2018). The specific conditions in a waterbody determine which of these two contrary processes determines phosphorus concentrations in the euphotic zone.

The concept of "carrying capacity" (section 4.3.5 and Figure 4.3) is helpful for assessing the impact of climate change *versus* the impact of trophic state: where there is not enough phosphorus and nitrogen to support substantial levels of phytoplankton biomass, warming is not likely to increase cyanobacterial proliferation. A number of modelling studies (reviewed by Elliott (2012)) show agreement that while cyanobacterial biomass is likely to increase in a warmer climate, the magnitude of this response will strongly depend on nutrient concentrations: where these are limiting, the annual amount of biomass is not likely to increase although the timing and proportional dominance of cyanobacteria may change. On the basis of data analysed from more than 1000 lakes, Rigosi et al. (2014) emphasise that the impacts of nutrients and climate are not synergistic; rather, "nutrients predominantly controlled cyanobacterial biovolume" and, although very eutrophic lakes are more sensitive to increased temperature, "ultimately nutrients are the more important predictor of cyanobacterial biovolume". Others, however, are convinced that a synergy between eutrophication and increased temperatures will produce even larger blooms in future (Moss et al., 2011; Havens & Jeppesen, 2018).

A further concern in climate change scenarios for some situations is the possible reduction in available water in lakes and reservoirs due to reduced rainfall. Water scarcity combined with increasing cyanobacterial blooms would create challenges for waterbody management in general and specifically for providing safe drinking-water. Yet, also for this aspect, direct human impact, for example, through large-scale depletion of groundwater for irrigation, could outweigh the possible long-term impact of changes in precipitation patterns in the course of climate change.

In conclusion, it is highly probable that a significant change in climatic conditions will affect the ecology of waterbodies. In some waterbodies, this could cause an increase in the frequency and the scale of cyanobacterial blooms, while not in others. Preparing today for possible effects of climate change on cyanobacterial blooms in the future should not divert the focus from measures to reduce cyanobacterial blooms that can be taken today: reducing eutrophication to levels unlikely to support blooms is an effective component of preparing for climate change.

4.6 WHICH FACTORS DETERMINE CYANOTOXIN CONCENTRATIONS?

In natural waters, complete absence of cyanobacteria is unlikely given the capability of cyanobacteria to thrive in the most extreme environments, and therefore, cyanotoxins are likely found in most waterbodies – yet often at very low concentrations and detectable only with highly sensitive analytical methods. Therefore, the qualitative detection of cyanobacterial toxins, for example, in samples concentrated with a plankton net, is only of limited relevance for public health management.

For health risk assessment, not mere occurrence but rather the concentration of cyanotoxins in waterbodies is important, and the concentration is tightly linked to the biomass of potentially toxigenic taxa. The most important ones include *Microcystis*, *Planktothrix*, *Dolichospermum*, *Raphidiopsis* and *Aphanizomenon* in the plankton of lakes and reservoirs, benthic taxa like *Phormidium* or *Microcoleus* in streams, (tychoplanktonic) *Tychonema* in lakes, and *Lyngbya* or *Nodularia* in brackish and coastal waters. Other taxa that have been reported to form blooms, such as *Limnothrix* or *Gomphosphaeria*, have not been unambiguously reported to produce known toxins, while, on the other hand, some toxigenic taxa such as *Hapalosiphon* or *Umezakia* have not been reported to form blooms.

As discussed in Chapters 2 and 3, the current understanding of cyanotoxin occurrence is incomplete, as the identification of further taxa producing specific cyanotoxins (and potentially yet unknown ones) is ongoing, and for most types of cyanotoxins, new producing taxa may be found in future. In particular, while it is well recognised that cyanobacterial populations consist of a mixture of clones, some of which produce toxins and others which do not, differentiation between toxic and nontoxic clones cannot be done reliably by microscopic examination. It became possible only with the advent of molecular methods (Kurmayer et al., 2004; section 13.4) or highly sensitive analytical methods that allow the detection of toxins in single colonies or filaments (Akçaalan et al., 2006; Welker & Erhard, 2007). Such methodology can characterise the clonal diversity of cyanobacterial communities, showing that a distinction between toxic and nontoxic clones is only a very rough classification (Janse et al., 2004). Rather, a multitude of genetically and metabolically distinct clones can be distinguished, with the production of cyanotoxins being one characteristic among others (Fastner et al., 2001a; Janse et al., 2003; Welker et al., 2004a).

Geographic variation in the genetic diversity of toxin-producing cyanobacteria is just beginning to be explored, and new insights into biogeographic patterns of toxin production are likely. Further surveys at a global scale will be particularly relevant for regions that have so far been poorly investigated.

4.6.1 Composition of blooms and cyanotoxin types

MCs appear to be the most widespread type of cyanotoxins occurring in health-relevant concentrations (Svirčev et al., 2019), and they are very likely to occur when species of *Microcystis* or *Planktothrix* are present. Field populations of these genera usually contain both clones with and without the genes for MC production (see below), yet it is very rare to find field populations of *Microcystis* without any MC-producing clones. In Europe, field populations of *Planktothrix agardhii* as well as of *P. rubescens* have never been observed without MC production (Kurmayer & Gumpenberger, 2006), although within these, some nonproducing clones occur frequently (Welker et al., 2004b). A number of studies have highlighted patterns of occurrence of particular MC congeners with particular taxa (e.g., Fastner et al., 1999). While for *Microcystis* blooms MC-LR, MC-RR and MC-YR typically are the most abundant congeners, in *P. agardhii* dominated blooms demethylated [Asp^3] congeners dominate and in *P. rubescens* the

[Asp^3, Dhb^7] congeners are typical. For risk assessment, specific toxicological data for congeners other than MC-LR are sparse or lacking, so as default or conservative estimate, these are regarded as being similarly toxic as MC-LR (see sections 2.1, 5.1 and 5.2).

Occurrence of cylindrospermopsin was first only expected when *R. raciborskii* was present in the plankton, but the number of taxa reported to produce CYNs is increasing (currently including species of *Aphanizomenon* and *Dolichospermum*). Further, among strains of *R. raciborskii*, only those originating from Australia or South East Asia were found to produce CYNs (Li et al., 2001b; Neilan et al., 2003; Piccini et al., 2011). Most CYN-producing species known so far belong to the order Nostocales (*sensu* Anagnostidis & Komárek (1985)) but evidence for CYN production in Stigonematales and Oscillatoriales indicates that the ability to produce CYN could be more widespread. However, CYN production appears to be more variable than MC production: while for MCs, occurrence can be reliably predicted when specific taxa are present with sufficiently high biomass (*Microcystis*, *Planktothrix*), no taxa have been reported which are almost always found together with CYNs. Thus, compared to MCs, for CYNs the frequency of nonproducing clones appears to be relatively high. This observation could, however, be biased to some degree by the poorly resolved taxonomy for some cyanobacteria (Chapter 3).

The same appears to apply to the cyanobacterial neurotoxins, although less data are available on their occurrence and producing strains. For neurotoxins, relationships between toxigenicity and taxonomy are less clear, as information on genes encoding their biosynthesis has become available later (see sections 2.3–2.5 and 13.4), but molecular methods are needed to unambiguously verify the potential to produce neurotoxins in large numbers of samples and strains (Wang et al., 2015). Like with MCs and CYNs, the ability to produce neurotoxins is not confined to a particular order of cyanobacteria, and within genera or even species, toxic and nontoxic clones exist (Beltran & Neilan, 2000). There are also possible biogeographic differences in neurotoxin production: for example, while for *Raphidiopsis* exclusively Brazilian strains seem to produce saxitoxins (Haande et al., 2008), for *D. circinale* only Australian strains have been shown to produce saxitoxins (Fergusson & Saint, 2000; Al-Tebrineh et al., 2010).

4.6.2 Toxin content in biomass

The term "concentration", according to IUPAC definitions, refers to an amount of toxin (generally expressed as mass or weight) per volume of water. This is best distinguished from toxin "content" referring to an amount of toxin per unit of cyanobacterial biomass, with the latter expressed as weight or as biovolume (Box 4.9). This distinction is not always made clear in the literature, rendering the comparison of data sometimes difficult.

BOX 4.9: UNITS TO REPORT CYANOTOXINS

The occurrence of cyanobacterial toxins is reported in a variety of units that are not always easily transformed from one to another. The most commonly used units are given in this box together with exemplary values that can serve as reference data to judge the plausibility of measured values.

Cell quota: amount of toxin per cell, expressed either in gravimetric or in molar units; this is femtogram (fg) per cell or femto- or attomol (fmol or amol, respectively) per cell. Cell quota are used primarily in laboratory experiments on MC production but also sometimes in field studies and generally for unicellular taxa like *Microcystis*. In individual toxigenic *Microcystis* strains, cell quota have been found to vary by a factor of about three, and the maximum values reported are 165 fg/cell (Orr & Jones, 1998; Long et al., 2001; Wiedner et al., 2003). From field samples, cell quota ranging from 1 to 144 fg/cell have been calculated (Okello et al., 2010). For CYNs, cell quota of 60 fg/cell (Orr et al., 2010) and 191 fg/cell (Vasas et al., 2013) have been reported for *R. raciborskii* and *Chrysosporum ovalisporum*, respectively. For saxitoxin, cell quota of 1.3 picogram (pg) per cell in a *Scytonema* strain have been reported together with a toxin content of 119 µg/g DW (Smith et al., 2011). Cells of *Scytonema* have a volume about fifty times that of a *Microcystis* cell. Any numbers considerably beyond these ranges may be the result of a conversion error or inaccuracies in the quantification of the cell number or the toxin concentration (see Box 4.10).

Toxin content: amount of toxin per mass of cells, expressed in µg or mg per gram dry weight (dw) in the older literature. A large number of values have been published for different cyanotoxins: in individual strains maximally 8 mg MC/g DW and 4.6 mg CYN/g dw (Sivonen, 1990; Long et al., 2001) and in field samples from below detection limit to a few mg/g dw. Toxin content may also be expressed as µg per mm^3 biovolume. As 1 mm^3 equals ~1 mg of fresh weight or about ~0.25 mg DW, and 1 µg toxin/mm^3 corresponds to about 4 µg toxin/mg dw. A third common unit for toxin content is µg toxin per µg chlorophyll-*a*, rarely exceeding a 1:1 ratio.

Toxin concentration: amount of toxin per volume of water, often expressed in µg/L or if concentrations are very low, in ng/L (= 0.001 µg/L). Two fractions need to be discriminated, particulate and dissolved. The particulate fraction comprises toxins contained in cells and adsorbed to particles or, from a practical point of view, the toxins that can be collected on a filter. Accordingly, the dissolved fraction is in the particle-free sample that passed the filter. Toxin concentration is the most important parameter for risk assessment.

Dry weight: A wealth of (published) data exists on toxin contents of bloom samples or isolated strains on a weight-to-weight basis (e.g., μg MC per g dry weight). Analytical methods can detect most toxins in the ppb range, that is, a few tens of nanograms per gram dry weight (for units and scales, see Box 4.10). The highest reported toxin contents for cultured strains are around 1.5% of dry weight for MCs and CYNs, corresponding to 15 μg toxin per mg dry weight (Chapter 2). The maximally achievable toxin contents of plankton net samples are thus reasonably expected in the same range. It follows that any reported toxin content exceeding 2% of dry weight is remarkably high and needs to be critically verified.

BOX 4.10: SCALES AND MAGNITUDES OF MASS AND VOLUME

"For thousands more years, the mighty ships tore across the empty wastes of space and finally dived screaming on the first planet they came across – which happened to be the earth – where due to a terrible miscalculation of scale the entire battle fleet was accidentally swallowed by a small dog"

Douglas Adams, The Hitchhiker's Guide to the Galaxy

What reads as funny side-kick in a novel conveys some deeper meaning: when dealing with masses and volumes in extremely low ranges, anyone who is not familiar with the units has to be very careful to avoid conversion errors that could lead to the reporting of false numbers.

For units of mass (often also termed "weight"), the relevant units for cyanotoxin research are gram (g), milligram (mg), microgram (μg), nanogram (ng), picogram (pg) and femtogram (fg). The conversion factor from one unit to the next is 1000; this means, for example, 1 μg equals 1,000,000,000 fg = 10^9 fg.

For units of volumes, in the context of toxic cyanobacteria, the units litre (L), millilitre (mL), cubic millimetre (mm^3; to report phytoplankton population biovolume) and cubic micrometre (μm^3; to report individual cells' volumes) are relevant. While conversion of L to mL to μL is done with respective factors of 1000, the conversion of mm^3 to μm^3 with a factor of 1 000 000 000 (10^9) requires particular attention. Conversion factors are summarised in the following table:

volume	*dm³*	*cm³*	*mm³*	*μm³*
dm³=L	1	1.0E+03	1.0E+06	1.0E+15
cm³=mL	1.0E−03	1	1.0E+03	1.0E+12
mm³=μL	1.0E−06	1.0E−03	1	1.0E+09
μm³=fL	1.0E−15	1.0E−12	1.0E−09	1

Assuming a specific density of ρ = 1 g/mL for fresh cyanobacterial cells – the density of water – biovolume converts to fresh weight (fw) as follows: 1 mm^3 → 1 mg and 1 μm^3 → 1 pg. For the conversion of fw to dry weight (dw), a factor of 0.25 is considered. Based on this and the assumptions of an MC content of 1% of dw, the amount of toxin per cell can be computed like in the following table (with three exemplary size classes of *Microcystis*):

	Dolichospermum	*Microcystis*	*Microcystis*	*Microcystis*
Cell diameter	10 µm	4 µm	4.5 µm	5 µm
Cell volume ($V = \pi/6 \times d^3$)	523 μm^3	34 μm^3	48 μm^3	65 μm^3
Cell mass (wet weight)	523 pg	34 pg	48 pg	65 pg
Cell dry weight (25%)	131 pg	8.5 pg	12 pg	16 pg
Calculated cell quota	1300 fg/cell	85 fg/cell	120 fg/cell	160 fg/cell

For *Microcystis*, the computed cell quota are in the range of reported cell quota estimated from direct cell counts and MC analyses; hence, the estimates are consistent with reported values. In turn, any cell quota substantially exceeding these values have to be critically evaluated. For example, cell quota reported for *Microcystis* of 1 pg/cell correspond to an improbable MC content of some 10% of dw, 5 pg/cell to very unrealistic 50%, and any value exceeding 10 pg/cell is close to the cells' dry weight or actually exceeds it – which is physically impossible.

It is thus highly advisable to perform basic plausibility tests based on known cell dimension, biovolumes, cell numbers, etc. to verify that calculated ratios are physically possible and in accordance with available data. Any value that is substantially beyond the values reported so far has to be carefully checked for possible errors in measurements and conversions – once erroneous data have been published, they can develop a life of their own through continuing citation.

Since most cyanobacterial blooms do not consist of a single clone with an extremely high toxin content, but comprise mixtures of clones with varying toxin contents, including nontoxigenic ones, field samples generally show toxin contents that are much lower than 1% of dry weight. A number of exemplary values for toxin contents of cultured strains are given in Chapter 2. However, dry weight is not practical for estimating toxin concentrations from biomass because it requires collecting a sufficiently large amount of biomass from a defined volume of water with a plankton net for drying and weighing. Water volumes are poorly defined if the sample is collected by hauling the net through the water column. With more sensitive analytical methods, small sample sizes suffice for accurate toxin analysis. The more practical approach therefore is to relate toxin content to biovolume (see Chapter 13 for the determination of biovolume by microscopy.)

Biovolumes: Data from field samples and laboratory experiments suggest that, for example, for MCs, a **ratio of 3 µg MC per mm³ biovolume** is rarely exceeded in the mixture of clones or genotypes occurring in waterbodies (Fastner et al., 2001b; Hesse & Kohl, 2001; Znachor et al., 2006) even though outliers of up to 13.8 µg/mm³ and 20.6 µg/mm³ have been found in cultured strains of *Microcystis* and *Planktothrix*, respectively (see Table 2.3). This small number of higher ratios reported in the literature should not be given too much weight unless the values have been verified with further values in the same range. One reason for outliers may partly be attributed to (unavoidable) inaccuracies in biovolume determination and analytical errors when the concentrations are close to the limit of quantification: any computing of ratios is sensitive to relative errors and errors in both values may potentiate the ratio's error (Chapter 13). For example, an underestimation of biovolume by setting the average cell diameter to 4 µm while the true value is 4.5 µm results in a biovolume with an error of 30%. Together with an overestimation of toxin concentration as 1 µg/L instead of truly 0.7 µg/L (error of 30%) already results in an overestimation of the toxin-to-biovolume ratio by a factor 2. Therefore, based on the bulk of the reported data, a value of 3 µg/mm³ of biovolume serves as a conservative assumption.

Cell quota: Not very many data are available on toxin contents per cell (generally expressed in femtogram (fg) per cell). Exemplary values are compiled in Box 4.9 and for cultured strains in Chapter 2; these suggest, for example, for **MCs, a content of 200 fg per cell for *Microcystis*** sp. is a conservative assumption. Again, outliers have to be critically verified before taken into consideration, for example, by rough estimates whether the cell quota are consistent with toxin contents per dry weight reported for specific toxins and taxa (Box 4.10). Based on assumed cell quota of 200 fg per cell of *Microcystis*, one million (10^6) cells in a bloom can account for maximally 0.2 µg MC but expectedly, in typically multiclonal populations, average cell quota are lower and hence the amount of toxin, too. Therefore, for an *in situ* MC concentration of 1 µg/L, *Microcystis* cell densities of at least 10^7 cells/L are expected. For other toxins and taxa, only few values are available (Chapter 2), and unfortunately erroneous toxin contents (even exceeding the cell's dry weight) have been published. Box 4.10 gives factors and an example for converting weights, biovolumes and cell quota to support checking for plausibility of values for toxin contents.

4.6.3 Toxin concentrations in water

The dynamic fluctuations in cyanobacterial occurrence discussed above impact on cyanotoxin concentrations. Studies of seasonal dynamics of cyanotoxin concentrations in relation to trophic and geographic gradients have mainly focused on MCs and cover a number of different types of waterbodies. A number of publications report significant correlations between MC concentrations and environmental variables – particularly total phosphorus (TP) (Kotak et al., 2000; Graham et al., 2006), N:P ratio (Kotak et al., 2000; Welker et al., 2003), dissolved inorganic and total nitrogen (TN) (Giani et al., 2005) or temperature (Albay et al., 2005). However, all of these conditions typically directly affect cyanobacterial proliferation and dominance (section 4.4), and the chief biological mechanism behind these correlations is the wax and wane of cyanobacterial populations rather than effects on toxin production by individual clones in these populations. Generally, the results show correlations of cyanotoxin concentration to a measure of phytoplankton biomass such as chlorophyll-*a* or – more specifically – biovolumes of potentially toxigenic cyanobacterial taxa (Znachor et al., 2006; Sitoki et al., 2012; Salmaso et al., 2014). The data in all of these studies show upper boundaries and below these a broad residual scatter, indicating that further variables, other than biomass, also influence cyanotoxin concentrations.

A key explanation for highly variable ratios of cyanotoxin to cyanobacterial biomass (e.g., expressed as biovolume) is shifts in clonal composition, that is, between clones of different toxin content (including nonproducers) which would influence the average cyanotoxin content of the cyanobacterial population and hence of the entire phytoplankton community (Kardinaal et al., 2007; Welker et al., 2007; Yepremian et al., 2007; Agha et al., 2014; Haruštiaková & Welker, 2017; Otten et al., 2017). Figure 4.8 illustrates this schematically: in (the improbable) scenario 1, all clones, each of which with a different toxin content, show homogenous dynamics and the relative clonal composition is stable. This would result in an equally stable average toxin content and a toxin concentration tightly following the total cell number. In scenario 2, individual clones show diverging dynamics. As a consequence, the average toxin content varies strongly, and the toxin concentration is partly uncoupled from the total cell number. In natural waters, dynamics of cyanobacterial populations are likely closer to scenario 2 than to scenario 1, although the number of respective studies is still limited. Coexisting toxigenic clones can also vary considerably in their toxin composition (i.e., in individual congeners produced) and in their (average) toxin content or toxin cell quota (Chapter 2 and Box 4.9) that may span ranges exceeding two orders of magnitude (Wilson et al., 2006b; Yepremian et al., 2007). The fact that cyanobacterial blooms and populations are generally composed of mixtures of toxin-producing and

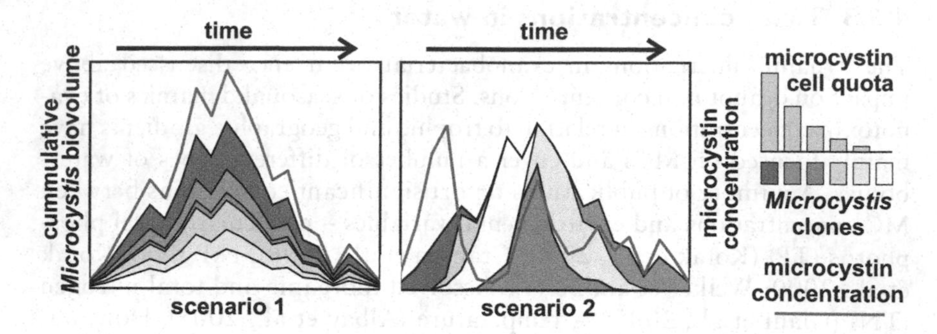

Figure 4.8 Schematic illustration of dynamics of *Microcystis* spp. clones with varying microcystin (MC) content (or cell quota) and the resulting MC concentration. In (unlikely) scenario 1, the temporal dynamic of all clones is homogeneous: that is, clone composition does not change over time, and the MC concentration follows *Microcystis* total biovolume. In scenario 2, the individual clones show variable dynamics resulting partly in a decoupling of *Microcystis* biovolume and MC concentration. For further explanations, see text.

nonproducing clones has also been demonstrated for CYN and for neurotoxins (Carmichael & Gorham, 1981; Fastner et al., 2001a; Wood & Puddick, 2017).

In some waterbodies, *Planktothrix*- and *Microcystis*-dominated cyanobacterial communities show seasonal patterns with the MC content of cyanobacterial biomass declining from a maximum at the onset of the bloom to lower values later in the season, coinciding with shifts in relative abundance of MC-producing clones (Kardinaal & Visser, 2005; Kardinaal et al., 2007; Welker et al., 2007; Davis et al., 2009), while in other waterbodies a reverse pattern was observed (see Chapter 5, Figure 5.1). In other waterbodies, the proportion of toxigenic cells in relation to total cell numbers proved rather stable for extended periods of time (Kurmayer et al., 2003; Salmaso et al., 2014). Also, MC concentrations may reach their maximum even at times when toxin content per cell or unit biovolume is relatively low if the total abundance of potentially toxigenic cyanobacteria is high.

In contrast to the decisive role of cyanobacterial biomass and clonal composition for toxin concentrations, laboratory studies showed that for a given clone, ratios of toxin per cell rarely change by more than a factor of 2–4. Moreover, the striking differences between laboratory studies addressing the impact of environmental conditions on toxin production are likely partially due to differences in study design (see Box 4.11). Compared to shifts in cyanobacterial biomass and/or clone composition, such shifts in toxin cell quota in individual clones have a relatively minor impact on toxin contents of field populations (Chapter 2).

BOX 4.11: PITFALLS WHEN DETERMINING CELLULAR CYANOTOXIN CONTENTS IN THE LABORATORY

Gary Jones

The literature is replete with laboratory studies attempting to correlate cellular cyanotoxin content with growth-limiting resources such as nitrogen, phosphorus, iron, light or temperature. Some of their apparently contradictory results may be due to inadequate experimental controls over the limiting resource and of how it regulates cyanobacterial growth rate and yield, if unequivocal single-resource limitation was not given. This is especially the case for batch cultures, where *in vitro* conditions (e.g., pH, light, nutrient concentration) and growth rate change continuously as the cells multiply (especially towards the end of exponential phase). Some of these batch cultures problems can be overcome through very carefully controlled design and sampling or, better still, by using continuous culture systems. An important consideration, often not fully accounted for, is that both batch and continuous culture experiments can be designed (intentionally or unintentionally!) so that the limiting resource reduces either one or both of (i) exponential growth rate and (ii) the cell concentration achieved in the plateau phase of a batch culture or the steady state of a continuous culture. Also often overlooked is the need for culture media which are chemically well defined, especially in the case of nutrients that have complex aqueous chemistry and speciation like phosphorus or iron.

Planning of experiments and interpretation of data requires proper training in microalgal culturing techniques, as well as a full understanding of the key theories behind resource limitation and growth (most notably the Monod and Droop equations), as well as the basic aqueous chemistry of common nutrient ions. Indeed, many of these considerations apply equally to the interpretation of field data as well.

Some excellent reference books to begin with are, for example, Stein et al. (1980) and Andersen (2005) on culture techniques and theory and Stumm & Morgan (1996) on water chemistry.

In conclusion, the average cyanotoxin concentration in a waterbody depends on

1. the combined biomass of potentially toxic cyanobacteria;
2. the clonal composition of populations and, in particular, on clone-specific toxin contents;
3. the physiological regulation of toxin biosynthesis in toxigenic clones.

While the clonal composition is crucial for the toxin concentration in the water, a distinction between only two types – toxigenic and nontoxigenic – is an undue simplification that may lead to invalid conclusions because both types likely are represented by multiple individual clones (Meyer et al., 2017). Other metabolites and physiological traits of individual clones may be decisive for the outcome of selection between clones. However, the driving forces regulating clonal composition of cyanobacterial populations are still poorly understood. For example, N limitation may play a role in selecting clones of *Microcystis* or *Planktothrix* that do not produce MCs (or other peptides) (Gobler et al., 2016), or parasite and phage infection may cause clone-specific mortality in cyanobacterial populations (Honjo et al., 2006; Rohrlack et al., 2013; Van Wichelen et al., 2016; Stough et al., 2017; Box 4.3).

Therefore, accurate predictions of cyanotoxin concentrations from cyanobacterial biomass are limited, even in an intensively studied waterbody. As an approximative approach, upper boundaries of possible toxin concentrations, based on conservative assumptions for toxin-to-biomass ratios, are used in the Alert Levels Frameworks in sections 5.1 and 5.2. As discussed in section 4.5, modelling is still struggling to predict cyanobacterial abundance as such, and forecasting cyanotoxin concentration adds a further dimension of complexity. However, some examples show that statistical models based on empirical data can predict cyanotoxin concentrations for a given waterbody with reasonable accuracy with respect to their timing as well as the magnitude of toxin occurrence (Jacoby et al., 2015; Recknagel et al., 2017). Although models cannot replace monitoring for the purpose of assessing cyanotoxin risks to human health, they can be used to streamline and focus monitoring, for example, by adjusting sampling frequency to expected bloom and toxin dynamics (Chapter 11). By using a default assumption for a ratio of toxin to cyanobacterial biomass, models to estimate cyanobacterial biomass can be expanded for a first, conservative estimate of toxin concentration. Such an approach likely results in an overestimation of the actual toxin concentration (Chapter 5 and section 4.6.5).

4.6.4 Spatial heterogeneity of toxin concentrations

As cyanobacterial biomass can show pronounced spatial variability, so do cyanotoxins: detectable cyanotoxin concentrations, reported as average concentrations from depth-integrated samples, generally range from just above detection limit to a few tens of µg/L. However, in individual samples taken in the same waterbodies on the same date, the toxin concentration can vary by several orders of magnitude, depending on where and how the sample is taken: surface layer *versus* hypolimnion or a central site *versus* the shoreline (see Figure 4.1, Figure 4.2, Box 4.8). In scums of *Microcystis*, for example, concentrations of MCs can reach several tens of mg/L (see section 2.1). In several multilake surveys, MC concentrations exceeding 50 µg/L were only found in scums where concentrations regularly reached the mg/L range, but not in integrated, averaged samples when cyanobacterial populations are suspended more or less homogeneously in the water column (Cook et al., 2004; Loftin et al., 2016). As concentrations of cyanotoxins can show pronounced horizontal gradients, data from a water sample taken at one site are unlikely to represent a lake-wide average (Welker et al., 2003; Ozawa et al., 2005; Dyble et al., 2008). It is thus of utmost importance to clearly define the objective and corresponding sampling strategy when designing monitoring programmes (Chapter 11).

Horizontal variability is mainly caused by two factors. One is differences in growth conditions that may occur between different parts of a lake, particularly in large lakes or reservoirs with complex morphology (e.g., Michalak et al., 2013; Chung et al., 2014). The other – often more important one – is the combination of surface bloom formation and horizontal dislocation by which cyanobacterial cells are generally concentrated at nearshore, downwind sites of a waterbody (section 4.5). However, the short-term fluctuations are not restricted to near-shore sites. Cyanotoxin concentrations in samples taken at high frequency at a fixed station showed fluctuations of several µg/L over the 6-h sampling interval (Miller et al., 2019).

Vertical variability in toxin concentration occurs mainly due to buoyancy regulation of the toxic cyanobacteria (section 4.2). Higher concentrations of cyanotoxins occur either in the topmost water layers, especially when buoyant taxa like *Microcystis* are dominant (Naselli-Flores et al., 2007), or in a metalimnetic layer in the case of *Planktothrix rubescens* (Ernst et al., 2009).

Since vertical and horizontal distributions depend to a large degree on current weather conditions such as wind direction and speed, dense cyanobacterial accumulations can occur within hours at one site of a lake and a few hours later at others. For approaches to early warning for surface bloom and scum occurrence, it is therefore effective to include the use of satellite images or continuous measurement of fluorescence, combined with numerical modelling (Ibelings et al., 2003; Chapter 11).

Two consequences are equally important for assessing the risk of cyanotoxin occurrence:

1. A cyanotoxin concentration of a few microgram per litre (that might be considered a low risk) at a central site in a waterbody can lead to cyanotoxin concentrations several orders of magnitude higher at downwind "hot spot" sites.
2. The timescales in which toxin concentrations can dramatically change at a particular (near-shore) site can be only a few hours.

4.6.5 Estimating cyanotoxin concentration from other limnological parameters

Where toxin data are not (or not readily) available for assessing and managing the health risks caused by cyanobacteria, measurements of cyanobacterial biomass are a useful point of departure for conservative estimates of **potential maximum cyanotoxin concentrations** in a waterbody. For practical purposes, such as the Alert Level Frameworks in sections 5.1 and 5.2, estimates therefore best use the maximum expectable concentrations. These will show where toxin analyses for the specific site are important in order to obtain more precise information to appropriately guide management responses, with significantly lower toxin concentrations being a likely outcome.

The limnological parameters allowing an estimation of cyanotoxin concentrations are primarily (cyanobacterial) biovolumes, cell counts, chlorophyll-*a* concentrations, water transparency (i.e., Secchi depth) and nutrient concentrations, in particular total phosphorus (TP). The uncertainty of cyanotoxin estimates increases in the same order. It is important to understand that estimates of maximum cyanotoxin concentrations based on these parameters will not be accurate; they merely serve as indicators to support decisions on where to focus efforts for monitoring and for further analyses, for example, of cyanotoxins (Chapters 13 and 14). In particular, it is important to realise that these indicators give conservative estimates gleaned from maximum likely ratios of toxin to the respective indicator. Due to their variability over time and between waterbodies, using any of them as an estimate for cyanotoxin concentration implies that follow-up by toxin analysis is most likely to result in a considerably lower rather than a higher human health risk.

While the ratios of toxin to indicator proposed here are primarily derived from data published for MCs, for other cyanotoxins, content per unit biomass does not appear higher than these maxima known for MCs (see Chapter 2); thus, the ratios of toxin to the concentration of the indicators discussed here may also be assumed as proxy for risk assessment. However – in contrast to MCs which largely occur cell-bound – applying this to CYNs may miss the higher dissolved fraction of this toxin.

Biovolumes can give a fairly accurate estimate of the maximum expectable cell-bound cyanotoxin concentrations, but yet bear uncertainty spanning more than an order of magnitude due to the variability of the toxin content of different clones of potentially toxigenic taxa, even within an individual waterbody and a single season. A key advantage of using biovolumes identified by microscopy as point of departure for risk assessment is that they include information on taxonomic composition, which provides a basis for understanding the potential further development of a bloom and thus of toxin occurrence. As discussed above in section 4.6.2, a ratio of 3 µg toxin per mm^3 biovolume serves as conservative estimate that is not likely to be exceeded in field samples.

Cell counts of potentially toxigenic cyanobacteria may likewise be used in combination with reported toxin cell quota of these taxa to estimate toxin concentrations. However, since reliable values for taxon and toxin-specific cell quota are scarce, this approach is limited to taxa for which cell quota estimates are available (see section 4.6.2). Furthermore, estimates based on cell counts and toxin cell quota are subject to the same limitations as estimates from biovolume.

Chlorophyll-*a* concentrations as a measure of biomass also include phytoplankton organisms other than cyanobacteria. Using them as indicator therefore requires a brief qualitative assessment by microscopy of whether or not the phytoplankton consists chiefly of cyanobacteria. While fluorescence methods, which include the cyanobacteria-specific pigment phycocyanin, can overcome this, they introduce a further dimension of variability (see section 13.5). Furthermore, the chlorophyll-*a* content of phytoplankton may vary in response to light and nutrient availability by a factor of up to 10 (Kruskopf & Flynn, 2006; Kasprzak et al., 2008). Also, if photometric analysis of chlorophyll-*a* includes a correction for phaeophytin, this may result in an underestimation of the concentration of chlorophyll-*a*, particularly in the low concentration range (Chapter 13.5), and thus in an overestimation of the toxin-to-chlorophyll-*a* ratio.

A number of reports on cyanotoxin concentrations (mainly MCs) suggest that the cell-bound toxin concentration (in µg/L) only very rarely exceeds the concentration of chlorophyll-*a* (also in µg/L) and if so, not by more than a factor of three (e.g., Fastner et al., 2001b; Carrasco et al., 2006; Sinang et al., 2013; Loftin et al., 2016; Mantzouki et al., 2018). For example, in Figure 4.9, the MC/chl-*a* ratio exceeded 1 only in a small share of samples, while the ratio for the majority of samples ranges between 0.1 and 0.5. An assumption of a **maximum ratio of 1 µg MC per µg chl-*a*** is hence a conservative approach, and in most cases, measured MC concentrations will be considerably lower than thus estimated.

Secchi depth is a parameter for water transparency which is easily measured on site. In many waterbodies, it correlates directly and inversely with phytoplankton abundance and chlorophyll-*a* concentration. However, respective correlations generally show considerable scatter because other water constituents (inorganic sediments suspended in the water, humic substances) and differences in optical properties of phytoplankton species affect water transparency. Nonetheless, waterbodies with consistently high Secchi depths are rarely dominated by cyanobacteria, and thus, high transparency indicates that potentially health-relevant cyanotoxin concentrations are unlikely. The transparency threshold value indicating cyanobacterial blooms is best established individually for a given waterbody or a region with a number of similar waterbodies, based on seasonal data for Secchi

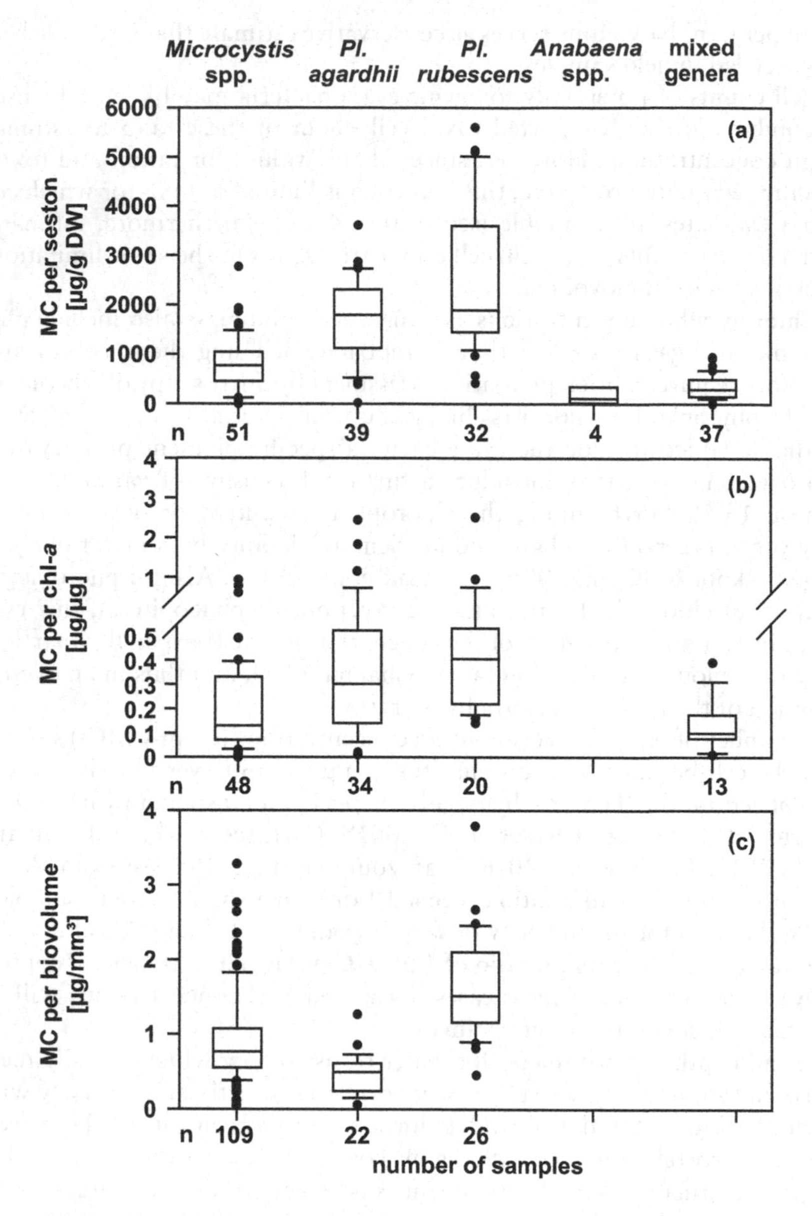

Figure 4.9 Distribution of microcystin (MCs) ratios in samples from German lakes: MC per dry weight of seston (a), MC per chlorophyll-*a* (b) and MC per biovolume (c). The samples are classified by types of dominance: that is, more than 50% of biovolume in a sample was due to a particular taxon. (Adapted by permission from Springer Nature, in: Cyanotoxins: occurrence, causes, consequences. Cyanotoxin occurrence in Germany: Microcystins and hepatocyte toxicity. Fastner J, Wirsing B, Wiedner C, Heinze R, Neumann U, Chorus I (2001b). Copyright 2001 Springer Nature. www.springernature.com/gp.)

depths and cyanobacterial biovolumes (or chlorophyll-*a* together with a semiquantitative check via microscopy as to whether cyanobacteria dominate). For small lakes of less than 50–100 ha, it may well be in the range of 2–3 m, while for larger waterbodies (due to the possibility of surface bloom recruitment from a large water volume; Figure 4.2), it may be higher, and scums may yet accumulate in a bay, while the Secchi disc reading taken at a central site may show several metres. Furthermore, Secchi disc readings cannot serve for assessing risks from benthic or tychoplanktonic cyanobacteria. Also, high Secchi depth readings will not indicate a cyanotoxin risk from *Planktothrix rubescens* in the metalimnion: these will not diminish transparency in the epilimnion (and may yet cause elevated toxin concentrations at drinking-water offtake depths).

In summary, while Secchi depth readings may be a very useful indicator of cyanotoxin risks, as for many other parameters their reliability depends on a sound understanding of the given waterbody.

Nutrient concentrations – primarily TP and total nitrogen (TN) – determine carrying capacity and hence the maximally achievable phytoplankton biomass (measured as biovolumes or chlorophyll-*a* concentrations) and hence the maximally achievable (average) cyanotoxin concentration (Dolman et al., 2012). Nutrient concentrations are useful as longer-term predictors whether blooms of cyanobacteria – and hence cyanotoxin concentrations at critical levels – are to be expected (Beaver et al., 2014; section 4.4 and Chapter 8).

REFERENCES

Abell JM, Özkundakci D, Hamilton DP (2010). Nitrogen and phosphorus limitation of phytoplankton growth in New Zealand lakes: implications for eutrophication control. Ecosystems. 13:966–977.

Agha R, Gross A, Gerphagnon M, Rohrlack T, Wolinska J (2018). Fitness and eco-physiological response of a chytrid fungal parasite infecting planktonic cya nobacteria to thermal and host genotype variation. Parasitology 145:1279–1286.

Agha R, Lezcano MA, del Mar Labrador M, Cirés S, Quesada A (2014). Seasonal dynamics and sedimentation patterns of *Microcystis* oligopeptide-based chemotypes reveal subpopulations with different ecological traits. Limnol Oceanogr 59:861–871.

Akçaalan R, Young FM, Metcalf JS, Morrison LF, Albay M, Codd GA (2006). Microcystin analysis in single filaments of *Planktothrix* spp. in laboratory cultures and environmental samples. Water Res. 40:1583–1590.

Al-Tebrineh J, Mihali TK, Pomati F, Neilan BA (2010). Detection of saxitoxin-producing cyanobacteria and *Anabaena circinalis* in environmental water blooms by quantitative PCR. Appl Environ Microbiol. 76:7836–7842.

Albay M, Matthiensen A, Codd GA (2005). Occurrence of toxic blue-green algae in the Kucukcekmece Lagoon (Istanbul, Turkey). Environ Toxicol. 20:277–284.

Anagnostidis K, Komárek J (1985). Modern approach to the classification system of Cyanophytes. 1-Introduction. Arch Hydrobiol Algol Stud. 38–39:291–302.

Andersen RA (2005). Algal culturing techniques. Amsterdam. Elsevier Academic Press:578 pp.

Anderson L, Sarmiento J (1994). Redfield ratios of remineralization determined by nutrient data analysis. Global Biogeochem Cy. 8:65–80.

Anneville O, Domaizon I, Kerimoglu O, Rimet F, Jacquet S (2015). Blue-green algae in a "Greenhouse Century"? New insights from field data on climate change impacts on cyanobacteria abundance. Ecosystems. 18:441–458.

Antunes JT, Leão PN, Vasconcelos VM (2015). *Cylindrospermopsis raciborskii*: review of the distribution, phylogeography, and ecophysiology of a global invasive species. Front Microbiol. 6:473.

Baker P, Humpage A (1994). Toxicity associated with commonly occurring cyanobacteria in surface waters of the Murray-Darling Basin, Australia. Mar Freshwater Res. 45:773–786.

Baker SM, Levinton JS, Kurdziel JP, Shumway SE (1998). Selective feeding and biodeposition by zebra mussels and their relation to changes in phytoplankton composition and seston load. J Shellfish Res. 17:1207–1213.

Barco M, Flores C, Rivera J, Caixach J (2004). Determination of microcystin variants and related peptides present in a water bloom of *Planktothrix* (*Oscillatoria*) *rubescens* in a Spanish drinking water reservoir by LC/ESI-MS. Toxicon. 44:881–886.

Barker GL, Konopka A, Handley BA, Hayes PK (2000). Genetic variation in Aphanizomenon (cyanobacteria) colonies from the Baltic Sea and North America. J Phycol. 36:947–950.

Beattie KA, Kaya K, Codd GA (2000). The cyanobacterium *Nodularia* in PCC 7804, of freshwater origin, produces [L-Har(2)]nodularin. Phytochemistry. 54:57–61.

Beaver JR, Manis EE, Loftin KA, Graham JL, Pollard AI, Mitchell RM (2014). Land use patterns, ecoregion, and microcystin relationships in US lakes and reservoirs: a preliminary evaluation. Harmful Algae. 36:57–62.

Begon M, Harper JL, Townsend CR (1996). Ecology: individuals, populations and communities, fourth edition. Oxford: Blackwell Scientific Publications.

Beltran EC, Neilan BA (2000). Geographical segregation of the neurotoxin-producing cyanobacterium *Anabaena circinalis*. Appl Environ Microbiol. 66:4468–4474.

Bianchi TS, Engelhaupt E, Westman P, Andren T, Rolff C, Elmgren R (2000). Cyanobacterial blooms in the Baltic Sea: Natural or human-induced? Limnol Oceanogr. 45:716–726.

Bolch CJS, Orr PT, Jones GJ, Blackburn SI (1999). Genetic, morphological, and toxicological variation among globally distributed strains of *Nodularia* (Cyanobacteria). J Phycol. 35:339–355.

Boström B, Pettersson A-K, Ahlgren I (1989). Seasonal dynamics of a cyanobacteria-dominated microbial community in surface sediments of a shallow, eutrophic lake. Aquat Sci. 51:153–178.

Bouma-Gregson K, Olm MR, Probst AJ, Anantharaman K, Power ME, Banfield JF (2019). Impacts of microbial assemblage and environmental conditions on the distribution of anatoxin-a producing cyanobacteria within a river network. ISME J. 13: 1618–1634.

Briand JF, Leboulanger C, Humbert JF, Bernard C, Dufour P (2004). *Cylindrospermopsis raciborskii* (Cyanobacteria) invasion at mid-latitudes: Selection, wide physiological tolerance, or global warming? J Phycol. 40:231–238.

Bright DI, Walsby AE (2000). The daily integral of growth by *Planktothrix rubescens* calculated from growth rate in culture and irradiance in Lake Zürich. New Phytol. 146:301–316.

Bulgakov N, Levich A (1999). The nitrogen: phosphorus ratio as a factor regulating phytoplankton community structure: nutrient ratios. Arch Hydrobiol. 146:3–22.

Burian A, Kainz MJ, Schagerl M, Yasindi A (2014). Species-specific separation of lake plankton reveals divergent food assimilation patterns in rotifers. Freshwat Biol. 59:1257–1265.

Cadel-Six S, Peyraud-Thomas C, Brient L, de Marsac NT, Rippka R, Méjean A (2007). Different genotypes of anatoxin-producing cyanobacteria coexist in the Tarn River, France. Appl Environ Microbiol. 73:7605–7614.

Cantoral Uriza E, Asencio A, Aboal M (2017). Are we underestimating benthic cyanotoxins? extensive sampling results from Spain. Toxins. 9:385.

Capone DG, Burns JA, Montoya JP, Subramaniam A, Mahaffey C, Gunderson T et al. (2005). Nitrogen fixation by *Trichodesmium* spp.: an important source of new nitrogen to the tropical and subtropical North Atlantic Ocean. Global Biogeochem Cy. 19.1–17.

Carey CC, Ibelings BW, Hoffmann EP, Hamilton DP, Brookes JD (2012). Ecophysiological adaptations that favour freshwater cyanobacteria in a changing climate. Water Res. 46:1394–1407.

Carlson RE (1977). A trophic state index for lakes. Limnol Oceanogr. 22:361–369.

Carmichael WW, Eschedor JT, Patterson GML, Moore RE (1988). Toxicity and partial structure of a hepatotoxic peptide produced by the cyanobacterium *Nodularia spumigena* Mertens emend. L575 from New Zealand. Appl Environ Microbiol. 54:2257–2263.

Carmichael WW, Gorham PR (1981). The mosaic nature of toxic blooms of cyanobacteria. In: Carmichael WW, editors: The water environment: algal toxins and health. New York: Plenum Press:161–172.

Carpenter SR, Ludwig D, Brock WA (1999). Management of eutrophication for lakes subject to potentially irreversible change. Ecol Appl. 9:751–771.

Carrasco D, Moreno E, Sanchis D, Wörmer L, Paniagua T, Del Cueto A et al. (2006). Cyanobacterial abundance and microcystin occurrence in Mediterranean water reservoirs in Central Spain: microcystins in the Madrid area. Europ J Phycol. 41:281–291.

Carvalho L, McDonald C, de Hoyos C, Mischke U, Phillips G, Borics G et al. (2013). Sustaining recreational quality of European lakes: minimizing the health risks from algal blooms through phosphorus control. J Appl Ecol. 50:315–323.

Chapman EJ, Byron CJ (2018). The flexible application of carrying capacity in ecology. Global Ecol Conserv. 13:e00365.

Cheng Y, Canuto V, Howard A (2002). An improved model for the turbulent PBL. J Atmos Sci. 59:1550–1565.

Chonudomkul D, Yongmanitchai W, Theeragool G, Kawachi M, Kasai F, Kaya K et al. (2004). Morphology, genetic diversity, temperature tolerance and toxicity of *Cylindrospermopsis raciborskii* (Nostocales, Cyanobacteria) strains from Thailand and Japan. FEMS Microbiol Ecol 48:345–355.

Chorus I, Dokulil M, Lammens E, Manca M, Naselli-Flores L, Nixdorf B et al (2011). Restoration responses of 19 lakes: are TP thresholds common? In: Chorus I, Schauser I, editors. Oligotrophication of Lake Tegel and Schlachtensee, Berlin Analysis of system components, causalities and response thresholds compared to responses of other waterbodies. Dessau: Umweltbundesamt:84–102.

Chorus I, Niesel V (2011). Steps towards a statistical model to predict phytoplankton responses to changes in trophic state. In: Chorus I, Schauser I, editors. Oligotrophication of Lake Tegel and Schlachtensee, Berlin Analysis of system components, causalities and response thresholds compared to responses of other waterbodies. Dessau: Umweltbundesamt:109–139.

Chorus I, Spijkerman E (2020). What Colin Reynolds could tell us about nutrient limitation, N:P ratios and eutrophication control. Hydrobiologia, in press

Chung S, Imberger J, Hipsey M, Lee H (2014). The influence of physical and physiological processes on the spatial heterogeneity of a *Microcystis* bloom in a stratified reservoir. Ecol Model. 289:133–149.

Cirés S, Wörmer L, Agha R, Quesada A (2013). Overwintering populations of *Anabaena*, *Aphanizomenon* and *Microcystis* as potential inocula for summer blooms. J Plankton Res. 35:1254–1266.

Conley DJ, Paerl HW, Howarth RW, Boesch DF, Seitzinger SP, Karl E et al. (2009). Controlling eutrophication: nitrogen and phosphorus. Science. 123:1014–1015.

Cook CM, Vardaka E, Lanaras T (2004). Toxic cyanobacteria in Greek freshwaters, 1987–2000: Occurrence, toxicity, and impacts in the Mediterranean region. Acta Hydrochim Hydrobiol. 32:107–124.

Cuypers Y, Vincon-Leite B, Groleau A, Tassin B, Humbert JF (2011). Impact of internal waves on the spatial distribution of *Planktothrix rubescens* (cyanobacteria) in an alpine lake. ISME J. 5:580–589.

Davis TW, Berry DL, Boyer GL, Gobler CJ (2009). The effects of temperature and nutrients on the growth and dynamics of toxic and non-toxic strains of *Microcystis* during cyanobacteria blooms. Harmful Algae. 8:715–725.

De Senerpont Domis LN, Elser JJ, Gsell AS, Huszar VL, Ibelings BW, Jeppesen E et al. (2013). Plankton dynamics under different climatic conditions in space and time. Freshwater Biol. 58:463–482.

Diaz M, Pedrozo F, Reynolds C, Temporetti P (2007). Chemical composition and the nitrogen-regulated trophic state of Patagonian lakes. Limnologica. 37:17–27.

Dionisio Pires LM, Bontes BM, van Donk E, Ibelings BW (2005). Grazing on colonial and filamentous, toxic and non-toxic cyanobacteria by the zebra mussel *Dreissena polymorpha*. J Plankton Res. 27:331–339.

Dokulil MT, Mayer J (1996). Population dynamics and photosynthetic rates of a *Cylindrospermopsis-Limnothrix* association in a highly eutrophic urban lake, Alte Donau, Vienna, Austria. Arch Hydrobiol Algol Stud. 83:179–195

Dokulil MT, Teubner K (2000). Cyanobacterial dominance in lakes. Hydrobiologia. 438:1–12.

Dolman AM, Rücker J, Pick FR, Fastner J, Rohrlack T, Mischke U et al. (2012). Cyanobacteria and cyanotoxins: the influence of nitrogen versus phosphorus. PLoS One. 7:e38757.

Downing JA, Watson SB, McCauley E (2001). Predicting Cyanobacteria dominance in lakes. Can J Fish Aquat Sci. 58:1905–1908.

Dyble J, Fahnenstiel GL, Litaker RW, Millie DF, Tester PA (2008). Microcystin concentrations and genetic diversity of *Microcystis* in the lower Great Lakes. Environ Toxicol. 23:507–516.

Dyble J, Paerl HW, Neilan BA (2002). Genetic characterization of *Cylindrospermopsis raciborskii* (Cyanobacteria) isolates from diverse geographic origins based on nifH and cpcBA-IGS nucleotide sequence analysis. Appl Environ Microbiol. 68:2567–2571.

Edmondson W, Litt AH (1982). *Daphnia* in lake Washington. Limnol Oceanogr. 27:272–293.

Elliott JA (2012). Is the future blue-green? A review of the current model predictions of how climate change could affect pelagic freshwater cyanobacteria. Water Res. 46:1364–1371.

Elser JJ, Bracken ME, Cleland EE, Gruner DS, Harpole WS, Hillebrand H et al. (2007). Global analysis of nitrogen and phosphorus limitation of primary producers in freshwater, marine and terrestrial ecosystems. Ecol Lett. 10:1135–1142.

Elser JJ, Urabe J (1999). The stoichiometry of consumer-driven nutrient recycling: Theory, observations, and consequences. Ecology. 80:735–751.

Ernst B, Hoeger SJ, O'brien E, Dietrich DR (2009). Abundance and toxicity of *Planktothrix rubescens* in the pre-alpine Lake Ammersee, Germany. Harmful Algae. 8:329–342.

Fastner J, Abella S, Litt A, Morabito G, Vörös L, Pálffy K et al. (2016). Combating cyanobacterial proliferation by avoiding or treating inflows with high P load – experiences from eight case studies. Aquat Ecol. 50:367–383.

Fastner J, Beulker C, Geiser B, Hoffmann A, Kröger R, Teske K et al. (2018). Fatal neurotoxicosis in dogs associated with tychoplanktic, anatoxin-a producing *Tychonema* sp. in mesotrophic Lake Tegel, Berlin. Toxins. 10:60.

Fastner J, Erhard M, Carmichael WW, Sun F, Rinehart KL, Rönicke H et al. (1999). Characterization and diversity of microcystins in natural blooms and strains of the genera *Microcystis* and *Planktothrix* from German freshwaters. Arch Hydrobiol. 145:147–163.

Fastner J, Erhard M, von Döhren H (2001a). Determination of oligopeptide diversity within a natural population of *Microcystis* spp. (Cyanobacteria) by typing single colonies by matrix-assisted laser desorption ionization-time of flight mass spectrometry. Appl Environ Microbiol. 67:5069–5076.

Fastner J, Wirsing B, Wiedner C, Heinze R, Neumann U, Chorus I (2001b). Cyanotoxin occurrence in Germany: microcystins and hepatocyte toxicity. In: Chorus I, editors: Cyanotoxins: occurrence, causes, consequences. Berlin: Springer:22–37.

Fergusson KM, Saint CP (2000). Molecular phylogeny of *Anabaena circinalis* and its identification in environmental samples by PCR. Appl Environ Microbiol. 66:4145–4148.

Fetscher AE, Howard MDA, Stancheva R, Kudela RM, Stein ED, Sutula MA et al. (2015). Wadeable streams as widespread sources of benthic cyanotoxins in California, USA. Harmful Algae. 49:105–116.

Feuillade J (1994). The cyanobacterium (blue-green alga) *Oscillatoria rubescens* D. C. Adv Limnol. 41:77–93.

Finlay BJ (2002). Global dispersal of free-living microbial eukaryote species. Science. 296:1061–1063.

Foy R, Gibson C, Smith R (1976). The influence of daylength, light intensity and temperature on the growth rates of planktonic blue-green algae. Brit Phycol J. 11:151–163.

Francis G (1878). Poisonous Australian lake. Nature. 18:11–12.

Frenken T, Wierenga J, Gsell AS, van Donk E, Rohrlack T, Van de Waal DB (2017). Changes in N: P supply ratios affect the ecological stoichiometry of a toxic cyanobacterium and its fungal parasite. Front Microbiol. 8:1015.

Gabriel J-P, Saucy F, Bersier L-F (2005). Paradoxes in the logistic equation? Ecol Model. 185:147–151.

Gallina N, Beniston M, Jacquet S (2017). Estimating future cyanobacterial occurrence and importance in lakes: a case study with *Planktothrix rubescens* in Lake Geneva. Aquat Sci. 79:249–263.

Gallon JR (2004). N-2 fixation by non-heterocystous cyanobacteria. In: Klipp W, Masepohl B, Gallon JR et al., editors: Genetics and regulation of nitrogen fixation in free-living bacteria. Dordrecht: Springer:111–139.

Gammeter S, Zimmermann U (2001). Changes in phytoplankton productivity and composition during re-oligotrophication in two Swiss lakes. Verh Int Verein Limnol. 27:2190–2193.

Garcia-Pichel F, Belnap J, Neuer S, Schanz F (2003). Estimates of global cyanobacterial biomass and its distribution. Algological Studies. 109:213–227.

Ger KA, Urrutia-Cordero P, Frost PC, Hansson L-A, Sarnelle O, Wilson AE et al. (2016). The interaction between cyanobacteria and zooplankton in a more eutrophic world. Harmful Algae. 54:128–144.

Giani A, Bird DF, Prairie YT, Lawrence JF (2005). Empirical study of cyanobacterial toxicity along a trophic gradient of lakes. Can J Fish Aquat Sci. 62:2100–2109.

Gobler CJ, Burkholder JM, Davis TW, Harke MJ, Johengen T, Stow CA et al. (2016). The dual role of nitrogen supply in controlling the growth and toxicity of cyanobacterial blooms. Harmful Algae. 54:87–97.

Gragnani A, Scheffer M, Rinaldi S (1999). Top-down control of cyanobacteria: a theoretical analysis. Am Nat. 153:59–72.

Graham JL, Jones JR, Jones SB, Clevenger TE (2006). Spatial and temporal dynamics of microcystin in a Missouri reservoir. Lake Reserv Manage. 22:59–68.

Gugger MF, Lenoir S, Berger C, Ledreux A, Druart JC, Humbert JF et al. (2005a). First report in a river in France of the benthic cyanobacterium *Phormidium favosum* producing anatoxin-a associated with dog neurotoxicosis. Toxicon. 45:919–928.

Gugger MF, Molica R, Le Berre B, Dufour P, Bernard C, Humbert JF (2005b). Genetic diversity of *Cylindrospermopsis* strains (Cyanobacteria) isolated from four continents. Appl Environ Microbiol. 71:1097–1100.

Haande S, Rohrlack T, Ballot A, Roberg K, Skulberg R, Beck M et al. (2008). Genetic characterisation of *Cylindrospermopsis raciborskii* (Nostocales, Cyanobacteria) isolates from Africa and Europe. Harmful Algae. 7:692–701.

Harris TD, Wilhelm FM, Graham JL, Loftin KA (2014). Experimental manipulation of TN:TP ratios suppress cyanobacterial biovolume and microcystin concentration in large-scale *in situ* mesocosms. Lake Reserv Manage. 30:72–83.

Haruštiaková D, Welker M (2017). Chemotype diversity in *Planktothrix rubescens* (cyanobacteria) populations is correlated to lake depth. Environ Microbiol Rep. 9:158–168.

Havens K, Jeppesen E (2018). Ecological responses of lakes to climate change. Water. 10:917.

Hesse K, Kohl J-G (2001). Effects of light and nutrients supply on growth and microcystin content of different strains of *Microcystis aeruginosa*. In: Chorus I, editors: Cyanotoxins: occurrence, causes, consequences. Berlin: Springer:104–115.

Higgins SN, Paterson MJ, Hecky RE, Schindler DW, Venkiteswaran JJ, Findlay DL (2018). Biological nitrogen fixation prevents the response of a eutrophic lake to reduced loading of nitrogen: evidence from a 46-year whole-lake experiment. Ecosystems. 21:1088–1100.

Honjo M, Matsui K, Ueki M, Nakamura R, Fuhrman JA, Kawabata Z (2006). Diversity of virus-like agents killing *Microcystis aeruginosa* in a hyper-eutrophic pond. J Plankton Res. 28:407–412.

Huber V, Wagner C, Gerten D, Adrian R (2012). To bloom or not to bloom: contrasting responses of cyanobacteria to recent heat waves explained by critical thresholds of abiotic drivers. Oecologia. 169:245–256.

Huisman J, Codd GA, Paerl HW, Ibelings BW, Verspagen JM, Visser PM (2018). Cyanobacterial blooms. Nat Rev Microbiol. 16:471.

Huisman J, Sharples J, Stroom JM, Visser PM, Kardinaal WEA, Verspagen JM et al. (2004). Changes in turbulent mixing shift competition for light between phytoplankton species. Ecology. 85:2960–2970.

Humphries SE, Lyne VD (1988). Cyanophyte blooms: the role of cell buoyancy. Limnol Oceanogr. 33:79–91.

Huszar V, Caraco N, Roland F, Cole J (2006). Nutrient-chlorophyll relationships in tropical-subtropical lakes: do temperate models fit? In: Martinelli LA, Howarth RW, editors: Nitrogen cycling in the Americas: natural and anthropogenic influences and controls. Netherlands: Springer:239–250.

Hutchinson PA, Webster IT (1994). On the distribution of blue-green algae in lakes: wind tunnel tank experiments. Limnol Oceanogr. 39:374–382.

Ibelings B, Kroon BM, Mur L (1994). Acclimation of Photosystem II in a cyanobacterium and a eukaryotic green alga to high and fluctuating photosynthetic photon flux densities, simulating light regimes induced by mixing in lakes. New Phytol. 128:407–424.

Ibelings BW, Maberly SC (1998). Photoinhibition and the availability of inorganic carybon restrict photosynthesis by surface blooms of cyanobacteria. Limnol Oceanogr. 43:408–419.

Ibelings BW, Mur LR, Kinsman R, Walsby AE (1991). *Microcystis* changes its buoyancy in response to the average irradiance in the surface mixed layer. Arch Hydrobiol. 120:385–401.

Ibelings BW, Portielje R, Lammens EH, Noordhuis R, van den Berg MS, Joosse W et al. (2007). Resilience of alternative stable states during the recovery of shallow lakes from eutrophication: Lake Veluwe as a case study. Ecosystems. 10:4–16.

Ibelings BW, Vonk M, Los HFJ, van der Molen DT, Mooij WM (2003). Fuzzy modeling of cyanobacterial surface waterblooms: validation with NOAA-AVHRR satellite images. Ecol Appl. 13:1456–1472.

Ihle T, Jähnichen S, Benndorf J (2005). Wax and wane of *Microcystis* (cyanophyceae) and microcystins in lake sediments: a case study in Quitzdorf reservoir (Germany). J Phycol. 41:479–488.

Imai H, Chang K-H, Nakano S-I (2009). Growth responses of harmful algal species *Microcystis* (Cyanophyceae) under various environmental conditions. In: Obayashi Y, Isobe T, Subramanian A et al., editors: Interdisciplinary studies on environmental chemistry—environmental research in Asia. Tokyo: Terrapub:269–275.

Jacoby JM, Burghdoff M, Williams G, Read L, Hardy J (2015). Dominant factors associated with microcystins in nine midlatitude, maritime lakes. Inland Waters. 5:187–202.

Jacquet S, Kerimoglu O, Rimet F, Paolini G, Anneville O (2014). Cyanobacterial bloom termination: the disappearance of *Planktothrix rubescens* from Lake Bourget (France) after restoration. Freshwater Biol. 59:2472–2487.

Janse I, Kardinaal WEA, Meima M, Fastner J, Visser PM, Zwart G (2004). Toxic and nontoxic *Microcystis* colonies in natural populations can be differentiated on the basis of rRNA gene internal transcribed spacer. Appl Environ Microbiol. 70:3979–3987.

Janse I, Meima M, Kardinaal WEA, Zwart G (2003). High-resolution differentiation of cyanobacteria by using rRNA-internal transcribed spacer denaturing gradient gel electrophoresis. Appl Environ Microbiol. 69:6634–6643.

Jensen J, Jeppesen E, Olrik K, Kristensen P (1994). Impact of nutrients and physical factors on the shift from cyanobacterial to chlorophyte dominance in shallow Danish lakes. Can J Fish Aquat Sci. 51:1692–1699.

Jeppesen E, Kristensen P, Jensen JP, Søndergaard M, Mortensen E, Lauridsen T (1991). Recovery resilience following a reduction in external phosphorus loading of shallow, eutrophic Danish lakes: duration, regulating factors and methods for overcoming resilience. Mem Ist Ital Idrobiol. 48:127–148.

Jeppesen E, Meerhoff M, Jacobsen B, Hansen R, Søndergaard M, Jensen J et al. (2007). Restoration of shallow lakes by nutrient control and biomanipulation—the successful strategy varies with lake size and climate. Hydrobiologia. 581:269–285.

Jeppesen E, Søndergaard M, Jensen JP (2003). Climatic warming and regime shifts in lake food webs—some comments. Limnol Oceanogr 48:1346–1349.

Jeppesen E, Søndergaard M, Jensen JP, Havens KE, Anneville O, Carvalho L et al. (2005). Lake responses to reduced nutrient loading–an analysis of contemporary long-term data from 35 case studies. Freshwater Biol. 50:1747–1771.

Jöhnk K, Huisman J, Sharples J, Sommeijer B, Visser P, Stroom J (2008). Summer heatwaves promote blooms of harmful cyanobacteria. Global Change Biol. 14:495–512.

Jones GJ, Blackburn SI, Parker NS (1994). A toxic bloom of *Nodularia spumigena* MERTENS in Orielton Lagoon, Tasmania. Aust J Mar Freshwater Res. 45:787–800.

Kardinaal WEA, Janse I, Kamst-van Agterveld MP, Meima M, Snoek J, Mur LR et al. (2007). *Microcystis* genotype succession in relation to microcystin concentrations in freshwater lakes. Aquat Microb Ecol. 48:1–12.

Kardinaal WEA, Visser PM (2005). Dynamics of cyanobacterial toxins: sources of variability in microcystin concentrations. In: Huisman J, Matthijs HCP, Visser PM, editors. Harmful cyanobacteria. Berlin: Springer:41–63.

Karlsson-Elfgren I, Rengefors K, Gustafsson S (2004). Factors regulating recruitment from the sediment to the water column in the bloom-forming cyanobacterium *Gloeotrichia echinulata*. Freshwater Biol. 49:265–273.

Karlsson-Elfgren I, Rydin E, Hyenstrand P, Pettersson K (2003). Recruitment and pelagic growth of *Gloeotrichia echinulata* (Cyanophyceae) in Lake Erken. J Phycol. 39:1050–1056.

Kasprzak P, Padisák J, Koschel R, Krienitz L, Gervais F (2008). Chlorophyll a concentration across a trophic gradient of lakes: An estimator of phytoplankton biomass? Limnologica. 38:327–338.

Kinsman R, Ibelings BW, Walsby A (1991). Gas vesicle collapse by turgor pressure and its role in buoyancy regulation by *Anabaena flos-aquae*. Microbiology. 137:1171–1178.

Kirk JTO (1994). Light and photosynthesis in aquatic ecosystems. Cambridge, UK: Cambridge University Press.

Klemer A, Feuillade J, Feuillade M (1982). Cyanobacterial blooms: carbon and nitrogen limitation have opposite effects on the buoyancy of *Oscillatoria*. Science. 215:1629–1631.

Kolzau S, Wiedner C, Rücker J, Köhler J, Köhler A, Dolman AM (2014). Seasonal patterns of nitrogen and phosphorus limitation in four German lakes and the predictability of limitation status from ambient nutrient concentrations. PLoS One. 9:e96065.

Konopka A (1989). Metalimnetic cyanobacteria in hard-water lakes: Buoyancy regulation and physiological state. Limnol Oceanogr. 34:1174–1184.

Konopka A, Kromkamp JC, Mur LR (1987). Buoyancy regulation in phosphate-limited cultures of *Microcystis aeruginosa*. FEMS Microbiol Lett. 45:135–142.

Kosten S, Huszar VL, Bécares E, Costa LS, Donk E, Hansson LA et al. (2012). Warmer climates boost cyanobacterial dominance in shallow lakes. Global Change Biol. 18:118–126.

Kosten S, Huszar VL, Mazzeo N, Scheffer M, da SL Sternberg L, Jeppesen E (2009). Lake and watershed characteristics rather than climate influence nutrient limitation in shallow lakes. Ecol App. 19:1791–1804.

Kotak BG, Lam AKY, Prepas EE, Hrudey SE (2000). Role of chemical and physical variables in regulating microcystin-LR concentration in phytoplankton of eutrophic lakes. Can J Fish Aquat Sci. 57:1584–1593.

Kromkamp J, Walsby AE (1990). A computer model of buoyancy and vertical migration in cyanobacteria. J Plankton Res. 12:161–183.

Kruk C, Mazzeo N, Lacerot G, Reynolds C (2002). Classification schemes for phytoplankton: a local validation of a functional approach to the analysis of species temporal replacement. J Plankton Res. 24:901–912.

Kruskopf M, Flynn KJ (2006). Chlorophyll content and fluorescence responses cannot be used to gauge reliably phytoplankton biomass, nutrient status or growth rate. New Phytol. 169:525–536.

Kuno S, Yoshida T, Kaneko T, Sako Y (2012). Intricate interactions between the bloom-forming cyanobacterium *Microcystis aeruginosa* and foreign genetic elements, revealed by diversified clustered regularly interspaced short palindromic repeat (CRISPR) signatures. Appl Environ Microbiol. 78:5353–5360.

Kurmayer R, Christiansen G, Chorus I (2003). The abundance of microcystin-producing genotypes correlates positively with colony size in *Microcystis* sp and determines its microcystin net production in Lake Wannsee. Appl Environ Microbiol. 69:787–795.

Kurmayer R, Christiansen G, Fastner J, Börner T (2004). Abundance of active and inactive microcystin genotypes in populations of the toxic cyanobacterium *Planktothrix* spp. Environ Microbiol. 6:831–841.

Kurmayer R, Gumpenberger M (2006). Diversity of microcystin genotypes among populations of the filamentous cyanobacteria *Planktothrix rubescens* and *Planktothrix agardhii*. Mol Ecol. 15:3849–3861.

Kurmayer R, Jüttner F (1999). Strategies for the co-existence of zooplankton with the toxic cyanobacterium *Planktothrix rubescens* in Lake Zurich. J Plankton Res. 21:659–683.

Lampert W, Fleckner W, Rai H, Taylor BE (1986). Phytoplankton control by grazing zooplankton: A study on the spring clear-water phase. Limnol Oceanogr. 31:478–490.

Lampert W, Sommer U (2007). Limnoecology: the ecology of lakes and streams. New York: Oxford University Press:336 pp.

Lemaire V, Brusciotti S, van Gremberghe I, Vyverman W, Vanoverbeke J, De Meester L (2012). Genotype×genotype interactions between the toxic cyanobacterium *Microcystis* and its grazer, the waterflea *Daphnia*. Evol Appl. 5:168–182.

Li H, Sherman DM, Bao SL, Sherman LA (2001a). Pattern of cyanophycin accumulation in nitrogen-fixing and non-nitrogen-fixing cyanobacteria. Arch Microbiol. 176:9–18.

Li J, Dittrich M (2019). Dynamic polyphosphate metabolism in cyanobacteria responding to phosphorus availability. Environ Microbiol. 21:572–583.

Li R, Carmichael WW, Brittain S, Eaglesham GK, Shaw GR, Mahakahant A et al. (2001b). Isolation and identification of the cyanotoxin cylindrospermopsin and deoxy-cylindrospermopsin from a Thailand strain of *Cylindrospermopsis raciborskii* (Cyanobacteria). Toxicon. 39:973–980.

Livingstone DM (2003). Impact of secular climate change on the thermal structure of a large temperate central European lake. Climatic Change. 57:205–225.

Loftin KA, Graham JL, Hilborn ED, Lehmann SC, Meyer MT, Dietze JE et al. (2016). Cyanotoxins in inland lakes of the United States: Occurrence and potential recreational health risks in the EPA National Lakes Assessment 2007. Harmful Algae. 56:77–90.

Loiselle SA, Cozar A, Dattilo A, Bracchini L, Galvez JA (2007). Light limitations to algal growth in tropical ecosystems. Freshwater Biol. 52:305–312.

Long BM, Jones GJ, Orr PT (2001). Cellular microcystin content in N-limited *Microcystis* aeruginosa can be predicted from growth rate. Appl Environ Microbiol. 67:278–283.

Lürling M, Eshetu F, Faassen EJ, Kosten S, Huszar VLM (2013). Comparison of cyanobacterial and green algal growth rates at different temperatures. Freshwater Biol. 58:552–559.

Mallet J (2012). The struggle for existence. how the notion of carrying capacity, K, obscures the links between demography, Darwinian evolution and speciation. Evol Ecol Res. 14:627–665.

Mamun M, An K-G (2017). Major nutrients and chlorophyll dynamics in Korean agricultural reservoirs along with an analysis of trophic state index deviation. J Asia-Pacific Biodiversity. 10:183–191.

Mantzouki E, Lürling M, Fastner J, de Senerpont Domis L, Wilk-Woźniak E, Koreivienė J et al. (2018). Temperature effects explain continental scale distribution of cyanobacterial toxins. Toxins. 10:156.

Mantzouki E, Visser PM, Bormans M, Ibelings BW (2016). Understanding the key ecological traits of cyanobacteria as a basis for their management and control in changing lakes. Aquat Ecol. 50:333–350.

Martiny JB, Bohannan BJ, Brown JH, Colwell RK, Fuhrman JA, Green JL et al. (2006). Microbial biogeography: putting microorganisms on the map. Nat Rev Microbiol. 4:102–112.

Matveev V, Matveeva L, Jones GJ (1994). Study of the ability of *Daphnia carinata* (King) to control phytoplankton and resist cyanobacterial toxicity: implications for biomanipulation in Australia. Mar Freshwater Res. 45:889–904.

McAllister TG, Wood SA, Hawes I (2016). The rise of toxic benthic *Phormidium* proliferations: a review of their taxonomy, distribution, toxin content and factors regulating prevalence and increased severity. Harmful Algae. 55:282–294.

Medrano EA, Uittenbogaard R, Pires LD, Van De Wiel B, Clercx H (2013). Coupling hydrodynamics and buoyancy regulation in *Microcystis aeruginosa* for its vertical distribution in lakes. Ecol Model. 248:41–56.

Mehnert G, Leunert F, Cirés S, Jöhnk KD, Rücker J, Nixdorf B et al. (2010). Competitiveness of invasive and native cyanobacteria from temperate freshwaters under various light and temperature conditions. J Plankton Res. 32:1009–1021.

Meyer KA, Davis TW, Watson SB, Denef VJ, Berry MA, Dick GJ (2017). Genome sequences of lower Great Lakes *Microcystis* sp. reveal strain-specific genes that are present and expressed in western Lake Erie blooms. PLoS One. 12. e0183859.

Michalak AM, Anderson EJ, Beletsky D, Boland S, Bosch NS, Bridgeman TB et al. (2013). Record-setting algal bloom in Lake Erie caused by agricultural and meteorological trends consistent with expected future conditions. Proc Natl Acad Sci USA. 110:6448–6452.

Miller TR, Bartlett S, Weirich CA, Hernandez J (2019). Automated sub-daily sampling of cyanobacterial toxins on a buoy reveals new temporal patterns in toxin dynamics. Environ Sci Technol. 53:5661–5670.

Mohamed ZA (2001). Accumulation of cyanobacterial hepatotoxins by *Daphnia* in some Egyptian irrigation canals. Ecotoxicol Environ Safety. 50:4–8.

Mooij WM, Hülsmann S, De Senerpont Domis LN, Nolet BA, Bodelier PL, Boers PC et al. (2005). The impact of climate change on lakes in the Netherlands: a review. Aquat Ecol. 39:381–400.

Mooij WM, Trolle D, Jeppesen E, Arhonditsis G, Belolipetsky PV, Chitamwebwa DB et al. (2010). Challenges and opportunities for integrating lake ecosystem modelling approaches. Aquat Ecol. 44:633–667.

Moss B, Jeppesen E, Søndergaard M, Lauridsen TL, Liu Z (2013). Nitrogen, macrophytes, shallow lakes and nutrient limitation: resolution of a current controversy? Hydrobiologia. 710:3–21.

Moss B, Kosten S, Meerhoff M, Battarbee RW, Jeppesen E, Mazzeo N et al. (2011). Allied attack: climate change and eutrophication. Inland Waters. 1:101–105.

Mur L (1983). Some aspects of the ecophysiology of cyanobacteria. Annales de l'Institut Pasteur/Microbiologie. 134B:61–72.

Nalewajko C, Murphy TP (2001). Effects of temperature, and availability of nitrogen and phosphorus on the abundance of *Anabaena* and *Microcystis* in Lake Biwa, Japan: an experimental approach. Limnology. 2:45–48.

Naselli-Flores L, Barone R, Chorus I, Kurmayer R (2007). Toxic cyanobacterial blooms in reservoirs under a semiarid mediterranean climate: the magnification of a problem. Environ Toxicol. 22:399–404.

Nehring S (1993). Mortality of dogs associated with a mass development of *Nodularia spumigena* (Cyanophyceae) in a brackish lake at the German North Sea coast. J Plankton Res. 15:867–872.

Neilan BA, Saker ML, Fastner J, Törökné AK, Burns BP (2003). Phylogeography of the invasive cyanobacterium *Cylindrospermopsis raciborskii*. Mol Ecol. 12:133–140.

Novak JT, Brune DE (1985). Inorganic carbon limited growth kinetics of some freshwater algae. Water Res. 19:215–225.

Nürnberg GK (1996). Trophic state of clear and colored, soft-and hardwater lakes with special consideration of nutrients, anoxia, phytoplankton and fish. Lake Reserv Manage. 12:432–447.

Nürnberg GK, LaZerte BD, Olding DD (2003). An artificially induced *Planktothrix rubescens* surface bloom in a small kettle lake in Southern Ontario compared to blooms world-wide. Lake Reserv Manage. 19:307–322.

Oberhaus L, Briand JF, Leboulanger C, Jacquet S, Humbert JF (2007). Comparative effects of the quality and quantity of light and temperature on the growth of *Planktothrix agardhii* and *P. rubescens*. J Phycol. 43:1191–1199.

Odum EP (1953). Fundamentals of ecology, first edition. Philadelphia (PA): WB Saunders.

Okello W, Ostermaier V, Portmann C, Gademann K, Kurmayer R (2010). Spatial isolation favours the divergence in microcystin net production by *Microcystis* in Ugandan freshwater lakes. Water Res. 44:2803–2814.

Oliver R, Ganf G (2000). Freshwater blooms. In: Whitton BA, Potts M, editors: The ecology of cyanobacteria: their diversity in time and space. Dordrecht: Springer:150–194.

Orr PT, Jones GJ (1998). Relationship between microcystin production and cell division rates in nitrogen-limited *Microcystis aeruginosa* cultures. Limnol Oceanogr. 43:1604–1614.

Orr PT, Rasmussen JP, Burford MA, Eaglesham GK, Lennox SM (2010). Evaluation of quantitative real-time PCR to characterise spatial and temporal variations in cyanobacteria, *Cylindrospermopsis raciborskii* (Woloszynska) Seenaya et Subba Raju and cylindrospermopsin concentrations in three subtropical Australian reservoirs. Harmful Algae. 9:243–254.

Otsuka S, Suda S, Li R, Watanabe M, Oyaizu H, Matsumoto S et al. (1999). Characterization of morphospecies and strains of the genus *Microcystis* (Cyanobacteria) for a reconsideration of species classification. Phycol Res. 47:189–197.

Otten TG, Paerl HW, Dreher TW, Kimmerer WJ, Parker AE (2017). The molecular ecology of *Microcystis* sp. blooms in the San Francisco Estuary. Environ Microbiol. 19:3619–3637.

Ozawa K, Fujioka H, Muranaka M, Yokoyama A, Katagami Y, Homma T et al. (2005). Spatial distribution and temporal variation of *Microcystis* species composition and microcystin concentration in Lake Biwa. Environ Toxicol. 20:270–276.

Padisák J (1997). *Cylindrospermopsis raciborskii* (Woloszynska) Seenayya et Stubba Raju, an expanding, highly adaptive cyanobacterium: worldwide distribution and review of its ecology. Arch Hydrobiol Suppl. 107:563–593.

Padisák J, Crossetti LO, Naselli-Flores L (2009). Use and misuse in the application of the phytoplankton functional classification: a critical review with updates. Hydrobiologia. 621:1–19.

Paerl HW, Huisman J (2009). Climate change: a catalyst for global expansion of harmful cyanobacterial blooms. Environ Microbiol Rep. 1:27–37.

Paerl HW, Scott JT, McCarthy MJ, Newell SE, Gardner WS, Havens KE et al. (2016). It takes two to tango: when and where dual nutrient (N & P) reductions are needed to protect lakes and downstream ecosystems. Environ Sci Technol. 50:10805–10813.

Paterson M, Schindler D, Hecky R, Findlay D, Rondeau K (2011). Comment: Lake 227 shows clearly that controlling inputs of nitrogen will not reduce or prevent eutrophication of lakes. Limnol Oceanogr. 56:1545–1547.

Phillips G, Pietiläinen O-P, Carvalho L, Solimini A, Solheim AL, Cardoso A (2008). Chlorophyll–nutrient relationships of different lake types using a large European dataset. Aquat Ecol. 42:213–226.

Piccini C, Aubriot L, Fabre A, Amaral V, González-Piana M, Giani A et al. (2011). Genetic and eco-physiological differences of South American *Cylindrospermopsis raciborskii* isolates support the hypothesis of multiple ecotypes. Harmful Algae. 10:644–653.

Porat R, Teltsch B, Perelman A, Dubinsky Z (2001). Diel buoyancy changes by the cyanobacterium *Aphanizomenon ovalisporum* from a shallow reservoir. J Plankton Res. 23:753–763.

Posch T, Köster O, Salcher MM, Pernthaler J (2012). Harmful filamentous cyanobacteria favoured by reduced water turnover with lake warming. Nat Clim Change. 2:809.

Preston T, Stewart WDP, Reynolds CS (1980). Bloom-forming cyanobacterium *Microcystis aeruginosa* overwinters on sediment surface. Nature. 288:365–367.

Price GD, Badger MR, Woodger FJ, Long BM (2007). Advances in understanding the cyanobacterial CO2-concentrating-mechanism (CCM): functional components, Ci transporters, diversity, genetic regulation and prospects for engineering into plants. J Exp Bot. 59:1441–1461.

Pridmore RD, Etheredge MK (1987). Planktonic cyanobacteria in New Zealand inland waters: distribution and population dynamics. New Zeal J Mar Freshwater Res. 21:491–502.

Ptacnik R, Lepistö L, Willén E, Brettum P, Andersen T, Rekolainen S et al. (2008). Quantitative responses of lake phytoplankton to eutrophication in Northern Europe. Aquat Ecol. 42:227–236.

Quiblier C, Wood SA, Echenique-Subiabre I, Heath M, Villeneuve A, Humbert JF (2013). A review of current knowledge on toxic benthic freshwater cyanobacteria–ecology, toxin production and risk management. Water Res. 47:5464–5479.

Rabouille S, Salençon M-J (2005). Functional analysis of *Microcystis* vertical migration: a dynamic model as a prospecting tool. II. Influence of mixing, thermal stratification and colony diameter on biomass production. Aquat Microb Ecol. 39:281–292.

Raikow DF, Sarnelle O, Wilson AE, Hamilton SK (2004). Dominance of the noxious cyanobacterium *Microcystis aeruginosa* in low-nutrient lakes is associated with exotic zebra mussels. Limnol Oceanogr. 49:482–487.

Recknagel F, Orr PT, Bartkow M, Swanepoel A, Cao H (2017). Early warning of limit-exceeding concentrations of cyanobacteria and cyanotoxins in drinking water reservoirs by inferential modelling. Harmful Algae. 69:18–27.
Redfield A (1934). On the proportions of organic derivations in seawater and their relation to the composition of plankton (reprint). In: Daniel R, editors: James Johnstone memorial volume. Liverpool: University Press of Liverpool: 176–192.
Reeders H, Boers P, Van der Molen D, Helmerhorst T (1998). Cyanobacterial dominance in the lakes Veluwemeer and Wolderwijd, The Netherlands. Wat Sci Technol. 37:85–92.
Reynolds C (1994). The ecological basis for the successful biomanipulation of aquatic communities. Arch Hydrobiol. 130:1–33.
Reynolds C (1999a). Modelling phytoplankton dynamics and its application to lake management. Hydrobiologia. 395:123–131.
Reynolds C, Maberly S (2002). A simple method for approximating the supportive capacities and metabolic constraints in lakes and reservoirs. Freshwater Biol. 47:1183–1188.
Reynolds CS (1984). The ecology of freshwater phytoplankton. Cambridge, UK: Cambridge University Press:396 pp.
Reynolds CS (1997). Vegetation processes in the pelagic: a model for ecosystem theory. Oldendorf/Luhe: Ecology Institute:371 pp.
Reynolds CS (1999b). Non-determinism to probability, or N: P in the community ecology of phytoplankton: nutrient ratios. Arch Hydrobiol. 146:23–35.
Reynolds CS (2006). The ecology of phytoplankton. Cambridge, UK: Cambridge University Press:552 pp.
Reynolds CS, Huszar V, Kruk C, Naselli-Flores L, Melo S (2002). Towards a functional classification of the freshwater phytoplankton. J Plankton Res. 24:417–428.
Reynolds CS, Jaworsky GHM, Cmiech HA, Leedale GF (1981). On the annual cycle of *Microcystis aeruginosa* Kütz emend. Elenkin. Philos Trans R Soc Lond B. 293:419–477.
Reynolds CS, Maberly SC, Parker JE, De Ville MM (2012). Forty years of monitoring water quality in Grasmere (English Lake District): separating the effects of enrichment by treated sewage and hydraulic flushing on phytoplankton ecology. Freshwater Biol. 57:384–399.
Rigosi A, Carey CC, Ibelings BW, Brookes JD (2014). The interaction between climate warming and eutrophication to promote cyanobacteria is dependent on trophic state and varies among taxa. Limnol Oceanogr. 59:99–114.
Rohrlack T (2018). Low temperatures can promote cyanobacterial bloom formation by providing refuge from microbial antagonists. AIMS Microbiol. 4:304–318.
Rohrlack T, Christiansen G, Kurmayer R (2013). Putative antiparasite defensive system involving ribosomal and nonribosomal oligopeptides in cyanobacteria of the genus *Planktothrix*. Appl Environ Microbiol. 79:2642–2647.
Romo S, Soria J, Fernandez F, Ouahid Y, Barón-Solá Á (2013). Water residence time and the dynamics of toxic cyanobacteria. Freshwater Biol. 58:513–522.
Rücker J, Wiedner C, Zippel P (1997). Factors controlling the dominance of *Planktothrix agardhii* and *Limnothrix redekei* in eutrophic shallow lakes. Hydrobiologia. 342/343:107–115.

Saker ML, Griffiths DJ (2000). The effect of temperature on growth and cylindrospermopsin content of seven isolates of *Cylindrospermopsis raciborskii* (Nostocales, Cyanophyceae) from water bodies in northern Australia. Phycologia. 39:349–354.

Salas HJ, Martino P (1991). A simplified phosphorus trophic state model for warm-water tropical lakes. Water Res. 25:341–350.

Salk KR, Bullerjahn GS, McKay RML, Chaffin JD, Ostrom NE (2018). Nitrogen cycling in Sandusky Bay, Lake Erie: oscillations between strong and weak export and implications for harmful algal blooms. Biogeosciences. 15:2891–2907.

Salmaso N, Boscaini A, Capelli C, Cerasino L (2018). Ongoing ecological shifts in a large lake are driven by climate change and eutrophication: evidences from a three-decade study in Lake Garda. Hydrobiologia. 824:177–195.

Salmaso N, Capelli C, Shams S, Cerasino L (2015). Expansion of bloom-forming *Dolichospermum lemmermannii* (Nostocales, Cyanobacteria) to the deep lakes south of the Alps: colonization patterns, driving forces and implications for water use. Harmful Algae. 50:76–87.

Salmaso N, Cerasino L, Boscaini A, Capelli C (2016). Planktic *Tychonema* (Cyanobacteria) in the large lakes south of the Alps: phylogenetic assessment and toxigenic potential. FEMS Microbiol Ecol. 92:fiw155.

Salmaso N, Copetti D, Cerasino L, Shams S, Capelli C, Boscaini A et al. (2014). Variability of microcystin cell quota in metapopulations of *Planktothrix rubescens*: causes and implications for water management. Toxicon. 90:82–96.

Sandrini G, Matthijs HC, Verspagen JM, Muyzer G, Huisman J (2014). Genetic diversity of inorganic carbon uptake systems causes variation in CO_2 response of the cyanobacterium *Microcystis*. ISME J. 8:589.

Sarnelle O (1993). Herbivore effects on phytoplankton succession in a eutrophic lake. Ecol Monogr. 63:129–149.

Sayre NF (2008). The genesis, history, and limits of carrying capacity. Ann Assoc Am Geogr. 98:120–134.

Schanz F (1994). Oligotrophication of Lake Zürich as reflected in Secchi depth measurements. Ann Limnol. 30:57–65.

Scheffer M, Rinaldi S, Gragnani A, Mur LR, van Nes EH (1997). On the dominance of filamentous cyanobacteria in shallow, turbid lakes. Ecology. 78:272–282.

Schindler DW (2006). Recent advances in the understanding and management of eutrophication. Limnol Oceanogr. 51:356–363.

Schindler DW, Carpenter SR, Chapra SC, Hecky RE, Orihel DM (2016). Reducing phosphorus to curb lake eutrophication is a success. Environ Sci Technol. 50:8923–8929.

Schindler DW, Hecky R, Findlay D, Stainton M, Parker B, Paterson M et al. (2008). Eutrophication of lakes cannot be controlled by reducing nitrogen input: results of a 37-year whole-ecosystem experiment. Proc Natl Acad Sci USA. 105:11254–11258.

Scott JT, Marcarelli AM (2012). Cyanobacteria in freshwater benthic environments. In: Whitton BA, editors: Ecology of Cyanobacteria II, second edition. Dordrecht: Springer:271–289.

Scott JT, McCarthy MJ (2010). Nitrogen fixation may not balance the nitrogen pool in lakes over timescales relevant to eutrophication management. Limnol Oceanogr. 55:1265–1270.

Scott JT, McCarthy MJ (2011). Response to comment: nitrogen fixation has not offset declines in the Lake 227 nitrogen pool and shows that nitrogen control deserves consideration in aquatic ecosystems. Limnol Oceanogr. 56:1548–1550.

Shams S, Capelli C, Cerasino L, Ballot A, Dietrich DR, Sivonen K et al. (2015). Anatoxin-a producing *Tychonema* (Cyanobacteria) in European waterbodies. Water Res 69:68–79.

Shapiro J (1997). The role of carbon dioxide in the initiation and maintenance of blue-green dominance in lakes. Freshwater Biol. 37:307–323.

Shatwell T, Köhler J (2019). Decreased nitrogen loading controls summer cyanobacterial blooms without promoting nitrogen-fixing taxa: long-term response of a shallow lake. Limnol Oceanogr. 64:S166–S178.

Simon RD (1971). Cyanophycin granules from the blue-green alga *Anabaena cylindrica*: a reserve material consisting of copolymers of aspartic acid and arginine. Proc Natl Acad Sci USA. 68:265–267.

Sinang SC, Reichwaldt ES, Ghadouani A (2013). Spatial and temporal variability in the relationship between cyanobacterial biomass and microcystins. Environ Monit Assess. 185:6379–6395.

Sitoki L, Kurmayer R, Rott E (2012). Spatial variation of phytoplankton composition, biovolume, and resulting microcystin concentrations in the Nyanza Gulf (Lake Victoria, Kenya). Hydrobiologia. 691:109–122.

Sivonen K (1990). Effects of light, temperature, nitrate, orthophosphate, and bacteria on growth of and hepatotoxin production by *Oscillatoria agardhii* strains. Appl Environ Microbiol. 56:2658–2666.

Skulberg OM, Skulberg R (1985). Planktic species of *Oscillatoria* (Cyanophyceae) from Norway. Characterization and classification. Arch Hydrobiol Algol Stud. 38–39:157–174.

Smith FM, Wood SA, van Ginkel R, Broady PA, Gaw S (2011). First report of saxitoxin production by a species of the freshwater benthic cyanobacterium, *Scytonema* Agardh. Toxicon 57:566–573.

Smith VH (1983). Low nitrogen to phosphorus ratios favor dominance by blue-green algae in lake phytoplankton. Science. 221:669–671.

Søndergaard M, Lauridsen TL, Johansson LS, Jeppesen E (2017). Nitrogen or phosphorus limitation in lakes and its impact on phytoplankton biomass and submerged macrophyte cover. Hydrobiologia. 795:35–48.

Soto D (2002). Oligotrophic patterns in southern Chilean lakes: the relevance of nutrients and mixing depth. Revista Chilena de Historia Natural. 75:377–393.

Steenhauer LM, Wierenga J, Carreira C, Limpens RW, Koster AJ, Pollard PC et al. (2016). Isolation of cyanophage CrV infecting *Cylindrospermopsis raciborskii* and the influence of temperature and irradiance on CrV proliferation. Aquatic Microbial Ecol. 78:11–23.

Steffen MM, Davis TW, McKay RML, Bullerjahn GS, Krausfeldt LE, Stough JM et al. (2017). Ecophysiological examination of the Lake Erie *Microcystis* bloom in 2014: linkages between biology and the water supply shutdown of Toledo, OH. Environ Sci Technol. 51:6745–6755.

Stein JR, Hellebust JA, Craigie J, editors (1980). Handbook of phycological methods: culture methods and growth measurements. Cambridge, UK: Cambridge University Press. Sterner RW (2008). On the phosphorus limitation paradigm for lakes. Int Rev Hydrobiol. 93:433–445.

Sterner RW, Andersen T, Elser JJ, Hessen DO, Hood JM, McCauley E et al. (2008). Scale-dependent carbon:nitrogen:phosphorus seston stoichiometry in marine and freshwaters. Limnol Oceanogr. 53:1169–1180.

Stough JM, Tang X, Krausfeldt LE, Steffen MM, Gao G, Boyer GL et al. (2017). Molecular prediction of lytic *vs* lysogenic states for *Microcystis* phage: Metatranscriptomic evidence of lysogeny during large bloom events. PLoS One 12:e0184146.

Stumm W, Morgan JJ (1996). Aquatic chemistry: chemical equilibria and rates in natural waters, third edition. Chichester: John Wiley & Sons:1040 pp.

Suda S, Watanabe MM, Otsuka S, Mahakahant A, Yongmanitchai W, Nopartnaraporn N et al. (2002). Taxonomic revision of water-bloom-forming species of oscillatoroid cyanobacteria. Int J Syst Evol Microbiol. 52:1577–1595.

Šulčius S, Mazur-Marzec H, Vitonytė I, Kvederavičiūtė K, Kuznecova J, Šimoliūnas E et al. (2018). Insights into cyanophage-mediated dynamics of nodularin and other non-ribosomal peptides in *Nodularia spumigena*. Harmful Algae 78:69–74.

Svirčev Z, Lalić D, Savić GB, Tokodi N, Backović DD, Chen L et al. (2019). Global geographical and historical overview of cyanotoxin distribution and cyanobacterial poisonings. Arch Toxicol. 93:2429–2481.

Takamura N, Yasuno M, Sugahara K (1984). Overwintering of *Microcystis aeruginosa* Kütz. in a shallow lake. J Plankton Res. 6:1019–1029.

Teubner K, Feyerabend R, Henning M, Nicklisch A, Woitke P, Kohl J-G (1999). Alternative blooming of *Aphanizomenon flos-aquae* or *Planktothrix agardhii* induced by the timing of the critical nitrogen:phosphorus ratio in hypertrophic riverine lakes. Arch Hydrobiol Spec Issues Advanc Limnol. 54:325–344.

Thomas MK, Litchman E (2016). Effects of temperature and nitrogen availability on the growth of invasive and native cyanobacteria. Hydrobiologia. 763:357–369.

Triest L, Stiers I, Van Onsem S (2016). Biomanipulation as a nature-based solution to reduce cyanobacterial blooms. Aquat Ecol. 50:461–483.

Uehlinger U (1981). Experimentelle Untersuchungen zur Autökologie von *Aphanizomenon flos-aquae*. Algol Stud/Arch Hydrobiol. 60:260–288.

Van Liere L, Mur LR (1980). Occurrence of *Oscillatoria agardhii* and some related species, a survey. SIL Workshop on Hypertrophic Ecosystems, Växjö, Sweden. Dordrecht:Springer:67–77.

Van Wichelen J, van Gremberghe I, Vanormelingen P, Debeer AE, Leporcq B, Menzel D et al. (2010). Strong effects of amoebae grazing on the biomass and genetic structure of a *Microcystis* bloom (Cyanobacteria). Environ Microbiol. 12:2797–2813.

Van Wichelen J, Vanormelingen P, Codd GA, Vyverman W (2016). The common bloom-forming cyanobacterium *Microcystis* is prone to a wide array of microbial antagonists. Harmful Algae. 55:97–111.

Vanderploeg HA, Liebig JR, Carmichael WW, Agy MA, Johengen TH, Fahnenstiel GL et al. (2001). Zebra mussel (*Dreissena polymorpha*) selective filtration promoted toxic *Microcystis* blooms in Saginaw Bay (Lake Huron) and Lake Erie. Can J Fish Aquat Sci. 58:1208–1221.

Vasas G, Surányi G, Bácsi I, Márta M, Máthé C, Gonda S et al. (2013). Alteration of cylindrospermopsin content of *Aphanizomenon ovalisporum* (Cyanobacteria, Nostocales) due to step-down from combined nitrogen to dinitrogen. Adv Microbiol. 3:557.

Verspagen JMH, Snelder EOFM, Visser PM, Huisman J, Mur LR, Ibelings BW (2004). Recruitment of benthic *Microcystis* (Cyanophyceae) to the water column: internal buoyancy changes or resuspension? J Phycol. 40:260–270.

Visser PM, Ibelings BW, Bormans M, Huisman J (2016a). Artificial mixing to control cyanobacterial blooms: a review. Aquat Ecol. 50:423–441.

Visser PM, Passarge J, Mur LR (1997). Modelling vertical migration of the cyanobacterium *Microcystis*. Hydrobiologia. 349:99–109.

Visser PM, Verspagen JM, Sandrini G, Stal LJ, Matthijs HC, Davis TW et al. (2016b). How rising CO_2 and global warming may stimulate harmful cyanobacterial blooms. Harmful Algae. 54:145–159.

Vollenweider RA (1968). Scientific fundamentals of the eutrophication of lakes and flowing waters, with particular reference to nitrogen and phosphorus as factors in eutrophication. Paris: OECD. 1–250.

Vollenweider RA (1976). Advances in defining critical loading levels for phosphorus in lake eutrophication. Mem Ist Ital Idrobiol. 33:53–83.

Vollenweider RA, Kerekes JJ (1982). Eutrophication of waters. Monitoring, assessment and control. Paris: Environment Directorate OECD:154 pp.

Wagner C, Adrian R (2009). Cyanobacteria dominance: quantifying the effects of climate change. Limnol Oceanogr. 54:2460–2468.

Wallace BB, Hamilton DP (2000). Simulation of water-bloom formation in the cyanobacterium *Microcystis aeruginosa*. J Plankton Res. 22:1127–1138.

Walsby A, Schanz F (2002). Light-dependent growth rate determines changes in the population of *Planktothrix rubescens* over the annual cycle in Lake Zürich, Switzerland. New Phytol. 154:671–687.

Walsby AE (2005). Stratification by cyanobacteria in lakes: a dynamic buoyancy model indicates size limitations met by *Planktothrix rubescens* filaments. New Phytol. 168:365–376.

Wang S, Zhu L, Li Q, Li G, Li L, Song L et al. (2015). Distribution and population dynamics of potential anatoxin-a-producing cyanobacteria in Lake Dianchi, China. Harmful Algae. 48:63–68.

Welker M, Brunke M, Preussel K, Lippert I, von Döhren H (2004a). Diversity and distribution of *Microcystis* (Cyanobacteria) oligopeptide chemotypes from natural communities studied by single colony mass spectrometry. Microbiology. 150:1785–1796.

Welker M, Christiansen G, von Döhren H (2004b). Diversity of coexisting *Planktothrix* (Cyanobacteria) chemotypes deduced by mass spectral analysis of microcystins and other oligopeptides. Arch Microbiol. 182:288–298.

Welker M, Erhard M (2007). Consistency between chemotyping of single filaments of *Planktothrix rubescens* (cyanobacteria) by MALDI-TOF and the peptide patterns of strains determined by HPLC-MS. J Mass Spectrom. 42:1062–1068.

Welker M, Šejnohová L, von Döhren H, Nemethova D, Jarkovsky J, Maršálek B (2007). Seasonal shifts in chemotype composition of *Microcystis* sp. communities in the pelagial and the sediment of a shallow reservoir. Limnol Oceanogr. 52:609–619.

Welker M, von Döhren H, Täuscher H, Steinberg CEW, Erhard M (2003). Toxic *Microcystis* in shallow lake Müggelsee (Germany) - dynamics, distribution, diversity. Arch Hydrobiol. 157:227–248.

Wetzel RG (2001). Limnology: lake and river ecosystems, third edition. San Diego (CA): Academic Press. 1005 pp.

Whitton BA (2012). Ecology of cyanobacteria II: their diversity in space and time. New York: Springer:760 pp.

Wiedner C, Rücker J, Brüggemann R, Nixdorf B (2007). Climate change affects timing and size of populations of an invasive cyanobacterium in temperate regions. Oecologia. 152:473–484.

Wiedner C, Visser PM, Fastner J, Metcalf JS, Codd GA, Mur LR (2003). Effects of light on the microcystin content of *Microcystis* strain PCC 7806. Appl Environ Microbiol. 69:1475–1481.

Wilson AE, Sarnelle O, Tillmanns AR (2006a). Effects of cyanobacterial toxicity and morphology on the population growth of freshwater zooplankton: meta-analyses of laboratory experiments. Limnol Oceanogr. 51:1915–1924.

Wilson AE, Wilson WA, Hay ME (2006b). Intraspecific variation in growth and morphology of the bloom-forming cyanobacterium *Microcystis aeruginosa*. Appl Environ Microbiol. 72:7383–7389.

Winder M, Sommer U (2012). Phytoplankton response to a changing climate. Hydrobiologia. 698:1–12.

Wood SA, Biessy L, Puddick J (2018). Anatoxins are consistently released into the water of streams with *Microcoleus autumnalis*-dominated (cyanobacteria) proliferations. Harmful Algae. 80:88–95.

Wood SA, Depree C, Brown L, McAllister T, Hawes I (2015). Entrapped sediments as a source of phosphorus in epilithic cyanobacterial proliferations in low nutrient rivers. PLoS One. 10:e0141063.

Wood SA, Kelly LT, Bouma-Gregson K, Humbert JF, Laughinghouse IV HD, Lazorchak J et al. (2020) Toxic benthic freshwater cyanobacterial proliferations: Challenges and solutions for enhancing knowledge and improving monitoring and mitigation. Freshwater Biol. 65:1824–1842.

Wood SA, Puddick J (2017). The abundance of toxic genotypes is a key contributor to anatoxin variability in *Phormidium*-dominated benthic mats. Marine Drugs. 15:307.

Wood SA, Selwood AI, Rueckert A, Holland PT, Milne JR, Smith KF et al. (2007). First report of homoanatoxin-a and associated dog neurotoxicosis in New Zealand. Toxicon. 50:292–301.

Yepremian C, Gugger MF, Briand E, Catherine A, Berger C, Quiblier C et al. (2007). Microcystin ecotypes in a perennial *Planktothrix agardhii* bloom. Water Res. 41:4446–4456.

Yoshida T, Takashima Y, Tomaru Y, Shirai Y, Takao Y, Hiroishi S et al. (2006). Isolation and characterization of a cyanophage infecting the toxic cyanobacterium *Microcystis aeruginosa*. Appl Environ Microbiol. 72:1239–1247.

Yoshida-Takashima Y, Yoshida M, Ogata H, Nagasaki K, Hiroishi S, Yoshida T (2012). Cyanophage infection in the bloom-forming cyanobacteria *Microcystis aeruginosa* in surface freshwater. Microbes Environ. 27:350–355.

Yuan LL, Jones JR (2020). Rethinking phosphorus–chlorophyll relationships in lakes. Limnol Oceanogr. 65:1847–1857.

Zhang X, Xie P, Huang X (2008). A review of nontraditional biomanipulation. Sci World J. 8:1184–1196.

Ziegler K, Diener A, Herpin C, Richter R, Deutzmann R, Lockau W (1998).

Molecular characterization of cyanophycin synthetase, the enzyme catalyzing the biosynthesis of the cyanobacterial reserve material multi-L-arginyl-poly-L-aspartate (cyanophycin). Eur J Biochem. 254:154–159.
Zimmermann U (1969). Ökologische und physiologische Untersuchungen an der planktischen Blaualge *Oscillatoria rubescens* DC unter besonderer Berücksichtigung von Licht und Temperatur. Schweiz Z Hydrol. 31:1–58.
Znachor P, Jurczak T, Komárková-Legnerová J, Jezberova J, Mankiewicz J, Kastovska K et al. (2006). Summer changes in cyanobacterial bloom composition and microcystin concentration in eutrophic Czech reservoirs. Environ Toxicol. 21:236–243.

Chapter 5

Exposure to cyanotoxins

Understanding it and short-term interventions to prevent it

CONTENTS

INTRODUCTION AND GENERAL CONSIDERATIONS

People may be exposed to cyanotoxins through oral, respiratory and dermal routes. Ingestion may occur through drinking-water (see section 5.1) or accidental uptake during water sports, recreational or occupational activity (see section 5.2). In some settings, contaminated food can be a source of dietary exposure, possibly significant (see section 5.3). This includes fish, shellfish and crustaceans collected from bloom-ridden waterbodies as well as leafy vegetable crops spray-irrigated with water containing cyanobacteria. A particularly high risk for specific subpopulations may be caused by exposure through haemodialysis (section 5.4): if dialysis centres do not take appropriate precautions and dialysate is contaminated with cyanotoxins, this can injure patients undergoing renal dialysis, because toxins from a large volume of water (>100 L per treatment) may gain direct access to the bloodstream via the intravenous route several times per week. Cyanobacterial dietary supplements may further be a potentially relevant route of oral intake for a small subpopulation using such products (see section 5.5).

While symptoms from cutaneous exposure to freshwater cyanobacteria have been quite widely reported (see section 5.2), these are usually mild and self-limiting. In contrast, marine cyanobacteria can cause severe skin lesions, for which, however, there is still a lack of dose–response information as a basis for estimating tolerable exposure levels (see section 2.6). Some people may experience allergic reactions to cyanobacteria, whereas others may be unaffected, and by the time of the publication of this book, it remains unclear which constituents of cyanobacterial cells – or associated microorganisms and compounds – actually cause allergic reactions.

The following sections 5.1 – 5.5 outline the specific exposure pathways and health risks through drinking-water, recreation and occupational use of water containing cyanobacteria and/or their toxins dissolved in water, food, renal dialysis and dietary supplements. They summarise available epidemiological knowledge as well as other indications of human exposure and relate these to the information on toxicity as discussed in Chapter 2 for the individual groups of cyanotoxins.

A caveat to keep in mind when assessing reports concerning human exposure to toxic cyanobacteria is that their estimates of exposure are almost always retrospective (it would not be ethically possible to conduct a prospective human study of a toxin at concentrations expected to show effects). That is, they provide information on human symptoms occurring at or just before the time of the study and try to explain these by looking into the past to make an "educated guess" as to what may have caused the observed symptoms. Even cyanotoxins detected in the tissues of people or animals do not solve this problem: while they provide absolute evidence of exposure, they do not necessarily demonstrate cyanotoxins to have been the sole cause of symptoms or elevated serum enzyme levels. Many of the reported symptoms in historical reports are quite general and cannot be considered in isolation as diagnostic of cyanotoxin poisoning. It is also not possible to know whether all potential causes and their interactions have been considered, nor whether the estimates of exposures are accurate. Thus, this type of study cannot prove that a cause–effect relationship exists, nor can it provide a quantitative dose–response estimate. This is why the guideline values (GVs) for all cyanotoxins except saxitoxins (STX) are based on animal studies, despite these also having many limitations. Saxitoxins are an exception due to the rapid onset of highly specific diagnostic symptoms following the consumption of contaminated seafood.

In spite of these limitations, however, it is highly useful to report incidents of suspected human and/or animal exposure, particularly for enabling direct interventions to prevent further exposure but also, in the longer term, to collate indicative evidence, particularly if reporting includes toxin concentrations observed in the field at the time of exposure or in the serum of those exposed, or cyanobacterial cells observed in stool samples.

Using concentrations of cyanobacterial biomass to trigger cyanotoxin alerts

Sections 5.1 and 5.2 propose Alert Level Frameworks (ALFs) to guide short-term interventions if cyanotoxins or cyanobacterial biomass are present in a waterbody in concentrations that may become or may already be relevant to human health. For triggering alerts, the ALFs offer different points of entry, ranging from visual assessment over microscopy and quantification of cyanobacterial biomass to toxin analysis. This allows the selection of parameters depending on national or local considerations, including the accessibility of analytical methods. Importantly, these ALFs are intended for national or even local adaptation: other parameters may also be used if these are more accessible or appropriate, provided their ratio to toxin concentrations can be determined periodically (see below), for example, cell numbers or turbidity readings in raw water entering a treatment plant. An advantage of defining the Alert Levels with a measure of cyanobacterial biomass (either biovolume or pigment concentrations; see Chapter 13) is that they are thus also protective against further unspecific health effects of blooms not attributable to the cyanotoxins.

The Alert Levels triggering interventions are based on concentrations of cyanobacterial biomass that correspond to the WHO health-based values for cyanotoxins (Table 5.1) – that is, depending on the Alert Level, those for drinking-water (lifetime, short-term or acute) or recreational exposure. Therefore, it is also possible to use the GVs in Table 5.1 directly to trigger alerts. Biomass is measured either as biovolume or as concentration of chlorophyll-*a* (the latter after a brief qualitative check by microscopy of whether chlorophyll-*a* is largely from cyanobacteria), and the Alert Levels for biovolume and chlorophyll-*a* proposed (Table 5.2) are

Table 5.1 Guideline values and health-based reference values for selected cyanotoxins and exposure scenarios (WHO, 2020)

Toxin	*Exposure*[a]	*Value (µg/L)*	*Value type* [b]
Microcystin-LR	Drinking-water, lifetime	1	Provisional guideline value
Microcystin-LR	Drinking-water, short term	12	Provisional guideline value
Microcystin-LR	Recreational	24	Provisional guideline value
Cylindrospermopsin	Drinking-water, lifetime	0.7	Provisional guideline value
Cylindrospermopsin	Drinking-water, short term	3	Provisional guideline value
Cylindrospermopsin	Recreational	6	Provisional guideline value
Anatoxin-a	Drinking-water, acute	30	Health-based reference value
Anatoxin-a	Recreational	60	Health-based reference value
Saxitoxin	Drinking-water, acute	3	Guideline value
Saxitoxin	Recreational	30	Guideline value

For details on derivation of individual values see sections 2.1–2.4.

[a] Note that short-term exposure refers to periods of about two weeks until enhanced drinking-water treatment or other measures can be implemented to achieve concentrations below the lifetime guideline value.

[b] Due to the overall quality of the database for their derivation and since the respective guideline values only cover specific congeners, the guideline values for microcystin-LR and for cylindrospermopsin are considered provisional.

In the absence of oral toxicity data for other congeners, itis recommended that the GVs be applied to total MCs, total CYNs and total STXs as gravimetric or molar equivalents, based on the worst-case assumption of the congeners having similar toxicity. For STX toxicity equivalents, see WHO 2020.

Furthermore, for ATX, the available toxicological information is not sufficient for deriving a formal guideline value (provisional or otherwise) for lifetime exposure, but it does show that health hazards are unlikely at levels above these health-based reference values (see sections 2.1–2.4 for details).

derived on the basis of conservative assumptions on ratios of microcystins (MCs) to either biovolume or chlorophyll-*a* found in publications covering a variety of waterbodies (reviewed in section 2.1 and discussed in section 4.6.2). Thus, if these Alert Levels are not exceeded, the concentrations of MCs are highly unlikely to exceed the respective GVs summarised in Table 5.1.

For the other cyanotoxins, less data are available to determine such ratios. The data available (see sections 2.2–2.4) show that their concentrations in the biomass of the producing cyanobacteria can attain the maximum levels similar to those attained by MCs, although this appears to occur less frequently. Thus, in many cases, the toxin/biomass ratios derived for MCs can be assumed as a conservative approach for these

Table 5.2 Conservative values for parameters of cyanobacterial biomass indicative of possible occurrence of cyanotoxin concentrations reaching guideline values

	Biovolume	*Chlorophyll-a*	
Alert Level	*MC/BV≤3/1* [µg/mm³]	*MC/Chl.a≤1:1* [µg/µg]	*Basis for conservative estimate*[a] *of toxin/biomass*
Alert Level 1 in drinking-water ALF	0.3 mm³/L	1 µg/L	$GV_{chronic}$ for MCs in drinking-water: 1 µg/L
Alert Level 2 in drinking-water ALF	4 mm³/L	12 µg/L	$GV_{short\text{-}term}$ for MCs in drinking-water: 12 µg/L
Alert Level 2 in recreational ALF	8 mm³/L	24 µg/L	$GV_{recreational}$ for MCs: 24 µg/L

For discussion of the biomass parameters and references, see text above as well as sections 2.1–2.4 and 4.6.2; for specifics of CYN, see Box 5.1.

Examples:

1. Observing 0.3 mm³/L biovolume or 1 µg/L chlorophyll-*a* (with dominance of cyanobacteria seen by brief visual assessment with microscopy) indicates that microcystin or cylindrospermopsin may occur at concentrations reaching the lifetime GV;
2. Observing>4 mm³/L biovolume or>12 µg/L chlorophyll-*a* (as above, with dominance of cyanobacteria) indicates that microcystins, cylindrospermopsins or saxitoxins may exceed the short-term GVs for these toxins.

[a] Note that in many cases, the ratio of toxin to either biomass parameter is likely to be substantially lower, often by up to a factor of 10. Periodically (i.e., 2–3 times during a cyanobacterial growing season) "calibrating" them with toxin analysis is likely to enable higher Alert Levels.

toxins as well. This is supported by the concentrations found in water for cylindrospermopsins (CYNs), saxitoxins (STXs) and anatoxins (ATXs), which are typically substantially lower than those of the MCs. For anatoxins, the biomass thresholds proposed in the ALFs are sufficiently protective because their health-based reference values are substantially higher than the GVs for MCs. For lifetime exposure to CYNs, the GV is in the same range as the corresponding GV for MCs, and therefore, if CYNs are monitored as described in Box 5.1, the biomass threshold for Alert Level 1 is considered sufficiently protective. However, for CYNs and STXs, there are some uncertainties about whether the Alert Level 2 biomass thresholds are sufficiently conservative to ensure that toxin levels are below the acute GV for STXs and the short-term GV for CYNs, as these values are fourfold lower than the corresponding value for MCs. For STXs, this applies particularly to *Dolichospermum* spp., for which high STX/biomass ratios have been reported (see Tables 2.6 and 2.7 in Chapter 2). Therefore, toxin concentrations should be determined for blooms when STX- or CYN-producing species are dominant, and there is evidence that Alert Level 1 may be exceeded.

BOX 5.1: CONSIDERATIONS FOR USING CYANOBACTERIAL BIOMASS AS INDICATOR OF CYLINDROSPERMOPSIN CONCENTRATIONS

As discussed in section 2.2, maximum CYN contents per unit biomass of the producing cells are in the same range as for MCs, and thus, the same biomass Alert Levels can be used. However, while MCs largely occur cell-bound, high proportions of CYNs can occur dissolved in water in concentrations exceeding the concentration of cell-bound CYNs and persist even after the producing cyanobacterial cells are no longer present. In consequence, levels of biovolume or chlorophyll-*a* at the time of sampling do not necessarily reliably indicate levels of the total concentration of CYNs.

Integrated samples taken in 2009 in Großer Plessower See illustrate this: Concentrations of cell-bound CYNs (combined cylindrospermopsin and deoxy-cylindrospermopsin) correlate to the biovolume of potentially CYN-producing species, summarized as Nostocales (*Rhaphidiopsis* (*Cylindrospermopsis*), *Aphanizomenon, Dolicospermum, Chrysosporum*). In contrast, dissolved CYNs reached its maximum concentration only once biovolumes of Nostocales and other cyanobacteria started to decline in September and remained on levels >1 µg/L until December (unpublished data, kindly provided by Karina Preussel, Robert-Koch-Institut, Berlin, and Jutta Fastner, Umweltbundesamt, Berlin).

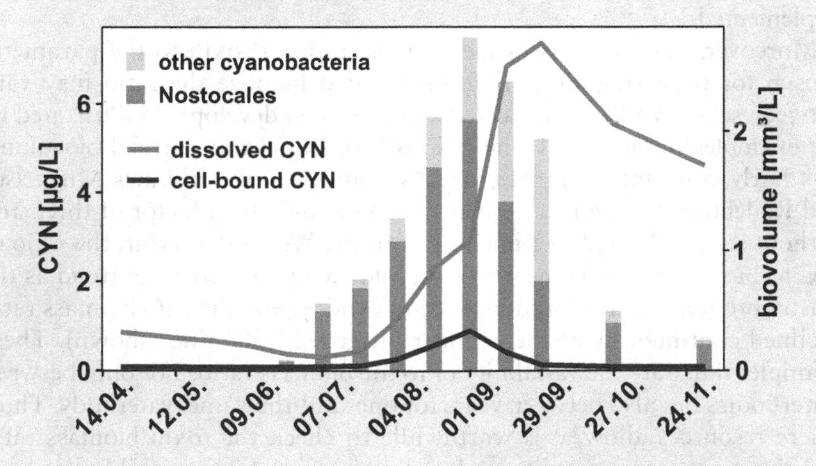

However, if monitoring on a regular, weekly or at least fortnightly basis has not identified any CYN-producing taxa (i.e., of the genera *Raphidiopsis* (*Cylindrospermopsis*), *Aphanizomenon*, or *Chrysosporum*) during the previous

4–6 weeks, the presence of CYNs is unlikely, in particular at concentrations above GVs.

If cyanobacteria of these genera have been found during previous weeks, but not at biovolume or chlorophyll-*a* levels exceeding the Alert Levels, the presence of CYNs exceeding the Alert Levels is also unlikely.

If, however, cyanobacteria of any of these genera have reached biomass levels corresponding to the Alert Levels during the 4–6 previous weeks, monitoring concentrations of dissolved and cell-bound CYNs is advised until concentrations of the sum of cell-bound and dissolved CYNs have declined below the guideline values (GVs).

It is generally useful to adapt the Alert Levels proposed in Table 5.1 to the toxin content of the locally prevalent cyanobacteria by occasional analyses of cyanotoxins together with the parameter used to trigger Alert Levels: periodically "calibrating" the trigger for alerts with cyanotoxin analyses will improve predictive power. As discussed in section 4.6, for any of the cyanotoxins, the ratio of toxin to biovolume or chlorophyll-*a* in a given waterbody may be substantially lower than the generally conservative assumption used in the Alert Level Frameworks, by an order of magnitude or more, and using a locally appropriate toxin/biomass ratio may serve to avoid undue restrictions of waterbody use or to lift restrictions previously implemented.

Moreover, periodic reassessment of the ratio of toxin to the parameter chosen for triggering alerts is recommended because the ratio may vary between seasons and within a season as a bloom develops, as illustrated by the examples in Figure 5.1: the ratio of MCs to cyanobacterial biovolume was fairly constant in the *Microcystis*-dominated waterbodies Müggelsee and Radeburg Reservoirs, Germany, varying only by a factor of three and without seasonal trend, but in contrast in the Weida Reservoir, the ratio of MC to biovolume varied nearly seven fold, with an increasing trend as the season progressed. Yet, in Müggelsee in other years, the MC/biomass ratio declined continuously as the summer progressed (data not shown). These examples illustrate the variability of toxin/biomass ratios not only between waterbodies but also between years for one-and-the-same waterbody. Thus, where resources allow, it is worthwhile to check the toxin/biomass ratio 2–3 times per season until a good understanding of its variability has been established in order to base management actions on the most appropriate information. Where access to capacity for cyanotoxin analysis is not readily possible, an option may be to send samples to regional laboratories or to seek support of research institutions.

For adapting the Alert Level Frameworks (ALFs) to national or local circumstances, the following further considerations are relevant:

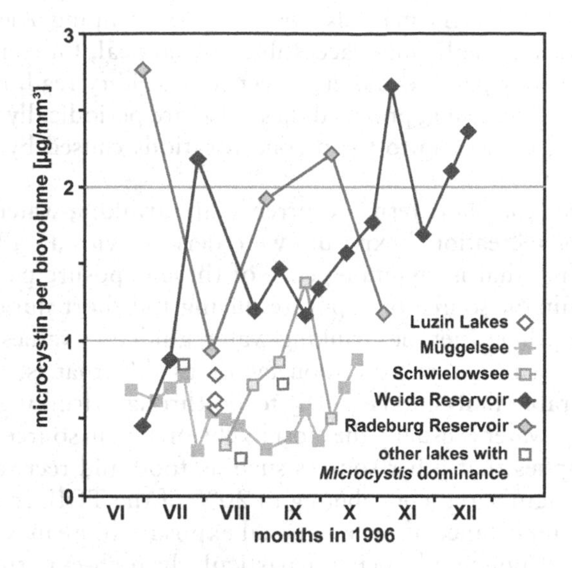

Figure 5.1 Ratios of microcystin (sum of all variants) to cyanobacterial biovolume over time in different lakes and reservoirs in Germany. Diamonds denote dominance of *Planktothrix rubescens* (Weida Reservoir and Luzin Lakes); the other waterbodies were dominated by *Microcystis* spp. (Jutta Fastner and Ingrid Chorus, unpublished data.)

1. If chlorophyll-*a* concentrations, Secchi depth readings or turbidity are used as triggers for alerts, a brief qualitative check by microscopy is important in order to assess whether chlorophyll-*a* or turbidity are largely due to cyanobacteria (and thus serve as effective indicators) or whether other phytoplankton, that is, eukaryotic algae (or in the case of turbidity, other particles), are causing elevated levels.
2. A reason to choose toxin concentrations rather than biomass indicators as parameters to define Alert Levels may be that the target is primarily to protect from cyanotoxins rather than from cyanobacterial cells as such; this may be appropriate particularly where drinking-water treatment reliably removes cells.
3. If cell counts are used to define Alert Levels, it is important to "calibrate" them against occasional toxin analyses because the cell quota data (i.e., toxin per cell) are available in the literature only for some taxa. However, cell sizes vary substantially, and as shown in section 4.6.2, cell size has a substantial impact on toxin quotas: if very small-sized cyanobacteria dominate, cell counts may be high and thus far too conservative, even if the water is very clear and toxin concentrations are negligible. "Calibration" of cell counts with toxin concentrations requires a significantly smaller number of samples over time as compared to regular monitoring.

4. Further parameters may also be used for defining Alert Levels, if locally or nationally more accessible or practical, for example, values for molecular parameters, fluorescence, turbidity readings or signals from remote sensing, provided these also are periodically "calibrated" regarding their ratio to toxin concentrations caused by the ambient cyanobacteria.
5. The GVs for short-term occurrence in drinking-water as well as those for recreational exposure were derived with an allocation factor of one, that is, assuming each of these exposure pathways to be the dominant source of exposure during the short duration of such exposure. The lifetime drinking-water guideline values (GVs) were derived assuming an allocation factor of 0.8, that is, that 80% of the tolerable daily intake (TDI) to be through drinking-water since drinking-water is usually the most likely long-term source of exposure. This implies that other sources such as food and recreational water are less significant (contributing to 20% of the TDI). In practice, the relative importance of each potential exposure route may be different, with food potentially being a particularly high-exposure source in some situations. When adapting an ALF to local circumstances, it is therefore important to assess the likelihood of simultaneous multiple routes of exposure – such as a population using bloom-ridden surface water with insufficient treatment for drinking *and* irrigation, perhaps also with freshwater fish as staple food. In such situations, it may be appropriate to consider reducing the allocation factors used in the derivation of the GVs. However, it is important to balance this with potential other negative consequences for the population's health and/or livelihoods that might result from severe restrictions of water use.

When using the information in this chapter as basis for developing locally appropriate guidance, it is further important to assess the patterns of bloom occurrence over time in the waterbodies of interest (see Chapter 8) and thus the likely duration of potential human exposure. This differs substantially between climates, regions and individual waterbodies: in temperate climates, some waterbodies dominated by surface scum-forming taxa such as *Microcystis* may have a bloom season of 3–5 months, and exposure then is typically seasonal. Other taxa, such as *Planktothrix agardhii*, may show perennial blooms even in cooler temperate climates, although generally with lower abundance during winter. In warmer climates, such as in some regions of Australia, South America, Asia and Africa, cyanobacteria may bloom for 6–10 months, and in relatively stable warm tropical climates, high numbers of cyanobacteria may occur year-round, potentially causing ongoing exposure. Importantly, however, in the same climates, other waterbodies may have no blooms at all or blooms occurring only sporadically and for only a few days or weeks.

5.1 DRINKING-WATER

Andrew Humpage and David Cunliffe

As outlined in the preceding chapters, toxigenic cyanobacteria are encountered in many waterbodies worldwide, including those from which water is abstracted for the production of drinking-water. The concentration of cyanotoxins in lakes and reservoirs can exceed the (provisional) GVs for lifetime daily exposure, as well as for short-term exposure, occasionally by orders of magnitude. To effectively remove cyanotoxins, drinking-water treatment needs to be optimised and validated for this target. Therefore, even when treatment is implemented, the possible breakthrough of cyanotoxins from raw water to the consumed drinking-water needs to be considered as a potential health risk and measures need to be validated to ensure that this risk is effectively controlled, or further measures be put in place to avert it.

A number of studies have concluded that cyanotoxins in drinking-water were the possible cause of documented cases of human illness. Further, even before the toxins were characterised in detail, there was compelling evidence of cyanobacterial toxicity from the deaths of animals following the consumption of water containing cyanobacteria. As discussed in section 5.0, historical literature about human illness after exposure to cyanotoxins must be treated with caution, however, because prior to their chemical characterisation neither the quantification of toxins nor the estimation of doses was possible. Further, other potential causes of the observed illnesses, such as viruses and protozoan pathogens, were not recognised, or could not be tested for, at the time. This does not, however, imply that the cases discussed in the next section are irrelevant with respect to the cyanotoxin risk.

Section 5.1.1 summarises evidence for the occurrence of toxigenic cyanobacteria in drinking-water sources and cyanotoxins in drinking-water distributed to consumers. It also provides data on human drinking-water-related poisoning events that have been documented adequately enough to provide reasonable indication that cyanotoxins were the causative agent of the poisonings. For further overview, readers are referred to the following publications: Harding and Paxton (2001); Chorus (2005); Codd et al. (2005); Falconer (2005); Falconer and Humpage (2005); Funari and Testai (2008); Hudnell (2008); Buratti et al. (2017). Section 5.1.2 gives guidance on assessing the risk of exposure to cyanobacteria or their toxins in drinking-water.

The cases discussed in the following sections demonstrate, firstly, that cyanotoxins in drinking-water sources and/or finished drinking-water are a worldwide phenomenon. Secondly, they also highlight that, depending on the level of contamination and the treatment processes employed, the toxins can contaminate treated drinking-water. Thirdly, they show that where treatment is insufficient or overwhelmed by a massive bloom, toxin concentrations can significantly exceed guideline values (GVs) in drinking-water.

5.1.1 Evidence of illness from exposure to cyanobacteria in drinking-water

All reports to date reporting symptoms have only been associated with the toxins MC or CYN. However, some of the human effects ascribed to the presence of cyanotoxins in drinking-water, such as gastrointestinal illness and pneumonia, may well be due to other, less well-described, cyanobacterial metabolites (see section 2.10) as well as bloom-associated pathogens or their metabolites. Furthermore, where blooms were treated with copper sulphate, high copper concentrations may be an explanation for symptoms such as diarrhoea, vomiting, stomach cramps and nausea; however, this would require concentrations above the range of 1–2mg/L at which it is used as an algicide (see WHO (2017) for a discussion of copper toxicity).

As discussed above, where cyanobacterial blooms in the source water and illnesses are observed at the same time, the obviousness of the bloom or scums makes it suggestive to presume cyanobacteria as the aetiologic agent. However, substantiating this with data is challenging as it requires analysing water samples taken when patients were exposed, and in the published case studies, this has very rarely been accomplished. Also, even if the cyanobacteria and their metabolites themselves are not the direct cause of the illness, the true aetiology may be closely linked to the bloom, for example, pathogens associated with the bloom (see, e.g., Berg et al., 2009). For effectively targeting measures to ensure or improve water quality, it is important to understand cause–effect relationships, particularly whether or not drinking-water was the actual cause. Illness suspected to be due to drinking-water requires a detailed investigation so that steps can be rationally applied to prevent such occurrences in future. While such follow-up investigations may fail to clearly identify a causative agent for observed illness, they will serve to identify water quality deficits and risks of events causing contamination hazardous to health.

5.1.1.1 Examples of potentially hazardous cyanotoxin concentrations in finished drinking-water

Local knowledge about the hazardous nature of cyanobacterial scums appears to have existed in some regions with eutrophic, bloom-ridden waterbodies for a long time, as discussed in Chapter 1. Scientific screening of occurrence began in the wake of emerging awareness of cyanobacterial toxicity in the 1980s. At this time, screening often followed deaths of farm animals and relied only on mouse bioassays to evaluate toxicity because methods for the chemical analysis of known toxins only became available from the later 1980s onwards. Since then, surveys have been conducted in many parts of the world, including in drinking-water. A summary of findings from a range of relatively detailed studies is provided below. Less extensive reports on cyanotoxin occurrence have come from Brazil, Europe, New Zealand, China, Thailand and Africa (Chorus, 2005; Codd et al., 2005). For exemplary data on cyanotoxin occurrence in a variety of waterbodies, see Chapter 2.

- One of the first surveys specifically targeting drinking-water sources was conducted in 1991 in the Murray–Darling Basin, Australia, which is a major agricultural region that relies on its rivers for irrigation and drinking-water supply (Baker & Humpage, 1994): Mouse bioassays were performed on 231 cyanobacterial grab samples from sites across the Basin. Approximately 60% of samples were from potential drinking-water sources (rivers, lakes, reservoirs). Mouse bioassays showed that 24% of samples were neurotoxic and a further 18% were hepatotoxic, thus demonstrating a need to ensure sufficiently effective drinking-water treatment.
- Low concentrations of MCs were detected in 15 finished drinking-water samples collected during the fall of 1992 from two Canadian water treatment plants (0.09–0.18 µg/L MC-LR equivalent in a protein phosphatase inhibition assay (Lambert et al., 1994).
- A survey of MCs in drinking-water utilities across the USA and Canada (June 1996 to January 1998; Carmichael, 2001) included over 24 utilities, and 677 samples were screened for MCs by ELISA. The samples were taken from blooms, plant intakes, plant influents (after preoxidation) and finished water. Although 80% of samples contained MC levels above the detection limit of 0.02 µg/L, only two finished water samples showed MC concentrations above 1 µg/L. These occurred in two of the three treatment plants that were facing significant MC challenges at the time of sampling in July 1997: at plant CM-1, MC concentrations at the intake were >1000 µg/L and 8 µg/L in the finished water, respectively. At plant IXC-3, the intake contained just over 2 µg/L MCs and the finished water contained about 1.3 µg/L. Plant CM-1 utilised prechlorination and granular activated carbon, whereas plant IXC-3 only added ammonium and chlorine to otherwise untreated source water (Carmichael, 2001).
- Cyanotoxin surveys in Florida in 1999 and 2000 (Burns, 2008) of surface water sources and finished waters collected 167 samples, of which 88 contained cyanotoxins (MCs, ATX, CYN). MCs were the most commonly found toxins, occurring in both pretreatment and posttreatment waters. Concentrations in the latter ranged from below detection to 12.5 µg/L. Three finished water samples contained ATX up to 8.46 µg/L, whereas nine finished water samples contained CYN at

concentrations of 8.1–97 μg/L. A survey of 52 source and finished water samples from two drinking-water treatment plants in Queensland, Australia, found that only two samples of finished water contained traces (<0.05 μg/L) of STX when the source waters contained up to 17 μg/L STX. The authors concluded that conventional drinking-water treatment (flocculation, sedimentation, PAC during high toxin load, sand filtration and chlorination) removed 99.9% of total STX (free and cell-bound) from water containing a toxic *Anabaena circinalis* bloom (Hoeger et al., 2004).

- During the summer of 2003, MCs were detected at low levels (0.15–0.36 μg/L) in 30 of 77 finished water samples from 33 US drinking-water treatment plants in Northeastern and Midwestern USA. However, only relatively low concentrations (0.15–5.6 μg/L) were detected in 87 of 206 raw water samples from the same plants (Haddix et al., 2007).
- CYN was detected (1.3 and 8.6 μg/L) during March 2007 in finished waters of two conventional treatment plants, as well as throughout the combined distribution system, on Kinmen Island, Taiwan (15 tap samples ranging from 0.7 to 2.2 μg/L), when the plants were challenged by high CYN levels in the raw water (0.7 and 36 μg/L; Yen et al., 2011).
- In another Canadian plant Zamyadi et al. (2012) detected up to 10 μg/L MCs in clarifier supernatants and up to 2.5 μg/L in the finished chlorinated drinking-water during the bloom seasons (June to October) of 2008, 2009 and 2010.
- MCs have also been detected in conventionally treated drinking-water (with flocculation, sedimentation, sand filtration, chlorination) in Saudi Arabia (range 0.33–1.6 μg/L over 8 monthly samples during May to December 2007; Mohamed & Al Shehri, 2009) and Egypt in May 2013 (up to 3.8 μg/L; Mohamed et al., 2015; Mohamed, 2016), and also in Algeria (up to 6.3 μg/L during a bloom in 2013, treatment process not reported; Saoudi et al., 2017).
- In Australia, during the summer of 2013–2014, a bloom of *Raphidiopsis* (*Cylindrospermopsis*) *raciborskii* occurred in the water supply of Mount Isa, Queensland. The water supply was treated by passage through a reed bed filtration lagoon before chlorination. *R. raciborskii* blooms were common in the supply reservoirs (Lake Moondara and Lake Julius), but this was the first time a bloom had occurred in the filtration lagoon. *R. raciborskii* numbers peaked at 425 000 cells/mL in the lagoon and 42 000 cells/mL in the finished water storage reservoir. The maximum toxin levels detected in treated water were 2 μg/L CYN in the storage reservoir and 0.5 μg/L CYN in the town reticulation. Chlorination was increased to maintain a residual and later a mobile ultrafiltration unit was installed. Cell counts and toxins in the treated water returned to safe levels after the ultrafiltration unit was installed (Janet Cumming, Queensland Department of Health, pers. comm., January 2017).

- In August 2014, the city of Toledo, Ohio, total MCs occurred in the city's finished drinking-water at levels up to 2.5 μg/L. A "do not drink or boil advisory" was issued to nearly 500 000 consumers. A cyanobacterial bloom near Toledo's drinking-water intake located on Lake Erie was the source of the MCs. The advisory was lifted 2 days later after optimisation of permanganate and PAC treatments led to the reduction of the MC concentrations to levels below 1 μg/L in all samples (US EPA, 2015).

Options for control of cyanobacterial occurrence and cyanotoxin removal through effective treatment are described in Chapters 7–10.

5.1.1.2 Case reports giving evidence of short-term health risks from acute exposure through drinking-water

Some case studies provide evidence that exposure to cyanobacterial toxins in drinking-water can lead to illness and even death. Due to the inability to identify the toxins at the time, the earliest reported cases offer only circumstantial evidence of a link between exposure to cyanotoxins and human illness.

- Gastroenteritis associated with cyanobacteria was observed in the population of a series of towns along the Ohio River in 1931. Low rainfall had allowed the water of a side branch of the river to develop a cyanobacterial bloom which was then washed into the main river. As this water moved downstream, a series of outbreaks of illness were reported (Tisdale, 1931).
- In Harare, Zimbabwe, children living in an area of the city supplied from a particular water reservoir developed gastroenteritis each year at the time when a natural bloom of *Microcystis* was decaying in the reservoir. Other children in the city with different water supplies were not affected (Zilberg, 1966).
- In an incident in Sewickley, Pennsylvania, 62% of the population connected to a filtered, chlorinated drinking-water supply developed symptoms of gastroenteritis within a period of five days. The water, sourced from groundwater contaminated by an intrusion from the Ohio River, was treated and then held in open holding reservoirs prior to distribution. One reservoir had over 100 000 cells/mL of *Schizothrix calcola*, *Plectonema*, *Phormidium* and *Lyngbya* in the open water. The reservoir had just been treated with copper sulphate when the poisoning event occurred (Lippy & Erb, 1976). Although not known to be toxic at the time, *Schizothrix*, *Phormidium* and *Lyngbya* have all since been shown to be toxin producers elsewhere (Falconer, 2005).

While these reports note that the health effects could not be attributed to infectious agents, a caveat on this conclusion is that many of the aetiologic agents leading to the described symptoms were unknown at the time (e.g., viruses) or not detectable with sufficient sensitivity by a standard laboratory (*Giardia*, *Cryptosporidium*). The following later study addressed many of these issues.

- An outbreak, with a high death rate attributed to cyanobacterial toxins in drinking-water, occurred in the Paulo Alfonso region of Bahia State in Brazil following the flooding of the newly constructed Itaparica Dam reservoir in 1988. Some 2000 gastroenteritis cases were reported over a 42-day period, and 88 deaths, mostly children, occurred (Teixera et al., 1993). Blood and faecal specimens from gastroenteritis patients were subjected to bacteriological, virological and toxicological testing, and drinking-water samples were examined for microorganisms and heavy metals. No infectious agent was identified, and cases occurred in patients who had been drinking only boiled water. The cases were restricted to areas supplied with drinking-water from the dam. Clinical data and water sample tests were reviewed, and it was concluded that the source of the outbreak was water from the dam and that a toxin produced by cyanobacteria (*Anabaena* and *Microcystis* in high densities) was the most likely responsible agent, although the toxin could not be identified.

A closer association between human illness and exposure to cyanotoxins is demonstrated when the cyanobacteria were shown to be toxin producers, as illustrated in the following examples:

- In Armidale, Australia, the water supply reservoir had been monitored for blooms of toxic *Microcystis* for several years, and MC-YM had been identified in these blooms. When a particularly dense bloom occurred, the water supply authority treated the reservoir with 1 mg/L of copper sulphate, which lysed the bloom, possibly causing a pulse of toxin release from the cells. An epidemiological study of the local population indicated subclinical liver damage occurring simultaneously with this treatment of the bloom (see Box 5.2).
- A more severe outbreak of cyanobacterial toxicity in a human population occurred on Palm Island, off the north-eastern coast of Australia in 1979. Complaints of bad taste and odour in the water supply were attributed to a cyanobacterial bloom, and the authorities therefore treated the reservoir with copper sulphate. Within a week, numerous children developed severe hepatoenteritis, and a total of 140 children and 10 adults required hospital treatment (Byth, 1980). A CYN-producing strain of *Raphidiopsis raciborskii* was later identified as the agent most likely to be responsible for this episode (see Box 5.3).

BOX 5.2: TOXIC *MICROCYSTIS* IN THE ARMIDALE WATER SUPPLY RESERVOIR AND PUBLIC HEALTH

At the time of this study, the city of Armidale, New South Wales, Australia, had a drinking-water supply from a eutrophic reservoir which had been experiencing repeated blooms of cyanobacteria since the early 1970s.

In 1981, a particularly extensive toxic bloom of *Microcystis aeruginosa* was monitored during its development. During the bloom, complaints of bad taste and odour in the drinking-water were received, leading to copper sulphate treatment of the reservoir. The toxicity of the bloom was monitored by mouse bioassay. A toxin had previously been isolated from Malpas Dam and partially described, which was later characterised as MC-YM (Botes et al., 1985). This event was used as the basis for a retrospective epidemiological study of liver function in the population consuming the water, compared with a population in the same region supplied from other reservoirs. The data for the activity of plasma enzymes reflecting liver function were obtained for patients having blood samples examined at the Regional Pathology Laboratory for the 5 weeks prior to the bloom, the 5 weeks of peak bloom and its termination and for 5 weeks after that. The data were then separated into those from patients having used the Malpas drinking-water supply and those using other supplies.

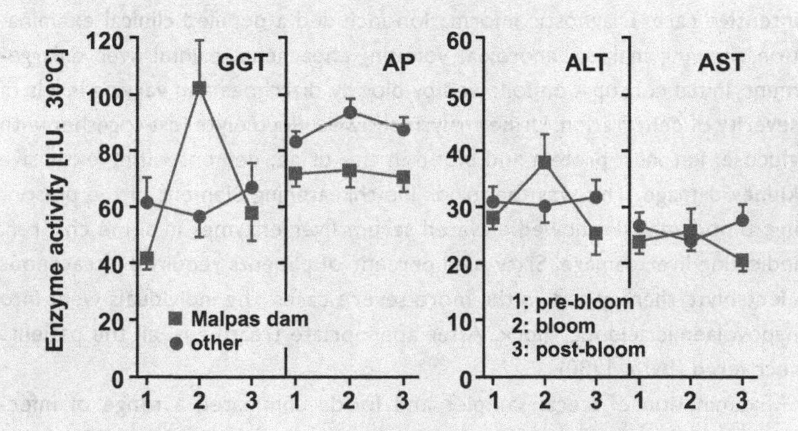

Serum enzymes reflecting liver function in patients consuming drinking-water from Malpas Dam or from other supplies included GGT=γ-glutamyl transferase; ALT=alanine aminotransferase; AST=aspartate aminotransferase and AP=alkaline phosphatase (Falconer et al., 1983). As shown in the figure above (redrawn from Falconer et al., 1983), γ-glutamyl transferase in the blood of the group using the Malpas Dam water supply during the peak of the bloom and its lysis with copper sulphate was significantly higher than that in the same population before and after the bloom, and higher than that in the other population served by different water supplies. The clinical record gave no evidence of an infectious hepatitis outbreak or disproportionate alcoholism (Falconer et al., 1983). While the mean increase in γ-glutamyl transferase activity was indicative of minor liver toxicity, some individuals within the population studied showed highly elevated enzyme activity, indicating substantial liver damage. This enzyme has also been shown to be elevated as a result of *Microcystis* toxicity in experimental studies with pigs and rodents, where it is used as an effective marker for liver injury (Fawell et al., 1993; Falconer et al., 1994).

BOX 5.3: PALM ISLAND MYSTERY DISEASE

In 1979, there was a major outbreak of hepatoenteritis among the children of an Aboriginal community living on a tropical island off the coast of Queensland, Australia. Altogether 140 children and 10 adults required treatment, which was provided by the local hospital for less severe cases and by the regional hospital on the mainland for severe cases possibly requiring intensive care. Diagnostic information included a detailed clinical examination showing malaise, anorexia, vomiting, headache, painful liver enlargement, initial constipation followed by bloody diarrhoea and varying levels of severity of dehydration. Urine analysis showed electrolyte loss together with glucose, ketones, protein and blood in the urine, demonstrating extensive kidney damage. This was the major life-threatening element of the poisoning. Blood analysis showed elevated serum liver enzymes in some children, indicating liver damage. Sixty-nine percent of patients required intravenous electrolyte therapy and, in the more severe cases, the individuals went into hypovolaemic/acidotic shock. After appropriate treatment, all the patients recovered (Byth, 1980).

Examination of faecal samples and foods eliminated a range of infectious organisms and toxins as possible causes for the outbreak and failed to identify the cause, hence the name "Palm Island Mystery Disease". The affected population, however, all received their drinking-water supply from one source, Solomon Dam. Families on alternative water supplies on the island were not affected by the disease. Prior to the outbreak of the illness, a bloom of cyanobacteria occurred in Solomon Dam. The bloom discoloured the water and gave it a disagreeable odour and taste. When the bloom became dense, the dam reservoir was treated with 1 ppm of copper sulphate (Bourke et al., 1983). Clinical injury among consumers on that water supply was reported the following week. In subsequent investigations, the organisms from the dam were cultured and administered to mice. Mice treated with *Raphidiopsis* (*Cylindrospermopsis*) *raciborskii* culture slowly developed (over several days) widespread tissue injury involving the gastrointestinal tract, the kidney and the liver (Hawkins et al., 1985). The widespread tissue damage and delayed effects are quite different to those following *Microcystis aeruginosa* administration (Falconer et al., 1981). Subsequent monitoring of the blooms in the dam – well after the outbreak – identified *R. raciborskii* as the cause of the blooms, with seasonal cell concentrations of up to 300 000 cells/mL of water. This organism did not form scums and has the highest cell concentrations well below the water surface. In order to reduce bloom formation, the responsible authorities later introduced destratification of the reservoir (Hawkins & Griffiths, 1993). Subsequent research on toxins produced by *R. raciborskii* has identified the cytotoxic alkaloid cylindrospermopsin.

5.1.1.3 Epidemiological studies addressing health risks from chronic, low-dose exposure through drinking-water

While a number of epidemiological studies of the possible association of MC exposure with cancer incidence are available, all of them have used retrospective estimates of MC exposure. However, as discussed at the beginning of this chapter, such retrospective approaches face pronounced uncertainty regarding both the concentrations of cyanotoxins and those of any other pollutants to which the population was exposed during the formative stages of their cancer. In fact, the occurrence of other pollutants in surface waterbodies with heavy cyanobacterial blooms is quite likely, as blooms are caused by heavy nutrient loads and these are often associated with substantial loads of pesticides and/or other contaminants from agriculture and/or poorly treated wastewater. In addition, demographic information was usually not provided so it is not clear whether dietary, genetic and/or lifestyle factors associated with cancer were adequately controlled in the analyses. It is therefore important that where an observed health impairment is connected to cyanobacterial blooms (as the most prominent and visible phenomenon), health authorities also look for other potential causative agents. In consequence, it is currently not possible to show causation or to derive concentration–response data from the epidemiological studies available to date. While for this reason they cannot serve as basis for deriving guideline values (GVs) (see above and Chapter 2), they are of some indicative value and are therefore summarised as follows:

- The possible link between chronic exposure to cyanotoxins and the incidence of human cancer has been studied in China and the USA. The incidence of hepatocellular carcinoma (HCC) in China has historically been one of the highest in the world, at least in part due to two proven risk factors: infection with hepatitis B virus (HBV), which increases the risk almost 10-fold (Yu et al., 2002), and intake of aflatoxin B1 from foods infected with moulds, which increases the risk in HBV-positive individuals by a further threefold (Lian et al., 2006). However, the uneven geographic distribution of HCC incidence in China could not be entirely explained by these factors and so other environmental factors were investigated (Yu, 1989; Yu, 1995; Yu et al., 2001). The source of a person's drinking-water was also found to be a significant risk factor with people drinking pond or ditch water having about 10-fold higher incidence of HCC when compared to those drinking deep well water. MCs were found to occur seasonally in water sources of Haimen city, China, with a summer survey detecting MCs in 17% of pond/ditch water samples, 32% of river water samples, 4% of shallow well and 0% of deep well water samples, with averages of 0.10, 0.16 and 0.07 μg/L for the first three, respectively (Ueno et al., 1996). Similar concentrations were found in a parallel study using different analytical methods (Harada et al., 1996). These concentrations seem quite low for untreated raw waters and are more similar to concentrations observed elsewhere in the world in finished waters (see examples given above). Nevertheless, based on the average MC contents of river and pond/ditch samples, Ueno et al. (1996) provide limited data that would lead to an estimated average daily exposure in the range of 0.2 μg/person during the summer months (note that the authors report 0.2 pg/person, but this is clearly a typographical error). Later studies from China have associated slightly higher exposure rates from food and water combined (0.36 to 2.03 μg/person per day) with detectable concentrations of serum MCs and increased levels of liver enzymes in the serum (Chen et al., 2009; Li et al., 2011), see below.
- A later case control study in Haimen city, China, did not find an association with drinking-water sources (Yu et al., 2002). However, this study did not analyse for the prevalence of aflatoxin-B1 antigens in the study population. There is evidence from animal studies that MC acts synergistically with aflatoxin tumour initiation to increase rates of liver cancer (Sekijima et al., 1999; Lian et al., 2006), whereas this may not be the case for HBV-related HCC (Lian et al., 2006).
- An increase in serum markers for hepatotoxicity (AST, ALP, ALT and lactate dehydrogenase, LDH) was observed in a cohort study of

Chinese fishermen exposed to MC-RR, MC-YR and MC-LR in Lake Chaohu through the consumption of contaminated water and food (Chen et al., 2009). The fishermen had a median serum MC concentration of about 0.2 ng/mL and an estimated daily intake of MC of 2.2–3.9 µg MC-LR equivalents (Chen et al., 2009). The relative proportion of the three variants in the fishermen's blood were similar to those in the carp and duck tissues used as typical food.

- Li et al. (2011) conducted a cross-sectional study assessing the relationship between liver damage in children (n>1000) and MC levels in drinking-water and aquatic food (carp and duck) in China. MC levels measured in three local sources of drinking-water were classified in three groups, as negative controls, low and high exposure, with children in the low-exposure group consuming an estimated 0.36 µg/day and high-exposure children consuming 2.03 µg/day. Mean serum levels of MC-LR equivalents in the groups were below the detection limit in the negative control, 0.4 in the low-exposure and 1.3 µg/L in the high-exposure groups, with mean detection rates of 1.9%, 84.2% and 91.9%, respectively (1.9% in the control group caused by 1 MC-positive among 54 serum samples). MC was associated with increases in aspartate aminotransferase (AST) and alkaline phosphatase (ALP), but not ALT or γ-glutamyl transferase (GGT). The odds ratio (OR) for liver damage associated with MC was 1.72 (95% CI: 1.05–2.76), after adjustment for HBV infection and use of hepatotoxic medicines as confounding factors. HBV infection was a greater risk for liver damage in children.

  Although these findings suggest a potential role of MCs in the high HCC incidences, they cannot be used, as was proposed by Ueno et al. (1996), to derive a guideline for MCs in drinking-water because (i) although the authors demonstrated an association between the type of water consumed by people living in high HCC areas and the presence of MCs in that water, they derive no quantitative relationship between MC exposure and cancer incidence; (ii) MC concentrations in similar waters in low HCC areas were not determined, so the association remains only suggestive; and (iii) as noted above, the high incidence of HCC in certain regions of China has also been linked to high hepatitis B infection rates and exposures to aflatoxin B1, so it would not be correct to extrapolate data from this population to other populations not exposed to these additional risk factors. These results about the possible, although not proven, higher HCC incidence are consistent with the activity of MC-LR as a tumour promoter, increasing the potency of known tumour initiators such as aflatoxin B1 (see section 2.1 and below).

- Another Chinese study has looked at the association between the incidence of colorectal cancer and drinking-water source (Zhou et al., 2002). In this case, 408 cases of colon or rectal cancer were retrospectively categorised by the source of drinking-water consumed by the patients (well, tap, river, pond). The relative risk of developing colorectal cancer was almost fourfold higher in consumers of pond or river water. Average and maximal concentrations of MCs were reported as follows: river waters (average 0.141 µg/L, maximum 1.083 µg/L, n=69), pond waters (0.106, 1.937 and 35), well waters (0.004, 0.009 and 12) and tap waters (0.005, 0.011 and 17). A positive association was found between MC concentration and colorectal cancer incidence, although as with the other studies, this association remains only suggestive.

- Svirčev et al. (2009; 2013) report an observational study that found an elevated incidence of primary liver cancer in regions served by drinking-water reservoirs that are subject to frequent summer blooms of cyanobacteria. However, no information on cyanotoxin exposures was presented.
- In the USA, the incidences of primary hepatocellular carcinoma (HCC) and colorectal cancer have been evaluated in relation to the study population's likely water source – surface water or ground water (Fleming et al., 2002). Only weak (HCC) or no (colorectal cancer) associations were found in these pilot studies.

As discussed above, such studies cannot be used for the derivation of GVs for safe levels in drinking-water. Because of the limitations of the human epidemiology studies, the best available animal studies have been used to derive the lowest, most protective GVs that are scientifically supported by robust quantitative evidence (see Chapter 2).

5.1.2 Assessing the risk of exposure to cyanotoxins through drinking-water and short-term responses to occurrence

A modern water treatment plant equipped with an effective filtration system for physical removal of cells as well as the removal of dissolved toxins should remove cyanotoxins to below hazardous levels, provided it is operated with attention to avoid disruption of cyanobacterial cells and release of dissolved toxin (see Chapter 10). However, this requires it to be validated for meeting this target. Also, many of the world's drinking-water supply systems and treatment plants are more rudimentary, and large populations may depend upon such vulnerable water supplies or on untreated surface waters for drinking and preparing food.

For exposure assessment, particularly for MCs and CYNs, it is important to differentiate between daily exposure for significant parts of a lifetime and short-term episodic exposure. If concentrations exceed the values intended for lifetime daily consumption of drinking-water, but are below the short-term guideline values (GVs) given in Table 5.1 (or nationally derived standards; Table 5.3), use of the water supply for drinking may continue, and action may first focus on assessing which measures are locally most appropriate to ensure better control of the cyanotoxin concentrations. These might include addressing the cause for waterbody conditions leading to

Table 5.3 Standards, guideline values, maximum acceptable concentrations or maximum values set by a number of countries for cyanotoxins in drinking-water

Cyanotoxin	*Type of value*	*Numerical value*	*Country*
Microcystins	Guideline value	1.3 µg/L MC-LR toxicity equivalents	Australia
	Standard	1 µg/L MCs	Brazil
	Maximum acceptable concentration	1.5 µg/L MC-LR	Canada
	Standard	1 µg/L MC-LR	Czech Republic
	Standard	1 µg/L sum of MCs	France
	Provisional maximum value	1.3 µg/L MC-LR equiv.	New Zealand
	Restrictions on water use	>1.0 µg/L sum of MCs	Finland
	Ban on water use	>10.0 µg/L sum of MCs	Finland
	Standard	1 µg/L MC-LR	Singapore
	Standard	1 µg/L sum of MCs	Spain
	Standard	1 µg/L MC-LR	Uruguay
	Standard	1 µg/L sum of MCs	Turkey
	Guideline value	1 µg/L MC-LR	South Africa
	Provisional maximum value	1 µg/L	New Zealand
Nodularin	Health Alert Level	1 µg/L	Australia
Cylindrospermopsin	Guideline value	1 µg/L	Brazil
	Provisional maximum value	1 µg/L	New Zealand
	Health Alert Level	3 µg/L	Australia
Saxitoxins (as saxitoxin toxicity equivalents)	Guideline value	3 µg/L	Brazil
	Provisional maximum value	3 µg/L	New Zealand
	Provisional maximum acceptable concentration	3.7 µg/L	Canada
Anatoxin-a	Provisional maximum value (valid also for homoanatoxin-a)	1 µg/L	New Zealand
	Provisional maximum value	1 µg/L	New Zealand

Source: Data from Ibelings et al. (2014).

blooms (which may or may not be feasible in the short term; see Chapters 7 and 8), shifting the raw water offtake to avoid blooms (Chapter 9) or implementing additional treatment steps (Chapter 10). Allowing such flexibility for the locally most effective response if a GV intended for lifetime daily exposure is exceeded is particularly pertinent to short-lived bloom situations if past experience shows that they are likely to disperse within a few days, thus no longer causing elevated cyanotoxin concentrations. The short-term GVs are intended for periods of about 2 weeks and are not intended to endorse repeated seasonal exceedances of the lifetime GV. Where water with concentrations ranging up to these values is distributed, it is important to inform the population about this situation so that specifically vulnerable groups may take specific measures, such as using bottled water. This may be relevant, for example, for hepatitis patients in the case of hepatotoxins and is particularly important for those responsible for bottle-fed infants because the short-term drinking-water GV is based on exposure of adults. Since infants and children can ingest a significantly larger volume of water per body weight (e.g., up to 5 times more drinking-water/kg bw for bottle-fed infants compared to an adult), as a precautionary measure WHO recommends that alternative water sources such as bottled water are provided for bottle-fed infants and small children when MC concentrations are greater than 3 μg/L for short periods (WHO, 2020).

5.1.2.1 Defining national or regional cyanotoxin levels requiring action

As discussed at the beginning of this chapter, when setting national standards or defining threshold concentrations that should trigger specific action, it is important to consider whether the WHO GVs given in Table 5.1 and used in the Alert Levels Framework (ALF) below are locally or nationally appropriate, or whether they would better be adapted to local or national circumstances. Besides differences in the ratios between toxin concentration and the indicators used to trigger the alert, such circumstances may include the amount of drinking-water consumed and the fraction of cyanotoxin allocated to uptake through drinking-water in relation to other exposure pathways (see sections 5.2–5.5). Further considerations include the extent and duration of cyanotoxin exposure in relation to other hazards: where public health impacts from exposure to other hazards (in particular pathogens) are substantial and toxic cyanobacterial blooms are short-lived events, a decision might be to tolerate somewhat higher concentrations (possibly only as an interim solution) in order to focus available capacity and resources on controlling exposure first to those hazards which are causing the highest risks for health. Such considerations are particularly important when setting national or local water quality regulations, because where other quality issues are likely to have a higher public health impact, enforcing a low cyanotoxin standard may distract funding from investments needed to remediate the more pressing public health problems.

A number of countries have implemented concentrations triggering action for a range of cyanotoxins (see examples in Table 5.3). Particularly for cyanotoxins other than MCs, they have typically not been set as standards in the legal sense of values that all water suppliers in the country need to meet in order to be in compliance with regulations but rather guideline values (GVs) or "Health Alert Levels" that are used to trigger a notification to the health authority, further assessment of the situation and/or other management responses.

5.1.2.2 Alert Levels for short-term responses to toxic cyanobacteria in drinking-water supplies

An Alert Levels Framework (ALF) is a monitoring and management action sequence, presented as a "decision tree" in Figure 5.2, which water treatment plant operators and managers can use to provide an immediate, graduated response to the onset and progress of a cyanobacterial bloom. An ALF was first developed in Australia in the 1990s and then introduced in the first edition of "Toxic Cyanobacteria in Water" in 1999 (Chorus & Bartram, 1999). Since then, this approach has been widely used, typically with some adaptation to local or national conditions (Ibelings et al., 2014). Circumstances and operational alternatives may vary depending upon the source of the water supply, as well as the analytical and water treatment facilities available. The ALF presented here is therefore intended as a general framework, recognising that it may be appropriate to adapt specific Alert Levels and actions to suit local conditions. This includes the choice of parameters used to trigger alerts: as discussed at the beginning of this chapter and in more detail below, other parameters such as cell numbers or turbidity readings may be used if they are periodically "calibrated" against toxin concentrations.

One important aspect of an ALF for potentially toxic cyanobacteria is that this specific hazard often occurs with some predictability. In many surface waters, cyanobacterial blooms (and phytoplankton blooms in general) follow a seasonal pattern, or they occur following distinct events such as drought or heavy rainfall (highly dependent on local circumstances). It is therefore important to keep any records that are taken when following the ALF. These data can serve to significantly refine the ALF for individual water supplies (see also Chapter 10). This applies equally to patterns of spatial heterogeneity (see Chapter 3) of cyanobacterial blooms in individual waterbodies. The formation and location of surface scum can potentially be anticipated, although with some uncertainty, for a given waterbody. Since accumulations of cyanobacteria next to sensitive sites, such as raw water offtakes, are highly relevant, these sites need to be included in the ALF.

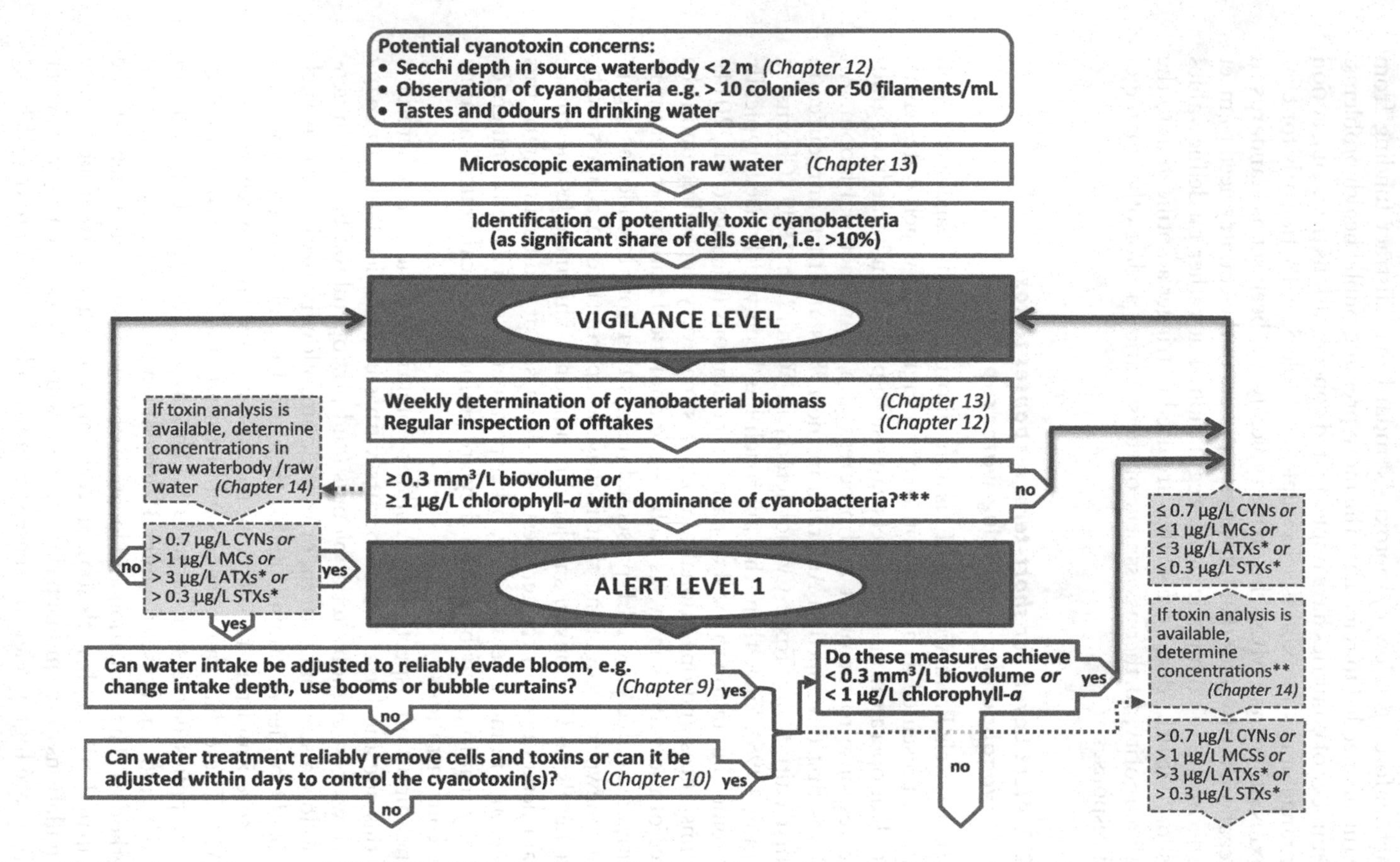

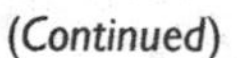

Figure 5.2 Alert Levels Framework (ALF): Decision tree for monitoring and managing cyanobacteria in drinking-water supplies (as template to be adapted to local conditions).

(Continued)

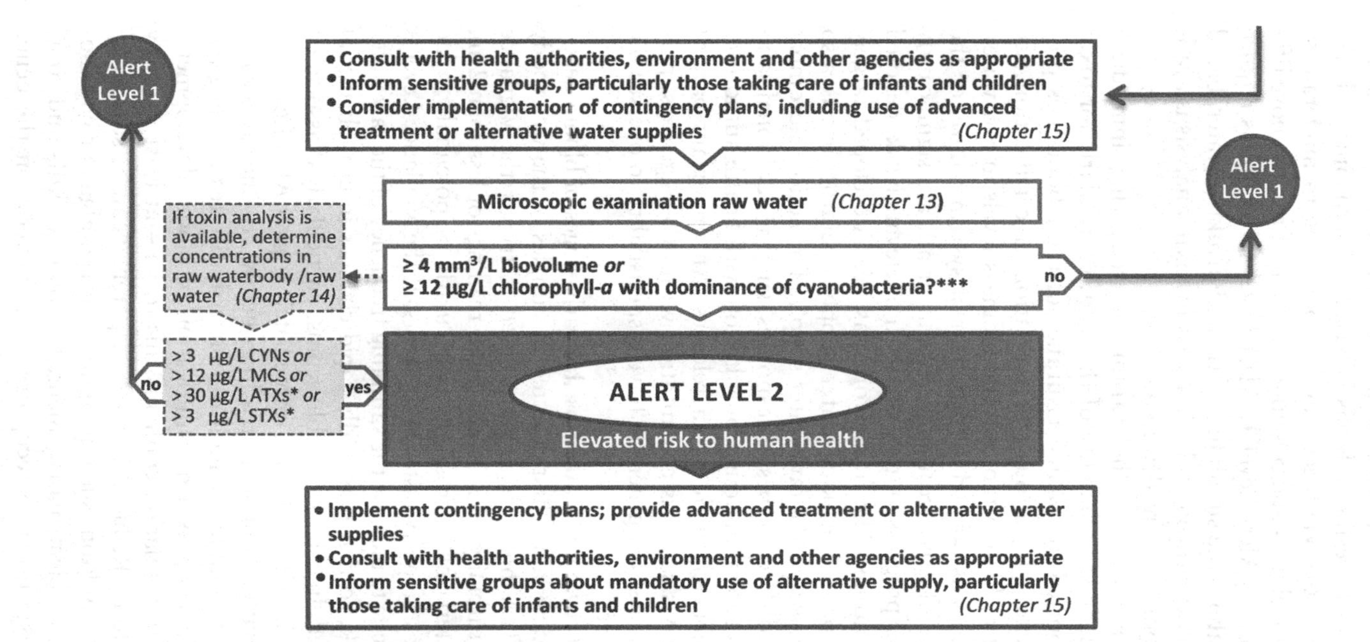

Alert Level 1
• Consult with health authorities, environment and other agencies as appropriate
• Inform sensitive groups, particularly those taking care of infants and children
• Consider implementation of contingency plans, including use of advanced treatment or alternative water supplies (Chapter 15)
Microscopic examination raw water (Chapter 13)
If toxin analysis is available, determine concentrations in raw waterbody /raw water (Chapter 14)
≥ 4 mm³/L biovolume or
≥ 12 µg/L chlorophyll-a with dominance of cyanobacteria?***
no
Alert Level 1
> 3 µg/L CYNs or
> 12 µg/L MCs or
> 30 µg/L ATXs* or
> 3 µg/L STXs*
no
yes
ALERT LEVEL 2
Elevated risk to human health
• Implement contingency plans; provide advanced treatment or alternative water supplies
• Consult with health authorities, environment and other agencies as appropriate
• Inform sensitive groups about mandatory use of alternative supply, particularly those taking care of infants and children (Chapter 15)

Note: As dissolved CYNs may be persistent, regular microscopy is particularly important to detect possible producer organisms and then trigger chemical analysis of CYNs (see Box 5.1); alternatively, monitor CYNs

* these values for STXs and ATXs are NOT Guideline values for lifetime exposure, but merely concentrations 10-fold below those for acute exposure (see text).

**analyse in raw water for checking effect of measures to avoid bloom intake and in treated water for checking the efficacy of treatment

*** if CYN- or STX-producers are dominant, analysing CYN/STX is recommended in order to determine whether Alert Levels for biovolume/chlorophyll *a* are sufficiently protective

Figure 5.2 (Continued) Alert Levels Framework (ALF): Decision tree for monitoring and managing cyanobacteria in drinking-water supplies (as template to be adapted to local conditions).

The ALF decision tree uses three "threshold" levels to guide the assessment of a potentially toxic cyanobacterial bloom, with appropriate actions and responses. The sequence of response levels is based upon the initial detection of cyanobacteria at the Vigilance Level, progressing to moderate to high cyanobacterial biovolumes and possible detection of toxins above lifetime GV concentrations at Alert Level 1. Alert Level 1 conditions require decisions to be made about the suitability of treated drinking-water, based on the efficacy of water treatment and – if access to toxin analysis is available – the concentrations of toxins detected.

An important issue regarding the parameters triggering immediate responses is confidence in the reliability of the data, particularly for toxin analyses. This is supported by quality assurance of the laboratories providing the data, for example, through accreditation or certification (see Chapters 11–14). At very high cyanobacterial biomass levels in raw water, the potential health risks associated with treatment system failure, or the inability to implement effective treatment systems at all, are significantly increased. This justifies progression to a situation of elevated risk, denoted by Alert Level 2 conditions. The framework has been developed largely from the perspective of the drinking-water supply operator, but is also important for the manager of the raw water supply. The actions accompanying each level cover different types of responses, such as additional sampling and testing, operational options, consultation with health authorities and informing the public through media releases. An important part of the framework at various stages is consultation with other agencies, particularly health authorities that generally have responsibility to oversee the safety of drinking-water.

The Vigilance Level encompasses the possible early stages of bloom development, when cyanobacteria are first detected in samples of the waterbody or raw water intake. Thresholds that may be used to trigger the Vigilance Level include elevated turbidity (e.g., Secchi depth readings of less than 2 m), detection of cyanobacteria by microscopy, particularly of potentially toxic species and, in some cases, musty tastes and odours. If the Vigilance Level is exceeded, it is appropriate to increase the sampling frequency of the raw water to at least once a week, so that potentially rapid changes in cyanobacterial biomass can be detected. In contrast, visible scums, particularly if associated with health complaints or animal deaths, immediately trigger Alert Level 1.

Elevated turbidity, with Secchi depth readings below 2 m due to greenish discoloration, or a correspondingly high online turbidity reading (e.g., at the raw water intake), serves as a first indication of bloom development, provided microscopic examination confirms this to be – at least partially – caused by cyanobacteria. Reduced water transparency can be seasonally caused by other phytoplankton, such as diatoms, green algae or euglenophytes. Therefore, for efficient management, microscopy skills and some taxonomic expertise (sufficient to recognise cyanobacteria on the genus level; see Chapter 13) are highly valuable. The detection of more than 10 colonies, or more than 50 filaments, of a cyanobacterium per 1 mL water sample is suggested as the trigger value for the Vigilance Level, although this threshold may be adapted according to local knowledge and prior history of occurrence. Taste and odour may become noticeable in the supply as the cyanobacterial population develops above the Vigilance Level and thus serve as a warning signal if they do occur, but their absence does not indicate the absence of toxic cyanobacteria (see section 2.8).

Alert Level 1 thresholds are defined in terms of cyanobacterial biomass, estimated as a biovolume of 0.3 mm^3/L or alternatively as a concentration of chlorophyll-*a* in the range of 1.0 µg/L, provided this chlorophyll is largely from cyanobacteria (for details see below). This can be ascertained by using probes which also detect phycocyanin – a pigment only found in cyanobacteria – or by qualitatively checking with microscopy. Qualitative microscopy is recommended in either case for obtaining visual information about the phytoplankton composition and the genera of cyanobacteria present.

These biomass indicators correspond to cyanotoxin concentrations possibly above the lifetime GVs but most likely well below the short-term GVs (i.e., for ATX the health-based reference value and for STX the acute GV). Biomass levels up to those corresponding to these short-term values may be tolerated in drinking-water for up to 2 weeks, provided the situation assessment and remediation steps taken show that the situation will not last longer, the public is informed and remediation measures are initiated. As discussed above, this approach provides important leeway for effective management: provided cyanotoxin concentrations stay below the Alert Level 2 thresholds, funds available may thus be focused on establishing remediation measures that avoid blooms or on bringing concentrations in finished waters back to below the lifetime guideline value (GV), rather than investing into short-term measures such as the provision of bottled water for the general population or expensive temporary technical remediation measures. Note that, as mentioned above, information to sensitive groups and those taking care of bottle-fed infants is important under Alert Level 1 conditions.

For CYN concentrations, cyanobacterial biomass can be a poor indicator, as (in contrast to MCs, ATXs and STXs) a large fraction of this toxin often occurs extracellularly and (in contrast to, e.g., ATXs) degradation in water may be slow, particularly at low temperatures (Chapter 2). Therefore, if CYN producers (e.g., *Raphidiopsis raciborskii* in the Americas and Australia and *Aphanizomenon* spp. in Europe) are, or have been, present, analysis of CYNs is recommended (see Box 5.1). Regular phytoplankton monitoring (visual, via qualitative microscopy) is important for identifying such situations.

Actions to take under Alert Level 1 include an assessment as to whether water treatment plant intakes can be adjusted or other physical actions can be implemented to reduce the cyanobacterial challenge; whether the water treatment system(s) available are effective in reducing toxin concentrations to acceptable levels (see Chapter 10) and whether waterbody conditions render a prolonged bloom likely or it is rather expected to be an occasional, short-lived event (Chapters 7 and 8). Cyanotoxin analysis of the raw and treated water (see Chapter 14) will allow a better assessment of the situation, potentially including adapting the biomass indicator values to the toxin content of the local bloom (see below). Alert Level 1 should further trigger an assessment of longer-term options to reduce the concentration of potentially toxic cyanobacteria in the raw water supply by measures in the catchment (see Chapter 7), in the waterbody (see Chapter 8) or in offtake management (Chapter 9).

Alert Level 1 conditions further require consultation with health authorities for ongoing assessment of the status of the bloom and of the suitability of treated water for human consumption. This consultation is best initiated early and should continue after the results of toxin analysis on drinking-water become available. Clearly, as the biomass of potentially toxic cyanobacteria increases in the raw water, so does the risk of adverse human health effects, particularly if water treatment systems are insufficient or other physical measures such as water treatment plant intake adjustments are not available or sufficiently effective. Therefore, ongoing monitoring for cyanobacterial biomass and, where possible, of toxin concentrations is important. It may also be appropriate to extend the monitoring programme, which should be at least weekly in frequency (in hot climates possibly more). Monitoring should be designed to establish the spatial variability of the cyanobacterial population and of toxin concentration (see Chapters 4 and 11).

An Alert Level 1 situation requires extensive public communication, particularly about the rationale for transiently tolerating levels above the lifetime GVs. Easing possible concerns of the public may be very important during phases with cyanobacterial biomass or toxin concentrations between the lifetime and short-term GVs. Media releases and even direct contact with consumers via letterbox delivery of leaflets with appropriate advice to householders may be appropriate (see Chapter 15 for further guidance). It may also be important to explicitly inform government departments, authorities and stakeholders with possible interests or legal responsibilities (beyond informing the health authority directly responsible for the surveillance of the water supply). Stakeholders may range from farmers needing information about possible impacts on livestock potentially exposed to blooms to organisations or facilities that treat or care for special "at-risk" members of the public (such as kidney dialysis patients, see section 5.4 or paediatricians and other health organisations advising parents of bottle-fed infants). Chapter 15 gives guidance on public communication.

If Alert Level 1 conditions continue, but toxins or toxicity are not detected in cyanobacterial or raw water samples, regular monitoring should nonetheless continue to ensure that toxic strains or species do not develop over ensuing weeks or months.

Alert Level 2 thresholds are defined as cyanobacterial biomass levels at ≥ 4 mm^3/L biovolume, or ≥ 12 μg/L chlorophyll-*a* (preferably with the presence of toxins confirmed by toxin analysis), and describe an established toxic bloom with rather high biomass and an elevated probability of scums. For CYNs, the caveat is – as for the Alert Level 1 threshold – the persistence of dissolved toxin, and regular microscopy is important to ensure that occurrence of possible producer organisms is detected on time to trigger chemical analysis of CYNs (see Box 5.1); alternatively, CYNs may be regularly included in the sampling programme.

In the Alert Level 2 situation, the sampling programme will have indicated that the bloom is widespread. Conditions in Alert level 2 correspond to cyanotoxin concentrations that may exceed even the short-term guidance values given in Table 5.1 and thus indicate an increased risk of adverse human health effects. Once the Alert Level 2 threshold is exceeded, an alternative water supply or effective water treatment system becomes urgent, as does ongoing monitoring of the performance of the system in place to control toxin concentrations.

Filtration systems (possibly combined with flocculation–coagulation) may remove cell-bound toxins, whereas dissolved toxin is likely to break through and require advanced treatment (see Chapter 10). If advanced treatment is not available or not sufficiently effective, Alert Level 2 conditions should result in the activation of a contingency water supply plan which is appropriate for the operator and the users or community. This may involve switching to an alternative supply for human consumption, the implementation of contingent treatment systems or, in some circumstances, the delivery of safe drinking-water to consumers by tanker or in bottles. While hydrophysical measures to reduce cyanobacterial growth or intake into the drinking-water system may still be attempted in this phase, application of algicides runs the risk of exacerbating the problem by causing high concentrations of dissolved toxins as a consequence of cell lysis (see Chapter 8).

Where advice is provided to the public not to drink water because of a cyanobacterial hazard to human health, it will usually emphasise that the water is still suitable for purposes such as washing, laundry and toilet flushing. Complete withdrawal of a piped drinking-water supply because of a cyanobacterial toxin hazard is not an option because the adverse health effects resulting from the disruption of supply (e.g., lack of water for toilet flushing, personal and household hygiene and in some situations also for firefighting) are likely to substantially outweigh the likely impact of the cyanobacterial toxin risk itself.

Monitoring of the bloom should continue in order to determine when the bloom starts to decline and normal supply can be resumed. The sequence at Alert Level 2 may follow through to deactivation of Alert Level conditions with media releases as well as advice to government departments and health authorities to confirm this. The collapse of a bloom, or a management action such as the flushing or mixing of a reservoir (Chapter 8), may lead to a rapid decline from Alert Level 2 back to Alert Level 1 or below.

Likewise, the sequence might escalate rapidly, bypassing Alert Level 1 to Alert Level 2, particularly if adequate monitoring and early warning information are not available. Cyanobacterial populations in natural waterbodies may increase by two- to threefold within 2 days (growth rate, μ=0.3/d; see Figure 5.3), especially in hot climates. Monitoring frequency needs to take such potentially rapid population growth rates into account.

The basis for deactivating Alert Level 2 and reverting back to Alert Level 1 or the Vigilance Level will depend on how it was triggered. If it was triggered by biovolumes or chlorophyll-*a* without cyanotoxin analyses, then deactivation can be based on biovolumes or chlorophyll-*a*. If cyanotoxin concentrations have been determined, these take precedence and Alert Level 2 should only be deactivated once the cyanotoxin concentrations have declined below the short-term guideline values (GVs).

5.1.2.3 Considerations for choosing parameters to trigger Alert Levels when adapting the Framework to local circumstances

The Alert Level Framework (ALF) proposed here focuses on indicators for which analytical methods are likely to be more readily accessible than for toxin analyses, that is, visual inspection (Secchi depth reading; scums) and cyanobacterial biomass. Their choice can be adapted as is nationally or locally practical: for example, measuring turbidity in the raw water entering

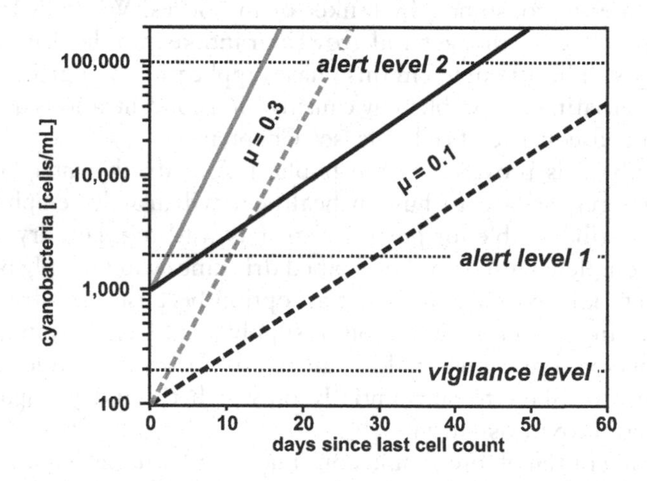

Figure 5.3 Predicted development of cyanobacterial population from initial concentration of 100 (dotted dashed lines) or 1000 (solid lines) cells per mL at exponential growth rates (μ) of 0.1 (dark lines) and 0.3 (light lines) per day. (Modified from Jones, 1997.)

a drinking-water treatment system (e.g., online) can replace measuring transparency in the waterbody with a Secchi disc. Cyanobacterial biomass is best determined as biovolume or, alternatively, as chlorophyll-*a* (in the latter case combined with qualitative microscopy); analysing both is not necessary and which of the two to choose will depend on locally available expertise and instrumentation. Also, techniques such as fluorescence probes (online or handheld), remote sensing (via satellite images, airplanes or drones), cell counts or molecular analyses for toxin-production genes may be used in a local adaptation of the ALF, provided the signals are "calibrated" with data from local sampling programmes and they depict local cyanobacterial biomass sufficiently well to be used as triggers in an ALF (see Chapter 13 for methods).

Alternatively, it is possible to analyse cyanotoxins directly if methods are accessible (see Chapter 14 for methods). However, including a biomass parameter to trigger action in the ALF offers the further advantage of encompassing any hazard caused by a cyanobacterial bloom: as discussed in section 2.10, while cyanotoxin concentrations below the trigger values given in the ALF imply a low health risk from exposure to cyanotoxins, blooms may contain further, yet unknown substances and/or organisms that may be hazardous. It is therefore prudent to avoid exposure to high concentrations of cyanobacterial biomass even if concentrations of the known cyanotoxins are low.

However, for any parameter used to trigger Alert Levels – including cyanotoxins – other than cyanobacterial biovolume, it is strongly recommended to include qualitative or semiquantitative microscopy in order to collect information on the dominant cyanobacterial genera in the waterbody. This is particularly important for timely recognition of possible CYN occurrence, as concentrations of dissolved CYNs do not relate to biovolume or chlorophyll-*a* (as measures of biomass) as immediately as do other cyanotoxins (see Box 5.1), but observing substantial amounts of potential CYN producers should trigger targeted analysis of CYNs. While identifying cyanobacterial species is often described as intimidating, as discussed in Chapter 12, identification on the genus level already provides highly valuable information, often quite sufficient for assessing the situation, and this is readily learnt by staff with some experience in microscopy. An understanding of the dominant cyanobacterial genera is also important for estimating their distribution in the waterbody as well as their likely responses to measures for control and remediation discussed in Chapters 7–9.

5.1.2.4 Considerations for setting the ALF thresholds and adjusting them to local circumstances

The value for chlorophyll-a at Alert Level 2 given in Figure 5.2 is now substantially lower than the values given for Alert Level 2 in the 1999 edition of this book. This is because the GVs for short-term exposure (Table 5.1) are now available (WHO, 2020), and Alert Level 2 should reflect the risk of exceeding these: at biomass concentrations up to Alert Level 2, that is, 4 mm^3/L biovolume or 12 µg/L chlorophyll-*a*, it is highly unlikely that concentrations of MCs can significantly exceed the provisional short-term guideline value (GV). Nor is it likely that concentrations of STXs can exceed the acute GV for STXs, or concentrations of ATXs exceed the health-based reference value for ATXs (see Chapter 2 for an explanation of these values). The same applies to the Alert Level 1 values of 0.3 mm^3/L biovolume or 1 µg/L chlorophyll-*a*: these are sufficiently conservative to maintain concentrations of MCs below the provisional lifetime GV, and the same applies to CYNs if monitored as described in Box 5.1.

For STXs and ATXs, the rationale for Alert Level 1 is different as no GVs for lifetime exposure are available. The Alert Level 1 value of 0.3 µg/L STX is merely 10-fold lower than the acute GV with the function of serving as a trigger for increased vigilance to avoid reaching the acute GV. This applies equally to the value of 3 µg/L proposed for ATX as a trigger in Alert Level 1: this is also not a toxicologically derived lifetime GV but merely a value set to be 10-fold lower than the proposed Health-based Reference Value as a trigger for increased vigilance (Table 5.1).

This is relevant because a further rationale for the thresholds proposed for Alert Level 1 is the potential for rapid exponential increase once cyanobacteria have been detected at this threshold level: even if the toxin content of the cells is substantially lower, concentrations in the water can increase exponentially as cells divide exponentially and thus reach levels exceeding Alert Level 1 within a few days: Figure 5.2 gives an indication of the rate of change of an exponentially dividing population at two growth rates typically observed in field studies of cyanobacteria (in the field, growth rates rarely exceed 0.3 per day).

Furthermore, as discussed at the beginning of this chapter, the Alert Levels proposed for biovolume and concentrations of chlorophyll-*a* are based on the upper range of cyanotoxin content typically found in cyanobacterial cells in the field (discussed in section 4.6.2), that is, on worst-case assumptions for the ratio of toxins to biovolume or chlorophyll-*a*. In many field situations, cyanotoxin concentrations will be lower, possibly by a factor of 10. It is therefore useful to support the assessment by analysing for the presence of cyanotoxins, and if their concentrations prove lower than the Alert Level values, this may revert the situation back to a lower level. Also, if the toxin content of the local cyanobacterial population is well understood, other, often higher Alert Levels may be set for biovolume or chlorophyll-*a*. In that case, checking the cyanotoxin content of the cyanobacterial population would remain necessary at larger intervals, for example, 2–3 times per season or monthly; however, for the more frequent monitoring between those occasions (e.g., weekly, daily or – with probes – continuously) biovolume or chlorophyll-*a* is likely to be sufficient.

REFERENCES

Baker P, Humpage A (1994). Toxicity associated with commonly occurring cyanobacteria in surface waters of the Murray-Darling Basin, Australia. Mar Freshwater Res. 45:773–786.

Berg KA, Lyra C, Sivonen K, Paulin L, Suomalainen S, Tuomi P et al. (2009). High diversity of cultivable heterotrophic bacteria in association with cyanobacterial water blooms. ISME J. 3:314–325.

Botes DP, Wessels PL, Kruger H, Runnegar MTC, Satikarn S, Smith RJ et al. (1985). Structural studies on cyanoginosins-LR, YR, YA, and YM, peptide toxins from *Microcystis aeruginosa*. J Chem Soc Perkin Trans. 1:2747–2748.

Bourke A, Hawes R, Neilson A, Stallman N (1983). An outbreak of hepato-enteritis (the Palm Island mystery disease) possibly caused by algal intoxication. Toxicon. 21:45–48.

Buratti FM, Manganelli M, Vichi S, Stefanelli M, Scardala S, Testai E et al. (2017). Cyanotoxins: producing organisms, occurrence, toxicity, mechanism of action and human health toxicological risk evaluation. Arch Toxicol. 91:1049–1130.

Burns J (2008). Toxic cyanobacteria in Florida waters. In: Hudnell HK, editors: Cyanobacterial harmful algal blooms: state of the science and research needs. New York: Springer:127–137.

Byth S (1980). Palm Island mystery disease. Med J Aust. 2:40–42.

Carmichael WW (2001). Assessment of blue-green algal toxins in raw and finished drinking water. Denver (CO): AWWA Research Foundation:179 pp.

Chen J, Xie P, Li L, Xu J (2009). First identification of the hepatotoxic microcystins in the serum of a chronically exposed human population together with indication of hepatocellular damage. Toxicol Sci. 108:81–89.

Chorus I (2005). Current appoaches to cyanotoxin risk asessment, risk management and regulations in different countries. Berlin: Federal Environmental Agency:117 pp. http://ww.umweltbundesamt.de.

Chorus I, Bartram J (1999). Toxic cyanobacteria in water: a guide to their public health consequences, monitoring and managemen. London: E & FN Spoon, on behalf of WHO:400 pp.

Codd GA, Morrison LF, Metcalf JS (2005). Cyanobacterial toxins: risk management for health protection. Toxicol Appl Pharmacol. 203:264–272.

Falconer I, Jackson R, Langley B, Runnegar M (1981). Liver pathology in mice in poisoning by the blue-green alga *Microcystis aeruginosa*. Aust J Biol Sci. 34:179–188.

Falconer IR (2005). Cyanobacterial toxins of drinking water supplies. Boca Raton (FL): CRC Press:279 pp.

Falconer IR, Beresford AM, Runnegar MT (1983). Evidence for liver damage by toxin from a bloom of the blue-green alge, *Microcystis aeruginosa*. Med J Aust. 1:511–514.

Falconer IR, Burch MD, Steffensen DA, Choice M, Coverdale OR (1994). Toxicity of the blue-green alga (cyanobacterium) *Microcystis aeruginosa* in drinking water to growing pigs, an an animal model for human injury and risk assessment. Environ Toxicol Wat Qual. 9:131–139.

Falconer IR, Humpage AR (2005). Health risk assessment of cyanobacterial (blue-green algal) toxins in drinking water. Int J Environ Res Public Health. 2:43–50.

Fawell J, Hart J, James H, Parr W (1993). Blue-green algae and their toxins-analysis, toxicity, treatment and environmental control. Water Supply. 11:109–109.

Fleming LE, Rivero C, Burns J, Williams C, Bean JA, Shea KA et al. (2002) Blue-green algal (cyanobacterial) toxins, surface drinking water, and liver cancer in Florida. Harmful Algae. 1:157–168.

Funari E, Testai E (2008). Human health risk assessment related to cyanotoxins exposure. Crit Rev Toxicol. 38:97–125.

Haddix PL, Hughley CJ, Lechevallier MW (2007). Occurrence of microcystins in 33 US water supplies. J - Am Water Works Assoc. 99:118–125.

Harada KI, Oshikata M, Uchida H, Suzuki M, Kondo F, Sato K et al. (1996). Detection and identification of microcystins in the drinking water of Haimen City, China. Nat Toxins. 4:277–283.

Harding WR, Paxton BR (2001). Cyanobacteria in South Africa: a review. Pretoria: Water Research Commission, Pretoria. 165 pp.

Hawkins P, Griffiths D (1993). Artificial destratification of a small tropical reservoir: effects upon the phytoplankton. Hydrobiologia. 254:169–181.

Hawkins PR, Runnegar MTC, Jackson ARB, Falconer IR (1985). Severe hepatotoxicity caused by the tropical cyanobacterium (blue-green alga) *Cylindrospermopsis raciborsckii* (Woloszynska) Seenaya and Subba Raju isolated from a domestic water supply reservoir. Appl Environ Microbiol. 50:1292–1295.

Hoeger SJ, Shaw G, Hitzfeld BC, Dietrich DR (2004). Occurrence and elimination of cyanobacterial toxins in two Australian drinking water treatment plants. Toxicon. 43:639–649.

Hudnell HK (2008). Cyanobacterial harmful algal blooms: state of the science and research needs. New York: Springer:950 pp.

Ibelings BW, Backer LC, Kardinaal WEA, Chorus I (2014). Current approaches to cyanotoxin risk assessment and risk management around the globe. Harmful Algae. 40:63–74.

Jones GJ (1997). Limnological study of cyanobacterial growth in three south-east Queensland reservoirs. In: Davis JR, editors: Managing algal blooms: outcomes from CSIRO's multi-divisional blue-green algal program. Canberra: CSIRO Land and Water Canberra:51–66.

Lambert TW, Boland MP, Holmes CF, Hrudey SE (1994). Quantitation of the microcystin hepatotoxins in water at environmentally relevant concentrations with the protein phosphatase bioassay. Environ Sci Technol. 28:753–755.

Li Y, Chen J-A, Zhao Q, Pu C, Qiu Z, Zhang R et al. (2011). A cross-sectional investigation of chronic exposure to microcystin in relationship to childhood liver damage in the Three Gorges Reservoir Region, China. Environ Health Persp. 119:1483.

Lian M, Liu Y, Yu S-Z, Qian G-S, Wan S-G, Dixon KR (2006). Hepatitis B virus x gene and cyanobacterial toxins promote aflatoxin B1-induced hepatotumorigenesis in mice. World J Gastroenterol. 12:3065.

Lippy EC, Erb J (1976). Gastrointestinal illness at Sewickley, PA. J Am Water Works Assoc. 68:606–610.

Mohamed ZA (2016). Breakthrough of *Oscillatoria limnetica* and microcystin toxins into drinking water treatment plants-examples from the Nile River, Egypt. Water SA. 42:161–165.
Mohamed ZA, Al Shehri AM (2009). Microcystin-producing blooms of *Anabaenopsis arnoldi* in a potable mountain lake in Saudi Arabia. FEMS Microbiol Ecol. 69:98–105.
Mohamed ZA, Deyab MA, Abou-Dobara MI, El-Sayed AK, El-Raghi WM (2015). Occurrence of cyanobacteria and microcystin toxins in raw and treated waters of the Nile River, Egypt: implication for water treatment and human health. Environ Sci Pollut Res. 22:11716–11727.
Saoudi A, Brient L, Boucetta S, Ouzrout R, Bormans M, Bensouilah M (2017). Management of toxic cyanobacteria for drinking water production of Ain Zada Dam. Environ Monit Assess. 189:361.
Sekijima M, Tsutsumi T, Yoshida T, Harada T, Tashiro F, Chen G et al. (1999). Enhancement of glutathione S-transferase placental-form positive liver cell foci development by microcystin-LR in aflatoxin B-1- initiated rats. Carcinogenesis. 20:161–165.
Svirčev Z, Drobac D, Tokodi N, Vidović M, Simeunović J, Miladinov-Mikov M et al. (2013). Epidemiology of primary liver cancer in Serbia and possible connection with cyanobacterial blooms. J Environ Sci Health Part C. 31:181–200.
Svirčev Z, Krstić S, Miladinov-Mikov M, Baltić V, Vidović M (2009). Freshwater cyanobacterial blooms and primary liver cancer epidemiological studies in Serbia. J Environ Sci Health Part C. 27:36–55.
Teixera MGLC, Costa MCN, Carvalho VLP, Pereira MS, Hage E (1993). Gastroenteritis epidemic in the area of the Itaparica Dam, Bahia, Brazil. Bull Pan Am Heal Organ. 27:244–253.
Tisdale ES (1931). The 1930–1931 drought and its effect upon public water supply. Am J Public Health Nations Health. 21:1203–1215.
Ueno Y, Nagata S, Tsutsumi T, Hasegawa A, Watanabe MF, Park HD et al. (1996). Detection of microcystins, a blue-green algal hepatotoxin, in drinking water sampled in Haimen and Fusui, endemic areas of primary liver cancer in China, by highly sensitive immunoassay. Carcinogenesis. 17:1317–1321.
US EPA (2015). Drinking water health advisory for the cyanobacterial microcystin toxins. Washington (DC): United States Environmental Protection Agency.
WHO (2017). Guidelines for drinking-water quality, fourth edition, incorporating the 1st addendum. Geneva: World Health Organization:631 pp. https://www.who.int/publications/i/item/9789241549950
WHO (2020). Cyanobacterial toxins: Anatoxin-a and analogues; Cylindrospermopsins; Microcystins; Saxitoxins. Background documents for development of WHO Guidelines for Drinking-water Quality and Guidelines for Safe Recreational Water Environments. Geneva: World Health Organization. https://www.who.int/teams/environment-climate-change-and-health/water-sanitation-and-health/water-safety-and-quality/publications
Yen H-K, Lin T-F, Liao P-C (2011). Simultaneous detection of nine cyanotoxins in drinking water using dual solid-phase extraction and liquid chromatography–mass spectrometry. Toxicon. 58:209–218.
Yu S-Z (1989). Drinking water and primary liver cancer. In: Tang ZY, Wu MC, Xia SS, editors: Primary liver cancer. Beijing: China Academic Publishers.

Yu S-Z (1995). Primary prevention of hepatocellular carcinoma. J Gastroenterol Hepatol. 10:674–682.
Yu S-Z, Huang XE, Koide T, Cheng G, Chen GC, Harada Ki et al. (2002). Hepatitis B and C viruses infection, lifestyle and genetic polymorphisms as risk factors for hepatocellular carcinoma in Haimen, China. Jpn J Cancer Res. 93:1287–1292.
Yu S-Z, Zhao N, Zi X (2001). The relationship between cyanotoxin (microcystin, MC) in pond-ditch water and primary liver cancer in China. Chin J Oncol. 23:96–99.
Zamyadi A, MacLeod SL, Fan Y, McQuaid N, Dorner S, Sauvé S et al. (2012). Toxic cyanobacterial breakthrough and accumulation in a drinking water plant: a monitoring and treatment challenge. Water Res. 46:1511–1523.
Zhou L, Yu H, Chen K (2002). Relationship between microcystin in drinking water and colorectal cancer. Biomed Environ Sci. 15:166–171.
Zilberg B (1966). Gastroenteritis in Salisbury European children – a five-year study. Cent African J Med. 12:164–168.

5.2 RECREATION AND OCCUPATIONAL ACTIVITIES

Ingrid Chorus and Emanuela Testai

Recreational activities may be a significant route of exposure to cyanotoxins. Throughout the world, the range and scope of recreational water activities vary as widely as does access to recreational waterbodies and their propensity to be impacted by cyanobacteria blooms. Where cyanobacterial blooms are pronounced and water sports are nonetheless popular, recreational activities are likely to be a major route of exposure to cyanotoxins. Occupational activities using cyanobacteria-affected waters may lead to similar patterns of cutaneous and inhalational exposures to cyanotoxins, though opportunities are available to reduce exposure through the use of personal protective equipment and other occupational management strategies. Understanding the usage patterns of untreated surface water is therefore fundamental for assessing exposure.

Scums of cyanobacteria in lakes and rivers used for recreational purposes have been well recognised as a public nuisance. Moreover, deaths of livestock, wild animals or pets have been observed after exposure to cyanobacteria. Such incidents raise the question whether affected waterbodies are safe for recreational use. Sometimes blooms are associated with unpleasant odours and a degraded appearance of lake shores, especially when scums aggregate and decay. Swimmers and other water users may avoid areas with extensive cyanobacterial scums or accumulated detached mats because of the obviously unpleasant environment, particularly when associated with related fish-kills.

However, sensory responses and reactions to cyanobacteria blooms vary. The smell of some blooms is not necessarily unpleasant, but more like freshly-mown grass, and some observers have described waters vividly coloured by blue-green cyanobacterial blooms as looking beautiful. Multiple anecdotal observations of children and adults playing with scum material have been reported (Figure 5.4). Where alternative recreational sites without cyanobacterial blooms are lacking and the demand for recreational water access is high, visual and olfactory amenity tend to be of lower priority, and people may tolerate water quality conditions that might otherwise discourage them from using the site. This has been observed in numerous countries, for example, in many parts of inland Australia that are subject to water scarcity, in arid regions of Hungary where few waterbodies are available for recreation, and in north-western Germany where for decades the majority of waterbodies were heavily eutrophic. In some regions in which cyanobacterial blooms have become a widespread phenomenon for more than a generation, site visitors have come to accept the degraded water quality as "natural" or "normal" for the region. In temperate climates, cyanobacterial dominance is most pronounced during the summer months, when the demand for recreational water is highest.

Figure 5.4 Playing children are particularly at risk to be exposed to critical quantities of cyanotoxins. (Kindly provided by Yora Tolman.)

Various enterprises may use untreated water from cyanobacteria-affected surface waters for a wide range of processes that can result in occupational exposure to cyanotoxins: for example, cooling in production processes, dust suppression by spraying, spray irrigation, workers exposed to raw water spray in waterworks or cooling of enclosed or semienclosed workspaces. Cell lysis and, in consequence, liberation of cell-bound toxins may be caused by pressure and shear stress during pumping. Occupational exposure may also occur through work directly in or on waterbodies affected by scums. Marine blooms of filamentous *Moorea* species (previously known as *Lyngbya majuscula*) can dry on fishing nets, and contact with fresh and dried material has caused severe skin reactions as well as breathing difficulties for workers in the fishing industry (Grauer & Arnold, 1961; Osborne et al., 2001).

Potential routes of occupational exposure to cyanotoxins include direct contact via exposed parts of the body and cell material trapped under clothing, accidental swallowing of contaminated water and inhalation. While some exposure pathways at workplaces are similar to those experienced during recreation, a difference may be longer and more frequently repeated exposure periods in occupational settings. Uncharacterised water supplies may contain further hazardous agents, and skin abrasion by protective clothing, potentially augmented in heat and by moist skin, may increase exposure.

Occupational settings may also involve a risk of exposure via drinking-water through cross-contamination of the potable water supply if this is not effectively separated from the process water or subject to poor labelling of pipework and fittings or poor process design and control. Where temperatures are high (e.g., >35 °C in some agricultural situations or in open-cast mines), poor access to potable water in sufficient quantity and proximity to the workplace may increase the risk of untreated water – potentially containing not only cyanotoxins but also pathogens and other hazards – being used for drinking.

Recreational and occupational exposure may be to whole cyanobacterial cells, lysates, dried cells or mixtures of these forms. Where blooms are of concern, water containing them may well contain further hazards, particularly microbial pathogens.

5.2.1 Evidence of health effects associated with exposure to cyanobacteria in water used for recreation or at workplaces

While reported concentrations of cyanotoxins in drinking-water are rarely found above the low microgram per litre range (see section 5.1), contact with scums through recreational activities more frequently results in exposure to cyanotoxin concentrations in a range of up to milligrams per litre (see Chapter 2), and acutely hazardous exposure is a realistic scenario if site users ingest scum. Evidence of health effects from recreational exposure has been published mainly as anecdotal reports, case studies and from epidemiological studies.

5.2.1.1 Case reports of short-term health effects from acute exposure

A number of published case reports of illness after exposure to cyanobacteria during recreation have been widely quoted to illustrate the relevance of this pathway. As discussed at the beginning of this chapter, in most of the published cases, the presence of infectious pathogens cannot be unambiguously excluded, and it is typically unclear whether the symptoms reported were caused by the known cyanotoxins or by other components of the bloom, including the possibility of yet unknown cyanobacterial metabolites. For example, enteritic viral or parasite pathogens may well have been present even where bacterial indicators were reported to have been absent. The case in Box 5.4 shows that later availability of new analytical methods can support or exclude cyanotoxins as cause if sample material is still available.

BOX 5.4: HUMAN MORTALITY FROM ACCIDENTAL INGESTION OF TOXIC CYANOBACTERIA – A CASE RE-EXAMINED

Wayne W. Carmichael

In July 2002, a 17 year-old male was taken to a local hospital emergency department in full cardiopulmonary arrest following an episode of vomiting and diarrhoea followed by seizure at his home. The patient, an athletic otherwise healthy individual, had no previous history of seizures, syncope or diarrhoeal illness. Extensive resuscitation efforts failed and the patient expired in the emergency department. An autopsy was performed the following day to determine the cause of death. After ruling out several possible aetiologies for death, including toxic chemicals and pathogenic microbes, the possible role of cyanotoxins was pursued since the youth was reported to have accidentally ingested water while swimming in a local golf course pond, about 2 h prior to symptoms, that was described as "dirty and scummy". Unfortunately, because cyanotoxins were considered as possible cause only late in the course of the investigation, no samples were taken from the pond. Samples of the youth's blood, liver and vitreous fluid were tested for MCs, STX, ATX and CYN. In addition, stool collected from autopsy was examined for the presence of cyanobacterial cells. ELISA was negative for microcystins and LC/MS analyses was negative for STX and CYN. ESI LC/MS did reveal a strong peak with *m/z* 166 with a retention time of 9.08 min, "similar"' to that of anatoxin-a, 8.51 min. This evidence allowed an initial listing of this cyanotoxin as a possible cause of death. Further analyses showed, however, that this peak with *m/z* 166 is not anatoxin-a but the ubiquitous amino acid phenylalanine.

In consequence, this example of a false-positive investigation of mortality from anatoxin-a should now be considered one of unknown cause.

A number of reports contain substantial evidence of the uptake of cyanobacteria and a likely connection to the symptoms observed:

- Dillenberg and Dehnel (1960) reported a case series of illness in 13 persons after swimming at various bloom-affected Canadian lakes (despite warnings posted following animal deaths); symptoms included headache, nausea, vomiting, painful diarrhoea, arthralgia and myalgia (i.e., pain in joints and muscles). Stool samples from two of the more severely affected individuals, one of whom was hospitalised overnight, were sent to the Saskatchewan public health laboratories, where *Microcystis* cells were identified in the specimens.
- Turner et al. (1990) reported that 10 out of 18 army recruits fell ill after training exercises involving canoeing – including practicing Eskimo rolls – in a waterbody affected by a *Microcystis* bloom, with two soldiers needing hospitalisation for a week because of severe atypical pneumonia and generalised illnesses. The authors suggested that inhalational exposure to cyanotoxins, especially to microcystin, may have been the probable cause, although that assertion has been challenged by others. This was the incident that first triggered wider attention to cyanobacterial toxicity in humans.

- In Argentina, a teenage jet-ski rider was hospitalised for several weeks, including an 8-day period in an intensive care unit during which time he required artificial ventilation. Acute respiratory symptoms were followed by hepatic insufficiency, which was essentially self-limiting. The presumed aetiologic agent was a microcystin-producing bloom of *Microcystis*, which was present as heavy scum in the dam at the time the young man spent several hours on and in the water (Giannuzzi et al., 2011).
- In Uruguay, a 20-month-old child suffered acute liver failure after repeated recreational activity at a beach of the Rio de La Plata River (Vidal et al., 2017) in January 2015. During this month, the river had a pronounced bloom of *Microcystis* sp. and microcystin concentrations up to 25 700 µg/L were reported in scum material. The child and her family first showed gastrointestinal symptoms a few hours after the final exposure, but she also developed jaundice and increased serum levels of liver enzymes as well as a need for mechanical respiratory support. A liver transplant was performed after 20 days, and microcystins were detected in the removed liver in concentrations up to 78 ng/g of tissue, which is in the range of the concentrations found in livers of the Caruaru victims (discussed in section 5.4). While the authors explicitly do not exclude other factors, for example, autoimmune hepatitis type II as cause (possibly triggered by the exposure to microcystins), they identify a high plausibility of direct damage through the repeated exposure to an estimated total of at least 1.78 L of microcystin containing water over a few days.
- In a review of CDC's Waterborne Disease and Outbreak Surveillance System in the USA in 2009–2010, 11 outbreaks were associated with cyanobacteria. In 70% of cases, health effects were associated with the major exposure route: rash, irritation, swelling or sores were reported in those outbreaks where exposure occurred mainly through dermal contact while gastrointestinal symptoms were reported after water ingestion. The outbreak with the more severe gastrointestinal and neurologic symptoms (one of the two hospitalisation cases) was characterised by the highest levels of MCs (>2000 µg MC-LR eq/L) and 9, 15 and 0.09 µg/L of CYN, ATX and STX. In the three cases in which ATX and STX were present, neurologic symptoms or confusion/visual disturbance were reported in addition to fever, headache and eye irritation. However, in all three cases, microcystins were also detected at often substantially higher concentrations (0.3–>2000 µg/L), and in one of them, CYN and STX were also present (Hilborn et al., 2014).

For assessing cases such as these, it is important that mere co-occurrence of cyanotoxins and unspecific symptoms (skin irritation, gastrointestinal, etc., see above) is not indicative of the known cyanotoxins having caused the symptoms; more likely the cyanobacterial biomass contains both toxins and other, yet unknown agents causing such general symptoms. In contrast, cause–effect relationships are likely if symptoms or analytical results are toxin-specific (e.g., for hepatotoxins elevated serum enzyme levels such as gamma glutamyl transferase; for neurotoxins respiratory difficulties, tingling of extremities, confusion or visual disturbance). While finding cyanotoxins in body fluids of patients and/or cyanobacterial cells in their stool confirms exposure, even this does not allow the conclusion that these were the cause of symptoms, as it is currently unknown how concentrations in serum relate to damage in the liver, for example.

Regarding occupational exposure, two studies have been undertaken by the mining industry in Australia. The Australian Coal Association Research Programme projects (Fabbro et al., 2008; Fabbro et al., 2010) investigated cyanobacteria and their toxicity in various waterbodies available to industry in Central Queensland, Australia, a semiarid region with a history of cyanobacterial blooms. Of the 180 samples tested for toxin, 17% contained CYN and 3% contained microcystin. Total CYN concentrations (CYN plus deoxycylindrospermopsin) ranged from 0.2 to 22.1 µg/L. Microcystin concentrations ranged from 1.7 to 3200 µg/L. Concentrations of toxin-producing cyanobacteria (*Dolichospermum circinale*) as high as 500 000 cells/mL were recorded from pit water (Fabbro et al., 2008). Workers can potentially have direct contact with pit water when installing pump facilities or when it is used for dust suppression, cooling or wash down. This research also provided the initial identification of novel toxicity associated with *Limnothrix/Geitlerinema* (Fabbro et al., 2010; Bernard et al., 2011; Humpage et al., 2012).

Other anecdotal and case reports of varying reliability describe acute gastrointestinal and respiratory illnesses associated with activities such as waterskiing (likely forming aerosols and spray) in recreational waters contaminated by cyanobacteria (reviewed in Stewart et al., 2006d), including a report of a windsurfer in the UK with hepatic dysfunction diagnosed by liver function tests and liver biopsy (Probert et al., 1995). In only a small proportion of such anecdotal reports documented in the biomedical literature were the subjects examined by medical practitioners. Anecdotal reports of illness are occasionally reported in local broadcast or print media, and some descriptions of the number and type of complaints received by public health authorities can be found in overview publications (see, e.g., Backer et al., 2015). A report from the US State of Nebraska recorded more than 50 complaints of skin eruptions, vomiting, diarrhoea and headache after swimming or waterskiing at a cyanobacteria-affected lake over a single summer weekend (Walker et al., 2008).

Severe skin reactions have been reported from contact with marine cyanobacteria, particularly with *Lyngbya majuscula* (now termed *Moorea producens*), which causes deep blistering particularly when trapped under bathing suits and where blooms have contained the toxins lyngbyatoxin A and debromoaplysiatoxin (see section 2.6). Severe dermatitis, resembling skin burns, has been reported from marine bathing in the presence of cyanobacteria dislodged from rocks, particularly after storms in tropical seas (Hashimoto et al., 1976; Moore et al., 1993). *Lyngbya/Moorea* has been recorded in many marine ecosystems worldwide, but is most common in tropical/subtropical locations. Intoxication events have been reported primarily in midsummer when both numbers of people engaged in recreational activities and the potential for bloom formation are high. Reports are chiefly from economically more developed countries, potentially due to a recording bias, and often include multiple morbidities.

Complaints of acute skin reactions have been associated with exposure to freshwater cyanobacteria as well as with eukaryotic microalgae; however, cyanobacteria are the focus of the majority of these reports (Stewart et al., 2006c) with clinical investigations suggesting allergic responses (Cohen & Reif, 1953; Stewart et al., 2006a; Stewart et al., 2006b; Geh et al., 2016). Two reports focus on the pigment phycocyanin as a suspect allergen (Cohen & Reif, 1953), and indeed a case investigation of anaphylaxis following consumption of *Spirulina* in tablet form (Petrus et al., 2009) and clinical laboratory allergy studies identified phycocyanin as an allergen (Geh et al., 2015; Lang-Yona et al., 2018). However, this requires further clarification as it would contradict other reports assigning antiallergic, anti-inflammatory and antioxidant properties to phycocyanins (Strasky et al., 2013; Liu et al., 2015; Wu et al., 2016). Investigators conducting epidemiological fieldwork at cyanobacteria-affected waters have received a small number of anecdotal reports from individuals with a history of allergy, though the association between anticipated symptom occurrence and cyanobacteria in such cases remains speculative. The possibility of serious anaphylactic reactions has been raised for some benthic cyanobacteria (Stewart et al., 2011). Thus, while allergic responses to some cyanobacteria are discussed in the literature, their relevance remains unclear.

A widespread problem that case studies, such as those discussed above, face is that in the course of steps taken to elucidate the possible cause of the observed symptoms, cyanobacteria are typically considered only rather late. If many days pass between symptom observation and sampling the water to which patients were exposed, a bloom may already have disappeared and the chance for establishing a causal connection is missed. This is true in particular for surface blooms or scums which can disperse within a few hours, for example, due to increased wind. Informing the medical community about toxic cyanobacteria may help to reduce the time between exposure and water sampling as well as to document the situation at the time of possible exposure, for example, with images taken with mobile phones.

5.2.1.2 *Epidemiological studies of acute health risks from short-term recreational exposure*

Several epidemiological studies investigating acute illness following recreational exposure to freshwater cyanobacteria have been conducted between 1990 and 2011. These studies utilised various retrospective and prospective designs capable of detecting relative differences in commonly reported symptoms between exposed and unexposed groups; however, levels of exposure were usually poorly characterised, and hence, these studies are inadequate for risk assessment purposes. Symptoms assessed included both cutaneous and systemic reactions – the statistical analyses of the studies do not differentiate between both.

- Philipp and coworkers conducted the first three formal epidemiological investigations into recreational exposure to cyanobacteria: These comprised a series of cross-sectional studies conducted in 1990 at inland waters in the UK, affected some weeks earlier by cyanobacteria blooms. They found only minor illnesses, with no statistically significant differences between symptoms reported by exposed and unexposed groups (Philipp, 1992; Philipp & Bates, 1992; Philipp et al., 1992).
- A retrospective study conducted in Australia in response to an extensive bloom of *Anabaena circinalis* in the River Murray in South Australia also did not detect any statistically significant increase in symptoms between those exposed to river water during recreational activities and nonexposed controls (El Saadi et al., 1995).
- Pilotto et al. (1997) conducted a prospective cohort study in 1995 at recreational waters in southern and south-east Australia and reported a statistically increased likelihood of symptom reporting compared to unexposed controls after 7 days (but not after 2 days) following exposure to low levels of cyanobacteria (5000 cells/mL) for more than 1 h. The cohort size for the statistically significant finding was small, comprising 93 exposed and 43 unexposed subjects.
- Stewart et al. (2006c) conducted a larger prospective cohort study in Australia and the USA and detected a statistically significant increase in symptom reporting, particularly respiratory symptoms, three days following exposure. These authors used multivariable analysis after adjusting for confounding variables such as age, smoking, geographic region and a prior history of allergic disease. Increased symptom reporting rates were seen only at higher cyanobacterial densities, using a biomass estimate of exposure, and symptom severity was rated as mild by most study subjects. These associations were linked to cyanobacterial cell densities higher than 100 000 cells/mL

- Two prospective cohort studies conducted in the USA by Backer et al. (2008; 2010) found no relationship to symptom reporting and exposure to microcystins, as measured by ELISA and LC-MS in lake water, aerosols and blood.
- Lévesque et al. (2014) conducted a prospective cohort study of residents living near three lakes in Quebec, Canada, which had a history of being impacted by cyanobacteria, one of which is also used as source for drinking-water. Exposure to cyanobacteria included a range of recreational water activities, drinking-water (for residents living near the lake with drinking-water abstraction from the lake) and consumption of fish from study lakes. Recreational exposure to cyanobacteria was associated with increased reporting of gastrointestinal symptoms; 466 individuals were enrolled in the study, although the number of subjects that engaged in recreational activities was not reported. The authors reported a strong statistically significant relationship between gastrointestinal illness and exposure to cyanobacterial cells above 100 000 cells/mL.

Most of the symptoms reported in these studies are mild and self-limiting. In contrast, the toxicological considerations discussed in section 5.2.3 show that serious morbidity or death through oral uptake of toxin is a realistic scenario in recreational water settings, if larger amounts of a highly toxic bloom are ingested. While the case study from Uruguay (Vidal et al., 2017) provides supporting evidence that they may occur, such events are, however, probably rare, and with the possible exception of the case–control design adopted by El Saadi et al. (1995), the prospective and retrospective epidemiological studies discussed above were not designed to detect the impact of massive oral exposure to high toxin concentrations.

The "gold standard" epidemiological design, a randomised controlled trial, could in theory be employed to investigate exposures and outcomes from oral consumption of cyanotoxin-contaminated recreational water, but this could not be done in practice on ethical grounds and would be logistically challenging. Future epidemiological investigations that seek to document events of severe acute illness following oral ingestion of cyanotoxin-contaminated waters would probably need to employ a case–control design. An advantage of these studies is that outcome data is ascertained by medical practitioners; however, disadvantages include exposure recall bias and recruitment of appropriate control groups (Stewart et al., 2006c). El Saadi et al. (1995) also alluded to difficulties in gaining cooperation of diagnosing practitioners.

In contrast to the limitations of field epidemiology, clinical studies overcome the reliance on self-reporting of symptom occurrence, severity and duration. The diagnosis and history of acute intoxication or allergic response to cyanobacteria and/or cyanotoxins is likely to be more reliable when conducted by expert clinicians, particularly when clinical histories and examinations can be supported by confirmatory or complementary diagnostic tests. Early clinical investigations, and in some cases desensitisation treatments, were concerned with allergic reactions to cyanobacteria in recreational waters (reviewed in Stewart et al., 2006c), and more recent clinical studies have addressed the topic of cutaneous and respiratory reactivity to cyanobacteria (Pilotto et al., 2004; Stewart et al., 2006a; Bernstein et al., 2011). The results of these clinical investigations confirm the case study reports discussed above that certain freshwater cyanobacteria can elicit hypersensitivity reactions in some individuals.

5.2.1.3 Responses to presumed cyanotoxin-related acute illness following exposure

With increasing public information and awareness of cyanotoxin occurrence, it is possible that more individuals will consult medical services if they develop symptoms after exposure – symptoms which not necessarily are caused by cyanobacteria and their toxins. However, particularly where symptoms set in rapidly, that is, within only a few hours after exposure, intoxication should be a diagnostic consideration. Medical consultation will primarily serve to clarify and treat symptoms. Although very few cases are known to date, patients may present with concerns of intoxication after exposure to scums or high concentrations of suspended cyanobacterial cells. For neurotoxins, these would be associated with symptoms of respiratory distress, and urgent respiratory support, including supplementary oxygen therapy, would be the appropriate response. Concerns about possible liver damage from microcystins or cylindrospermopsin after exposure can be met by surveillance of serum parameters reflecting liver function, particularly markers of acute injury such as hepatic transaminases.

Beyond this primary function, however, reporting such cases to public health authorities is helpful for promoting the understanding of the public health impact of recreational exposure to (toxic) cyanobacteria. As discussed above, analysis of water samples for cyanobacteria and cyanotoxins very soon after exposure would be most useful, and to make this happen, it is important that medical services or public health authorities trigger such action. Specific biomarkers of exposure to cyanotoxins are not routinely available, but a range of diagnostic criteria may be applied to support the identification of possible cyanobacterial intoxication (Box 5.5).

BOX 5.5: DIAGNOSTIC CRITERIA TO SUPPORT THE IDENTIFICATION OF POSSIBLE CYANOBACTERIAL INTOXICATION

- Routine diagnostic tests used by clinicians in fields such as clinical microbiology and clinical biochemistry, to investigate whether other causes may explain presenting signs and symptoms;
- a recent history of engaging in recreational water activity, with ingestion of water at a site contaminated by a planktonic bloom, scum material or detached benthic mats of cyanobacteria;
- the confirmation of cyanotoxins and/or cyanotoxin-producing cyanobacteria in water samples or benthic mats collected at or close to the time and location of exposure;
- signs and symptoms of acute hepatic toxicity, supported by findings of hepatic impairment at clinical examination and abnormal liver function tests;
- signs and symptoms of motor nerve deficit, which may or may not manifest in acute respiratory insufficiency, seen at clinical examination where the clinical history indicates recent exposure to cyanobacteria;
- cyanobacterial cells and trichomes in vomitus and stool samples identified by microscopy; although this procedure is a simple, low-tech method for identifying a biomarker of exposure to cyanobacteria, it seems to have been scarcely reported in human case investigations since the 1960s (Dillenberg & Dehnel, 1960; Schwimmer & Schwimmer, 1964).

When allocating symptoms to cyanotoxins, it is important to realise that mere co-occurrence is insufficient for establishing a causal connection: even if cyanotoxins are found in patients' serum, it remains possible that other components of the bloom caused the symptoms, particularly if symptoms are unspecific. If, however, they relate to the mode of action and exposure to high toxin concentrations, this is indicative of the respective toxin to be a likely cause. To support diagnosis, awareness and networking of laboratories involved in microbiological and chemical analyses is important so that they too can trigger a timely sampling campaign at the site where patients were exposed – within a short reaction time to capture the situation *in situ* as close to the potential exposure event as possible.

5.2.2 Pathways for exposure through recreational or occupational water activities

With the exception of the toxins from marine cyanobacteria (see below), the water-soluble cyanotoxins known to date are highly unlikely to be able to disrupt the normal protective barrier function of the skin. Thus, cutaneous exposure will not cause access to the bloodstream in concentrations sufficient to cause generalised organ system dysfunction. Activities involving full immersion (e.g., jumping from diving boards, sailboarding, canoe capsizing, competitive swimming) or potential exposure to spray and aerosols (e.g., jet skiing, spray irrigation, cooling of mining drills) may facilitate the entry of cyanotoxins into the systemic circulation, both through ingestion and through inhalation (these are sometimes termed "primary exposure"). Powered watercraft activities such as tube skiing and wakeboarding are likely to cause more frequent and forceful immersions than, for example, sailing or fishing from a dinghy. Other recreational or occupational activities present low risks of ingesting cyanotoxin-contaminated water, for example, shoreline or jetty fishing, wading, low-speed boating, operating irrigation channels. Exposure to cyanotoxins is potentially through the following routes:

- unintentional ingestion of water through reflex swallowing, or in the case of infants "intentionally" during playing;
- water entering the nasopharynx which is subsequently swallowed;
- inhalation when respirable aerosol or spray is formed and droplets/particles enter the nasopharynx and are subsequently swallowed or when dried scums present on the shore are raised as respirable dust;
- for marine cyanotoxins skin and mucous membrane contact.

Of these exposure routes, the one understood best from numerous animal studies is ingestion (see Chapter 2), and dose–effect relationships will follow the patterns assumed for other oral exposure routes, for example, through drinking-water or food. Moreover, toxin concentrations in water can be measured and amounts ingested be estimated from this.

In contrast, while inhalation has frequently been flagged as a concern, quantitative information on exposure is scant: while the formation of spray through fast power boats, jet skis and water skiing appears likely and exposure may well be enhanced by wind, the dynamics of spray formation are poorly understood and the amount of water to which a person is thus exposed is difficult to quantify. Data on toxin concentrations in spray are limited to microcystins for which concentrations were mostly in the low pg/m³ range but occasionally up to a few 2.89 ng/m³ when the toxin concentration in water was high (Backer et al., 2010; Wood & Dietrich, 2011; Gambaro et al., 2012). The particle size of the contaminated aerosols or spray droplets will determine their ability to reach the alveoli, but information on cyanotoxin uptake through the respiratory tract is limited. The following information is available:

- Benson et al. (2005) exposed male BALB/c mice with the nose-only modality to purified MC-LR and described slight to moderate multifocal degeneration and necrosis in the respiratory epithelium and atrophy of the olfactory epithelium at doses up to 265 µg/m³ after 7 days of daily exposure up to 180 min/d. The authors identified a no observed adverse effect level (NOAEL) for nasal lesions after inhalation of 3 µg/kg bw or 20 ng/cm² of nasal epithelium.
- Fitzgeorge et al. (1994) performed an acute study with MC-LR administered via intratracheal instillation to guinea pigs and determined an LD_{50} of 250 µg/kg (similar to the i.p. lethal dose), with necrosis starting in the high airways, progressing to alveoli and resulting in liver damage, but this route of exposure with purified toxin is poorly representative of human exposure via inhalation (Buratti et al., 2017).
- Backer et al. (2008; 2010) detected microcystins in environmental air samples (0.052–2.89 ng MC/m³ in aerosol with MC-LA as the dominant variant in water at 15–350 µg/L) and at lower levels in nasal swabs (from below the limit of detection to 5 ng) of 81 individuals practising recreational activities in lakes during cyanobacterial blooms. However, MCs were undetectable (<1 µg/L) in the blood of those exposed. This can suggest that the aerosol had a limited systemic bioavailability after inhalation, but no conclusion can be drawn due to the small size of control group (n=7), the variability of aerosol particle size and some analytical problems with the detection of microcystins (matrix effects with ELISA detection in blood; Buratti et al. (2017)).
- Wood and Dietrich (2011) give theoretical considerations for protection from systemic effects of microcystins in spray: from the tolerable daily intake (TDI) of 0.04 µg/kg bw per day and considering an average ventilation volume of 30.3 L/min, typical of sustained activity, and a high bioavailability of inhaled toxin (similar to that after i.p. administration, based on the similar lethal dose as proposed by Fitzgeorge et al., 1994), they estimate that people should not be exposed to more than 4.58 ng/m³ of air. This is higher than the levels so far detected in air or in spray.

In consequence, available data are not sufficient to derive cell densities specifically associated with local or systemic symptoms due to inhalation of contaminated water (Funari et al., 2017). A further possible effect is local irritation of the upper airway mucosa through other substances in cyanobacteria. Also, in many recreational activities, multiple exposure scenarios will occur simultaneously, rendering discrimination between them difficult.

With respect to cutaneous, ocular and respiratory tract symptoms, as discussed above, there is strong evidence, both experimentally and from field observations, of marine toxic *Moorea* species (*Lyngbya majuscula*) containing lyngbyatoxins and/or debromoaplysiatoxin causing such symptoms in a high proportion of exposed individuals. However, no comparable body of evidence exists to support a similar clinical profile and symptomatology for exposure to freshwater planktonic cyanobacteria. Cutaneous exposure may be aggravated by bathing and diving suits, as these may trap and accumulate cyanobacterial cells, enhance their disruption and hence the liberation of cell contents onto the wearer's skin. Disruption by bathing costumes of *L. majuscula* filaments has been reported (Osborne et al., 2001).

5.2.3 Assessing the risk of exposure to planktonic cyanotoxins through recreational or occupational activities and short-term responses to occurrence

In contrast to dogs or to livestock lacking access to scum-free water, humans will rarely swallow a cupful of thick scum intentionally, but bolus-type exposure can occur, for example, in the context of accidents such as capsizing boats or sailboards. Exposure scenario estimates from scum concentrations of cyanotoxins in the range of mg/L show that an acutely hazardous cyanotoxin dose is rarely likely, but cannot be dismissed as a possibility if fairly large water volumes containing highly toxic scum are ingested. If a toddler of 10 kg body weight swallows 100–200 ml of scum containing 25 mg/L, it would reach an exposure of 2.5–5 mg for microcystin-LR, sufficient to cause liver damage, and the case report from Uruguay mentioned above (Vidal et al., 2017) highlights that such scenarios may be realistic. Thus, even a life-threatening dose cannot be totally excluded, particularly for sensitive individuals, if scums are thick and highly toxic. A possibly more relevant concern, however, is injury through frequently repeated exposure to a subacute dose, most likely for microcystins in face of their high concentrations in surface scums.

The extent to which public authorities are able to conduct surveillance and to respond to blooms with temporary warnings or bans may be limited by the number of sites to monitor in relation to their institutional capacity. For example, north-western Germany faces the challenge of a high number of eutrophic, frequently cyanobacteria-ridden lakes used with varying intensity for recreation by the local population, regardless as to whether or not sites are officially designated as recreational sites and are accordingly monitored. Another common scenario is that of densely populated lowland regions with slowly flowing, nutrient-rich and bloom-affected rivers which are nonetheless intensively used for sailboarding, swimming and other water sports even though sites are not explicitly designated for recreational use and are monitored accordingly. In Australia, blooms may affect over 1000 km of continuous stretches of inland river systems (Al-Tebrineh et al., 2012). Unless regular monitoring is in place for other reasons such as drinking-water abstraction, such situations may pose considerable challenges for the monitoring of recreational water and for interventions to protect public health.

Where people use waterbodies for recreation or irrigation, technical barriers against exposure to water potentially containing toxic cyanobacteria are typically lacking. They may also be lacking for other occupational uses of surface water, even where, in principle, some treatment would be possible, for example, for water used to cool drills in mining. Control options to avoid exposure to cyanobacteria and cyanotoxins include catchment and waterbody management geared towards reducing the potential for blooms, as discussed in Chapters 7 and 8. They also include the considerations discussed in section 9.1 for the assessment and choice of drinking-water offtake sites to avoid scums: where the shoreline geography of a given waterbody and/or other already established usages allow a choice, similar considerations may serve to optimise the choice of sites for recreational use. Where none of these management approaches are successful, the option that remains is to guide and influence the behaviour of site users. Options for this range from informing users, that is, creating awareness and enabling individual responses to bloom situations, to temporarily banning waterbody use for the duration of the bloom.

Site users differ in their risk perception, in how receptive they are to information and how willing they are to adapt their behaviour in order to avoid contact. As discussed above, areas with extensive cyanobacterial scums or accumulated detached mats on bathing beaches may be avoided by swimmers and other water users, but as cyanobacterial dominance is typically most pronounced in climates and seasons in which the demand for recreational water is high, scums may also be ignored. Differences in usage patterns, user perception and willingness to engage are depicted in Table 5.4. It is valuable to consider likely behaviour of site users when deciding how intensive monitoring should be at a specific site and whether temporary usage bans are necessary or whether information and warning are sufficient.

Understanding the potential for blooms in a waterbody is a further important basis for prioritising monitoring. It depends on a few key conditions, in particular the concentration of total phosphorus, turbidity, water exchange rate and for lakes or reservoirs also on thermal stratification. For example, if total phosphorus concentrations in a waterbody do not exceed 20 µg/L and the water is clear, with Secchi depths above 2–3 m, blooms are very unlikely. Waterbody conditions that render blooms likely are discussed in detail in Chapters 7 and 8, which also give checklists for assessing the risk of bloom occurrence. The advantage of understanding the potential for blooms is that usually this potential does not change quickly in a given

Table 5.4 Usage patterns of waterbodies prone to blooms as criteria for monitoring and intervention

Appropriate intensity of monitoring and intervention	*Waterbody usage pattern*	
	Almost daily exposure during the bloom season, for example, at lakeside holiday homes and campsites or at a workplace	Recreational sites used by a high number of people Occupational exposure to aerosol likely for a high number of workers and/or regularly over several weeks
	Water sports with high probability of immersion of the head and/or oral uptake of bloom material; lakeshore bathing sites with diving boards or rafts, water slides or other attractions likely to increase the probability of incidental oral uptake	
	Sites used only by a small number of people and only occasionally, discontinuously	
	Occupational exposure only occasionally, intermittedly and/or to a small number of workers	
	Site users/workers receptive to information on blooms, how to recognise them and how to respond to them	Site users/workers willing to engage in initiatives to assist surveillance, for example, by scum scouting and checking turbidity, reporting observations to the responsible authority and thus triggering targeted surveillance

waterbody, and after once having assessed such baseline data throughout one to three bloom seasons and their patterns over time, such analyses may not need to be repeated frequently; occasional checking whether the situation has substantially changed may be sufficient.

Longer-term data on cyanobacteria and toxin concentrations help to understand their variability in the given waterbody and are therefore highly valuable for prioritising waterbodies of concern: for example, if data covering 2–3 years or seasons of cyanobacterial dominance regularly showed high amounts of toxic cyanobacteria occurrence, this would indicate a high priority for the monitoring of cyanobacteria at recreational sites or in water used at a workplace. By contrast, if data with sufficient resolution over time (i.e., at least monthly, preferably fortnightly sampling) show that over a period of 2–3 years, cyanobacteria were never dominant or exceeding the biovolumes given in the Alert Levels Framework (ALF; see Figure 5.5), monitoring of such a waterbody could be a lower priority (see also Chapter 11).

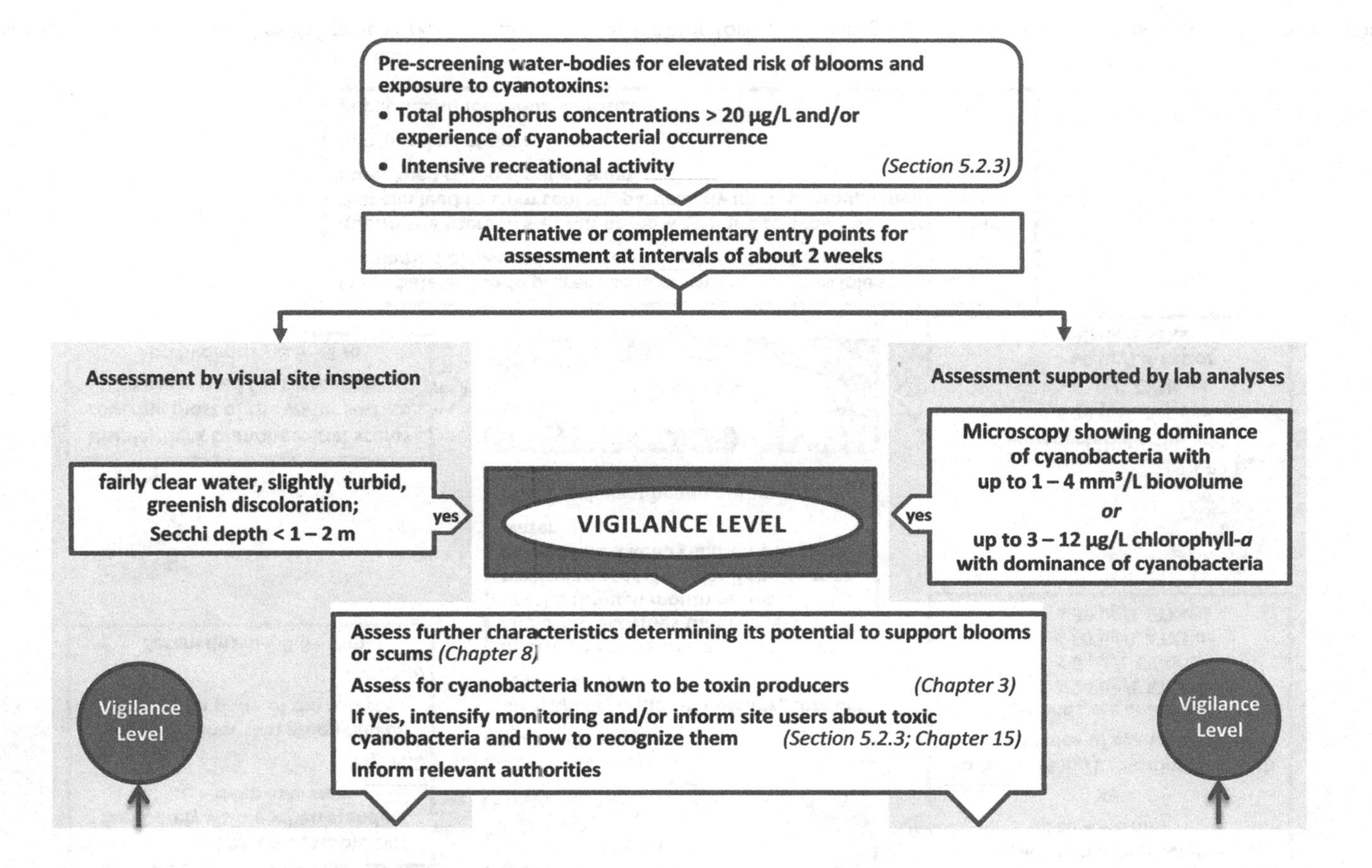

Figure 5.5 Alert Levels Framework (ALF): Decision tree for monitoring and managing cyanobacteria in waterbodies used for recreation.

(*Continued*)

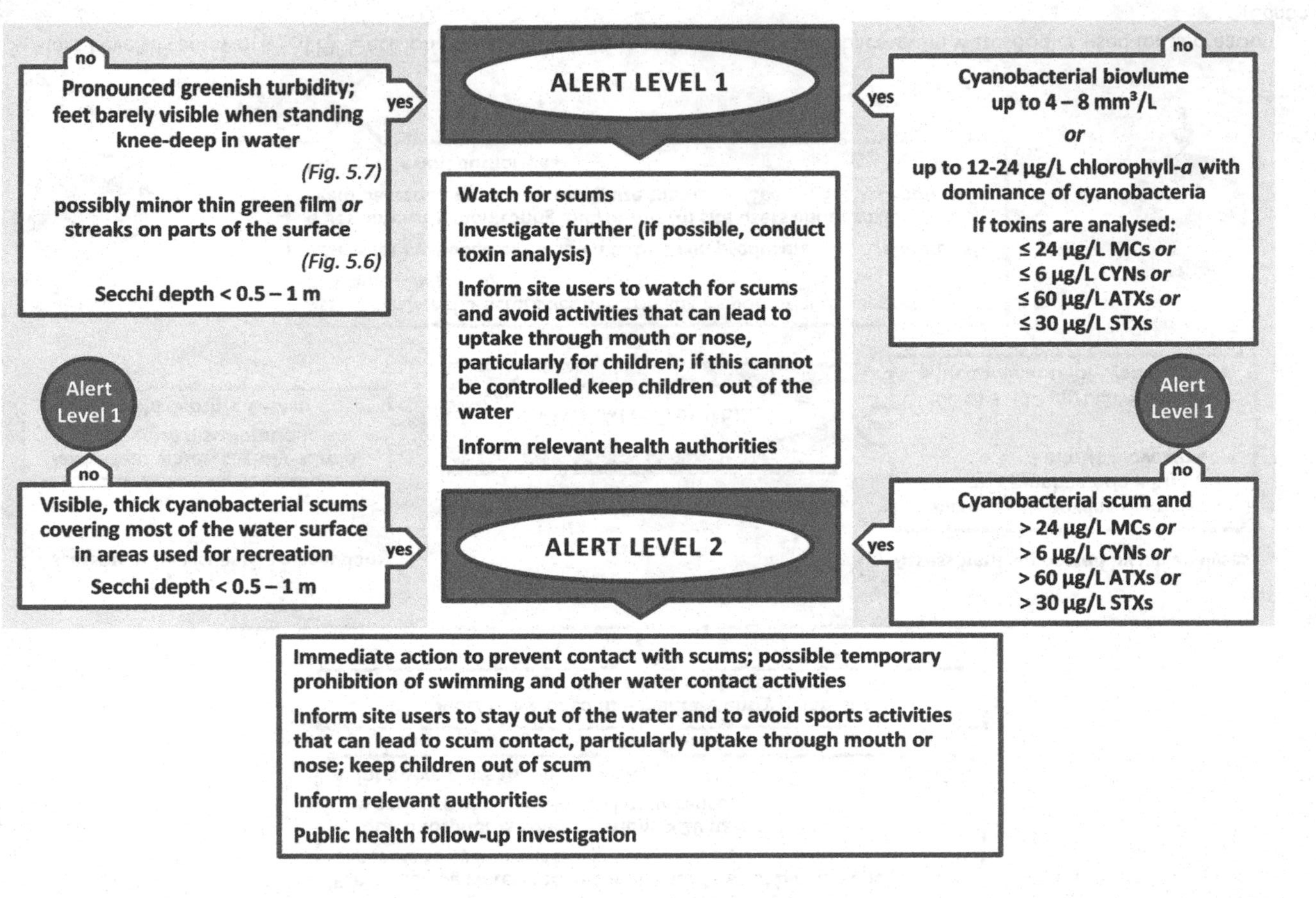

Figure 5.5 (Continued) Alert Levels Framework (ALF): Decision tree for monitoring and managing cyanobacteria in waterbodies used for recreation.

Checklist 5.1 summarises aspects to consider when designing the overall approach to assessing exposure risks through direct contact to cyanobacteria in recreational or occupational settings.

CHECKLIST 5.1 FOR ASSESSING THE LIKELIHOOD OF EXPOSURE TO CYANOTOXINS THROUGH RECREATIONAL AND OCCUPATIONAL USE OF A WATERBODY

- Is information available to indicate the likelihood of bloom occurrence, that is, from catchment characteristics and land use governing nutrient loads (see checklists in Chapter 7) or from direct observations of cyanobacteria and/or waterbody characteristics (see checklists in Chapter 8)?
- If scums occur, are there bays and shorelines where they chiefly tend to accumulate (see section 4.1.2, Figure 8.1 and the checklist in section 9.1.5), and if so, how does the location of the site used (e.g., for a beach or for the offtake of water for production purposes) relate to these?
- How intensively is the site used (see Table 5.4)? Is individual use occasional, or are the same people exposed frequently, for example, almost daily?
- Are the majority of users receptive to information and likely to adapt their behaviour accordingly?
- Are site operators or users potentially willing to engage in initiatives to assist surveillance, for example, by scum scouting and/or checking turbidity and reporting observations?

5.2.3.1 Defining national cyanotoxin levels that trigger action

The edition of "Toxic Cyanobacteria in Water" published in 1999 proposed two points of entry for assessing the "guidance level or situation" – either the concentration of chlorophyll-*a* (with dominance of cyanobacteria) as measure for biomass, or cell numbers, with Table 5.2 and Figure 6.5 in that edition differentiating three "Guidance levels" for recreational exposure.

A number of countries have since used this guidance as basis for implementing guidelines or action levels for assessing health risks from cyanobacteria through recreational usage of waterbodies (see Chorus, 2012; Ibelings et al., 2014; Funari et al., 2017 for overviews). While the actions taken in these countries at each of the three levels are similar (ranging from information and the issuing of warnings to temporary site closure), they vary considerably in the cell count levels triggering them and in their assessments of the health risk arising from exposure: the distinction between risks categorised as "low" and those categorised as "moderate" varies 20-fold including one extreme case even 200-fold, that is, from 500 cells/mL in New Zealand to 100 000 cells/mL in Canada (with intermediate values such as 5000 cells/mL in Australia and 20 000 cells/mL in the Czech Republic, Italy and France).

In contrast, where countries base their Alert Levels on cyanobacterial biovolumes, differences between levels triggering alerts are less pronounced, varying only by a factor of less than 10 and ranging from 1.8 mm^3/L in New Zealand to 15 mm^3/L in the Netherlands, and almost all countries place the presence of scums in the high-risk category. The range of variation is similar where countries include microcystins in their risk assessment: levels considered as high risk range from 10 to 20 µg/L.

Such variation reflects differences not only in the assessment of the uncertainty of the toxicological data used but also in the estimates of exposure, particularly regarding the water volumes assumed for oral uptake and the duration of exposure. The background of local experience with the recreational use of waterbodies affected by blooms is also relevant: in countries with a long history of recreational use, despite waterbodies frequently suffering visible discoloration, high turbidity and blooms (such as the Netherlands and Germany), authorities tend to set triggers for warning and for closure at higher levels. For example, in two provinces of the Netherlands warnings, discouraging of bathing and even prohibition occurred in numerous waterbodies even though the thresholds triggering these actions are set quite high, and Ibelings et al. (2014) analyse the situation as follows:

> *"Setting the alert levels in the Netherlands is the outcome of intensive discussions between scientists, lake managers and policy makers, in a country known for the highly eutrophic state of its lakes (despite successful restoration efforts ...), where stricter alert levels might result in extended closure of many lakes. Safety clearly must come first, but the protocol used in the Netherlands – in addition to health risks – takes into account the promotion of outdoor activities, feasibility, complexity and costs of monitoring or risk control of cyanobacteria, as well as the ease of communication to the public. Given the large uncertainty in the derivation of TDI for cyanotoxins it is not possible to say whether the higher alert levels in the Netherlands truly result in decreased protection. We merely know it is uncertain." (p. 68)*

The WHO guidance level of 100 000 cells/mL potentially triggering restrictions of site use, published in 1999, was based on the potential for health impacts of cyanotoxins through ingestion and systemic intoxication inferred from toxicological considerations, using the provisional WHO TDI for microcystin-LR. The lower guidance level of 20 000 cell/mL triggering information to site users was also based on the complementary criterion of a potential for irritant and/or allergic reactions, inferred as outcomes of the epidemiological study on health effects from recreational exposure to cyanobacteria conducted by Pilotto et al. (1997). The subsequent studies by Stewart et al. and Backer et al. discussed above did not find increased symptoms at the low levels of cyanobacteria that Pilotto et al. (1997) concluded to be associated with illness, thus casting doubt about the validity of this complementary criterion. In addition, Pilotto et al. (2004) in a subsequent study found no direct dose–response to a wide range of cell densities during a study with skin patches to assess skin effects after exposure to cyanobacteria. These new results question the basis for the criterion of 20 000 cells/mL.

Meanwhile, WHO has derived guideline values (GVs) for recreational exposure to microcystins, cylindrospemopsin and saxitoxins as well as a health-based reference value for anatoxin-a (Table 5.1). This has opened a new rationale for setting guidance or Alert Levels for recreational exposure. The values for the Alert Levels are now based on the minimum amount of biomass that is likely to contain the toxin concentrations amounting to the recreational GVs. Also, an Alert Levels Framework (ALF) for recreational exposure (Figure 5.5) now replaces the guidance levels described in 1999, while maintaining the differentiation of three levels. A further change reflects experience with cell numbers leading to undue restrictions of recreational use if the dominant cyanobacteria are species with very small cells: as toxin concentrations relate to biomass rather than numbers, even at high cell numbers of very small cells water is clear and toxin concentrations are negligible (see discussion in section 4.6). Therefore, while the ALF continues the use of concentrations of chlorophyll-*a* as indicator for possible cyanotoxin occurrence, it now uses biovolumes instead of cell numbers as a further parameter for defining Alert Levels.

Importantly, as for the drinking-water ALF described in section 5.1.2.2, this ALF for recreational water use is intended as template for adaptation to national circumstances or even to local conditions: where appropriate, regulators can modify both the choice of parameters as indicators of possible cyanotoxin concentrations and the levels at which alerts are set. The Alert Levels on the basis of biomass are defined from experience regarding the maximum ratio between microcystins and cyanobacterial biovolume or chlorophyll-*a* typically found in the field, based on the assumption that maximum ratios of toxin to biomass will not be higher for the other cyanotoxins (see section 4.6). Note that for CYNs, due to the high share of this toxin potentially dissolved in water, the biomass at the time of sampling does not necessarily indicate occurrence of dissolved CYNs; this needs to be inferred from the biomass of CYN producers observed in the 4–6 preceding weeks (see Box 5.1 for details).

As discussed at the beginning of this chapter, the ratio used is in the upper range of ratios from field data reported in the literature and thus is highly conservative: if these Alert Levels are met, it is highly likely that the WHO GVs for recreational exposure to cyanotoxins summarised in Table 5.1 are not exceeded. However, the cyanobacteria in a given waterbody may well contain far less toxin per unit biovolume or chlorophyll-*a*, and determining such ratios specifically for the waterbody may well lead to setting different Alert Levels, that is, tolerating higher amounts of cyanobacteria before moving on to the next Alert Level. Monitoring can then continue based on cyanobacteria; however, during the course of a bloom or if its composition changes, it may be important to repeat toxin analysis to check whether the toxin/biomass ratio is still appropriate, as this may change as a bloom develops.

If other parameters that serve as indicators of toxin concentrations are chosen to define the Alert Levels (e.g., cell numbers, pigment concentration measured by hand-held probes, remote sensing signals or molecular tools), it is important to periodically "calibrate" these locally against toxin concentrations in order to ensure that they adequately reflect these. When using such parameters or when using data from toxin analyses to define Alert Levels, qualitative microscopy is recommended in order to assess which cyanobacterial genera dominate, as this information is important both for understanding scum behaviour and for waterbody management. An advantage of using cyanobacterial biomass for defining Alert Levels is that this encompasses any further hazards potentially associated with cyanobacteria, including those from yet unknown substances they may contain (section 2.10) or pathogenic organisms associated with their mucilage. Although health risks from such poorly understood agents cannot be quantified and thus no health-based limits for cyanobacteria can be derived, meeting the biomass-based Alert Levels is likely to provide level of protection from such agents as well.

Furthermore, water-use patterns may determine the ALF thresholds used locally. The WHO recreational guideline values (GVs) are calculated on the basis of a 15 kg child ingesting 250 mL of water, and an adult swimmer, sailboard rider or water skier would need to ingest 1 L to reach the WHO GV.

5.2.3.2 Alert Levels for short-term responses to toxic cyanobacteria in waterbodies used for recreation

As for drinking-water in section 5.1, Figure 5.5 provides an Alert Levels Framework (ALF), that is, monitoring and management action decision tree as for planning immediate short-term responses to cyanobacterial occurrence in waterbodies used for recreation (for longer-term measures addressing the causes of cyanobacterial proliferation, see Chapters 6–9). Depending on local circumstances, this ALF may be adapted for water used at workplaces as well.

As for drinking-water, the basis for this ALF is an assessment of the likelihood for the waterbody to harbour health-relevant amounts of cyanobacteria. While for drinking-water supplies such an assessment may be driven by the water supplier for the specific waterbody used as raw water, for recreational sites, in some cases, a private operator may be responsible for a site, but often it is a public authority. Sometimes these carry the responsibility for assessing the safety of a larger number of sites. Prescreening waterbodies to set priorities for the surveillance of bathing sites may then be important, and criteria include data from previous monitoring (if available) as discussed above in section 5.2.3 as well as the

- likelihood of cyanotoxin risks to occur, to be assessed from characteristics of the waterbody and/or previous observations of their occurrence as described in Chapter 6;
- the potential public health impact which is influenced by the intensity of site use, as depicted in Table 5.4.

Furthermore, in many settings, waterbodies for recreational use are not monitored as intensively as drinking-water, and in consequence, monitoring is less likely to follow the onset and progress of a cyanobacterial bloom. Inspection conducted only occasionally may well find a pronounced bloom, and the outcome will then lead straight to Alert Level 1 or 2.

Differently from the ALF for drinking-water (Figure 5.1), the next step after prescreening in this ALF offers two points of entry, of which the simpler approach of visual inspection may well be used alone. The supplementary point of entry is analyses of cyanobacterial biomass – depending on the equipment and expertise available either by microscopy as biovolume or by chemical analysis as the concentration of chlorophyll-*a* (combined with qualitative microscopy to assess whether or not the biomass primarily consists of cyanobacteria or other phytoplankton). This adds a more objective trigger for action which may be important particularly where warnings or temporary site closure is likely to lead to concern or opposition, for example, because of substantial restrictions of site use and economic consequences for operators. Biomass should, however, be used in addition to – and not instead of – the visual assessment described on the left-hand side of Figure 5.5.

As discussed at the beginning of this chapter (and more specifically for drinking-water in section 5.1.2.2), for MCs, STXs and ATXs the toxin/biomass ratios proposed for triggering Alert Levels are quite conservative and thus highly protective of cyanotoxin occurrence, while for CYNs, with its low GV, they are somewhat more uncertain. Also note for CYNs that dissolved CYNs can persist well after CYN-producing taxa (i.e., particularly species of *Aphanizomenon*; in Australia and the Americas also *Raphidiopsis raciborskii*) are no longer conspicuously present in a sample, and where these have been observed, sampling should follow the guidance given in

Box 5.1, and/or CYN concentrations should be determined. In contrast, the WHO recreational GV for STXs and the health-based reference value for anatoxin-a are substantially higher, and it is unlikely that the Alert Levels for biovolume and chlorophyll-*a* will not be sufficiently protective to ensure that meeting these also ensures not exceeding the 30 µg/L for STX or 60 µg/L for ATX.

The Alert Levels Framework (ALF) given in Figure 5.5 does not address the concerns of individuals with an allergic predisposition, for example, atopy, who may experience acute cutaneous or respiratory allergies at quite low concentrations of cyanobacteria, whereas those without such predisposition will likely be unaffected. Also, the aggravation of cutaneous reactions due to accumulation of cyanobacterial material and enhanced disruption of cells under bathing suits and wet suits discussed above may occur even at densities below the Alert Level values used in the ALF.

The Vigilance Level addresses a situation with dominance of cyanobacteria in the phytoplankton, but at biomass levels too low to contain hazardous toxin levels and thus with fairly clear water that might show slight turbidity with greenish discoloration; transparency determined with a Secchi disc will usually be in the range of 1–2 m. However, because of the potential for rapid increase or even scum formation, it is appropriate to intensify surveillance and inform site users about their potential to increase to higher levels. In this range, that is, at a maximum of 4 mm^3/L cyanobacterial biovolume or 12 µg/L of chlorophyll-*a*, microcystins can reach concentrations in the range of 12 µg/L provided that the cyanobacteria present have a high content of this toxin. However, generally concentrations of toxin will be lower than this (see section 4.6).

Because of the potential for rapid increase of cyanobacterial biomass and thus toxin levels between monitoring occasions, dominance of cyanobacteria even at low levels should not be a cause for complacency, particularly if recreational site monitoring occurs only at intervals longer than once per week. Concentrations in this range are a cause for alertness and locally appropriate responses with a focus on improving the understanding of the specific situation.

Vigilance is particularly relevant for waterbodies with total phosphorus concentrations well above 20 µg/L (provided N is not reliably limiting) because cyanobacteria, once dominant, may reach a higher biomass within a few days. It is also particularly relevant for very large waterbodies because they have a potential for scum formation even at these rather low biomass levels, as scums can accumulate from very large water volumes. However, lakes and reservoirs with low phytoplankton density rarely show prolonged dominance of cyanobacteria, and such scums tend to be short-lived minor events.

Alert Level 1 addresses a situation in which cyanobacteria are clearly visible when inspecting the site, particularly as greenish turbidity or discoloration and possibly also as minor green streaks or specs floating on parts of the water surface, but not as scum covering major parts of the surface area, with Secchi depth in the range of 0.5 – 1 m or even less (Figure 5.6). In such a situation, cyanotoxin concentrations can reach potentially hazardous levels even without scums, but typically they do not, and recreational use may be continued without exposure to cyanotoxins exceeding the recreational guideline values (GVs). This is particularly the case for scum-forming microcystin producers such as *Microcystis* or *Dolichospermum* which may be visible as slight streaks or small specs between which water is fairly clear.

Determining biomass and possibly toxin concentrations provides more precise information and is important in waterbodies with a history of supporting the growth of non-scum-forming species of cyanobacteria: for example, *Planktothrix agardhii* can reach very high densities, particularly in shallow waterbodies, up to 70 mm^3/L biovolume or 200 µg/L of chlorophyll-*a*. Secchi depth in such situations will be less than 0.5 m. *P. agardhii* may contain particularly high cell contents of microcystins (up to or exceeding 1 µg toxin per µg chlorophyll-*a*; see section 2.1).

Figure 5.6 Streaks, specs and Secchi disc reading depicting Alert Level 1 conditions.

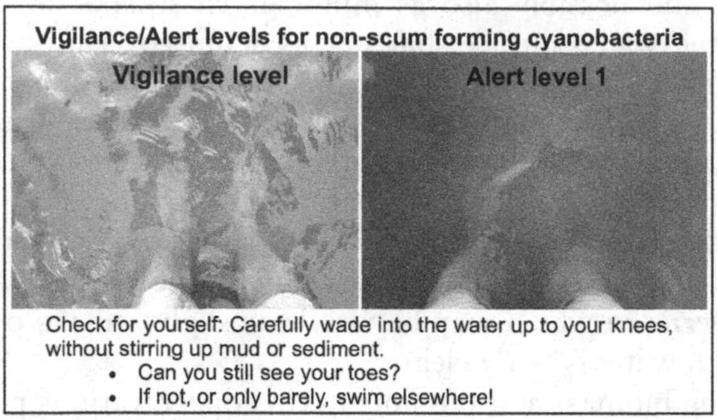

Vigilance/Alert levels for non-scum forming cyanobacteria

Vigilance level

Alert level 1

Check for yourself: Carefully wade into the water up to your knees, without stirring up mud or sediment.

- Can you still see your toes?
- If not, or only barely, swim elsewhere!

Figure 5.7 Simple and efficient guidance for recreational site users for checking whether non-scum-forming cyanobacteria are present at potentially unhealthy concentrations.

Though such extreme situations are rarely observed, in such situations, it is possible that microcystin concentrations range up to several hundred µg/L or more, without scum formation.

Informing site users to avoid exposure to high densities of such evenly dispersed cyanobacteria is less straightforward than informing them to avoid scums because the situation is harder to describe. Figure 5.7 shows one option for visualising a criterion for self-assessment of the situation.

Where data from visual inspection and quantifying cyanobacterial biomass can be supported by cyanotoxin analyses, this can serve to avoid undue restrictions of recreational site use in situations where cyanobacterial biomass is high but toxin content is low, rendering toxin concentrations in the water below Alert Level 1.

At Alert Level 1, the cyanobacteria present may well increase to a heavy bloom within a few days if conducive conditions prevail in the waterbody. Watching out for scums is therefore recommended, and increasing surveillance may therefore be appropriate, particularly for heavily used recreational sites, in order to rapidly detect if the situation escalates to Alert Level 2.

Alert Level 2 describes a situation with scums or very high cell density leading to substantial turbidity (Figure 5.8). While scums can be thick in parts of the waterbody, other parts may still show a Secchi depth ranging up to about 1 m. Whilst in such a situation the recreational GVs for cyanotoxins are not necessarily exceeded, this is quite likely. Cyanotoxin analysis can be used to confirm or downgrade the Alert Level status. As discussed above, if scum material is both very thick and highly toxic, the ingestion of 100–200 mL by a toddler can contain an acutely hazardous dose. The presence of cyanobacterial scums is a readily observable indicator of a high risk of adverse health effects.

Alert Level 2 situations call for immediate action to avoid scum contact and, in particular, oral uptake. Temporary banning of use may be

Figure 5.8 Examples of *Microcystis* scums depicting Alert Level 2 conditions.

appropriate, and intensified monitoring may be important to either confirm or downgrade the Alert Level status in order to not unnecessarily restrict use. Providing information to the site users is important to achieve an understanding of the hazard and thus compliance. Measures to reduce exposure that can be implemented quickly may include the installation of floating physical barriers to prevent the scum from being driven into the swimming area, provided that surface scums are the key issue (rather than dispersed suspended cells or colonies). If scums typically accumulate at certain sites while others largely remain unaffected, directing recreational use to another site may be an option. Removing drying scum accumulated on beaches may be necessary to avoid the development of dust (using personal protecting equipment if scum is already dry).

Misconceptions of what constitutes a scum are common for large, deep and usually clear lakes with low nutrient concentrations: in such lakes, cyanobacteria may become transiently dominant in the phytoplankton but only at low concentrations. Cells from the large water volume may rise to the surface and be swept into a downwind bay where they may form a surface film, typically thin and with cyanotoxin concentrations well below hazardous levels. Site users not accustomed to any visible phytoplankton on the surface may interpret even a very thin and locally limited film as "scum" and be unduly concerned, and advisories may need to explain what amounts to a sufficiently pronounced scum to cause concern. Local information may be appropriate to dispel such concerns.

Rescinding warnings after a bloom, when recreational use is safe again, is important in order to avoid undue discouragement of healthy outdoor recreational activity as well as "warning fatigue": if warning signs remain posted even though the water is clear, it is likely that site users will tend to ignore them in the future. This is discussed in more detail in Chapter 15.

5.2.4 Assessing risks from recreational exposure to cyanobacteria on benthic and other surfaces

Death of dogs and livestock has been observed even where the water was clear and toxins from cyanobacteria growing attached to submerged aquatic plants ("periphyton"; including species of *Microcoleus* (*Phormidium*), *Tychonema* and possibly of other genera) or the sediment ("benthic cyanobacteria") have been identified as the cause (e.g., Puschner et al., 2010; Wood et al., 2017; Fastner et al., 2018; see also section 4.2.2). Some countries, for example New Zealand, report widespread occurrence of benthic mats of *Phormidium* found under a wide range of water qualities and proliferating during stable stream flow (Wood & Williamson, 2012). *Microcoleus* (*Phormidium*) is known to produce anatoxins. Benthic mats have become a concern because of the frequency of dog deaths, attributed to anatoxin-a, homoanatoxin-a and saxitoxins and in one case also to microcystins (Wood & Williamson, 2012). Likewise, ATX-producing *Tychonema* species may colonise submerged macrophytes such as *Fontinalis* (Fastner et al., 2018).

Such mats of *Microcoleus* or lumps of macrophytes with attached *Tychonema* may cause a lethal dose for animals ingesting substantial amounts. While dogs sometimes appear to be attracted to decaying material and to ingest substantial amounts, this behaviour is unlikely for humans, including small children. However, swimmers may also be in direct contact with such material after a storm breaks off clumps of material or it naturally detaches from the sediment and is accumulated in shallow water or on the shore.

Implications for managers deciding on whether to restrict recreational use of such waterbodies are challenging: while dead animals are a cause for concern, negligibly low cyanotoxin concentrations in the water would not be. Where water is very clear, that is, with Secchi depths of 2 m or more, concentrations of concern are unlikely, as dissolved toxin leaching from detached mats or macrophytes will dilute and/or degrade quickly. Such situations require a rapid assessment of the risk and its communication to site users – for example, assurance that concentrations in the water are indeed low. Confirmation through cyanotoxin analysis is the most convincing way forward for hazard analysis in such situations. Box 5.6 gives a case example.

BOX 5.6: DOG DEATHS ATTRIBUTED TO *TYCHONEMA* GROWING ON *FONTINALIS* IN A CLEAR LAKE

Lake Tegel is an important suburban resource for the city of Berlin, Germany, both for abstracting drinking-water via bank filtration and for intensive recreational use. Lake Tegel went through a history of eutrophication in the 1980s followed by restoration efforts resulting in the re-estabishment of mesotrophic conditions. Since the begin of the millenium, the lake has become clear, with Secchi depths rarely less than several metres, and heavy cyanobacterial blooms no longer occur. Stands of submerged macrophytes now cover large areas of the lake bottom, and as these bind nutrients that thus are no longer available for phytoplankton, they contribute to keeping the water clear.

Against the background of this success story, in May 2017, the acute neurotoxicosis of 12 dogs, several of which died, after playing and swimming in Lake Tegel caused considerable concern. Intensive investigation detected high biomass of anatoxin-a-producing *Tychonema* spp. in detached and floating water moss (*Fontinalis*), which led to concentrations of anatoxin-a of up to 8700 µg/L detected in dog stomach content (Fastner et al., 2018). Interestingly, while the aqueous fraction of some samples of floating *Fontinalis* with *Tychonema* contained up to 1870 µg/L of anatoxin-a, concentrations in other *Fontinalis* samples were low, that is, in the range of 10–20 µg/L. Moreover, water samples – even those taken only a few metres away from the floating *Fontinalis*/*Tychonema* clumps – contained 1 µg/L or less of anatoxin-a. The available data indicated the occurrence of *Tychonema* on *Fontinalis* to be highly variable and the extremely high concentrations detected in May 2017 appear to be an isolated event.

This situation demonstrates a typical challenge for risk assessment: should recreational use of a lake in such a situation be discouraged or is it safe to continue activities? In this specific case, the public authority responsible for site surveillance initiated a monitoring campaign focusing on *Tychonema* and anatoxin-a for Lake Tegel and later also for other Berlin lakes, the removal of macrophyte clumps on the shoreline and information of the public advising against contact with the clumps. Based on the low toxin concentrations found outside the areas with dislodged macrophyte clumps, the lake's importance for recreation and its otherwise high water quality (as compared to the other, more eutrophic and bloom-ridden waterbodies available in and around the city), the authority did not discourage or ban recreational use.

New Zealand has introduced a three-tiered Alert Level Framework for benthic cyanobacteria which is similar to the guidelines for planktonic cyanobacteria, but based on the percent coverage of the waterbody's sediment as well as on detached material from benthic mats accumulating along the shore (Wood & Williamson, 2012).

5.2.5 Assessing risks from recreational exposure to marine dermatotoxic cyanobacterial

As discussed above, some marine beaches have been reported to cause widespread health problems due to the benthic marine cyanobacterium, *Moorea* sp., growing on rocks and in shallow embayments in tropical and subtropical seas. *Moorea producens* and possibly other species of *Moorea* can cause severe blistering when people swimming in affected coastal waters come into contact with strands of these filamentous cyanobacteria, particularly if trapped and macerated under bathing suits (section 2.6). This response may be due to acute toxicity, as *Moorea* can produce irritant toxins. The dermatotoxic alkaloids produced by *Moorea* are not considered in Table 5.4 because exposure patterns to them are different – that is, not unintentional ingestion of planktonic cells or colonies, but cutaneous contact with clusters of filaments (each 10–30 cm in length). Measures to protect site users include providing information about avoiding skin contact, removing bathing suits and showering after immersion to ensure removal of any *Moorea* from the skin (Osborne et al., 2007).

For example, the Moreton Bay Regional Council, Queensland, Australia, has established a three-level approach. Where *Moorea* deposits on beach and adjacent waters are small to moderate and away from built-up areas, they monitor and install warning signs for the public (level 1). Where large quantities of *Moorea* are washing ashore or beginning to form rafts adjacent to built-up areas, they advocate removal from beaches with tractors and excavators (level 2) and notify relevant stakeholders including other government authorities and media. Where very large quantities are washing ashore, in addition to the level 2 procedures, the beach will be closed to the public to safeguard against associated risk of wading or swimming (level 3) (Moreton Bay Regional Council, 2018).

5.2.6 Research to improve our understanding of recreational exposure

As discussed above, symptoms clearly caused by microcystins, cylindrospermopsins, anatoxins or saxitoxins following recreational exposure are not very likely; however, as compared to exposure through drinking-water uptake, recreational activities are more likely to lead to exposure to higher concentrations, possibly causing detectable symptoms. A larger body of thoroughly investigated cases is therefore valuable to improve our understanding of the hazards that cyanotoxin exposure imply for human health. A key issue for this aim is the quantification of exposure. While rapid (preferably within hours) site inspection and bloom sampling mentioned above would be the best approach, this is often hampered by limited institutional capacity and communication between the institutions responsible for public health versus environmental monitoring. Continuous online monitoring of cyanobacterial biomass development with *in situ* fluorescence probes can greatly improve the understanding of the wax and wane of blooms, as can remote sensing if data can be obtained with sufficient frequency.

Biomarkers are a further helpful tool to assess exposure. Notable advances have occurred in the analytical detection and quantification of cyanotoxins in physiological fluids such as serum, blood, vomitus and urine from exposed groups using chemical and antibody-based methods, although for human blood these findings so far have only been reported for microcystins (Hilborn et al., 2007; Chen et al., 2009; Li et al., 2011). However, for other cyanotoxins such as anatoxin-a, similar advances have been reported from veterinary researchers investigating dog poisonings. While such investigations are usually conducted on necropsied tissues, particularly liver in the case of microcystins or nodularin, analytical chemists have confirmed the presence of anatoxin-a in dog urine (Puschner et al., 2010) and stomach contents (Hoff et al., 2007; Fastner et al., 2018). Such methods are useful in order to support or exclude diagnoses of cyanotoxin exposure and possible intoxication. Many laboratories can also identify cyanobacterial cells and trichomes in vomitus and stool samples or at least have the capacity to capture photomicrographs of stool or vomitus, which can be referred to expert phycologists for confirmation or exclusion of cyanobacterial cells.

The "ideal" case investigation would be triggered by one or several individuals exposed to significant levels of cyanobacteria in recreational waters presenting soon after symptom onset for medical assessment, providing samples of blood, stool, urine and potentially vomitus, a good estimate of the amount of water ingested and the time and location of exposure, from where water samples would be immediately collected for cyanobacterial and cyanotoxin analysis. The putative case would then be rapidly assessed by either an expert hepatologist for a comprehensive assessment of liver function, or, in the case of exposure to a cyanobacterial neurotoxin, for nerve conduction studies and detailed assessment of neuromuscular function. Likewise, the ideal assessment for those presenting with anaphylaxis or other allergic reaction, possibly due to cyanobacterial exposure, would be rapid referral (after recovery) to a clinical immunologist, asthma specialist or dermatologist for confirmatory challenge testing. Furthermore, ideal patients would be willing to consent to publication of their case history, and the attending clinicians will be keen to publish. Substantial public health benefits would arise from a better scientific understanding gleaned from a series of studies employing various subsets of the aforementioned "ideal" case criteria.

REFERENCES

Al-Tebrineh J, Merrick C, Ryan D, Humpage A, Bowling L, Neilan BA (2012). Community composition, toxigenicity, and environmental conditions during a cyanobacterial bloom occurrence along 1100km of the Murray River. Appl Environ Microbiol. 78:263–272.

Backer L, Manassaram-Baptiste D, LePrell R, Bolton B (2015). Cyanobacteria and algae blooms: review of health and environmental data from the harmful algal bloom-related illness surveillance system (HABISS) 2007–2011. Toxins. 7:1048–1064.

Backer LC, Carmichael W, Kirkpatrick B, Williams C, Irvin M, Zhou Y et al. (2008). Recreational exposure to low concentrations of microcystins during an algal bloom in a small lake. Mar Drugs. 6:389–406.

Backer LC, McNeel SV, Barber T, Kirkpatrick B, Williams C, Irvin M et al. (2010). Recreational exposure to microcystins during algal blooms in two California lakes. Toxicon. 55:909–921.

Benson JM, Hutt JA, Rein K, Boggs SE, Barr EB, Fleming LE (2005). The toxicity of microcystin LR in mice following 7days of inhalation exposure. Toxicon. 45:691–698.

Bernard C, Froscio S, Campbell R, Monis P, Humpage A, Fabbro L (2011). Novel toxic effects associated with a tropical *Limnothrix/Geitlerinema*-like cyanobacterium. Environ Toxicol. 26:260–270.

Bernstein JA, Ghosh D, Levin LS, Zheng S, Carmichael W, Lummus Z et al. (2011). Cyanobacteria: an unrecognized ubiquitous sensitizing allergen? Allergy Asthma Proc. 32:106–110.

Buratti FM, Manganelli M, Vichi S, Stefanelli M, Scardala S, Testai E et al. (2017). Cyanotoxins: producing organisms, occurrence, toxicity, mechanism of action and human health toxicological risk evaluation. Arch Toxicol. 91:1049–1130.

Chen J, Xie P, Li L, Xu J (2009). First identification of the hepatotoxic microcystins in the serum of a chronically exposed human population together with indication of hepatocellular damage. Toxicol Sci. 108:81–89.

Chorus I (2012). Current approaches to cyanotoxin risk assessment, risk management and regulations in different countries. , Dessau: Federal Environment Agency.

Cohen SG, Reif CB (1953). Cutaneous sensitization to blue-green algae. J Allergy. 24:452–457.

Dillenberg HO, Dehnel MK (1960). Toxic waterbloom in Saskatchewan, 1959. Can Med Assoc J. 83:1151–1154.

El Saadi O, Esterman A, Cameron S, Roder DM (1995). Murray River water, raised cyanobacterial cell counts, and gastrointestinal and dermatological symptoms. Med J Australia. 162:122–125.

Fabbro L, Unwin L, Barnett L, Young L, Orr N (2008). Mine water quality - spread of blue-green algae. Brisbane: Australian Coal Association Research Program (ACARP). Report C14051.

Fabbro LD, Bernard C, Monis PT (2010). Improved morphometric and genetic tools for better identification and management of blue green algae. Brisbane: Australian Coal Association Research Program (ACARP). Report C16033.

Fastner J, Beulker C, Geiser B, Hoffmann A, Kröger R, Teske K et al. (2018). Fatal neurotoxicosis in dogs associated with tychoplanktic, anatoxin-a producing *Tychonema* sp. in mesotrophic Lake Tegel, Berlin. Toxins. 10:60.

Fitzgeorge RB, Clark SA, Keevil CW (1994). Routes of exposure. In: Codd GA, Jefferies TM, Keevil CW et al., editors: Detection methods for cyanobacterial toxins. Cambridge, UK: The Royal Society of Chemistry:69–74.

Funari E, Manganelli M, Buratti FM, Testai E (2017). Cyanobacteria blooms in water: Italian guidelines to assess and manage the risk associated to bathing and recreational activities. Sci Tot Environ. 598:867–880.

Gambaro A, Barbaro E, Zangrando R, Barbante C (2012). Simultaneous quantification of microcystins and nodularin in aerosol samples using high-performance liquid chromatography/negative electrospray ionization tandem mass spectrometry. Rapid Commun Mass Spectrom. 26:1497–1506.

Geh EN, Armah A, Ghosh D, Stelma G, Bernstein JA (2016). Sensitization of a child to Cyanobacteria after recreational swimming in a lake. J Allergy Clin Immunol. 137:1902–1904. e1903.

Geh EN, Ghosh D, McKell M, de la Cruz AA, Stelma G, Bernstein JA (2015). Identification of *Microcystis aeruginosa* peptides responsible for allergic sensitization and characterization of functional interactions between cyanobacterial toxins and immunogenic peptides. Environ Health Persp. 123:1159.

Giannuzzi L, Sedan D, Echenique R, Andrinolo D (2011). An acute case of intoxication with cyanobacteria and cyanotoxins in recreational water in Salto Grande Dam, Argentina. Mar Drugs. 9:2164–2175.

Grauer FH, Arnold HL (1961). Seaweed dermatitis: first report of dermatitis-producing marine algae. Arch Dermatol. 84:720–732.

Hashimoto Y, Kamiya H, Yamazato K, Nozawa K (1976). Occurrence of a toxic blue-green alga inducing skin dermatitis in Okinawa. In: Ohsaka A, Hayashi K, Sawai Y, editors: Animal, plant, and microbial toxins. New York: Plenum:333–338.

Hilborn E, Carmichael W, Soares R, Yuan M, Servaites J, Barton H et al. (2007). Serologic evaluation of human microcystin exposure. Environ Toxicol. 22:459–463.

Hilborn ED, Roberts VA, Backer L, DeConno E, Egan JS, Hyde JB et al. (2014). Algal bloom-associated disease outbreaks among users of freshwater lakes—United States, 2009–2010. Morb Mortal Wkly Rep. 63:11–15.

Hoff B, Thomson G, Graham K (2007). Neurotoxic cyanobacterium (blue-green alga) toxicosis in Ontario. Can Vet J. 48:147.

Humpage A, Falconer I, Bernard C, Froscio S, Fabbro L (2012). Toxicity of the cyanobacterium *Limnothrix* AC0243 to male Balb/c mice. Water Res. 46:1576–1583.

Ibelings BW, Backer LC, Kardinaal WEA, Chorus I (2014). Current approaches to cyanotoxin risk assessment and risk management around the globe. Harmful Algae. 40:63–74.

Lang-Yona N, Kunert AT, Vogel L, Kampf CJ, Bellinghausen I, Saloga J et al. (2018). Fresh water, marine and terrestrial cyanobacteria display distinct allergen characteristics. Sci Tot Environ. 612:767–774.

Lévesque B, Gervais M-C, Chevalier P, Gauvin D, Anassour-Laouan-Sidi E, Gingras S et al. (2014). Prospective study of acute health effects in relation to exposure to cyanobacteria. Sci Tot Environ. 466:397–403.

Li Y, Chen J-A, Zhao Q, Pu C, Qiu Z, Zhang R et al. (2011). A cross-sectional investigation of chronic exposure to microcystin in relationship to childhood liver damage in the Three Gorges Reservoir Region, China. Environ Health Persp. 119:1483.

Liu Q, Wang Y, Cao M, Pan T, Yang Y, Mao H et al. (2015). Anti-allergic activity of R-phycocyanin from *Porphyra haitanensis* in antigen-sensitized mice and mast cells. Int Immunopharmacol. 25:465–473.

Moore RE, Ohtani I, Moore BS, De Koning CB, Yoshida WY, Runnegar MTC et al. (1993). Cyanobacterial toxins. Gazz Chim Ital. 123:329–336.

Moreton Bay Regional Council (2018). Harmful algal bloom response plan. Caboolture, Queensland: Moreton Bay Regional Council. RIO Reference: A17098456. 10 pp. https://www.moretonbay.qld.gov.au/files/assets/public/services/environment/harmful-algal-bloom-response-plan.pdf.

Osborne NJ, Webb PM, Shaw GR (2001). The toxins of *Lyngbya majuscula* and their human and ecological health effects. Environ Int. 27:381–392.

Osborne NJT, Shaw GR, Webb PM (2007). Health effects of recreational exposure to Moreton Bay, Australia waters during a *Lyngbya majuscula* bloom. Environ Int. 27:309–314.

Petrus M, Culerrier R, Campistron M, Barre A, Rougé P (2009). First case report of spirulin anaphylaxis caused by the photosynthetic pigment phycocyanin. Allergy. 65:924–925.

Philipp R (1992). Health risks associated with recreational exposure to blue-green algae (cyanobacteria) when dinghy sailing. Health Hygiene. 13:110–114.

Philipp R, Bates A (1992). Health-risks assessment of dinghy sailing in Avon and exposure to cyanobacteria (blue-green algae). Water Environ J. 6:613–617.

Philipp R, Brown M, Bell R, Francis F (1992). Health risks associated with recreational exposure to blue-green algae (cyanobacteria) when windsurfing and fishing. Health Hygiene. 13:115–119.

Pilotto LS, Hobson P, Burch MD, Ranmuthugala G, Attewell R, Weightman W (2004). Acute skin irritant effects of cyanobacteria (blue-green algae) in healthy volunteers. Aust New Zeal J Public Health. 28:220–224.

Pilotto LS, Douglas RM, Burch MD, Cameron S, Beers M, Rouch GJ et al. (1997). Health effects of exposure to cyanobacteria (blue-green algae) during recreational water-related activities. Aust N Z J Public Health. 21:562–566.

Probert CS, Robinson RJ, Jayanthi V, Mayberry JF (1995). Microcystin hepatitis. Arq Gastroenterol. 32:199.

Puschner B, Pratt C, Tor ER (2010). Treatment and diagnosis of a dog with fulminant neurological deterioration due to anatoxin-a intoxication. J Vet Emerg Crit Care. 20:518–522.

Schwimmer D, Schwimmer M (1964) Algae and medicine. In: Jackson DF, editor: Algae and man. Boston (MA): Springer:368–412.

Stewart I, Carmichael WW, Backer LC, Fleming LE, Shaw GR (2011). Recreational exposure to cyanobacteria. In: Nriagu JO, editor: Encyclopedia of environmental health. Amsterdam: Elsevier:776–788.

Stewart I, Robertson IM, Webb PM, Schluter PJ, Shaw GR (2006a). Cutaneous hypersensitivity reactions to freshwater cyanobacteria–human volunteer studies. BMC Dermatol. 6:6.

Stewart I, Seawright AA, Schluter PJ, Shaw GR (2006b). Primary irritant and delayed-contact hypersensitivity reactions to the freshwater cyanobacterium *Cylindrospermopsis raciborskii* and its associated toxin cylindrospermopsin. BMC Dermatol 6:5.

Stewart I, Webb PM, Schluter PJ, Fleming LE, Burns JW, Gantar M et al. (2006c). Epidemiology of recreational exposure to freshwater cyanobacteria - an international prospective cohort study. BMC Public Health. 6:93.

Stewart I, Webb PM, Schluter PJ, Shaw GR (2006d). Recreational and occupational field exposure to freshwater cyanobacteria–a review of anecdotal and case reports, epidemiological studies and the challenges for epidemiologic assessment. Environ Health. 5:6.

Strasky Z, Zemankova L, Nemeckova I, Rathouska J, Wong RJ, Muchova L et al. (2013). *Spirulina platensis* and phycocyanobilin activate atheroprotective heme oxygenase-1: a possible implication for atherogenesis. Food & Funct. 4:1586–1594.

Turner PC, Gammie AJ, Hollinrake K, Codd GA (1990). Pneumonia asociated with contact with cyanobacteria. Brit Med J. 300:1440–1441.

Vidal F, Sedan D, D'Agostino D, Cavalieri ML, Mullen E, Parot Varela MM et al. (2017). Recreational exposure during algal bloom in Carrasco Beach, Uruguay: a liver failure case report. Toxins. 9:267.

Walker SR, Lund JC, Schumacher DG, Brakhage PA, McManus BC, Miller JD et al. (2008). Nebraska experience. Adv Exp Med Biol. 619:139–152.

Wood S, Williamson W (2012). New Zealand: regulation and management of cyanobacteria. In: Chorus I, editor: Current approaches to cyanotoxin risk assessment, risk management and regulations in different countries. Dessau: Umweltbundesamt:97–108.

Wood SA, Dietrich DR (2011). Quantitative assessment of aerosolized cyanobacterial toxins at two New Zealand lakes. J Environ Monitor. 13:1617–1624.

Wood SA, Puddick J, Fleming RC, Heussner AH (2017). Detection of anatoxin-producing *Phormidium* in a New Zealand farm pond and an associated dog death. New Zeal J Botany 55:36–46.

Wu H-L, Wang G-H, Xiang W-Z, Li T, He H (2016). Stability and antioxidant activity of food-grade phycocyanin isolated from *Spirulina platensis*. Inter J Food Properties. 19:2349–2362.

5.3 FOOD

Bastiaan W. Ibelings, Amanda Foss and Ingrid Chorus

Four chief sources of exposure to cyanotoxins through food for which data have been published include: (i) animals grown in aquaculture or harvested as food in brackish or freshwater containing cyanobacteria (for examples, see Table 5.5), (ii) so-called blue-green algal food supplements (BGAS, see section 5.5), (iii) food prepared using water contaminated with cyanotoxins (ineffectively treated or untreated) and (iv) crops irrigated with water from waterbodies with toxic blooms. Key mechanisms include toxin adsorption to the surface of plants or translocated to leaves and fruits after root uptake and trophic transfer to animals along food chains. Further sources for cyanotoxins and conceivable pathways into food for which, however, published data are largely lacking, include soil amended with sediment dredged from waterbodies with blooms and the use of algae, including cyanobacteria, as a cheap source of food for poultry or other farm animals.

5.3.1 General considerations on risk assessment and risk management

For assessing and managing health risks from food, the Codex Alimentarius provides the HACCP concept – Hazard Analysis Critical Control Points – which is very similar to the Water Safety Plan (WSP; see Chapter 6) approach. The WSP approach draws on many of the principles and concepts from other risk management approaches, including HACCP. Both approaches emphasise that monitoring the end product alone will not ensure safety. Rather, they focus on controlling the processes that are crucial for the safety of food or drinking-water. Both call on the managers and technical operators of a given facility to conduct a comprehensive analysis of the hazards that could occur in their system, to assess the human health risks they cause, to identify the key measures that are critical for safety ("*Control Measures*" in WSP terminology or "*Critical Control Points*", that is, CCPs in HACCP terminology), and to develop management plans to ensure that these measures are in place and properly functioning at all times.

Where the production of drinking-water and food draw on the same waterbody, the WSP for drinking-water and the HACCP plan for food may interface with respect to assessing and managing the waterbody and its catchment. Naturally, a close collaboration between the teams developing respective plans is desirable. This is important for risk assessment since exposure to cyanotoxins in water and food would add up. HACPP is being implemented in marine fisheries and shellfish harvesting, and it is interesting to note that even here, in the marine environment, one argument

Table 5.5 Examples of cyanotoxin concentrations in foods

A	*B*	*C*	*D*	*E*	*F*	*G*	*H*
Organism	*Organs or tissue*	*Cyanotoxin*	*Content [µg/kg ww]*	*Max. [µg] ingested with serving of 0.1 kg*	*% of TI for DW short term (MC; CYN) or acute (STX) exposure*	*Study type*	*Reference*
Crops							
Lactuca sativa	Leaf	MC-LR	2.4–147	14.7	61	Lab	Cordeiro-Araújo et al. (2017)
Lactuca sativa	Unknown	MC-YR, MC-RR	ND–108	10.8	45	Field	Li et al. (2014)
Ipomoea aquatica	Unknown	MC-RR, MC-LR	ND–68	6.8	29	Field	Li et al. (2014)
Brassica oleracea	Unknown	MC-RR	ND–20	2.0	9	Field	Li et al. (2014)
Brassica rapa var. *Parachinensis*	Unknown	MC-RR	ND–40	4.0	17	Field	Li et al. (2014)
Brassica oleracea var. Sabellica	Leaf	CYN	3.1	0.3	5	Lab	Kittler et al. (2012)
Brassica juncea	Leaf	CYN	4.0	0.4	7	Lab	Kittler et al. (2012)
Lactuca sativa	Leaf	CYN	3.1–8.2	0.8	13	Lab	Cordeiro-Araújo et al. (2017)
Eruca sativa	Leaf	CYN	5.5–11.5	1.1	18	Lab	Cordeiro-Araújo et al. (2017)
Spinacea oleracea	Leaf	CYN	9.5–120	12.0	200	Lab	Llana-Ruiz-Cabelo et al. (2019)[a]
Lactuca sativa	Leaf	CYN	2.4–42	4.2	70	Lab	Llana-Ruiz-Cabello et al. (2019)[a]

(*Continued*)

Table 5.5 (Continued) Examples of cyanotoxin concentrations in foods

A	*B*	*C*	*D*	*E*	*F*	*G*	*H*
Organism	*Organs or tissue*	*Cyanotoxin*	*Content [µg/kg ww]*	*Max. [µg] ingested with serving of 0.1 kg*	*% of TI for DW short term (MC; CYN) or acute (STX) exposure*	*Study type*	*Reference*
Molluscs							
Mytilus galloprovincialis	Whole	dmMC-RR[b]	ND–39	3.9	16	Field	Rita et al. (2014)
Patinopecten yessoensis	Whole	MC-LR	ND–4.3	0.43	2	Field	Cui et al. (2018)
Crassostrea virginica	Whole	MC-RR, MC-LR	ND–9.8	1.0	4	Field	Cui et al. (2018)
Mytilus galloprovincialis	Whole	CYN	28.1–41.6	4.2	70	Lab	Freitas et al. (2016)
Anodonta cygnea	Whole	CYN	247	24.7	412	Lab	Saker et al. (2004)
Anodonta cygnea	Whole	$STX_{t.eq}$	160–220	22	73	Lab	Pereira et al. (2004)
Alathyria condola	Muscle	$STX_{t.eq}$	144–179	18	60	Lab	Negri & Jones (1995)
Crustaceans							
Astacus astacus	Head/ thorax	[Asp^3,Dhb^7]MC-LR, [Asp^3]MC-RR, [Asp^3,Dhb^7]MC-RR	10	1.0	4	Field	Miles et al. (2013)
Astacus astacus	Tail	[Asp^3,Dhb^7]MC-LR, [Asp^3]MC-RR, [Asp^3,Dhb^7]MC-RR	<1	<0.1	<1	Field	Miles et al. (2013)
Procambarus clarkii	Abdomen	MC-RR, MC-LR	1.4–17.1	1.7	7	Field	Ríos et al. (2013)
Cherax quadricarinatus	Muscle	CYN	205	20.5	342	Field	Saker & Eaglesham (1999)

(Continued)

Table 5.5 (Continued) Examples of cyanotoxin concentrations in foods

A	*B*	*C*	*D*	*E*	*F*	*G*	*H*
Organism	*Organs or tissue*	*Cyanotoxin*	*Content [µg/kg ww]*	*Max. [µg] ingested with serving of 0.1 kg*	*% of TI for DW short term (MC; CYN) or acute (STX) exposure*	*Study type*	*Reference*
Fish							
Aristichthys nobilis	Muscle	MC-RR, MC-YR, MC-LR	177[2]	17.7	74	Field	Chen et al. (2007)
Hypophthalmichthys molitrix	Muscle	MC-RR	0.4[2]	0.0	0.2	Field	Chen et al. (2009b)
Hypophthalmichthys molitrix	Muscle	MC-RR, MC-LR	ND–249[c] (mean: 39.4)	24.9	104	Field	Chen et al. (2006)
Pomoxis nigromaculatus	Muscle	MC-LR	ND–70	7.0	29	Field	Schmidt et al. (2013)
Cyprinus carpio	Muscle	MC-LR	ND–4	0.4	2	Fleld	Schmidt et al. (2013)
Cyprinus carpio	Muscle	MC-RR	0.6[b]	0.1	0.3	Field	Chen et al. (2009b)
Oreochromis niloticus	Muscle	MC-LR[b]	4.2–5.2	0.5	2	Field	Greer et al. (2017)
Geophagus brasiliensis	Muscle	$STX_{t\text{-}eq}$	12–20	2.0	7	Field	Clemente et al. (2010)

For contextualising the health risk due to these concentrations, column "F" relates to the amount of toxin ingested with a serving of 0.1 kg to the percentage of the *tolerable intake for short-term (MCs, CYN) or acute (STX) exposure* (TI calculated from the NOAEL, UF and bodyweight of 60 kg as given in Chapter 2). Note that these short-term GVs were derived assuming drinking-water to be the major source of exposure to these cyanotoxins, not leaving a proportion to other sources, and that this comparison is intended merely to give a rough estimate of the health relevance of concentrations found in food, not for defining safe levels for food. This table does not take into account the relative toxicities of microcystin congeners.

ND = not detected above the detection limit. $STX_{t,eq}$ = saxitoxin toxicity equivalents.

[a] When applied with equal amounts of MC; without MC accumulation was fourfold lower.
[b] Secondary technique (i.e., MMPB, ELISA) indicated higher levels may have been present.
[c] Converted to wet weight using a wet-to-dry weight ratio of 5.

for implementing HACPP is the risk of microcystins present in mussels (Tzouros & Arvanitoyannis, 2000). HACPP in the seafood industry, as elsewhere, is based upon seven principles: (i) hazard analysis, (ii) identification of the critical points in the process, (iii) establishment of critical limits, (iv) requirements for CCP monitoring, (v) corrective actions, (vi) record keeping procedures and (vii) verification. Implementation of HCAPP based upon these principles in freshwater fisheries and harvesting greatly enhances the protection of the consumers.

5.3.2 Sources of exposure

Both plants and animals have shown highly variable accumulation of cyanotoxins. Table 5.5 shows results of concentrations measured in organisms collected in the field, or exposed experimentally in the laboratory, focusing on experiments using concentrations in a realistic range. To enable an estimate of the health implications of these concentrations, in face of the lack of guideline values (GVs) for concentrations in foods, Table 5.5 relates the exposure from a serving of 0.1 kg of these foods to the tolerable intake (calculated from the no observed adverse effect level (NOAEL), UF and bodyweight of 60 kg as given in Chapter 2) for short-term (MCs, CYNs) or acute (STXs) exposure. The data in Table 5.5 show that while in many cases concentrations in foods are low, some field observations and laboratory experiments found concentrations that would lead to a dose in the range of – or above – that which would be acceptable for up to 2 weeks for an adult consuming 2 L of drinking-water per day, using the short-term WHO GVs for drinking-water. Trends that can be discerned are that consuming molluscs and crustaceans collected from environments with blooms might cause higher risks, particularly because they are eaten with the viscera which can contain large amounts of toxic cyanobacteria. In contrast, the edible portions of higher trophic-level organisms (e.g., muscle tissue of fish), excluding viscera, have less chance of containing a large amount of free toxin. Livestock reports are, however, based on very few animals, rendering results uncertain.

A key problem in using published literature is the uncertainty of many results: The extensive literature survey on cyanobacterial toxins in food by Testai et al. (2016) concluded that the majority of publications had significant flaws in toxin extraction, sample cleanup and/or the analytical methods which undermine the confidence in the data on toxin levels in food. These impair the quality of the analytical data, due to inefficient extraction, poor quality controls and downstream matrix effects resulting in a loss of sensitivity and inaccurate quantification, as well as missing reporting of how quantification was achieved (Testai et al., 2016). For accumulation in plants, information on how toxins were applied, that is, via the soil (enabling only root uptake) or irrigated also on leaves (thus possibly adhering to leaf surfaces), is important but often not clearly described. A further challenge is accounting for metabolised or protein-bound toxins, as it is yet unclear whether they represent a potential reservoir of toxin that may be released in the gut. Strategies to account for metabolised or protein bound toxins, such as Lemieux oxidation or protein deconjugation techniques, require careful calibration and intimate knowledge of analyte chemistry to avoid producing data that may lead to overestimating the hazard. Dionisio Pires et al. (2004) used Lemieux oxidation to extract microcystins from mussels and found that the bound fraction was always smaller than free microcystin. They contrast this with earlier mussel studies which reported a 10 000-fold in the bound fraction compared to the free fraction. Also, beyond experimental data, more data on levels of cyanotoxins found in food items in markets are needed in order to assess actual rather than perceived risk, with testing geared towards capturing the "total" cyanotoxin pool in order to remain conservative.

5.3.2.1 Microcystins

Information on microcystins detected in crops is limited, with a few studies conducted in the laboratory showing accumulation via uptake through roots and/or leaves (Table 5.5). None of these data indicate a dose substantially above that which can be tolerated from drinking-water for up to 2 weeks unless a serving size significantly above 100 g is assumed. Interestingly, one study did not result in detectable microcystin in spinach and lettuce even though it confirmed cylindrospermopsin uptake (Llana-Ruiz-Cabello et al., 2019) and another resulted in detections in lettuce leaves but not in arugula (Cordeiro-Araújo et al., 2017). These results indicate that other factors (e.g., plant variety, physiology, morphology) influence accumulation and require consideration when monitoring. Only one field study was identified confirming microcystins in crop irrigated with water (Dianchi Lake, China; Li et al. (2014). None of these studies addressed bound microcystin (e.g., conjugated) content in crops.

Among the studies with mussels or crustaceans assessed by Testai et al. (2016) as having been performed with reliable methods, sufficiently comprehensively reported, maximum concentrations in wet weight ranged up to 3400 (±1000) µg/kg for *Mytilus edulis* and up to 329±95 µg/kg for crayfish, while other authors found only low concentrations of MCs (Table 5.5). For fish muscle, the review by Testai et al. (2016) includes a study that found up to 2860 µg/kg in silver carp from China and one with up to 340 µg/kg for *Odontesthes bonariensis* from Argentina. However, most analyses targeting specific microcystin congeners via LC-MS/MS have rarely found MCs in fish muscle, even though they reported microcystins in the source water and other organs (e.g., liver; Kohoutek et al., 2010; Hardy et al., 2015), and if MCs were found, in most cases, concentrations were low. This is likely due to rapid microcystin elimination as well as the remaining fractions in muscle tissue being unextractable (bound to proteins) or, to a lesser degree, extractable but inactivated by metabolism (e.g., conjugated with thiols; Williams et al., 1997a; Williams et al., 1997b). While extractable conjugated microcystins are detectable with ELISA and MMPB ((2S, 3R)-2-methyl-3-methoxy-4-phenylbutanoic acid), which is also able to detect protein bound fractions), many other methods are too specific to detect these fractions. This partially explains why methods targeting free microcystins (i.e., not bound or degraded) such as LC-MS/MS tend to report lower values than ELISA and MMPB (Li et al., 2014; Foss et al., 2017; Greer et al., 2017).

Studies reporting high microcystin levels in fish muscle (>12 µg/kg wet weight) frequently employed ELISA (Freitas de Magalhães et al., 2001; Berry et al., 2011; Poste et al., 2011). Not only do some of these assays react with microcystin conjugates, but they are also prone to nonspecific binding resulting in overestimation (Hardy et al., 2015; Foss et al., 2017). This could be considered a welcome conservative approach for the case that protein-bound microcystins become released (Smith et al., 2010) and/or metabolised MCs become deconjugated (Miles et al., 2016), regaining some toxicity. However, little is known about these processes, and undue restrictions of food use may also impair health (see section 5.3.4). Research addressing these potential reservoirs, geared to resolving the disparity between fractions of total MC burden (bound, free and conjugated), is therefore important, particularly for molluscs and crustaceans which are a relevant protein source in some regions.

Testai et al. (2016) assess the few results available for livestock as insufficient because they rely on a very limited number of animals and insufficient analytical method that does not include bound MCs; thus, the possible transfer of MCs to the milk or meat cannot be assessed. Chen et al. (2009b) found low levels of MC-RR, MC-YR, MC-LR in muscle tissue of the common duck (*Anus platyrhynchos*) and Chinese softshell turtle (*Pelodiscus sinensis*), that is, 3 and 0.6 µg/kg ww, respectively.

In spite of the analytical limitations of many studies, a general trend in literature indicates higher microcystin content in liver and viscera than in muscle. Further, the available evidence supports microcystins to be biodiluted rather than biomagnified in the aquatic foodweb (Ibelings et al., 2005; Ibelings & Havens, 2008), indicating that properly cleaning meat in higher trophic level animals reduces the risk of exposure to microcystins.

5.3.2.2 Cylindrospermopsin

As for microcystins, plant studies have addressed the uptake of cylindrospermopsin into leafy vegetables such as lettuce, arugula and mustard, sometimes at higher levels than microcystins (Table 5.5), with one study finding CYN in spinach at levels higher than other leafy greens, possibly resulting in exposure at levels relevant to health. Llana-Ruiz-Cabello et al. (2019) found high concentrations in spinach and lettuce, but only when applying CYN in concentrations of 25 µg/L together with 25 µg/L of MCs; applying 25 µg/L CYN alone resulted in fourfold lower concentrations in the plant material. However, such levels have not been reported from field studies or market acquired vegetables. In animals, CYN has been studied less than MCs, with accumulation reported from bivalves and crustaceans, in one case at levels potentially relevant to health (Table 5.5). For fish, studies with sufficiently selective methods (e.g., LC-MS/MS) are largely lacking; the review by Testai et al. (2016) includes three studies that found no or only very low concentrations in fish.

5.3.2.3 Saxitoxins

Saxitoxins (STXs) in food products are well documented in the marine environment, including numerous cases of human illness and death. To date, there have not been any reports of paralytic shellfish poisoning caused by freshwater cyanobacteria even though STX accumulation in freshwater mussels has been demonstrated (Negri & Jones, 1995). Freshwater fish, *Oreochromis niloticus* and *Geophagus brasiliensis*, were found to accumulate STXs from the environment, but not in concentrations that would lead to exposure in a health-relevant range (Table 5.5; Galvão et al., 2009). Testai et al. (2016) include one study finding up to 30.6±14.5 µg/kg of PSP toxins in Cichlidae. Interestingly, intraperitoneal dosing of the tropical freshwater fish *Hoplias malabaricus* four times with STX at 800 µg/kg did not result in accumulation in muscle tissue (da Silva et al., 2011).

5.3.2.4 Anatoxins

Very little is known regarding the accumulation of anatoxin-a and/or homoanatoxin-a, with studies lacking on crops or invertebrates. One study has shown anatoxin-a to bioaccumulate in fish (Osswald et al., 2011), but others have shown it to rapidly eliminate from fish and mussels (Osswald et al., 2008; Colas et al., 2020).

5.3.2.5 Conclusions on exposure via food

In summary, as preliminary assessment considerering all types of cyanobacterial toxins, the data available by 2019 do not point to a high level of short-term exposure to cyanotoxins in crops or muscle tissue of fish and crayfish, whereas exposure may be more significant if viscera are eaten, as is the case for small fish, crustaceans and mussels. If for instance crops are sprayed or irrigated with lake water containing scums or high levels of cyanotoxins and in particular if foods are not sufficiently washed or prepared, risks may be higher. However, data obtained with reliable methods are insufficient for drawing clear conclusions.

Where crop irrigation with scum material is widespread (as described, e.g., in Li et al., 2014) or fish, mussels and crayfish from bloom-ridden waterbodies constitute staple foods, screening cyanotoxin concentrations in such foods is recommended, with attention to the methodological requirements described in section 5.3.4.

Hazard analysis for any of the above settings may indicate that when cyanotoxins in foods cannot be excluded because of – substantial – cyanobacterial blooms in the waterbody used for the production of the food, a more detailed analysis becomes important. Checklist 5.2 provides guidance for conducting such an analysis.

CHECKLIST 5.2 FOR ASSESSING THE RISK OF CYANOTOXIN EXPOSURE THROUGH FOOD

1. Are blooms of potentially toxic cyanobacteria present in the waterbodies used for collecting, producing or preparing food (see Chapters 4 and 8)?
   - 1.1. Inspect these waterbodies to collect information on the presence of surface blooms or scums, strong greenish discoloration and turbidity.
   - 1.2. Collect samples for species identification and quantification, particularly if these observations indicate cyanobacteria could be present.
   - 1.3. Particularly if potentially toxic cyanobacteria are found and if feasible, have toxin content of the cells and bloom analysed (see point 2.3).
   - 1.4. If cyanotoxins are present currently or were present during the previous month, further risk analysis in food becomes relevant. Clarify the time pattern of toxin occurrence – is it sporadic for a few days, or continuous for many weeks or months?
2. Are organisms (e.g., fish, shellfish, snails, bivalves) harvested for food from the impacted waterbodies? If so,
   - 2.1. Find out whether these species are likely to filter-feed particles, including cyanobacteria, and whether they have been reported to contain cyanotoxins.
   - 2.2. Find out whether viscera and gonads are removed prior to consumption or whether the organisms are consumed whole.
   - 2.3. Check whether analyses of their cyanotoxin content are feasible, and if so, together with experts derive a plan for sampling and analyses.
3. Are crops irrigated with water containing high amounts of cyanobacteria?
   - 3.1. If so, check whether the use of alternative water sources, free of blooms, is feasible or run a programme of sampling and analyses to assess whether the practices used lead to cyanotoxins in the crop.

3.2. Investigate whether substantial amounts of cells cling to the surface of fruits or vegetables which are potentially consumed without sufficient treatment to remove them.

4. Are soils augmented with sediment dredged from systems containing high amounts of cyanobacteria? If so, check dredged material for cyanotoxins and – depending on the results – also the crop.
5. Find out whether other exposure pathways to these cyanotoxins are likely (drinking-water or recreation)? If so, estimate the dose from these and determine the proportion from food which is most appropriate for your setting.
6. Estimate the contribution of the affected foods to the local diet and the time spans of their contamination with cyanotoxins.

6.1. Is it consumed seasonally or year-round? On a daily basis, or occasionally? Are exposure patterns likely to be short term and occasional (justifying assessing exposure in relation to a short-term tolerable daily intake, TDI) or more likely to be continuous for many weeks on end and several times a week (necessitating application of a TDI for chronic lifetime exposure)?

6.2. Estimate the amounts consumed and the impact of local traditions for collecting and preparing these foods on exposure pathways.

7. Clarify the tolerable cyanotoxin dose from food in the local setting together with toxicologists, taking points 5 and 6 into account. Note that in deriving the WHO guideline values for chronic exposure via drinking-water, WHO apportioned 20% of intake to other sources, including food, while the short-term values are based on exposure only to drinking-water. As discussed above (see point 5 of this checklist), this apportionment may need to be adjusted locally, depending on other exposure routes and the contribution of foods containing cyanotoxins to the local diet.

7.1. From the results of local analyses and/or published data on the potential toxin content of these foods (see Table 5.5 and section 5.3.2) and the dose found tolerable for food in your setting, estimate how likely the cyanotoxin contents in the edible parts of these organisms are to exceed that tolerable dose and by how much.

7.2. If restricting access to fish, mussels and shellfish is considered, what are the consequences for overall local diet? Are suitable alternatives available, accessible and accepted?

7.3. If restricting access to fish, mussels and shellfish is considered and access to alternative protein food sources is poor or in question, how high is the uncertainty of the information base on cyanotoxin

5.3.3 Assessing and managing exposure via food

Cyanobacterial metabolites may also cause a musty or earthy taste of fish ("tainting"; see section 2.9). While this is a mere quality issue with no direct health relevance, it does indicate that cyanobacteria – and thus cyanotoxins – may be present. This may serve as a warning signal, but it is not a reliable one: cyanotoxins may well occur without the presence of taste-and-odour compounds, and other organisms such as Actinomycetes may also cause tainting. Therefore, the absence of a musty taste is not a reliable indicator of the safety from cyanotoxins.

The use of waterbodies for aquaculture or fisheries usually is not the primary cause of excessive nutrient concentrations leading to cyanobacterial blooms and cyanotoxin occurrence. However, these activities may augment nutrient loading to the waterbody they use, particularly where aquaculture or fisheries are intensive (Rickert et al., 2016). Flow-through aquaculture systems may drain into the waterbody which they also tap to feed their basins or ponds, thus contributing to the waterbody's nutrient load. Fisheries may involve fertilising ponds and lakes (including with manure, organic wastes or agricultural byproducts) in order to augment fish production. Feeding may significantly contribute to the nutrient load to the waterbody, thus enhancing cyanobacterial blooms. Cage culture ("net-pen") systems rear animals in cages or nets floating within the waterbody, thus adding feed directly into the waterbody.

For commercial food production, control measures can be taken in planning, design and during operation (Table 5.6). In planning, they may involve land-use and waterbody management to avoid cyanobacterial proliferation (see Chapters 7 and 8), or, where this is not sufficiently successful, (re)locating aquaculture to sites where cyanotoxin levels are low. Where fisheries or aquaculture are a major cause of eutrophication, permits limiting size of stock and amounts of feeding may be appropriate in order to control eutrophication of the waterbody. Fish rearing systems may be designed to recirculate the used water back to the fish rearing unit through a treatment system which removes nutrients (and other harmful substances such as antibiotics). Control measures for the operation of fish rearing systems include regular removal of sludge from basins, ponds and water treatment units in order to remove nutrients which otherwise would supply cyanobacterial growth (see also Rickert et al., 2016).

Where food production is continued even though cyanobacteria occur, further control measures may be required in order to keep toxin concentrations in food below hazardous levels. Typically, they involve public information and creating awareness, particularly for subsistence fisheries. Keeping the live animals in clear water for a depuration period of a few days may be a Critical Control Point in a producer's HACCP plan. Depuration of microcystins in various marine and freshwater mussels has been shown to occur within days to a few weeks – although small amounts may remain for periods longer than this (Dionisio Pires et al., 2004). Thus, possibly a couple of weeks after blooms have disappeared, eating shellfish may be safe again, but the fragmented knowledge about depuration does not allow this to be generally assumed, and it is therefore important to verify that concentrations are safe with appropriate analytical techniques.

Another control measure may be to remove the body parts of the animals which contain high cyanotoxin concentrations, that is, viscera and liver of fish or the guts and hepatopancreas of crayfish and mussels, before they are sold on the market, or to inform consumers of the need to do so. This control measure, however, cannot always be applied, so that some animals will be eaten whole (e.g., bivalves, snails, small fish such as smelt). In such cases, where cyanobacteria occur seasonally, harvesting animals can be restricted to seasons with low cyanobacterial occurrence, and operational monitoring will check that they are not marketed during these seasons. Where seasonal patterns are less reliable, harvesting may be restricted when simple indicators show that levels of cyanobacteria – or cyanotoxins – have exceeded a predefined limit. Monitoring of cyanotoxin concentrations in the animals harvested may also be an option to control exposure, as is the case in Victoria, Australia, where authorities advise to refrain from consumption when concentrations exceed Alert Levels (Van Buynder et al., 2001; Saker et al., 2004).

Not many countries have regulated cyanotoxins in food from freshwaters and between those that do, values vary considerably: For MCs, five authorities give values ranging from 5.6 μg/kg for fish (France) to 51 μg/kg for molluscs; for CYN, three authorities give values from 18 μg/kg (two states in Australia) to 70 μg/kg (California, USA); for STX in fish, prawns and shellfish, the value of 800 μg/kg is the same in these two Australian states and in Canada and for ATX, only California, USA, gives a value set at 5000 μg/kg for fish (see Table 3.1.3.3 in Testai et al., 2016). Choices of control measures can be optimised depending on the conditions in the specific aquatic setting that lead to blooms, the local patterns of consumption potentially leading to exposure and the available institutional capacity for operational monitoring.

Table 5.6 shows some examples of measures to control cyanotoxin levels in food collected or farmed in waterbodies. It is important to emphasise that cooking (e.g., boiling, frying, microwave) offers no reliable protection against cyanobacterial toxins in food. Contrasting results published for MC vary from a decrease to no effect or even an increase after cooking (Testai et al., 2016). Table 5.6 also gives options for monitoring to ensure that the intended control measures are being implemented and that they are functioning during day-to-day operations, as required both for HACCP and Water Safety Plan (WSP). Further information can be found in Rees et al. (2010).

Table 5.6 Examples of control measures for the commercial production of fish, crayfish and mussels and of options for monitoring their implementation and functioning

Process step	***Examples** of control measures for food*	*Options for their operational monitoring and/or verification 3.4*
Planning	For measures to control cyanobacteria through catchment management, land-use planning and waterbody management, see Chapters 7–9	
	Designate sites with low levels of cyanobacteria for harvesting and/or farming aquatic organisms as well as for abstracting water for irrigation	Conduct periodic site inspections during the cyanobacterial growing season
	Require permits for location, design and operation of aquatic farming operations (e.g., net-pens) and fish stocking	Review (application for) permit with respect to adequacy of choice of site, planning and operation
	Plan intensive land-based aquaculture systems with treatment of the outflow (e.g., in a wetland) to avoid eutrophication	Inspect outflow and check for illicit direct flow to the waterbody
	Plan irrigation schemes to avoid direct contact between water containing cyanobacteria and the crop to be consumed	Review plans
Design, Construction and Maintenance	Design aquaculture as closed recirculation system with treatment, aeration, sustainable stocking rates and controlled feeding rates	Conduct visual site inspection; review management plan for stocking and feeding rates (require development, if nonexistent)
	Avoid discharge of untreated effluent – treat it or use it as liquid fertiliser on crops	Monitor effluent flow; review information about its designation
	Construct and maintain particle traps in tanks (with separate sludge outlet) and collect waste from cages	Inspect structures; require records of waste collection and review them regularly
	Design irrigation systems as drip or ditch systems without direct crop contact; abstract water for irrigation outside of scum areas and depths as discussed in Chapter 9	Inspect abstraction points for irrigation water

(*Continued*)

Table 5.6 (Continued) Examples of control measures for the commercial production of fish, crayfish and mussels and of options for monitoring their implementation and functioning

Process step	***Examples*** *of control measures for food*	*Options for their operational monitoring and/or verification 3.4*
Operation	For cultures of aquatic organisms (e.g., net-pens), limit stock density and feeding to levels not likely to enhance eutrophication and thus cyanobacterial development	Inspect sites and enterprises for compliance with permits, for example, farm records for fish stock and food application
	Use low-polluting feed, high levels of lipid, lowered protein content, typically with high digestibility value, low in phosphorus	Inspect feed used; discuss criteria for its choice with operators
	If manure, fertilisers or wastewater are applied, base amounts on nutrient budget and optimise application times in relation to animal demand	Inspect materials applied; discuss practices with operator; if available, inspect records of application
	Remove viscera, liver or guts from organisms before marketing	Inspect products marketed
	Allow for depuration times of animals after exposure to toxin-containing cyanobacteria	Inspect enterprises for availability and functioning of facilities for depuration in clear water and for records of their use
	Restrict food collection during specific seasons for which cyanotoxin contamination is known and/or when Alert Levels (for cyanobacteria or for cyanotoxins) are exceeded	Monitor compliance with seasonal marketing restrictions or an indicator of cyanobacterial biomass or cyanotoxins in the water

Source: Modified from Rickert et al. (2016).

Which of these control measures – or others – are to be implemented in a given setting needs to be determined locally, depending on the specific natural and socioeconomic conditions. Implementation is most effective if the stakeholders involved collaboratively develop their specific management plans (e.g., WSP or HACCP Plans or a combination of both) in which they define the control measures and how their performance is to be monitored, as well as responsibilities, lines of communication and documentation requirements. For situations in which operational monitoring shows that a control measure is not operating adequately (i.e., within its predefined limits), management plans should include a description of the corrective action to take. Note that the options for monitoring suggested in Table 5.6 focus on the functioning of the control measures rather than on cyanotoxin levels in food.

5.3.4 Verification monitoring of cyanotoxin levels in food from aquatic systems versus operational monitoring

As mentioned above, monitoring cyanotoxin levels in food from aquatic systems is most useful for risk assessment, that is, to inform planning and to adapt management strategies in the medium to longer term. It is also valuable for verifying whether the whole set of control measures implemented in a given situation is meeting its target. As discussed above, some countries have regulated cyanotoxin concentrations in food (Testai et al., 2016) which trigger immediate action: for example, food exceeding them may be banned from further marketing. Such consequences of violating limits may be useful to enforce improved control measures. However, the basis for day-to-day management is operational monitoring that checks the functioning of the control measures by methods such as regular inspection, where HACCP is implemented, in the context of HACCP management plans. This allows quick responses if it shows a measure not to be functioning within its boundaries, and many operational monitoring approaches are possible at low costs.

As discussed above, monitoring food products for cyanotoxin levels is more challenging than water, with most analytical techniques compromised by matrix (see Testai et al. (2016) for an extensive literature survey describing the chemico-analytical and biological methods available for sample preparation and detection in detail). Although readily available and easy to use, the ELISA format is inadequate for food testing without, at minimum, proper cleanup, quality controls (e.g., spiking) and confirmatory testing, particularly for commercially available ELISAs specifically intended to be used for water testing. ELISA should be considered a screening tool, requiring confirmation of identity, and quantity, with strategies employed to address extraction efficiency and bound analytes. Spiking subsets of material prior to extraction allows for an assessment of extraction efficiency. Adding analyte to aliquots immediately prior to testing (after extraction) will help determine if the extract matrix causes inhibition or nonspecific binding. If sample cleanup using solid-phase extraction and/or liquid–liquid extraction does not prevent matrix effects, a dilution series can be employed to assess such effects (although dilution can compromise the detection limit). When using ELISAs, the following need to be addressed, at minimum:

- inhibition resulting in underestimation and false negatives;
- nonspecific binding of matrix components to antibodies or antigen resulting in overestimations and false positives;

- varying reactivity to nontarget (but related) analytes, such as degradation products and metabolites resulting in overestimation;
- varying reactivity to target analytes of similar structure (e.g., microcystin congeners) resulting in overestimation or underestimation.

Other methods for the analysis of cyanotoxins (discussed in more detail in Chapter 14) include liquid chromatography (HPLC, LC, UPLC) coupled with various detectors, such as photodiode array/ultraviolet (PDA; UV), mass spectrometer (MS) or fluorescence detector (FL). LC-UV has been employed for microcystin (223–238 nm), cylindrospermopsin (262 nm) and anatoxin-a (227 nm), but detection limits may be insufficient, and if identification is based solely on peak retention time (without another in-line detector such as MS), this increases the chance of misidentification in complex matrices such as food. Higher interferences from matrix also hinder quantification, making LC-UV techniques inadequate for monitoring most food items. Even the use of single-quadrupole mass spectrometry (LC-MS) is prone to over-reporting microcystin in matrices such as fish tissues (Kohoutek et al., 2010). Highly specific LC-MS/MS methods are useful for the analysis of complicated matrices, which, with proper calibrations, are recommended or anatoxin-a/homoanatoxin-a and cylindrospermopsin. It is more difficult to fully account for all saxitoxins (STXs) (>57 analogues) using a targeted LC-MS/MS approach, with the sum of toxins detected possibly underrepresenting totals, although improvements to this analysis have been made (Turner et al., 2019). Therefore, until methods have been adequately developed to address STXs in freshwater-related food contamination, it is recommended that accepted methods for monitoring PSP in shellfish be used (e.g., Lawrence et al., 2005; AOAC, 2011a; b). Similar to STX, as discussed in Chapter 14, targeting microcystin congeners with MS is limited by the availability of standard reference material, unless the water source has been thoroughly characterised and the microcystin congeners are known. In order to assess fractions bound to either proteins or thiols, thiol-deconjugation (Miles et al., 2016) or the MMPB technique (see above) can be used. However, careful calibration for the MMPB method requires preoxidation spiking with intact microcystin to properly account for oxidation efficiency and recovery, increasing the time needed for preparation and analysis.

In summary, while monitoring cyanotoxins with relatively few structures (e.g., cylindrospermopsin) can be easily achieved, monitoring the microcystins and saxitoxins is significantly more complicated. A useful approach is to screen for food items with, for example, ELISA (which are available for most cyanotoxins) and if this indicates levels of concern in food items and quality assurance controls indicate the test is performing properly, to confirm the data with a further test. These can be related to toxicity (e.g., receptor-binding assay or protein phosphatase inhibition assay; see Chapter 14) or more targeted analyses of cyanotoxins (e.g., LC-MS/MS). Appropriate calibration standards (including internal standards) should be utilised, with certified reference materials used where available.

5.3.5 Balancing cyanotoxin risks against the risk of malnutrition

A critically important public health aspect when deciding which control measures to implement is their possible impact on the nutritional status of the population which may depend on fisheries and/or aquaculture as key protein source. This needs to be balanced against the health risks through cyanotoxins possibly contained in these foods. Restriction may well prove scarcely feasible where intensive aquafarming or angling is needed as basis for the population's protein supply. On the other hand, the published information on cyanotoxin concentrations in food (Table 5.5) indicates a fair likelihood that these may well be below concentrations of potential concern, although in exceptional cases like fishermen on Lake Chaohu exposure clearly is likely to be elevated (Chen et al., 2009a). In consequence, before taking measures with a potentially major impact on peoples' livelihoods and nutritional status, it may be worthwhile to assess the relevance of such foods as staple protein source for a population and to invest in a survey to sample and analyse cyanotoxin concentrations in the local produce from fisheries and aquaculture in order to avoid undue restrictions causing more harm than good. A critical issue to consider here is that cyanotoxin concentrations in food produce varies greatly between points in time. A further point to consider, albeit challenging, is the risk of exposure to multiple toxins as well as to multiple sources, that is, exposure via food augmented by toxins in drinking-water and/or recreational use of waterbodies with blooms.

5.3.6 Public awareness and information

For small-scale commercial and particularly for recreational, noncommercial angling and harvesting of invertebrates from aquatic systems, effective controls are difficult to implement, and creating public awareness of potential risks may be a more effective or the only feasible approach. In contrast to cyanotoxins in freshwater, for the marine environment, public awareness of "algal toxins" is well developed in many regions: for example, native Americans already warned early settlers in the USA not to eat shellfish in the summer months. Today, among tourists or other non-natives, marine bivalves cause disproportionately high numbers of cases of paralytic shellfish poisoning, and this is attributed to tourists' disregard for either official quarantines or traditions of safe consumption, both of which tend to protect the local population (see Ibelings & Chorus, 2007). Many of the states in the USA and Australia and countries in Europe host hotlines with information for shellfish collectors. South Australia classifies collecting sites for shellfish in four categories: approved, conditionally approved, restricted and fully restricted. This approach is familiar from other contaminants: for example, banning fishing in certain waterbodies to avoid consumption of pathogen-contaminated or of mercury-contaminated fish. A further option is issuing quantitative advisories on the amount that may be safely consumed or the frequencies at which fish may be eaten (e.g., US EPA, 2017).

Public awareness approaches that have been successful for seafood from marine environments can be similarly applied to cyanotoxin risks from freshwater environments, from collecting shellfish and snails or catching fish where water is visibly greenish or covered by scums. Information campaigns successful elsewhere are best adapted locally or regionally, since the type of food varies greatly between different geographic regions. Information particularly needs to reach specifically sensitive subpopulations, for example, in the case of cyanobacterial hepatotoxins persons with chronic hepatitis or other liver disorders. Also, information campaigns about using food from waterbodies with cyanobacteria may be effectively combined with information on their recreational use. See Chapter 15 for more information on public communication and participation targeting toxic cyanobacteria.

REFERENCES

AOAC (2011a). Official method 2011.02 determination of paralytic shellfish poisoning toxins in mussels, clams, oysters and scallops. Rockville (MD): Association of Official Analytical Chemists International. http://www.eoma.aoac.org/methods/.

AOAC (2011b). Official Method 2011.27: Paralytic shellfish toxins (PSTs) in shellfish, receptor binding assay. Rockville (MD): Association of Official Analytical Chemists International. http://www.eoma.aoac.org/methods/.

Berry JP, Lee E, Walton K, Wilson AE, Bernal-Brooks F (2011). Bioaccumulation of microcystins by fish associated with a persistent cyanobacterial bloom in Lago de Patzcuaro (Michoacan, Mexico). Environ Toxicol Chem. 30:1621–1628.

Chen J, Xie P, Li L, Xu J (2009a). First identification of the hepatotoxic microcystins in the serum of a chronically exposed human population together with indication of hepatocellular damage. Toxicol Sci. 108:81–89.

Chen J, Xie P, Zhang D, Ke Z, Yang H (2006). *In situ* studies on the bioaccumulation of microcystins in the phytoplanktivorous silver carp (*Hypophthalmichthys molitrix*) stocked in Lake Taihu with dense toxic *Microcystis* blooms. Aquaculture. 261:1026–1038.

Chen J, Xie P, Zhang D, Lei H (2007). *In situ* studies on the distribution patterns and dynamics of microcystins in a biomanipulation fish–bighead carp (*Aristichthys nobilis*). Environ Pollut. 147:150–157.

Chen J, Zhang D, Xie P, Wang Q, Ma Z (2009b). Simultaneous determination of microcystin contaminations in various vertebrates (fish, turtle, duck and water bird) from a large eutrophic Chinese lake, Lake Taihu, with toxic *Microcystis* blooms. Sci Tot Environ. 407:3317–3322.

Clemente Z, Busato RH, Ribeiro CAO, Cestari MM, Ramsdorf WA, Magalhaes VF et al. (2010). Analyses of paralytic shellfish toxins and biomarkers in a southern Brazilian reservoir. Toxicon. 55:396–406.

Colas S, Duval C, Marie B (2020). Toxicity, transfer and depuration of anatoxin-a (cyanobacterial neurotoxin) in medaka fish exposed by single-dose gavage. Aquatic Toxicol. 222:105422.

Cordeiro-Araújo MK, Chia MA, do Carmo Bittencourt-Oliveira M (2017). Potential human health risk assessment of cylindrospermopsin accumulation and depuration in lettuce and arugula. Harmful Algae. 68:217–223.

Cui Y, Li S, Yang X, Wang Y, Dai Z, Shen Q (2018). HLB/PDMS-coated stir bar sorptive extraction of microcystins in shellfish followed by high-performance liquid chromatography and mass spectrometry analysis. Food Anal Meth. 11:1748–1756.

da Silva CA, Oba ET, Ramsdorf WA, Magalhães VF, Cestari MM, Ribeiro CAO et al. (2011). First report about saxitoxins in freshwater fish *Hoplias malabaricus* through trophic exposure. Toxicon. 57:141–147.

Dionisio Pires LM, Karlsson KM, Meriluoto JA, Kardinaal WEA, Visser PM, Siewertsen K et al. (2004). Assimilation and depuration of microcystin-LR by the zebra mussel, *Dreissena polymorpha*. Aquat Toxicol. 69:385–396.

Foss AJ, Butt J, Fuller S, Cieslik K, Aubel MT, Wertz T (2017). Nodularin from benthic freshwater periphyton and implications for trophic transfer. Toxicon. 140:45–59.

Freitas de Magalhães V, Soares RM, Azevedo SM (2001). Microcystin contamination in fish from the Jacarepaguá Lagoon (Rio de Janeiro, Brazil): ecological implication and human health risk. Toxicon. 39:1077–1085.

Freitas M, Azevedo J, Carvalho AP, Mendes VM, Manadas B, Campos A et al. (2016). Bioaccessibility and changes on cylindrospermopsin concentration in edible mussels with storage and processing time. Food Control. 59:567–574.

Galvão JA, Oetterer M, Bittencourt-Oliveira MD, Gouvêa-Barros S, Hiller S, Erler K et al. (2009). Saxitoxins accumulation by freshwater tilapia (*Oreochromis niloticus*) for human consumption. Toxicon. 54:891–894.

Greer B, Maul R, Campbell K, Elliott CT (2017). Detection of freshwater cyanotoxins and measurement of masked microcystins in tilapia from Southeast Asian aquaculture farms. Anal Bioanal Chem. 409:4057–4069.

Hardy FJ, Johnson A, Hamel K, Preece E (2015). Cyanotoxin bioaccumulation in freshwater fish, Washington State, USA. Environ Monit Assess. 187:667.

Ibelings BW, Bruning K, de Jonge J, Wolfstein K, Dionisio Pires LM, Postma J et al. (2005). Distribution of microcystins in a lake foodweb: No evidence for biomagnification. Microb Ecol. 49:487–500.

Ibelings BW, Chorus I (2007). Accumulation of cyanobacterial toxins in freshwater "seafood" and its consequences for public health: A review. Environ Pollut. 150:177–192.

Ibelings BW, Havens KE (2008). Cyanobacterial toxins: a qualitative meta–analysis of concentrations, dosage and effects in freshwater, estuarine and marine biota. In: Hudnell HK, editors: Cyanobacterial harmful algal blooms: state of the science and research needs. New York: Springer:675–732.

Kittler K, Schreiner M, Krumbein A, Manzei S, Koch M, Rohn S et al. (2012). Uptake of the cyanobacterial toxin cylindrospermopsin in *Brassica* vegetables. Food Chem. 133:875–879.

Kohoutek J, Adamovský O, Oravec M, Šimek Z, Palíková M, Kopp R et al. (2010). LC-MS analyses of microcystins in fish tissues overestimate toxin levels – critical comparison with LC-MS/MS. Anal Bioanal Chem. 398:1231–1237.

Lawrence JF, Niedzwiadek B, Menard C (2005). Quantitative determination of paralytic shellfish poisoning toxins in shellfish using prechromatographic oxidation and liquid chromatography with fluorescence detection: collaborative study. J AOAC Int. 88:1714–1732.

Li Y-W, Zhan X-J, Xiang L, Deng Z-S, Huang B-H, Wen H-F et al. (2014). Analysis of trace microcystins in vegetables using solid-phase extraction followed by high performance liquid chromatography triple-quadrupole mass spectrometry. J Agric Food Chem. 62:11831–11839.

Llana-Ruiz-Cabello M, Jos A, Cameán A, Oliveira F, Barreiro A, Machado J et al. (2019). Analysis of the use of cylindrospermopsin and/or microcystin-contaminated water in the growth, mineral content, and contamination of *Spinacia oleracea* and *Lactuca sativa*. Toxins. 11:624.

Miles CO, Sandvik M, Haande S, Nonga H, Ballot A (2013). LC-MS analysis with thiol derivatization to differentiate [Dhb^7]-from [Mdha7]-microcystins: analysis of cyanobacterial blooms, *Planktothrix* cultures and European crayfish from Lake Steinsfjorden, Norway. Environ Sci Technol. 47:4080–4087.

Miles CO, Sandvik M, Nonga HE, Ballot A, Wilkins AL, Rise F et al. (2016). Conjugation of microcystins with thiols is reversible: Base-catalyzed deconjugation for chemical analysis. Chem Res Toxicol. 29:860–870.

Negri AP, Jones GJ (1995). Bioaccumulation of paralytic shellfish poisoning (PSP) toxins from the cyanobacterium *Anabaena circinalis* by the freshwater mussel *Alathyria condola*. Toxicon. 33:667–678.

Osswald J, Azevedo J, Vasconcelos V, Guilhermino L (2011). Experimental determination of the bioconcentration factors for anatoxin-a in juvenile rainbow trout (*Oncorhynchus mykiss*). Proc Int Acad Ecol Environ Sci. 1:77.

Osswald J, Rellan S, Gago A, Vasconcelos V (2008). Uptake and depuration of anatoxin-a by the mussel *Mytilus galloprovincialis* (Lamarck, 1819) under laboratory conditions. Chemosphere. 72:1235–1241.

Pereira P, Dias E, Franca S, Pereira E, Carolino M, Vasconcelos V (2004). Accumulation and depuration of cyanobacterial paralytic shellfish toxins by the freshwater mussel *Anodonta cygnea*. Aquat Toxicol. 68:339–350.

Poste AE, Hecky RE, Guildford SJ (2011). Evaluating microcystin exposure risk through fish consumption. Environ Sci Technol. 45:5806–5811.

Rees G, Pond K, Kay D, Bartram J, Santo Domingo J, editors (2010). Safe management of shellfish and harvest waters. London: IWA Publishing on behalf of World Health Organization. https://apps.who.int/iris/handle/10665/44101

Rickert B, Chorus I, Schmoll O (2016). Protecting surface water for health. Identifying, assessing and managing drinking-water quality risks in surface-water catchments. Geneva: World Health Organization:178 pp. https://apps.who.int/iris/handle/10665/246196

Ríos V, Moreno I, Prieto AI, Puerto M, Gutiérrez-Praena D, Soria-Díaz ME et al. (2013). Analysis of MC-LR and MC-RR in tissue from freshwater fish (*Tinca tinca*) and crayfish (*Procambarus clarkii*) in tench ponds (Cáceres, Spain) by liquid chromatography–mass spectrometry (LC–MS). Food Chem Toxicol. 57:170–178.

Rita DP, Valeria V, Silvia BM, Pasquale G, Milena B (2014). Microcystin contamination in sea mussel farms from the Italian southern Adriatic coast following cyanobacterial blooms in an artificial reservoir. J Ecosystems. 2014.

Saker ML, Eaglesham GK (1999). The accumulation of cylindrospermopsin from the cyanobacterium *Cylindrospermopsis raciborskii* in tissues of the Redclaw crayfish *Cherax quandricarinatus*. Toxicon. 37:1065–1077.

Saker ML, Metcalf JS, Codd GA, Vasconcelos VM (2004). Accumulation and depuration of the cyanobacterial toxin cylindrospermopsin in the freshwater mussel *Anodonta cygnea*. Toxicon. 43:185–194.

Schmidt JR, Shaskus M, Estenik JF, Oesch C, Khidekel R, Boyer GL (2013). Variations in the microcystin content of different fish species collected from a eutrophic lake. Toxins. 5:992–1009.

Smith JL, Schulz KL, Zimba PV, Boyer GL (2010). Possible mechanism for the foodweb transfer of covalently bound microcystins. Ecotox Environ Safe. 73:757–761.

Testai E, Buratti FM, Funari E, Manganelli M, Vichi S, Arnich N et al. (2016). Review and analysis of occurrence, exposure and toxicity of cyanobacteria toxins in food. Parma: EFSA Supporting Publications:EN-998. 309 pp.

Turner AD, Dhanji-Rapkova M, Fong SY, Hungerford J, McNabb PS, Boundy MJ et al. (2019). Ultrahigh-performance hydrophilic interaction liquid chromatography with tandem mass spectrometry method for the determination of paralytic shellfish toxins and tetrodotoxin in mussels, oysters, clams, cockles, and scallops: Collaborative study. Journal of AOAC International. 103:533–562.

Tzouros N, Arvanitoyannis I (2000). Implementation of hazard analysis critical control point (HACCP) system to the fish/seafood industry: A review. Food Rev Int. 16:273–325.

US EPA (2017). Eating fish: what pregnant women and parents should know. Washington (DC): United States Environmental Protection Agency. https://www.fda.gov/downloads/Food/FoodborneIllnessContaminants/Metals/UCM537120.pdf.

Van Buynder PG, Oughtred T, Kirkby B, Phillips S, Eaglesham G, Thomas K et al. (2001). Nodularin uptake by seafood during a cyanobacterial bloom. Environ Toxicol. 16:468–471.

Williams DE, Craig M, Dawe SC, Kent ML, Andersen RJ, Holmes CFB (1997a). 14C-Labeled microcystin-LR administered to Atlantic salmon via intraperitoneal injection provides in vivo evidence for covalent binding of microcystin-LR in salmon livers. Toxicon. 35:985–989.

Williams DE, Craig M, Dawe SC, Kent ML, Holmes CF, Andersen RJ (1997b). Evidence for a covalently bound form of microcystin-LR in salmon liver and dungeness crab larvae. Chem Res Toxicol. 10:463–469.

5.4 RENAL DIALYSIS

Sandra M. F. O. Azevedo

Renal dialysis patients are a group of the population with a specific and increased risk of cyanotoxin poisoning. The exposure pathway through haemodialysis is intravenous and to a large water volume – approximately 120 L are used in each treatment, three times per week. Hence, this group can be affected even by cyanotoxin concentrations far below the lifetime guideline values (GVs) for drinking-water.

According to Couser et al. (2011), approximately two million people are receiving haemodialysis worldwide, of which 90% live in North America, Japan and Europe. Dialysis is not regularly available in low-income countries, mainly due to a limited access to medical assistance.

In a disastrous incident early in 1996 in Caruaru, Brazil, 131 dialysis patients were exposed to cyanotoxin-contaminated water. Of these, 116 people experienced symptoms, including visual disturbances, nausea and vomiting, 110 developed acute liver failure, and 60 deaths were attributed to acute intoxication by cyanotoxins (microcystins and cylindrospermopsin) from water used for haemodialysis treatment (Jochimsen et al., 1998; Carmichael et al., 2001; Azevedo et al., 2002).

In a second episode of human microcystin exposure by the intravenous route documented among patients undergoing dialysis (Soares et al., 2006), a complete water treatment system including reverse osmosis, operating according specific procedures for dialysis use, proved insufficiently safe to prevent microcystin exposure. Notably, in this case, the microcystin concentration in drinking-water distribution system of the city was below the provisional WHO drinking-water guideline value (GV) of 1 µg/L.

In face of the 100-fold higher water volume to which dialysis patients are exposed, tolerable concentrations in dialysis water would correspondingly need to be at least 100-fold lower. Additionally, however, with oral exposure, only a fraction of the cyanotoxins is efficiently absorbed by the gastrointestinal tract through an active transport involving organic anion transporting polypeptides (OATP; Shitara et al., 2013). This process is saturable and affected by the presence of other chemicals and dependent on the relative affinity of individual compounds (Fischer et al., 2005; Fischer et al., 2010). In contrast, if exposure is intravenous, the systemic bioavailability is close to 100%. Therefore, in face of the present sparse quantitative understanding of the kinetics of sublethal doses of cyanotoxins in humans, especially for renal disease patients, it is not possible to establish threshold values for the induced adverse effects, and thus, no GVs for cyanotoxins in water used for dialysis can be derived. Certainly, however, the GVs for cyanotoxins in drinking-water are not sufficiently protective.

Hazard analysis for cyanotoxins in water used for hemodialysis therefore needs to assess the source of the raw water. Surface water potentially containing even traces of cyanotoxins needs to be avoided whenever possible.

5.4.1 Assessing and controlling the risk of cyanotoxin exposure

The WHO guidelines for Drinking-Water Quality (WHO, 2017) do not consider the especially high quality of water needed for dialysis treatment, intravenous therapy or other clinical uses. The treatment processes used at conventional surface water treatment plants (such as coagulation, clarification and sand filtration) are effective in removing cyanobacterial cells, but may not be sufficiently effective in removing or destroying dissolved cyanotoxin concentrations to below GVs, especially from water supplies with a high organic content and cyanobacterial dominance (see Chapter 10). Consequently, clinics and hospitals with special water needs, such as for dialysis treatment or for transfusions (intravenous administration), often apply additional water treatment, for example, for the removal of cyanotoxins. Such treatment ranges from granular activated carbon filtration, followed by reverse osmosis, to more elaborate treatment, including membrane filtration. The extent of treatment necessary depends on the quality of the municipal water supply.

Continuous monitoring of performance and equipment is essential to ensure adequate quality of the water. On-site water treatment systems in clinics and hospitals require rigorous monitoring and regular maintenance, including back-flushing of filters and recharge of activated carbon, according to manufacturer's specifications. It is important that manufacturer specifications should be assessed under local conditions for their adequacy in maintaining performance. Activated carbon, for example, may be exhausted for its ability to remove cyanotoxins long before it reaches saturation for the removal of other organic compounds, and some manufacturers may be unaware of this.

As emphasised above, the present knowledge about toxicity of different cyanotoxins does not allow establishment of any safe concentration for intravenous exposure. Therefore, a monitoring programme for water quality used for dialysis procedure needs to be performed with methods of utmost sensitivity (see Chapter 14).

Contingency plans and actions for prevention or management of health hazards from cyanotoxins for this specifically susceptible subpopulation are usually developed and managed at local or regional level. Additionally, national authorities may have important roles in organising, supporting and facilitating plan formulation, particularly after an event of suspected or proven intoxication. Some key actions for preventive management of these special water uses include the following:

- Establishment of a multiagency and multidisciplinary regional committee with participation of public health authorities, water supply managers, hospital and dialysis clinic technically responsible for elaborating an effective plan of communication about incidents of cyanobacteria blooms in water supplies and cyanotoxins levels in drinking-water system used to supply health units. This communication plan needs to guarantee information about cyanotoxins concentration in drinking-water distributed to hospital and dialysis clinics within less than 24 h.
- A compilation of information about reservoirs or rivers used as water supplies to each community and a comprehensive map of the water distribution system, including location of hospitals and dialysis clinics needs to be available to health authorities to support any contingency plan when it is needed. The data of basic limnological parameters monitored in water supplies, including phytoplankton density (with special emphasis on cyanobacterial biovolume or cell numbers per litre) should be up to date. Interagency cooperation, especially between the drinking-water supplier and the health authority, is crucial to prevent an incident.
- If cyanotoxins are detected in drinking-water used to directly prepare water for dialysis or infusions, even in concentration well below lifetime GVs, a contingency plan to supply alternative safe water to health units needs to be implemented immediately. It should be previously developed and established, including specific actions and responsibilities of different actors. It needs to include previous identification of potential alternative water supplies, preferably from uncontaminated groundwater; plans for transporting safe water from other areas or deploying portable water treatment systems.
- In case an alternative safe water supply is not available, the dialysis service should be interrupted and patients should be transferred to other health units with no risk of exposure from the dialysis water. In this situation, the dialysis unit potentially exposing patients to cyanotoxins needs to be thoroughly cleaned, including exchange of activated carbon and the cleanup of all filters and membrane systems used.
- A regular monitoring programme for cyanotoxin analysis in the in-house water treatment systems of a dialysis unit should be implemented in regions where cyanobacterial blooms occurrence in water supplies cannot be excluded because no source water without potential contamination is available, particularly if blooms were detected during the past 12 months. This analysis needs to include sampling of

water before and after the treatment steps in order to assess treatment performance. It requires a highly sensitive methodology which can detect cyanotoxins in the nanogram per litre range.

- Preparing a standardised press release (previously agreed between the authorities which need to be involved) and an agreement on the triggers for its publication can help inform patients early in an incident, if one occurs.

Guidelines for quality assurance of dialysis equipment and fluids generally are more focused on (heterotrophic) microbial and chemical contaminations (e.g., Kawanishi et al. (2009); Penne et al. (2009)). However, best practice standards for the production of pure or ultrapure water for renal dialysis do apply to all chemicals (Ledebo, 2007).With respect to cyanobacterial toxins, dialysis units need to inquire from the water supplier whether there is a risk of cyanotoxin contamination in drinking-water, either seasonally or for extended periods. In this case, periodic use of an alternative water source may be a way forward if the water source cannot permanently be altered.

More information on quality control for dialysis, including fluid quality, is available on the websites of the US National Kidney Foundation (https://www.kidney.org) or the European Renal Association – European Dialysis and Transplant Association (https://www.era-edta.org). Guidelines of the latter can be found in a supplement issue of "Nephrology Dialysis Transplantation" (ERA-EDTA, 2002). Further information and guidelines are given in ISO 11663 (ISO, 2009) and the standards cited therein.

REFERENCES

Azevedo SM, Carmichael WW, Jochimsen EM, Rinehart KL, Lau S, Shaw GR et al. (2002). Human intoxication by microcystins during renal dialysis treatment in Caruaru-Brazil. Toxicology. 181–182:441–446.

Carmichael WW, Azevedo SMFO, An JS, Molica RJR, Jochimsen EM, Lau S et al. (2001). Human fatalities from cyanobacteria: Chemical and biological evidence for cyanotoxins. Environ Health Persp. 109:663–668.

Couser WG, Remuzzi G, Mendis S, Tonelli M (2011). The contribution of chronic kidney disease to the global burden of major noncommunicable diseases. Kidney Int. 80:1258–1270.

ERA-EDTA (2002). European Renal Association: European best practice guidelines for haemodialysis. Nephrol Dial Transplant. 17(supplement 7).

Fischer A, Höger SJ, Stemmer K, Feurstein D, Knobeloch D, Nussler A et al. (2010). The role of organic anion transporting polypeptides (OATPs/SLCOs) in the toxicity of different microcystin congeners in vitro: a comparison of primary human hepatocytes and OATP-transfected HEK293 cells. Toxicol Appl Pharmacol. 245:9–20.

Fischer WJ, Altheimer S, Cattori V, Meier PJ, Dietrich DR, Hagenbuch B (2005). Organic anion transporting polypeptides expressed in liver and brain mediate uptake of microcystin. Toxicol Appl Pharmacol. 203:257–263.
ISO (2009). ISO 11663. quality of dialysis fluid for haemodialysis and related therapies. Geneva: International Organization for Standardization.
Jochimsen EM, Carmichael WW, An J, Cardo DM, Cookson ST, Holmes CEM et al. (1998). Liver failure and death after exposure to microcystins at a hemodialysis center in Brazil. New England J Med. 338:873–878.
Kawanishi H, Akiba T, Masakane I, Tomo T, Mineshima M, Kawasaki T et al. (2009). Standard on microbiological management of fluids for hemodialysis and related therapies by the Japanese Society for Dialysis Therapy 2008. Therapeutic Apheresis and Dialysis. 13:161–166.
Ledebo I (2007) Ultrapure dialysis fluid—how pure is it and do we need it? Nephrol, Dial, Transplant. 22:20–23.
Penne EL, Visser L, Van Den Dorpel MA, Van Der Weerd NC, Mazairac AH, Van Jaarsveld BC et al. (2009). Microbiological quality and quality control of purified water and ultrapure dialysis fluids for online hemodiafiltration in routine clinical practice. Kidney Int. 76:665–672.
Shitara Y, Maeda K, Ikejiri K, Yoshida K, Horie T, Sugiyama Y (2013). Clinical significance of organic anion transporting polypeptides (OATPs) in drug disposition: their roles in hepatic clearance and intestinal absorption. Biopharm Drug Dispos. 34:45–78.
Soares RM, Yuan M, Servaites JC, Delgado A, Maglhaes VF, Hilborn ED et al. (2006). Sublethal exposure from microcystins to renal insufficiency patients in Rio de Janeiro, Brazil. Environ Toxicol. 21:95–103.
WHO (2017). Guidelines for drinking-water quality, fourth edition, incorporating the 1st addendum. Geneva: World Health Organization:631 pp. https://www.who.int/publications/i/item/9789241549950

5.5 CYANOBACTERIA AS DIETARY SUPPLEMENTS

Daniel Dietrich

Cyanobacteria, specifically *Arthrospira* sp. (previously classified as *Spirulina* sp.; see Chapter 3), were used as a food staple by indigenous people in Central America and in the Rift Valley of Africa. Large-scale production of cyanobacteria and microalgae for marketing in western society started about 50 years ago. Much of the early research work dealt with the basic photosynthetic properties of microalgae, their possible therapeutic, antibiotic and toxic properties and their potential as an agricultural commodity for human consumption. The microalgae biomass industry now provides biomass for pigments and speciality chemicals used primarily in the food industry and, more recently, as food supplements, also termed health foods, nutraceuticals, esoteric foods or simply blue-green algal supplement (BGAS). These mostly originate from three filamentous genera of cyanobacteria: *Arthrospira* (*Spirulina*), including *A. platensis* and *A. maxima* (Belay & Ota, 1994), *Nostoc* (*N. commune* and *N. flagelliforme*) and *Aphanizomenon flosaquae*.

While *Arthrospira* is grown in cultures, often in outdoor ponds, mainly in the USA (southern California and Hawaii), Chad, France, Mexico, Myanmar, Thailand, Taiwan and Japan, *Nostoc* (*N. commune*) is either grown by indigenous people as food staples, also known as *llullucha* (Johnson et al., 2008) or as dietary food supplements in South-East Asia (Saker et al., 2007) and China (Gao, 1998) while *Aphanizomenon* is primarily harvested from a dammed natural lake (Klamath Lake, Oregon, USA; Carmichael et al., 2000). Production of food-grade "*Spirulina*" largely depends on the production region, for example, the UN estimates approximately 250 tons/year to be produced in Chad for sale on local markets, while Henrikson (2011) estimated the internationally oriented commercial enterprises to produce more than 500 tons/year. The production volumes of *Nostoc* are presently unknown and cannot be extrapolated from sales or consumption data, as these are missing as well.

Aphanizomenon production is also substantial; however, data on production volumes have not been possible to obtain. The only indication of the amounts of *Aphanizomenon flosaquae*-based dietary supplements is their annual sales, which range in the tens of millions US dollars (ODA, 2017).

5.5.1 Cyanotoxins potentially present in cyanobacterial food supplements

Cyanobacteria used as dietary supplements can be a source of cyanotoxins even when the main ingredient is considered nontoxic, such as *Arthrospira maxima*. Nonetheless, some studies suggest a potential for "*Spirulina*" products to contain cyanotoxins, possibly via contamination of cultures with other, toxigenic cyanobacteria: the anatoxin-a analogs epoxyanatoxin-a and dihydrohomoanatoxin-a have been identified at concentrations ranging from nondetectable to 19 µg/g dry weight in "*Spirulina*"-based dietary supplements (Salazar et al., 1996; Salazar et al., 1998; Draisci et al., 2001). A market analysis demonstrated concentrations of anatoxin-a ranging between 2.50 and 33 µg/g, whereby these included products intended for human and animal consumption (Rellán et al., 2009). In alkaline crater lakes in Kenya, *Arthrospira fusiformis* was found to produce small amounts of both microcystins and anatoxin-a (Ballot et al., 2004; Ballot et al., 2005), and ELISA results were positive for microcystins in "*Spirulina*" food supplements, suggesting a contamination with a microcystin producer (Gilroy et al., 2000). There are no proven cases of human injury as a result of ingesting "*Spirulina*"-based food supplements, although these were proposed as the cause of liver injury of a 52-year-old Japanese (Iwasa et al., 2002). However, consumption of "*Spirulina*" as well as other cyanobacteria-based food supplements are frequently accompanied by massive diarrhoea, nausea, abdominal pain and skin rash (Rzymski & Jaśkiewicz, 2017).

Nostoc commune produced by indigenous people of Peru were found to contain β-methyl-amino-alanine (BMAA; Johnson et al., 2008). However, the analytical method used is now known to substantially overestimate BMAA concentrations, and the toxic potential of BMAA is debated highly controversially. The conclusion of section 2.7 of the present volume is that, at present, the weight of evidence suggests that BMAA is present in insufficiently high concentrations to cause neurogenerative diseases.

Aph. flosaquae can contain cylindrospermopsins, anatoxin-a and saxitoxins as well as toxicity not attributable to any of the known cyanotoxins (see Heussner et al., 2012, and Chapter 2). Although microcystin production has not been observed for *Aphanizomenon* sp., in natural blooms, *Aphanizomenon* sp. is often found associated with other cyanobacteria which are known to be toxigenic.

Common cyanobacteria associated with blooms of *Aphanizomenon* sp. are *Microcystis* sp. and *Dolichospermum* sp., that is, species that potentially produce microcystins (Ekman-Ekebom et al., 1992; Teubner et al., 1999; Wood et al., 2011; Shams et al., 2015; Chapter 4). Analysis of *Aph. flosaquae* samples taken from Lake Klamath for dietary supplement production demonstrated that approximately 80% of the samples taken between 1994 and 1998 contained >1 µg MC-LR equivalents per gram dry weight, which is the maximum acceptable content established by the state of Oregon in the USA (Gilroy et al., 2000). Further studies showed higher as well as lower microcystin contents (Table 5.7), which is partly attributed to shifts in taxonomic composition within the blooms in Lake Klamath dominated by *Aph. flosaquae*, in particular, the variable share of toxigenic *Microcystis* sp. in bulk phytoplankton biomass. The studies summarised in Table 5.7 show a trend to lower maximum microcystin contents over time.

Table 5.7 Microcystin concentration in *Aphanizomenon* sp. dietary supplements from the market

Number of Samples	*% samples exceeding 1.0 µg/g DW*	*Microcystin content µg/g DW*	*Detection method*	*Reference*
87	72	2.2–10.9	ELISA	Gilroy et al. (2000)
52	50	0–35.7 0–49.0 0–35.7	ELISA cPPA LC-MS/MS	Lawrence et al. (2001)
6	100	11–24.7	ELISA, cPPA HPLC	Schaeffer et al. (1999)
18	80	0.3–8.3 0.5–5.9	Adda-ELISA cPPA	Hoeger & Dietrich (2004)
12	33	0.1–4.7	ELISA	Saker et al. (2005) Saker et al. (2007)
26	35	<LoD–5.2	LC-MS/MS	Vichi et al. (2012)
10	60 50 40	<LoD–6.1 <LoD–11.0 <LoD–5.8	Adda-ELISA cPPA LC-MS/MS	Heussner et al. (2012)
60	6 7	0–3.0 <0.25–2.8	LC-MS/MS PPA	Marsan et al. (2018)

DW: dry weight; LoD: limit of detection; ELISA: enzyme-linked immunosorbent assay; cPPA: colorimetric protein phosphatase inhibition assay, HPLC: high-pressure liquid chromatography; LC-MS/MS: liquid chromatography–mass spectrometry; Adda-ELISA: enzyme-linked immunosorbent assay with a recognition antibody specifically directed against the Adda-moiety of microcystins.

5.5.2 Assessing and managing the risk of cyanotoxin exposure through food supplements

In the studies summarised in Table 5.7, maximum contents of microcystin per gram dry weight range between 3.0 and 49 µg/g, and therefore, a risk of exposure to cyanotoxins cannot be ignored. A detailed assessment, however, is difficult, firstly, because the manufacturer's recommendations for daily consumption vary widely from 0.5 to 15 g/day with some products indicating no maximum limit (Marsan et al., 2018) and, secondly, because individual consumption also varies and may largely exceed recommendations. However, based on reported possible toxin contents and a consumption of a few grams per day, exposure may well be at levels exceeding the provisional tolerable daily intake (TDI) of 0.04 µg/kg (see section 2.1) for adults and especially for children. Further, in deriving its drinking-water guideline values (GVs) for lifetime exposure, 20% of intake are allocated to sources other than drinking-water, which may not be appropriate for persons consuming cyanobacterial products on a regular basis (see sections 2.1 and 2.2). Dietrich and Hoeger (2005) discuss these aspects for varying levels of microcystin contamination of food supplements and propose corresponding maximum amounts that can be safely consumed by infants, children and adults.

As with other health risks, animal poisoning indicate potential adverse health effects in humans (Hilborn & Beasley, 2015). The case of an 11-year-old female spayed pug dog, weighing 8.95 kg and presenting with abnormally high alanine aminotransferase (ALT), alkaline phosphatase (ALP) and aspartate aminotransferase (AST) activities and serious liver dysfunction, indicates uptake of a hepatotoxin. This dog was fed single to multiple daily rations of 1 gram of 100% certified organic *Aph. flosaquae* for approximately three and a half weeks. The analysis of the powder via LC-MS/MS revealed 0.166 µg/g of MC-LR and 0.962 µg/g of MC-LA, while no other MCs were reported (Bautista et al., 2015). Thus, the MC content would approximate the Oregon provisional guidance value of 1 µg/g dw (Gilroy et al., 2000). However, with an analytical method including more microcystin variants, as suggested in section 14.3, a higher actual total MC content may have been found. Further, neither the number of daily rations nor any further potential source of the dog's exposure – such as cyanobacterial blooms in a waterbody – are known, making it difficult to estimate retrospectively whether the undoubted exposure to microcystins through dietary supplements was enough to explain the observed symptoms in this single study on one animal.

A further issue in this context is the as of yet very incomplete understanding of the bioactivity of cyanobacterial metabolites beyond the known toxins. Underdal et al. (1999) found protracted toxic response in test animals exposed to extracts of *Aph. flosaquae* but could not identify any toxins. Similarly, Heussner et al. (2012) found cytotoxicity in *Aph. flosaquae* product extracts that were not associated with any of the known cyanobacterial toxins. Indeed, particularly *Aphanizomenon* species are known for inducing effects not yet explained by any identified cyanobacterial metabolite, for example, malformation of fish embryos (Oberemm et al., 1997; Berry et al., 2009). While such effects cannot be quantitatively used for a human health risk assessment, they do indicate potential presence of further hazards to clarify.

Further, field collections of cyanobacteria and, possibly to a lesser extent, cyanobacteria harvested from open tanks contain a high diversity of heterotrophic bacteria, including human pathogens (Berg et al., 2009) that may cause further health hazards.

5.5.3 Approaches to assessing and controlling the potential cyanotoxin hazards

The regulation of dietary supplements is generally less strict compared to regulations for food, pharmaceutical or drinking-water, and only few regulatory schemes are in place. For example, since 1994, dietary supplements have been regulated in the USA under the Dietary Supplement Health and Education Act (DSHEA; FDA, 2017). Because cyanobacteria are capable of producing toxins and their presence has been confirmed in some dietary supplements, it is appropriate to regulate and monitor these toxins in dietary supplements, including the provision of adequate information to consumers. Considerations include the following:

Testing for cyanotoxin content: Biomass collected from natural blooms or open tank incubators should be tested, lot by lot as recommended by the regulatory authority, for possible contamination with potentially toxigenic cyanobacteria, for example, *Microcystis* sp. in blooms dominated by *Aph. flosaquae*. Production lots should be managed by unique identifying numbers and production dates. For potential subsequent reanalysis by regulatory authorities, producers should be mandated to retain representative samples of each charge produced and to make these available upon official request.

Testing for other contaminants: Dietary supplement products should be tested for other potential contaminants, including indicators for pathogenic bacteria and protozoa, where and when contamination is expected. This is best based on an assessment of contamination risks from the catchment or the culture conditions. Examples of contamination sources include excreta of migrating birds or surface runoff following rainfall.

Claims on possible effects: The proposed beneficial effects of the consumption of cyanobacterial food supplements have not been demonstrated in scientifically sound studies; only subjective and anecdotal evidence is proposed by the vendors. Therefore, product information should not suggest that consumption of larger amounts would produce more positive effects.

Consumer information: Producers should clearly inform the consumers which quality control procedures are in place and give access to the test results. Further they should give a clear maximum daily doses, specified for infants, children and adults. None of these measures, however, can serve to protect from negative effects of known and yet unknown bioactive substances in cyanobacteria, as discussed in section 2.10.

REFERENCES

Ballot A, Krienitz L, Kotut K, Wiegand C, Metcalf JS, Codd GA et al. (2004). Cyanobacteria and cyanobacterial toxins in three alkaline Rift Valley lakes of Kenya—Lakes Bogoria, Nakuru and Elmenteita. J Plankton Res. 26:925–935.

Ballot A, Krienitz L, Kotut K, Wiegand C, Pflugmacher S (2005). Cyanobacteria and cyanobacterial toxins in the alkaline crater lakes Sonachi and Simbi, Kenya. Harmful Algae. 4:139–150.

Bautista AC, Moore CE, Lin Y, Cline MG, Benitah N, Puschner B (2015). Hepatopathy following consumption of a commercially available blue-green algae dietary supplement in a dog. BMC Vet Res. 11:136.

Belay A, Ota Y (1994). Production of high quality spirulina at Earth Rise Farms. 2nd Asia Pacific Conference on Algal Biotech. Kuala Kumpur, Malaysia.

Berg KA, Lyra C, Sivonen K, Paulin L, Suomalainen S, Tuomi P et al. (2009). High diversity of cultivable heterotrophic bacteria in association with cyanobacterial water blooms. ISME J. 3:314–325.

Berry JP, Gibbs PDL, Schmale MC, Saker ML (2009). Toxicity of cylindrospermopsin, and other apparent metabolites from *Cylindrospermopsis raciborskii* and *Aphanizomenon ovalisporum*, to the zebrafish (*Danio rerio*) embryo. Toxicon. 53:289–299.

Carmichael WW, Drapeau C, Anderson DM (2000). Harvesting of *Aphanizomenon flos-aquae* Ralfs ex Born. & Flah. var. *flos-aquae* (Cyanobacteria) from Klamath Lake for human dietary use. J Appl Phycol. 12:585–595.

Dietrich D, Hoeger SJ (2005). Guidance values for microcystins in water and cyanobacterial supplement products (blue-green algal supplements): a reasonable or misguided approach? Toxicol Appl Pharmacol. 203:273–289.

Draisci R, Ferretti E, Palleschi L, Marchiafava C (2001). Identification of anatoxins in blue-green algae food supplements using liquid chromatography-tandem mass spectrometry. Food Addit Contam. 18:525–531.

Ekman-Ekebom M, Kauppi M, Sivonen K, Niemi M, Lepistö L (1992). Toxic cyanobacteria in some finnish lakes. Environ Toxicol Wat Qual. 7:201–213.

FDA (2017). US Food and Drug Administration dietary supplements. Silver Spring, MD: Food and Drug Administration United States of America. Available at: https://www.fda.gov/Food/DietarySupplements/default.htm.

Gao K (1998). Chinese studies on the edible blue-green alga, *Nostoc flagelliforme*: a review. J Appl Phycol. 10:37–49.

Gilroy DJ, Kauffman KW, Hall RA, Huang X, Chu FS (2000). Assessing potential health risks from microcystin toxins in blue-green algae dietary supplements. Environ Health Persp. 108:435–439.

Henrikson R (2011). Development of a *Spirulina* Industry – Production. Algae Industry Magazine.

Heussner AH, Mazija L, Fastner J, Dietrich DR (2012). Toxin content and cytotoxicity of algal dietary supplements. Toxicol Appl Pharmacol 265:263–271.

Hilborn E, Beasley V (2015). One health and cyanobacteria in freshwater systems: animal illnesses and deaths are sentinel events for human health risks. Toxins. 7:1374–1395.

Hoeger S, Dietrich DR (2004). Possible health risks arising from consumption of blue-green algae food supplements. 6th International Conference on Toxic Cyanobacteria. Bergen, Norway.

Iwasa M, Yamamoto M, Tanaka Y, Kaito M, Adachi Y (2002). *Spirulina*-associated hepatotoxicity. Am J Gastroenterol. 97:3212–3213.

Johnson HE, King SR, Banack SA, Webster C, Callanaupa WJ, Cox PA (2008). Cyanobacteria (*Nostoc commune*) used as a dietary item in the Peruvian highlands produce the neurotoxic amino acid BMAA. J Ethnopharmacol. 118:159–165.

Lawrence JF, Niedzwiadek B, Menard C, Lau BPY, Lewis D, Kuper-Goodman T et al. (2001). Comparison of liquid chromatography/mass spectrometry, ELISA, and phosphatase assay for the determination of microcystins in blue-green algae products. J AOAC Int. 84:1035–1044.

Marsan DW, Conrad SM, Stutts WL, Parker CH, Deeds JR (2018). Evaluation of microcystin contamination in blue-green algal dietary supplements using a protein phosphatase inhibition-based test kit. Heliyon. 4:e00573.

Oberemm A, Fastner J, Steinberg CEW (1997). Effects of microcystin-LR and cyanobacterial crude extracts on embryo-larval development of zebrafish (*Danio rerio*). Water Res. 31:2918–2921.

ODA (2017). Klamath headwaters agricultural water quality management area plan. Salem (OR): Oregon Department of Agriculture:81 pp. https://www.oregon.gov/ODA/shared/Documents/Publications/NaturalResources/KlamathAWQMAreaPlan.pdf.

Rellán S, Osswald J, Saker M, Gago-Martinez A, Vasconcelos V (2009). First detection of anatoxin-a in human and animal dietary supplements containing cyanobacteria. Food Chem Toxicol. 47:2189–2195.

Rzymski P, Jaśkiewicz M (2017). Microalgal food supplements from the perspective of Polish consumers: patterns of use, adverse events, and beneficial effects. J Appl Phycol. 29:1841–1850.

Saker ML, Jungblut AD, Neilan BA, Rawn DFK, Vasconcelos VM (2005). Detection of microcystin synthetase genes in health food supplements containing the freshwater cyanobacterium *Aphanizomenon flos-aquae*. Toxicon. 46:555–562.

Saker ML, Welker M, Vasconcelos VM (2007). Multiplex PCR for the detection toxigenic cyanobacteria in dietary supplements produced for human consumption. Appl Microbiol Biotechnol. 73:1136–1142.

Salazar M, Chamorro GA, Salazar S, Steele CE (1996). Effect of *Spirulina maxima* consumption on reproduction and peri- and postnatal development in rats. Food Chem Toxicol. 34:353–359.

Salazar M, Martínez E, Madrigal E, Ruiz LE, Chamorro GA (1998). Subchronic toxicity study in mice fed *Spirulina maxima*. J Ethnopharmacol. 62:235–241.

Schaeffer DJ, Malpas PB, Barton LL (1999). Risk assessment of microcystin in dietary *Aphanizomenon flos-aquae*. Ecotoxicol Environ Safety. 44:73–80.

Shams S, Capelli C, Cerasino L, Ballot A, Dietrich DR, Sivonen K et al. (2015). Anatoxin-a producing *Tychonema* (Cyanobacteria) in European waterbodies. Water Res. 69:68–79.

Teubner K, Feyerabend R, Henning M, Nicklisch A, Woitke P, Kohl J-G (1999). Alternative blooming of *Aphanizomenon flos-aquae* or *Planktothrix agardhii* induced by the timing of the critical nitrogen:phosphorus ratio in hypertrophic riverine lakes. Arch Hydrobiol Spec Issues Advanc Limnol. 54:325–344.

Underdal B, Nordstoga K, Skulberg OM (1999). Protracted toxic effects caused by saline extracts of *Aphanizomenon flos-aquae* (Cyanophyceae/Cyanobacteria). Aquat Toxicol. 46:269–278.

Vichi S, Lavorini P, Funari E, Scardala S, Testai E (2012). Contamination by *Microcystis* and microcystins of blue–green algae food supplements (BGAS) on the italian market and possible risk for the exposed population. Food Chem Toxicol. 50:4493–4499.

Wood SA, Rueckert A, Hamilton DP, Cary SC, Dietrich DR (2011). Switching toxin production on and off: intermittent microcystin synthesis in a *Microcystis* bloom. Environ Microbiol Rep. 3:118–124.

Chapter 6

Assessing and managing cyanobacterial risks in water-use systems

Ingrid Chorus and Rory Moses McKeown

CONTENTS

INTRODUCTION

Cyanotoxin occurrence in water to which people may be exposed depends on the extent to which conditions in the respective waterbody favour the proliferation of cyanobacteria. Where barriers (or "control measures") are in place (e.g., natural, as well as technical treatment or engineered barriers), as well as managerial and planning measures, human exposure will depend on how effectively these measures are working to limit cyanobacterial growth and/or to prevent exposure. Assessing a given system's efficacy in controlling this risk requires understanding the entire water-use systems – from the catchment or source right through to the point of contact with the end user (e.g., consumers of drinking-water or fish/shellfish, or end users such as those involved in recreation or occupational exposures). Assessing the efficacy of the barriers in place is an essential basis for identifying, planning and implementing priority measures to control the conditions that may cause cyanobacterial blooms, thus limiting human exposure. Further elements essential for planning include time spans expected for control measures to take effect as well as expertise, investments and regulatory frameworks necessary for their implementation.

This chapter presents a proactive risk assessment and management framework that can be applied to identify and manage threats to public health from water-use systems, including cyanotoxins, namely, water safety planning. Water safety planning is advocated for by the World Health Organization (WHO) as the most effective means of ensuring the safety of drinking-water supplies (Rickert et al., 2016). In the context of toxic cyanobacteria, this chapter provides guidance not only on how the Water Safety Plan (WSP) approach can be used to manage the risk of cyanotoxin occurrence in drinking-water, but also on how the framework can be adapted and expanded to consider cyanotoxin exposure from other routes relevant for a given context (e.g., exposure from recreational or occupational contact or food consumption). The approach is illustrated with worked examples from three different scenarios, ranging from larger water-use systems to smaller private supplies.

6.1 LEVELS FOR EXERTING CONTROL OVER CYANOTOXIN OCCURRENCE AND EXPOSURE

As depicted in Figure 6.1, the most fundamental control level is catchment management to prevent or reduce nutrient loads to the waterbody, particularly those of phosphorus (Chapter 7), and control measures at this level can be supported by measures at the level of waterbody management (Chapter 8). If control measures on these levels are not in place or fail to meet their targets and toxic cyanobacteria proliferate, the remaining management option is to control human exposure. In some situations, this is possible by shifting sites for drinking-water offtake or recreation to where cyanobacteria do not accumulate. A further option may be to implement management measures such as mechanical mixing within the waterbody to reduce cyanobacterial biomass (Chapters 8 and 9). For drinking-water, the removal of cyanobacterial cells and/or toxins dissolved in water in the drinking-water treatment plant is an additional important control measure (Chapter 10). For other water uses (e.g., swimming and other recreational contact), the consumption of fish and shellfish or spray irrigation – temporarily limiting or banning use as an emergency response (as discussed in sections 5.2–5.5) – may be the only option if cyanotoxin levels are in a range causing exposure to inacceptable concentrations.

A cornerstone of the WSP philosophy is the promotion of a "multiple-barrier approach". This approach advocates for the use of more than one type of barrier or control measure (Figure 6.1) throughout the water-use system (i.e., from the source to the point of use/contact) to minimise risks from cyanotoxin exposure. Through this approach, in the event that an upstream control measure fails (e.g., in the case of a drinking-water supply system, failure of a multilevel raw water offtake), the presence of downstream barriers may still limit the risk from cyanotoxin exposure (e.g., drinking-water treatment optimised to remove cyanotoxins). Exceptions may include settings where risks from cyanotoxin occurrence are considered to be very low, such as pristine and protected catchments that are under the direct control of a single management entity (e.g., water supplier or a catchment management authority).

Targets can be set at each of these levels to achieve the target set for cyanotoxins: a target for cyanobacterial biomass may be based on ratios of toxin to biomass either expected from the literature (section 4.2.6) or determined locally from data for the specific waterbody. A target for limiting the concentration of a key nutrient may be set, for example, in the range of 10–50 μg/L of total phosphorus, depending on the specific characteristics of the waterbody and on how stringently cyanobacterial biomass is to be limited (Chapter 7). The corresponding target for the nutrient load from the catchment to the waterbody depends on further waterbody characteristics, and models are available to determine which load is likely to achieve which concentration in a given waterbody (Chapter 7).

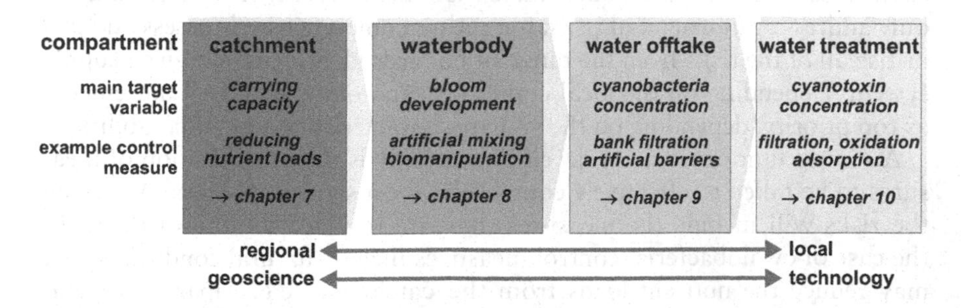

Figure 6.1 Levels and scales of measures for controlling cyanotoxin occurrence

The challenge for management is to set targets based on an assessment of the specific situation to determine which control options are realistically achievable. Depending on the local conditions, this may focus on the activities in the catchment that cause nutrient loads from different effluents and surfaces. Additionally or alternatively, it may involve setting targets for the performance of drinking-water treatment in cyanotoxin removal (Chapter 10), or for public outreach campaigns to inform recreational site users of potential cyanotoxin risks and of prudent behaviour for avoiding exposure (Chapter 15). Thus, decisions on setting targets for a given setting at each of these control levels will depend on the control options available, and setting them effectively requires an understanding of the specific system.

Such an in-depth understanding of the specific setting is most effectively achieved by following the steps of developing a WSP as described in the WHO Guidelines for Drinking-water Quality (WHO, 2017), in the WHO/IWA Water Safety Plan Manual (Bartram et al., 2009; Figure 6.2) or the guidelines for safe recreational environments (WHO, 2003), and extending the approach by considering exposure routes in addition to drinking-water and recreation if these are relevant for the specific system. Further, WHO also provides guidance tailored specifically for water safety planning in small systems (e.g., rural communities; WHO, 2012).

In contexts where a full WSP is not developed, many of the steps and elements of this approach are highly useful for assessing and managing cyanotoxin risks.

6.2 WATER SAFETY PLANNING AS A FRAMEWORK FOR ASSESSING AND MANAGING CYANOBACTERIAL RISKS

A comprehensive Water Safety Plan (WSP) should consider the potential risks from all of the threats or "hazards" (i.e., typically microbial, chemical or physical agents that can impact public health or disrupt system operations and service delivery) within the entire water supply system (i.e., at catchment/source, treatment, storage, distribution and consumer levels), as well as the hazardous events that may introduce them. As such, a WSP should not only address cyanobacterial risks, but rather comprehensively assess and prioritise all of the risks from the range of hazards identified for a given supply system. Depending on the local context, cyanotoxins may well not emerge as top priority, depending on the system characteristics and vulnerabilities.

An important outcome of developing a WSP is the prioritisation of measures to be taken to effectively control the most significant risks. Assessing the risks will include the measures that are in place to control them. In the case of cyanobacteria, control measures include natural conditions that may reduce the nutrient loads from the catchment (e.g., riparian vegetation buffer strips) as well as engineered control measures (e.g., mechanical

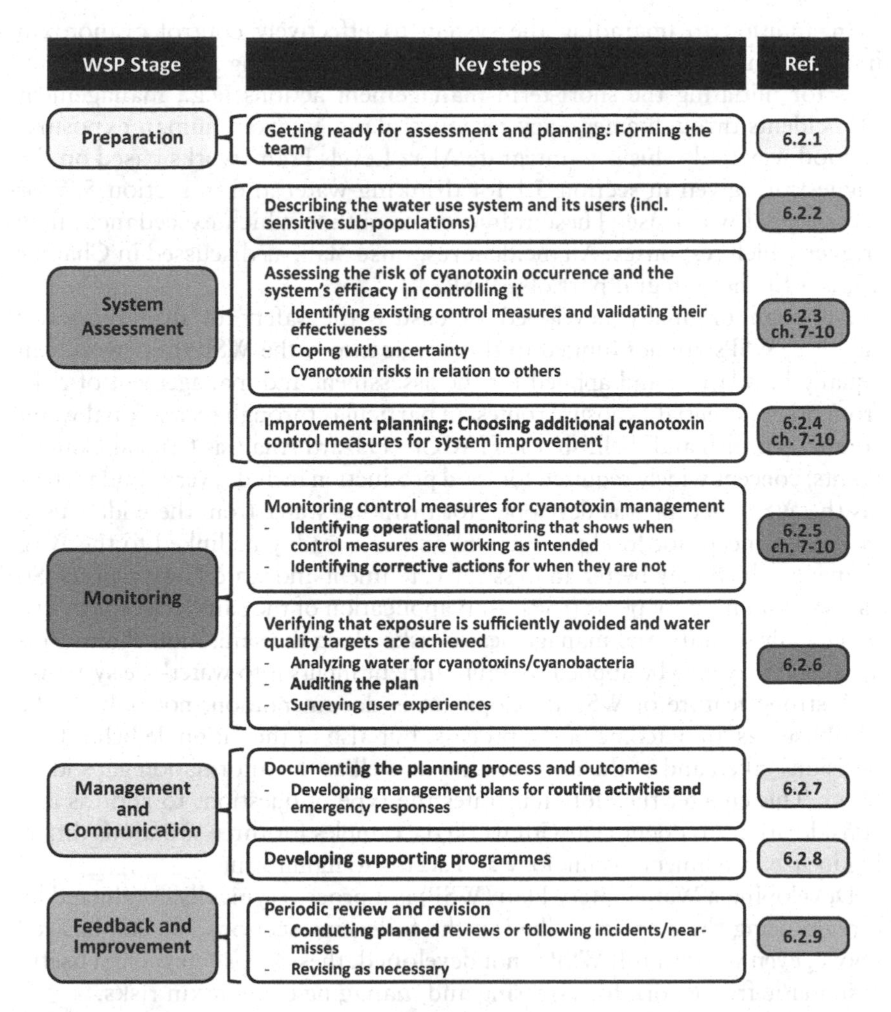

WSP Stage
Key steps
Ref.
Preparation
Getting ready for assessment and planning: Forming the team
6.2.1
System Assessment
Describing the water use system and its users (incl. sensitive sub-populations)
6.2.2
Assessing the risk of cyanotoxin occurrence and the system's efficacy in controlling it
- Identifying existing control measures and validating their effectiveness
- Coping with uncertainty
- Cyanotoxin risks in relation to others
6.2.3 ch. 7-10
Improvement planning: Choosing additional cyanotoxin control measures for system improvement
6.2.4 ch. 7-10
Monitoring
Monitoring control measures for cyanotoxin management
- Identifying operational monitoring that shows when control measures are working as intended
- Identifying corrective actions for when they are not
6.2.5 ch. 7-10
Verifying that exposure is sufficiently avoided and water quality targets are achieved
- Analyzing water for cyanotoxins/cyanobacteria
- Auditing the plan
- Surveying user experiences
6.2.6
Management and Communication
Documenting the planning process and outcomes
- Developing management plans for routine activities and emergency responses
6.2.7
Developing supporting programmes
6.2.8
Feedback and Improvement
Periodic review and revision
- Conducting planned reviews or following incidents/near-misses
- Revising as necessary
6.2.9

Figure 6.2 Selected steps of developing a Water Safety Plan (WSP) for a specific water-use system, and corresponding subsections within this chapter. Following these steps is effective for managing cyanotoxin control also when no full WSP is developed.

waterbody mixing) to limit stratification. An important principle of water safety planning is the ongoing routine monitoring of control measures (referred to as "operational monitoring"), which shows that control measures are working within acceptable operational limits, showing that the hazard continues to be managed effectively on an ongoing basis. For control measures to be effective in the longer term (e.g., existing control measures or new, additional control measures implemented in consequence of the risk assessment process), it is prudent to include considerations of how climate change may impact on the given water-use system.

In addition to upgrading the system to effectively control cyanotoxin hazards, an important element of water safety planning is to include triggers for initiating the short-term management actions (e.g., management of incidents through emergency response plans) to avoid human exposure. A good way to do this is to integrate Alert Levels Frameworks based on the suggestions given in section 5.1 for drinking-water and in section 5.2 for recreational water use. These frameworks outline which exceedances may trigger which responses. An incident response plan, as discussed in Chapter 15, is a further integral part of a WSP.

Although originally developed to ensure the safety of drinking-water supplies, WSPs are not limited to this application – the WSP framework can equally be adapted and applied for the assessment and management of risks from other potential exposure routes, in particular through recreational water contact. For fish and shellfish, the HACCP (Hazard Analysis Critical Control Points) concept widely required for food production includes very similar steps (as the WSP concept was developed for drinking-water from the widely used HACCP concept for food production) and can readily be linked to the WSP elements which may better address the catchment and waterbody aspects. So can sanitation safety plans (SSPs) – an application of the same concept to systematically identify and manage health risks along the sanitation chain. This approach may also be applied to limit nutrient inputs into water-use systems.

A strong feature of WSP development is documentation, not only of the WSP itself as an outcome of the process, but also of the rationale behind the decisions taken and of the uncertainties as well as the information gaps identified. This chapter therefore illustrates the type of questions to address and considerations to document with worked examples for three situations differing in size, catchment, technology and access to monitoring.

Developing a Water Safety Plan (WSP) is a process typically conducted by a team using the steps described in the following sections. As emphasised above, even where a full WSP is not developed, these steps represent a useful systematic framework for assessing and managing cyanotoxin risks.

6.2.1 Getting ready for assessment and planning: forming the team

The first step is to establish an experienced, multidisciplinary team whose role is to develop and drive the day-to-day implementation of the plan. Assessing and controlling the risk of cyanotoxin occurrence tends to require a broad range of expertise: for example, setting targets in terms of concentrations and loads of phosphorus requires an in-depth understanding of the specific waterbody's hydrological conditions, the land uses and nutrient dynamics. In contrast, setting performance targets in water treatment requires engineering and operational knowledge. Such expertise is spread across different institutions and stakeholders. Moreover, the stakeholders to involve for effective cyanotoxin control tend to span quite a range of responsibilities and areas of influence, that is, for activities and hydrological management in the catchment, for managing the waterbody as well as for drinking-water abstraction and treatment. They may also include those using a waterbody for recreation, irrigation or other workplace exposure to spray, as well as fisheries. A drinking-water supplier alone cannot make decisions on measures in the catchment or waterbody (unless the supplier owns the catchment). Rather, the implementation of measures beyond the water supply depends on good cooperation with the other stakeholders. For example, for the catchment aspects of a WSP, an authority responsible for catchment management may already have a leading role in this area. Collaboration in system assessment is also essential in order to gain access to the information needed from the different stakeholders. Furthermore, including stakeholders – particularly for land use in the catchment – early on in the process can develop a sense of ownership and involvement that will facilitate the subsequent implementation of management measures.

This is why developing a WSP typically begins with forming a team of experts. Whether a full WSP is developed or only an approach to controlling cyanotoxin occurrence is sought, such a team is an excellent platform for bringing together stakeholders and information for interdisciplinary and intersectoral collaboration. The team needs the participation of technical operators as well as that of senior managers. The full support of the leading management is essential for allocating staff time and resources. This is particularly relevant when developing a WSP but also applies when the scope of planning is limited to controlling only cyanotoxins. It is also important for later acceptance of control measures and subsequent day-to-day practices, including the implementation and monitoring of system improvements.

Such a team is most effective if it includes people with the competence needed to analyse the factors leading to cyanotoxin risks and the efficacy of the measures in place to control these risks, as well as staff with the authority to implement any further measures decided upon. However, it is useful to limit the core team to those needed throughout the process of developing the WSP, particularly to ensure that all key stakeholders identify themselves with the process and its outcome. In contrast, those needed for the clarification of specific aspects are best included on an *ad hoc* basis, that is, only when these aspects are discussed. Such specific expertise may include the fields of:

- phytoplankton ecology to understand the likelihood of bloom occurrence;
- nutrient dynamics to set adequate targets for nutrient concentrations and nutrient loading and to propose measures to achieve these targets;
- drinking-water treatment to set performance targets that ensure cyanotoxin removal and – if necessary – to propose further measures to better achieve these targets;

- analytical skills, ranging from cyanobacterial identification and quantification to cyanotoxin analysis, depending on programmes to be implemented;
- public health and/or water quality who can advise on the health impacts of cyanotoxins and support risk assessment;
- integrated water resource management;
- emergency response planning;
- recreational water management;
- food safety management;
- occupational health and safety management;
- integration of climate change considerations (e.g., climatologist, hydrologist, strategic planners, climate change and public health risk specialist).

Representatives from the relevant end-user groups should also be involved in the process at key stages (e.g., drinking-water users associations, community groups, recreational groups). This can provide important user perspectives from "on the ground", particularly in relation to catchment activities. Such involvement also serves to ensure that the relevant end users are informed and support the process – and thus the longer-term effectiveness and sustainability of its outcomes.

A team leader should be designated who drives the process of Water Safety Plan (WSP) development. If the most important or most sensitive use of a waterbody is the provision of drinking-water, it is usually most effective for the team to be led by the water supplier, while including relevant experts and decision-makers from the catchment and waterbody. However, in certain contexts, and depending on the most sensitive use of water, the WSP may be driven by the authority responsible for public health or for management and protection of surface water. Table 6.1 shows three examples of how teams may vary in size and expertise, depending on the requirements and available options in the respective setting.

It is further useful to define and record the roles and responsibilities of the team members, potentially differentiating between core team members (i.e., those who are responsible for the more day-to-day aspects of WSP implementation) and those who support specific parts of the WSP development. Challenges include finding stakeholders in the catchment who are willing to be involved, potentially with the consequence of changing their way of doing things in order to reduce nutrient loads to the waterbody, holding regular meetings with team members from different organisations over a longer period of time, and finding and involving individuals with sufficient expertise. Chapters 7–10 therefore include guidance on the scientific and technical expertise that may be required. Generic aspects of team formation are discussed – with examples of challenges and benefits – in the Water Safety Plan Manual (Bartram et al., 2009) and Water Safety Planning for Small Community Water Supplies (WHO, 2012).

Table 6.1 Three example settings: team composition for assessing and managing the risk of cyanotoxin occurrence

Examples of settings	*Team composition for each of the three settings*
1: Slow-flowing large river serving as raw water source for drinking-water for a town of 500 000 inhabitants	*Core team*: the water supply's technical manager (team leader); two operators responsible for abstraction and treatment; an officer from the local public health authority; a senior officer from the water board; a representative from the catchment land-use association. *Expertise consulted on an ad hoc basis*: a limnologist, a hydrogeologist with experience in modelling nutrient loads, a microbiologist, a climate expert. *Initial management decision*: thorough system assessment to be undertaken; full support to be provided by staff with dedicated allocation of staff working time for this purpose; regular presentation of interim results at staff meetings to be undertaken.
2: Reservoir serving about 7000 people (three villages and a number of farms)	*Core team*: one engineer (leader), an officer from the local public health authority, an officer from the environmental authority and a representative from the local boating club. No funding is available for external support, but a cooperation with the hydrobiological faculty of a nearby university will be undertaken for scientific support; there will be participation in some relevant scientific and management meetings for knowledge transfer.
3: Farm dugout serving as drinking-water source for 20–50 people	The health authority has identified that farm dugouts pose a risk to human health, both because incidents of diarrhoea are attributed to *Cryptosporidium* and because cyanobacterial blooms are generally frequent in the region. It has therefore invited farm owners in the region to a series of workshops for assessing the situation and finding appropriate management solutions, and the owner of this farm is developing a WSP for her water supply on the basis of this support. *Core team*: the farmer has formed a three-person team for her farm with herself as the designated team leader, with support from an officer from the health authority and the farm manager. An engineer has been contracted for consultation to join relevant team meetings.

6.2.2 Describing the water-use system and its users

A thorough understanding of the system – from the catchment to user, that is, the point of exposure – is the basis for identifying and assessing hazards/hazardous events, existing control measures and risks. To facilitate this, an accurate and up-to-date system description should be prepared, which can contribute to system understanding and support the identification of system vulnerabilities and informs the subsequent hazard analysis and risk assessment (Table 6.2). A flow diagram is a helpful tool for visualisation, which can help support the identification of system vulnerabilities in subsequent steps.

A comprehensive description of the water-use system should begin with an inventory of conditions in the catchment that determine water flow, that is, the water budget, as well as an estimate of potential pathways for nutrient loading through erosion, seepage, inflows and tributaries. The target is to document geographical and hydrogeological conditions as well as land use that may affect nutrient loads, for example, agriculture, direct discharge of wastewater (including information on wastewater treatment efficacy for removing phosphorus and nitrogen), indirect discharge of wastewater (e.g., seepage/overflow of on-site sanitation systems such as septic tanks or latrines) or manure, drainage from roofs and roads. (For further information and examples of catchment inventories, refer to "Protecting Surface Waters for Health"; Rickert et al., 2016.) The description further includes the morphological, hydrophysical, chemical and biological characteristics of the waterbody. If drinking-water is abstracted, the description covers offtake site(s) and patterns as well as the steps of the treatment train. Where recreational use may lead to cyanotoxin

Table 6.2 Examples of basic system descriptions as a basis for assessing the risk of cyanotoxin occurrence

Example 1: *Slow-flowing large river serving as raw water source for drinking-water for a town of 500 000 inhabitants – summary system description.*
90% of the (mostly cross-border) catchment is used for agriculture.
Sewage from smaller settlements is discharged up to 50 km upstream (this is treated, but without nutrient removal [i.e., phosphorous or nitrogen]).
Larger cities are located further upstream.
Heavy cyanobacterial blooms occur every summer with lower biomass persisting in winter; no cyanotoxin data available; climate projections indicate that warmer water temperatures might exacerbate seasonal blooms during the summer.
Drinking-water treatment is in place (system is >30 years old) with pre-oxidation, coagulation/ flocculation, filtration and disinfection; powdered activated carbon is stored but rarely used (operators consider the black dust too messy to handle).
Water is distributed via three intermittent storage tanks and piped to households.

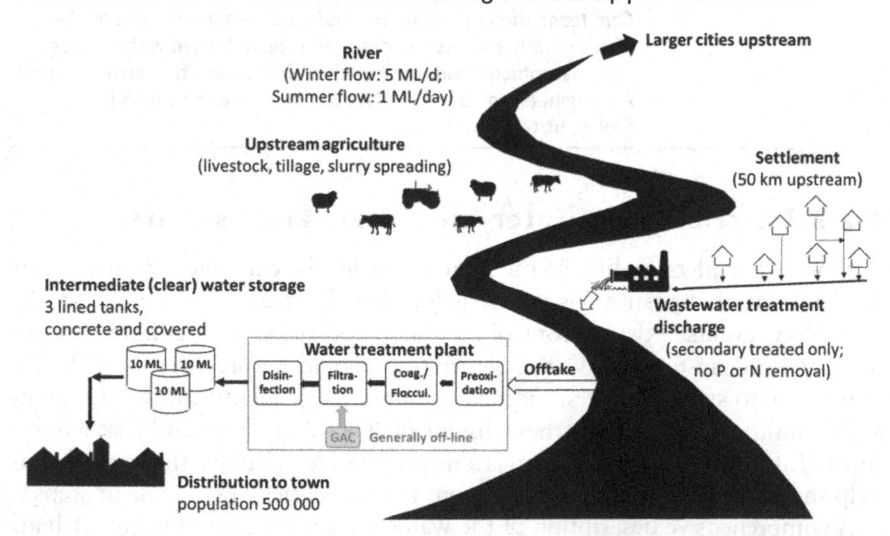

(Continued)

Table 6.2 (Continued) Examples of basic system descriptions as a basis for assessing the risk of cyanotoxin occurrence

Example 2: *Reservoir serving about 7000 people, that is, three villages and a number of farms – summary system description*

Reservoir is located uphill from the villages and farms in a middle-range mountain area (see diagram for hydrological details); 70% of the catchment is used for forestry and some hunting; 20% is rocky and not used; meadow area (≈5%) is used for extensive sheep farming.

Ecotourism has developed over the past 10–15 years, including a small hotel ("Ecolodge") with 60 beds, a restaurant, boating club and up to 500 week-end restaurant visitors; the drinking-water supply is from its own spring; wastewater effluent discharges into a septic system (no documentation on the system or effluent volumes discharged is available).

The meadow ends in a small beach with a steep rock which is very popular for diving (no official bathing site, not monitored).

No data on phytoplankton or cyanobacteria, but locals and tourists occasionally report either reddish discoloration (particularly in autumn and winter) or thin bright-green scums in the bay used for swimming.

Drinking-water treatment for the villages and connected farms is limited to flocculation, filtration and disinfection, which may be insufficient during blooms. Piped distribution system is on-premises. Recreational use is banned (to avoid human pathogens in the reservoir) but enforcement is poor.

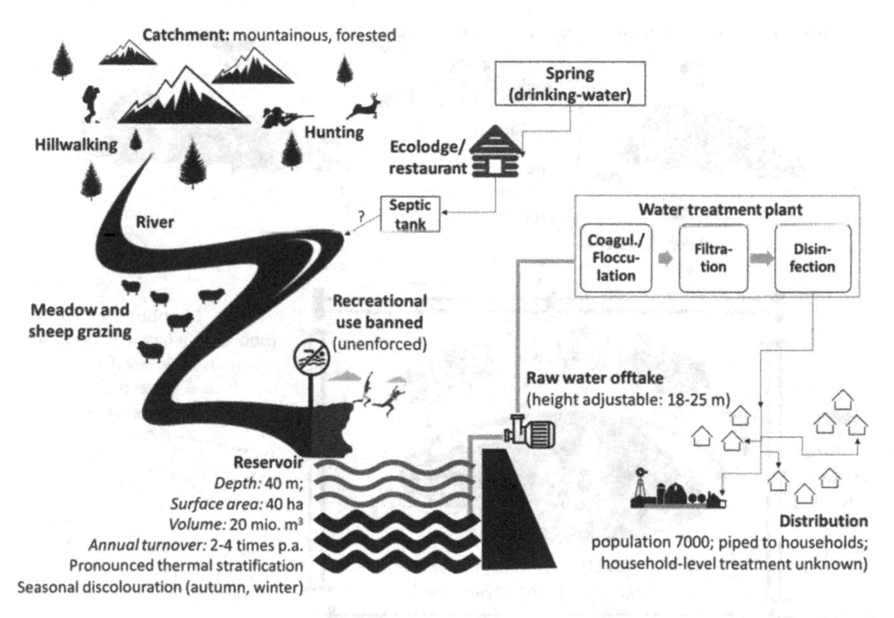

(Continued)

Table 6.2 (Continued) Examples of basic system descriptions as a basis for assessing the risk of cyanotoxin occurrence

Example 3: *Farm dugout serving as water source for 20–50 people – summary of system description*

Farm dugout is located at the foot of a hill that is used for cattle grazing; summer cyanobacterial blooms are common.

The extent of groundwater *versus* surface run-off is unclear but slope and the traces of erosion suggest a fair amount of direct run-off; vegetation cover is only grass and appears ineffective in intercepting run-off carrying phosphorus or nitrogen (or pathogens like *Cryptosporidium*) from cattle manure; future climate projections indicate more intense rainfall to be expected.

The fence around the dugout is intact and effective in keeping cattle out.

No alternative drinking-water source is available within reasonable distance.

Drilling a deeper well is not an option in this rocky area; funding for expensive interventions is not feasible.

Water for household use is piped directly at 1–1.5 m depth; filtration device of unclear design is in place for food preparation.

Site inspection showed a small self-made diving board, with about 15 children in the water (sailboard, swimming, diving, playing in moderate scum).

Impact on local farm produce marketed (e.g., meat, milk and grains) is unlikely; impact on vegetables grown for own use is a possibility, the risk of which is to be included in the assessment.

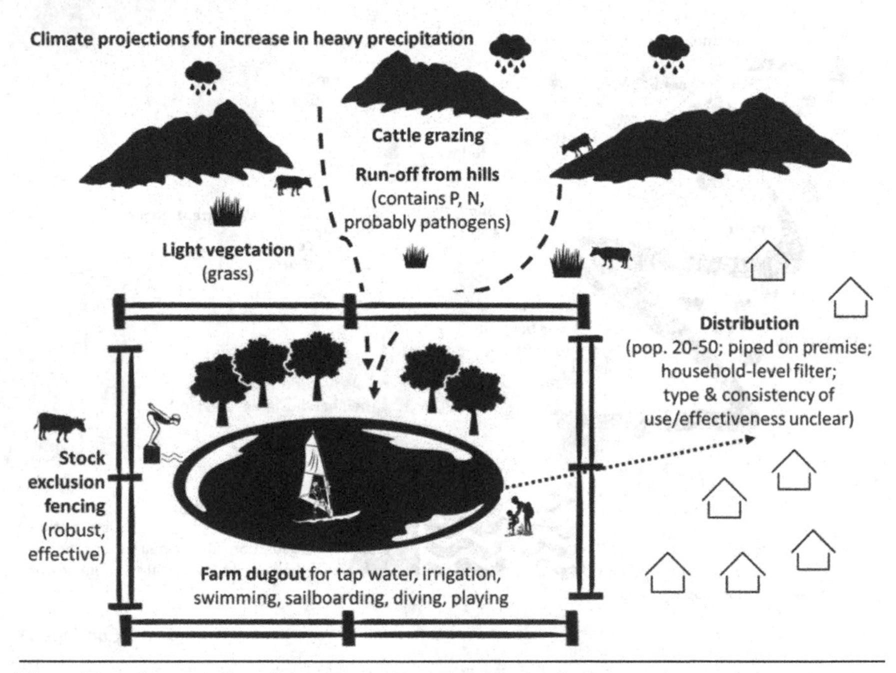

Note: Typically, system flow diagrams would include more detailed information to support subsequent hazard identification and risk assessment, including quantitative or relative relevance information for specific loads and pathways, detailed flow diagrams for water treatment plants.

exposure, information on bathing sites and other water sport activities relative to prevailing wind directions is important, including time patterns of such activities in order to assess critical periods for exposure. This also applies to other direct water uses potentially causing human exposure, like spray irrigation. Where exposure through food (irrigated vegetables, recreational angling, professional fisheries, aquaculture and mussel harvesting) may be quantitatively relevant, information on the affected crops, fish or mussel species, amounts typically consumed and time patterns of consumption may be important to estimate a potential cyanotoxin dose via food.

Visual inspection of the waterbody and its catchment is a highly valuable basis both for collecting information on the aspects listed above and for validating information gleaned from documents and interviews. This process of "walking the system" can help identify potential threats to the waterbody which cannot be identified through desk-based assessment alone (e.g., activities that are not permitted, but are occurring nevertheless). Such an inspection is best prepared by collating documented information, for example, from authorities managing the waterbody and its tributaries, authorities responsible for issuing permits for discharges into water courses or activities in the catchment, operators of enterprises and activities in the catchment as well as from site users. The latter may include professionals as well as the public, for example, people observing scums and greenish turbidity. Including any available (semi)quantitative information about loads of phosphorus or nitrogen to the waterbody, expected and/or measured concentrations within the waterbody, cyanobacterial biomass or cyanotoxin concentrations is highly useful for the steps of identifying hazards and conducting a risk assessment. Examples of key information to support the description of the water-use system include the following:

- *for the catchment*: Activities and conditions likely to lead to erosion and nutrient input, particularly during storm events;
- *for the waterbody and the catchment*: Where available, impacts anticipated in the wake of developments such as land use or climate change, both those currently known and those anticipated for the future;
- *for the waterbody*: Water quality data as available, in particular nutrient concentrations, Secchi depth readings, phytoplankton data (biomass and extent of cyanobacterial dominance); potentially also data on organisms at higher trophic levels as these may impact on phytoplankton biomass and species composition; any climate change scenario projections available;
- where available, data on cyanobacterial and cyanotoxin occurrence and any indication of human or animal illness suspected to have been caused by these bacteria;
- *if the waterbody is used as a drinking-water resource*: A description of
  - the drinking-water treatment train (e.g., offstream storage reservoirs before or after treatment, pretreatment [e.g., addition of powdered activated carbon, oxidation], coagulation/flocculation,

sedimentation, dissolved air floatation, filtration, ozonation, granulated activated carbon (GAC) filtration, slow sand or riverbank filtration, disinfection [e.g., chlorination, UV irradiation]);
  - the amount of water produced and the households it serves (potentially as map);
  - a map of the mains, including reservoirs in the distribution system and their condition, whether these are covered or open (relevant also for other hazards potentially introduced during distribution) and retention time (relevant for cyanotoxin degradation in the mains);
- information on the water users (including any sensitive population groups) and for which purposes it is being used (e.g., drinking, other nondrinking household uses [such as washing, bathing], or sensitive applications [like dialysis]; see below).

Chapters 7–10 give detailed checklists on information to collect and evaluate about the catchment, the waterbody, the location of abstraction points for drinking-water and for bathing sites, and drinking-water treatment. These checklists are intended to support both the description of the system and subsequent risk assessment. Table 6.2 introduces the three examples used in this chapter with a short summary of system descriptions that highlight the potential variability in coverage and amount of information available. Table 6.2 includes basic flow diagrams that provide an illustrative example of how these tools can support the identification of system vulnerabilities.

The typical situation is indeed that not all of the desirable information will be available. The Water Safety Plan (WSP) concept promotes "incremental improvement", that is, encouraging to get started in the first instance, and improve the WSP stepwise over time as capacity and system knowledge build and resources become available. As such, a first iteration of assessing the system and the risk of cyanotoxin occurrence should begin even in the absence of all of the necessary information, to find out which information gaps are the most crucial for the decisions that need to be made. If these gaps prove relevant for the assessment, they will be the first ones to address with targeted programmes. It is important to validate the description of the system through site inspection, particularly from the perspective of identifying any illegal/unauthorised activities in the catchment, and to accurately document the system description.

6.2.2.1 Identifying water users and uses (including sensitive subpopulations)

Assessing who uses the water and for which purposes helps evaluate the public health risk arising from the exposure of the respective population. Information about any groups of the population with specific exposure risks or particularly susceptibility to specific hazards provides a basis for specifically targeted information and warning. In the case of toxic cyanobacterial blooms, such groups might include the following:

- those preparing dialysis water as well as dialysis users, as dialysis directly exposes people intravenously to large amounts of water (approximately 120 L per treatment), thus increasing toxicity by at least an order of magnitude (section 5.4). Early information if cyanobacteria build up in the waterbody and cyanotoxins may occur is important, as even trace amounts are of concern for dialysis;
- users of private drinking-water supplies using surface water or shallow wells strongly influenced by a waterbody with blooms (e.g., supplies for holiday houses);
- people potentially exposed via aerosol and small water droplets from the waterbody, for example, through spray irrigation, decorative fountains, or water use for cooling;
- recreational user groups, particularly where exposure is frequently repeated (e.g., sports clubs, lakeside campsites) or where direct contact with bloom material (e.g., water skiing, wind surfing) is likely;
- operators of fisheries and consumers of fish/shellfish.

Using the three examples in this chapter, Table 6.3 highlights how such information may be documented, including existing information gaps to close with high priority.

6.2.3 Assessing the risk of cyanotoxin occurrence and the system's efficacy in controlling it

The next step aims to identify:

- Conditions (in particular eutrophication) potentially causing cyanotoxin occurrence.
- Events augmenting this occurrence (e.g., extended spells of warm weather).
- The efficacy of the existing control measures in place (if present) to control the occurrence of cyanobacteria.
- The likelihood of occurrence and severity of the consequences (or impact), resulting in assessment and prioritisation of the risks.

This assessment requires an understanding of potential nutrient sources in the catchment causing eutrophication, conditions causing them to reach the water source, how these pathways are best controlled, whether these existing controls are effective, and if controls at this level fail, how effective downstream control measures are at minimising exposure (e.g., where present, how effectively can downstream drinking-water treatment remove cyanotoxins).

The key steps of risk assessment in the context of Water Safety Plan (WSP) development are summarised briefly here – for detailed information

Table 6.3 Three example settings: water users/uses documented in the system description for assessing cyanotoxin exposure risk

Examples of settings	*Overview of water uses/users in each of the three settings*
1: Slow-flowing large river serving as raw water source for drinking-water for a town of 500 000 inhabitants	• Drinking-water for the population of the town, used also for standard household activities (washing, bathing, food preparation, etc.). • A hospital with a dialysis unit (information gap: find out whether it buys water specifically designated for dialysis or uses tap water and conducts its own treatment). • Water for irrigation pumped from the river, but currently not for vegetable or fruit crop with direct water contact. • Private angling – unclear how widespread.
2: Reservoir serving about 7000 people (three villages and a number of farms)	• Drinking-water for the villages and farms, used also for standard household activities. • Some recreational uses at the beach at the bay near the ecolodge, boating.
3: Farm dugout serving as water source for 20–50 people	• For standard household activities, including food preparation for the 20 persons living on the farm and for up to 30 workers commuting daily; inhabitants and workers emphasise that they drink bottled water. • Intensively for swimming by the children of the family, the farm workers and their friends (information gap: clarify awareness of the need to avoid ingesting water or inhaling aerosol and spay). • Irrigation of the farm's vegetable garden for own use; no produce sold.

on the theory and practical application of risk assessment through the WSP process, refer to the Water Safety Plan Manual (Bartram et al., 2009) or Water Safety Planning for Small Community Water Supplies (WHO, 2012).

For assessing the risk of cyanotoxins to occur, it is important to firstly identify hazardous events from the source through to the point of contact with the end user that may result in cyanotoxin occurrence and exposure. Such hazardous events may range from stormwater run-off and tributaries introducing nutrients resulting in bloom occurrence to failure of a water treatment plant component to remove toxins.

To assess the likelihood of cyanotoxin occurrence, any observations on the patterns of cyanobacterial occurrence in the waterbody are valuable, and people living near the waterbody or regularly visiting it may provide important information. Checklist 8.1 in Chapter 8 shows the type of questions to ask regarding direct indication of the occurrence of potentially toxic cyanobacteria. However, blooms can be short-lived periodic events and may well be missed unless observations are quite frequent (e.g., twice a month or even weekly during seasons in which blooms are most likely). Therefore, and to address the potential *causes* of blooms, it is important to understand which activities in the catchment (in particular intensive farming and wastewater inflow) might lead to nutrient loads causing eutrophication and whether conditions within the waterbody are conducive to bloom formation. To support this assessment, Chapter 7 discusses key causes of nutrient loads to the waterbody and provides checklists for assessing conditions and activities in the catchment that are likely to contribute them, and Chapter 8 discusses the conditions within it that determine phytoplankton biomass and species dominance.

The severity of the impact of cyanotoxin occurrence may be determined from the toxin concentrations in relation to the guideline values discussed in Chapter 2 and summarised in Table 5.1, from the size of the population affected, and the duration of the exposure (as depicted for recreational exposure in Table 5.4 in section 5.2).

When determining the likelihood and severity of the consequences of a particular hazard or hazardous event, it is important that the risk assessment include the identification and assessment (or "validation") of the existing control measures that are already in place within the water-use system to determine

- whether the control measures in place are fundamentally capable of effectively controlling the hazard/hazardous event;
- any information gaps and uncertainties of this assessment.

Risk assessment matrices relating the likelihood of hazards/hazardous events to occur against the severity of their impact are frequently used to help understand and make transparent the underlying assumptions leading the assessment. Note that such an assessment can not only be made for a drinking-water system, as is typically done when developing a Water Safety Plan (WSP), but can also be adapted for the whole of the relevant exposure pathways to water containing cyanobacteria, as is the case in the three examples in Tables 6.1–6.7. While such assessments are inevitably somewhat subjective and their value does not lie in finding "absolute truth", they prove valuable for stepwise, systematic and consistent identification, assessment and prioritisation of risks, particularly if the team agrees on definitions for likelihood and severity prior to the assessment. Their value particularly lies in making transparent the (otherwise merely implicit) assumptions that drive decisions on implementing control measures. This transparency makes the assumptions accessible to debate and the decisions accessible to potential improvement.

If the risk assessment identifies that the risk is not adequately managed (e.g., there are no control measures in place, or the existing control measures in place are insufficient to effectively manage the risk), the next step is to suggest upgrading the control measures or to propose new ones for implementation (typically documented in an "improvement plan"; see section 6.2.4).

If the outcome of the risk assessment is that the risk is adequately controlled, typically no further improvement actions (or additional control measures) are required. This would be a very important outcome of the risk assessment: if it shows, for example, that nutrient loads to the waterbody are low and not likely to lead to concentrations supporting a substantial biomass of cyanobacteria or that conditions within the waterbody are not conducive to cyanobacterial blooms, it is worthwhile to understand which conditions and control measures are currently maintaining the good situation. An important outcome of the assessment then is that these beneficial conditions should be maintained. For example, if the assessment shows land in the catchment to be chiefly covered by pristine vegetation and in parts used for forestry, a decision may be to maintain this for sustainable future use of the water resource rather than to re-designate it for farming or urban development. It is also important to ensure proper documentation of the assessment as basis for any future planning of land use and permits for new activities in the catchment.

For the three case examples used in this chapter, Table 6.4 shows how the respective teams assessed the health risks due to cyanotoxin occurrence.

Table 6.4 Three example settings: outcomes from assessments of the risk of cyanotoxins to occur in health-relevant concentrations and the reasoning leading to the assessments

Example 1: Slow-flowing large river serving as raw water source for drinking-water for a town of 500 000 inhabitants

		Severity of public health impact		
		Minor impact	*Moderate impact*	*Major impact*
Likelihood of occurrence	*Often*		Exposure to cyanotoxins in drinking-water (chiefly microcystins) due to ineffective water treatment	
	Occasional			
	Rarely	Exposure to cyanotoxins due to recreational contact (bathing)[a] Exposure to cyanotoxins due to the consumption of contaminated foods[a]		

Dark grey = high risk; Medium grey = medium risk; Light grey = low risk

Rationale:

Blooms, particularly of *Microcystis* and *Planktothrix agardhii*, documented by the water board and the drinking-water supplier, last up to 3 months during summer; they were extreme during unusually extended periods of drought 1 and 3 years ago.

Phosphorus concentrations of 60 to >250 µg/L sustain blooms; very slow river flow promotes their persistence; sources are primarily from two other upstream countries

(Continued)

and reduction only possible through agreements established in the international river basin commission, possible only in the longer term.
Shifting the site for the drinking-water offtake is not an option, as cyanobacterial biomass is evenly distributed across the river profile.
Drinking-water treatment will currently remove only cell-bound cyanotoxins and reliability of operations for this is not certain.
Resulting exposure risks are as follows:

- high for drinking-water, as current treatment system is inadequate for the removal of dissolved cyanotoxins;
- low for recreation; this use is banned because of heavy ship traffic;
- low for food – no professional fishery, almost no private angling; crops directly irrigated from the river are limited to grain and fruit trees (none that are eaten directly such as lettuce or strawberries).

Example 2: Reservoir serving about 7000 people – three villages and a number of farms

		Severity of public health impact		
		Minor impact	*Moderate impact*	*Major impact*
Likelihood of occurrence	*Often*			
	Occasional	Exposure to cyanotoxins (including potentially neurotoxic ones) due to recreational contact (bathing)[a]	Exposure to microcystins (chiefly from *P. rubescens*) in drinking-water due to ineffective water treatment, probably at low concentrations based on likely low cell densities (<10 μg/L)	
	Rarely			

Dark grey = high risk; Medium grey = medium risk; Light grey = low risk

Rationale:
High levels of turbidity at the thermocline but rarely in raw water; visual reports of sometimes wine-red and sometimes bright-green thin surface films covering a small part of the shoreline water; this indicates *Planktothrix rubescens* (red) and *Dolichospermum* sp. (green) likely at low cell density.
Assessment is uncertain; improvement would be through data on phytoplankton species, toxins and nutrient concentrations, but nutrients are likely low because of the catchment conditions.
Resulting exposure risks are as follows:

- drinking-water risk from microcystins most likely low, but classified as medium as default assumption until data available;
- low for recreational exposure: even if the occasional scums consist of neurotoxic *Dolichospermum*, scum is too limited to cause substantial exposure;
- not given for food as no exposure pathways were identified.

Population potentially affected is small; financial resources are very limited; data gaps to be reduced where cooperation with the university allows, but not as high priority of the public surveillance authority.

(Continued)

Example 3: Farm dugout serving as water source for 20–50 people

		Severity of public health impact		
		Minor impact	*Moderate impact*	*Major impact*
Likelihood of occurrence	*Often*			
	Occasional	Exposure to cyanotoxins due to the consumption of contaminated foods (i.e., cyanotoxins on irrigated vegetables and fruit)[a]	Possible exposure to cyanotoxins in drinking-water due to uncertain reliability of household-level treatment Exposure (including young children) to cyanotoxins due to recreational contact (bathing)[a]	
	Rarely			

Dark grey = high risk; Medium grey = medium risk; Light grey = low risk

Rationale:

Local residents describe observations on bloom intensity (including some photographs) and duration typically lasting for about one week, sometimes more.

Results of three summer sampling campaigns: total phosphorus 40 – 55 µg/L, high turbidity (possibly due in part to suspended clay particles), and *Microcystis* biovolume up to 10 mm³/L, slight scum, microcystins 0.3 – 1.7 µg/L (determined by ELISA).

Phosphorus loads are uncertain; site inspection suggests primary source is run-off from pasture (mostly cattle) around the dugout.

Awareness of residents and workers is well developed: all individuals addressed during site inspection emphasised bad taste from the dugout water and drinking only bottled water (although the reliability of this appears uncertain for children).

Tap water appears to be filtered for preparing tea and coffee (a filter is installed under the kitchen sink); efficacy of the filter is unclear; it is also unclear to which extent this is used for preparing food.

Exposure of children due to recreational activities is evident.

Resulting exposure risks are as follows:

- likely low for drinking-water but provisionally documented as medium given that the uncertainty of the assessment is substantial;
- moderate for recreational use: exposure of children, including small ones, is repeated at almost daily intervals for many successive weeks on end, but often without the presence of blooms;
- low for food, while not totally to be excluded if lettuce and strawberries (grown for own use) are irrigated with scum (thorough washing should be recommended).

Note: The above tables include a number of risks (denoted by [a]) that are not typically considered under a conventional WSP for drinking-water, but would be assessed in a risk assessment/management plan for recreational water safety, or for food under the similar principles of HACCP.

6.2.3.1 Coping with uncertainty

Collating in-depth information for risk assessment can be an extensively time-consuming exercise, and often information will not be readily available. Estimating nutrient loads from the catchment, understanding the ecological interactions within the waterbody and collecting information on socio-economic aspects of water-use patterns may be particularly challenging. Therefore, it is necessary to consider the detail of information required for making decisions – possibly only preliminary ones, documenting uncertainties and information gaps along with the information. Risk assessment will then show which information gaps and uncertainties most urgently need to be closed, that is, those for which uncertainty precludes decision-making.

For example, reducing uncertainties in estimating a nutrient load may be critically important in a setting where controlling nutrient loading is the key measure to avoid cyanobacterial proliferation and thus cyanotoxin occurrence (e.g., an upstream reservoir in a fairly pristine catchment). In contrast, in a setting where nutrient loads and concentrations in the waterbody are already excessively high and not readily amenable to local control (e.g., a downstream river reach with complex transnational nutrient sources), identifying key nutrient sources and taking action to reduce loads remain important in face of long-term benefits, but in the shorter term, other control measures may need to take priority.

Therefore, risk assessment is an iterative rather than a linear process: it should incorporate the WSP principle of "incremental improvement" as described above, with an emphasis on the importance of getting started and improving over time as information becomes available.

For the three case examples used in this chapter, Table 6.5 shows how the respective teams described the uncertainties of their risk assessments.

Documenting uncertainties and making them transparent, including information gaps to close, are important. This will inform decisions on which measures to take first – whether these should be interventions to reduce exposure or rather programmes to collect data and information before decisions on any investments into measures are to be made.

6.2.3.2 Cyanotoxin risks in relation to other public health risks from exposure to water

A key purpose of risk assessment in water safety planning is to determine priorities for maintaining, upgrading or implementing measures to control public health impacts from the hazards identified. For the overall target of protecting public health, it is important to assess the public health risk from cyanotoxins in relation to that from other hazards/hazardous events potentially occurring in the water. This is also useful because some of the events causing other hazards will also cause cyanobacteria – for example, sewage loads carry both nutrients that support blooms and pathogens. A comprehensive risk assessment would be developed in the context of developing a full WSP, but it is also valuable to contextualise the potential cyanotoxin risk even without developing a complete WSP. Generally, public health risks from pathogens in the water are likely to be of higher priority

Table 6.5 Three example settings: uncertainties arising during the assessments of the risk of cyanotoxins to occur in health-relevant concentrations

Examples of settings	*Uncertainties of the risk assessment for each of the three settings introduced in Table 6.1*
1: Slow-flowing large river serving as raw water source for drinking-water for a town of 500 000 inhabitants	Uncertainty due to the lack of cyanotoxin data, but the dominant cyanobacterial taxa almost certainly contain microcystins, and steps available in the drinking-water treatment system – while probably removing cells containing toxin – cannot remove dissolved toxins and may well lead to the lysis of some cells, thus releasing further toxins. Therefore, even without toxin data, exposure to microcystin concentrations in the range of a few µg/L is likely, particularly as occurrence is ongoing for periods of several months on end. While uncertainty regarding the extent of private angling is high, it is clear that this lead to relevant exposure only for a small population, possibly for some people with low income frequently relying on fish from the river as a relevant source of protein.
2: Reservoir serving about 7000 people (three villages and a number of farms)	Uncertainty is considerable: although the visual reports suggest cyanobacteria to be the cause, it is unclear whether they indeed cause the discoloration. While catchment conditions do not indicate sufficiently high nutrient loads to support substantial biomass, this cannot be totally excluded, particularly in face of increasing tourism.
3: Farm dugout serving as water source for 20–50 people	Uncertainty regarding concentrations of phosphorus as well as cyanobacterial taxa and their concentrations is relevant, as the data from the three sampling occasions suggest them to be only slightly above thresholds for interventions to prevent human exposure, thus indicating that interventions to reduce phosphorus loads from erosion may be effective. As finances for a more intensive monitoring programme are lacking, uncertainty will be addressed by intensified visual inspection for blooms. Uncertainty also exists with regard to the efficacy of the point of use filters and whether filtered water is reliably used for food preparation or not.

due to the potential for severe acute illness, even death, and also because even a small number of people infected through exposure to water containing pathogens can communicate the infection to a potentially exponentially increasing number of others.

For the three case examples used in this chapter, Table 6.6 shows how the respective teams related the health risks due to cyanotoxins to other health risks in the respective setting.

Table 6.6 Three example settings: health risk assessments due to cyanotoxins in relation to risks from other hazards

Example 1: Slow-flowing large river serving as raw water source for a town of 500 000 inhabitants

		Severity of public health impact		
		Minor impact	*Moderate impact*	*Major impact*
Likelihood of occurrence	*Often*	Contamination of drinking-water with musty taste and odour (of unclear origin) possibly due to ineffective water treatment	Exposure to cyanotoxins in drinking-water (chiefly microcystins) due to insufficiently effective water treatment	Microbial contamination of the drinking-water due to failure of filtration in water treatment
	Occasional			Microbial contamination (i.e., *Legionella*) in household installations due to inappropriate management of internal plumbing systems
	Rarely	Exposure to cyanotoxins due to recreational contact (bathing)[a] Exposure to cyanotoxins due to the consumption of contaminated foods[a]	Chemical contamination of source water due to spills of hazardous chemicals	Drowning, injuries due to illegal swimming in boating channels *

Dark grey=high risk; Medium grey=medium risk; Light grey=low risk

Rationale: high risk from cyanotoxins (see Table 6.4); high risk of pathogen breakthrough, particularly of *Cryptosporidium* (which are resistant to disinfection). *Legionella* are known to have caused numerous cases of serious pneumonia and two deaths in two hotels and one senior citizens' residence. Public concern is high for spills of hazardous chemicals but actual risks are low, due to the lack of industry in nearer catchment (if they occur, concentrations would be low). Slight risk from musty taste and odour with increased risk if people use other less safe water sources. Two known incidents of teenagers severely injured by boats when swimming in spite of warning notices.

(Continued)

Example 2: Reservoir serving about 7000 people – three villages and a number of farms

		Severity of public health impact		
		Minor impact	*Moderate impact*	*Major impact*
Likelihood of occurrence	*Often*	Sunburn due to excessive exposure to sun as a result of underestimating the impact at 1000 m altitude[a]		
	Occasional	Exposure to cyanotoxins (including potentially neurotoxic ones) due to recreational contact (bathing)[a]	Exposure to microcystins (chiefly from *P. rubescens*) in drinking-water due to ineffective water treatment, probably at low concentrations based on likely low cell densities (<10 µg/L)	Cranial and spinal injury due to unsafe diving[a]
	Rarely		Microbial contamination of drinking-water due to the presence of inadequately treated human effluent from the bathing area	

Dark grey=high risk; Medium grey=medium risk; Light grey=low risk

Rationale: risks from cyanotoxins provisionally moderate for drinking-water and low for recreation (see Table 6.4). Diving injuries reported more than once a year; sunburn frequently. Pathogens from sewage seeping through rock fissures and from people using the beach are not likely to reach the waterworks (inactivation by long travel times in the reservoir).

(Continued)

Example 3: Farm dugout serving as water source for 20–50 people

		Severity of public health impact		
		Minor impact	*Moderate impact*	*Major impact*
Likelihood of occurrence	*Often*			Microbial contamination of farm reservoir used for drinking-water due to defecation and run-off from livestock
	Occasional	Exposure to cyanotoxins due to the consumption of contaminated foods (i.e., cyanotoxins on irrigated vegetables and fruit)[a]	Possible exposure to cyanotoxins in drinking-water due to uncertain reliability of household-level treatment Exposure (including young children) to cyanotoxins due to recreational contact (bathing)[a]	
	Rarely			

Dark grey=high risk; Medium grey=medium risk; Light grey=low risk

Rationale: moderate risks exists from exposure to cyanotoxins through recreational contact, low for foods (see Table 6.4). *Cryptosporidium* likely, due to cattle uphill of the dugout, which may cause severe illness, so risk is high.

Note: The above tables include a number of risks (denoted by [a]) that are not typically considered under a conventional WSP for drinking-water, but would be assessed in a risk assessment/ management plan for recreational water safety or for food under the similar principles of HACCP.

6.2.4 Improvement planning: choosing additional cyanotoxin control measures for system improvement

If the outcome of the risk assessment identifies that high-priority risks are not adequately managed, then upgrade of exiting controls and/or additional control measures are needed. These actions are typically documented in an "improvement plan", which should capture which improvement is needed, who is responsible for doing it, by when should it be done (i.e., reflecting its priority) and how the improvement will be funded.

Measures to control the risk of human exposure to cyanobacterial blooms range from simple physical interventions like vegetation buffer strips around a waterbody, or behavioural ones like banning recreational use of a waterbody, to more complex technical interventions like the implementation and use of appropriate drinking-water treatment trains. Examples of additional control measures to consider at the different stages of the water-use system are provided in Chapters 7–10, and for the three scenarios, they are presented in Table 6.7.

6.2.5 Monitoring the functioning of control measures for cyanotoxin management and developing a management plan

Validation determines that a control measure is fundamentally capable of controlling a hazard/hazardous event (see section 6.2.3). However, to determine that the control measure actually does continue to function effectively over time, routine monitoring is required (referred to as "operational monitoring"). This will show whether the control measure is reliably managed/operated such that it continues to provide effective protection. Ideally, operational monitoring should use quick and simple monitoring parameters (see below) that provide a rapid result so the performance of a control measure can be continuously determined, and if necessary, corrective action can be taken in an efficient and timely manner.

Operational monitoring also requires setting performance criteria for the respective control measure and critical limits which indicate if the measure is working within the established acceptable performance criteria. Furthermore, it is useful to define corrective action(s) to be taken if the monitoring shows that the control measure is no longer working within the critical limits. For example, for filtration to remove cyanobacterial cells in drinking-water treatment, turbidity, measured continuously at the outflow of each individual filter, is a simple operational monitoring parameter that indicates whether filtration is working optimally. Critical turbidity limits can be set, and if they are exceeded, this would indicate that the filtration processes are not operating optimally, triggering, for example, filter backwashing as the corrective action to restore optimal operation of the control measure.

This approach can be similarly applied to control measures in catchment or offtake management; for example, vegetation cover to prevent erosion from catchment areas identified as critical for the nutrient load to the waterbody can be defined as control measure, compliance to which can be monitored either by remote sensing or by periodic site inspection. If such monitoring detects violation, corrective action would be an immediate enforcement of revegetation and compliance to the dedicated land use. Likewise, adjusting the drinking-water offtake depth to avoid cyanobacterial intake can be defined as a control measure with online monitoring of

Table 6.7 Three example settings: additional measures to control cyanotoxin risks and their operational monitoring identified through WSP development

Examples of settings	*Additional control measures and their operational monitoring identified for each of the three settings*
1: Slow-flowing large river serving as raw water source for a town of 500 000 inhabitants	Implement Alert Levels Framework; install an online fluorescence analyser to indicate when cyanobacterial levels are >1 µg/L at raw water intake to trigger microscopy for cyanobacteria; incident response plans to be developed as part of an emergency response. Upgrade the drinking-water filtration system in the treatment train (see technical specification for details) to ensure an effective cell removal avoiding rupture and lysis (note: this will also reduce risks of breakthrough of disinfection-resistant pathogens). For operational monitoring: install online turbidity analyser (with corresponding "auto dial" alarming for operator notification) at the outlet of each filter.
2: Reservoir serving about 7000 people (three villages and a number of farms)	Any investment into treatment targeting cyanotoxin removal may well prove futile; as a first step, gain the necessary data via the university collaboration described in Table 6.1; decide on appropriate control measures only after the data are available.
3: Farm dugout serving as water source for 20–50 people	Plant a vegetation buffer strip of 10 m between the uphill pasture and the dugout (this likely represents a sufficient intervention to reduce loads from erosion; note: this will also intercept particles like pathogens, reducing infection risks). Encourage farm inhabitants and farm workers to continue to drink bottled water, to ensure children understand this, and to use packaged water for food preparation. Replace the filtration device in the kitchen by one with a carbon cartridge with regular renewal following the manufacturer's instructions. Ensure children understand the need to avoid swallowing water when using the dugout for recreation and to keep out of scum. Advise to water the vegetable garden via the soil rather than causing direct water contact with produce. Operational monitoring of the vegetation buffer strip through visual inspection – annually by the public authority responsible for oversight, by the farmer herself at monthly intervals as well as during and after stormwater events to look for traces of erosion and for immediate repair of any damage. Operational monitoring of behaviour by spontaneous random household surveys of people on the premises during inspections to check their awareness.

a characteristic cyanobacterial pigment, phycocyanin, with a specific fluorescence probe as a means of operational monitoring. Critical fluorescence limits can be set, and if they are exceeded, this would inform managers that they need to take corrective action by switching the offtake to a different depth or site, or temporarily ceasing raw water harvesting.

Operational monitoring aims to ensure that the water-use system is "proactively" managed to avoid human exposure to unsafe water (e.g., containing cyanotoxin concentrations exceeding the guideline values or prevailing national standards). Proactive management can thus be far more effective (and less costly) than reacting to water quality issues after they have arisen. Additionally, operational monitoring is more practical and cost-effective than relying primarily on cyanotoxin monitoring. Evidently, by the time violation of a land-use plan has led to cyanobacterial blooms that show up in cyanotoxin monitoring data, "fixing the problem" has become far more difficult. Similarly, by the time cyanotoxin monitoring data show that the water quality target for finished drinking-water is exceeded, the water has already reached the consumer, whereas routine process monitoring would indicate the development of the problem (e.g., declining filter performance) with time to fix it before it leads to high levels of toxin concentrations. Chapters 7–10 therefore include text and tables suggesting the selected examples of control measures that can be implemented for the respective targets as well as operational monitoring parameters that indicate whether the measure is working as it should.

Beyond their use for day-to-day operation, the data documented from operational monitoring of control measures can be highly valuable for system and risk assessment, as they may also indicate/validate how effectively a control measure is working. Documentation also supports the identification of trends over time and of conditions that may impact the efficacy of control measures (such as patterns of precipitation or drought).

Furthermore, a management plan should be developed which defines how the performance of key control measures is monitored and which corrective action should be taken if monitoring indicates poor performance, or if incidents occur (typically referred to as "operational monitoring plans", which may be part of standard operating procedures, SOPs). Operational monitoring plans for key control measures are important to ascertain their reliable performance at all times. These specify:

- **Operational monitoring parameters** for key control measures. An important criterion for the choice of the monitoring parameter is that it gives a result with sufficient time for taking corrective action before failure leads to cyanobacterial proliferation or cyanotoxin breakthrough and exposure.
- **Documentation of data from operational monitoring**: For each operational monitoring parameter, it is important to keep records of the monitoring data collected in order (i) to be able to trace what went

wrong and why in cases of incidents or to validate that the system was working well even when excessively challenged, for example, by a bloom, (ii) to allow the recognition of trends in the data which may indicate a decline in the performance of the control measure (e.g., gradual reduction in filter runtimes at a water treatment plant over time may indicate that the filter media needs replacing) and (iii) to demonstrate due diligence in managing the system.

- **Critical limits** for each of the monitoring parameters that show operators when the system is "out of bounds" and corrective action needs to be taken on time.
- **Corrective action**(s) to take immediately in case monitoring shows a process to be outside of the critical limits, that is, performance criteria are not being met, including lines of responsibility and communication.

6.2.6 Verifying that exposure is sufficiently avoided and water quality targets are achieved

Verification in Water Safety Plan (WSP) terms refers to obtaining evidence that the WSP is working as whole to deliver safe drinking-water. In the context of toxic cyanobacteria, verification may involve:

- **Compliance monitoring**, that is, water quality testing to ensure that water quality objectives (e.g., national standards) are being achieved; this may be concentrations of the toxins themselves (e.g., against the guideline values given in Chapter 2 and Table 5.1 or prevailing national standards) or of cyanobacteria in the waterbody: if sufficiently frequent monitoring of cyanobacteria (or measures indicating their levels of biomass such as the concentration of chlorophyll-*a* or even turbidity) shows that cyanobacteria are absent or only present at low concentrations, verification does not require monitoring toxin concentrations.
- **WSP audits** to ensure the WSP is up-to-date, is complete, is being implemented and is effective; depending on the local context, this may be carried out by internal or external bodies and may be a supportive assessment or more formal audit (which may or may not include penalties for noncompliance).
- **Surveying user satisfaction** may yield important information on, for example, taste/odour issues for drinking-water quality, observation of discoloration or odour/scum issues for recreational water use.

6.2.7 Documenting the planning process and outcomes

For an assessment as described above (whether or not it is conducted in the context of developing a WSP), it is important to document the considerations involved. This begins with a description of a water-use system. Maps of the catchment as well as water flow diagrams are useful not only for documentation, but also for understanding land-use patterns and how critical they are for the water budget and for nutrient loading. Flow diagrams can help conceptualise and visualise the points at which control can be exerted upon factors that affect cyanobacterial proliferation, toxin removal and water-use patterns. For each control measure, documentation should include the reasons for its choice and the targets it should achieve as well as how its adequacy for achieving the targets was validated. For control measures to be upgraded or newly implemented, documenting the rationale for such investments is important and provides reasoning for mobilising the necessary investments.

Documentation of the risk assessment and the criteria that led to its results is a necessary basis for successively further developing this assessment and for improving it. Therefore, this documentation should explicitly include information gaps and an assessment of how critical they are for making management decisions.

The target of documentation is a comprehensive overview rather than an extensive document. Where more in-depth information is needed, the overview best refers to further in-depth documents, like records of operational monitoring data.

6.2.7.1 Documenting management procedures

Management procedures should include the documentation of how to perform key operational activities (including operation of control measures) for normal operating conditions and incidents, as well as for emergency situations. For normal operating conditions and incidents, particularly for key control measures, typically this may take the form of SOPs (standard operating procedures), for example, day-to-day operation and monitoring of a water treatment filter, and what to do in an incident situation when this control measure fails. For these control measures, the level of detail useful in documentation will vary between settings. In general, such documents should be concise and readily available (including to technical staff). General experience is that after initial reluctance to document SOPs, they are found to be highly useful, particularly for maintaining "institutional memory", for keeping information accessible and for training new staff.

For emergency situations, emergency response plans should be developed (see also Chapter 15). They typically include the following example information:

- triggers for activating emergency response (e.g., detection of cyanotoxin levels above guideline values or prevailing national standards, or threshold values as given in the Alert Levels Frameworks in sections 5.1 and 5.2);

- steps to protect water quality/public health (e.g., initiate pretreatment step; issue "do not drink" advisory and provision of alternative water supplies; issue "do not swim/fish" advisory);
- general roles and responsibilities (both within the water supply or waterbody management entity as well as external stakeholders);
- communication protocols (internal to the management entity and external, for example, to stakeholders such as users, regulators, health authorities, environmental agencies, recreational groups, community groups);
- in the case of drinking-water, alternative/emergency drinking-water supplies (e.g., emergency provision of bottled water, water tankers and public collection points).

6.2.8 Developing supporting programmes

Supporting programmes are actions that contribute to drinking-water safety but do not directly affect water quality. Such programmes can develop capacity (e.g., water treatment plant operator training), can strengthen relationships (outreach and awareness raising for recreational user groups), and can create enthusiasm and buy-in to the process from key stakeholders. Figure 6.3 shows examples of supporting programmes relevant to cyanobacterial management.

6.2.9 Periodic review and revision

Land uses and population densities in catchments undergo change, resulting in changes in the nutrient load to the waterbody. The climate is changing, resulting in changes in hydrodynamics, precipitation and seasonal patterns

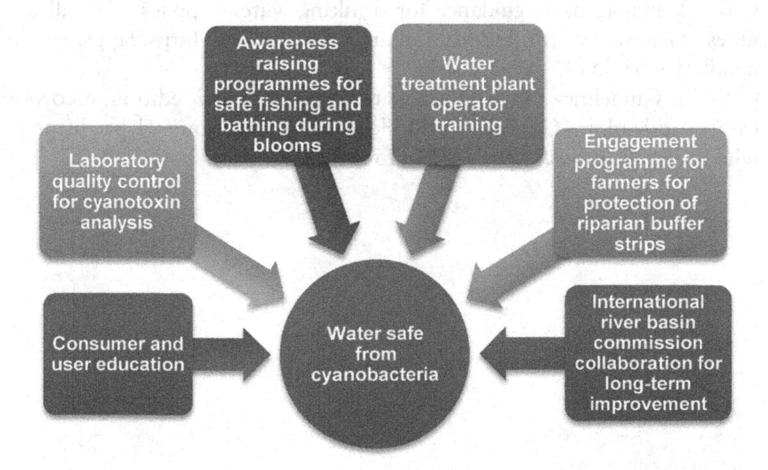

Figure 6.3 Examples of programmes to support the management of cyanotoxins in water-use systems.

of thermal stratification. Such changes may significantly shift phytoplankton species occurrence and lead to increases or decreases in cyanobacterial blooms. Also, conditions in drinking-water treatment plants undergo changes. Therefore, periodic revision is necessary to ensure that the assessments and measures are still appropriate. It is additionally important to incrementally improve the WSP, incorporating experience gained as well as further expertise, capacities and resources. Revision is also recommended after incidents/near-misses to document key lessons learned which may inform the review and strengthen key aspects of the plan: for example, a heavy cyanobacterial bloom is a useful opportunity to study whether the control measures in place have proven valid even in an extreme event, or whether an upgrade is advisable, or if the emergency response plans implemented were effective in protecting human exposure.

REFERENCES

Bartram J, Correales L, Davison A, Deere D, Drury D, Gordon B et al. (2009). Water safety plan manual: step-by-step risk management for drinking-water suppliers. Geneva: World Health Organization:103 pp. https://apps.who.int/iris/handle/10665/75141

Rickert B, Chorus I, Schmoll O (2016). Protecting surface water for health. Identifying, assessing and managing drinking-water quality risks in surface-water catchments. Geneva: World Health Organization:178 pp. https://apps.who.int/iris/handle/10665/246196

WHO (2003). Guidelines for safe recreational water environments. Vol. 1: Coastal and fresh waters. Geneva: World Health Organization:251 pp. https://apps.who.int/iris/handle/10665/42591

WHO (2012). Water safety planning for small community water supplies: step-by-step risk management guidance for drinking-water supplies in small communities. Geneva: World Health Organization:55 pp. https://apps.who.int/iris/handle/10665/75145

WHO (2017). Guidelines for drinking-water quality, fourth edition, incorporating the 1st addendum. Geneva: World Health Organization:631 pp. https://www.who.int/publications/i/item/9789241549950

Chapter 7

Assessing and controlling the risk of cyanobacterial blooms

Nutrient loads from the catchment

Ingrid Chorus and Matthias Zessner

CONTENTS

INTRODUCTION AND GENERAL CONSIDERATIONS

As discussed in Chapter 5, cyanobacterial blooms in surface waters are most effectively and sustainably controlled by limiting nutrient concentrations in the waterbody, and this requires sufficiently limiting the nutrient loads that it receives from its **catchment** (for terminology, see Box 7.1). These loads enter a waterbody from point sources such as discharges and sewage outfalls and from nonpoint sources (also termed "diffuse sources") such as surface run-off or drainage from fields. In some cases, inflow of groundwater may also carry significant nutrient loads. Furthermore, sediments may release nutrients, particularly phosphorus (P), into the waterbody. These releases are termed "internal loads", and they delay the decline of concentrations in the water after the external load has been reduced. However, already Vollenweider and Kerekes (1980) showed that in many cases, sediments are – on an annual scale – a sink rather than a source for phosphorus; thus, if the external load reduction is effective and water exchange rates are sufficiently high, the sediments will become a sink again, typically several years after load reduction. While such time spans may be of concern, particularly if a rapid remediation is necessary or water exchange rates are low, the first step for the target of reducing nutrient concentrations in the waterbody is to reduce the external load; otherwise, measures to reduce the internal load have little chance of being sustainably effective. Assessing the role of sediment nutrient release and

BOX 7.1: TERMINOLOGY

A **catchment** is the entire land area from which rain, snowmelt or groundwater drain into a waterbody, typically delineated by the crests of the hills or mountains that form water divides. Synonyms include "watershed", "river basin" and "drainage area". Catchments span a very wide range from fairly small, for example, for the close surroundings of hydrologically isolated ponds, to a continental scale for large rivers.

Point sources release nutrients to a waterbody at a single localised point of discharge, such as a sewage outfall.

Diffuse sources, also termed "nonpoint sources", are many smaller or scattered sources from which nutrients may be released to a waterbody, for example, from the land surface through rainwater runoff, through groundwater or from scattered rural dwellings. The combined impact of diffuse sources on the waterbody may be significant.

Riverine loads are the mass of a contaminant transported per unit of time, typically expressed as kg or tons per year. The nutrient load in the river reflects the sum of inputs upstream of the monitoring point at which these loads are calculated minus the possible retention in the river sediment. As such, these loads provide a first check: the sum of inputs from individual and separate sources should broadly equate to the total riverine load if retention is neglected for very rough estimation. More detailed investigations include retention.

Riparian buffer strips are the areas around a waterbody of about 10–30 m width or more, covered with dense vegetation which can effectively intercept surface run-off carrying phosphorus-rich soil eroded from arable land and pastures.

Tile drainage is a term used for draining land that would otherwise be too saturated with water for crops to grow. The term derives from installing drainage in a grid pattern covering the otherwise too moist field.

options for controlling it are discussed in section 8.6. This chapter focuses on assessing and managing external nutrient loads to a waterbody.

As outlined in Figure 7.1, the first step for this purpose is to estimate the maximum nutrient concentration in the waterbody that can be tolerated to effectively control (toxic) cyanobacterial blooms (section 7.1) and the corresponding load to the waterbody that may be tolerated to avoid exceedance of this target concentration (section 7.2). The next steps are to identify the main pathways and sources of nutrients (section 7.3) and to estimate the respective loads they contribute (section 7.4). The approach to estimating these loads may range from qualitative expert judgement (including that of local stakeholders) to quantitative load modelling (see tiers in

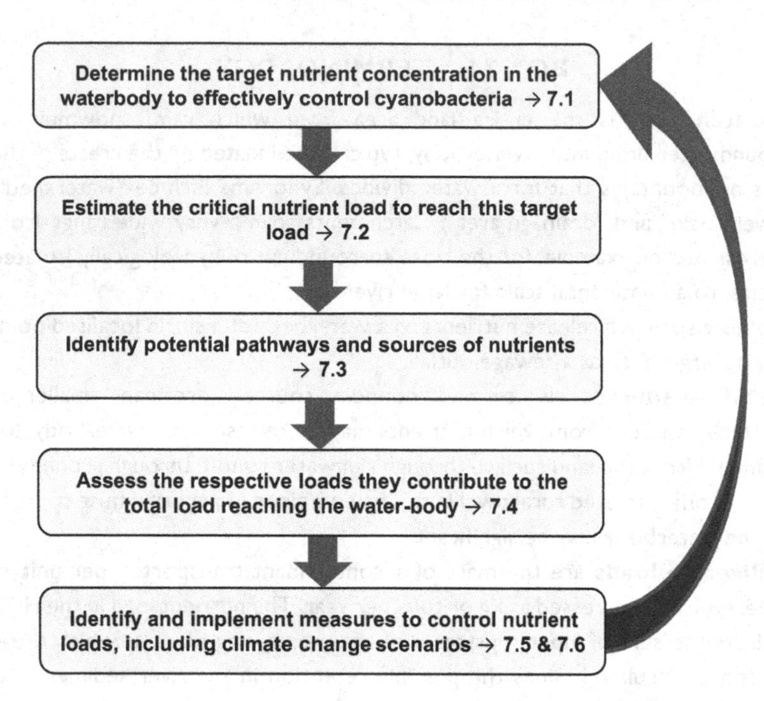

Figure 7.1 Steps in the selection of measures for controlling nutrient loads from the catchment.

section 7.4), depending on the available information. Once the loads from key pathways and sources are clear, the next step is to identify the most promising and most cost-effective measures to control them, to implement these measures, and to ensure they are operating effectively (section 7.5). After the implementation of measures, it is important to monitor whether they are taking effect as planned. This involves going back to assessing the nutrient load in order to validate that it has been sufficiently reduced, and it involves monitoring the nutrient concentration in the waterbody.

Implementing effective measures to control nutrient loads takes time, often several years, and particularly for P, it takes further time for concentrations in the waterbody to decline and for biota to respond to lower concentrations: Jeppesen et al. (2005) reviewed data on the responses to a nutrient load reduction of 35 lakes and found that at least 3 retention times (i.e., exchanges of the lake's volume) were necessary to dilute 95% of the excess P out of the waterbody and <10–15 years for total phosphorus (TP) to reach a new equilibrium between water and sediment; deep lakes tend to take longer. Although time scales of years or even decades may seem prohibitive, in the longer term controlling cyanobacterial blooms through keeping the nutrient load sufficiently low is the most sustainable approach, often rendering further (usually costly and continuously necessary) measures within the waterbody (discussed in Chapter 8) unnecessary.

The guidance given in this chapter for assessing and controlling nutrient loads from catchments to waterbodies is valid independently of the size of the catchment. Estimating nutrient loads from large catchments with a range of possible nutrient sources can be particularly challenging. For rivers crossing municipal, state or national borders, planning and management require collaboration across jurisdictions, and for transnational basins, international commissions have proven useful. For example, the Water Framework Directive of the European Union requires transboundary river basin management plans. Nutrient loads are more readily assessed and controlled for smaller, more readily controllable catchments, for example, those of reservoirs in middle-range mountains.

For further information and guidance on managing waterbodies and their catchments, readers are referred to the WHO guidebook *Protecting Surface Water for Health* (Rickert et al., 2016).

7.1 DETERMINING TARGETS FOR NUTRIENT CONCENTRATIONS IN THE WATERBODY

Which nutrient to address – P or both N and P – is a key question for planning measures to reduce the maximum possible amount of phytoplankton biomass – and thus of cyanobacteria – in a given waterbody. As discussed in Chapter 4, while in theory any nutrient could be limiting, in practice the macronutrients phosphorus (P) and in some cases nitrogen (N) are decisive for the amount of biomass that can occur. Moreover, if the concentration of one nutrient is sufficiently low, reducing those of others will not contribute to controlling cyanobacteria (see section 4.3.2 and Box 4.5). Reducing P loads to waterbodies has been widely successful, provided the measures taken achieved sufficiently low concentrations within the waterbody (Jeppesen et al., 2005; Phillips et al., 2008; Evans et al., 2011; Carvalho et al., 2013; Søndergaard et al., 2017), while there is very little experience with exerting control by reducing N. However, in many eutrophic shallow waterbodies, N limits phytoplankton biomass during the later summer months (Søndergaard et al., 2017). Shatwell and Köhler (2019) show the example of a shallow lake in which phosphorus cycled between water and sediment during summer perpetuated high concentrations of P even after substantial load reduction, and reduced N loads were therefore decisive for controlling summer phytoplankton biomass. Such situations may be particularly relevant for waterbodies with low rates of water exchange in which the gradual dilution of P takes many years (Conley et al., 2009).

In contrast to P, which is removed from a waterbody only by dilution and adsorption to particles with which it is deposited in the sediment, N is lost to the atmosphere through the bacteria-driven process of denitrification. At elevated summertime temperatures, this process can significantly reduce the concentrations of N within days, and in face of this quick response time, N load reduction may quickly render concentrations in the waterbody sufficiently limiting to control blooms during their peak season. As P release from sediments may also be particularly pronounced during later summer, Shatwell and Köhler (2019) propose to assess whether activities causing N loads (fertilisation, spreading of manure) can be timed to avoid loads specifically during the critical summer weeks in which keeping N limiting can control cyanobacterial blooms. In consequence, in situations in which the target concentration for TP cannot readily be reached, it may be effective to also control N loading. This may also serve to protect the macrophyte cover that would otherwise support the improvement of water quality `

N may also be relevant in the wider context of environmental targets for aquatic ecosystem protection. Conley et al. (2009) discuss the negative ecological impacts of reducing only P on some coastal waters and estuaries, such as parts of the Baltic Sea, Wadden Sea, and Gulf of Mexico. In such situations, excessive N may also lead to coastal harmful algal (including cyanobacterial) blooms exposing people during recreational use. Furthermore, N emissions into aquatic environments may be directly relevant to human health where they not only reach surface waterbodies but also reach groundwater, causing elevated nitrate concentrations in drinking-water (WHO, 2017b). As much of the N and P that reach waterbodies originate from the same sources, that is, human and animal excreta and/or fertiliser, some measures for controlling P loads can be readily designed to also reduce N loads, in particular reducing excessive application of fertilisers or manure on land. However, techniques for their removal in sewage treatment tend to be more expensive for N than for P. Also, as discussed below, the transport pathways of N and P to waterbodies are different, and intercepting also those of N may therefore require additional measures to those for intercepting P. Where prioritising investments is necessary, focusing on P is likely to be more effective for the target of controlling cyanobacteria.

The following considerations may serve to assess whether to focus measures on controlling P loads or to also address those of N:

1. Is the waterbody shallow and mixed (with thermal stratification at most lasting for a few days)?
2. Is P clearly too high to be limiting for extended periods during the cyanobacterial growth season, that is, total phosphorus (TP)>25–50 μg/L (depending on the waterbody) or even soluble P "left over" by the phytoplankton, that is, in concentrations>5–10 μg/L? Do concentrations of P increase during summer, indicating release from the sediment?
3. How do concentrations of N relate to those that can realistically be achieved by load reduction measures – that is, is a target of 200–500 μg/L of total nitrogen (TN) and<100 μg/L for dissolved N achievable?

Importantly, because of the possibility of N limitation shifting phytoplankton to N-fixing cyanobacteria, controlling N is not an alternative to measures reducing P loads, but rather an additional approach, focusing on specific summer situations.

Setting target nutrient concentrations: How low must the concentration of phosphorus or nitrogen be to effectively limit cyanobacterial biomass? Sections 4.3 and 4.4 summarise information and references showing that TP scarcely limits the biomass of phytoplankton – including that of cyanobacteria – if concentrations are above 100 µg/L. It can limit biomass to some extent in the concentration range of 50–100 µg/L and more effectively below 20–50 µg/L, while at less than 10 µg/L TP cyanobacteria scarcely occur and if so, health-relevant levels are unlikely in most situations. For nitrogen, section 4.3 shows that the limitation of biomass occurs at 7- to 10-fold higher concentrations as compared to those of TP.

Within this range, the target to set for a specific waterbody depends on both its intended use and specific conditions in it, particularly on its hydrological and morphological features. For example, in some shallow lakes with extensive macrophyte cover, cyanobacteria have only rarely developed blooms even at TP concentrations in the range of 100 µg/L (Jeppesen et al., 2007), and for the purpose of recreational use, this level may be sufficient as target nutrient concentration. At the other extreme, in a large deep lake or reservoir, cyanobacteria may develop and accumulate to scums on leeward shores at 20 µg/L TP, or *Planktothrix rubescens* may form metalimnetic maxima in the depths of the drinking-water offtake, and controlling these cyanobacteria may necessitate a TP target of 10 µg/L or even slightly lower.

For setting a target TP concentration, a general orientation can be gleaned from the experience with lake and reservoir restoration discussed in section 4.4: that is, lower TP concentrations in the range of 20–30 µg/L are typically necessary for thermally stratified waterbodies, yet lower ones closer to 10 µg/L may be needed to control *P. rubescens* in deep reservoirs, whereas shallow lakes with dense macrophyte stands may remain clear at TP concentrations even in the range of 100 µg/L. While much less experience exists for target N concentrations, multiplying these values for P by 7 may serve for a rough estimate. Beyond these rules of thumb, setting nutrient targets for a specific waterbody requires a good understanding of its ecology and the conditions that favour cyanobacterial blooms, and this is best done in collaboration with experts in limnology. It is further important to collaborate with authorities and stakeholders in waterbody and catchment management to identify overlap between targets for human health protection and aquatic ecosystem protection in order to efficiently coordinate measures within this larger context. Models outlined in section 4.4 can support setting target nutrient concentrations.

The following guidance focuses on assessing and controlling phosphorus loads as the nutrient, which is most frequently decisive for controlling cyanobacterial blooms. However, many aspects can likewise be used for developing measures to control nitrogen loads.

7.2 DETERMINING CRITICAL NUTRIENT LOADS TO THE WATERBODY

Once the target for the nutrient concentration is clear, it is possible to estimate the maximum nutrient load that must not be exceeded in order to meet this target. This is termed "critical load". Determining the critical load does not yet differentiate by nutrient sources and pathways but merely focuses on the total amount that should not be exceeded.

The critical load L_{crit} is given in mass per time (e.g., in tons per year). If there were no loss processes removing the nutrient, its critical load could be calculated from the target nutrient concentration (given in mg/m^3, which is equal to µg/L) multiplied by the amount of water flowing through the system. The latter is given in water volume per unit time: for rivers, this is discharge, Q_{river}, (given in m^3/s); for lakes and reservoirs, it is the water exchange rate or flushing rate ρ, given as the number of times the total waterbody volume is exchanged per year. For a river, the critical load of total phosphorus (TP) can be estimated as

$$L_{crit} = TP_{target} \times Q_{river}$$

- *Example for a river*: If TP_{target} is set to 25 mg/m^3 (=25 µg/L) and the average Q_{river} is 2 m^3/s, this gives a critical load L_{crit} of 50 mg/s, which is 1 576 800 000 mg/yr=1.58 t/yr.

As Q_{river} varies over time, Q_{river} can be defined as mean flow or low flow of the river, with low flow providing the higher level of protection, particularly as cyanobacteria tend to develop during periods of low flow. A more detailed level of emission modelling would include instream retention, as it is presented in section 7.3 as tier 3 approach.

For a lake or reservoir, the critical load of TP would be estimated as

$$L_{crit} = TP_{target} \times \rho \times z_{mean}$$

which is conceptually the same as the approach for rivers, but it is an established practice to use a different dimension for discharge, that is, the flushing rate ρ multiplied by the waterbody's mean depth (z_{mean}) which together describe the volume of water exchanged under one m^2 of waterbody surface.

- *Example for a lake or reservoir*: If TP_{target} is 25 mg/m³, the flushing rate is twice per year and the mean depth is 15 m (and thus the water volume under 1 m² of surface is 15 m³), this gives a critical load L_{crit} of 10×2×15=300 mg/m² per year.

  Multiplying this critical load by the total lake area then gives the critical load for the entire waterbody – that is, if the lake area were 1 km², L_{crit} would amount to 300 kg/km² or 0.3 t/yr.

However, for lakes and reservoirs, the role of P exchange with the sediments is usually far more significant than for rivers, particularly if the flushing rate is low, that is, less than 2–3 times per year. Sediment influence therefore needs to be included in the calculation of the critical load. For this purpose, Vollenweider (1976) (modified by Cooke et al., 2005) empirically developed a term relating ($\rho^{0.5}$) the interaction of TP with the sediment to the flushing rate. Adding this term to the equation gives

$$L_{Crit} = TP_{target} \times \left(\rho + \rho^{0.5}\right) \times z_{mean}$$

- *Example for a lake or reservoir*: If TP_{target} is 25 mg/m³, the flushing rate is twice per year and the mean depth is 15 m (and thus the water volume under 1 m² of surface is 15 m³), this gives a critical load L_{crit} of 10×(2+1.41)×15=512 mg/m² per year.

  Multiplying this critical load by the total lake area then gives the critical load for the entire waterbody – that is, if the lake area were 1 km², L_{crit} would amount to 512 kg/km² or 0.5 t/yr.

The difference between both approaches highlights that the role of sediments as sink for phosphorus strongly depends on flushing rates: recalculating these two examples with 10-fold higher flushing rates results in a much lower difference between the approach including the term for interaction with the sediment and the approach without ithat term – that is, 3671 mg/m² as compared to 3000 mg/m². The higher the flushing rate, the lower the role of losses of phosphorus via sedimentation.

However, the addition of $\rho^{0.5}$ to ρ introduced by Vollenweider (1976) was empirically derived from the OECD data set and thus is a rough approximation across a range of different waterbodies. A range of factors other than the flushing rate will influence P sedimentation or release, including lake morphometry and patterns of thermal stratification. In particular, if load reduction is pronounced, during the first years after load reduction the sediments are likely to release phosphorus through mineralisation or through desorption of redox sensitively bound P, depending on temperature and redox conditions (see section 8.6). For such situations, the equation will overestimate the acceptable external load or underestimate the time it takes to reach the target TP concentration. For further models that incorporate P release from sediments, including lake-specific approaches, see the discussion in Cooke et al. (2005) or literature therein, for example, Nürnberg (1998). Further complexity results from a substantial variation of loads in time, multiple inflows and/or heterogeneous distribution of the inflow within the waterbody.

The advantage of the loading equation above is its simplicity and reliance on only two terms which tend to be known for reservoirs, that is, flushing rate and mean depth. While it may serve for preliminary orientation, limnological expertise is important when setting a target for the TP load. This includes assessing the quality of the available data for flushing (ρ) and mean depth (z_{mean}) and the applicability of this simple approach to the specific waterbody.

The basic hydrological information needed for determining critical loads with the equation above may be available from authorities responsible for the management of the waterbody and its catchment. If not, mean depth (z_{mean}) and water volume can be determined if topographic maps are available and these maps include bathymetric contour lines showing depths. The rate of flushing (ρ) or its inverse, the water residence time (also termed "retention time"), can be derived from a water budget, which is calculated from flows and water volume – that is, from the balance between inflows (tributaries and in some cases also groundwater), run-off from surfaces and rainfall versus outflow, and – if relevant – amounts lost to seepage and evaporation. Besides outflow (which is often easiest to measure, particularly for reservoirs), further water losses relevant for the budget can include recharge to groundwater as well as evaporation. Measuring inflows requires determining stream flow of tributaries, which can be done by measuring the water level of the tributaries at river gauges (for instance continuously by pressure sensors) and transforming it by a rating curve to flow values. Rating curves quantify the relationship between water level and flow, and they need to be regularly controlled and adapted to changing conditions at the river gauge.

Once a water budget is available, it is further useful to estimate a nutrient budget, that is, the waterbody's total nutrient content (usually given in tons) compared to the total amounts that flow in and out of it. With sufficient resolution in time and space, a nutrient budget provides a valuable indication of nutrient sources as well as sinks: where it shows imbalances, this implies that there are further sources or sinks, for example, surface run-off not sufficiently well quantified, P losses through sedimentation or gains from sediment release. The nutrient budget can be derived from the water budget, the respective nutrient concentrations of the waterbody and the relevant in- and outflows. The nutrient budget may also vary considerably over time.

For determining the nutrient content of a thermally stratified waterbody, sampling should include depth profiles because concentrations may show pronounced depth gradients. Nutrient concentrations may also show pronounced seasonal patterns, and therefore, sampling and analysis monthly or even twice per month may be necessary for a sufficiently accurate assessment of the waterbody's nutrient content. Where this is not feasible, in some situations a good first estimate may be possible from one sample obtained during spring overturn, that is, when the waterbody is well mixed, rendering one sample quite representative for the whole lake or reservoir and the growing season (Reynolds & Maberly, 2002). However, this is only meaningful for waterbodies with fairly low water exchange rate and only moderate variation in stream flow of its tributaries, conditions more commonly found in temperate than in tropical climates. For example, tributaries often carry the greatest phosphorus loads during rain-event inflows when tributary streams and rivers are swollen (Zessner et al., 2005; Zoboli et al., 2015). Further challenges to establishing water and nutrient budgets include multiple inflows, for example, with small tributaries that run water only after major precipitation events or snowmelt, or significant groundwater flows which typically are difficult to measure. A comprehensive introduction to approaches to assessing nutrient budgets and critical phosphorus loads is given by Cooke et al. (2005).

7.3 IDENTIFYING KEY NUTRIENT SOURCES AND PATHWAYS CAUSING LOADS

Once the critical load has been determined, this needs to be compared to the current load to the waterbody in order to assess by how much the load needs to be reduced in order to remain below the critical load. The current load then needs to be differentiated according to the locally relevant sources and pathways in order to identify measures for reducing or controlling loads from these sources. This is also useful in situations in which the critical load is not exceeded: this serves to identify situations and measures worth maintaining in order to ensure that a currently good situation does not deteriorate.

7.3.1 Background information

Figure 7.2 shows principal sources, pathways and internal processes of nutrient loads to a waterbody. In this conceptual framework, all processes and activities that are likely to contribute to the input of nutrients are defined as sources. The most important **point sources** for nutrients are settlements which dispose wastewaters to surface waters via sewage without or after treatment and, depending on processes, also industrial facilities (the latter being typical point sources also for specific other pollutants). Relevant **diffuse or nonpoint sources** most frequently originate from agriculture, but they may include other fertilising activities, some urban emissions (including

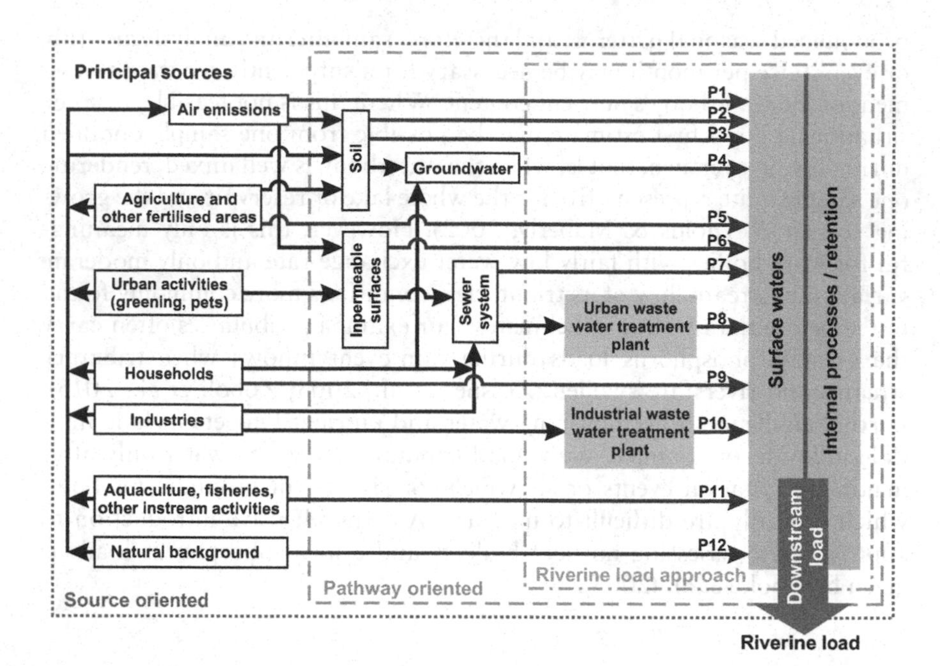

Pathways

P1	Atmospheric deposition directly to surface water	P7	Stormwater outlets and combines sewer overflows+unconnected sewers
P2	Erosion	P8	Urban wastewater treated
P3	Surface run-off from unsealed areas	P9	Individual – treated and untreated – household discharges
P4	Interflow, drainage and groundwater	P10	Individual wastewater treated
P5	Direct discharges and drifting	P11	Direct discharges from aquaculture, fisheries and other instream activities
P6	Surface run-off from sealed areas	P12	Natural background

Figure 7.2 Sources and pathways of nutrients and different levels (tiers) of their quantitative assessment in the context of emission inventories (Adapted from European Commission (EC), 2012.)

into air and then precipitating on the water surface, contribution to water pollution via atmospheric deposition), and wastewater from rural dwellings not connected to central sewage treatment. Typically, diffuse sources are more variable in space and time than point sources, and quantifying them may be more challenging.

Pathways are the means or routes by which nutrients can migrate or are transported from their various sources to the waterbody. Following release, they may be directly emitted to a waterbody or reach it after being transferred to and stored within environmental media, including soil and impermeable surfaces. Typical pathways of wastewater from industrial or urban sources to a waterbody are sewer systems and wastewater treatment plant effluents or groundwater in unsewered areas. Pathways transporting nutrients from agricultural areas and other surfaces follow the hydrological pathways as surface run-off, interflow (a subsurface run-off component that does not reach the groundwater), **tile drainage** (artificial pipe installations that drain agricultural areas to avoid soil being too wet) and groundwater. While nitrogen in form of nitrate is very soluble and readily reaches waterbodies via drainage, phosphorus supplied to soil in higher amount than needed by the crop is usually adsorbed to a high extent to soil particles. Erosion transports such particles over the land surface to waterbodies. Aerial emission is an important pathway for nitrogen and can result in subsequent direct deposition on the surface of a waterbody or indirect entry via soil or a sewer system.

The differentiation between sources and pathways is useful because measures to reduce nutrient emissions may either directly address the sources of nutrients (e.g., reduced fertilisation or livestock, improvement of industrial production processes, P-free detergents) or intercept the pathways of nutrients to the waterbody (as, for instance, erosion abatement by **riparian buffer strips**), and because, as discussed above, some pathways differ for N and P.

Besides external loads, processes within surface waters determine the nutrient concentrations in the water. These processes include a wide range, for example, sorption onto suspended particles, plant uptake, desorption or – for nitrate and ammonium – denitrification. Retention is a broad term used to describe the outcome if loads entering surface water remain there, without, for example, being discharged to coastal waters or – in case of nitrogen be lost to the atmosphere through denitrification (see section 4.3.2), a process relevant particularly in shallow lakes at elevated temperatures. The fractions that are retained by sedimentation in the river, along riverbanks or in sediments of lakes and reservoirs, can potentially be mobilised in future; however, this is not always the case. The extent of their retention depends on the nutrient (N or P) as well as hydromorphological conditions of the waterbody (Behrendt & Opitz, 1999; EC, 2012).

While nitrogen largely reaches waterbodies as dissolved inorganic N, for phosphorus, loads can occur in different binding forms. As discussed in Chapter 4, for limiting cyanobacterial biomass in the waterbody, it is important to assess not only the concentration of soluble reactive phosphorus (SRP) but rather that of total phosphorus (TP). P binding forms are also relevant for assessing P transport: some of the pathways discussed below transport a high share of P as SRP (groundwater, treated wastewater). Via other pathways, P is transported primarily in particulate forms, that is, P adsorbed to soil particles from erosion or P in organic material from raw wastewater. Whether particulate P may become available for the growth of cyanobacteria and algae depends on P forms in the particulate matter and the physiochemical conditions in the respective waterbody, which determine the fate of the respective P forms: for instance, P in apatite (as part of soil material) will rapidly settle to the sediment and not become available even over long periods of time, while P bound in organic matter will become available as organic matter decomposes, and P bound to iron salts may dissolve in anaerobic zones of the sediment (Psenner et al., 1988). On the other hand, if potential binding partners for phosphorus, such as iron- and aluminium oxides and hydroxides as well as certain clay minerals, are available in a waterbody or reach it together with the P load, dissolved phosphorus may adsorb to these binding partners, and if these complexes settle to the sediments, they will contribute to removing phosphorus from the productive water layers. Consequently, either they may be buried under younger sediment layers and thus be permanently removed from the system, or they may be mobilised again later on by desorption, particularly during events of sediment resuspension, increasing the concentration of dissolved P forms in the water system. Therefore, availability of P is not only a question of its emission pathway but also a question of complex biological and chemical processes of the P cycle within the waterbodies.

Similar processes of interaction between nutrients and soil also apply on land. If agricultural soils with increased P concentrations erode, P is transported together with soil particles and eventually emitted to surface waters. Depending on soil properties and soil saturation with P, P might be transported in soluble form and reach surface waters with surface run-off, tile drainages (i.e., drainage from fields and meadows), interflow or groundwater. In most settings, transport with erosion dominates. Losses of P from agricultural soil are impacted by many factors. Fox et al. (2016) give a review of these processes, including a discussion of "legacy P" accumulated in soils on land with literature indicating that this may be released for years or even centuries after it has been deposited.

7.3.2 Identifying nutrient sources and pathways

A good way to get started is to establish a qualitative overview of potential nutrient sources to the waterbody, that is, to compile an inventory of activities in the catchment, to collect the information available on their potential nutrient discharge and to map where in the catchment they are occurring in relation to the hydrophysical conditions that determine their pathway to the waterbody (for relevant activities, see Figure 7.2). Geographical Information Systems (GIS; see Box 7.2) are highly useful up-to-date tools for organising such spatial data. Such an inventory best begins with a detailed topographical map and with available, documented data, particularly data that can be obtained from public authorities, for example, from permits issued for discharges or for land use. Such data may be spread across a number of authorities, depending on responsibilities for the respective activity in the catchment. Some data may also be available from research institutes in the region.

BOX 7.2: USING GEOGRAPHICAL INFORMATION SYSTEMS (GIS)

GIS is a system designed to capture, store, manipulate, analyse, manage and present spatial or geographic data. GIS applications are tools that allow users to create interactive queries (user-created searches), analyse spatial information, edit data in maps and present the results of all these operations. An example for a typical application is the creation of maps that show the distribution of land–use types in relation to water courses. A more advanced application would be the implementation of the universal soil loss equation (Wischmeier & Smith, 1960) on a regional scale: spatial data on slope (from digital elevation model), slope length, rainfall intensity, soil erosivity and cultivation of crop types are merged to derive data in order to calculate the spatial distribution of erosive soil loss in a catchment. Practically, all advanced methods for modelling emissions rely on more or less comprehensive GIS applications.

For example, in the Action Plan for the Santa Lucia River Basin (see Box 7.4), GIS tools were used at the step of assessing the loads discharged from non-point sources and to develop an environmental information platform of open access ("Observatorio Ambiental Nacional") that centralises and organises the environmental information generated in various areas of the state. This includes a geo-integrator that provides access to georeferenced information and interactive maps, allows territorial analysis of information and makes files available for downloading.

It is useful to include an inventory of control measures that are already in place as well as information on how well they are currently managed (see section 7.5). The WHO guidebook "Protecting Surface Water for Health" (Rickert et al., 2016) gives an introduction into identifying sources and pathways for hazardous contaminants in general, including pathogens, harmful chemicals and also nutrients causing eutrophication and cyanobacterial blooms. This guidebook includes guidance on developing an inventory of activities potentially releasing contaminants hazardous for health and on conducting a catchment inspection, with checklists addressing loads from, for example, wastewater, agriculture, aquaculture and fisheries that can be downloaded and adapted to one's specific situation and needs.

Who should conduct the assessment?

Typically, compiling information on nutrient sources and pathways to the waterbody is a multisectoral exercise for which no one single public authority has the competence and possibility to enforce cooperation or compliance. Success chances therefore increase substantially if good will and motivation can be established among the stakeholders in the catchment. A Water Safety Plan team (see Chapter 6) can be an effective platform for bringing together staff from the public authorities involved; stakeholders from activities in the catchment of the waterbody (e.g., from agricultural or wastewater sector); and technical experts in the fields of, for example, hydrology, catchment management, geography, soil science and wastewater treatment. Together, they can compile information on potential nutrient sources and pathways, and develop proposals for the most effective way of controlling these sources and pathways. The lead for developing the catchment management aspects of a Water Safety Plan may best be taken by those responsible for water or environmental management. However, either the water supplier or the health authority responsible for the quality of drinking-water and/or the safety of recreational water use can take an active role in initiating the assessment and bringing together the key actors.

The relevance of catchment inspection

Regardless of the sources of information thus collated, validating it on site is important as conditions often change without proper notification to authorities. Also, a number of discharges as well as activities relevant to nutrient loading are often not notified, known and documented. Thus, while data available in documents provide a good point of departure for assessing nutrient loads from the catchment, they may not provide a sufficiently comprehensive picture, and visual inspection will reveal which activities relevant to nutrient loading are going on and which pathways for nutrients are evident. In smaller watersheds, catchment inspection is an applicable and valuable tool for the validation of information. In larger ones, inspection may only be partially possible, and it may be necessary to organise a review of the data available through intersectoral collaboration between a range of stakeholders and authorities.

Catchment inspection can be a time-intensive undertaking even in smaller catchments. Good preparation the therefore important, that is, to collect and evaluate as much information as possible prior to the inspection in order to focus on things to look for, which questions to clarify, which experts to ask to participate. Catchment inspection also provides an opportunity to identify owners and operators who may need to be interviewed afterwards (e.g., about discharge amounts, fertiliser application or records of manure application), and contact with them may be established directly during the inspection. It is generally useful to seek contact with locals during catchment inspection, as their information and observations can provide a valuable indication of factors otherwise overlooked. Catchment inspection usually provides a considerable amount of information to follow up afterwards, and this in turn improves and facilitates the next inspection. It is an iterative process, to be well documented and to be repeated at intervals. Rickert et al. (2016) give more guidance on catchment inspection, including checklists for this purpose to download and adapt to local circumstances.

The role of monitoring nutrient loads

Information on the relative contribution of different sources to the total load of a nutrient to the waterbody evidently is valuable for planning measures to control it. Sampling and analysing concentrations of nutrient concentrations may be fairly inexpensive, particularly if a water-quality monitoring programme is in place anyhow. However, capturing events causing peak loads may be more challenging, possibly requiring automated sampling triggered by some signal reflecting changes in discharge or precipitation.

Monitoring nutrient concentrations is valuable to assess the impact of implementing a new control measure if it is done before and after the intervention. For point sources along a water course, this may be quite straightforward. For the overall response of a waterbody to reduced loading, it may take a resolution in space and time (e.g., depth profiles if the waterbody stratifies and monthly or even weekly sampling intervals) for a year before and a few years after implementation and then at larger intervals in the scale of several years. While a cause–effect relationship may well be clouded by other changes in the catchment, such a "try and see" approach may be effective particularly where major loads to control are quite evident. Waterbody data are important for load modelling: they provide the empirical basis to test whether the model correctly depicts developments.

Nutrients and cyanobacteria in the broader context of health hazards

In many cases, it will be effective to assess loads and pathways of nutrients in the broader context of preventing water pollution causing health risks, as one aspect of developing a Water Safety Plan (see Chapter 6). The WHO guidebook "Protecting Surface Water for Health" (Rickert et al., 2016) gives guidance on estimating the health risks caused by the whole range of different hazards from the catchment, based on estimates of their likelihood to occur and their significance for human health. This broader context is important when assessing risks from potentially toxic cyanobacteria, as some sources of nutrients as well as pathways to the waterbody may be identical – for example, pathogens from human excreta – and therefore, one-and-the-same control measure may be significant for both. Recognising and highlighting such combinations may facilitate mobilising funding for implementing control measures.

Events causing loading

The control of chemical pollution is commonly based on monitoring at regular, predefined intervals. This approach to control risks missing major emissions causing peak loads and concentrations that occur during specific events, such as heavy precipitation bypassing wastewater treatment and thus causing sewer overflow directly into the waterbody, spreading of manure on frozen ground, stormwater run-off shortly after the application of fertiliser or manure, or illegal discharges conducted after the sampling team has left the premises. When planning the assessment of loads, it is therefore important to consider which events – from regular continuous emissions to sporadic intermittent extreme ones – are likely to cause relevant emissions and how such loads can be captured in the assessment.

Information to compile about the catchment
Checklist 7.1 (in part adapted from Rickert et al., 2016) outlines the broader information needed as a basis for characterising conditions and activities in the catchment area of the waterbody with respect to their relevance for nutrient loading. More detailed checklists for assessing nutrient loads from individual activities are given in the following sections of this chapter. Important expertise for this initial assessment includes geography, hydrology, local l–nd-use planning as well as wastewater management and agriculture. For later quantification of loads, it is useful to include expertise in catchment modelling when planning the assessment and inspecting the catchment.

CHECKLIST 7.1: ASSESSING WHICH ACTIVITIES IN THE CATCHMENT ARE LIKELY TO CONTRIBUTE MAJOR FRACTIONS OF THE NUTRIENT LOAD TO THE WATERBODY

Which basic information is available for assessing the relevance of different sources?

- Is a detailed topographical (digital) map available? When was it last updated? What topographical data are available on drainage areas, slopes and lengths?
- Which natural conditions in the catchment enhance nutrient pathways from the land to the waterbody, that is, topography (slopes), precipitation patterns, frost and snowmelt, soil types, erosion potential and drainage? Can areas in the catchment be identified which are most vulnerable to nutrient losses from the land to the waterbody?
- Are data available for discharge volumes of key point sources? Which fraction of the total discharge is from such inflows? Are data available for the nutrient loads these carry?

What activities are going on in the catchment of the waterbody and where are they located in relation to it? (See the template for site inspection in Rickert et al. (2016).)

Which areas in the catchment are covered by

- agriculture?
- aquaculture?
- suburban settlements?
- urban housing?
- informal settlements?
- industry?
- paved areas or otherwise impermeable surfaces draining to the waterbody?
- wilderness?
- forest, including use for logging?
- areas for recreational activities?
- other uses which potentially release nutrients to the waterbody or its tributaries?

Based on documentation available, what are the locations, spatial distribution and scale of the activities identified (generate map if possible)?

Are there trends or changes in land use, including population forecast studies?

What is the linear and hydrological distance (i.e., travel time of the run-off or seepage) to the waterbody from these activity points?

How are activities that potentially release nutrients managed, controlled and regulated?

- What national, regional, local or catchment-specific legislation, rules, recommendations, voluntary cooperation agreements or common codes of good practice are in place? How effectively are they enforced?
- Are there regulations for drinking-water protection zones or riparian buffer strips?
- Is land use subject to planning and permission? If so, do criteria for issuing permits include an assessment of the potential nutrient loads to the waterbody? How effectively are land-use regulations being enforced?
- Who are the main stakeholders to involve in the assessment?

7.3.3 Nutrient loads from wastewater, stormwater and commercial wastewater

Wastewater and stormwater inflows chiefly reach waterbodies as point sources and can cause significant nutrient loads. As point sources are more readily identified than diffuse sources a range of approaches is available to control them (see section 7.5.1). A fairly complete inventory of them is therefore an important basis for assessing and managing loads.

7.3.3.1 Sources

In municipal wastewater, human excreta are the dominating source of P. The average person-specific amount of P in excreta depends to some extent on the populations' nutrition but varies only within relative small boundaries of 1.3 – 1.7g P per person and day (Zessner & Lindtner, 2005) with the lower end of this range reflecting the emissions of populations with a low consumption of meat (Thaler et al., 2015). Where detergents containing P are used, the daily P load discharged to wastewater may vary between 2.0 and 3.0g per person. Thus, the density of the population living in the catchment of a sewage system determines the P load of municipal wastewater and its potential impact on waterbodies. Other significant point sources may be P from commercial and industrial wastewater, particularly from food processing enterprises or fertiliser industry.

While conventional wastewater treatment with biological organic carbon removes 30–40% of the P load, this is relatively easily improved by simultaneous precipitation techniques and/or biological P removal as "tertiary treatment", which typically achieve 80–90% P removal, leading to effluent concentrations in the range of 500–1000 µg/L. With an additional post-treatment step (e.g., post-precipitation and filtration), effluent P concentrations may even be reduced to < 200 µg/L. Where sewage effluent constitutes a major fraction of river flow, even concentrations in the range of 200 µg/L may cause too high a load of phosphorus to prevent cyanobacterial blooms, and specific filtration steps may be necessary to reduce effluent concentrations yet further. Also, attention to the storage of sewage sludge from the treatment process is important: if it is stored or disposed of inadequately close to the waterbody, run-off or seepage from this may be a further source of nutrient loading.

For assessing nutrient loads from industrial discharges, it is important to check whether the production line involves phosphorus (or nitrogen) compounds which are discharged and if so, whether data on the amounts are available or can be estimated from their content in substances purchased by the company for its production line. Enterprises that do not use phosphorus for their production may be adding it to their wastewater treatment system because the bacteria biodegrading organic substances in wastewater treatment require a minimum amount of P to work effectively. If the industry's wastewater does not contain enough P for an efficient biodegradation of organic substance, it may be dosing this to the biological treatment step. Dosing needs to be precise to avoid excessive P in the effluent, and when identifying P sources, including such enterprises in the assessment may be relevant.

A further nutrient source is rainwater run-off ("stormwater") from impervious areas, that is, roofs, roads, sidewalks and parking lots. It can contain significant nutrient loads particularly after extended periods of "dry deposition" from the air, garbage and excreta (from livestock, pets and where open defecation is practised, also from humans), particularly in the first flush after extended periods of dry weather. Where stormwater is collected by sewers that discharge directly into the waterbody, this will be an intermittent point source in the event of rainfall or snowmelt. While this source is generally less relevant than sewered wastewater, no generic concentration ranges can be given, as amounts depend entirely on local conditions.

7.3.3.2 Pathways

Outfalls of sewers carrying untreated wastewater directly to the waterbody and those of wastewater treatment plants are obvious point discharges in which nutrient concentrations can be directly analysed and/or loads be estimated from the size of the populations served and/or the type of enterprise emitting the wastewater (see section 7.4). Sewer systems carrying both domestic wastewater and stormwater from surfaces, that is, so-called combined sewers systems, can protect waterbodies during precipitation events that do not exceed the sewer capacity: these systems treat stormwater together with the domestic wastewater, and if treatment includes nutrient removal, this will reduce loads from run-off that would otherwise reach the receiving waterbody. However, they can be significant intermittent point discharges during heavy rainfall causing stormwater volumes beyond the capacity of sewerage and/or the sewage treatment system, and then sewage overflows allow this mixture of untreated domestic wastewater and stormwater to flow directly into the waterbodies, bypassing the treatment facility. Even where capacities of stormwater retention basins are large, they can rarely be built large enough to totally avoid such overflow events.

Diffuse pathways originate where wastewater from households and/or commercial activities is not sewered or where many small sewers discharge untreated wastewater directly to the waterbody. While in such situations diffuse loads from agriculture (see below) may be the major nutrient source for surface waterbodies, diffuse wastewater loads can also be significant if, for example,

- A number of dwellings or enterprises located sufficiently close to a waterbody lead wastewater pipes directly into it (possibly undocumented and informal) – a situation commonly causing diffuse loading from dispersed settlements along river courses or lakeshores.
- Open defecation is widely practised close to a waterbody and/or latrines close to the waterbody are poorly managed so that rain can wash excreta directly into a waterbody.
- The underground is very porous and soil filtration is poor so that seepage from latrines and septic system can reach the waterbody.

Otherwise, for phosphorus, even short distances of filtration through soil will achieve quite effective retention. Nitrogen may be less well retained if ammonium or nitrate from unsewered wastewater reaches shallow groundwater that drains into a surface waterbody (for an overview, see MacDonald et al., 2011).

7.3.3.3 What to look for when compiling an inventory of loads from sewage, stormwater and commercial wastewater sources

For assessing loads from wastewater and stormwater, Checklist 7.2 suggests a range of questions to address, depending on which appear locally relevant. These questions can support developing a checklist for catchment inspection. The data thus collected can be the basis for calculating not only the current loads from different sources but also the expected impacts of measures to reduce them (see section 7.5). Checklist 6 in "Protecting Surface Water for Health" suggests further questions that may be useful particularly if the purpose of inventory is to address not only nutrient loads, but also health hazards in general, including pathogens (Rickert et al., 2016).

Not all of these questions will be important in all situations, and information for answering all of them may not be available. Nonetheless, it is important to make a beginning with the information available while identifying the gaps and estimating how important it is to fill them in order to plan catchment management measures which are effective for meeting the nutrient load targeted.

Important specific expertise for assessing nutrient loads from sewage and stormwater includes environmental engineering with a focus on wastewater management.

CHECKLIST 7.2: COMPILING AN INVENTORY OF NUTRIENT LOADS FROM SEWAGE, STORMWATER AND COMMERCIAL WASTEWATER

GENERAL:

- Is the catchment primarily urban or rural, or a combination of both?
- Is there a relevant use of detergents containing P in the catchment? If so, could the use of P-free detergents be implemented? Would the P load from wastewater nonetheless remain in a range requiring removal in sewage treatment?
- Are there any enterprises which process food or nutrient-rich materials (fertilisers) in the region? Or any which are adding P to their wastewater treatment system to enhance its performance?
- Are enterprises operating at up-to-date technologies, for example, according to BAT (best available technique) requirements? Are improvements conceivable?

WASTEWATER SEWERAGE AND TREATMENT:

- Is the population density moderate or high and is a high share of people connected to public sewer systems? Is this share known?
- Is sewage treated? If so, with which steps? Are data on nutrient concentrations in treatment plant effluents as well as discharge rates available?
- Are there treatment plants with the removal of organic carbon loads in operation which could be upgraded with simultaneous P precipitation? With the removal of N?
- Are new treatment plants planned? Will they include nutrient removal applying P precipitation and/or biological P removal? If not, are there options to implement such a treatment step?
- Are effluents of treatment plants with P removal nonetheless a significant source of P for the catchment? Can a post-treatment step for additional P removal be implemented?
- Are industrial wastewater discharges significantly contributing to the nutrient load from the catchment? Can their effluent quality be improved by enhanced treatment (or where P-dosage is practised, by better control of the dose)?
- Is there any regulation in place that requires nutrient removal at treatment plants? Is this regulation considering the target concentration in the waterbody necessary to avoid cyanobacteria blooms?

UNSEWERED AREAS:

- How many households are not connected to sewer systems? Which type of disposal do they have? Are there direct discharges into the surface water? Does rainfall rinse the content of poorly managed latrines, open defecation or septic systems directly into surface water?
- If answers to question 12 indicate potential for a significant nutrient load to the waterbody, which options are available to prevent this (e.g., implementing improved on-site sanitation systems, including collecting and transporting the content of septic tanks to wastewater treatment plants or safe use in agriculture)?
- Can safe dry systems for collecting and treating human excreta be promoted as alternative to developing sewerage?
- Is a sewer development planned, and should this approach be further pursued? If so, go back to points 4–11.

SEWER SYSTEMS:

- Are there separate sewer systems in place? Are stormwater sewers draining areas with heavy nutrient pollution (e.g., farmyards, excreta from pets)? If so, could emissions be reduced by installing infiltration ponds?
- Are there connections between stormwater and wastewater sewers, discharging untreated wastewater continuously?
- Are combined sewer systems in place? If so, how frequently and at what type of rain events to the overflow, bypassing treatment?
- Are combined sewer systems equipped with retention tanks or basins? If yes, to which extent? Are regulations in place and if so, how stringently are they implemented?

SEWAGE SLUDGE:

- Is sludge used as fertiliser? (If yes, see Checklist 7.3 for agricultural activities in section 7.3.4.)
- If sludge is disposed, is the site and method adequate to avoid nutrients reaching the waterbody?

For estimating how these nutrient loads relate to loads targeted for the waterbody as discussed in section 7.1, see section 7.4.

7.3.4 Nutrient loads from agriculture and other fertilised areas

While in some regions of the world, agricultural productivity is low due to a lack of fertiliser, in other regions, fertilisation is excessive and the primary cause of diffuse nutrient loads to waterbodies where they cause eutrophication and cyanobacterial blooms. For phosphorus, MacDonald et al. (2011) give an overview of this global imbalance, and Withers et al. (2014) describe agriculture as prominent and persistent cause of diffuse nutrient loads in many parts of the world. However, the latter authors also emphasise the importance of farming and food production, in consequence of which measures potentially imposing restrictions must be reasonable and effective. This requires a sound identification of nutrient loads and an assessment of their relevance for eutrophication of the specific waterbody.

7.3.4.1 Sources

Sources of nutrient loads from agricultural activities are fertilisers and manure or slurry spread on fields as well as excreta from free-range animal herds on pastures. Animal husbandry is typically relevant where feedlots, large stables or manure piles are located close to a waterbody. Nutrient loads from fertiliser and manure/slurry can range from almost negligible to extremely high, depending on how much is applied in excess of that which the crop can take up and convert to biomass. The excessive application of fertilisers and manure in many of the world's more affluent agricultural communities has been based on policies actively encouraging and subsidising intensive fertilisation (Withers et al., 2014) and on the widespread concept of soils serving as storage for P, adsorbing it and releasing it when needed by the crop. Where this has led to amounts applied to soils that exceed their binding capacity, soluble P will leach to the waterbody (Behrendt et al., 2000). Also, as discussed in section 7.3.1, soil particles to which P is adsorbed can release it once erosion carries them to the waterbody (Novotny, 2003). Fox et al. (2016) show that streambank soils can contain from nondetectable to more than 1000 mg P per kg soil. From their evaluation of the modest number of studies available from Europe and North America, these authors conclude that where catchments are impacted by excessive nutrient application, soils are likely to contain more than 250 mg/kg. The fraction of this which becomes available for algae and bacteria when erosion carries such soils into a waterbody strongly depends on the physical and chemical conditions in the waterbody. Where the "legacy phosphorus" from excessive fertilisation in agricultural soils is high and/or the time span for which it is likely to cause loads to a waterbody is difficult to assess, it is particularly important to assess the erosion pathways to the waterbody (Sharpley et al., 2015).

While excessive fertilisation also increases nitrogen (N) loads, these are often due to the large size of intensive animal husbandry operations: where these produce amounts of manure and slurry that cannot be spread on nearby fields and pasture without exceeding the uptake capacity of crops and meadows, this causes loads of both N and P – possibly significant or the predominant source of eutrophication and cyanobacterial blooms (additionally, excessive manure and slurry spread on land can cause elevated nitrate concentrations in groundwater used as source of drinking-water; see Schmoll et al., 2006).

In some situations, substantial fertilisation of other land, such as golf courses or lakeside lawns, may also be a relevant source of loads to a waterbody.

7.3.4.2 Pathways

Where erosion occurs, soil particles will be carried towards the waterbody, thus transporting the P adsorbed to them. There is consensus that streambank erosion is a highly relevant pathway for phosphorus loading: Fox et al. (2016) review case studies of P loads from streambanks and conclude that 7–92% of the total P loads could be accounted for by streambank and gully erosion. Peacher et al. (2018) also review streambank erosion as a major source of nutrient loads transported with sediment and report own results for P loss rates with soil eroded from riverbanks in the range of 38–49 g/m and year for the riverbanks of two streams in Missouri; these loads amounted to 67% of the P transported in these creeks. Although there are examples of situations in which plant root growth contributes to riverbank erosion, in general an intact cover of vegetation ("riparian buffer strip") stabilises the riverbank and can serve to intercept soil in surface run-off (see discussion in Fox et al., 2016).

In tributaries and drainage ditches, P thus transported will interact with the channel bed sediments, leading either to a reduction (through adsorption and sedimentation) or to an increase of P (through resuspension and desorption) transported in the stream – or to periodic alternation of both processes, depending on river flow and stormwater events. Often only a fraction of such a P load will become relevant for eutrophication of the downstream waterbodies, and the size of this fraction depends on a range of chemical and physical variables. Quantifying these variables is still challenging, and Fox et al. (2016) review publications on methods and models for this purpose.

In contrast to pathways for P, excessive nitrogen (N) from fertiliser or animal excreta scarcely binds to soil. Animals release N as urea which rapidly degrades to ammonium, some of which is lost to the atmosphere by volatilisation (where in the form of N_2O, it acts as greenhouse gas, enhancing climate warming) and some of which is oxidised to nitrate by bacteria in the soil (nitrification). Nitrate is also the form in which fertilisers contain N. As nitrate is very well soluble, excessive N readily leaches from soils and reaches waterbodies not only by surface run-off, but also via tile drainage (Novotny, 2003).

Nutrient loads not only depend on the amounts of fertilisers and manure applied, but also depend on timing of the application as well as on conditions determining pathways to the waterbody. These include natural geographical conditions such as the slope of the land as well as agricultural practices: untimely application of manure (e.g., on frozen ground or before strong rainfall) may cause major nutrient loading. This not only pollutes water, but also loses potentially valuable fertiliser from the farmland. Methods of ploughing have a strong impact on the extent of erosion, and so does leaving fields barren, without vegetation cover. Access of cattle and other farm animals to a waterbody or its tributaries can cause loading through a direct input of faecal material when animals wallow in the water or defecate near it. In particular, cattle can cause massive erosion of shorelines saturated with the animals' faeces: the trampling of larger herds can destroy the vegetation cover and also create pathways for erosion farther into the catchment, as reviewed by Wilson and Everard (2018).

Pathways for both nutrients are also created by clear-cutting of forests and woodland by logging on steep slopes or burning of woodland to convert it into farmland. Without vegetation cover or through trampling by herds of livestock, steep slopes become unstable and susceptible to heavy erosion. Particularly in climates with heavy rainfall, such practices may massively promote erosion and thus nutrient loads.

7.3.4.3 What to look for when compiling an inventory of loads from agricultural activities

As pathways for nutrients from land to water depend so strongly on the general geophysical characteristics of the land, for assessing nutrient loads collecting information about these characteristics is as important as developing the inventory of the activities that potentially release nutrients. Checklist 2 in "Protecting Surface Water for Health" (Rickert et al., 2016) supports this with questions on local climatic and hydrological characteristics, tributaries and their discharges, topographical data and soil types, signs of erosion and flooding.

The following checklist 7.3 is intended as point of departure when planning the assessment of nutrient loads from agriculture. As for the checklist above for loads from wastewater and stormwater, aspects of this may feed into developing a checklist for inspecting a specific catchment, and the information thus serves to estimate both current loads and the impact of measures to control the loads. Checklist 4 in Rickert et al. (2016) adds further aspects.

CHECKLIST 7.3: COMPILING AN INVENTORY OF NUTRIENT LOAD FROM AGRICULTURAL ACTIVITIES, GOLF COURSES, LAWNS AND OTHER FERTILISED AREAS

LAND USE AND REGULATIONS:

- What types of land use are being conducted that could cause nutrient emission, for example, arable land, pasture, irrigated or drained agriculture, horticulture, market gardening, golf courses, lawns and parks reaching all the way to the shoreline? Which types are being conducted on land with steep slopes (more than 8% grade)?
- Which regulatory frameworks (specific legislation, regulations, recommendations, voluntary cooperation agreements, codes of good practices, restrictions, bans) exist, particularly for the application of fertiliser and manure? How well are they known to the farmers? Could their implementation being enforced?
- Are drinking-water protection zones established around the reservoir and/or its tributaries? If so, is a map of their delineation available? Which limitations do they involve, and how stringently are these implemented? If not, could they be established?
- Are policy instruments in place such as financial incentives (e.g., subsidies, low-interest loans or compensation for lost income during transition to more environmentally friendly practices) or financial disincentives (e.g., penalties for nutrient loads caused by poor agricultural practice) that can be used to initiate agricultural practices with low nutrient emissions? Are any future incentives reasonable and realistic?

- Are agricultural advisory services in place, and what practices do they recommend to farmers, particularly regarding fertilisation, stock size and practices of animal husbandry? What have they recommended in the past, possibly having led to legacy P in soils?

APPLICATION OF NUTRIENTS:

- Is manure or sewage sludge being applied to fields or lawns? If so, are amounts and dates of application documented? What information is available about the storage conditions and handling practices for manure or sewage sludge? Are application rates on farms mostly based on farm nutrient budgets (see Box 7.3) and crop uptake rates, or are they roughly estimated, or are they based on the need of getting rid of manure or sewage sludge, for example, in areas with high livestock densities, or in proximity to a wastewater treatment facility? Are incentives operating (e.g., expert consultations) to improve practice and achieve a balanced soil nutrient budget?
- Is application timed in relation to hydrological events and seasonal aspects, for example, presence/absence of vegetation cover, frozen ground? How adequate are spreading methods and timing in relation to weather conditions? Are there any incentives to improve current practice?
- If fertilisers are applied, which types and products with which composition (e.g., nitrogen and phosphorus contents) and in which amounts? Is information available on amounts applied? On concentrations in soils? Are application rates based on plant needs and up-to-date information? Are there any guidelines to support the calculation of appropriate fertilisation? If not, can they be provided?
- Are arrangements in place that limit the amounts to be applied? For example, agreements between farmers and drinking-water suppliers or managers of waterbodies used for recreation?

NUTRIENT LOSSES DUE TO AGRICULTURAL PRACTICES:

- What main crops are cultivated currently and during the past seasons? What trends or changes are anticipated? Which of the main crops have a low vegetation cover especially during rainy seasons? Are they cultivated on steep slopes? Can this be avoided?
- What ploughing practices are being applied? To which extent does ploughing promote soil erosion? Are any guidelines on best practice in place? Is any consultancy to farmers in place?

- Are winter cropping, mulch seed or any other practices to avoid soil loss from steep fields in place? Are there financial incentives on regional, national or catchment scale in place to support these practices?
- Is there indication of gullying, soil scouring and land slipping in steep areas in the catchment (including changes over time)? If so, what are possible causes (trampling of herds, ploughing practices, barren fields)? Is there an awareness of possible causes?
- Is the land drained, and do drainage ditches or pipes carry dissolved nutrients to the waterbody?
- Is there any other indication of fertiliser, manure or nutrient-rich soils being lost from land to the waterbody, such as periodic heavy loads of suspended solids in the tributaries?

LIVESTOCK AS NUTRIENT SOURCE AND CAUSE OF PATHWAYS TO THE WATERBODY:

- What are the livestock densities, animal species and amount of manures produced? Are they exceeding the nutrient needs on farm level? Can they be better utilised by better distribution?
- Are stables and/or feedlots close to the waterbody or its tributaries? If so, are there run-off pathways (gullies) to the waterbody? If yes, could manure collection and storage be improved?
- Is there sufficient storage volume for manure and slurry? Is the storage time long enough for seasons where applications are not favourable (e.g., winter)?
- Are pastures fenced, or can livestock access the waterbody or its tributaries? Are fences intact and regularly inspected?
- If there is an indication of direct impact of livestock excreta on the waterbody or its tributaries, how many heads of stock are there in the area? How much nutrient input can their excreta cause at maximum? Is this relevant in respect to P loading to the waterbody?
- Is there an indication of erosion damage from livestock?

INTERCEPTING TRANSPORT OF SOIL NUTRIENTS FROM LAND TO WATER:

- Can specific areas or practices be identified as likely main causes of nutrient loading, particularly of phosphorus? Can areas be identified which could be used as buffers to interrupt the transport of soil particles into surface waters or to allow denitrification to take place before the inflow of drainage to the waterbody?

- Are vegetation-covered buffer strips in place between fields or pastures and the waterbody or its tributaries? If so, are they properly located in respect to erosion transport? How wide are they, and are they intact or are they frequently interrupted? Are they managed and/or well maintained?

As for Checklist 7.2 on nutrient loads from point sources, neither will all of these questions be equally important in all situations, nor is information for answering all of them likely to be available, and identifying the gaps relevant for planning catchment management measures is an important element of an initial assessment of diffuse nutrient loads.

BOX 7.3: AGRICULTURAL NUTRIENT BUDGETS

A nutrient budget estimates the nutrient surplus as the difference between nutrient inputs and nutrient outputs for a certain boundary, for example, the amount of nutrient that enters a farm with fertilisers and feedstuff for animals minus the amount that leaves the farm with the produce. Nutrient budgets for agriculture can be distinguished by the definition of the boundary (farm, soil or land) they refer to.

A soil nutrient budget estimates nutrient surplus from nutrient inputs to the soil (e.g., fertilisers) and nutrient outputs from the soil (e.g., harvest).

Nutrients accumulate in soils as nutrient stock, changes of which are difficult to quantify. Therefore, they are frequently accounted in the surplus.

Nutrient budgets provide a valuable information about the link between agricultural activities and environmental impacts of nutrient use and management in agriculture. Nutrient budgets can be used to determine areas at risk of releasing nutrients to waterbodies (when estimated at low regional levels), to identify driving factors behind nutrient pollution resulting from agriculture and to follow trends over time. For further information, see Eurostat (2013).

Important specific expertise for assessing nutrient loads from agriculture particularly includes agricultural practitioners (preferably from the region and thus familiar with local practices, habits and attitudes), soil scientists, hydrologists and – if modelling loads is intended – also catchment modellers.

For estimating how these nutrient loads relate to loads targeted for the waterbody as discussed in section 7.1, see section 7.4.

7.3.5 Nutrient loads from aquaculture and fisheries

Aquaculture and fishponds in a catchment may be a major point source where their effluent reaches a waterbody. They are not typically the main source of nutrient loading, but in many regions, aquaculture and fish production are increasing. In particular, cage cultures ("net pens") within waterbodies may also introduce substantial nutrient loads directly into waterbodies. So may fisheries involving feeding or even fertilisation of the waterbody to enhance its productivity. Fisheries management within the waterbody can further impact water quality through its internal effects on the food chain and/or through bottom-dwelling fish which resuspend sediment, as discussed in Chapter 8 in the context of waterbody management.

The maximum amount of nutrients introduced can be calculated from the amount in the applied feed minus the amount in the fish biomass harvested from the system. While this approach disregards potential losses through sedimentation occurring in the tributary between the fishpond and the waterbody of concern, it provides a useful worst-case estimate for assessing the relative importance of loads from aquaculture and fisheries. An estimate is also possible from the biomass of fish produced, using factors that describe the amount of feed necessary for this growth and the efficiency of the conversion of fish food into fish biomass.

Nutrient concentrations in a fishpond effluent can vary widely over time, depending on current operations such as the cleaning of tanks, backwashing filters or emptying ponds. Alabaster (1982) showed that a 30-min cleaning operation discharged 75% of the total phosphorus and 10% of the total nitrogen (TN) from a fishpond, highlighting that such events may be important when estimating nutrient loads. Such short-lived nutrient pulses in the receiving waterbody may be rapidly utilised for the growth of cyanobacterial and/or algal biomass, causing sudden increases and triggering blooms.

7.3.5.1 What to look for when including aquaculture and fisheries in the inventory of activities causing nutrient loads

Checklist 7.4 suggests questions to address when assessing the contribution of aquaculture and fisheries to the nutrient load of a waterbody as well as for assessing the expected impacts of measures to reduce these loads. When using this for catchment inspection, adaptation to the questions which appear locally relevant is recommended. Including specific expertise from aquaculture and fisheries is valuable for estimating the nutrient loads (e.g., from feeding rates) as well as for assessing the efficacy of control measures in place or to be implemented. See also Checklist no. 5 in Rickert et al. (2016) for further information, particularly if the assessment is to include the wider context of health hazards from aquaculture and fisheries in water.

CHECKLIST 7.4: IDENTIFYING SOURCES FROM AQUACULTURE AND FISHERIES

- If aquaculture is practised in the catchment or waterbody, where are which operations located, and how much fish do they produce per year or season?
- If cage culture ("net pens") or fisheries are practised within the waterbody, where are they located, and how much fish do they produce per year or season?
- Are data on feeding and the source, amount and type of feed applied available, and on its phosphorus and nitrogen contents? If not, can nutrient loads be estimated from fish production rates and conversion factors?
- Are fertilisers applied? If so, what amounts, types, products and composition of fertilisers are used?
- Are manures applied? If so, what are the source, amount, composition and application patterns for the manure?
- Is wastewater or sewage sludge applied? If so, what information is available on the wastewater (e.g., amount; is it raw or has it undergone some treatment or ageing; is it pure domestic wastewater or might it contain commercial effluents?)
- Do regulations (specific legislation, recommendations, voluntary cooperation agreements, codes of good practices, restrictions, bans) for these activities exist, and if so, how well are they being enforced?
- Are flow-through or recirculating systems being used?
- Is effluent discharged directly to the waterbody, or is it treated? If it is treated, how? Are data available on nutrient concentrations in the effluent and on the water volume of the effluent?
- Are data on nutrient concentrations in the effluent available from the aquaculture operator? Can nutrient loads from their discharge be estimated, for example, from fish food consumption and the amount of fish produced?

- What time patterns of these operations (e.g., emptying and cleaning of tanks) might be relevant to nutrient loading? Which nutrient sources may be relevant that are not readily detected through inspection of the operations (e.g., in sludge and sediment at the bottom of the tanks)?

For estimating how these nutrient loads relate to loads targeted for the waterbody as discussed in section 7.1, see section 7.4.

7.4 APPROACHES TO QUANTIFYING THE RELEVANCE OF SOURCES AND PATHWAYS

Obviously quantifying the loads introduced to the waterbody from the major sources via major pathways will provide the best basis for identifying the most effective measures to control nutrient loads. However, key parameters necessary for quantitative models are often unknown and not readily measured. Qualitative assessments then are a valuable beginning. Rickert et al. (2016) describe a qualitative approach to assessing loads, roughly categorising their relevance from negligible to extreme and highlighting uncertainties necessitating more in-depth assessment.

In some settings, a qualitative assessment of potential sources of nutrient loading can provide a sufficiently clear basis for planning control measures even without quantifying the relative relevance of different sources and pathways of loads. This is possible if conditions are quite clear. For example, in a densely settled, largely urbanised catchment, much of the nutrient load will originate from sewerage, and focusing on measures that reduce sewage nutrient emission loads will directly impact concentrations in the waterbody. For such a setting, it will be clear that without control of such substantial point sources, further load reduction measures cannot achieve sufficiently low loads to the waterbody to reach target nutrient concentrations. Similarly, for a largely agricultural catchment, identifying the steepest slopes for implementation of the most stringent management of fertilisation and tillage may be a sufficiently effective basis for achieving a substantial load reduction, even if the reduction achieved can hardly be quantitatively predicted and only assessed by subsequent monitoring of the change in concentrations in the waterbody. A merely qualitative approach – based on "getting the job done" for obvious measures – can provide substantial load reductions, particularly in situations in which resources for more elaborate approaches are lacking. Qualitative or semiquantitative approaches may suffice for estimating loads that are either self-evidently major or likely to be low.

Quantification becomes important where the load that a source contributes may be relevant but uncertainties are too significant to make decisions on investments or regulations to reduce it. Approaches to quantification cover a wide range, requiring different levels of information, staff capacity and resources. The European Union summarises some in its Guidance Document No. 28 (EC, 2012) in the context of its Common Implementation Strategy for the EU Water Framework Directive. This document has the advantage not only of being harmonised as outcome of discussions between a number of countries, but also of giving guidance at 4–5 different levels of complexity and data requirements. Thus, the technical guidance on the development of an inventory of emissions, discharges and losses of substances described in this document can be used for a range of situations with different resources. While this document focuses on priority hazardous substances, it outlines general principles which can be applied for nutrients as well. It therefore uses some key considerations from the Guidance Document No. 28 (EC, 2012), giving specific attention to situations where data availability is lower than can be expected in EU countries.

In face of the complexity of systems and the challenges associated with data collection, three broad quantitative approaches in the establishment of inventories can be distinguished, which are shown in Figure 7.2 with their scope indicated by the dashed boxes in diagram and their complexity increasing from right to left:

- the riverine load-oriented approach, which estimates the observed total load that a river carries into a lake or reservoir. This information can be used together with a quantification of point source inputs to calculate an estimate of the diffuse inputs (green dashed line in Figure 7.2);
- the pathway-oriented approach (POA), also called "regionalised pathway analysis" (RPA), which models the different transport phenomena for the final input routes to the river system starting from the "interface media" as soil, groundwater or wastewater treatment plants. This approach calculates regionalised emissions for small catchments (termed "analytical units"), which can be subsequently aggregated to river basins or subunits (yellow dashed line in Figure 7.2);
- the source-oriented approach, which addresses the whole system starting from the principal sources of substance release. Such an approach includes substance flow analysis (SFA; red dashed line in Figure 7.2).

As situations differ strongly in the range of information and data sources available, the following introduces a tiered (or level) approach whereby the complexity increases with each progressive tier, beginning with purely qualitative (tier 0) or semiquantitative (tier 1) assessment. Quantitative tiers (2–4) require further data as well as more in-depth understanding of sources and pathways, resolution and detail. On this basis, they allow a better discrimination of the relevance of sources, for example, the relative contribution of those emitting nutrients to sewers and wastewater treatment plants rather than the (lower tier) lumped treated effluent discharge which does not allow for discrimination of the original source. Thus, the different tiers support a progressively improved understanding of the emission situation and, therefore, the ability to effectively allocate financial resources and evaluate (cost-)effective measures for emission reduction.

The approach to select for a given catchment will depend on its size, the availability of data and resources as well as the relevance of the problem. Five levels or "tiers" (one qualitative, one semiquantitative and the three quantitative approaches outlined in Figure 7.2) of emission estimation methods are summarised in Table 7.1 and explained in the following.

Table 7.1 Five tiers for the elaboration of emission inventories in catchments – overview

Tier	*Required information*	*Expected output*	*Results from the inventory*
0. Qualitative assessment	• Catchment inspection and/or qualitative description of main activities in the catchment	• Overview over catchment characteristics	• Identification of potentially relevant sources and pathways
1. Emission factors (semiquantitative)	• Data on population, land use and wastewater disposal • Population and area-specific emissions	• Availability of data • Assessment of the quality of data • Identification of information gaps	• First rough estimate of point source emissions in relation to diffuse emissions • List of identified data gaps
2. Riverine load approach	In addition to tier 1: • Data on point discharge • River concentration • Data on river discharge • In-stream processes	• Riverine load • Trend information • Proportion of diffuse and point sources • Identification of information gaps	• Rough estimation of total lumped diffuse emissions • Verification data for emission estimates and for results from tier 3 and 4 • Listing of identified data gaps

(Continued)

Table 7.1 (Continued) Five tiers for the elaboration of emission inventories in catchments – overview

Tier	*Required information*	*Expected output*	*Results from the inventory*
3. Pathway-orientated approach	In addition to tier 2: • Agricultural statistics (fertiliser, crops…) • Soil data • Data on hydrology • Others depending on the applied model	• Quantification and proportion of pathways • Identification of hotspots • Information on adequacy of pathway-oriented protection measures (scenario calculations)	• Pathway-specific emissions • Additional spatial information on emissions
4. Source-orientated approach	In addition to tier 3 • Production and use data (nutrition statistics) • Substance flow analyses • Others depending on the applied model	• Quantification of primary sources • Complete overview about substance cycle • Information on adequacy of source protection measures	• Source-specific emissions • Total emissions to environment and proportion to surface waters

Source: Adapted from European Commission (EC) (2012).

7.4.1 Tier 1: Assessment using emission factors

In more heterogeneous and larger catchments or in situations with enhanced wastewater disposal, the main sources of nutrients in a catchment might not be obvious, nor may be the pathways by which they enter the river system. In such a setting, the tier 1 approach based on emission factors (also called "export coefficient method") is helpful to obtain a first semiquantitative overview of the contribution of main sources of nutrient emissions to a waterbody – namely, municipalities (including households and industries) and agriculture.

7.4.1.1 Municipalities

As discussed in section 7.3.3, phosphorous is an essential nutrient in human nutrition and human excreta, which therefore are the dominating source of P in municipal wastewater, and other significant sources may be P from detergents or commercial and industrial wastewater, particularly from food processing enterprises or fertiliser industry. An estimation of P emissions from municipalities (which includes wastewater from households, commerce and industries) is possible on the basis of the following information:

- the number of people in the catchment, with 1.3–1.7 g P emitted per person and day (Zessner & Lindtner, 2005);
- whether or not detergents used locally are P free (i.e., which type is typically sold on local markets), with emissions per person amounting to 2–3 g if detergents contain P;
- the share of the population connected to sewer systems;
- the estimated amount of commerce and industry likely to emit P;
- the level of wastewater treatment (no treatment, biological treatment with removal only of organic substance ["C removal"], or treatment with P removal/P precipitation).

In municipalities without any significant commercial and industrial activity, P in wastewater predominantly originates form households. In case of high commercial and industrial activities, experience shows that those activities may increase P loads in wastewater by up to 1.5 g per person and day (Zessner & Lindtner, 2005), and this may serve as first rough estimate. Any large-scale fertiliser or food processing industries are not included in this number and would have to be accounted separately.

The impact of populations not connected to sewer systems on P emissions is highly dependent on their sanitation system. Where this is through subsurface or soil treatment, the impact will usually be small and can be neglected for a first estimate. Where household wastewater is directly discharged into a river or its tributary, the total load from the population would have to be accounted as emission into the surface waters.

Wastewater collected in sewers is usually discharged to surface waters, with P emissions depending on the level of treatment. Emissions are:

- 100% of P in raw wastewater in case of no treatment;
- 60–70% of the P concentration in raw wastewater in case of biological treatment without P removal;
- 10–20% of the P concentration in raw wastewater in case of biological treatment with P precipitation or biological P removal (with further reduction by a factor of up to 10 if a combination of post-precipitation and filtration step is added after conventional treatment) (Heinzmann & Chorus, 1994).

Example: There are four municipalities in a catchment. The first is a town and has 50 000 inhabitants (inh), all connected to the public sewer system, an average amount of commercial and industrial activity and a wastewater treatment plant, including biological C removal without P removal. The markets in town sell P-free detergents for washing laundry and dishes. The other three municipalities are small settlements with all together 1500 inhabitants not connected to sewer systems and buying in the markets of the town.

P emissions from the town to surface waters=(1.5 g P/(inh×d) [from households]+0.75 g P/(inh×d) [from commercial activity])×50 000 inh×0.65 (P in effluent after treatment)=73 kg P/d=2.7 t P/yr.

P in wastewater from settlements=1.5 g P/(inh×d)×1500 inh=2.3 kg P/d=0.8 t P/yr. The fate of this P load is not known. Catchment inspection would serve to collect evidence whether wastewater disposal from these settlements could directly seep into surface waters of the catchment.

7.4.1.2 Agriculture

The main external pathway by which nutrients reach a farm is the input of external (mineral) fertiliser and the input of feedstuff for livestock. Farm animals process their feed, excreting faeces and urine which is then spread as manure and slurry on fields. Nutrient exports from the farm as loads to a waterbody can ideally be avoided if feed and manure are kept in an internal on-farm cycle, with the phosphorus content of the agricultural products (as output of P from the agricultural production process) approximately balancing the P input through externally imported fertiliser and feedstuff. While this situation appears idealistic, agricultural nutrient budgets (see Box 7.3) have indeed proven to be a highly effective approach to controlling nutrient loads (see section 7.5.2).

Soil is the essential medium in this production process as it provides the nutrients to the plants for their growth. As discussed above, during the production process, nutrients not taken up by the plants may be transported from agricultural areas to the waterbody, and while nitrogen in form of nitrate is very soluble, P is usually adsorbed to a high extent to soil particles, if supplied to soil in higher amount as needed by the plants. This may lead to increased concentrations of P in agricultural soils, and if erosion occurs, this can transport P together with soil particles to surface waters. Depending on soil properties and soil saturation with P, it might also be transported in soluble form and reach surface waters with surface run-off, tile drainages, interflow or groundwater, but in most settings, particulate transport with erosion dominates. Losses of P from agricultural soil are impacted by many factors, and the load emitted into surface waters is determined by the concentrations of P in soils and the amount of soil mobilised from the field that reaches the waterbody. Because of the high numbers of factors that determine this process, even rough estimates of P emissions from agricultural soils to waterbodies are less straightforward than for wastewater from municipalities. Nonetheless, a first rough tier 1 estimate based on emission factors is possible for this source as well. Where more precise quantification is needed, more elaborated quantifications (tier 3) are recommended in cooperation with modelling specialists. For this tier 1 approach, the following information needs to be known:

- arable area in the catchment;
- basic information on slopes of land used for agriculture;
- connectivity of arable land to the waterbody or its tributaries (ranging from "well connected" to "poorly connected", e.g., due to the interception of erosion through natural vegetation or buffer strips);
- plant cover of crops during rainy seasons (missing to high);
- soil properties (clay, silt, loam, sand);
- density of livestock, particularly cattle, and fertilisation level (high to none);
- erosion abatement in place (high to none).

Inputs of P from the catchment to the waterbody range from about 0.1–0.2 kg P/(ha×yr) from areas with dense perennial vegetation cover up to 5.0 kg P/(ha×yr) from arable land (Franke et al., 2013). The highest values can be expected if high P concentrations in soils occur in situations with pronounced soil erosion and the eroded soil is easily transported to the surface waters. High P concentrations can be expected if agricultural management is characterised by high livestock densities and/or fertilisation levels exceeding plant requirements resulting in high P surpluses. P surpluses in soils can be estimated via soil nutrient budgets from the amount of nutrient in fertiliser, manure and slurry spread on the fields and pastures in relation to the amount in the harvests leaving the farm (for further information and data, see EUROSTAT (2019) and FAOSTAT (2019)). With clay/silty soils, the concentrations in eroded soil material may be further enriched as fine particles usually have the highest concentrations and are predominately transported.

Several local factors determine the levels of soil erosion. Firstly, soil erosion is impacted by the energy with which raindrops mobilise soil particles when they reach the surface. This is especially high if the crops grown have a low plant cover during rainy season (which is often the case for, e.g., maize, soya bean) and in regions with high rainfall intensity (volume per area and time). Secondly, the slope of a field and its length determine the transport capacity of water during surface run-off. Therefore, soil erosion increases at fields with steep and long slopes. Clayey/silty soils are especially vulnerable against soil erosion as small particles are more readily mobilised and transported as larger particles from sandy soils. Further, high organic (humus) content of the soil and improved soil structure reduce erodibility. If specific erosion abatement measures are in place, erosion is reduced.

Winter crops and mulch seed, for instance, increase the coverage of soils and thus hinder rain to mobilise and transport soil particles. High transport to surface waters of eroded material can be expected if the fields where erosion takes place are well connected to surface waters (high connectivity): run-off from such fields may directly enter surface water because there are no other types of land uses (e.g., buffer stripes) between them and the river is not able to hinder the transport of soil particles.

Example: A catchment of 50 km^2 has a share of 30% arable land. The amount of fertiliser applied is in a medium range; the region is hilly with a significant share of steep slopes and pronounced connections between arable land and creeks of the catchment. Silty soils prevail; no specific erosion abatement measures are in place. About 35 km^2 in the catchment is covered with natural vegetation or grassland. Input into surface waters can be assumed to be at the low end of the range given above, that is, 0.1–0.2 kg P/(ha×yr) from these areas as soil loss from these areas and P content of soil material are usually low:

- P emissions form naturally covered land and grassland = 0.1–0.2 kg P/(ha×yr)×3500 ha=0.35–0.70 t P/yr.

About 15 km^2 are covered with arable land. We assume relatively high levels of P emissions due to unfavourable conditions with respect to erosion (high soil loss) but only average fertilisation levels (moderate P concentrations in soils) of 1.5–3.0 kg P/(ha×yr):

- P emissions form arable land=1.5–3.0 kg P/(ha×yr)×1500 ha= 2.25–4.50 t P/yr.

Franke et al. (2013) present a more elaborate tier 1 approach for nutrient emissions from agricultural fields in the context of grey water footprint calculations. This can be applied if fertilisation levels are known. If requirements for the quantitative assessment of nutrient emissions are higher, the higher-level tiers discussed below should be applied. These tiers require including experts in the field of nutrient monitoring and nutrient emission modelling in the planning team.

7.4.2 Tier 2: Assessment using the Riverine Load Approach

The riverine load approach (RLA) as presented in EC (2012) is based on data measured on site, that is, for the water, the suspended solids, river discharge as well as monitoring data from relevant point sources, and it calculates the basic processes of transport, storage or temporary storage and degradation of substances. The resulting riverine load provides quantitative information about the recent status of loading and, provided long-term information is available, also about trends over time (see Figure 7.3). In particular, it allows the allocation of observed river loads to point and diffuse sources (i.e., a basic source apportionment). If a reservoir or a lake is fed by different rivers, RLA needs to be implemented for each one significantly contributing to the nutrient inputs into the reservoir or lake. High nutrient concentrations, an increasing trend, or a high relevance of diffuse sources indicate a need for a more detailed analysis using the approaches in tiers 3 and 4.

The nutrient load transported by a river is estimated by taking the product of the mean flow-weighted concentration and the total river flow, expressed by the following formula (OSPAR, 2004a):

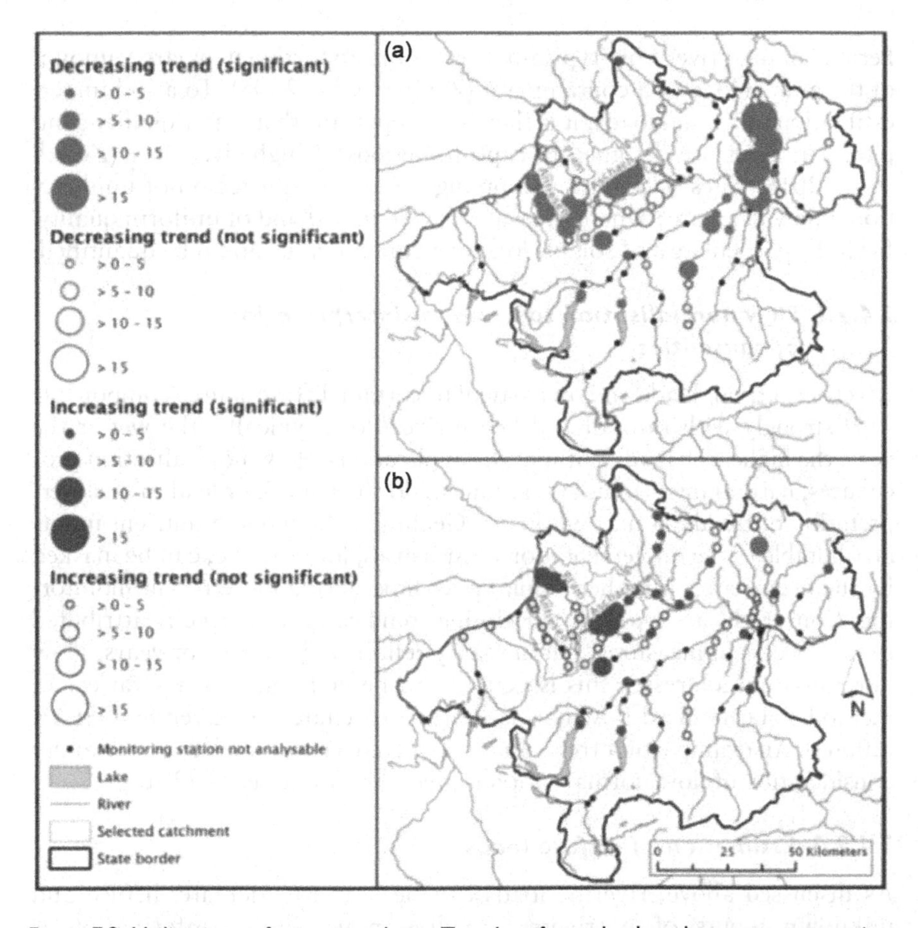

Figure 7.3 Utilisation of riverine data: Trends of total phosphorus concentration (flow adjusted in μg TP/L×year) in surface waters in Upper Austria for the periods: (a) 1990–2000 and (b) 2001–2004. (From Zessner et al., 2016.)

$$L_Y = \frac{Q_d}{Q_{Meas}} \times \left(\frac{1}{n} \sum_{i=1}^{n} C_i \times Q_i \times U_f \right)$$

L_Y=annual load (t/yr)
Q_d=arithmetic mean of daily flow (m^3/s)
Q_{Meas}=arithmetic mean of all daily flow data with concentration measurement (m^3/s)
C_i=concentration (mg/L)
Q_i=measurement of daily flow (m^3/s)
U_f=correction factor for the different locations of flow and water quality monitoring station
n=number of data with measured concentrations within the investigation period.

Periods of high river flow typically carry a disproportionately large amount of the annual load of a contaminant (Zessner et al., 2005). To avoid underestimation of annual loads, it is therefore important that water quality sampling strategies are designed to capture periods of high river flow (Zoboli et al., 2015). Sites selected for sampling should be in a region of unidirectional flow in an area where the water is well mixed and of uniform quality. Both the particulate and soluble load of a contaminant should be quantified.

7.4.2.1 Flow normalisation to avoid misinterpretation of causalities

Riverine nutrient loads and, in particular, certain diffuse source components vary strongly with rainfall and hence river flow; typically, the wetter the year, the higher the load. Without the application of flow normalisation procedures, natural interannual variations in flow can mask or lead to misinterpretation of trends in nutrient loads. Genuine reductions in nutrient inputs attributable to the implementation of measures, for example, can be masked by the occurrence of higher annual river flow during more recent monitoring. Conversely, an apparently declining trend can be incorrectly attributed to the success of measures, but in reality reflects a drier year or years. Flow normalisation addresses this issue and can be undertaken via a variety of methods. Harmonised flow normalisation procedures are given by OSPAR (2004a). An example of a trend analysis of P concentrations in a river under consideration of flow normalisation is given by Zessner et al. (2016).

7.4.2.2 Estimation of diffuse loads

As discussed above, riverine loads can be used to calculate diffuse and unknown inputs of nutrients providing point source information is available. In the most basic approach, the diffuse load can be estimated as the difference between the total load (measured from river discharge multiplied by concentrations; see above) and the load discharged from point sources, as follows:

$$L_{Diff} = L_{yr} - D_P$$

where, for a given contaminant, L_{Diff} is the anthropogenic diffuse load, L_{yr} is the total annual riverine load, and D_P is the total point source discharge. Such an approach ignores any potential in-river processes such as sedimentation and remobilisation, but provides a useful approximate estimate of the diffuse load of a given substance.

A more detailed formulation will be necessary where processes in the river or stream and natural background loads are thought to be significant. The following formula is based on an approach established by OSPAR (2004b) for the calculation of diffuse nutrient loads; in-river nutrient processing is typically significant:

$$L_{Diff} = L_Y - D_P - L_B + N_P$$

where, for a given contaminant, L_B is the natural background load of the contaminant, and N_P is the net outcome of in-river processes upstream of the monitoring point. There are several methods to estimate N_P on a catchment scale. For example, Vollenweider and Kerekes (1982) derived a formula which described the relationship between the nutrient concentration at the inflow of a lake or reservoir and the concentration within it based on the water residence time. This formula can be used to calculate the retention of nutrients by in-lake processes (N_P for lakes). Behrendt and Opitz (1999) proposed something similar for rivers at a catchment-scale level. They derived a relationship between area-specific run-off (river flow subdivided by the area of the catchment) and nutrient retention induced by processes within the stream or river as well as a relationship between hydraulic load of a river (river flow subdivided by the surface of the waterbodies in the catchment) and retention. If the flow of a river, the nutrient load, the catchment area and/or the surface of waterbodies in the catchment are known, this approach can be used to estimate N_P for rivers on a catchment scale (OSPAR, 2004c).

The riverine load approach (RLA) provides a useful means of estimating diffuse inputs and/or validating modelled predictions. However, diffuse inputs from different sources are merged into a single value and are not, for example, distinguished between inputs arising from agriculture and those arising from urban run-off.

7.4.3 Tier 3: Pathway-Oriented Approach

The pathway-oriented approach (POA) (see EC (2012) for more detail) uses more specific information about the land use, hydrology and basic transport processes involved. It adds an estimate of the impact of key processes of transformation, removal and temporary storage taking place between the source of emission and the receiving waterbody to the assessment. Therefore, data requirements are higher than for the lower tiers, but so is the level of information available for the inventory and for deciding on priorities for load control because this approach quantifies specific emissions (e.g., area-specific loads, stormwater run-off loads). It thus allows the identification of the main nutrient pathways and regional emission hotspots as well as providing a holistic overview of the emission status. POAs are well established and applied, for example, in many European River Basin Districts (RBDs) for the quantification of nutrients and heavy metal inputs.

As defined above, inputs can be caused by point and diffuse sources. Accordingly, point source pathways are defined by being discrete, having distinct locations and in many cases a quasi-continuous discharge, for example, the discharge of municipal wastewater treatment plants and industrial plants. Diffuse source inputs use different pathways and are discharged via different run-off components into surface waters, often driven or augmented by extreme events. A differentiation of the run-off components is necessary as substance concentrations as well as the underlying processes may differ significantly for the considered substances and localities. Actually 12 potential pathways for inputs into surface waters are identified for nutrients. This is summarised in the general working scheme (Figure 7.2). The pathways can be classified into three blocks:

1. pathways transporting nutrients from point sources;
2. pathways transporting nutrients from diffuse nonurban sources;
3. pathways transporting nutrients from diffuse urban sources.

The calculation of emissions from point sources can be straightforward if data on effluent concentration and the amount of water are available, or can be derived from statistical data with the required accuracy.

The inputs caused by diffuse sources are the result of more or less complex interactions with different interfaces, including temporary storage, transformation and losses. These processes have to be adequately integrated into the approaches.

As outlined above, pathways from agricultural diffuse sources include erosion, surface run-off, interflow, tile drainage and groundwater as well as direct discharges and wind drifting (e.g., of slurry sprayed on fields). The principles of POAs are best illustrated using the example of erosion, particularly as P can readily attach to soil and eroded sediment (Figure 7.4).

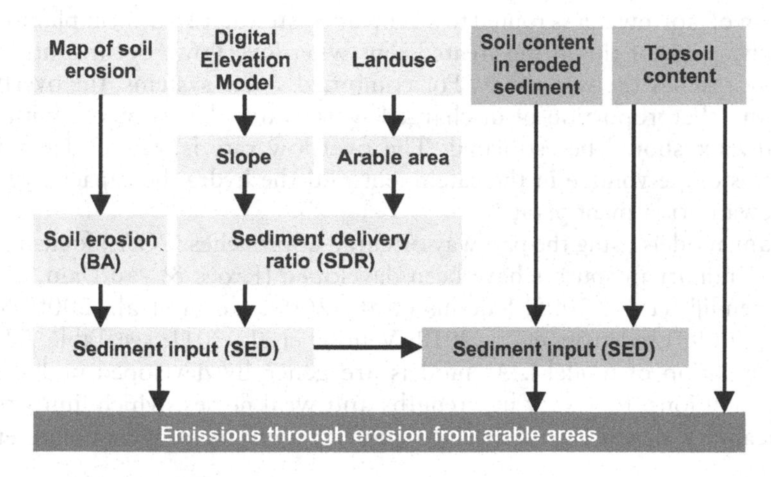

Figure 7.4 Input data to quantify the emissions from erosion. (Adapted from Fuchs et al., 2010.)

As discussed above, erosion begins with the mobilisation of top soil caused by heavy rainfall. At a river basin scale, the soil loss from arable land is commonly calculated using an adapted version of the universal soil loss equation (Wischmeier & Smith, 1960), which considers the slope, rainfall (energy input), soil characteristics, land cover and cultivation as well as erosion protection measures in place. The proportion of eroded soil entering the surface water is called "sediment delivery ratio". Different approaches can be applied for its calculation. As an example, individual areas within a catchment can be identified where eroded soil reaches a waterbody based on a Geographical Information System (GIS)-supported submodel, giving a relationship between sediment delivery and catchment characteristics (Behrendt et al., 2000).

During the erosion process, fine particles accumulate in the transported sediment. Phosphorus is predominantly bound to finer grains which accumulate during the transport process. The enrichment of a substance in the erosion material is described by the enrichment ratio (EnR), which is the ratio between the substance concentration in the topsoil and that in the sediment reaching the waterbody. Beyond the initial substance concentration, the grain size distribution of the topsoil and the intensity of the classification process are the most important factors influencing sediment concentrations.

As discussed above, in urbanised parts of a river basin, the important diffuse pathways for nutrients are overflows from stormwater sewers (i.e., those carrying rainwater run-off from paved or otherwise sealed surfaces) and in particular overflows from sewers carrying both stormwater and raw sewage. Their relevance is highly variable as overflows are caused by heavy precipitation when run-off volumes exceed the storage capacity of the sewer system, depending very much on local conditions. In general, a more complex situation can be assumed in combined sewer systems where a certain portion of stormwater is routed to a central wastewater treatment plant with the advantage of this being treated, but when overflows occur, untreated sewage reaches the waterbody. For combined sewer systems, the overflow rate and the proportion of discharged wastewater that is mixed with the stormwater should be estimated. The overflow rate is strictly dependent on the storage volume in the catchment and the hydraulic capacity of the wastewater treatment plant.

Many models using the pathway oriented approaches (POAs) focusing on diffuse nonurban sources have been developed (Kroes & van Dam, 2003; Groenendijk et al., 2005; Siderius et al., 2008; Gebel et al., 2009; Smit et al., 2009; Lindström et al., 2010; Venohr et al., 2011; see Table 7.2 for a compilation of models). As models are generally developed under specific conditions, they vary in strengths and weaknesses, which limits their applicability, depending on specific regional requirements. Schoumans et al.

Table 7.2 Examples of models for watershed-scale distributed simulation of nutrient transport in river basins (in alphabetical order)

Model	*Temporal scale*	*Description*	*Reference*
AnnAGNPS	Day or less	Annual-scale agricultural nonpoint-source pollution model, annualised version of AGNPS for continuous simulation of hydrology, erosion, transport of nutrients, sediment and pesticides	Young et al. (1995) Bingner & Theurer (2001)
ANSWERS-continuous	Day or less	Areal Nonpoint Source Watershed Environment Response Simulation, expanded with elements from other models (GLEAMS, EPIC) for nutrient transport and inputs	Bouraoui et al. (2002)
Hydrological Simulation program – Fortran	Hour	Continuous watershed simulation of water quantity and quality at any point in a watershed, developed for US Environmental Protection Agency (EPA).	US EPA (2011) Skahill (2004)
IBIS-HYDRA	Variable, 1 day to 1 year	Land surface and terrestrial ecosystem model IBIS with hydrology model HYDRA, used for modelling dissolved inorganic nitrogen fluxes and removal	Donner et al. (2002) Donner et al. (2004)
IMAGE DGNM	Month	Same as above, but with mechanistic instream model for C, Si, N and P, including sediment–water exchange	Vilmin et al. (2018)
IMAGE-GNM	Year	Detailed description of delivery of nutrients, at annual scale globally and by river basin; includes aquifer transport and processing	Liu et al. (2018)

(*Continued*)

Table 7.2 (Continued) Examples of models for watershed-scale distributed simulation of nutrient transport in river basins (in alphabetical order)

Model	*Temporal scale*	*Description*	*Reference*
INCA	Day	Integrated flow and nitrogen model for multiple-source assessment in catchments	Wade et al. (2002) Whitehead et al. (1998b) Whitehead et al. (1998a)
MIKE-SHE	Variable, depending on numerical stability	Comprehensive, distributed, physically based model to simulate sediment and water quality parameters in two-dimensional overland grids, one-dimensional channels, and one-dimensional unsaturated and three-dimensional saturated flow layers, with both continuous and single-event simulation capabilities	Refsgaard & Storm (1995)
MONERIS	Month or year	Empirically derived nutrient emission model, considering all relevant input pathways and instream retention processes on subcatchment level with a size of> 100 km^2.	Venohr et al. (2011)
NL_{CAT}		A combination of the models ANIMO/ SWAP/SWQN/SWQL. Based on the representation of system processes, nutrient concentrations can be calculated (inorganic and organic components). Furthermore, water flow and overland particulate and nutrient flow are modelled (run-off, erosion, subsurface run-off/leaching) in order to assess the total nutrient load to surface waters.	Groenendijk et al. (2005) Kroes & van Dam (2003) Smit et al. (2009) Siderius et al. (2008)
Riverstrahler	Reach, decade	Riverstrahler allows for analysing, apart from other disturbances, the impact of changing nutrient load and changing nutrient ratios, and potential saturation of retention processes such as denitrification and P retention by sediment. While in-stream processes are modelled with a mechanistic model, the delivery processes are described with coefficients, lumping soils, aquifers and riparian zones	Garnier et al. (1995) Billen & Garnier (2000)
SWAT	Day	Soil Water Assessment Tool to predict the impact of management on water, sediment and agricultural chemical losses in large ungauged river basins	Arnold & Fohrer (2005)

(2009) give an overview of some of the emission models and their specific applicability, and Moriasi et al. (2007) give guidelines for the estimation of accuracy in catchment models.

Monitoring → Modelling → Management are the key steps in a successful strategy for the development of sound policies for nutrient load control. Well-developed and appropriate emission modelling is central for this target. In addition to being necessary for assessing the relevance of different emission pathways, modelling helps to extrapolate information to locations or situations where monitoring has not been done or is not possible. Therefore, models use process relations or empirically derived relations expressed in quantitative formulas. Technically sound model applications, validated against measurements, implemented in a river basin or a watershed provide many potential applications, beginning with improving system understanding, increasing insights into cause–effect relationships and helping to identify emission hotspots.

For locations in a basin where monitoring is missing, modelling can provide an assessment of the risk to exceed water quality targets, and this can support the development of future monitoring schemes. Models are particularly useful to calculate scenarios in order to estimate the impact of a measure under consideration or – if the model structure allows – other potential future developments such as trends in population, land use or climate (Schönhart et al., 2018). The ability of calculating scenarios depends on the specific model structure and to which extent it depicts the complexity of the system. Whether models are able to assess scenarios quantitatively can be checked by running them with older data to see how well they are able to depict developments of the past.

The information derived from these investigations further provides a basis for cost-effectiveness or cost–benefit analyses. However, even technically sound models, the plausibility of which is validated by data from monitoring, will never provide an exact information. Therefore, uncertainty considerations are an important element of good practice in modelling.

7.4.4 Tier 4: The Source-Oriented Approach

This tier is based on substance-specific information on production, sales and consumption. It provides a comprehensive picture of the life cycle of a substance, for example, a nutrient. The benefit of this approach is that the information gained on the relative contribution of a source to the total nutrient load is far more precise than that gleaned from tiers 0 to 3, and thus, this provides a better basis for prioritising control measures addressing the primary sources of the nutrients. This level of precision may be relevant when advocating for control measures that require substantial investments (e.g., larger storage volume for stormwater to avoid overflow) or substantial changes of practice that may impact on people's livelihoods. (e.g., reducing fertiliser use or stock density) or require more elaborate management practices (e.g., introducing farm nutrient budgets; see Box 7.3).

Substance Flow Analysis (SFA), a source-oriented approach, is a method of analysing the flows of a substance in a well-defined system, including through industries producing and using it, households, wastewater treatment plants and all connected media such as soil, air and water. All the applications and uses of a substance are collated, enabling the development of strategies to reduce the impact of the substance. Such measures can also encompass source controls such as changes in consumption patterns (e.g., nutrition). SFA is applied in connection with the early recognition of potentially harmful or beneficial accumulations and depletions in stocks, as well as the prediction of future environmental loads. SFA methods, as we know them today, were first applied by Wolman (1965) in the wake of introducing metabolism studies for cities. Later, Baccini and Brunner (2012) developed a more specific method for the evaluation of the metabolism of the anthroposphere.

For nutrients, many SFAs have been performed on a supranational, national or district scale. Examples are the P balances of EU countries by Ott and Rechberger (2012) or the N and P balance of Busia District, Uganda, which was performed to identify the potential of improved waste uses for agricultural balances and productivity (Lederer et al., 2015). Zoboli et al. (2016a; 2016b) developed a time series in phosphorus flow analysis for the Austrian P-budget from 1990 to 2011 to assess drivers for changes in the national P metabolism and analysed the potential of different strategies for optimising the national P budget (Figure 7.5). This work goes far beyond just addressing P releases into waterbodies and possibilities for their reduction, but additionally includes measures addressing resource-efficient management strategies such as recycling technologies and changes in nutrition behaviour of people, which in turn affects agricultural practices.

One drawback to SFA is that applicable data tend to be limited to specific spatial or temporary solutions. Data sets are often only available on a country level. If the perspective is limited to a river basin, proxies may have to be used to illustrate the regional situation. And even though national data may be of high quality because they were compiled accurately, downsizing to the regional level can incorporate errors. On the other hand, it is also possible to combine a source-oriented approach of tier 4 with a pathway-oriented tier 3 approach by subdividing a country or region into subcatchments which form the basis for pathway-oriented emission modelling and integrate the aggregated subcatchment results for agricultural nutrient turnover and emissions into river systems into a regional or national SFA. This would consider the overall nutrient turnover, including imports, exports, production and consumption. Thaler et al. (2013; 2015), for instance, implemented such an approach, showing that in Austria P imports and emissions into the environment on country level could be reduced by about 20% if the population would change from its actual (meat-rich) diet to a healthy balanced diet (reduction of meat consumption by 50%) as recommended from nutritional experts. A further example is the reduction of P loads to a country's waterbodies merely by banning P from laundry detergents: this has reduced loads by about 50%, as has been shown for the Upper Danube in Germany and Austria by Zessner (1999).

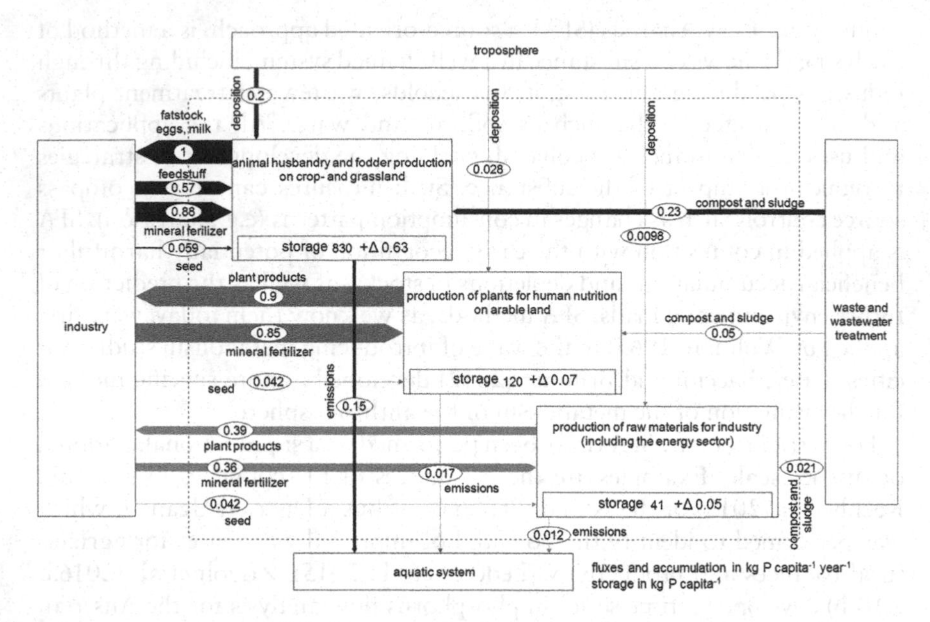

Figure 7.5 Phosphorus fluxes in the agricultural system of Austria differentiated by production for animal products, production for plant products and production for industrial raw materials (Thaler et al., 2013).

7.5 MANAGING NUTRIENT LOADS

The next step after identifying the relevant nutrient sources is to develop a management plan to mitigate them to the level targeted in order to reach the concentration targeted for the waterbody. Even if quantification of loads is only rudimentary or can only be roughly estimated, major source may be evident, and it will be important to get started with the implementation of measures to control them. The case study of the Santa Lucia River Basin (SLRB) in Box 7.4 shows that addressing major sources can lead to a substantial load reduction within a few years.

BOX 7.4: REDUCING EUTROPHICATION OF THE SANTA LUCIA RIVER BASIN IN URUGUAY

Juan Pablo Peregalli, Carolina Michelena and Giannina Pinotti

The Santa Lucia River Basin (SLRB) is the drinking-water source for 60% of Uruguay's population. Although it is not the biggest catchment in the country, it concentrates a major portion of Uruguay's industrial and agricultural activities. From 2004 to 2011, studies for assessing the water quality in the basin were conducted by the municipalities. The national authority responsible for the environment, in cooperation with external support, is implementing a water quality monitoring plan that includes 25 monitoring stations in different waterbodies of the basin. Most of the results show eutrophic to hyper-eutrophic conditions, and during March 2013, taste and odour events in supplied drinking-water were related to cyanobacterial blooms in the Santa Lucia River.

The initial assessment of the catchment showed that an average of 80% of the organic matter and nutrient load (N and P) received by the waterbodies originated from diffuse sources, including soil erosion and agricultural activities such as forage areas, fruit and vegetable plantations, dairy farms and feedlots. Point source contamination accounts for 20% of the total load and is mainly from nonsewered or insufficiently sewered settlements and from industrial facilities, mainly slaughterhouses, dairy, tannery and solid waste processing plants.

NUTRIENT LOADS MANAGEMENT STRATEGY

Based on these monitoring results, in 2013, the National Authority of Environment launched the "Action Plan for the protection of environmental quality and drinking - water sources in the SLRB". This plan included a series of measures to "control, stop and reverse the deterioration of the water quality and ensure the quantity and quality of water resources for a sustainable use of water in the river basin" (Table 1).

Table 1 Control measures in the Action Plan for the protection of the SLRB

MEASURE 1	• **Reduce the impact of effluents dicharge from industrial activities**
MEASURE 2	• **Reduce the impact of municipal wastewater discharge**
MEASURE 3	• **Control the excessive use of fertilizers**
MEASURE 4	• **Control the load discharge from feedlots**
MEASURE 5	• **Control the load discharge from dairy farms**
MEASURE 6	• **Management of sludge from the drinking-water treatment plant.**
MEASURE 7	• **Limit the access of animals to water in the waterbodies of the catchment.**
MEASURE 8	• **Establishments of riparian buffer zones**
MEASURE 9	• **Require licences for surface water and groundwater extraction.**
MEASURE 10	• **Declare the catchment of Casupá stream as drinking-water reservoir.**
MEASURE 11	• **Involve the different actors in the management of the basin.**

PRELIMINARY RESULTS AFTER 6 YEARS OF WORK IN PROGRESS

Industries were already under governmental surveillance, and the quality of the wastewater discharge was regulated by the Decree 253 from 1979, which includes targets for total phosphorus (TP) and ammonium. However, most of the wastewater treatment plants focus on organic matter removal, mainly stabilisation ponds, and did not achieve the levels targeted for nutrient emissions (5 mg/L for total phosphorous and ammonium). Measure 1 therefore focused on 24 industries that were responsible for 90% of the organic matter and 95% of nutrients discharged to the river basin from the industrial sector. For the emissions of those industries, additional standards were set and they were urged to build wastewater treatment systems with nutrient removal processes. As a result, discharge loads were reduced by 18% for biological oxygen demand (BOD5), by 52% for TN and 30% for TP, from 2014 to 2018 (Figure 1a).

Regarding municipal wastewater, measure 2 focused mainly on settlements of more than 2000 inhabitants. The existing wastewater treatment plants were urged to include nutrient removal, two new plants were built, the wastewater of two major settlements was transferred to the Rio de la Plata Basin (a much larger and less vulnerable catchment), and two plants were relocated from flooding areas. Because of the construction works that these measures require, most of these changes are still in progress; however, some results are already visible in the discharge data (Figure 1b).

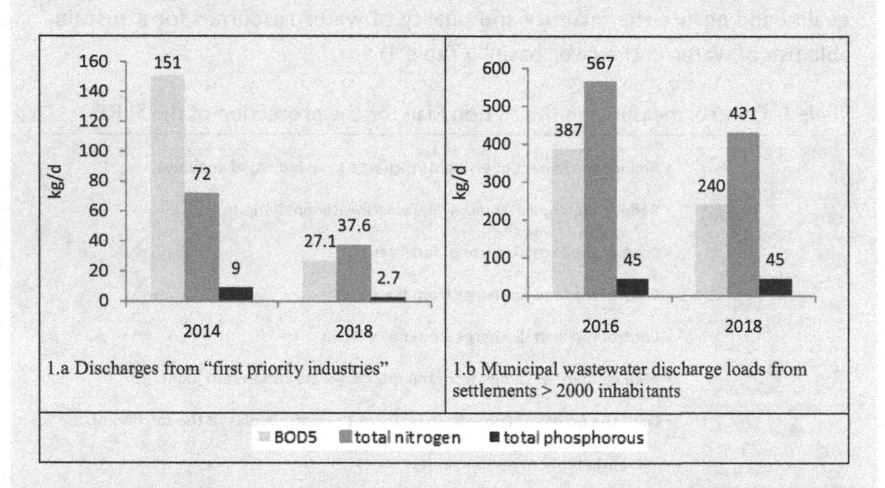

Figure 1 Results of measures 1 and 2 to control point source loads. (a) Discharges from "first-priority industries". (b) Municipal wastewater discharge loads from settlements > 2000 inhabitants.

Measure 4 banned the installation of new feedlots on most of the catchment and also the extension of existing ones. Also, in 2014, Decree 162/014 regulated environmental aspects of feedlots. Since then, national authorities regulating environment and livestock production and the association of feedlot owners are collaborating to establish the application of good environmental practices. In particular, this includes recommending the use of wastewater from feedlots and farms for irrigation instead of discharging it directly to waterbodies. For nine of the 20 feedlots in the catchment, that is, those with more than 500 animals, wastewater management systems were implemented. This has considerably reduced the loads of organic matter, N and P that otherwise would potentially be discharged to the basin (Figure 1a).

SLRB concentrates a major portion of the dairy farms in the country with a total of 1200 establishments, 92% of which are small (<300 animals), 5% are medium (300–500 animals) and 3% are big farms with more than 500 animals. Control measures first focused on the large farms, with a similar approach to the one used for feedlots. An interagency activity was launched to train farmers about sustainable wastewater management and to encourage its use for controlled irrigation. Additionally, a project was launched to provide economical support to more than 50% of the small farms to build wastewater management systems (drainage, accumulation ponds and irrigation systems). The rest of the small farms and medium-sized establishments still needs to be addressed but so far, about 50% of the organic matter and nutrients load discharged from dairy farms is being managed by the aforementioned initiatives (Figure 2).

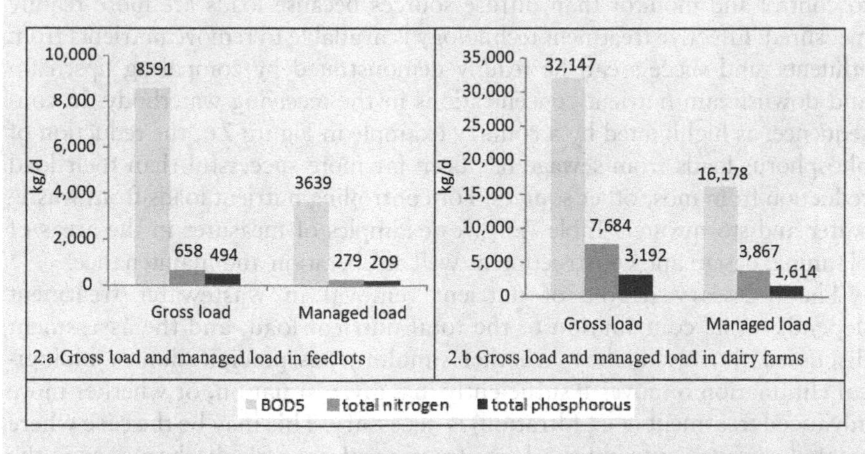

Figure 2 Results of measures 4 and 5 to control feedlots and dairy farms. (a) Gross load and managed load in feedlots. (b) Gross load and managed load in dairy farms.

In order to mitigate nutrient loads from soil erosion and run-off that reach the waterbodies of the catchment, measure 8 defined buffer strips on which agriculture and agrochemicals are banned to preserve and restore the riparian vegetation. Their width was set as 100 m for water reservoirs, 40 m for main rivers and 20 m for main tributaries. Analyses of satellite images show that most of the areas were already in compliance with the restriction (>98%), without considering urban areas that cover 14% of the buffer zones. Additionally, work with the local community to re-grow native riparian vegetation around the Paso Severino Reservoir is ongoing.

FUTURE STEPS

The action plan is still in progress, and it will take time to see the effects of the control measures on the water quality of the catchment. In December 2018, an update of the plan, based on the experience gained until then, implemented the so-called second-generation measures which are structured in four strategic lines: to ensure water quality, to reduce discharge loads, to protect and restore the ecosystems and to improve the understanding of the river system's dynamics.

7.5.1 Measures to control nutrient loads from sewage, stormwater and commercial wastewater

Wastewater is a key point source of nutrient emission, and it is usually easier to control and monitor than diffuse sources because loads are more readily measured. Effective treatment technology is available to remove nutrients from effluents, and success can be readily demonstrated by comparing upstream and downstream nutrient concentrations in the receiving waterbody. In consequence, as highlighted by a country example in Figure 7.6, the reduction of phosphorus loads from sewage has been far more successful than their load reduction from most other sources. For controlling nutrient loads from wastewater and stormwater, Table 7.3 gives examples of measures in the areas of planning, design and construction as well as operation and maintenance.

The necessary degree of nutrient removal in wastewater treatment depends on its contribution to the total nutrient load, and the assessment discussed above will show whether simultaneous precipitation or biological elimination removes P sufficiently in a given situation, or whether more advanced treatment (e.g., filtration) is necessary. This may be the case where treated sewage contributes a large fraction of a river's discharge, as is the case for, for example, River Spree in Berlin with 20–50% of the discharge being treated sewage (Fritz et al., 2004) but also for numerous other densely populated lowland river basins.

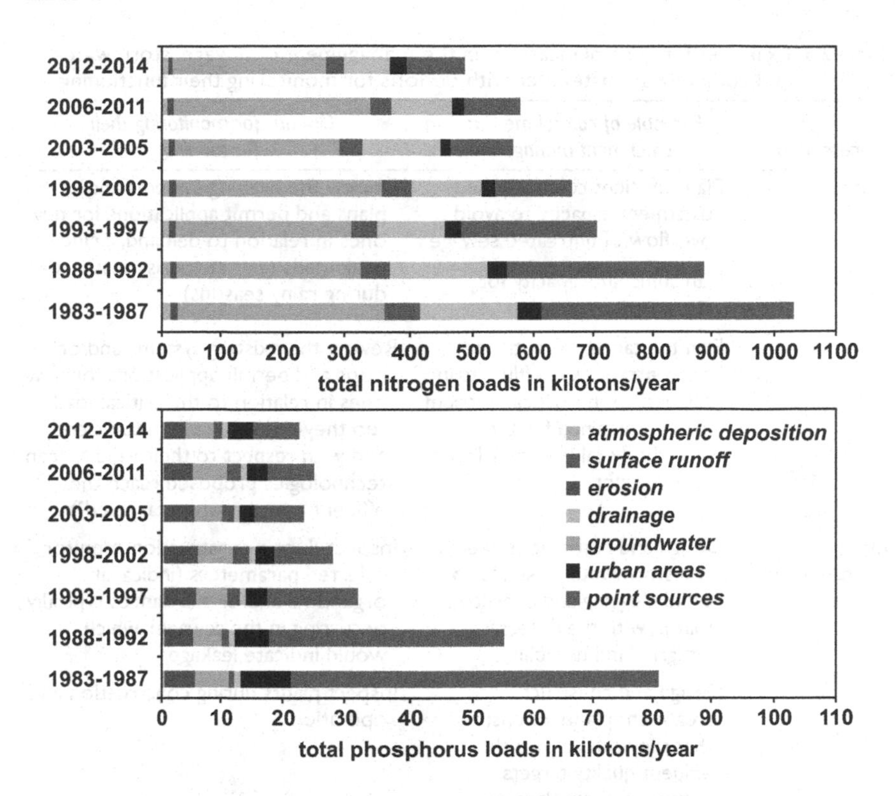

Figure 7.6 Loads of nitrogen and phosphorus from variable sources to surface waterbodies in Germany in kilotons per year (kt/yr). (Modified from UBA, 2017.)

A pronounced contribution to source abatement of P from municipal sewage, which affects all related emission pathways, is using P-free detergents. P loads and concentrations have been substantially and immediately reduced in countries introducing them. However, where wastewater causes substantial P loads, this may not suffice to achieve the target concentration for P in the waterbody, and the remaining P load from excreta nonetheless renders measures to reduce the P load from wastewater necessary.

Where loads from sewer systems carrying stormwater need to be reduced, an effective measure can be to intercept this on its pathway to the waterbody by constructing sufficiently large storages for flushes of rainwater: in such retention basins, a fraction of the suspended solids will settle to the sediment (which, however, needs to be removed periodically). From these storages, stormwater can also be gradually fed to a wastewater treatment, as capacity allows. A further option to retain stormwater is to create wetlands or to construct depressions for on-the-spot infiltration into the underground during storm events (e.g., areas in parks, covered

Table 7.3 Examples for control measures in the management of sewage, stormwater and commercial wastewater with options for monitoring their functioning

Process Step	*Example of control measures in catchment management*	*Options for monitoring their functioning*
Planning	Plan sufficient collection and treatment capacity to avoid overflow of untreated sewage Plan sufficient capacity for stormwater retention	Review the existing systems and/or plans and permit applications for new ones in relation to demand, including peak loads (e.g., at tourist season or during rainy seasons)
	Plan the target for nutrient concentrations in effluents in relation to the critical nutrient load determined for the waterbody, taking other loads into account	Review the existing systems and/or plans and permit applications for new ones in relation to the critical load (do they reduce the load sufficiently?) and with respect to their validity (can technologies proposed reach the effluent concentration targeted?)
Design, construction	Design and construct sewer and stormwater systems to avoid clogging and overflow, that is, with the necessary integrity and durability	Inspect during construction; monitor selected parameters (indicator organisms and/or substances typically occurring in the sewage), which would indicate leakage
	Design and construct treatment plants to ensure that they can achieve the effluent quality targets defined in their planning	Inspect plants during construction and operation
	Design and construct stormwater retention and infiltration systems to meet integrity and durability criteria	Inspect during construction and integrity during operation and when emptied for maintenance
Operation and maintenance	Clean sewers and drains at intervals necessary to avoid clogging	Inspect conditions; review records of cleaning and maintenance
	Keep wastewater treatment plants operating effectively	Monitor process parameters (see above) indicating process functioning; monitor nutrient concentrations in the discharge; monitor nutrient concentrations in effluent at regular intervals and during events (e.g., drought, heavy rainfall)
	Remove sludge from stormwater retention basins at appropriate intervals	Inspect condition of retention basins; monitor basin effluent suspended solids levels; monitor sludge levels in basin periodically; review records of sludge removal and maintenance
	Minimise nutrient accumulation on surfaces flushed by stormwater	Inspect street-sweeping and garbage-collection operations and records

with scenic vegetation). Figure 7.6 shows an example on the nation-wide level demonstrating the potential efficacy of the reduction of loads from combined sewer overflows and separate sewer systems (= urban systems in the figure).

For controlling diffuse nutrient loads from wastewater emitted by dispersed settlements along a river course or lakeshore, there are several options. One is the introduction of sewerage. Communities often construct sewers for the undisputable benefits of reducing infections spread through inappropriate disposal of excreta, and if this is not accompanied by sufficient treatment, it may dramatically increase phosphorus loads as compared to – for example – latrines from which a lesser fraction of these loads will reach surface waters. To prevent cyanobacterial blooms, it is therefore essential to introduce sewerage together with introducing appropriate wastewater treatment, including sufficient nutrient removal, to avoid exceeding critical loads to the receiving waterbody. An effective measure to ensure that no insufficiently treated sewage reaches the waterbody is to intercept its pathway by sewage diversion, for example, through installing a sewage channel between the settlements and the shore, that is, a channel or pipe which collects all of the wastewater and carries it to a treatment plant.

An alternative to sewerage is to implement sufficiently effective on-site treatment. A further option in dispersed settlements is to avoid sewerage and rather install dry sanitation systems, which, however, need to be safely designed and managed in order to avoid the contamination of groundwater or surface water (particularly with pathogens) and to achieve acceptance by users. In rural areas, when safely operated, such systems can have the benefit of providing fertiliser to use in agriculture (for safe use of wastewater, excreta and greywater, see WHO (2006); specifically for groundwater, see Howard et al. (2006); and for surface water, see Rickert et al. (2016)).

7.5.1.1 Operational monitoring for control measures in wastewater management

As discussed in Chapter 6, complementary to water quality monitoring, the purpose of operational monitoring is to continuously check whether or not a control measure is working as it should. For technical controls in wastewater treatment, effective parameters indicating treatment performance include, for example, flow rates and detention times in the treatment plant, suspended solids, dissolved oxygen, pH and chemical oxygen demand. These parameters can be measured with continuous, online recording by operators of the wastewater facilities. However, as Table 7.4 shows, many control measures which are important to ensure that nutrients from these point sources do not reach the waterbody are best monitored by inspection and reviewing of records. Time patterns for such operational monitoring can be at larger, irregular intervals, and it may be important to monitor when events occur that affect the process, or when operations change.

7.5.1.2 Validation of control measures in sewage and stormwater management

Where advanced wastewater treatment technology is implemented, the validation of whether or not these measures are actually achieving the targeted effluent nutrient concentration is quite straightforward: it requires a regular monitoring of effluent discharge and nutrient concentrations in it, that is, of the nutrient load leaving the plant. This applies equally to most alternative treatment technologies such as artificial wetlands: it is often feasible to run a focused programme to validate whether the loads discharged are within the targeted limits. Even where such monitoring programmes appear expensive, their costs need to be viewed in relation to the investment and operation costs, and such considerations typically show that it is worthwhile to validate whether or not a measure is sufficiently effective before taking decisions, for example, on continuing its operation or even upgrading it.

In contrast, it is challenging to validate whether the load management from overflows of mixed sewage systems and from stormwater in separate systems meets its targets: estimating nutrient loads during stormwater run-off requires sampling effluents and measuring their flows under conditions of heavy precipitation, and these samples are needed with a tight resolution over time (both water volumes and nutrient concentrations may change within minutes, and possibly also in space if numerous (usually dry) overflows come into operation more or less simultaneously). This can be achieved with automated sampling and/or a sufficiently large team of highly motivated staff.

Waterbody data on nutrient concentrations may support validation if they have a high resolution in time and space so that sudden concentration peaks can reflect loading patterns.

7.5.2 Measures to control nutrient loads from agriculture and other fertilised areas

Agriculture is a key diffuse source of nutrients, and in many cases, reducing the loads it causes has been less successful than reducing the point sources from wastewater discussed above. Improving practices of fertilisation to optimise the balance between crop yield and nutrient load control requires not only engagement of farmers and managers for the target of protecting the water source, but also expertise and training. Furthermore, the overall socioeconomic context strongly influences the locally realistic options for agricultural practice. Agriculture needs to remain productive

Table 7.4 Examples for control measures in agriculture and fertilised land use with options for monitoring their functioning

Process Step	*Example of control measures in catchment management*	*Options for monitoring their functioning*
Planning	Define criteria for exclusion or restriction of activities (e.g., stock density, type of crop) in vulnerable catchments (e.g., implement protection zones or riparian buffer strips)	Monitor land use in vulnerable areas/protection zones and ensure that restrictions are implemented (site inspection)
	Require permits for the location, design and operation of feedlots in vulnerable drinking-water catchments	Review plans and applications for permits for agricultural activities in relation to the nutrient load expected from the area
	Set financial incentives (subsidies, credit, low-interest loans to fund changes, compensation for lost income during transition periods to new practices) and/or disincentives such as penalties for nutrient loading caused by poor management practice	Check compliance with practices negotiated before granting financial incentives or applying penalties; check compliance to restrictions set in regulations
	Regulate operations (e.g., types of crop; stock density on fields and in stables; use of fertiliser and/or manure) in vulnerable catchments	Review considerations as to whether they can achieve the target load estimated to be tolerable for the waterbody
Design and construction	Apply best management practices for treating wastewater from feeding operations	Check compliance of treatment structures with best management practices
	Construct fencing to protect waterbodies from livestock	Inspect integrity regularly
Operation and maintenance	Implement regulations for operations (e.g., types of crop; stock density on fields and in stables; use of fertiliser and/or manure) in vulnerable catchments	Inspect records of crops grown, fertiliser and manure application; count heads of stock
	Require soil tillage methods that minimise erosion	Visual site inspection
	Grow winter crop cover to reduce erosion	Visual site inspection
	Match irrigation and fertilisation (mineral fertiliser and/or manure) to the needs of crops or lawns; implement farm nutrient management plans budgeting fertiliser purchased against nutrient content of crop leaving the farm	Inspect drainage and monitor its nutrient concentrations; inspect farm records for nutrient budgets, amounts and timing of fertiliser/manure application

Source: Adapted from Rickert et al. (2016).

and economically viable, and measures to control eutrophication need to be sustainable also in this respect. Thus, aspects to consider when planning measures may therefore include a wide range, for example, prices achievable for produce, habits and demands of consumers, education status of farmers, messages conveyed by agricultural advisory services, access to sites for watering animals and costs of fertiliser or material for fences to keep stock away from water courses.

A range of measures to control nutrient emissions from fertilised land to surface waters is available. They can be summarised in three groups: (i) to avoid to high nutrient surpluses and enrichment in soils by crop choices and fertilisation limited to actual plant needs and uptake; (ii) to avoid nutrient (soil) loss from land by, for example, maintaining a vegetation cover on fields or other erosion abatement measures; and (iii) to intercept the transport of nutrients (soil) from agricultural land into surface water (e.g., with densely vegetated riparian buffer strips). They cover the areas of planning, design and construction as well as operation and maintenance (Table 7.4).

Good agricultural practice as implemented in some larger enterprises with accordingly trained staff involves balancing a farm's nutrient budget to where the farm does not import more N and P than the amount that leaves the farm with the produce (see Box 7.3). Regulations may require a farm operator to keep records of the amounts of fertiliser purchased and spread on the fields, or financial support from the water supplier or a government authority can be made dependent on demonstrating that the farm maintains this balance. Where large-scale intensive animal husbandry can lead to an imbalance between farm fertiliser needs and manure production, measures to avoid excessive spreading of manure on land are important.

Implementing farm advisory systems (or strengthening the existing ones) and providing information on appropriate or innovative approaches are important to optimise practices. Information campaigns can communicate best management practices such as fertilisation on demand. Such campaigns are most effective if combined with the analysis of soil nutrient content at the end of the growing season as a sound basis for assessing the seasonal fertiliser application needs for next year's crop. Successful multiple-stakeholder approaches have shown that funding soil analyses and information campaigns, for example, by state authorities or even the water supplier are cost-effective: the money invested for better raw water quality saves investment in treatment technology and is a more sustainable approach. Stock density and fertiliser application rates may also be limited by law, for example, by banning manure application during specific seasons with heavy rainfalls or snow and ice-cover or limiting stock density. Where large-scale animal husbandry operations cause amounts of manure and slurry significantly exceeding the nutrient demand of fields and pastures, loads may be controlled by organising manure export to other farms and regions with a need for nutrients (MacDonald et al., 2011). Further control measures may include requiring sufficiently large storage volumes for manure and slurry and banning manure application during specific seasons with heavy rainfalls or snow and ice-cover. Nutrient loads from agricultural activities, golf courses, lawns or other land uses involving fertilisation can be avoided if these loads are sited sufficiently far from the waterbody in order to avoid a direct input of fertiliser or manure during application or from grazing animals.

For mitigating phosphorus loads transported via erosion, it is generally important to establish and manage water retention measures that minimise the concentrated run-off. Such measures include avoiding devegetated land where slopes will allow erosion by heavy rainfall. On sloping land, it is particularly important to control run-off (e.g., by water retention measures such as retention ponds, vegetated buffer strips) and to avoid practices that enhance erosion, that is, planting of specific crops with low vegetation cover (e.g., maize), deep ploughing, ploughing groves perpendicular to the slope, intensive grazing or slash-and-burn agriculture. Winter cropping and mulch seed increase vegetation cover of arable land and therefore may significantly reduce erosion if professionally implemented. Maintaining the soil organic content and soil structure reduces soil erodability. Other measures for erosion abatement are contour ploughing, conservational tillage (see examples and discussion in Tiessen et al. (2010)), avoiding tillage altogether or creating terraces (Novotny, 2003). Intensive grazing and animal access to watering points near rivers can cause severe erosion of riverbanks, and such destruction of riparian vegetation as well as faecal loads should be avoided. Peacher et al. (2018) give an overview of publications discussing the benefits of fencing to exclude cattle from streams, thus preventing riverbank erosion; they also review publications on the relevance of vegetation cover directly on the areas on top of riverbanks to prevent erosion.

Riparian buffer strips covered with dense vegetation can effectively intercept surface run-off carrying phosphorus-rich soil eroded from arable land and pastures. If they are applied at sufficient width (i.e., up to 30 m) and at the right locations (where they effectively interfere to relevant P transport – this can be along the riverbank but also somewhere else in the catchment), they may effectively reduce particle-bound P transported by soil erosion. Nevertheless, in contrast to erosion abatement measures mentioned above, buffer strips may retain soil particles by being a sink and avoid their input into surface waters but they do not avoid soil and nutrient losses from productive agricultural areas. Furthermore, to maintain their longer-term efficiency, proper management of buffer strips is important. Other potential nutrient transport control measures include grassed waterways, grass filters, constructed ponds, reservoirs and wetlands as well as connected or reconnected floodplains.

7.5.2.1 Operational monitoring of control measures in agriculture and land use involving fertilisation

In contrast to monitoring effects on water quality, monitoring whether or not a measure is operating as intended can be simple and be performed by those managing the agricultural activities. Operational monitoring should provide a timely signal if the specified control measure is not operating within the acceptable limits, so that operators can take corrective action before nutrients are washed from the land to the waterbody. Monitoring the operation of many measures that can control nutrient losses from agricultural land or from fertilised lawns to water needs to be done by farmers or gardeners themselves, and motivation for such measures can effectively be developed by involving these stakeholders in planning the protection of the waterbody, as highlighted in Table 7.4. Time intervals for the monitoring of control measures in agriculture can often be much larger than for operational monitoring of technical control measures (such as wastewater treatment, which may even involve continuous reading and recording of parameters). For many measures, operational monitoring can simply be visual, that is, through inspection, for example, of the integrity of erosion control structures, fences or riparian buffer strips. It may also include desk work to assess whether reports required by regulations (e.g., on stock density or farm nutrient budgets) have been submitted and the information reported complies with the requirements. This is most effective if sporadically checked personally on site, for example, in the context of catchment inspection. Satellite data can support visual inspection. Some continuous, online recording is feasible for some agricultural control measures as well, for example, using the interruption of an electrical current as a monitoring parameter for the integrity of an electrical fence.

While enforcing compliance particularly in agriculture and of home owners has been experienced as notoriously difficult in many countries, experience is also that it tends to improve as the education level of farmers (and home or golf course owners) increases. It is also likely to improve where these stakeholders are successfully involved in the planning process or obtain support in implementing measures, for example, in the context of co-operation agreements with water suppliers.

7.5.2.2 Validation of control measures in agriculture and for land use involving fertilisation

In contrast to the rather straightforward options for validating the efficacy of measures to control nutrient loads from effluents discussed in section 7.5.2, validating whether or not the measures taken actually achieve the target of retaining nutrients on the land rather than losing them to the waterbody is more challenging: it requires quantitative approaches to estimating diffuse loads as discussed in section 7.4. Validation of measures controlling diffuse loads is typically a discontinuous activity to be repeated at intervals, particularly after changes in the system, with time scales for changes in nutrient losses from land to water usually being in the range of seasons or years.

Where improvement is substantial and visibly evident – for example, the reduction of soil erosion – validation may be possible by inspection. Another approach to validation is trend analysis of the development of agricultural practices which impact the nutrient emissions to surface waters, that is, of livestock densities, fertiliser applications, farm nutrient balances, cultivated crops and tillage practices.

7.5.3 Measures to control nutrient loads from aquaculture and fisheries

Where cage cultures and fisheries within a waterbody cause nutrient loads in excess of the load which is acceptable in order to meet the phosphorus concentration target for the waterbody, the only option available to control nutrient loads from feeding is to restrict or totally ban these activities from the waterbody. In practice, clear water with no or only low cyanobacterial biomass and productive fisheries are conflicting, scarcely compatible targets that require a decision on priorities.

In contrast, for aquaculture operations in the catchment, effluent treatment can be an option to reduce nutrient loads. Treatment can include technical methods as well as treatment through a wetland or even – provided concentrations of other contaminants such as medication are not too high – application of effluent on farmland, combining irrigation with fertilisation. However, the latter requires a tight control in order to avoid application in excess of demand, as it may risk "getting rid of the waste" driving the effluent amounts applied, which can in turn cause run-off to the waterbody. It further requires assessing whether such effluent causes inacceptable loads with treatment chemicals and pharmaceuticals (Table 7.5).

7.5.3.1 Operational monitoring of control measures in aquaculture and fisheries

Operational monitoring of control measures in aquaculture and fisheries may be difficult to implement particularly in small-scale operations that typically have poor recording and documentation. Specific motivation and training of operators can be important, and this may be facilitated by involving them or their representatives in planning (e.g., in the team for developing a Water Safety Plan). Where the target of clear water without cyanobacterial blooms is in conflict with fisheries and cage cultures as the basis for people's livelihoods, resolving this may require substantial discussion, potentially resulting either in an alternative drinking-water source or in alternative sources of income and potentially also of the local population's protein.

Table 7.5 Examples for control measures in the management of aquaculture and fisheries with options for monitoring their functioning

Process Step	*Example of control measures in catchment management*	*Options for monitoring their functioning*
Planning	Plan operations *within a waterbody* (such as cage cultures or feeding of fish) in relation to targets for the maximum acceptable nutrient load (see section 7.2)	Review the existing systems and/or plans and permits for the nutrient load; the operations are likely to introduce in relation to the nutrient load acceptable for the waterbody
	Where targets are in conflict, that is, between a TP concentration in the waterbody and aquaculture or fisheries, decide on the water-use priority	If the decision is against aquaculture or fish stocking, inspect the waterbody or catchment to check for compliance
	Plan aquaculture operations *in the catchment* with respect to the nutrient load acceptable for the waterbody, potentially introducing effluent treatment (settling ponds, wetlands)	Review the removal efficiency treatment that can potentially achieve the resulting load in relation to the nutrient load acceptable for the waterbody
Design, construction	Line or reline fishponds from which water may seep to a waterbody with impervious material; protect from storm and flood damage, for example, through stormwater bypasses	Inspect structures as to suitability of design for the purpose and their integrity. Monitor water balance in ponds to determine if seepage is occurring
	Use closed re-circulation system with treatment, aeration, sustainable stocking rates and controlled feeding rates	Inspect design and construction; review the management plan for stocking and feeding rates
	Avoid discharge of untreated effluent – treat it or use it as liquid fertiliser on land areas that are not susceptible to run-off and leaching	Review information about its designation
	Construct and maintain particle traps in tanks (with separate sludge outlet)	Inspect structures
	Use removed sludge as fertiliser on land areas that are not susceptible to run-off and leaching	Inspect storage and application sites; review records of sludge application

(*Continued*)

Table 7.5 (Continued) Examples for control measures in the management of aquaculture and fisheries with options for monitoring their functioning

Process Step	*Example of control measures in catchment management*	*Options for monitoring their functioning*
Operation and maintenance	Match amount of feed to intake by the fish, using feeding methods and patterns adapted to satiation time, transit rate and subsequent return of appetite	Inspect feed used; discuss practices (e.g., timing and amounts) with operator; if available, inspect records of feed purchasing and application Estimate fish stock density; discuss practices and use of specific diets with operators and feed supplier
	Use low-polluting feed, high levels of lipid, lowered protein content, typically with high digestibility value, low in phosphorus	...
	Collect waste from tanks and cages	Inspect records of waste collection and cleaning activities, with waste volume estimated and disposal practice (sites) noted
	When emptying and cleaning basins, ponds and tanks, avoid discharge of untreated water	Discuss attention to this point with operators
	Keep fish stock density below a threshold defined as acceptable in relation to the nutrient loading target for the waterbody	Inspect records of fish stock density

7.6 INCLUDING CLIMATE CHANGE SCENARIOS WHEN PLANNING MEASURES

Climate change is expected to increase the frequency of extreme events such as rainfall patterns, floods or drought, and such events will impact on nutrient loads to rivers, lakes and reservoirs (WHO, 2017a). The extent to which such events are likely to change nutrient loads to a given waterbody, however, strongly depends on local conditions. For example, a heavy rainfall event may enhance erosion and thus increase the phosphorus load. In contrast, depending on the soils eroded, this may be accompanied by increased loads of silt that binds phosphorus with which it settles to the sediment. Also, if previous drought has rendered the soil surface hard and almost impermeable, such an event may dilute the concentration in the waterbody rather than increasing it. For example, Schindler (2006) found that for lakes in the Canadian Experimental Lakes Area, dry periods with less inflow actually reduced the phosphorus load, and this effect more than compensated that of a reduced flushing rate, actually reducing P concentrations in the lakes.

Generalised predictions of the impact of climate change on nutrient loading on a specific waterbody are therefore not possible; rather, site-specific scenario considerations are a prerequisite for including potential climate change impacts when planning measures to control nutrient loads. Past observations during extreme events are useful for such considerations. Where models can be constructed to depict nutrient loads relative to weather events, these are a highly useful tool to test climate change scenarios. Schönhart et al. (2018), for instance, include climate scenarios to assess climate-driven impacts on land use and nutrient emission in an integrated impact modelling framework (IIMF) for the whole Austrian territory. To address the uncertainty of predictions for precipitation, these authors tested two scenarios, one with increasing precipitation in future and another with decreasing precipitation. Results show that drier conditions could increase the pressure on freely flowing river stretches because reduced dilution of permanent emissions would cause higher concentrations, while increasing precipitation would, in contrast, increase the pressure on stagnant waterbodies because of increasing transport of loads from diffuse sources.

As discussed in the World Health Organization's guide on "managing health risks associated with climate variability and change" (WHO, 2017a), developing a Water Safety Plan provides a good platform to include the experts and specialists "*to understand potential climate change impacts in the context of their water supply*" and thus integrate aspects of climate resilience into planning improved management of a catchment.

REFERENCES

Alabaster JS (1982). Report of the EIFAC workshop on fish-farm effluents, Silkeborg, Denmark, 26–28 May 1981. Roma: Food and Agriculture Organization of the United Nations.

Arnold JG, Fohrer N (2005). SWAT2000: current capabilities and research opportunities in applied watershed modelling. Hydrol Process. 19:563–572.

Baccini P, Brunner PH (2012). Metabolism of the anthroposphere: analysis, evaluation, design, 2nd edition. Cambridge (MA): MIT Press:391 pp.

Behrendt H, Huber P, Opitz D, Schmoll O, Scholz G, Uebe R (2000). Nutrient emissions into river basins of Germany. Umweltbundesamt (Federal Environmental Agency). 75:350 pp.

Behrendt H, Opitz D (1999). Retention of nutrients in river systems: dependence on specific runoff and hydraulic load. Hydrobiologia. 410:111–122.

Billen G, Garnier J (2000). Nitrogen transfers through the Seine drainage network: a budget based on the application of the 'Riverstrahler'model. Hydrobiologia. 410:139–150.
Bingner R, Theurer F (2001). AnnAGNPS: estimating sediment yield by particle size for sheet and rill erosion. Proceedings of the 7th Interagency Sedimentation Conference, Reno, Volume 1:1–7.
Bouraoui F, Braud I, Dillaha T (2002). ANSWERS: a nonpoint source pollution model for water, sediment and nutrient losses. In: Singh VP, Frevert DK, editors: Mathematical models of small watershed hydrology and applications. Littleton (CO): Water Resources Publications:833–882.
Carvalho L, McDonald C, de Hoyos C, Mischke U, Phillips G, Borics G et al. (2013). Sustaining recreational quality of European lakes: minimizing the health risks from algal blooms through phosphorus control. J Appl Ecol. 50:315–323.
Conley DJ, Paerl HW, Howarth RW, Boesch DF, Seitzinger SP, Karl E et al. (2009). Controlling eutrophication: nitrogen and phosphorus. Science. 123:1014–1015.
Cooke GD, Welch EB, Peterson S, Nichols SA (2005). Restoration and management of lakes and reservoirs. Boca Raton (FL): CRC Press:616 pp.
Donner SD, Coe MT, Lenters JD, Twine TE, Foley JA (2002). Modeling the impact of hydrological changes on nitrate transport in the Mississippi River Basin from 1955 to 1994. Global Biogeochem Cy. 16:16–11–16–19.
Donner SD, Kucharik CJ, Oppenheimer M (2004). The influence of climate on in-stream removal of nitrogen. Geophys Res Lett. 31:L20509.
EC (2012). Common Implementation Strategy for the Water Framework Directive (2000/60/EG), Guidance document No. 28: Technical Guidance on the Preparation of an Inventory of Emissions, Discharges and Losses of Priority and Priority Hazardous Substances. European Commission.
Eurostat (2013). Nutrient budgets – methodology and handbook. Version 1.02. Luxembourg: Eurostat and OECD.
EUROSTAT (2019). Agri-environmental indicator – risk of pollution by phosphorus. Available at: https://ec.europa.eu/eurostat/statistics-explained/index.php/Agri-environmental_indicator__risk_of_pollution_by_phosphorus.
Evans MA, Fahnenstiel G, Scavia D (2011). Incidental oligotrophication of North American great lakes. Environ Sci Technol. 45:3297–3303.
FAOSTAT (2019). Available at: http://www.fao.org/faostat/en/#data.
Fox GA, Purvis RA, Penn CJ (2016). Streambanks: a net source of sediment and phosphorus to streams and rivers. J Environ Manage. 181:602–614.
Franke N, Boyacioglu H, Hoekstra A (2013). Grey water footprint accounting: tier 1 supporting guidelines. Delft: UNESCO-IHE Institute for Water Education.
Fritz B, Rinck-Pfeiffer S, Nuetzmann G, Heinzmann B (2004). Conservation of water resources in Berlin, Germany, through different re-use of water. IAHS Publications-Series of Proceedings and Reports. 285:48–52.
Fuchs S, Scherer U, Wander R, Behrendt H, Venohr M, Opitz D et al. (2010). Calculation of Emissions into Rivers in Germany using the MONERIS Model. Nutrients, heavy metals and polycyclic aromatic hydrocarbons. Dessau-Roßlau: Federal Environment Agency (Umweltbundesamt).
Garnier J, Billen G, Coste M (1995). Seasonal succession of diatoms and Chlorophyceae in the drainage network of the Seine River: observation and modeling. Limnol Oceanogr. 40:750–765.

Gebel M, Halbfaß S, Bürger S (2009). Stoffbilanz – Modellerläuterung (in German). Dresden: Gesellschaft für Angewandte Landschaftsforschung.

Groenendijk P, Renaud L, Roelsma J (2005). Prediction of nitrogen and phosphorus leaching to groundwater and surface waters; process descriptions of the ANIMO4. 0 model. Wageningen: Alterra – Green World Research.

Heinzmann B, Chorus I (1994). Restoration concept for Lake Tegel, a major drinking and bathing water resource in a densely populated area. Environ Sci Technol. 28:1410–1416.

Howard G, Reed B, McChesney D, Taylor R, Schmoll O, Chilton J (2006). Human excreta and sanitation: control and protection. In: Schmoll O, Howard G, Chilton J et al., editors: Protecting groundwater for health: managing the quality of drinking-water sources. Geneva: World Health Organization:677 pp. https://apps.who.int/iris/handle/10665/43186

Jeppesen E, Meerhoff M, Jacobsen B, Hansen R, Søndergaard M, Jensen J et al. (2007). Restoration of shallow lakes by nutrient control and biomanipulation—the successful strategy varies with lake size and climate. Hydrobiologia. 581:269–285.

Jeppesen E, Søndergaard M, Jensen JP, Havens KE, Anneville O, Carvalho L et al. (2005). Lake responses to reduced nutrient loading–an analysis of contemporary long-term data from 35 case studies. Freshwater Biol. 50:1747–1771.

Kroes J, van Dam J (2003). Reference Manual SWAP; version 3.0. Wageningen: Alterra – Green World Research:211 pp.

Lederer J, Karungi J, Ogwang F (2015). The potential of wastes to improve nutrient levels in agricultural soils: a material flow analysis case study from Busia District, Uganda. Agric, Ecosyst Environ. 207:26–39.

Lindström G, Pers C, Rosberg J, Strömqvist J, Arheimer B (2010). Development and testing of the HYPE (Hydrological Predictions for the Environment) water quality model for different spatial scales. Hydrol Res. 41:295–319.

Liu X, Beusen AH, Van Beek LP, Mogollón JM, Ran X, Bouwman AF (2018). Exploring spatiotemporal changes of the Yangtze River (Changjiang) nitrogen and phosphorus sources, retention and export to the East China Sea and Yellow Sea. Water Res. 142:246–255.

MacDonald GK, Bennett EM, Potter PA, Ramankutty N (2011). Agronomic phosphorus imbalances across the world's croplands. Proc Nat Acad Sci USA. 108:3086–3091.

Moriasi DN, Arnold JG, Van Liew MW, Bingner RL, Harmel RD, Veith TL (2007). Model evaluation guidelines for systematic quantification of accuracy in watershed simulations. Trans ASABE. 50:885–900.

Novotny V (2003). Water quality: diffuse pollution and watershed management, second edition. New York: John Wiley & Sons:865 pp.

Nürnberg GK (1998). Prediction of annual and seasonal phosphorus concentrations in stratified and polymictic lakes. Limnol Oceanogr. 43:1544–1552.

OSPAR (2004a). OSPAR Guidelines for Harmonised Quantification and Reporting Procedures for Nutrients (HARP-NUT), Guideline 7: quantification and reporting of the monitored riverine load of nitrogen and phosphorus, including flow normalisation procedures. London: OSPAR Commission.

OSPAR (2004b). OSPAR Guidelines for Harmonised Quantification and Reporting Procedures for Nutrients (HARP-NUT), Guideline 8: quantification of nitrogen and phosphorus losses from diffuse sources by riverine load apportionment. London: OSPAR Commission.

OSPAR (2004c). OSPAR Guidelines for Harmonised Quantification and Reporting Procedures for Nutrients (HARP-NUT), Guideline 9: quantification and reporting of the retention of nitrogen and phosphorus in river catchments. London: OSPAR Commission.

Ott C, Rechberger H (2012). The European phosphorus balance. Resour Conserv Recy. 60:159–172.

Peacher R, Lerch R, Schultz R, Willett C, Isenhart T (2018). Factors controlling streambank erosion and phosphorus loss in claypan watersheds. J Soil Water Conserv. 73:189–199.

Phillips G, Pietiläinen O-P, Carvalho L, Solimini A, Solheim AL, Cardoso A (2008). Chlorophyll–nutrient relationships of different lake types using a large European dataset. Aquat Ecol. 42:213–226.

Psenner R, Boström B, Dinka M, Pettersson K, Pucsko R (1988). Fractionation of phosphorus in suspended matter and sediment. Arch Hydrobiol Supl. 30:98–110.

Refsgaard JC, Storm B (1995). MIKE SHE. In: Singh VP, editors: Computer models of watershed hydrology. Littleton (CO): Water Resources Publications.

Reynolds C, Maberly S (2002). A simple method for approximating the supportive capacities and metabolic constraints in lakes and reservoirs. Freshwater Biol. 47:1183–1188.

Rickert B, Chorus I, Schmoll O (2016). Protecting surface water for health. Identifying, assessing and managing drinking-water quality risks in surface-water catchments. Geneva: World Health Organization:178 pp. https://apps.who.int/iris/handle/10665/246196

Schindler DW (2006). Recent advances in the understanding and management of eutrophication. Limnol Oceanogr. 51:356–363.

Schmoll O, Howard G, Chilton J, Chorus I (2006). Protecting groundwater for health: managing the quality of drinking-water sources. Geneva: World Health Organization. https://apps.who.int/iris/handle/10665/43186

Schönhart M, Trautvetter H, Parajka J, Blaschke AP, Hepp G, Kirchner M et al. (2018). Modelled impacts of policies and climate change on land use and water quality in Austria. Land Use Policy. 76:500–514.

Schoumans O, Silgram M, Groenendijk P, Bouraoui F, Andersen HE, Kronvang B et al. (2009). Description of nine nutrient loss models: capabilities and suitability based on their characteristics. J Environ Monit. 11:506–514.

Sharpley AN, Bergström L, Aronsson H, Bechmann M, Bolster CH, Börling K et al. (2015). Future agriculture with minimized phosphorus losses to waters: research needs and direction. AMBIO. 44:163–179.

Shatwell T, Köhler J (2019). Decreased nitrogen loading controls summer cyanobacterial blooms without promoting nitrogen-fixing taxa: long-term response of a shallow lake. Limnol Oceanogr. 64:S166–S178.

Siderius C, Groenendijk P, Jeuken MHJL, Smit AAMFR (2008). Process description of NUSWALITE: a simplified model for the fate of nutrients in surface waters. Wageningen: Alterra – Green World Research:71 pp.

Skahill BE (2004). Use of the hydrological simulation program-FORTRAN (HSPF) model for watershed studies. Vicksburg (MS): U.S. Army Engineer Research and Development Center:26 pp.

Smit AAMFR, Siderius C, van Gerven LPA (2009). Process description of SWQN: a simplified hydraulic model. Wageningen: Alterra – Green World Research:52 pp.

Søndergaard M, Lauridsen TL, Johansson LS, Jeppesen E (2017). Nitrogen or phosphorus limitation in lakes and its impact on phytoplankton biomass and submerged macrophyte cover. Hydrobiologia. 795:35–48.

Thaler S, Zessner M, Mayr MM, Haider T, Kroiss H, Rechberger H (2013). Impacts of human nutrition on land use, nutrient balances and water consumption in Austria. Sustain Water Qual Ecol. 1:24–39.

Thaler S, Zessner M, Weigl M, Rechberger H, Schilling K, Kroiss H (2015). Possible implications of dietary changes on nutrient fluxes, environment and land use in Austria. Agr Syst. 136:14–29.

Tiessen K, Elliott J, Yarotski J, Lobb D, Flaten D, Glozier N (2010). Conventional and conservation tillage: influence on seasonal runoff, sediment, and nutrient losses in the Canadian prairies. J Environ Qual. 39:964–980.

UBA (2017). Einträge von Nähr- und Schadstoffen in die Oberflächengewässer. Available at: https://www.umweltbundesamt.de/daten/wasser/fliessgewaesser/eintraege-von-naehr-schadstoffen-in-die.

US EPA (2011). Hydrological Simulation Program - FORTRAN (HSPF). Available at: http://www.epa.gov/ceampubl/swater/hspf/.

Venohr M, Hirt U, Hofmann J, Opitz D, Gericke A, Wetzig A et al. (2011). Modelling of nutrient emissions in river systems – MONERIS – methods and background. Int Rev Hydrobiol. 96:435–483.

Vilmin L, Mogollón JM, Beusen AH, Bouwman AF (2018). Forms and subannual variability of nitrogen and phosphorus loading to global river networks over the 20th century. Global Planet Change. 163:67–85.

Vollenweider R, Kerekes J (1980). Loading concept as basis for controlling eutrophication philosophy and preliminary results of the OECD programme on eutrophication. Progr Water Technol. 12:5–38.

Vollenweider RA (1976). Advances in defining critical loading levels for phosphorus in lake eutrophication. Mem Ist Ital Idrobiol. 33:53–83.

Vollenweider RA, Kerekes JJ (1982). Eutrophication of waters. Monitoring, assessment and control. Paris: Environment Directorate OECD:154 pp.

Wade AJ, Durand P, Beaujouan V, Wessel WW, Raat KJ, Whitehead PG et al. (2002). A nitrogen model for European catchments: INCA, new model structure and equations. Hydrol Earth Syst Sci. 6:559–582.

Whitehead P, Wilson E, Butterfield D, Seed K (1998a). A semi-distributed integrated flow and nitrogen model for multiple source assessment in catchments (INCA): part II – application to large river basins in south Wales and eastern England. Sci Tot Environ. 210:559–583.

Whitehead PG, Wilson E, Butterfield D (1998b). A semi-distributed integrated nitrogen model for multiple source assessment in catchments (INCA): part I – model structure and process equations. Sci Tot Environ. 210:547–558.

WHO (2006). Guidelines for the safe use of wastewater, excreta and greywater. Geneva: World Health Organization. https://apps.who.int/iris/handle/10665/78265

WHO (2017a). Climate-resilient water safety plans: managing health risks associated with climate variability and change. Geneva: World Health Organization: 82 pp. https://apps.who.int/iris/handle/10665/258722

WHO (2017b). Guidelines for drinking-water quality, fourth edition, incorporating the 1st addendum. Geneva: World Health Organization:631 pp. https://www.who.int/publications/i/item/9789241549950

Wilson JL, Everard M (2018). Real-time consequences of riparian cattle trampling for mobilization of sediment, nutrients and bacteria in a British lowland river. Int J River Basin Manage. 16:231–244.

Wischmeier WH, Smith DD (1960). A universal soil-loss equation to guide conservation farm planning. Trans 7th Int Cong Soil Sci. 1:418–425.

Withers PJ, Neal C, Jarvie HP, Doody DG (2014). Agriculture and eutrophication: where do we go from here? Sustainability. 6:5853–5875.

Wolman A (1965). The metabolism of cities. Sci Am. 213:178–193.

Young RA, Onstad C, Bosch D, Singh V (1995). AGNPS: an agricultural nonpoint source model. In: Singh VP, editors: Computer models of watershed hydrology. Littleton (CO): Water Resources Publications:1001–1020.

Zessner M (1999). Bedeutung und Steuerung von Nährstoff und Schwermetallflüssen des Abwassers. Vienna: Institution.

Zessner M, Lindtner S (2005). Estimations of municipal point source pollution in the context of river basin management. Toxicol Lett. 52:175–182.

Zessner M, Postolache C, Clement A, Kovacs A, Strauss P (2005). Considerations on the influence of extreme events on the phosphorus transport from river catchments to the sea. Water Sci Technol. 51:193–204.

Zessner M, Zoboli O, Hepp G, Kuderna M, Weinberger C, Gabriel O (2016). Shedding Light on Increasing trends of phosphorus concentration in upper Austrian rivers. Water. 8:404.

Zoboli O, Laner D, Zessner M, Rechberger H (2016a). Added values of time series in material flow analysis: the Austrian phosphorus budget from 1990 to 2011. J Ind Ecol. 20:1334–1348.

Zoboli O, Viglione A, Rechberger H, Zessner M (2015). Impact of reduced anthropogenic emissions and century flood on the phosphorus stock, concentrations and loads in the Upper Danube. Sci Tot Environ. 518:117–129.

Zoboli O, Zessner M, Rechberger H (2016b). Supporting phosphorus management in Austria: potential, priorities and limitations. Sci Tot Environ. 565:313–323.

Chapter 8

Assessing and controlling the risk of cyanobacterial blooms

Waterbody conditions

Mike Burch, Justin Brookes, and Ingrid Chorus

CONTENTS

INTRODUCTION

While site inspection is highly valuable for assessing the risk of cyanobacterial proliferation or blooms, taken alone this may not be sufficient to assess this risk because of the inherent variability of cyanobacterial occurrence. As explained in Chapter 4, certain waterbody conditions – particularly high levels of phosphorus and high turbidity – are particularly favourable for cyanobacterial growth and enable them to outcompete many planktonic microalgae. An understanding of the growth conditions in the respective waterbody therefore helps to predict whether blooms are likely. It is also a basis for planning measures to control them. Among these measures, the reduction of phosphorus loads from the catchment (Chapter 7) is the method of choice to address the root of the problem, but in some situations, load reduction is not possible to an extent which is sufficiently effective within the required time frame. For such situations, waterbody management options are available to shift growth conditions, making them less favourable for cyanobacterial proliferation. These are termed "internal measures" (in contrast to the management of external nutrient loads to a waterbody). The feasibility of internal measures depends on waterbody characteristics. While Chapter 4 discusses growth conditions from the cyanobacterial perspective, this chapter takes the complementary perspective, focusing on using the assessment of waterbody conditions to estimate the likelihood of cyanobacterial mass development. This includes guidance on estimating whether or not internal measures are both necessary and feasible.

If internal waterbody management measures are to be planned, implemented or validated, a comprehensive knowledge and understanding of the hydrophysical and hydrochemical characteristics of the waterbody as well as of its biota and their interaction (with a focus on phytoplankton ecology) is important as basis for choosing measures that are likely to be successful. This requires involvement of limnological expertise. Where a Water Safety Plan (WSP; see Chapter 6) is developed, it is effective to invite corresponding experts to advise and support the WSP team. Further, particularly where cyanotoxin risks are assessed for the first time, it is valuable to check with waterboards, health inspection and environmental authorities as well as with research institutions in the region whether data from monitoring or research are available, for example, from specific programmes addressing levels of eutrophication, phytoplankton (and specifically cyanobacterial) biomass or cyanotoxin concentrations. Historical data and results from longer-term monitoring, if available, are particularly useful for understanding the cyanobacterial development of a waterbody.

This chapter intends to support a preliminary assessment of the waterbody situation, to give an overview of measures that may be taken within a waterbody to control cyanobacterial growth and bloom formation, and to give guidance on the questions to address when considering the implementation of internal measures. Sections 8.2–8.11 address different internal control measures that can be implemented alone or in combination, depending on the conditions in the respective waterbody. They provide checklists for assessing the situation and the possible benefits from control measures considered. For each control measure, they also propose options for its operational monitoring, that is, of parameters that indicate failure long before blooms re-appear, and they propose an approach to validating whether the measure is likely to be sufficiently effective.

8.1 DIRECT INDICATION OF THE OCCURRENCE OF (POTENTIALLY TOXIC) CYANOBACTERIA

For the purpose of hazard assessment, regular monitoring of the occurrence of cyanobacteria or cyanotoxins is most effective if it is focused on waterbodies in which cyanobacteria are likely to occur in quantities potentially relevant to health through the exposure pathways caused by the way these waterbodies are used.

Site inspection is extremely valuable: in contrast to many other chemical and biotic hazards, cyanobacteria are often readily visible if they occur in potentially hazardous concentrations. Site inspection provides information on the overall situation of the waterbody and its surroundings beyond that which can be gleaned from data and documents, including activities and conditions in the vicinity of the waterbody. Site inspection may miss bloom situations, particularly if it occurs at long time intervals (i.e., greater than weekly), but such observations may be available from authorities (waterboards, health inspection, environmental authorities), from members of the local community ("citizen scientists"; see Chapters 11 and 15), operators of campsites, boat rental operators, restaurants and from scientific organisations conducting research on the waterbody. Asking these stakeholders to report the occurrence of surface streaks, scums, pronounced greenish or reddish discoloration or greenish turbidity may help to identify circumstances for further investigation.

A range of instruments and sensors, in particular fluorescence probes, allow real-time monitoring of cyanobacteria and algae for management purposes, as does remote sensing (see Chapter 13). Water utilities increasingly employ water quality monitoring systems with probes that measure temperature, conductivity, dissolved oxygen profiles, etc., together with fluorescence of chlorophyll and that of accessory pigments to monitor phytoplankton distribution and/or concentrations at raw water intakes. However, along with the opportunities given through these *in situ* fluorometers, it is important to be aware of uncertainties and technical limitations of this potentially complex technology (see Chapter 13, Zamyadi et al., 2016, and Bertone et al., 2018).

Checklist 8.1 is intended as a template for questions to address in preparation of site inspection and to ask residents of the area and local stakeholders during site inspection.

CHECKLIST 8.1: ASSESSING DIRECT INDICATION OF THE OCCURRENCE OF (POTENTIALLY TOXIC) CYANOBACTERIA

- Have scums been observed? (Note: sometimes duckweed is taken for cyanobacterial scum, so ask whether individual tiny plants of a few mm diameter were recognisable – this is likely to be duckweed, i.e., *Lemna minor*.)
- Has conspicuous colour (greenish or wine-red) been observed with low transparency (i.e., <1 m)?
- Has the water been investigated for the occurrence of cyanobacteria? If so, were any quantitative data obtained? If so, do they include biovolumes >0.3–1 mm^3/L or concentrations of chlorophyll-*a* (during dominance of cyanobacteria) >1 µg/L?
- Have cyanotoxins ever been detected? If so, at what concentrations?
- Are there any reports of animal or even human illness associated with exposure to blooms? (Note: while a clear association of human symptoms with cyanobacteria is not very likely, such concern may well have been voiced.)

8.2 ASSESSING A WATERBODY'S POTENTIAL FOR CYANOBACTERIAL BLOOMS

The key prerequisite for a high biomass density of any phytoplankton, including cyanobacteria, is elevated concentration of nutrients. In most cases, the concentration of total phosphorus (TP) can be used to estimate the potential of bloom development in a waterbody (i.e., the "carrying capacity"): in general, high concentrations of phytoplankton biomass – typically dominated by cyanobacterial blooms – occur at TP concentrations above 20–50 µg/L, depending on hydrophysical conditions of the waterbody, particularly mixing depth (see section 4.3).

Exceptions to this widespread pattern include some very large deep lakes and reservoirs which develop thin and transient surface scums even at TP concentrations in the range of 10–20 µg/L, and while these can support only a low cyanobacterial cell density and biomass, scums are nevertheless possible because they can be recruited out of a large volume of water. Figure 4.6 in Chapter 4 illustrates how this can concentrate toxins by several orders of magnitude. Also, metalimnetic populations of *Planktothrix rubescens*, typical for deep, stably stratified mesotrophic waterbodies, may be relevant for human exposure if drinking-water is abstracted from that level. They typically stay at depth until autumn or winter when mixing entrains them throughout the waterbody and up to the surface where they then can form significant and spectacular surface blooms which may pose a potential toxin hazard (see Boscaini et al. (2017) for an example). Benthic cyanobacteria are a further exception as these typically occur in very clear, shallow water with low nutrient concentrations (section 4.2.2); however, while they result in high toxin concentrations in detached mats or agglomerations of macrophyte material, they are not known to cause high concentrations in the water.

If the concentration of TP in a given waterbody can be sufficiently controlled in order to suppress cyanobacterial dominance and biomass, excessive nitrogen concentrations will not be relevant for cyanotoxin occurrence. However, where reaching this target is not feasible, for certain periods of the seasonal cycle or year-round, assessing whether nitrogen is limiting may be relevant. This is most likely in mesotrophic and eutrophic shallow lakes which can be N-limited during summer if organic substance on the sediment surface and high temperatures promote denitrification. While low N concentrations have been proposed to risk shifting cyanobacterial species composition to those that can fix atmospheric nitrogen, field data show this to rarely be the case because nitrogen fixation requires high amounts of light energy which are usually not available in turbid eutrophic waterbodies. Nitrogen limitation can be assumed at concentrations of dissolved inorganic N below 100 µg/L (see section 4.2.1 and Kolzau et al., 2014.). While there are methods to reduce phosphorus cycling within a waterbody discussed below, for nitrogen the only options for reduction are natural denitrification and to control nitrogen loading to the waterbody (see Chapter 7; for a broader discussion of the role of N see Chorus and Spijkerman, 2020).

While a high concentration of total phosphorus (TP) and TN (total nitrogen) is a prerequisite for high cyanobacterial biovolumes, this is not the only one.

A low water exchange rate is usually a further condition for bloom formation. Given that cyanobacteria grow rather slowly, they need sufficient residence time in the waterbody to establish large populations – often in the range of weeks. High river flows will dilute and wash out cyanobacteria faster than they can grow, and blooms are unlikely in lakes and reservoirs with water retention times of less than one month. For temperate climate settings, a further consequence of low growth rates is that it often takes till late summer for cyanobacteria to outcompete the more rapidly growing springtime algal phytoplankton and become dominant; that is, the typical "cyanobacterial season" is from mid-summer to (late) autumn. Exceptions include the non-scum-forming *Limnothrix* spp., often associated with other fine filamentous cyanobacteria and or *Planktothrix agardhii*; these may occur with high biomass in summer, particularly if they survive in winter and dominate already in spring.

Low water transparency is often a further condition linked to the dominance of a range of cyanobacterial species, not because it is a precondition for their growth, but because it is a consequence of bloom occurrence. Also, for some cyanobacterial species, this may function as positive feedback loop: for those that grow well at rather low light intensity (better than many phytoplanktonic algae), the likelihood of their dominance is higher in turbid water (Table 8.1).

Table 8.1 Conditions affecting or indicating the likelihood of high cyanobacterial biomass (see section 5.3 for references)

Total phosphorus	*Mixing conditions*		*Transparency*	*pH*
>50 µg/L	Stagnant, depth >5–10 m, with stable thermal gradients: Favours scum-forming taxa, i.e., *Microcystis, Dolichospermum, Aphanizomenon*	Stagnant, shallow and well mixed: Favours non-scum forming taxa, i.e., *Planktothrix agardhii* and other fine filamentous forms, e.g., *Limnothrix*	Low; Secchi depth often <1 m	pH >7 (often >8 or possibly >9 due to high rates of photosynthesis associated with high biomass)
20–50 µg/L	Stagnant, deeper than 10 m, stratified: potential for mass development of *Planktothrix rubescens* which accumulates at the metalimnion		Moderate; Secchi depth ~1–3 m	pH ≥7
10–20 µg/L	Fast flowing river	Lake or reservoir with water residence time <1 month	High; Secchi depth ~3–7 m	pH 6–7
≤10 µg/L	Mountain stream or brook		Very High – Clear water; Secchi depth often >7 m	pH<6
	Exception: mats of cyanobacteria attached to surfaces			

Waterbody mixing is well tolerated by many cyanobacteria because of their tolerance of low light availability, as mixing entrains cells into deeper, low-light layers. Cyanobacterial taxa with tolerance of low light intensity are widespread in relatively shallow, well-mixed waterbodies. Deep mixing of a waterbody can, however, suppress the proliferation of scum-forming cyanobacteria such as *Microcystis* spp. and *Dolichospermum* spp.: these cyanobacteria are less effective competitors for light, and they compensate for this by regulating their buoyancy and thus their vertical position in the waterbody (see section 4.3). If they are entrained by deep and sufficiently strong mixing, they lose this competitive advantage.

High temperature is often assumed to enhance cyanobacterial growth. As discussed in section 4.4.2, this may chiefly act indirectly through the stabilisation of thermal stratification. Thus, high temperature may be an indicator for the increased likelihood of cyanobacterial blooms.

A further indicator is high pH. It is well established that cyanobacteria are rarely found at levels that represent a health hazard if pH is <6–7 (note that high pH is not a cause, but rather a consequence of high phytoplankton biomass, often due to cyanobacterial blooms (section 4.3.6 and Tables 8.1 and 4.2); nonetheless, high pH indicates that blooms may be occurring).

In summary, cyanobacteria are likely to occur at concentrations that represent a health hazard for periods longer than a few days if the waterbody meets most of the conditions in Checklist 8.2.

CHECKLIST 8.2: ASSESSING WHETHER WATERBODY CONDITIONS FAVOUR CYANOBACTERIAL GROWTH AND PROLIFERATION, LEADING TO THE OCCURRENCE OF BLOOMS

- Is the waterbody eutrophic, that is, are concentrations of TP>20–50 µg/L and those of TN about 10-fold higher?
- If yes, are concentrations of dissolved inorganic phosphorus detectable (that is, above 5–10 µg/L) and those of inorganic nitrogen above 100 µg/L?
- Is the water retention time>one month?
- If transparency is low, is this due to phytoplankton (and not to suspended minerals), that is, are Secchi disc readings<1–2 m (during the bloom season)?
- Is the waterbody either shallow and well mixed or deep with stable thermal stratification?
- Is the water alkaline rather than acidic, that is, pH>7?
- For temperate climates, is the season late summer and early autumn?
- For rivers, are there tributaries that import blooms (e.g., from impoundments in which they may develop)?

If a waterbody meets all of these conditions, this does not mean that cyanobacterial blooms will necessarily occur – it merely indicates an elevated likelihood of their occurrence. *Vice versa*, if it does not meet these conditions, cyanotoxin concentrations exceeding the guideline levels discussed in Chapter 2 are not likely but they cannot be totally excluded – for example, from metalimnetic maxima entering a drinking-water offtake (Chapter 9) or from detached benthic cyanobacteria.

8.3 ESTIMATING THE IMPACT OF CLIMATE CHANGE ON CYANOBACTERIAL PROLIFERATION AND BLOOMS

As discussed in section 4.5, understanding how climate warming will affect cyanobacteria is far more complex than the direct impact of higher water temperature (by, e.g., 2–3 degrees Celsius) on their growth rate: rather, it requires an understanding of the shifts in the aquatic ecosystem's functioning that may occur and how such shifts affect the dominance of cyanobacteria relative to that of other species. This includes changes in duration of thermal stratification, ice cover and patterns of drought versus heavy rainfall, all of which in turn can have significant impacts on waterbody mixing patterns, water levels, oxygen concentrations, nutrient loads and concentrations, and fish and zooplankton populations (section 4.5 and overviews by Moss et al., 2011; Winder & Sommer, 2012, and Hamilton et al., 2016).

Whether climate change will enhance cyanobacterial proliferation in a specific waterbody depends on which way the impacts of warming and the associated changes in the aquatic ecosystem will interact – resulting in a potential increase or decrease in the dominance and biomass of cyanobacteria. For example, while warmer weather and the resulting increased stability of thermal stratification may favour cyanobacteria blooms, reduced mixing may also reduce the nutrient supply from deeper layers to upper layers where cyanobacteria can grow (Salmaso et al., 2018). Less runoff in dry warm periods may also work both ways, decreasing or increasing phosphorus input with the corresponding impact on the biomass of cyanobacteria that may develop.

Predicting the potential impacts of climate change on cyanobacteria in a given waterbody requires assessing locally how climate changes are most likely to impact on the conditions listed in Table 8.2. While this list incorporates our conventional understanding of cyanobacterial ecology and the drivers for cyanobacterial growth, using these categories for predicting whether blooms will increase or decrease involves considerable uncertainty, particularly for the hydrological drivers of bloom development. Uncertainty

Table 8.2 Possible effects of global warming that can increase or decrease cyanobacteria

Impacts of global warming that can	
Increase cyanobacteria	*Decrease cyanobacteria*
In waterbodies with higher trophic state, **higher temperature** may increase nutrient release from sediments, fertilising growth, particularly in nonstratifying waterbodies where these nutrients reach the upper water layers	**Higher temperature** may lead to more stable thermal stratification, thus reducing nutrient transport from deep water layers into upper layers where they would fertilise growth (relevant in waterbodies in which biomass in the surface layers is nutrient limited)
More frequent storms can reduce thermal stratification, thus transporting nutrient-rich water from deeper layers into upper layers where it fertilises growth	**More frequent storms** can disrupt cyanobacterial blooms, giving other taxa a chance to outcompete them after the storm
Stronger and/or more frequent storms can increase flow and thus nutrient loads	**Stronger and/or more frequent storms** can increase flow and thus water exchange rates, potentially carrying cyanobacteria out of the system
Longer periods of drought reduce water exchange rates; thus there is less dilution of nutrient discharges to the waterbody, increasing concentrations and fertilising cyanobacteria	**Longer periods of drought** reduce inflow and thus nutrient loads from erosion and other diffuse sources in the catchment, potentially limiting cyanobacterial biomass
Longer periods of drought reduce water exchange rates, allowing blooms to last longer	
Longer periods of drought can reduce water levels in stratified reservoirs to where no "clean" layer without blooms is available to abstract drinking-water	
Longer periods of drought can reduce water levels in stratified reservoirs and lead to warmer water over the sediment. As a result, more phosphorus is released from the sediment	
Longer periods of stable thermal stratification can give cyanobacterial dominance more time to form and to last, thus increasing the duration of the season with toxic cyanobacteria	
In lakes with higher trophic state, anoxia above the sediments and redox-sensitively bound phosphorus, **longer periods of stable thermal stratification** can enhance the consumption of oxygen in the deep water, leading to more release of phosphorus from the sediment	

is lower regarding the impact of nutrient concentrations because these set clear limits: if concentrations are too low to support a substantial planktonic biomass, it cannot develop even as water gets warmer or stratification more stable (see discussion in section 4.5).

Prediction for a given waterbody therefore requires a comprehensive understanding of local conditions as well as expertise in hydrology, limnology and possibly in modelling phytoplankton occurrence (see section 4.4.1).

8.4 INTERNAL MEASURES AGAINST CYANOBACTERIA: WHAT, WHY AND WHEN?

The experience across the globe of several decades of lake and reservoir restoration – that is, reversing eutrophication – shows that the benefit of internal measures mostly lies in speeding up the ecosystem response once external nutrient loads have been significantly reduced. Internal measures are rarely successful if external loads of nutrients, particularly phosphorus, remain high. There are, however, settings in which cyanobacteria are a problem requiring a rapid solution while sufficient reduction of external loads is unlikely to be rapidly feasible or sufficiently effective to solve the problem. Lowland stretches of trans-boundary rivers are a typical example: some are naturally eutrophic. Also, typically many stakeholders in the catchment may need to engage in nutrient control. It takes time to establish an effective policy in catchment management to reduce river phosphorus concentrations to sufficiently low levels along the entire river reach. Some internal measures can work in such situations, but they need to be continuously or repeatedly applied until the external load will have been sufficiently reduced.

In other situations, external load reduction may be effective, but internal phosphorus loads, released from sediments, are high and will sustain cyanobacterial biomass above target levels for quite some time. Particularly in lakes with very low flushing rates, the phosphorus supply in the lake may remain high for many years, even after substantial load reduction. In such settings, internal measures may give the ecosystem a kick to interrupt the prevalent ecosystem feedback mechanisms that stabilise the eutrophic or hypertrophic situation. Such measures can target a transition in sediment chemistry to reduce phosphorus release rates from the sediment (section 8.7) or they can target other conditions to make them less favourable for cyanobacteria, for example,

i. by suppressing the dominance of cyanobacteria in favour of other (nontoxic, non–scum-forming) phytoplankton species and – in lakes with substantial shallow areas – of aquatic macrophytes;
ii. by increasing loss rates of the cyanobacterial populations.

Table 8.3 gives an overview of internal measures.

Monitoring the system's response to an internal measure is important to assess whether the measure needs to be repeated or whether the system self-stabilises with a less eutrophic biotic community structure in which cyanobacteria only play a minor role.

The most important first step in planning is to make sure that internal measures are necessary. Two reasons can stand against this: (i) it may be more sustainable and effective to invest in reduction of external nutrient loads, and this needs to have been sufficiently explored before deciding to invest in internal measures, and (ii) external load reductions may already be leading to a decline in internal phosphorus concentrations, and it is worthwhile monitoring this decline to see if it will be sufficient to reach target phosphorus concentrations within the necessary time span.

Time lags between the reduction of external nutrient loads and the desired results achieved in the waterbody may be substantial, that is, in the range of a decade, and response times strongly depend on water exchange rates. Resilience effects are not uncommon, even after substantial reduction of inputs below thresholds calculated to be effective. Sas (1989) discusses two (partially connected) resilience mechanisms

Table 8.3 Overview of measures to suppress cyanobacterial growth by influencing internal waterbody processes

Intervention target	*Intervention type*	*Technique*
Suppress dominance of cyanobacteria, potentially in favour of other phytoplankton	Hydrophysical control of growth conditions	Mixing – artificial destratification Decreasing water retention time Maintaining sufficient flow and thus a rapid change of hydrophysical conditions, that is, avoiding or removing impoundments
Suppress internal phosphorus (P) load released from the sediment *Note*: this is only likely to be successful if sediments are a major P source relative to the external P load	Internal phosphorus control	Sediment removal Sediment treatment with P-binding agents, for example, lime, alum, modified clay, zeolite Suppressing redox-sensitive P release by oxidising the sediment surface (through hypolimnetic aeration or oxygenation)
Enhance loss rates of phytoplankton, including cyanobacteria, or support their competitors	Biological control	Biomanipulation Barley straw Viruses, bacteria
Induce rapid lysis of cyanobacterial cells or inhibition of their growth	Chemical control	Algicides, algistats

for phosphorus concentrations: (i) a delayed response of in-lake total phosphorus (TP) concentrations to a reduction of input, due to the time required for flushing phosphorus out of the waterbody, and (ii) phosphorus release from the sediments ("internal loading") until a new sediment–water equilibrium is established.

Additionally, as mentioned above, the community structure of the organisms in a waterbody may show resilience to change – through several mechanisms. For many shallow mesotrophic to slightly eutrophic lakes, two alternate states are possible at the same external nutrient load: either clear water due to dominance of macrophytes (i.e., submersed aquatic plants and reed belts) or turbid water due to dominance of phytoplankton, often cyanobacteria (Scheffer et al., 1993; Scheffer et al., 1997). Once phytoplankton dominates at high density, this causes turbidity which shades the macrophytes to the extent where they are unable to grow, and thus, phytoplankton dominance self-stabilises. Such a lake can be returned to the clear water macrophyte-dominated state if nutrient concentrations can be reduced to the point where phytoplankton are nutrient–limited and thus less dense; the water therefore becomes clearer and the previously light–limited macrophytes are then able to ecolonize.

Also, phytoplankton species composition may resist change: once high cell densities of cyanobacteria are established in a waterbody, some of these will survive, for example, on the sediment surface even when conditions become less favourable (e.g., during winter), and these cells are available as inoculum to seed the population in the next growing season. Furthermore, fish populations affect zooplankton populations, which in turn feed on phytoplankton, thus affecting its species-specific loss rates, and such food-chain mechanisms can cause resilience to change or be manipulated to enhance change (e.g., through fish stock management; see below). Nature may overcome such biological resilience phenomena by itself with time, but change can be accelerated by the interventions described below.

Waterbody ecosystem processes are highly complex and quite specific to an individual waterbody (Chapter 4). Consequently, predictions of their response to interventions have a higher uncertainty than predictions of responses to measures in most technical systems (e.g., drinking-water treatment). A prerequisite for success of internal measures is that they are planned on the basis of comprehensive understanding of the waterbody ecosystem – of its hydrological regime, its biota and its sediment chemistry (see Checklist 8.3). Involving limnology experts in planning and designing such measures is therefore fundamental for success. Additionally, expert review of the plans as well as later validation of the measures taken (preferably by experts independent of those contracted for planning and implementing the measure) is important to understand factors leading to success or failure, and if necessary, to readjust measures. Neglecting the need to understand the individual waterbody for planning, validating and adjusting internal measures risks failed investments. The data to collect should cover at least one growing season.

CHECKLIST 8.3: ASSESSING THE POTENTIAL BENEFITS AND THE PROSPECT OF SUCCESS OF IMPLEMENTING AN INTERNAL MEASURE

- Before planning any specific internal measure, answering the following questions will help to clarify whether or not to take an internal measure at all, and if so, to decide on which option(s) would be best for the given waterbody:
- Is an internal measure necessary because
  - the external phosphorus (P) load cannot be sufficiently (quickly) reduced?
  - the waterbody ecosystem is responding slower than necessary to the external load reduction?
  - no alternative water sources are available for an important use of the waterbody, making rapid improvement necessary?
  - the fraction of internal phosphorus loading from the sediments is high in relation to external loads and will likely stay high for many years or even decades?
- What exactly should the internal measure target – a suppression of cyanobacterial dominance within the phytoplankton, a shift from phytoplankton to macrophytes (in shallow lakes), a reduction of internal phosphorus loading or an increase of overall phytoplankton loss rates and thus a decrease of biomass – including that of the cyanobacteria?
- Is the internal measure needed once (maybe with one repeat) to help to overcome ecosystem resilience and trigger the waterbody's shift to a lower trophic state, or will it be needed continuously because external P loads remain too high?
- Are investment and operation costs adequate in relation to the waterbody's priority for human use and/or environmental targets?
- Is the necessary expertise available for this decision, that is, for planning the measure in more detail and for monitoring success?
- Is the necessary data and information on P sources and trends available to make a decision on the most effective approach, and is capacity available to monitor the response of the waterbody?

8.5 HYDROPHYSICAL CONTROL OF GROWTH CONDITIONS

Thermal stratification of a waterbody influences the depth at which cyanobacteria occur and the amount of light they receive. It may also affect the nutrient concentrations available in the upper layers in which there is enough light for growth (section 4.3.3). Water exchange rates determine whether cyanobacterial populations are diluted faster than they can grow. Changing hydrophysical conditions, for example, through impoundment, may suddenly turn a waterbody previously scarcely affected by cyanobacteria into one with heavy blooms. *Vice versa*, some hydrophysical measures may specifically reduce cyanobacterial growth, allowing other phytoplankton species (planktonic microalgae) to become dominant. Given that freshwater phytoplankton species other than cyanobacteria are not known to be toxic, they may be less of a problem, particularly for nonpotable water use.

A caveat with all hydrophysical control approaches is that if nutrient levels remain high enough to support substantial amounts of phytoplankton biomass (i.e., >20–50 µg/L total phosphorus (TP) and >100–150 µg/L dissolved nitrogen; see section 4.3), their success in suppressing cyanobacterial biomass can be somewhat uncertain. Also, other phytoplankton, that is, eukaryotic algae, may continue to reach substantial levels of biomass which can cause problems, for example, challenge drinking-water treatment with organic matter. Thus, depending on the specific circumstances, hydrophysical measures may not be sufficient to achieve water quality targets if overall phytoplankton biomass remains high even if they succeed in suppressing cyanobacteria.

Because of the inherent uncertainty involved in the manipulation of complex biological interactions, for hydrophysical interventions validation is particularly important, that is, checking whether the measure chosen proves effective for suppressing cyanobacterial blooms. As discussed above, this will require observing whether cyanobacterial blooms still occur – preferably for several growing seasons (see Chapters 11–13).

8.5.1 Artificial destratification

While artificial mixing or circulation is one of the most commonly recommended and employed management interventions to attempt to reduce the growth of cyanobacteria in lakes and reservoirs, it requires continuous operation, involving costs for energy and maintenance. Chances of success depend very much on a thorough prior baseline assessment of the conditions in the specific waterbody: it will not work in every case.

Its application is based upon the aim to eliminate the stratification in the system and affect the balance of growth between different species of phytoplankton, including cyanobacteria, by changing the physical conditions in the waterbody and thereby leading to a change in the phytoplankton composition and/or biomass. This is specifically based on the fact that the composition of the phytoplankton community in hypertrophic and eutrophic systems is often strongly driven by competition for light, and light availability is impacted by waterbody mixing and the vertical distribution of phytoplankton cells (Huisman et al., 2004). A comprehensive review of artificial mixing explains this: Visser et al. (2016) examined the mechanism of action of mixing upon the phytoplankton community. Artificial mixing works through increasing mixing depth, which affects the competition between cyanobacteria and eukaryotic algae in two ways. Firstly, mixing reduces the sedimentation losses of phytoplanktonic algae which are not buoyant (e.g., green algae and diatoms), and hence, their net growth rates tend to increase. Secondly, the buoyant cyanobacteria with a tendency to float are entrained into the deep, artificially induced turbulence and experience a lower light dose and stronger light fluctuations, and hence, their net growth rate tends to decrease. Based upon these principles, artificial mixing has been applied with the aim to disrupt the suitability of stable growth conditions favourable for cyanobacteria and prevent buoyant cyanobacteria from forming scums.

Visser et al. (2016) also reviewed the evidence from an extensive number of studies of mixing as to their success or failure to draw conclusions about possible reasons for this. They concluded that artificial mixing was not successful if the system was: (i) not sufficiently well mixed vertically, (ii) too shallow or (iii) if the horizontal distribution and position of mixing devices was not adequate to cover the entire lake to induce turbulent flow fully across the lake. If the mixing rate is not high enough to entrain the cyanobacteria and decrease their light exposure, buoyant colony-forming cyanobacteria can "escape" from the turbulent flow due to their high flotation velocity. In addition, artificial mixing will generally only be effective in relatively deep lakes. Their review of studies resulted in a minimum depth of >15 m at which mixing was successful to control *Microcystis*, which has a high flotation velocity; however, mixing depth could be less and mixing could work in shallower lakes for other cyanobacteria such as filamentous types which have a much lower flotation velocity (e.g., *Planktothrix* spp.).

Other studies have shown that artificial destratification has been successful in a number of cases with less critical examination of the principles (Reynolds et al., 1983; Hawkins & Griffiths, 1993; Heo & Kim, 2004; Lewis, 2004; Becker et al., 2006), or less successful at controlling cyanobacteria while effectively controlling other problems such as the release of iron, manganese and nutrients from sediments by maintaining oxidising conditions (McAuliffe & Rosich, 1989). The reasons for lack of success are typically not well documented; however, it most likely relates to insufficient mixing to counteract the effect of surface heating causing the formation of a warm surface layer which favours cyanobacteria. It is important to recognise that artificial mixing and destratification may not be practical and feasible in very large and very deep lakes.

A number of techniques have been proposed to directly disrupt the thermal stratification of a waterbody, including aeration systems, mechanical devices (like pumps) and solar-powered water mixers (Visser et al., 2016). Hydrodynamic modelling by Antenucci et al. (2003) suggests that for aeration systems, a combination of a deep diffuser to disturb the seasonal thermocline and a shallow diffuser to enhance vertical mixing of the epilimnion could be more effective. Surface-mounted mechanical mixers proved effective at promoting circulation and improving dissolved oxygen concentrations but not more successful than bubble plume aerators, and they have higher maintenance costs (Lewis, 2004). For solar-powered upflow, water circulators Upadhyay et al. (2013) found that these mixers tend to circulate only the epilimnetic water and stratification is maintained; thus, these mixers have a limited zone of influence and are unable to adequately mix entire lakes to effectively control cyanobacteria.

The most common and effective destratification devices are bubble plume aerators. They require a compressor to pump air to a diffuser line in the reservoir. They work by releasing a series of fine air bubbles from a pipe or line near the bottom of the lake. As the bubbles rise, they entrain water from different depths into a plume. When the plume reaches the surface, the air dissipates and the plume plunges to a depth of equivalent density and moves through the reservoir as an intrusion. Return currents flow on each side of the intrusion and generate basin-scale circulation (Schladow, 1993; Whittington et al., 2000).

Often, some stratification will still be evident outside of the immediate influence of the aerator (Sherman et al., 2000), and while this stratification is weaker than without mixing, cyanobacterial cells near the surface may not be entrained (Visser et al., 1996). This means there is still a habitat for buoyant cyanobacteria to exploit (Sherman et al., 2000). In warm climates and where night-time temperatures are high, cyanobacterial growth may still be observed in artificially destratified reservoirs, for example, in Chaffey Dam, Tamworth Australia (Sherman et al., 2000), and in North Pine Dam, Queensland Australia (Burford & O'Donohue, 2006).

Many mixing and circulating systems are available commercially and are actively marketed. Engineering expertise is sufficiently developed to design systems that can meet the specific local mixing requirements. Care must be taken, however, to engage competent companies and to plan the management and ecological targets set in combating cyanobacterial blooms. Particularly in tropical and subtropical countries with high and prolonged insolation, the energy costs of systems to maintain mixing can be significant. Poorly designed aerators may transport nutrients from sediment-near layers to the epilimnion without reducing stratification sufficiently to meet the targets, and this will favour the growth of cyanobacteria (Tsujimura, 2004). Given the initial capital expenditure and the ongoing energy and maintenance costs, it is important to properly size the aerator. Hydrodynamic modelling is recommended to predict the likely changes in the stability of stratification, potentially refining the aerator design in response to the results. A design methodology for the design of aerators for destratification of lakes is given by Schladow (1993).

Four notes of warning need to be considered when considering and planning destratification:

1. In some (primarily shallower and naturally weakly stratified) lakes, artificial destratification may promote the growth of cyanobacteria that are favoured by mixed conditions such as *Planktothrix agardhii, Planktothrix rubescens* or *Raphidiopsis* (*Cylindrospermopsis*) *raciborskii* (see section 4.2.1).
2. Artificial destratification may not only increase P concentrations in the epilimnion as discussed above. It will also increase temperatures above the sediment and thus biodegradation rates of organic matter, which further increases P release (see section 8.3.4). If phosphorus is the limiting nutrient, this could increase overall phytoplankton biomass, particularly that of cyanobacteria.
3. Mixing of mesotrophic and otherwise stably stratified lakes can induce growth and entrainment of *Planktothrix rubescens* during autumn or winter and in spring (Nürnberg et al., 2003).
4. Drinking-water is preferably extracted from deeper, cooler water layers (to reduce microbial growth in the distribution network), and artificial mixing of deep waterbodies causes warming of that water layer.

Aerators can have additional benefits or primary uses other than controlling cyanobacteria. They have also been proposed as mechanisms to control cold water pollution to downstream fisheries (Sherman et al., 2000), iron and manganese dissolution (Raman & Arbuckle, 1989; Ismail et al., 2002), hydrogen sulphide release (Cowell et al., 1987) and oxygenation of the hypolimnion to expand fish habitats. Checklist 8.4 suggests questions to clarify before deciding on introducing artificial destratification or selecting a specific scheme.

CHECKLIST 8.4: ASSESSING THE PROSPECT OF SUCCESS OF ARTIFICIAL DESTRATIFICATION

- How strongly is the waterbody stratified, and during which months of the year?
- Are the dominant cyanobacterial taxa potentially scum-forming?
- Can mixing be designed to entrain cyanobacteria into deep, dark layers for long enough to substantially reduce their growth rate and to counteract their buoyancy?

- Is the mixing design proposed technically adequate to meet the target?
- Is mixing (of shallower, weakly stratified waterbodies) likely to induce a shift to non-scum-forming filamentous cyanobacteria (e.g., *Planktothrix agardhii*) or (for mesotrophic stably stratified waterbodies) a shift to *Planktothrix rubescens* or *Raphidiopsis raciborskii*?
- Is there a risk that artificial mixing will substantially increase temperatures at the sediment surface, thus enhancing phosphorus release from the biodegradation of organic matter?
- For which part of the year must aerators be operated to effectively suppress cyanobacterial dominance?
- Is the required infrastructure and funding (reliable power supply, maintenance, monitoring) available for continuous operation?

Operational monitoring for artificial destratification

Options for parameters that indicate failure of devices installed for destratification are straightforward, including the electrical power consumption of devices installed for this purpose. For aerator designs causing a bubble plume, daily visual checks of the size of bubble plumes seen at the surface are a further option for operational monitoring.

Validation of artificial destratification designs

Whether or not the destratification system that is installed actually achieves the mixing target can readily be validated by measurements of temperature at different depths and locations, and in different weather situations or even seasons (in reservoirs, e.g., the simplest option can be at the surface and in the bottom outlet). One option is to collect data continuously or at short intervals using thermistor chains that relay these data back to the operator (these can also be used as online monitoring systems for operational monitoring of aerator function). Alternatively, temperature depth profiles can be monitored at specific occasions, focusing on weather conditions when thermal stratification has the best chances to develop in spite of the mixing, for example, during extended periods of sunny, warm and nonwindy weather. Such monitoring would demonstrate that the destratification scheme is sufficient to ensure mixing and is therefore fit for purpose.

If the outcome shows that further fine-tuning is necessary, improving the design may require more information on the response of the waterbody's thermal stratification as well as the phytoplankton populations. Data on wind, solar irradiation, temperature and precipitation from meteorological stations near to the waterbody are valuable for hydrodynamic modelling. Together with phytoplankton cell counts and nutrient data, information on reservoir or lake hydrodynamics is very useful in determining and confirming the conditions that promoted cyanobacterial growth. This serves to validate the mixing concept and enables predictive capacity for forecasting future cyanobacterial growth – and for risk assessment.

Such in-depth validation typically requires experts in limnology with a focus on phytoplankton ecology.

8.5.2 Managing river flow regimes to suppress cyanobacterial growth

Planktonic cyanobacteria generally do not develop blooms in rapidly flowing rivers. Possible reason for this may be increased turbidity due to high loads of inorganic particles and hence limited light availability, losses due to benthic grazing or highly fluctuating conditions lowering growth rates (Dokulil, 1994; Reynolds et al., 1994; Welker & Walz, 1998; Caraco et al., 2006) – or combinations of these and other factors that prevent cyanobacterial (and other phytoplankton) blooms from developing within the limited time water flows towards the sea, in most cases within days or weeks.

In rivers with long stretches of slow flow, like the lowland Murray–Darling River in Australia, this is different: hydrophysical conditions remain fairly constant over long stretches of such rivers. If nutrient concentrations are also high, the cyanobacteria which are typically found in well-mixed shallow waterbodies – for example, *Planktothrix agardhii* and other fine filamentous species – may become dominant and reach high population densities. To break their dominance, hydrophysical interventions would need to introduce pronounced changes to flow or mixing conditions at time intervals in the range of 1–2 doubling times of the cyanobacteria, that is, within several days or one to two weeks.

Impoundments or constructed barriers markedly reduce both turbulent kinetic energy and river flow and increase residence times. For example, where, without impoundment, the water would take one week to travel from the foothills to the river's mouth (i.e., 1–2 doubling times of the cyanobacteria) impoundments can reduce this travelling time to many weeks. This gives cyanobacterial populations sufficient time for many cell divisions and thus for the formation of blooms. Where impoundments are being planned, this potential impact on water quality should be assessed. Where impoundments already exist and have been identified as one cause of cyanobacterial proliferation, managing them differently or even restoring natural flow regimes may be an option, depending on other management targets.

Low flow conditions in lowland rivers can even lead to stratification. The correlation observed between buoyant species of *Dolichospermum* and low flow in some large rivers suggests that the manipulation of flow may be used to control cyanobacteria (Baker et al., 2000; Maier et al., 2004). In regulated rivers, the magnitude and timing of discharge can be manipulated to disrupt stratification every few days, thereby controlling cyanobacterial growth. Bormans and Webster (1997) developed a mixing criterion for turbid rivers that can be used to determine the flow required to disrupt stratification.

River management strategies to generate higher flow and reduce the risk of cyanobacterial blooms depend upon the availability, cost and ability to deliver enough water to provide that flow. It is also important to weigh the likelihood of success and cost–benefit of such interventions against further socioeconomic criteria (i.e., the need for an impoundment to store water or enable shipping; loss of provision of water for irrigation) and ecological criteria (i.e., the implications of flow regime changes on the riverine ecosystem that is adapted to the slow flow or impoundment regime).

Checklist 8.5 suggests questions to address when considering changes in river flow management with respect to their impact on cyanobacterial growth.

CHECKLIST 8.5: ASSESSING RIVER FLOW REGIMES AND OPTIONS FOR THEIR MANAGEMENT

- Are data on flow rates available? If not, can they be collected?
- What is the goal for the flow management or manipulation regime: to reduce residence time and dilute cyanobacteria? To disrupt stratification and reduce growth through altering mixing and light availability?
- What changes in flow regime are required to reach this goal?
- Is enough water available stored upstream in the catchment for the flow rate targeted?
- Which other sectors need to be involved in developing a flow management strategy?

Operational monitoring for flow regime management

Operational monitoring will record whether river flows are as planned, that is, through measuring flow rates. For major rivers, data on river flow are often available from water resource authorities who generate them for other purposes.

Validation of flow regime management

Validation of the flow regime management involves monitoring whether the intended flow rates are achieved (for measuring them, see section 7.2).

8.5.3 Managing water retention time in lakes and reservoirs to suppress cyanobacterial growth

Management interventions reducing water retention times in a waterbody may successfully reduce cyanobacterial biomass, if dilution rates can be achieved that are higher than their growth rates – that is, retention times of one a month or less. Retention time is the quotient of the basin volume divided by the inflow. For many waterbodies, particularly lakes, retention times are not well known, and they may be difficult to measure directly particularly if there is more than one inflow or if a lake is strongly connected to groundwater flows (see Chapter 7). Water budgets may be calculated from concentration changes of a conservative tracer substance like chloride analysed in the inflow(s), in the lake and in the outflow. As for managing river flow, a caveat may be the lack of water availability to increase the water exchange rate, particularly during seasons with little precipitation in the catchment.

CHECKLIST 8.6: ASSESSING WATER RETENTION TIMES AND OPTIONS FOR THEIR MANAGEMENT

- Are data on water retention times available? If not, can inflow and outflow rates be established or inferred from concentration differences of a tracer substance (like chloride)?
- Can a water retention time target of approximately one month or less be achieved in the lake or reservoir during the growing season?
- Are sufficient water volumes of suitable quality available in the catchment for this target?
- Are there conflicting interests for the use of this additional water or for environmental targets affected by diverting water to increase exchange rates?

Operational monitoring of water retention time management

Operational monitoring will serve to ensure that inflows to the waterbody remain in the predefined range, and thus, it will require monitoring inflow or outflow to determine whether the intended retention time is actually achieved. For reservoirs, data on outflow from the dam are usually available, and data on drinking-water abstraction may need to be included in the budget, if this volume amounts to more than a few percent of the river flow.

Validation of water retention time management

Validation of the water retention time will focus on checking whether it is indeed short enough in all parts of the waterbody to achieve its target of reducing cyanobacterial biomass. Particularly for waterbodies with many bays, as is typical for reservoirs, retention time may not be homogenous and inflows may find a preferential flow path through them, with much lower water exchange rates in the bays. Monitoring cyanobacterial occurrence also at such locations is particularly important so that further measures can be taken if the management of retention time proves insufficient.

8.6 ASSESSING AND CONTROLLING INTERNAL PHOSPHORUS RELEASE FROM THE SEDIMENTS

The amounts of phosphorus (P) released from sediments vary from negligible to being a substantial load for many years after external inputs have been reduced (Orihel et al., 2017). However, once a waterbody has reached a new equilibrium, the sediments may well become a sink rather than a source for phosphorus. Whether and when sediments are a source or a sink for phosphorus is governed by a range of conditions, in particular by

- water exchange rates across the sediment water interface;
- chemical and physical processes of precipitation and resolution as well as adsorption and desorption, which depend on the availability of binding partners for phosphorus in the sediment (e.g., silt from river inflow);
- biotic processes, that is, mineralisation of organic matter as well as bioturbation through fish and invertebrates which resuspend sediment;
- redox conditions and pH influencing the binding potential for phosphorus;
- temperature, with higher temperatures enhancing release through biodegradation of organic matter.

Chemical conditions in the upper sediment layers, temperature and waterbody mixing affect these processes. Iron-bound phosphorus is highly sensitive to oxygen concentrations and redox conditions: when such sediment surfaces become anoxic during summer stratification, phosphorus concentrations in the water above the sediment may increase significantly, and if some of this water is mixed into upper layers during the optimum growing season of cyanobacteria, it may provide nutrients for their further growth. In some situations, this may act as feedback loop, with more cyanobacterial growth increasing pH and the amount of organic material which consumes oxygen when it degrades, thus triggering more phosphorus release.

Such processes can have a significant impact on the seasonal or interannual changes of phosphorus concentrations (Xie & Xie, 2002).

Aerobic phosphorus—release mechanisms may also be significant, that is, through microbial degradation of organic material, enhanced through bioturbation by feeding fish and invertebrates (Gardner et al., 1981; Søndergaard et al., 2003; Hölker et al., 2015).

Phosphorus budgets can be estimated by balancing phosphorus loads to the waterbody against losses from the waterbody. This requires data on water inflow and outflow (preferably from continuous recording) and on phosphorus concentrations in this water (at least from monthly sampling, ideally supplemented by flow-based monitoring and sampling to capture events likely to change the load, i.e., pronounced changes in flow). Determining P concentrations requires capacity for sampling and laboratory analysis. Determining all relevant inflows and outflows can be challenging, particularly where there are many small and variable streams and/or influence from groundwater (see section 7.2). For thermally stratified lakes and reservoirs, it is relevant to measure depth profiles of P and temperature in order to differentiate between the P content of the epilimnion and the hypolimnion and to assess the stability of stratification: this allows an understanding of the potential for P-rich water from deep layers to reach the surface layer. One possible outcome of a P budget may be that sediments act as a sink for P on an annual basis and nonetheless as a P source during part of the summer, thus making P available for phytoplankton growth.

While such P budgets help clarify the role of P from the sediments for the overall P concentration in the waterbody, they scarcely contribute information for differentiating between the two most important processes of P release: desorption of iron-bound P and mineralisation of organically bound P. If measures are to be taken to reduce P release from the sediment, this differentiation is important for choosing an effective method. The potential for the release of redox-sensitively iron-bound P and organic-bound P may be estimated from chemical analyses of sediment cores following Psenner et al. (1984), and this requires access to expertise and capacity in sediment chemistry. Further indication may be derived from the analysis of time patterns of phosphorus peaks in the water above the sediment in relation to temperature (which strongly governs mineralisation) and redox conditions (indicated, e.g., by the concentrations of oxygen or nitrate, or measured directly with a probe; an example is given by Chorus & Schauser, 2011). Although such analyses require time and expertise, measures to reduce internal phosphorus loads require investment, and this information is important to estimate their chance of success and thus avoid failed investment.

If a thorough analysis reveals P release from the sediment to be a major source of P in the waterbody, and if demands for water use do not allow waiting for years until measures to control the external load have taken effect and sediments turn into a sink rather than a source for P, it may be necessary to reduce the in-lake phosphorus pool. This can particularly be the case if water exchange rates are low. Sometimes this situation also applies to lakes that are naturally eutrophic, like lakes in western Canada situated on phosphorus-rich glacial till (Prepas et al., 1997). A comprehensive review by Bormans et al. (2016) of opportunities and methods for controlling internal phosphorus loading gives many examples and case studies of techniques and an analysis of their success. These examples include dredging (see section 8.6.2), hypolimnetic aeration or oxygenation (see section 8.6.3) and hypolimnetic withdrawal (see section 8.6.4).

Checklist 8.7 serves for a first assessment of whether or not internal phosphorus control measures should be considered.

CHECKLIST 8.7: ASSESSING POTENTIAL BENEFITS AND THE PROSPECT OF SUCCESS OF INTERNAL PHOSPHORUS CONTROL MEASURES

- Have external inputs been sufficiently reduced so that these loads do not override the effect of the internal measure?
- Has a comprehensive phosphorus budget been calculated and an analysis of phosphorus release from the sediments been conducted, as outlined above, to clarify the relevance of the sediments as phosphorus source to the waterbody?
- Is the control of internal phosphorus loading necessary because external measures are not likely to reduce phosphorus concentrations in the waterbody sufficiently within the targeted time frame?

Several options for internal phosphorus control are introduced in the following sections. Their best choice depends on hydrological conditions and in particular on sediment chemistry. Investing in sediment analysis to determine phosphorus-binding forms and binding partners before investing in sediment treatment is strongly recommended.

8.6.1 P reduction by in-lake phosphorus precipitation and capping

The phosphorus-binding capacity of sediments depends upon the sediment type, primarily on the sediment's adsorption capacity. If the sediment has received phosphorus-rich waters for a considerable period of time, then the adsorption capacity may be saturated. Adsorbents naturally reach waterbodies with the load of silt, mineral and clay particles eroding from the catchment. If measures to reduce phosphorus loading in the catchment shift the balance between P and adsorbents reaching the waterbody, the new sediment forming in the waterbody may attain a sufficient binding capacity without further measures. Adding P-adsorbing substances (flocculants) to the waterbody can accelerate this process.

Conversely, waterbodies with naturally high concentrations of iron or calcium compounds in the inflow and thus sufficient adsorption capacity to induce natural phosphorus precipitation may lose this ability through measures in the catchment area or changes in inflow regime.

Precipitation of phosphorus from the waterbody to the sediment, and sediment "capping", can be successful if phosphorus then remains permanently bound in the sediment. This is either done by binding P to insoluble iron compounds (which, however, remain insoluble only under oxidative conditions), aluminium sulphate (alum) or by adsorption onto calcium carbonate or clay particles modified to enhance adsorption. Experience with both failures and successes has shown that effective treatment requires a careful design on the basis of comprehensive understanding of the sediment chemistry and hydrology of the individual waterbody (Cooke et al., 2005).

Prerequisites for lasting success include low external P loading; a sufficient oxygen supply of the deep water if capping is done with redox-sensitive compounds; sufficient depth to prevent sediment resuspension; and appropriate choice of P-adsorptive materials used as flocculants such as ferric salts (chlorides, sulphates), ferric aluminium sulphate, zeolites, lanthanum-modified clay, clay particles and lime (as $Ca(OH)_2$ and as $CaCO_3$); and by-products of mining, mineral processing or industries (Akhurst et al., 2004; Douglas et al., 2016). Douglas et al. (2016) describe the mode of action of P adsorbents and report that those with substantial uptake capacity are generally enriched in Ca, Fe and/or Al; they may also incorporate the rare earth element lanthanum (La). These compounds may all have some degree of undesirable side-effects, from toxicity to disrupting food resource supply to zooplankton; therefore, care needs to be taken with their application.

Ferric salts are effective in precipitating phosphorus, but difficult to handle because of their aggressive acidity. In particular, the iron–phosphorus complex is stable only under oxic conditions. Therefore, the application of ferric salts usually requires oxic conditions to be ensured down to the sediment, for example, through continuous aeration (which may remain necessary for many years, until P concentrations and plankton biomass are so low that the waterbody remains oxic down to the sediments including during summer). In practice, oxic conditions are unlikely to be reached quickly in waterbodies that have experienced years or decades of hypertrophic conditions accumulating a thick anoxic sediment rich in P and biodegradable organic matter. This will continue to degrade, consuming oxygen and thus reducing the iron–phosphorus complex, releasing the bound phosphorus.

In addition, Prepas et al. (1997) pointed out that iron may be a limiting micronutrient in some systems, and in such situations, treatment with ferric salts may actually stimulate growth of cyanobacteria and algae. This stimulation cannot occur if sufficiently stringent phosphorus limitation can be achieved.

Aluminium sulphate treatment does not require oxic conditions as it is poorly soluble, provided pH conditions remain neutral or high pH. However, it may decrease pH in waters with low buffering capacity, and this leads to solubilisation and problems of alum toxicity.

Although lime (both $Ca(OH)_2$ and $CaCO_3$) has been used primarily to coagulate and precipitate phytoplankton cells (see also section 8.4.9), it has also been successfully applied to precipitate phosphorus from the water, for example, in farm dugouts (dams) in Alberta, Canada (Murphy & Prepas, 1990; Zhang & Prepas, 1996) and in other natural waters and wastewaters (Douglas et al., 2016). Hart et al. (2003) assessed the effectiveness of three forms of $CaCO_3$ (crushed limestone and two forms of precipitate calcite) on phosphorus binding in lake sediments. Limestone was found to be ineffective, but the precipitated calcite products reduced phosphorus release by up to 100 times under anoxic conditions. It appears that $Ca(OH)_2$ is more effective than $CaCO_3$ in precipitating phosphorus (Murphy & Prepas, 1990), and it is possible that the technique may be more effective in these conditions than in soft water.

Another approach is to bind P in sediments with lanthanum-modified clay (Douglas et al., 2016). Robb et al. (2003) reported that modified clay successfully bound phosphorus in the rivers of Canning and Vasse in Australia. Akhurst et al. (2004) investigated the ability of modified clay to reduce the levels of phosphorus released from the sediments of Lake Ainsworth, Australia. They found the bentonite clay highly effective at reducing P under both anoxic and oxic conditions. However, levels of dissolved Fe were enhanced with its use, and this may result in a water quality issue (see above). Ross and Cloete (2006) showed a significant reduction in filterable reactive phosphorus (FRP) and a drop in the amounts of phytoplankton after the addition of modified clay to Hartbeespoort Dam, South Africa.

A natural phenomenon potentially useful in specific circumstances is the effect of desiccation/oxidation of sediments on P adsorption as this can significantly reduce the release of phosphorus from lake sediments upon rewetting attributable to a number of interrelated factors (Baldwin, 1996; Mitchell & Baldwin, 1998; Baldwin et al., 2000). While it is not known how long the effects of desiccation will last, drying the lake sediments may be a suitable strategy to reduce P release in some shallow lakes or reservoirs, if, for example, natural seasonal wetting and drying cycles can be (re)introduced.

As already mentioned above, some precipitation and/or capping agents have an impact (adverse or toxic effect) upon lake biota. Douglas et al. (2016) describe the need to establish the ecotoxicological profile of P-adsorptive material applied to natural waters; these authors also provide references to available ecotoxicological test that can be employed for this purpose.

CHECKLIST 8.8: ASSESSING THE PROSPECT OF SUCCESSOF P REDUCTION BY IN-LAKE PHOSPHORUS PRECIPITATION OR SEDIMENT CAPPING

- Has the P-binding capacity of the sediments been assessed?
- How large is the "inventory" of phosphorus likely to be released from the sediments?
- Have the phosphorus loads (inputs) been assessed?
- Has the oxygen consumption of the sediment during thermal stratification been assessed?
- What substances are to be precipitated for sediment capping?
- Do the compounds have appropriate local regulatory approval for application to waterbodies?
- Do they have side-effects? If so, what are they and to which extent?
- Do further stakeholders need to be involved to decide on their acceptability?

Operational Monitoring for in-Lake Phosphorus Precipitation and Capping

For these techniques, it is important to monitor the application with respect to sufficient amounts and even distribution of the substance applied throughout the waterbody during application. After the measure, inspecting the sediment surface from time to time may be important to ensure that the capping cover is still sufficiently thick.

Validation of in-Lake Phosphorus Precipitation and Capping

As for P removal by flushing, validation of precipitation and capping measures requires a monitoring programme of the P content of the waterbody, that is, concentrations in different depths. For lakes and reservoirs with a very heterogeneous shape, possibly this needs to include bays and subsystems. P concentration monitoring may need to be repeated at intervals (e.g., annually or in each season) to ensure that the measure was sufficient and further interventions are not necessary. If concentrations decline to the level targeted, this is sufficient, and the P-response monitoring programme can be reduced to periodic reviews (e.g., once per year in spring). If not, this indicates either insufficient capping or external phosphorus sources that require reduction before capping can be successful.

8.6.2 P reduction by sediment dredging

Dredging to remove P-saturated sediments intuitively appears to be an attractive solution. However, it is costly and will reduce release rates only if

1. (as for all internal measures to reduce P release), the internal phosphorus load from the sediment is significant in relation to the external load;
2. it is carried out all the way down to sediment layers with a lower or less mobile phosphorus content;
3. phosphorus-rich interstitial water from the sediments dredged is handled in such a fashion that it does not reach the waterbody and cause additional inputs during the dredging operation;
4. dredged sludge can be deposited where it does not create a new external load through runoff into the waterbody.

In some urban and industrial regions, dredging is not possible or complicated by high concentrations of heavy metals and organic contaminants in the sediments which would then require disposal as hazardous waste. Dredging may be a good solution for smaller waterbodies, the trophic state of which can additionally be improved by increasing their depths, or which also need to be cleared of dumped rubbish. A review of dredging case studies by Bormans et al. (2016) found both successful and unsuccessful examples and that dredging is costly. The authors concluded that where it has been unsuccessful, in most cases the reason for this was that external loading was not sufficiently controlled, and that it is more successful when combined with other restoration techniques.

CHECKLIST 8.9: ASSESSING THE PROSPECT OF SUCCESS OF P REDUCTION BY DREDGING

- Is the input of phosphorus from the catchment area low enough to ensure that sediment removal will have a lasting effect?
- Will dredging remove P-saturated sediments down to layers that are not likely to cause a continuing internal P load?
- Is a suitable site available for depositing the sludge, so that P-rich sludge water can be kept away from the waterbody?

Operational monitoring for removal of in-lake phosphorus by dredging

Operational monitoring of dredging involves checking, for example, by visual monitoring of the amounts dredged and the sites at which dredging is performed, that the amount of sediment intended for dredging is actually removed from the sediment surface areas at the depths designated for removal. Visual inspection can serve to monitor that dredged sludge is not deposited so close to the waterbody that will reintroduce part of the phosphorus load through seepage, erosion and stormwater runoff.

Validation of removal of in-lake phosphorus by dredging

Dredging may easily miss the target of removing all of the sediment layers which release phosphorus, and as for other P removal methods, validation requires monitoring whether the P content of the waterbody responds sufficiently to the measure, that is, whether concentrations decrease to the target level. If concentrations decline to the target level, this is sufficient, and the P-response monitoring programme can be reduced to periodic reviews (e.g., once each year in spring). If not, this indicates either insufficient dredging or external phosphorus loads that require reduction before dredging can be successful.

8.6.3 Binding phosphorus through hypolimnetic aeration

Hypolimnetic aeration or oxygenation (with pure oxygen) aims at providing oxygen to the hypolimnion without disrupting thermal stratification to enhance the binding of phosphorus to iron (Bormans et al., 2016). The prerequisite for this technique is that the waterbody is thermally stratified during relevant parts of the year and that the sediment contains enough redox-sensitive P-binding material, that is, iron. If the ratio between iron and phosphorus is low, oxidising the sediment will not help. Schauser et al. (2006) and Schauser and Chorus (2007) highlight this with a case study of Lake Tegel, Germany: budget calculations demonstrated that P accumulating in the hypolimnion did not primarily result from anoxic release of iron-bound P, but largely originated from the mineralisation of recently sedimented biomass, and later investigations showed that indeed the sediment's iron concentrations were rather low in relation to P concentrations. In this case, the increase of sediment temperatures caused by poorly managed aeration actually enhanced this biodegradation-driven P release and scarcely served to bind P.

Other researchers discuss further examples of marginal or lack of success of hypolimnetic aeration where the chief target of aeration was reducing phosphorus release from sediments (Gächter & Wehrli, 1998; Gächter & Müller, 2003; Hupfer & Lewandowski, 2008; Bormans et al., 2016). These examples show that planning hypolimnetic aeration requires a particularly good understanding of the sediment chemistry for success of this intervention, and it is therefore not recommended without this knowledge.

While hypolimnetic aeration may use aerators like those used for artificial destratification (section 8.2.1), these aerators must be designed and operated in a way that minimises the disruption of thermal stratification for two reasons: (i) to avoid accelerating P release from the sediment by from biodegradation and (ii) because stable stratification resists deep mixing and entrainment of nutrients from the hypolimnion into epilimnion where they can be used for phytoplankton growth. Designs to avoid destratification have been used for fisheries management where warm surface waters and cool bottom water are both required for fish habitat (Moore et al., 2014), but it also has the advantage that nutrients can remain limiting in the surface water layers. The challenge is to inject oxygen into the hypolimnion in such a way that it does not form rising air bubbles that can form a plume and disrupt stratification.

Hypolimnetic oxygenation can be achieved with a number of methods, including airlift pumps, side-stream oxygenation and direct injection of air or oxygen using a bubble contact chamber like a Speece cone (Beutel & Horne, 1999; Cooke et al., 2005; Singleton & Little, 2006). These techniques vary in cost, but they are all relatively expensive. The technique is also used and well understood for controlling soluble iron and manganese release in drinking-water reservoirs (Gantzer et al., 2009).

Hypolimnetic aeration can easily be a waste of money if selecting this management option is not carefully designed and operated and not based on a good understanding of the waterbody's hydrodynamics, sediment oxygen demand, iron concentration and nutrient release rates as well as the relative contribution of internal versus external nutrient load (see Singleton & Little, 2006).

CHECKLIST 8.10: ASSESSING THE PROSPECT OF SUCCESS OF BINDING P THROUGH HYPOLIMNETIC AERATION

- Does the sediment contain enough redox-sensitive binding sites (i.e., iron) for a success chance of this strategy?
- Is the system designed to keep destabilisation of thermal stratification minor?
- As aeration will continue to be necessary for a prolonged time period, is funding for maintenance and operation secure for several years to come?

Operational monitoring for binding P through hypolimnetic aeration

Operational monitoring of hypolimnetic aerators can be performed in a similar way to that for artificial mixing described above, that is, by recording electrical power consumption of the pumps, the oxygen concentration in the water or some other indicator for their continuous operation.

Validation of binding P through hypolimnetic aeration

As for artificial mixing, recording temperature and dissolved oxygen profiles over depth is also important for validation, in this case to demonstrate that the function of the aerators is actually limited to the hypolimnion, without significant impact on the layers above. This can either be done continuously using a thermistor chain, or at regular intervals (e.g., weekly), or under selected sensitive conditions where stratification is less stable than normal, for example, in spring when it begins developing, after storms or in late summer when it begins eroding. Validation should also check for (unintentional) entrainment of dissolved substances (particularly phosphorus) from the hypolimnion into the surface water, particularly if temperature profiles indicate that aeration does cause some intensification of water exchange between the hypolimnion and surface-near water layers.

In particular, the further primary parameter to monitor is the oxygen concentration in the hypolimnion in order to validate that the measure actually achieves the targeted oxygen concentration. This can be done periodically, for example, every two weeks during thermal stratification or continuously if sensors are installed. The second important level of validation is to show that this target oxygen concentration – if it is met – is actually successful in suppressing phosphorus release. As for the other internal measures for reducing P in a waterbody, this requires monitoring phosphorus concentrations. If concentrations decline to the target level, this is sufficient, and the P-response monitoring programme can be reduced to periodic reviews (e.g., once per year in spring)

Possibly, if target levels are not met, either the sediment is less of a P source than assumed (indicating the presence of further sources) or P release from the sediments is not as redox-sensitive as assumed and rather driven by mineralisation, which is more effective if oxygen is present and therefore may even be enhanced by aeration, particularly if this measure increases the water temperature at the sediment surface.

8.6.4 Withdrawal of P with the bottom water (hypolimnetic withdrawal)

In thermally stratified eutrophic lakes, phosphorus accumulates in the hypolimnion during conditions of stable stratification during summer. While during summer this layer is largely separated from the epilimnion in which phytoplankton grows, autumn and winter mixing will redistribute this phosphorus throughout the waterbody, making it available for phytoplankton for spring growth. Although most natural outflows drain surface water, it may also be possible to abstract hypolimnetic water instead, by installing a pipe that reaches down into the hypolimnion and has an outlet positioned lower into the outlet (like an "upside-down U"; Olszewski, 1961). For reservoirs with multiple outlets, it is often possible to use a lower outlet of the dam to withdraw hypolimnetic water.

Where P concentrations in the hypolimnion are high and the hypolimnion depth and volume are considerable, this method can reduce in-lake concentrations significantly. For example, in the Swiss Mauensee, the biomass of *Planktothrix rubescens* was reduced from 152 to 42 g m^3 (corresponding to biovolume of 152 and 42 mm^3/L) using this approach (Gächter, 1976). With data from numerous case studies, Nürnberg (1997; 2007) compiled the advantages of hypolimnetic withdrawal during summer stratification as a method based on the selective outflow of P-rich water. Further case studies are given by Bormans et al. (2016). Advantages of the method are as follows:

- It addresses the cause of eutrophication.
- It does not introduce chemicals.
- It can be used without changing the water budget.
- It can break the cycle of enhanced sediment accumulation of total phosphorus (TP).
- If hydrological regimes allow, it can flush more phosphorus out of the system than the sediments accumulate and/or release each year.
- Costs are relatively low, and where operation is by gravity, it does not require energy.

Hypolimnetic withdrawal is effective only if enough water flows into the lake to balance consumptive needs, that is, amounts abstracted for use (e.g., drinking-water supply, irrigation). It is possible only if the depth and volume of the hypolimnion are sufficiently large to have an overall impact on the phosphorus budget of the waterbody without removing the entire hypolimnion: if drawdown is too pronounced, this may destabilise thermal stratification, causing entrainment of P-rich hypolimnion to surface layers and increasing temperatures at the sediment surface, which in turn enhances the release of P through mineralisation of organic matter.

In addition, impairment of water quality downstream may require attention if oxygen demand and phosphorus concentrations of the water removed are high in relation to the total flow of the receiving waterbody. Downstream phosphorus pollution may be avoided by treatment of the hypolimnion outlet with chemical phosphorus precipitation. In addition, the low temperatures of the hypolimnetic water may have an impact upon downstream biological processes, like fish breeding. Use of this phosphate-rich water for agricultural purposes may be an option, depending upon local needs for irrigation in addition to considering the implications of other contaminants which can be dissolved in water with a low oxygen concentration (e.g., iron and manganese).

CHECKLIST 8.11: ASSESSING THE PROSPECT OF SUCCESS OF P REDUCTION THROUGH HYPOLIMNETIC WITHDRAWAL

- Are the depth and volume of the hypolimnion sufficient to allow major removal without destabilising thermal stratification?
- Does enough water flow into the waterbody to balance the hypolimnion offtake, or is some drawdown of the water level acceptable?
- Have downstream effects been assessed and stakeholders possibly affected been involved?

Operational monitoring for removing P through hypolimnetic withdrawal

Operational monitoring of hypolimnetic withdrawal should include monitoring of both the reservoir water and the outlet (e.g., temperature or conductivity) to ensure the desired target layer is being removed.

Validation of P removal through hypolimnetic withdrawal

As for artificial mixing, measuring temperature profiles is also important for validation. This can either be done continuously using a thermistor chain, or at regular intervals (e.g., weekly) or under selected sensitive conditions where stratification is less stable than normal, for example, in spring when it begins developing, after storms or in late summer when it begins eroding.

In particular, the additional parameter to monitor is the oxygen concentration in the hypolimnion in order to validate that low oxygen or anoxic water is being removed as an additional surrogate for phosphorus content.

8.6.5 Reducing the P content of the waterbody by flushing

Flushing with water of low phosphorus concentrations can substantially accelerate recovery from internal loading by removing in-lake phosphorus that would otherwise be recycled for several seasons. If suitable water is available in sufficient quantity, flushing can be a very effective tool for reduction of cyanobacterial proliferation. Successful examples include Veluwemeer in the Netherlands (Reeders et al., 1998), Moses Lake in the USA (Welch et al., 1972) and two lakes in Germany (Figure 8.1). The latter show that if P concentrations in the inflow are sufficiently low (in these cases < 10–20 µg/L TP), exchanging the lake's water volume 2–4 times per year can reduce TP concentrations from several hundred µg/L to 20–50 µg/L within less than 10 years despite a continued internal load from the sediments and some remaining external load. These examples show that while diluting cyanobacteria out of a waterbody as discussed in section 8.5.3 requires extremely high flushing rates – that is, in the range of an exchange of the water volume once a month – diluting out phosphorus can be successfully done more slowly. The time it takes to achieve the target phosphorus concentration depends chiefly upon the initial P content (i.e., the waterbody's volume multiplied with its mean concentration of total P) and the P load of the inflow(s) (i.e., concentration multiplied with the water volume entering the waterbody) plus the internal P load (i.e., the amount released from the sediments, which can be estimated by budget calculations as outlined above and described in Schauser and Chorus (2009)).

A flushing option for reservoirs is to use heavy rainfall events to fill the reservoir with water of lower P concentration while releasing water with higher concentration downstream. For thermally stratified waterbodies,

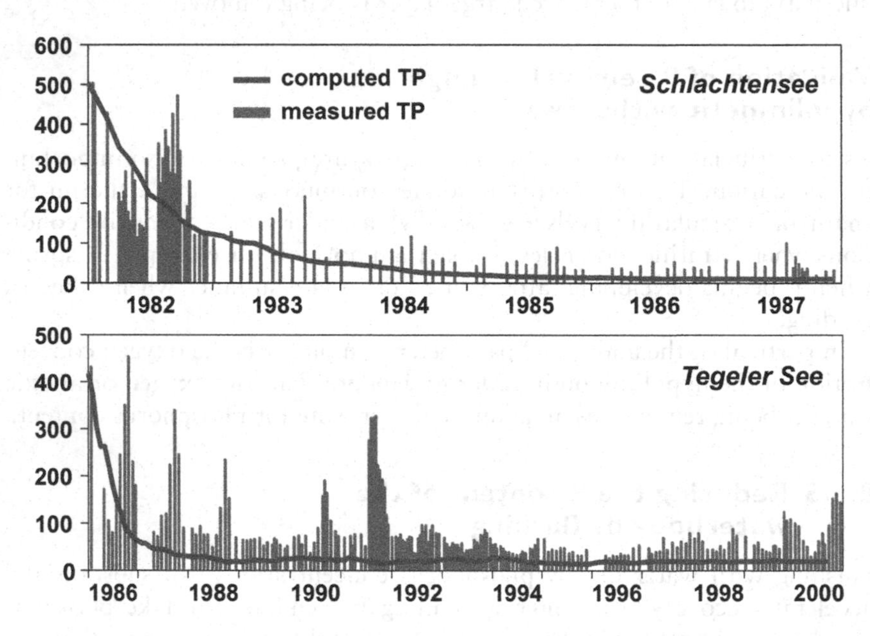

Figure 8.1 Theoretical dilution of concentration of total phosphorus (TP) in two Berlin lakes after flushing with water of low TP concentrations began (curves) compared to concentrations actually measured (bars): for Schlachtensee (a; flushing rate of 2 times per year with water containing ~8 µg/L TP), the impact of other TP sources was much lower than in Tegel (see b; flushing rate of 2–3 times per year with water containing ~20 µg/L TP) where the marked seasonal peaks of the concentrations measured in the lake indicate substantial further TP sources. (Data from Chorus & Schauser, 2011.)

this can be particularly effective if water can be released from P-rich bottom layers, as described above. However, this measure causes a relocation of the phosphorus to a downstream waterbody, and this requires impact assessment with the stakeholders responsible for that waterbody. If the amount released and its P load is small in relation to the river flow into which it is released, this may not cause a substantial change of P concentrations in that river.

CHECKLIST 8.12: ASSESSING THE PROSPECT OF SUCCESS OF P REDUCTION BY FLUSHING

- Has a P budget been established, and have water volumes needed for flushing been estimated? Which P concentration can be achieved by flushing?
- Is sufficient water of suitable quality available in the catchment with significantly lower P concentration to achieve targets for the waterbody by flushing?
- What are the oxygen and phosphorus concentrations in the water to be released, and what potential impact will that have downstream of the release?

Operational monitoring for P reduction by flushing

Similar to operational monitoring for managing river flow or water retention times, operational monitoring for P reduction by flushing should also record whether flows are occurring as planned, that is, through measuring flow rates or other indicators of water exchange rates, for example, concentrations of tracers such as chloride. Where a specific selected water layer is to be drained from a thermally stratified waterbody, it is advisable to monitor the water temperature both in the waterbody and in the released flow to ensure that the layer actually being drained is indeed the one targeted.

Validation of P reduction by flushing

Validation of the flow regime management to flush P out of the systems requires monitoring the P content of the waterbody, that is, concentrations in different depths and for lakes and reservoirs with very heterogeneous shape, possibly also in different bays and subsystems. If concentrations decline to the target level, this is sufficient, and the P-response monitoring programme can be reduced to periodic reviews (e.g., once a year in spring). If not, more detailed sampling and analyses of P concentrations in tributaries and at the outflow may be needed in order to improve the P budget calculation as a basis for identifying the remaining P sources (e.g., surface runoff, minor tributaries or sediment release).

8.7 BIOLOGICAL CONTROL OF CYANOBACTERIA

The term "biomanipulation" describes a range of techniques that influence phytoplankton community composition and growth by influencing parts of the food web of a lake. One approach aims to stimulate the growth of zooplankton that graze phytoplankton and is termed the "top-down" approach for phytoplankton control, as opposed to "bottom-up control" by nutrient reduction. The other approach aims at stimulating the growth of submerged aquatic plants ("macrophytes") or reeds that can serve two functions: they compete with phytoplankton for nutrients, and they provide refuges for zooplankton (thus supporting the top-down approach). Both approaches do not specifically address cyanobacteria, but phytoplankton in general. If these measures successfully decreased phytoplankton biomass and thus turbidity, this tends to favour species other than cyanobacteria, thus shifting species dominance away from cyanobacteria.

Biomanipulation as a management tool to reduce algal or cyanobacterial growth is most likely to be successful in situations of moderate nutrient concentrations (i.e., total phosphorus (TP) <50 µg/L) and in many situations also requires the reduction of nutrient loads. Experience shows that as long nutrient concentrations remain high, the risk that the ecosystem switches back into its original state is also higher. Also, stimulating zooplankton grazing without reducing concentrations of nutrients may stimulate dominance of grazing-resistant phytoplankton species, such as colony-forming (*Microcystis, Aphanizomenon*) or filamentous cyanobacteria (*Planktothrix agardhii*) (see section 4.1.5). For scientific reviews on biomanipulation, see Triest et al. (2016), DeMelo et al. (1992), Kitchell (1992), Carpenter & Kitchell (1996), Moss et al. (1994) and Jeppesen et al. (2007a).

A striking example of how biomanipulation can complement and enhance nutrient control measures to return lakes back to a clear state is documented by Ibelings et al. (2007) for the shallow lake Veluwemeer in the Netherlands: as nutrient concentrations in the lake increased in the 1960s, biota shifted from a macrophyte-dominated state to a turbid phytoplankton-dominated state. Even though nutrient inputs and in-lake concentrations were significantly reduced and overall phytoplankton biomass declined, this state was resistant to change and macrophytes did not reappear; that is, the lake remained in a stable turbid state (Scheffer & Carpenter, 2003). This state was maintained by resuspension of sediments both through wind and through benthivorous fish. A significant reduction in the bream population allowed zebra mussels to recolonise and clear the water with their high filtration capacity. The clearing of the water then enabled macrophytes to recolonise, and these macrophytes in turn support and contribute to keeping the water clear by binding nutrients in their biomass.

These examples demonstrate the complexity of aquatic ecosystems and highlight the potential for different management options that may need to be tested in practice to find the approach which is most effective in the specific waterbody. Regardless of whether biological controls are being considered via fish stock management or through the introduction of macrophytes – or both – the following checklist helps assess the success chances of a biological approach.

CHECKLIST 8.13: ASSESSING THE PROSPECT OF SUCCESS OF BIOLOGICAL CONTROL MEASURES

- Is the waterbody only slightly eutrophic, thus rendering the food web susceptible to a "switch" in species composition away from cyanobacterial dominance, or are total phosphorus concentrations high (>50 µg/L), with stable dominance of filamentous or colony-forming cyanobacteria, which are poorly edible by zooplankton and likely to cause too much turbidity for macrophyte growth?
- Has the ecosystem of the waterbody been intensively studied for at least 1–2 years, thus providing an in-depth understanding of trophic interactions as a basis for planning biomanipulation measures?
- Is funding available to continue these studies in order to monitor the ecosystem response, to possibly fine-tune the management of the measures taken and/or to repeat them if necessary?

The following two sections provide an introduction into the two biological approaches – that is, supporting zooplankton grazing by managing fish which would otherwise decimate the zooplankton and supporting recolonisation of shallow areas of a waterbody with macrophytes.

8.7.1 Suppressing cyanobacteria through increasing grazing pressure by fish stock management

Interventions into established hypertrophic ecosystem structures by fish stock management techniques have been successful, particularly in small ponds and lakes over shorter periods of time (Hrbáček et al., 1978). The target to increase zooplankton grazing pressure on phytoplankton can be achieved by reducing the populations of fish that feed on zooplankton – either by regularly removing such planktivorous fish manually, that is through net hauls, or by introducing predatory fish that feed on the planktivorous fish. Numerous examples and potential challenges of this strategy are discussed extensively by Triest et al. (2016). These authors include examples of insufficient removal or suppression of planktivorous fish, insufficient macrophyte coverage to provide refuges for zooplankton as well as for the fry of predatory fish (which are intended to reduce the planktivorous fish) as well as unintentional promotion of the dominance of poorly edible cyanobacterial species. These potential pitfalls highlight the need for comprehensive previous studies of ecosystem as well as for ongoing monitoring after measures have been undertaken (see also Reynolds, 1997).

A successful contrasting example from a hypereutrophic subtropical lake in China suggests that cyanobacteria could be controlled directly by fish grazing upon them (Xie & Liu, 2001). This is a different approach from traditional biomanipulation through enhancing zooplanktonic grazers, and it is reported to have been effective for over 30 years.

The monitoring of fish populations is time-intensive and requires substantial expertise, and consequently, this is often a major cost factor for such biomanipulation schemes, while the fish stock itself may be comparatively cheap. Biomanipulation requires continued monitoring of the development of the plankton as well as regular stocking with predatory fish. Where fishing or angling occur to the extent that it impacts the fish population, it is important to involve anglers and fishermen in planning and operating the measure. They need to be encouraged not to decimate the predatory fish but to remove planktivorous fish (which tend to be unattractive commercially or as trophy), as removing these may support the desired outcome. Biomanipulation through fish stock management may not continue to work naturally and unaided unless nutrient concentrations are also sufficiently reduced.

CHECKLIST 8.14: ASSESSING THE PROSPECT OF SUCCESS OF BIOLOGICAL CONTROL OF CYANOBACTERIA THROUGH INCREASING GRAZING PRESSURE BY FISH STOCK MANAGEMENT

- Is the biotic structure of the waterbody ecosystem described and understood – that is, are data available on fish stock, zooplankton population sizes and biomass of phytoplankton taxa?
- Do these results suggest that biological control will be successful?
- Does the trophic state also imply that success of biological control of cyanobacteria can be successful?
- Can stocking measures of predatory fish or removal of planktivorous fish be repeated, for example, annually?
- Can anglers be prevented from removing the predatory fish introduced for biomanipulation?

Operational monitoring of grazing pressure management by fish stock management

Fish stock management interventions tend to be discontinuous, and accordingly so is operational monitoring. It addresses whether planktivorous fish have been removed as intended and/or predatory fish fry have been stocked as intended. This can be through, for example, surveillance of the documentation of the hatcheries and operators involved in stocking operations as well as through monitoring compliance of angling activities to permits.

Validation of grazing pressure management by fish stock management

It is challenging to validate fish stock management as to whether or not it is effectively changing the food web of a waterbody. This validation requires monitoring the population changes of fish, zooplankton and phytoplankton, including cyanobacteria, and interpreting whether changes observed are likely to be responses to the interventions, or whether they are due to normal fluctuations or other changes in the ecosystem. While assessing changes in cyanobacterial density alone will indicate success or failure, it misses providing an in-depth understanding of the mechanisms leading to any observed changes, and therefore, zooplankton data are also important. Willmitzer (2010) showed a simple approach to validating whether planktivorous fish have been sufficiently reduced to allow zooplankton to reduce phytoplankton via biofiltration, that is, through monitoring the presence of large zooplankton, particularly *Daphnia* ("water fleas") which have high filtration rates. For this, the proportion of large and small zoo plankton is a useful indicator (CSI=Cladocera Size Index), and for this purpose, sampling with simple plankton net hauls with large and small mesh sizes is sufficient. Evaluation of the zooplankton does not need to be taxonomically rigorous, as the data needed are size distribution rather than species. However, this does not necessarily indicate whether or not the large *Daphnia* are able to exert sufficient grazing pressure upon the cyanobacterial species responsible for the bloom, because many bloom species are indeed poorly edible.

Validating the success of fish stock management is therefore most effective through a comprehensive limnological assessment of the aquatic ecosystem on the basis of quantifying both cyanobacteria and zooplankton conducted sufficiently often. While fish stock studies are needed at intervals of several years, plankton usually needs to be monitored at least monthly. Misinterpretations include that where the necessary limnological expertise and/or data on the prior condition of the waterbody were lacking, seasonal fluctuations like spring clearwater phases (typical for many temperate lakes and reservoirs) have been misinterpreted as success of measures.

8.7.2 Enhancing competition of macrophytes against cyanobacteria

The introduction of macrophytes (totally or partially submerged aquatic plants) has the greatest chance of success in waterbodies with a relatively large shallow littoral area of lakes. Macrophytes are mostly unable to colonise reservoirs with pronounced fluctuations of water level. Submerged macrophytes also require relatively high light penetration, and therefore, introducing them works best at moderate concentrations of total phosphorus (TP), that is, less than 50 µg/L. This is because under field conditions, 1 µg of TP can support about 1 µg of chlorophyll-*a* (see sections 4.3.2 and 4.4), and if the spring phytoplankton reaches a biomass containing 50–100 µg/L of chlorophyll, this renders the water very turbid, thus suppressing the growth of submerged macrophytes. In contrast, if it is nutrient-limited and therefore clearer, macrophytes have an increased chance to begin to grow. If further nutrient loads during summer are not too high, they can then incorporate enough of the available phosphorus with their growth to achieve substantial phosphorus limitation of phytoplankton biomass for the remaining growing season. Thus, measures to support macrophytes may switch a slightly eutrophic aquatic ecosystem – particularly a shallow one with large areas covered by macrophytes – into a different, potentially more stable aquatic community, resulting in clear water and low cyanobacterial biomass. This effect can be enhanced by also managing fish stock, as discussed above.

Excessive nutrient concentrations tend to be detrimental to aquatic macrophytes: for example, a comprehensive study based upon 97 shallow lakes on the Yangtze Plain, China, found that macrophytes begin to collapse and degrade when the TP concentration increased to more than 60–80 µg/L, and the authors suggest that the collapse of particular macrophyte species can serve as an early warning signal for the regime shifts from clear to turbid state with increasing phosphorus levels (Su et al., 2019). While Jeppesen et al. (2007b) suggested the growth of submerged aquatic plants to diminish if nitrogen concentrations were above 1000–2000 µg/L, Søndergard et al. (2015) showed this to be relevant only if TP concentrations are also high (>100 µg/L) and that for numerous lakes in Denmark, reducing TP has been more effective in promoting re-colonisation with macrophytes than reducing only N (see discussion in Chorus and Spijkerman, 2020).

Where boat traffic damages reed belts both, directly and through wave action, protecting the reed belt may be a further measure to foster macrophytes as competitors against cyanobacteria. This can be achieved by constructing structures (e.g., wooden palisades) in the water at a depth of about 2 m in front of the remnants of reed stands or newly planted ones. Such structures should reach about 50 cm above the water surface in order to effectively intercept even larger waves. They can be occasionally interrupted by small gaps to allow water birds to move between the lake and the quiescent protected water behind the structures. Where recreational use pressure on a lake shore is high, protecting the reed belt with a fence on its landward side may also be important to prevent erosion and mechanical damage and to allow recovery.

CHECKLIST 8.15: ASSESSING THE PROSPECT OF SUCCESS FOR BIOLOGICAL CONTROL OF CYANOBACTERIA THROUGH INTRODUCING OR SUPPORTING MACROPHYTES

- Have existing macrophyte stands been assessed, particularly in spring?
- Does the waterbody have large shallow areas that could potentially harbour major macrophyte stands?
- Are concentrations of total phosphorus (TP) usually below 50–100 µg/L, so that there is a chance of sufficient P reduction by P binding in macrophyte stands?
- Are reasons for poor macrophyte development understood or at least working hypotheses available that merit testing?
- Would macrophytes freshly introduced into the waterbody have a fair chance of developing, or is the water too turbid or recreational pressure on shorelines too high?
- Are shorelines protected from wave erosion and plant shoots likely to be sufficiently protected from bird grazing?

Operational monitoring of enhancing competition against cyanobacteria by introducing or supporting macrophytes

Operational monitoring of macrophyte management measures can be relatively straightforward by regular reviews of measures for introducing macrophytes, by observing macrophyte growth and/or by inspecting the integrity of structures to protect, for example, reed belts.

Validation of enhancing competition against cyanobacteria by introducing or supporting macrophytes

Whether macrophytes successfully compete against cyanobacteria can be readily validated by observation and recording the sediment area covered by macrophytes as well as visual indicators of cyanobacterial blooms, that is, a decrease in turbidity and scum occurrence. Additionally, a limnological monitoring programme involving reductions in concentrations of TP and cyanobacterial biomass as well as an increase in water transparency is useful to obtain a better in-depth understanding of the mechanisms determining success or failure.

Surveying the increase of macrophyte stands can be conducted on an annual basis and thus requires low frequency monitoring. In contrast, as for validating biomanipulation through managing the fish stock (section 8.7.1), monitoring of transparency, cyanobacterial biomass and concentrations of total phosphorus needs to be conducted sufficiently often to be able to distinguish between success of the measure and natural fluctuations or those due to other changes in the system – that is, at intervals of weekly to monthly. In both cases, several years of observation will be necessary to assess success or failure.

8.8 INTRODUCING BARLEY STRAW

The effect of rotting or decomposing barley straw in reducing the growth of algae was first demonstrated by Welch et al. (1990) who showed the reduction of the growth of green filamentous algae, *Cladophora glomerata*, in a canal. Since then, the use of decomposing barley straw for the control of algae and cyanobacteria has been the subject of considerable interest and investigation, with numerous publications showing some effect. Numerous other studies have not supported barley straw's algicidal activity. It has been even suggested that anaerobic decomposition of straw produces chemicals which actually stimulate the growth of algae, because the algae can use them as a source of carbon (Martin & Ridge, 1999; Terlizzi et al., 2002).

Mechanisms postulated include the production of antibiotics by fungal flora or the release of phenolic compounds such as ferulic acid and *p*-coumaric acid from the decomposition of straw cell walls (Newman & Barrett, 1993); the possibility that the straw acts as a carbon source for carbon limited microbial growth which then uses available phosphorus preventing its use by cyanobacteria (Anhorn, 2005); compounds that chelate with essential metals, thus making them unavailable (Geiger et al., 2005); or antialgal activity of fungi present in the straw (Pillinger et al., 1992). Other studies have indicated a large number and range of compounds, including phenolic, quinone compounds and flavolignans, extracted from straw to have a significant cyanocidal toxicity (Murray et al., 2010; Xiao et al., 2014). One of the first theories proposed for cyanocidal action of barley straw was the generation of hydrogen peroxide during photooxidation of constituents in the straw (Everall & Lees, 1997). Iredale et al. (2012) confirmed that hydrogen peroxide does form during decomposition, but many variables may determine its cyanocidal effects, including the cyanobacterial strains treated, the amount of UV-supplemented visible light, the temperature and the form of straw used and its state of decomposition. The activity of barley straw is usually described as being algaestatic (prevents new growth of algae) rather than algicidal (kills already existing algae).

The contradictory findings, the inconclusive understanding of the mechanisms where effects have been observed and the unknown identity of the potentially phytotoxic compound(s) in rotting barley straw indicate that this technique is still too poorly understood to recommend it for widespread use as a measure to control cyanobacteria, particularly for potable water supply.

8.9 CHEMICAL CONTROL OF CYANOBACTERIA

Algicides have been used rather widely in some regions to rapidly eliminate cyanobacterial bloom outbreaks, to avoid cyanotoxins as well as off-flavour problems caused by cyanobacteria. Algicide treatment has been proposed as being more cost-effective than toxin and/or off-flavour removal in drinking-water treatment, because an extended period of persistent blooms greatly enhances the need for additional treatment for removal of dissolved organic carbon, off-flavours and toxins. Environmental concerns have been raised, also because the most commonly used algicide, copper sulphate, has broad ecological impact and the copper may accumulate in the lake sediments (Prepas & Murphy, 1988). Copper and other heavy metals differ from some other toxic contaminants in that they are not biodegradable, and once they have entered the environment, their potential toxicity is controlled largely by their speciation or physicochemical form (Florence, 1982; Mastin & Rodgers, 2000).

These concerns tend to limit algicide treatment to special circumstances for reservoirs relevant for water supply, as an emergency measure applied at one point in time, particularly where alternative drinking-water sources are not available and preventive measures are not feasible or not yet effective. As a long-term solution, algicide treatment is unsatisfactory, and wherever possible, control measures which address the factors that promote cyanobacteria are preferable. In many countries, national or local environmental regulations prohibit or limit the use of algicides because of their adverse environmental impact. Legal requirements therefore need to be clarified prior to considering the use of algicides.

A major limitation of any agent which disrupts cyanobacterial cells is the release of toxins and of taste and odour compounds from the cells to which they are normally confined (Lahti et al., 1996). Toxin release upon treatment can be quite rapid, and different studies have shown it to occur within 3–24 h (Kenefick et al., 1993; Jones & Orr, 1994). These dissolved toxins will then disperse and be diluted throughout the waterbody, and they may not be removed by conventional flocculation and filtration procedures in drinking-water treatment. Installation of additional treatment for removing dissolved cyanotoxins may be costly. The risk of treating dense blooms with algicides was demonstrated in an incident which occurred on tropical Palm Island, Australia, where members of the community became ill with hepato-enteritis following treatment of a bloom in the water supply reservoir with copper sulphate (Bourke et al., 1983) (see Box 5.3). If algicide treatment is used, this is therefore better done early at the beginning of bloom development to prevent further cyanobacterial growth (Cameron, 1989), thus limiting the amount of toxin that can be released.

Algicides, like all management techniques, must be applied correctly to work effectively. Application at the early stages of bloom development when cell densities are low not only reduces the potential for liberation of intracellular toxin but also will enhance the effectiveness of treatment because cyanobacterial cells can form a major part of the copper demand along with other organic matter in natural water.

If algicides are used to control toxic cyanobacteria, the reservoir should be isolated from the drinking-water supply for a sufficiently long time period to allow the toxins and odours to degrade. Unfortunately, very little data exist on the withholding period in relation to toxin loss specifically after algicide treatment. However, as discussed in general for biodegradation of microcystins in section 2.1, these cyanotoxins are likely to be degraded within a few days if conditions are favourable. However, this needs to be checked on site, as an early study of Jones et al. (1994) showed degradation after algicide treatment took more than 14 days. Furthermore, cylindrospermopsin appears to be particularly poorly biodegradable (section 2.2).

In some cases, algicide treatment may be unsuccessful or only partially successful. This can be due to inadequate dispersal and contact with the target organisms, variable sensitivity of cyanobacteria, and reduced efficacy due to complexation of the copper (Burch et al., 1998). The form of copper compound which is most bioavailable and toxic to aquatic organisms is influenced by pH, organic carbon, alkalinity, ionic strength or conductivity (McKnight et al., 1983; Mastin & Rodgers, 2000).

8.9.1 Copper sulphate

Records of the use of copper sulphate date from 1890 in Europe (Sawyer & Hazzard, 1962), from 1904 in the USA (Moore & Kellerman, 1905) and at least back to the mid-1940s in Australia (Burch et al., 1998). Copper sulphate has been regarded as the algicide of choice because it is economic, effective, relatively safe and easy to apply. It is also considered to be of limited significance to human health at the doses commonly used (WHO, 2017) and has been considered not to cause extensive environmental damage (Elder & Horne, 1978; McKnight et al., 1983). The latter point has been an issue of debate for some time (see Mackenthun & Cooley, 1952) because of the abovementioned tendency of copper to accumulate in lake sediments (Hanson & Stefan, 1984). In some cases, it appears not to be remobilised and is bound permanently to the sediments (Elder & Horne, 1978). However, in a study of 10 drinking-water dugouts (small reservoirs) in Canada, sediment copper (previously accumulated from copper sulphate treatments) was released back into the open water under conditions of low dissolved oxygen in the hypolimnion in summer (Prepas & Murphy, 1988). It has also been suggested that sediment-bound copper could have an impact on the benthic macroinvertebrate community (Hanson & Stefan, 1984).

The effect of copper as an algicide is generally short-lived due to rapid loss via precipitation as insoluble salts and hydroxylates, depending upon the chemical conditions in the receiving water (Cooke et al., 1993; Fan et al., 2013). Copper sulphate treatment has been shown to cause short-term changes in phytoplankton abundance and species succession (Effler et al., 1980; McKnight, 1981). An additional consideration is that the chronic application of copper algicides may encourage cyanobacteria to become resistant to it, and it thus may cause shifts in the composition of the phytoplankton community with prevalence of copper-resistant green algae (Qian et al., 2010; Rouco et al., 2014). Fish kill has also been reported following copper sulphate treatment, although it is not clear whether this was a result of copper toxicity or oxygen depletion caused by the decaying bloom (Hanson & Stefan, 1984).

8.9.2 Copper chelates

Chelated copper algicides were developed to overcome the problems of the complexation and precipitation loss of toxic copper and thus reduced effectiveness of copper sulphate treatment in hard alkaline water. The different copper formulations include copper oxychloride, organocopper complexes like copper ethanolamine complex or copper citrate used in commercial preparations (Murray-Gulde et al., 2002; Calomeni et al., 2014). The effectiveness of these formulations depends upon a range of factors, including both the target organism and the effect of chemical conditions in the water on bioavailability and toxicity (Mastin & Rodgers, 2000; Calomeni et al., 2014).

8.9.3 Hydrogen Peroxide

A range of both stabilised compounds and liquid hydrogen peroxide (H_2O_2) have been developed and used more recently in a desire to overcome the environmental issues associated with copper algicides (Matthijs et al., 2016). Hydrogen peroxide is a strong candidate due to its apparently selective toxicity to cyanobacteria combined with the rapid degradation of the chemical to water and oxygen with no residual. Matthijs et al. (2016) describe it as potentially a more specific and sustainable "cyanocide" compared to copper chemicals, herbicides, natural compounds from plant extracts (e.g., barley straw) and other organic chemicals compared for cyanobacterial control.

The effects of hydrogen peroxide on photosynthesis are reported to be relatively rapid – that is, within 3–5 h (Matthijs et al., 2016), and loss of membrane integrity with some evidence of release of cyanotoxins follows over several days after treatment (Matthijs et al., 2012; Lürling et al., 2014). Cyanobacteria appear to be more sensitive to hydrogen peroxide than eukaryotic algae as they do not possess the defensive enzyme and substrate mechanisms of green algae to convert reactive oxygen species and to neutralise their toxic effects upon photosynthesis and subsequent growth of cells (Drábková et al., 2007a; Drábková et al., 2007b). This mechanism and mode of action as a specific cyanocide has been tested successfully in both the laboratory (Drábková et al., 2007a; Weenink et al., 2015) and in the field with natural phytoplankton populations (Matthijs et al., 2012). Hydrogen peroxide has been applied in a range of lakes, and the effective dose rates for control of cyanobacteria vary widely from <5 mg/L up to 100 mg/L. They depend upon the target strain type and its density and also on the presence of eukaryotic algae. While in the Netherlands the dose is restricted to <5 mg/L to avoid killing non-target species, concentrations in the higher end of the range may be required for colony-forming taxa such as *Microcystis aeruginosa*, where mucilage may protect cells against the oxidising effects of H_2O_2 (Lürling & Tolman, 2014b). At the time of publication of this document, adverse and potentially toxic effects of hydrogen peroxide on nontarget organisms in natural treatment situations have not been widely reported. Nonetheless, monitoring the impacts on non-target organisms is advised on a case-by-case basis, particularly when utilising dose rates approaching up to 100 mg/L.

Addressing the questions in Checklist 8.16 will help decide whether to apply algicide treatment. If this is intended, choosing the best method for the respective waterbody will require expertise particularly in aquatic ecology and limnology and a good understanding of the waterbody's chemical condition and biota.

CHECKLIST 8.16: ASSESSING THE BENEFITS AND THE PROSPECT OF SUCCESS OF CONTROL OF CYANOBACTERIA THROUGH ALGICIDE TREATMENT

- Have other options been sufficiently assessed, and is algicide treatment the only feasible short-term option?
- Which cyanotoxins are expected to be the predominant problem – cell-bound microcystins and neurotoxins or rather cylindrospermopsin with a potentially high extracellular fraction?
- How likely is a low cyanobacterial population to grow into a major bloom? Are data available from the waterbody for previous years – is this a typical pattern for the waterbody? Are nutrient concentrations

available to infer the carrying capacity for biomass (see section 4.3.2), and is this potentially limiting, so that major blooms are unlikely?
- Will the impact on cyanobacteria and on released toxins be sufficiently monitored to prevent exposure to higher concentrations of dissolved cyanotoxins? How will the cyanobacterial response be monitored?
- If the waterbody is used as drinking-water supply, can it be isolated from the supply until the lysed algal biomass, toxins and possibly occurring off-flavour substances have degraded? How long is this expected to take and what criterion will be used as basis for the decision to take the waterbody back into the drinking-water supply (e.g., concentrations of dissolved organic carbon (DOC) or cyanotoxin)?
- If the waterbody remains online for drinking-water supply, is treatment available to remove dissolved toxins, and how will it respond to the challenge of the DOC pulse expected from algicide treatment?
- Which algicide will be used (see below), what dose is necessary, how will it be applied and at which time interval is repeated treatment likely to be necessary?
- Is the chosen algicide likely to be sufficiently effective in the specific water (e.g., copper may be less effective in waters with high dissolved carbonate or at alkaline pH) at the dosing regime planned?
- Is algicide treatment compliant with local regulations?
- Do further stakeholders need to be involved (e.g., environmental and health authorities, water boards)?

Operational monitoring of chemical control of cyanobacteria

Operational monitoring will check whether chemicals are dosed as planned, whether they reach the target concentration in the waterbody and whether measures to protect humans (like isolating the reservoir from a drinking-water supply until dissolved toxins have been degraded) are in place.
Validation of chemical control of cyanobacteria
Validation of plans for chemical control best begins with reviewing whether these measures are necessary or whether other options are available that should be tested first. It will then review whether the intended chemical cyanobacterial control measure is likely to be effective and feasible for the specific waterbody and its specific blooms, whether potentially detrimental impacts on other biota can be accepted and whether human exposure to dissolved toxins can effectively be prevented. This typically needs expertise in phytoplankton ecology and waterbody management, but also an understanding of water chemistry. Post-treatment monitoring should include the

development of the biota to both ensure the efficacy of treatment and better understand the ecological effects. In addition, analyses of dissolved cyanotoxins should be included where drinking-water is being abstracted.

In the interest of long-term sustainability, it is advisable to initiate the implementation of other controls which are more sustainable than algicide treatment, even if these measures are effective only in the longer term – that is, the reduction of nutrient loads and/or internal measures as discussed above.

8.10 ULTRASONICATION FOR CONTROL OF CYANOBACTERIA

The use of ultrasound for the control of cyanobacteria in both freshwater and wastewater storages has long been proposed as a non-chemical technique and thus as an attractive prospect. It has also become popular across a range of countries since about the turn of the millennium, due to the availability and promotion of relatively low-cost commercial ultrasound systems for cyanobacterial control. Ultrasound is sound energy with frequencies that are higher than can be detected by the human ear (i.e., approximately>20 kHz). It has been reported to reduce cyanobacterial growth through a range of mechanisms that affect the structural integrity of cells (membrane damage or gas vesicle collapse and destruction) or impact cell physiology (photosynthetic activity and growth reduction; Wu et al., 2011; Rajasekhar et al., 2012; Park et al., 2017).

A reduction in cyanobacterial growth through the application of sound energy to cells and colonies is likely to be due to a combination of many effects and will depend upon the target organism, the power or energy applied, the frequency and the design of application of ultrasound. Key operational parameters in the application of ultrasound for algal control are frequency, intensity (or power) and duration of exposure (Park et al., 2017), and the more successful ones among the studies available have been able to explore and optimise these parameters.

Although research results to support the effectiveness and mechanism of action of ultrasound in aquatic systems have been accumulating, objective assessments and comparisons of findings are not straightforward. The studies and reviews of the mechanisms and effectiveness of ultrasound for cyanobacterial control describe a wide range of test conditions applied in the laboratory or field, both with cultures and with natural material (Pavagadhi et al., 2013; Lürling & Tolman, 2014a; Lürling et al., 2016; Park et al., 2017). Published studies have mostly used custom-made or experimental equipment with a range of frequencies and power intensity which are different to commercially available units. Most of the studies that have shown clear effects upon cyanobacteria have used high-power purpose-built experimental devices while the commercial devices sold for algal control are relatively low power (Lürling et al., 2016). The low-energy ultrasound devices are all very similar in design and application; however, they use a range of different ultrasound frequencies in the lower range and various configurations of pumps to circulate water past ultrasonic transducers. The range of conflicting reports from the literature on the effectiveness of ultrasound in natural waterbodies with cyanobacterial contamination is most likely due to difference in both design and configuration of the sonication equipment in relation to the size of the waterbodies. In addition, designing studies to account for confounding ecological effects that influence growth is challenging.

Several examples from the published studies highlight the conflicting results: field trials with ultrasound devices conducted in the Netherlands in 2007 concluded that ultrasound was not effective in reducing cyanobacteria (Kardinaal et al., 2008). Similarly, in the UK, Purcell et al. (2013) tested two models of commercial ultrasound devices with inconsistent results. By contrast, an evaluation of four commercial ultrasound devices in a pair of reservoirs in New Jersey (USA) found positive results (Schneider et al., 2015). A further comprehensive study in the Netherlands based upon both experiments and a critical literature review concluded that the commercially available ultrasound transducers tested would not control cyanobacteria *in situ* (Lürling et al., 2014).

In conclusion, the application of ultrasound for cyanobacterial control remains under question because of the limited number of both validated field and pilot tests and the lack of information on the feasibility of both commercial and experimental devices for use in larger waterbodies. In particular, a set of criteria for successful application appears to be lacking.

8.11 MEASURES TO CONTROL THE PROLIFERATION OF BENTHIC CYANOBACTERIA

Mitigation strategies to remove or reduce toxic benthic cyanobacteria are very limited because for the large majority of situations, there is little knowledge about the environmental variables that regulate their proliferation. There are very few reports on the management of benthic cyanobacteria, and anecdotally, they are regarded as more difficult to control or remove (Liu et al., 2019). As discussed in section 4.2.2, benthic cyanobacteria may grow at low nutrient concentrations and be less sensitive to nutrient reduction than phytoplankton (Bonilla et al., 2005), so that nutrient management is less likely to limit their growth. Where nutrients and fine sediment have been implicated as a cause (e.g., McAllister et al., 2016), actions such as planting riparian margins, regulation of fertiliser application in the catchment and preventing stock access to river/streams edges may ultimately reduce proliferation, but this has not been demonstrated.

In flow-regulated rivers and streams, flushing flows have been used to remove other nuisance periphyton growth, but this is not known to have been successfully applied to cyanobacteria, probably because toxic benthic cyanobacteria have not been reported from rivers or streams where this would be feasible, particularly where they occur naturally even in pristine streams and oligotrophic lakes (see section 4.2.2). Techniques that are recommended for the management of odour-producing benthic cyanobacteria are discussed by Liu et al. (2019), including the application of algicides, physical removal by disturbance treatments and reducing water levels to dry out or desiccate the benthic cyanobacteria. These techniques all have limitations, and the more destructive techniques (algicides and physical removal) may have potential adverse impacts such as lysing cells or dislodging mats which may then release toxins.

Mitigation of the growth of toxic benthic cyanobacteria may neither be practical nor necessary for the target of protecting human health, as they are not known to cause health-relevant toxin concentrations in the surrounding water. Therefore, the most feasible management option may be to provide information on avoiding direct contact with dislodged and floating mats (including preventing pets from coming into contact; see Section 5.2 and Chapter 15).

REFERENCES

Akhurst D, Jones GB, McConchie DM (2004). The application of sediment capping agents on phosphorus speciation and mobility in a sub-tropical dunal lake. Mar Freshwater Res. 55:715–725.

Anhorn R (2005). A study of the water quality of 145 Metropolitan area lakes. Saint Paul (MN): Metropolitan Council:451 pp.

Antenucci JP, Alexander R, Romero J, Imberger J (2003). Management strategies for a eutrophic water supply reservoir - San Roque, Argentina. Wat Sci Technol. 47:149–155.

Baker PD, Brookes JD, Burch MD, Maier HR, Ganf GG (2000). Advection, growth and nutrient status of phytoplankton populations in the lower River Murray, South Australia. Reg Riv Res Man. 16:327–344.

Baldwin DS (1996). The phosphorus composition of a diverse series of Australian sediments. Hydrobiologia. 335:63–73.

Baldwin DS, Mitchell A, Rees G (2000). The effects of *in situ* drying on sediment–phosphate interactions in sediments from an old wetland. Hydrobiologia. 431:3–12.

Becker A, Herschel A, Wilhelm C (2006). Biological effects of incomplete destratification of hypertrophic freshwater reservoir. Hydrobiologia. 559:85–100.

Bertone E, Burford MA, Hamilton DP (2018). Fluorescence probes for real-time remote cyanobacteria monitoring: a review of challenges and opportunities. Water Res. 141:152–162.

Beutel MW, Horne AJ (1999). A review of the effects of hypolimnetic oxygenation on lake and reservoir water quality. Lake Reserv Manage. 15:285–297.

Bonilla S, Villeneuve V, Vincent WF (2005). Benthic and planktonic algal communities in a high arctic lake: pigment structure and contrasting responses to nutrient enrichment. J Phycol. 41:1120–1130.

Bormans M, Maršálek B, Jančula D (2016). Controlling internal phosphorus loading in lakes by physical methods to reduce cyanobacterial blooms: a review. Aquat Ecol. 50:407–422.

Bormans M, Webster IT (1997). A mixing criterion for turbid rivers. Environ Modell Softw. 12:329–333.

Boscaini A, Brescancin F, Cerasino L, Fedrigotti C, Fano EA, Salmaso N (2017). Vertical and horizontal distribution of the microcystin producer *Planktothrix rubescens* (Cyanobacteria) in a small perialpine reservoir. Adv Oceanogr Limnol. 8:208–221.

Bourke A, Hawes R, Neilson A, Stallman N (1983). An outbreak of hepato-enteritis (the Palm Island mystery disease) possibly caused by algal intoxication. Toxicon. 21:45–48.

Burch M, Velzeboer R, Chow C, Stevens H, Bee C, House J (1998). Evaluation of copper algaecides for the control of algae and cyanobacteria. Urban Water Research Association of Australia, Melbourne. Research Report No. 130. pp.

Burford MA, O'Donohue MJ (2006). A comparison of phytoplankton community assemblages in artificially and naturally mixed subtropical water reservoirs. Freshwater Biol. 51:973–982.

Calọmeni A, Rodgers JH, Kinley CM (2014). Responses of *Planktothrix agardhii* and *Pseudokirchneriella subcapitata* to Copper Sulfate ($CuSO_4 \cdot 5H_2O$) and a Chelated Copper Compound (Cutrine®-Ultra). Water Air Soil Poll. 225:2231.

Cameron CD (1989). Is this a way to run a reservoir? In: Practical lake management for water quality control. Denver (CO): American Water Works Association:63–83.

Caraco N, Cole J, Strayer D (2006). Top down control from the bottom: regulation of eutrophication in a large river by benthic grazing. Limnol Oceanogr. 51:664–670.

Carpenter SR, Kitchell JF (1996). The trophic cascade in lakes. Cambridge, UK: Cambridge University Press:400 pp.

Chorus I, Schauser I (2011). Oligotrophication of Lake Tegel and Schlachtensee, Berlin-Analysis of system components, causalities and response thresholds compared to responses of other waterbodies. Dessau: Umweltbundesamt:157 pp.

Chorus I, Spijkerman E (2020). What Colin Reynolds could tell us about nutrient limitation, N: P ratios and eutrophication control. Hydrobiologia. 1–17.

Cooke GD, Welch EB, Peterson SA, Newroth PR (1993). Artificial circulation. In: Cooke GD, Welch EB, Peterson SA et al., editors: Restoration and management of lakes and reservoirs. Boca Raton (FL): CRC press:419–449.

Cooke GD, Welch EB, Peterson S, Nichols SA (2005). Restoration and management of lakes and reservoirs. Boca Raton (FL): CRC press:616 pp.

Cowell BC, Dawes CJ, Gardiner WE, Scheda SM (1987). The influence of whole lake aeration on the limnology of a hypereutrophic lake in central Florida. Hydrobiologia. 148:3–24.

DeMelo R, France R, McQueen DJ (1992). Biomanipulation: hit or myth? Limnol Oceanogr. 37:192–207.

Dokulil MT (1994). Environmental control of phytoplankton productivity in turbulent turbid systems. In: Descy JP, Reynolds CS, Padisák J, editors: Phytoplankton in turbid environments: rivers and shallow lakes. Dordrecht: Kluwer Academic Publishers:65–72.

Douglas G, Hamilton D, Robb M, Pan G, Spears B, Lurling M (2016). Guiding principles for the development and application of solid-phase phosphorus adsorbents for freshwater ecosystems. Aquat Ecol. 50:385–405.
Drábková M, Admiraal W, Maršálek B (2007a). Combined exposure to hydrogen peroxide and light selective effects on cyanobacteria, green algae, and diatoms. Environ Sci Technol. 41:309–314.
Drábková M, Matthijs H, Admiraal W, Maršálek B (2007b). Selective effects of H_2O_2 on cyanobacterial photosynthesis. Photosynthetica. 45:363–369.
Effler S, Litten S, Field S, Tong-Ngork T, Hale F, Meyer M et al. (1980). Whole lake responses to low level copper sulfate treatment. Water Res. 14:1489–1499.
Elder JF, Horne AJ (1978). Copper cycles and $CuSO_4$ algicidal capacity in two California lakes. Environ Manage. 2:17–30.
Everall N, Lees D (1997). The identification and significance of chemicals released from decomposing barley straw during reservoir algal control. Water Res. 31:614–620.
Fan J, Ho L, Hobson P, Brookes J (2013). Evaluating the effectiveness of copper sulphate, chlorine, potassium permanganate, hydrogen peroxide and ozone on cyanobacterial cell integrity. Water Res. 47:5153–5164.
Florence T (1982). The speciation of trace elements in waters. Talanta. 29:345–364.
Gächter R (1976). Die Tiefenwasserableitung, ein Weg zur Sanierung von Seen. Schweizerische Zeitschrift für Hydrologie. 38:1–28.
Gächter R, Müller B (2003). Why the phosphorus retention of lakes does not necessarily depend on the oxygen supply to their sediment surface. Limnol Oceanogr. 48:929–933.
Gächter R, Wehrli B (1998). Ten years of artificial mixing and oxygenation: no effect on the internal phosphorus loading of two eutrophic lakes. Environ Sci Technol. 32:3659–3665.
Gantzer PA, Bryant LD, Little JC (2009). Controlling soluble iron and manganese in a water-supply reservoir using hypolimnetic oxygenation. Water Res. 43:1285–1294.
Gardner WS, Nalepa TF, Quigley MA, Malczyk JM (1981). Release of phosphorus by certain benthic invertebrates. Can J Fish Aquat Sci. 38:978–981.
Geiger N, Gearheart R, Henry E, Rueter J, Pan Y (2005). Preliminary research on *Aphanizomenon flos-aquae* at Upper Klamath Lake, Oregon. Klamath Falls (OR): Klamath Falls Fish and Wildlife Office:158 pp.
Hamilton DP, Salmaso N, Paerl HW (2016). Mitigating harmful cyanobacterial blooms: strategies for control of nitrogen and phosphorus loads. Aquat Ecol. 50:351–366.
Hanson MJ, Stefan HG (1984). Side effects of 58 years of copper sulfate treatment of the Fairmont Lakes, Minnesota. J Am Water Resour Ass. 20:889–900.
Hart BT, Roberts S, James R, O'Donohue M, Taylor J, Donnert D et al. (2003). Active barriers to reduce phosphorus release from sediments: effectiveness of three forms of $CaCO_3$. Aust J Chem. 56:207–217.
Hawkins P, Griffiths D (1993). Artificial destratification of a small tropical reservoir: effects upon the phytoplankton. Hydrobiologia. 254:169–181.
Heo W-M, Kim B (2004). The effect of artificial destratification on phytoplankton in a reservoir. Hydrobiologia. 524:229–239.
Hölker F, Vanni MJ, Kuiper JJ, Meile C, Grossart H-P, Stief P et al. (2015). Tube-dwelling invertebrates: tiny ecosystem engineers have large effects in lake ecosystems. Ecol Monogr. 85:333–351.

Hrbáček J, Desortova B, Popovský J (1978). Influence of the fishstock on the phosphorus—chlorophyll ratio. Verh Int Ver Limnol. 20:1624–1628.
Huisman J, Sharples J, Stroom JM, Visser PM, Kardinaal WEA, Verspagen JM et al. (2004). Changes in turbulent mixing shift competition for light between phytoplankton species. Ecology. 85:2960–2970.
Hupfer M, Lewandowski J (2008). Oxygen controls the phosphorus release from lake sediments – a long-lasting paradigm in limnology. Int Rev Hydrobiol. 93:415–432.
Ibelings BW, Portielje R, Lammens EH, Noordhuis R, van den Berg MS, Joosse W et al. (2007). Resilience of alternative stable states during the recovery of shallow lakes from eutrophication: Lake Veluwe as a case study. Ecosystems. 10:4–16.
Iredale RS, McDonald AT, Adams DG (2012). A series of experiments aimed at clarifying the mode of action of barley straw in cyanobacterial growth control. Water Res. 46:6095–6103.
Ismail R, Kassim M, Inman M, Baharim N, Azman S (2002). Removal of iron and manganese by artificial destratification in a tropical climate (Upper Layang Reservoir, Malaysia). Wat Sci Technol. 46:179–183.
Jeppesen E, Meerhoff M, Jacobsen B, Hansen R, Søndergaard M, Jensen J et al. (2007a). Restoration of shallow lakes by nutrient control and biomanipulation—the successful strategy varies with lake size and climate. Hydrobiologia. 581:269–285.
Jeppesen E, Søndergaard M, Meerhoff M, Lauridsen TL, Jensen JP (2007b). Shallow lake restoration by nutrient loading reduction—some recent findings and challenges ahead. Hydrobiologia. 584:239–252.
Jones GJ, Bourne DG, Blakeley RL, Doelle H (1994). Degradation of cyanobacterial hepatotoxin microcystin by aquatic bacteria. Nat Toxins. 2:228–235.
Jones GJ, Orr PT (1994). Release and degradation of microcystin following algicide treatment of a *Microcystis aeruginosa* bloom in a recreational lake, as determined by HPLC and protein phosphatase inhibition assay. Water Res. 28:871–876.
Kardinaal E, De Haan M, Ruiter H (2008). Maatregelen ter voorkoming blauwalgen werken onvoldoende. H_2O. 41:4–7.
Kenefick SL, Hrudey SE, Peterson HG, Prepas EE (1993). Toxin release from *Microcystis aeruginosa* after chemical treatment. Wat Sci Technol. 27:433–440.
Kitchell JF (1992). Food web management: a case study of Lake Mendota. New York: Springer:556 pp.
Kolzau S, Wiedner C, Rücker J, Köhler J, Köhler A, Dolman AM (2014). Seasonal patterns of nitrogen and phosphorus limitation in four German lakes and the predictability of limitation status from ambient nutrient concentrations. PLoS One. 9:e96065.
Lahti K, Kilponen J, Kivimaeki A (1996). Removal of cyanobacteria and their hepatotoxins from raw water in soil and sediment columns. In: Kivimki A-L, Suokko T, editors: Artificial recharge of groundwater NHP/Report No 38. Helsinki:187–195.
Lewis DM (2004). Surface mixers for destratification and management of *Anabaena circinalis*. School of Civil and Environmental Engineering, University of Adelaide:249 pp.
Liu Z, Lin T-F, Burch M (2019). Management of T&O in source water. In: Lin T-F, Watson S, Dietrich AM et al., editors: Taste and odour in source and drinking water: causes, controls, and consequences. London: International Water Association:304.
Lürling M, Meng D, Faassen EJ (2014). Effects of hydrogen peroxide and ultrasound on biomass reduction and toxin release in the cyanobacterium, *Microcystis aeruginosa*. Toxins. 6:3260–3280.

Lürling M, Tolman Y (2014a). Beating the blues: is there any music in fighting cyanobacteria with ultrasound? Water Res. 66:361–373.

Lürling M, Tolman Y (2014b). Effects of commercially available ultrasound on the zooplankton grazer Daphnia and consequent water greening in laboratory experiments. Water. 6:3247–3263.

Lürling M, Waajen G, de Senerpont Domis LN (2016). Evaluation of several end-of-pipe measures proposed to control cyanobacteria. Aquat Ecol. 50:499–519.

Mackenthun K, Cooley H (1952). The biological effect of copper sulphate treatment on lake ecology. Trans Wisconsin Acad Sci, Arts Lett. 41:177–187.

Maier HR, Kingston GB, Clark T, Frazer A, Sanderson A (2004). Risk-based approach for assessing the effectiveness of flow management in controlling cyanobacterial blooms in rivers. Riv Res Appl. 20:459–471.

Martin D, Ridge I (1999). The relative sensitivity of algae to decomposing barley straw. J Appl Phycol. 11:285–291.

Mastin B, Rodgers J (2000). Toxicity and bioavailability of copper herbicides (Clearigate, Cutrine-Plus, and copper sulfate) to freshwater animals. Arch Environ Contam Toxicol. 39:445–451.

Matthijs HC, Jančula D, Visser PM, Maršálek B (2016). Existing and emerging cyanocidal compounds: new perspectives for cyanobacterial bloom mitigation. Aquat Ecol. 50:443–460.

Matthijs HC, Visser PM, Reeze B, Meeuse J, Slot PC, Wijn G et al. (2012). Selective suppression of harmful cyanobacteria in an entire lake with hydrogen peroxide. Water Res. 46:1460–1472.

McAllister TG, Wood SA, Hawes I (2016). The rise of toxic benthic *Phormidium* proliferations: a review of their taxonomy, distribution, toxin content and factors regulating prevalence and increased severity. Harmful Algae. 55:282–294.

McAuliffe T, Rosich RS (1989). Review of artificial destratification of water storages in Australia. Water Research Association of Australia, Sydney. Research Report No. 9. 233 pp.

McKnight D (1981). Chemical and biological processes controlling the response of a freshwater ecosystem to copper stress: a field study of the $CuS0_4$ treatment of Mill Pond Reservoir, Burlington, Massachusetts. Limnol Oceanogr. 26:518–531.

McKnight DM, Chisholm SW, Harleman DR (1983). $CuSO_4$ treatment of nuisance algal blooms in drinking water reservoirs. Environ Manage. 7:311–320.

Mitchell A, Baldwin DS (1998). Effects of desiccation/oxidation on the potential for bacterially mediated P release from sediments. Limnol Oceanogr. 43:481–487.

Moore BC, Cross BK, Clegg EM, Lanouette BP, Skinner M, Preece E et al. (2014). Hypolimnetic oxygenation in Twin Lakes, WA. Part I: distribution and movement of trout. Lake Reserv Manage. 30:226–239.

Moore GT, Kellerman KF (1905). Copper as an algicide and disinfectant in water supplies. US Government Printing Office. Bulletin Nr. 76. 55 pp.

Moss B, Kosten S, Meerhoff M, Battarbee RW, Jeppesen E, Mazzeo N et al. (2011). Allied attack: climate change and eutrophication. Inland Waters. 1:101–105.

Moss B, McGowan S, Carvalho L (1994). Determination of phytoplankton crops by top-down and bottom-up mechanisms in a group of English lakes, the West Midland meres. Limnol Oceanogr. 39:1020–1029.

Murphy T, Prepas EE (1990). Lime treatment of hardwater lakes to reduce eutrophication. Verh Int Ver Limnol. 24:327–334.
Murray D, Jefferson B, Jarvis P, Parsons S (2010). Inhibition of three algae species using chemicals released from barley straw. Environ Technol. 31:455–466.
Murray-Gulde C, Heatley J, Schwartzman A, Rodgers Jr J (2002). Algicidal effectiveness of clearigate, cutrine-plus, and copper sulfate and margins of safety associated with their use. Arch Environ Contam Toxicol. 43:19–27.
Newman JR, Barrett P (1993). Control of *Microcystis aeruginosa* by decomposing barley straw. J Aquat Plant Manage. 31:203.
Nürnberg GK (1997). Coping with water quality problems due to hypolimnetic anoxia in Central Ontario Lakes. Water Qual Res J Can. 32:391–405.
Nürnberg GK (2007). Lake responses to long-term hypolimnetic withdrawal treatments. Lake Reserv Manage. 23:388–409.
Nürnberg GK, LaZerte BD, Olding DD (2003). An artificially induced *Planktothrix rubescens* surface bloom in a small kettle lake in Southern Ontario compared to blooms world-wide. Lake Reserv Manage. 19:307–322.
Olszewski P (1961). Versuch einer Ableitung des hypolimnischen Wassers an einem See. Ergebnisse des ersten Versuchsjahres. Verh Int Ver Limnol. 18:1792–1797.
Orihel DM, Baulch HM, Casson NJ, North RL, Parsons CT, Seckar DC et al. (2017). Internal phosphorus loading in Canadian fresh waters: a critical review and data analysis. Can J Fish Aquat Sci. 74:2005–2029.
Park J, Church J, Son Y, Kim K-T, Lee WH (2017). Recent advances in ultrasonic treatment: challenges and field applications for controlling harmful algal blooms (HABs). Ultrason Sonochem. 38:326–334.
Pavagadhi S, Tang ALL, Sathishkumar M, Loh KP, Balasubramanian R (2013). Removal of microcystin-LR and microcystin-RR by graphene oxide: adsorption and kinetic experiments. Water Res. 47:4621–4629.
Pillinger J, Cooper J, Ridges I, Barrett P (1992). Barley straw as an inhibitor of algal growth III: the role of fungal decomposition. J Appl Phycol. 4:353–355.
Prepas E, Murphy T (1988). Sediment–water interactions in farm dugouts previously treated with copper sulfate. Lake Reserv Manage. 4:161–168.
Prepas E, Murphy T, Dinsmore W, Burke J, Chambers P, Reedyk S (1997). Lake management based on lime application and hypolimnetic oxygenation: the experience in eutrophic hardwater lakes in Alberta. Water Qual Res J Can. 32:273–294.
Psenner R, Pucsko R, Sager M (1984). Die Fraktionierung organischer und anorganischer Phosphorverbindungen von Sedimenten–Versuch einer Definition ökologisch wichtiger Fraktionen. Arch Hydrobiol Suppl. 70:111–155.
Purcell D, Parsons SA, Jefferson B, Holden S, Campbell A, Wallen A et al. (2013). Experiences of algal bloom control using green solutions barley straw and ultrasound, an industry perspective. Water Environ J. 27:148–156.
Qian H, Yu S, Sun Z, Xie X, Liu W, Fu Z (2010). Effects of copper sulfate, hydrogen peroxide and N-phenyl-2-naphthylamine on oxidative stress and the expression of genes involved photosynthesis and microcystin disposition in *Microcystis aeruginosa*. Aquat Toxicol. 99:405–412.
Rajasekhar P, Fan L, Nguyen T, Roddick FA (2012). A review of the use of sonication to control cyanobacterial blooms. Water Res. 46:4319–4329.

Raman RK, Arbuckle BR (1989). Long-term destratification in an Illinois Lake. J - Am Water Works Assoc. 81:66–71.

Reeders H, Boers P, Van der Molen D, Helmerhorst T (1998). Cyanobacterial dominance in the lakes Veluwemeer and Wolderwijd, The Netherlands. Wat Sci Technol. 37:85–92.

Reynolds C, Descy J-P, Padisák J (1994). Are phytoplankton dynamics in rivers so different from those in shallow lakes? Hydrobiologia. 289:1–7.

Reynolds C, Wiseman S, Godfrey B, Butterwick C (1983). Some effects of artificial mixing on the dynamics of phytoplankton populations in large limnetic enclosures. J Plankton Res. 5:203–234.

Reynolds CS (1997). Vegetation processes in the pelagic: a model for ecosystem theory. Oldendorf/Luhe: Ecology Institute:371 pp.

Robb M, Greenop B, Goss Z, Douglas G, Adeney J (2003). Application of Phoslock TM, an innovative phosphorus binding clay, to two Western Australian waterways: preliminary findings. Hydrobiologia. 494:237–243.

Ross G, Cloete T (2006). Phoslock® field trial at K'shani Lake Lodge, Hartbeespoort Dam. Pretoria: University of Pretoria:13 pp.

Rouco M, López-Rodas V, González R, Huertas IE, García-Sánchez MJ, Flores-Moya A et al. (2014). The limit of the genetic adaptation to copper in freshwater phytoplankton. Oecologia. 175:1179–1188.

Salmaso N, Boscaini A, Capelli C, Cerasino L (2018). Ongoing ecological shifts in a large lake are driven by climate change and eutrophication: evidences from a three-decade study in Lake Garda. Hydrobiologia. 824:177–195.

Sas H (1989). Lake restoration by nutrient control: expectations, experiences, extrapolations. Sankt Augustin: Academia Verlag Richardz:520 pp.

Sawyer CN, Hazzard AS (1962). Causes, effects, and control of aquatic growths. J Water Pollut Control Fed. 34: 279–290.

Schauser I, Chorus I (2007). Assessment of internal and external lake restoration measures for two Berlin lakes. Lake Reserv Manage. 23:366–376.

Schauser I, Chorus I (2009). Water and phosphorus mass balance of Lake Tegel and Schlachtensee–A modelling approach. Water Res. 43:1788–1800.

Schauser I, Chorus I, Lewandowski J (2006). Effects of nitrate on phosphorus release: comparison of two Berlin lakes. Acta Hydrochim Hydrobiol. 34:325–332.

Scheffer M, Carpenter SR (2003). Catastrophic regime shifts in ecosystems: linking theory to observation. Trends Ecol Evol. 18:648–656.

Scheffer M, Hosper SH, Meijer ML, Moss B, Jeppesen E (1993). Alternative equilibria in shallow lakes. Trends Ecol Evol. 8:275–279.

Scheffer M, Rinaldi S, Gragnani A, Mur LR, van Nes EH (1997). On the dominance of filamentous cyanobacteria in shallow, turbid lakes. Ecology. 78:272–282.

Schladow SG (1993). Lake destratification by bubble-plume systems: design methodology. J Hydraul Eng. 119:350–368.

Schneider OD, Weinrich LA, Brezinski S (2015). Ultrasonic treatment of algae in a New Jersey reservoir. J - Am Water Works Assoc. 107:E533–E542.

Sherman B, Whittington J, Oliver R (2000). The impact of artificial destratification on water quality in Chaffey Reservoir. Arch Hydrobiol Spec Issues Advanc Limnol. 55:15–29.

Singleton VL, Little JC (2006). Designing hypolimnetic aeration and oxygenation systems– a review. Environ Sci Technol. 40:7512–7520.

Søndergaard M, Jensen JP, Jeppesen E (2003). Role of sediment and internal loading of phosphorus in shallow lakes. Hydrobiologia. 506:135–145.
Søndergaard M, Lauridsen TL, Johansson LS, Jeppesen E (2017). Nitrogen or phosphorus limitation in lakes and its impact on phytoplankton biomass and submerged macrophyte cover. Hydrobiologia 795:35–48.
Su H, Wu Y, Xia W, Yang L, Chen J, Han W et al. (2019). Stoichiometric mechanisms of regime shifts in freshwater ecosystem. Water Res. 149:302–310.
Terlizzi DE, Ferrier MD, Armbrester EA, Anlauf KA (2002). Inhibition of dinoflagellate growth by extracts of barley straw (*Hordeum vulgare*). J Appl Phycol. 14:275–280.
Triest L, Stiers I, Van Onsem S (2016). Biomanipulation as a nature-based solution to reduce cyanobacterial blooms. Aquat Ecol. 50:461–483.
Tsujimura S (2004). Water management of Lake Yogo targeting internal phosphorus loading. Lakes Reserv Res Manage. 9:171–179.
Upadhyay S, Bierlein KA, Little JC, Burch MD, Elam KP, Brookes JD (2013). Mixing potential of a surface-mounted solar-powered water mixer (SWM) for controlling cyanobacterial blooms. Ecol Eng. 61:245–250.
Visser PM, Ibelings BW, Bormans M, Huisman J (2016). Artificial mixing to control cyanobacterial blooms: a review. Aquat Ecol. 50:423–441.
Visser PM, Ibelings BW, van der Veer B, Koedood J, Mur LR (1996). Artificial mixing prevents nuisance blooms of the cyanobacterium *Microcystis aeruginosa* in Lake Nieuwe Meer, the Netherlands. Freshwater Biol. 36:435–450.
Weenink EF, Luimstra VM, Schuurmans JM, Van Herk MJ, Visser PM, Matthijs HC (2015). Combatting cyanobacteria with hydrogen peroxide: a laboratory study on the consequences for phytoplankton community and diversity. Front Microbiol. 6:714.
Welch EB, Buckley JA, Bush RM (1972). Dilution as an algal bloom control. J Water Pollut Cont Fed. 44:2245–2265.
Welch I, Barrett P, Gibson M, Ridge I (1990). Barley straw as an inhibitor of algal growth I: studies in the Chesterfield Canal. J Appl Phycol. 2:231–239.
Welker M, Walz N (1998). Can mussels control the plankton in rivers? - A planktological approach applying a Lagrangian sampling strategy. Limnol Oceanogr. 43:753–762.
Whittington J, Sherman B, Green D, Oliver RL (2000). Growth of *Ceratium hirundinella* in a subtropical Australian reservoir: the role of vertical migration. J Plankton Res. 22:1025–1045.
WHO (2017). Guidelines for drinking-water quality, fourth edition, incorporating the 1st addendum. Geneva: World Health Organization:631 pp. https://www.who.int/publications/i/item/9789241549950
Willmitzer H (2010). Bewertung und Bedeutung der Biofiltration des Zooplanktons zur Verbesserung der Wasserqualität in Talsperren. GWF Wasser Abwasser. 151:1070.
Winder M, Sommer U (2012). Phytoplankton response to a changing climate. Hydrobiologia. 698:1–12.
Wu X, Joyce EM, Mason TJ (2011). The effects of ultrasound on cyanobacteria. Harmful Algae. 10:738–743.
Xiao X, Huang H, Ge Z, Rounge TB, Shi J, Xu X et al. (2014). A pair of chiral flavonolignans as novel anti-cyanobacterial allelochemicals derived from barley straw (*Hordeum vulgare*): characterization and comparison of their anti-cyanobacterial activities. Environ Microbiol. 16:1238–1251.

Xie L, Xie P (2002). Long-term (1956–1999) dynamics of phosphorus in a shallow, subtropical Chinese lake with the possible effects of cyanobacterial blooms. Water Res. 36:343–349.

Xie P, Liu J (2001). Practical success of biomanipulation using filter-feeding fish to control cyanobacteria blooms: a synthesis of decades of research and application in a subtropical hypereutrophic lake. Sci World J. 1:337–356.

Zamyadi A, Choo F, Newcombe G, Stuetz R, Henderson RK (2016). A review of monitoring technologies for real-time management of cyanobacteria: recent advances and future direction. Trends Anal Chem. 85:83–96.

Zhang Y, Prepas E (1996). Short-term effects of $Ca(OH)_2$ additions on phytoplankton biomass: a comparison of laboratory and in situ experiments. Water Res. 30:1285–1294.

Chapter 9

Managing cyanotoxin risks at the drinking-water offtake

Justin Brookes, Mike Burch, Gesche Grützmacher, and Sondra Klitzke

CONTENTS

INTRODUCTION

The characteristics of cyanobacteria to form surface scums or subsurface accumulation in deeper layers (see Chapter 4) mean there can be patchy and variable vertical and horizontal distribution of cells. This is important to consider for abstracting drinking-water from surface waterbodies. The intake of cyanobacterial biomass and cyanotoxins in the raw water can potentially be greatly reduced – sometimes by orders of magnitude – if site and depth of drinking-water abstraction are chosen to avoid these accumulations. The water offtake is one of the few control points available to managers to improve the quality of abstracted water and thereby ensure product water quality following treatment. Where control of abstraction is possible, the challenge to drinking-water treatment from high levels of biomass, DOC, TOC (dissolved and total organic carbon) and cyanotoxins can be managed in order to reduce the reliance and pressure on water treatment options (see Chapter 10).

Where surface water can be filtered through sediment, that is, by abstraction through bank filtration or artificial groundwater recharge, or if slow sand filtration can be installed, if operated with sufficiently low and long filtration rates, this can also be highly effective in removing both cyanobacterial cells (as well as other particles, including pathogens) and dissolved toxins.

9.1 OPTIMISING THE LOCATION AND DEPTH FOR THE OFFTAKE

9.1.1 Vertical variability of cyanobacterial occurrence

Spatial and vertical variability in the concentration of contaminants in lakes and reservoirs, as described in section 4.6.4, is common. This applies not only to cyanobacteria but also to pathogens, iron, manganese and other contaminants. Thermal stratification of the waterbody leads to warmer water layered above cooler, denser water. This provides suitable conditions for cyanobacterial growth as the common bloom-forming types are buoyant and can avoid sedimentation losses during the periods of stratification. Under some conditions, the cyanobacteria can accumulate to scums at the surface, and in mesotrophic waterbodies, the maximum concentration can occur as a band in a deeper layer, as discussed in section 4.2.1.

Many modern reservoir offtake structures (often towers but also some dams) have multiple offtake depths as part of good design. If multiple offtakes are not available, in small systems it may be possible to siphon water from a specific depth using large pipes as a temporary management measure. The depth at which cyanobacterial cell densities are greatest – at the surface or at specific depths – may show diurnal and seasonal patterns, and the range of passive diurnal sinking and rising of cells is due to light- and photosynthetic-driven changes in cell buoyancy (section 4.1.2). While the high variability in surface concentrations can be a relevant risk in all waterbodies, the formation of pronounced layers or maxima at deeper and often variable depths may only be relevant in thermally stratified (see section 4.3.4) mesotrophic reservoirs. The best-known example of this occurs in temperate regions with the formation of distinctive deep metalimnetic maxima of low-light-adapted *Planktothrix rubescens*. In subtropical and tropical systems, *Raphidiopsis* (*Cylindrospermopsis*) *raciborskii* can also reach high numbers deeper in the epilimnion, but these rarely form surface scums.

Figure 9.1 provides a good example of spatial variability and vertical patterns for buoyant cyanobacteria in a medium-sized reservoir. These

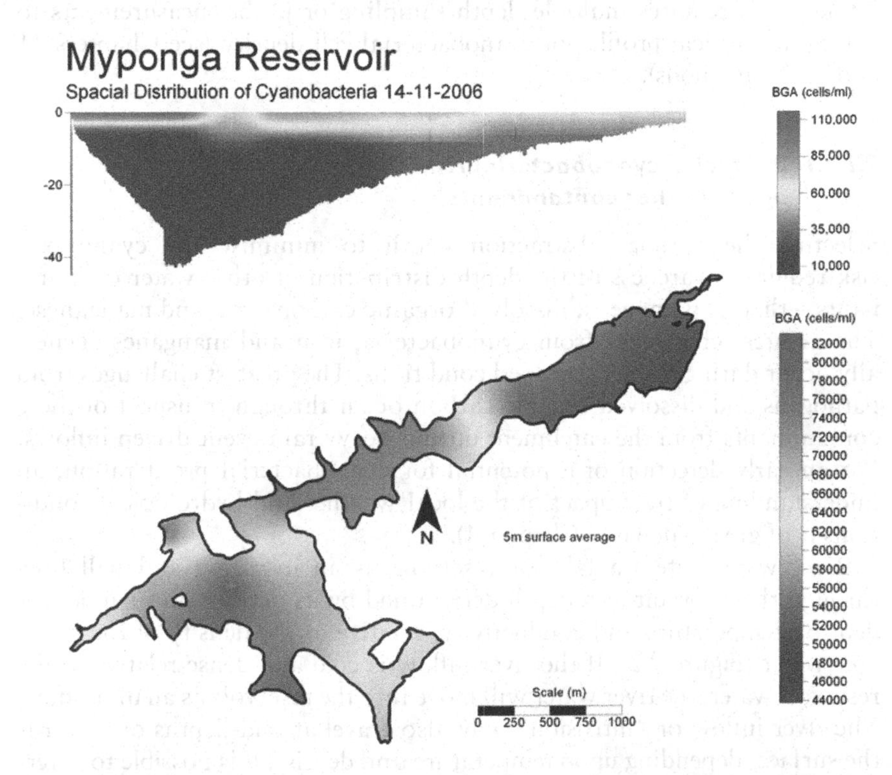

Figure 9.1 Vertical profiles and horizontal variability of a population of *Dolichospermum circinale* in a horizontal transect across a reservoir in South Australia measured by phycocyanin fluorescence, converted to cells per mL.

cyanobacteria, predominantly *Dolichospermum circinale*, develop surface and subsurface maxima at several depths at the deepest site (Site 1), while more mixing occurs at shallower locations in the reservoir (Sites 6–8). This accumulation of buoyant cyanobacteria is associated with temperature stability in the growing season and shows how strong surface maxima can develop near the reservoir surface while very low numbers occur in deeper water. This drinking-water reservoir has the option for water offtake at the surface, 5, 15 and 25 m at Site 1. The operational practice at this reservoir is to draw water from a depth of 25 or 15 m when cyanobacteria are present, which leads to much lower intake of both cells and potentially toxin and odour metabolites into the water treatment plant. A further case study is given by Fastner et al. (2001) for the Deesbach Reservoir, where at that time surface concentrations ranged up to 570 µg/L, while at the offtake depth of 17 m, they were in the range of less than 1 µg/L.

In conclusion, it is important for operators to obtain information about the range of vertical movement of local cyanobacterial populations and also to be aware of the potential for the formation of metalimnetic (deep-depth) maxima in order to avoid drawing high cell densities into the raw water intake. This requires multiple depth sampling or probe measurements to determine vertical profiles of cyanobacterial cell density (see Chapters 11 and 12 for methods).

9.1.1.1 Balancing cyanobacterial risk against other contaminants

Selecting the optimal abstraction depth to minimise the cyanotoxin risk requires awareness of the depth distribution of other water contaminants – that is, pathogens, dissolved organic carbon, iron and manganese. The greatest challenges from cyanobacteria, iron and manganese generally occur during stable, stratified conditions. The greatest challenges from pathogens and dissolved organic carbon occur through transport of these contaminants from the catchment during heavy rain event-driven inflows. For an early detection of a potential for cyanobacterial proliferation, an understanding of the impact of the local weather and hydrological conditions is of great value (see Chapter 4).

River water enters a lake or reservoir as an intrusion and will flow through the reservoir at a depth determined by its density (in turn dependent on temperature and conductivity) relative to the density of the reservoir water (Figure 9.2). If the river inflow is cold and dense relative to the reservoir water, the river water will move into the reservoir as an underflow. The river inflow or "intrusion" may also travel at mid-depths or towards the surface, depending upon temperature and density. It is possible to determine the depth of the riverine intrusion using a simple online tool which considers temperature detailed in section 9.1.4.

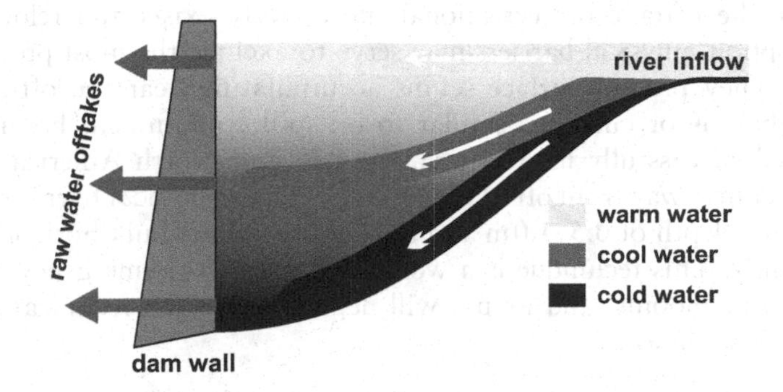

Figure 9.2 Formation of river water intrusions during inflow events. The darker the shading, the colder and denser the water. The depth of formation of the riverine intrusion will correspond to the depth in the reservoir at which water has equivalent density.

9.1.1.2 Releasing poor-quality water downstream

The water abstraction depth can also be utilised to release water of poorer quality downstream and optimise the quality of water remaining in the reservoir. Riverine intrusions into a reservoir often have a higher concentration of phosphorus and nitrogen, so releasing this water downstream can result in lower reservoir nutrient concentrations in the reservoir, thus reducing the maximum achievable cyanobacterial biomass. This method is generally only suitable in areas without water shortage, while in arid regions, water harvesting is often maximised, therefore avoiding downstream release. Also, the quality impact of the water released to the downstream river reach may be an issue to clarify with stakeholders and authorities responsible for the river water quality (often, this impact is minor and only relevant for a very short downstream part of the river).

9.1.2 Horizontal variability of cyanobacterial occurrence

The horizontal variation in the distribution of cyanobacterial populations can also be considerable. Figure 9.1 shows the horizontal and vertical heterogeneity of cyanobacterial distribution. Observing seasonal patterns of cyanobacterial scum location and/or predicting them from the main wind direction may be useful for planning a drinking-water offtake or the location of a recreational site or beach (see below).

Substantial contamination of raw water can be avoided by locating offtakes away from sheltered bays where scums may accumulate (usually downwind of the prevailing winds during the critical summer growth period). If this is not practical, it may be possible to employ temporary extensions to pipe intake points.

Where the offtake or recreational site already exists and relocation is not an option, physical barriers may serve to exclude the most pronounced blooms. They prevent surface scums accumulating near the offtake site. Surface booms or curtains, similar to oil-spill containment booms, have been used successfully in Australia, the UK and North America to keep surface scums away from offtake structures. These physical barriers usually extend to a depth of 0.5–1.0 m and thus do not affect bulk horizontal flow significantly. This technique is a worthwhile practical emergency measure for transient blooms, and its use will depend upon the technical requirements of installation.

9.1.3 Data collection for optimising offtake sites

Collecting information as well as building knowledge and understanding of the local ecology and conditions can greatly increase flexibility in responses to blooms. When collecting data to optimise offtake depth, it is important to include relevant hazards or indicators for their occurrence. Vertical profiles of temperature provide a basis for assessing thermal stratification, and profiles of oxygen concentration and redox potential indicate the likelihood of higher iron and manganese concentrations. Data from turbidity profiles may also be useful to indicate the location of a bloom. Many water quality probes that measure temperature, conductivity, dissolved oxygen, etc can also be equipped with a fluorometer, which measures chlorophyll, including the specific or other pigment fluorescence – a useful surrogate for phytoplankton biomass. Fluorometers can specifically differentiate between cyanobacteria and other phytoplankton if they measure phycocyanin fluorescence, an accessory pigment only of the cyanobacteria. While the ratio between cyanobacterial biomass and the fluorescence signal may vary to some extent, fluorescence signals have proven highly useful as a relative measure of the distribution of blooms, both vertically and horizontally. These probes may be installed online at offtakes as part of Alert Level Frameworks (see section 5.1.2) to adjust water treatment responses. The fluorometry signal can be calibrated by comparing phycocyanin fluorescence with the monitored cell counts (section 13.6.1). However, it is important to be aware of the issues and technical limitations of *in situ* fluorometers for monitoring of cyanobacteria (see, e.g., Zamyadi et al., 2016 or Bertone et al., 2018).

The following checklist outlines information needed to assess cyanotoxin intake from raw water offtake systems. It is neither complete nor designed as a template for direct use, and should be adapted to specific local conditions. Support from lake experts or limnologists with hydrodynamic expertise is highly valuable for this assessment. For measures to minimise the occurrence of cyanobacteria through reducing nutrient loading or waterbody management, see Chapters 7 and 8.

CHECKLIST 9.1: COLLECTING INFORMATION ON THE RISK OF INTAKE OF CYANOBACTERIA OR DISSOLVED CYANOTOXINS WITH THE DRINKING-WATER OFFTAKE

What meteorological, hydromorphological and hydrodynamics characteristics of the waterbody could affect cyanobacterial distribution?

- Compile information on depth, volume, bathymetry and thermal stratification;
- Collect information on the prevailing wind speed and direction;
- Determine the location of raw water offtake.

What information is available on cyanobacterial or cyanotoxin occurrence, and where they tend to accumulate in relation to the drinking-water intake?

- Collate historical information on the occurrence of cyanobacteria and consider initiating a new programme or adapting an ongoing programme to survey cyanobacterial occurrence and to determine differences in their distribution.
- Evaluate the local cyanobacterial species with regard to their buoyancy characteristics and their potential vertical distribution.

What data are available and/or necessary to inform selection of an offtake site?

- Determine the scale of monitoring and expertise necessary and available to effectively manage the offtake.
- Assess whether continuous online monitoring of temperature profiles could be installed to better understand both thermal stratification and flow regime, and whether the information gained is likely to justify the costs.
- Evaluate whether sensors could be installed to effectively monitor the offtake for the indication of cyanobacterial biomass, for example, phycocyanin fluorescence, chlorophyll fluorescence and turbidity.

Are hazardous events likely to cause cyanobacteria to concentrate near the offtake?

- How heavy are cyanobacterial blooms? How long do they persist, and what seasonal patterns do they show? Which toxins do they contain? Are substantial extracellular amounts of toxin likely?

- Assess whether scum-forming cyanobacteria are likely to accumulate at the surface and to concentrate on leeward shores due to wind action.
- Assess whether storm event inflows will transport high concentrations of cells to the offtake from upstream of the waterbody.

Where blooms occur at the offtake, are cells likely to die and lyse through pumping the water from the offtake site to the waterworks?
What management options are available for the drinking-water offtake?

- For thermally stratifying lakes and reservoirs, is a multilevel offtake available? If so, can it be readily operated based upon current monitoring information? If not, can effective options for monitoring indicators for cyanobacterial biomass be installed?
- For thermally stratifying lakes and reservoirs with heavy cyanobacterial blooms, is artificial mixing an option? That is, can aeration sufficiently suppress their development (e.g., through intermittent operation) or shift cell distribution to reduce concentrations at the intake (see also Chapter 8)?
- For near-surface intakes (particularly from shallow waterbodies), determine whether their sites are optimal in relation to chief areas of bloom accumulation and if not, whether relocation (e.g., through pipe extension) is possible.
- For bloom-affected near-surface intakes which cannot be relocated, consider installing physical barriers as discussed above to reduce bloom intake.

What other water quality hazards should be considered when changing an offtake for cyanobacterial management?

- Check whether avoiding the intake of pathogens is an important target for reservoir offtake management.
- Assess whether intake needs to periodically avoid depth layers which have low dissolved oxygen or with high iron or manganese concentrations.
- Develop a strategy for balancing cyanobacterial hazards against other hazards associated with low dissolved oxygen, iron and manganese, or pathogens in inflows.

What regulatory framework exists for abstracting drinking-water?

- Are there surface water abstraction regulations that need to be considered when planning offtake sites and amounts?

Document and visualise the information on the abstraction scheme and regime:

- Compile a summary report and consolidate information from your checklist.
- Map the spatial distribution of existing or potential sites for raw water intake.

Outcome of system assessment:

- *Estimate the risk of cyanotoxin intake*: What maximum levels do you expect to find, and for what time periods?
- *Estimate the uncertainty of this assessment*: Is the information base sufficient to make management decisions? If not, which information gaps should be closed and with which priority?

9.1.4 Operational monitoring of control measures in raw water abstraction

The most effective way to ensure that control measures are operating as intended is to monitor readily observable indicators that show whether structures are intact and processes operating as intended – that is, operational monitoring (Chapter 6). For the measures proposed above to control cyanotoxin concentrations in raw water intakes, operational monitoring will largely address the integrity of structures and whether flexible choice of offtake is operated adequately in relation to bloom occurrence. Surveillance will check the records of this monitoring as well as the adequacy of planning and design.

9.1.5 Validation of control measures for raw water offtake

Validation to ensure that the drinking-water offtake is appropriately sited and performing optimally is best achieved by a cyanobacterial monitoring programme during a bloom period (Table 9.1). The validation programme would ideally include several different bloom events, consider how wind direction influences the accumulation of scums and consider different types of cyanobacteria with different properties that may occur (e.g., scum-forming *Microcystis* versus dispersed filamentous *Planktothrix agardhii*). Several parameters for cyanobacterial biomass measurement may be used for these investigations, including biovolume, pigment analysis either by *in situ* measurement of fluorescence by fluoroprobe

Table 9.1 Examples for control measures in drinking-water offtake with options for monitoring their functioning

Process step	*Examples of control measures for intake of contaminants*	*Options for monitoring their functioning*
Planning	Appropriate selection of offtake site and depth in relation to cyanobacterial and cyanotoxin accumulation, including the occurrence of other detrimental or hazardous contaminants (e.g., iron, manganese, pathogens)	Review plans/applications for permits in relation to information on bloom accumulation and occurrence of other contaminants
	Downstream release of nutrient-rich water, artificial mixing or physical barriers against surface scums	Review plans and applications for permits
Design, construction, maintenance	Ensure that offtake structures (e.g., extension pipes, depth-variable offtakes) are constructed according to good practice and that they are withdrawing water at the location intended	Inspect structures during construction and at intervals during operation; monitor their integrity
	If measures such as downstream release of nutrient-rich water, artificial mixing or physical barriers against surface scums are implemented, ensure proper design, construction, maintenance	Inspect structures during construction and at intervals during operation; inspect records of maintenance
	Adapt variable offtake depth to stratification of cyanobacterial accumulation, taking other contaminants (iron, manganese) into account	As above, but for deep-layer offtakes include redox measurement or oxygen concentration to detect high levels of iron/manganese
	If artificial destratification is used, adapt the mixing intensity to stability of thermal stratification	Monitor temperature profiles over depth in reservoir or lake
	If artificial mixing is intermittent, adapt periodicity to phytoplankton development	Monitor phytoplankton development, for example, by fluorescence (differentiating between cyanobacteria and algae); verify qualitatively by microscopy
	If physical barriers to deflect blooms are in place, monitor their integrity and proper positioning	Visual inspection, possibly supported by on-site probing of fluorescence or turbidity

(preferably distinguishing between chlorophyll-*a* and phycocyanin specific for cyanobacteria) or chlorophyll extraction and photometric determination (in conjunction with qualitative microscopy to determine the proportion of chlorophyll that is likely to originate from cyanobacteria). Other parameters indicating the amount of cyanobacterial biomass may be used as well, including cell counts, molecular information, remote sensing data or turbidity measurements (as with chlorophyll-*a*, in conjunction with qualitative microscopy to check whether a major fraction of the turbidity is likely to originate from cyanobacteria). While these are less precise, they can be very suitable surrogate monitoring parameters particularly if locally "calibrated" against periodic toxin analyses as described in Chapter 5.

Validation would also include other contaminants to determine differences in spatial distribution compared to that of the cyanobacteria, for example, iron and manganese or pathogens potentially accumulating in the hypolimnion. This is to ensure that optimising for cyanotoxin control does conflict with optimising for the control of other contaminants. Validation should be repeated from time to time, particularly if bloom patterns show conspicuous changes or if the hydrodynamics of the waterbody have changed.

9.2 SEDIMENT PASSAGE: MANAGED AQUIFER RECHARGE VIA SOIL AQUIFER TREATMENT OR POND INFILTRATION, SLOW (SAND) FILTRATION AND BANK FILTRATION

In many settings, managed aquifer recharge (MAR; Tufenkji et al., 2002; Maeng et al., 2010; Romero et al., 2014) or slow sand filtration (Ellis & Wood, 1985; Haig et al., 2011) has proven to be highly effective and low-cost options for the removal of cyanobacteria and dissolved cyanotoxins (Grützmacher et al., 2002). A prerequisite for MAR is the suitability of the given sediment (which is not always given, e.g., in rocky mountainous or karstic regions) or the availability of space for constructing and operating slow sand filters (which may be a constraint where large volumes of water need to be treated). For many scenarios of cyanotoxin occurrence and conditions for MAR or slow sand filtration, substantial toxin removal is very likely. It is thus a very valuable option worth considering when assessing a given drinking-water supply at risk of cyanotoxin contamination.

Managed aquifer recharge utilizes subsurface passage through porous media (mostly gravel or sandy material) to achieve purification of surface water for drinking water production. Types of MAR are e.g., bank filtration (Figure 9.3), pond infiltration or aquifer recharge via trenches or injection wells (an overview of techniques applied in Europe is given in Sprenger et al. (2017). Slow sand filtration (Figure 9.4) uses the same principle, though on a smaller scale, i.e., through a less thick layer of porous medium. All these methods are applied in a wide variety of cases with different results and can therefore be seen as abstraction methods, pre-treatment or treatment steps.

Managed aquifer recharge comprises both systems saturated with water (i.e., riverbank filtration, groundwater and (slow) sand filtration) and unsaturated systems such as soils, in which not all pores are filled with water, leading to different transport dynamics, especially to higher temporal variability. Therefore, some of the processes eliminating cyanotoxins differ between both, and hence, results from saturated systems cannot always be transferred to unsaturated systems. The following sections give an overview on the removal of different cyanotoxins in various managed aquifer recharge systems, as well as information on different aspects of managing these systems.

9.2.1 Background information to assess the subsurface conditions relevant for performance in retaining cells and dissolved cyanotoxins

Soil aquifer treatment is one type of managed aquifer recharge, in which irrigation water that may contain toxins is applied onto soil. The soil is meant to act as a filter to retain toxins and to prevent them from further transport to groundwater.

Bank filtration is characterised by drinking-water wells in the vicinity of a lake or river that are fed mainly by water infiltrating from the surface water supply (Figure 9.3). For other types of MAR, surface water is first conveyed into infiltration ponds, trenches or wells from which it infiltrates into the subsurface aquifer and is reclaimed in nearby wells. Due to different hydrogeological settings, residence times may vary between a few days to several months. The decisive parameters for the residence time are the hydraulic conductivity of the aquifer (governed by the grain size distribution as well as the existence of a clogging layer) and the distance between point of infiltration and well. Usually, infiltrated surface water blends in the well with ambient groundwater, diluting possible contaminants originating from the surface water (and *vice versa*; however, groundwater contaminants may also be diluted by uncontaminated surface water).

An important mechanism of sediment passage is the reduction of peak contaminant concentrations by dispersion (i.e., mixing of water with different residence times due to varying flow paths). This may help avoid hazardous concentrations, though it does not serve to reduce the total load of the contaminant (this is also true for the decrease in concentration by mixing with groundwater). The main purification processes during infiltration and subsurface passage are straining of particles as well as adsorption onto aquifer material and biological or chemical degradation of dissolved substances.

For predominantly cell-bound cyanotoxins, the main elimination process is the straining of cells on the sediment surface. Phytoplankton removal through sand filtration is very efficient and may eliminate up to 97% (Pereira et al., 2012). Efficacy may be limited if the grain size distribution of the

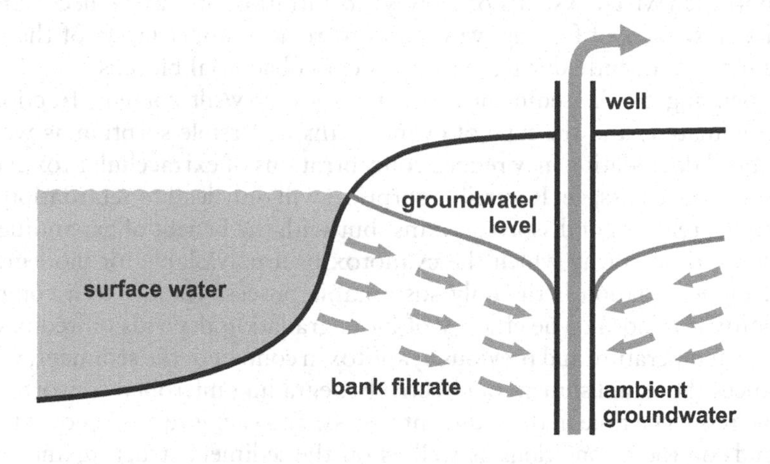

Figure 9.3 Bank filtration scheme. (Adapted by permission from Springer Nature, Hydrogeology Journal, Future management of aquifer recharge. Dillon P. Copyright Springer Nature 2005. www.springernature.com/gp.)

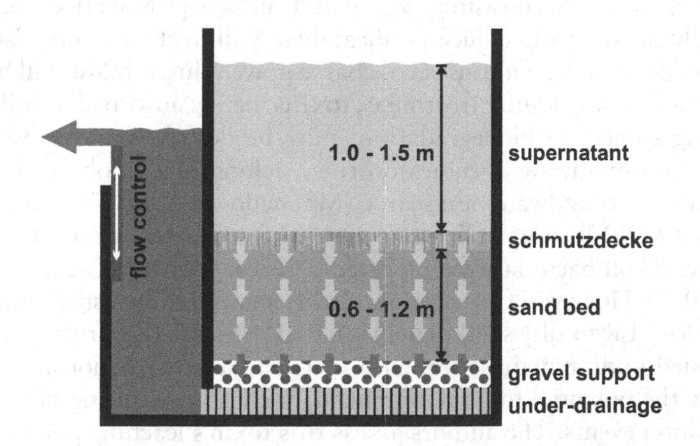

Figure 9.4 Slow sand filtration scheme. Raw water passes the sand filter through gravity. The hydrostatic pressure is controlled by a flow control.

sediment is coarse (e.g., gravel) and if there is no clogging layer (the clogging layer is defined as the uppermost part of a sediment or filter in which hydraulic conductivity is reduced due to the accumulation of organic and inorganic debris as well as biofilms resulting from high biological activity). An additional parameter that may limit cell removal is the morphotype of the cyanobacterial species (see review by Romero et al., 2014): removal of filamentous cyanobacteria may be more effective than that of single cells or small colonies. Where cell removal is effective, cyanotoxin transport is limited to the migration of dissolved (extracellular) toxins through the subsurface.

As cell lysis may release high amounts of toxin (see Chapter 4), the accumulation of cells on the sediment surface or filter is best avoided. Therefore, in controlled MAR systems or slow sand filtration, it may be necessary to reduce flow rates of feeding water, and to remove upper layers of the filter sand more frequently during periods of cyanobacterial blooms.

Depending on the sediment characteristics (clay/silt content, fraction of organic matter) and the type of cyanotoxins, reversible sorption as well as biological degradation may reduce concentrations of extracellular toxin during subsurface passage. Reversible sorption will only lead to retardation and not to the removal of dissolved toxins, but with the benefit of expanding the residence time during which the cyanotoxins are available for biodegradation. Biodegradation is the only sustainable process leading to a complete cyanotoxin removal. The efficacy of biodegradation depends on redox conditions, temperature and previous cyanotoxin contact of the sediment, which enhances the establishment of effectively degrading microbial consortia.

The residence time in the sediment necessary to achieve the removal target depends on these conditions as well as on the sediment structure and material: a sufficiently large fraction of fine particles is necessary to provide surface area not only for sorption, but also for the establishment of biofilms that harbour the toxin-degrading microorganisms. A fraction of >1% fines (i.e., particle size <63 μm, consisting of clay and silt) has proven effective, though this might significantly reduce permeability. Sediment structure also determines residence time: if it consists of coarse gravel, throughflow will be rapid.

However, in soil aquifer treatment, toxins merely adsorbed to soils, without being subject to biodegradation, may be desorbed and subsequently leached to groundwater. Studies on the leaching of cyanobacterial toxins from soils to groundwater are scarce (Machado et al., 2017). Batch studies show that MC-LR sorbs well to clay and organic matter (Miller et al., 2001) and several soil bacteria were reported to break down MC-LR (Machado et al., 2017). However, Corbel et al. (2014) showed only a small amount of applied MC-LR to silty sand to be degraded, whereas the largest fraction of the applied toxin remained sorbed in extractible soil fractions. This toxin thus has the potential to be remobilized, for instance during precipitation or irrigation events. The authors assess this toxin's leaching potential from soil to groundwater to be high. So far, there are no studies which investigated the leaching potential of CYN in soils, but due to its low sorption potential (Klitzke et al., 2011b) its mobility, i.e. leaching potential, is likely to be high.

The leaching potential of cyanobacterial toxins strongly depends on site conditions such as soil texture, irrigation frequency with polluted water and toxin concentrations in the water as well as time elapsed after toxin application. Due to the lack of data, general management recommencations cannot be derived, and site-specific validation is necessary. Because of the potential risk of leaching, where water with high toxin concentrations is applied in soil aquifer treatment, validation of removal efficacy is particularly important.

Knowledge on sediment passage elimination varies between types of cyanotoxins, with a large share of publications reporting on microcystin and relatively few on cylindrospermopsin, saxitoxin and anatoxin.

9.2.2 Degradation of microcystin during bank filtration

Biodegradation as opposed to sorption was found to be the dominating process for microcystin elimination in sandy porous material (Grützmacher et al., 2010). While sorption on sandy aquifer material was very low (Grützmacher et al., 2010), clay (Wu et al., 2011) and silt were found to contribute to microcystin adsorption (Miller et al., 2005). Organic carbon did not enhance microcystin adsorption. Miller et al. (2005) and Wu et al. (2011) even observed a decrease in microcystin sorption at low organic carbon contents (i.e., <8%). The authors explain this decrease by sorption competition between dissolved organic substances and microcystin.

Grützmacher et al. (2002) demonstrated an efficient removal of microcystins by slow sand filtration, simulating the first few decimetres of subsurface passage also for sandy bank filtration and aquifer recharge sites. Under optimal conditions (aerobic, moderate to high temperatures and previous microcystin contact), degradation is rapid, with rates of around 1/d (assuming exponential first-order decay). The residence time in the sediment needed for sufficient toxin removal (<1 µg/L) under optimal conditions even for very high microcystin concentrations amounts to 10 days. At less-than-optimal conditions, microcystins will also be degraded, though at a slower rate (Grützmacher et al., 2007).

Each of the critical conditions described in the following increases the probability that the minimum residence times necessary for sufficient toxin removal will be achieved or exceeded.

Anaerobic/anoxic degradation of microcystins: It may be much slower than degradation under aerobic conditions. Lag phases may occur but rarely last more than 1–2 days (Grützmacher et al., 2010). Redox conditions in the subsurface may be monitored by regular sampling of observation wells in the flow path, analysis of redox-sensitive parameters (e.g., iron and ammonium) and measurement of oxygen content.

Lack of previous contact of the system with microcystins: Previous microcystin contact will be given for most settings as water reservoirs and rivers with toxic cyanobacterial mass occurrences tend to show these regularly. Even for surface waters without a history of cyanobacterial blooms, the time usually needed for a mass occurrence to develop (i.e., a few weeks) is likely to be sufficient for adaptation of bacteria in the subsurface. This may, however, be a crucial point for managed aquifer recharge (MAR) sites and slow filtration if sand/sediment has been exchanged just before or during a massive cyanobacterial bloom. Checking records for sediment exchange may therefore be important when increasing amounts of cyanobacteria or blooms are observed.

Low temperature: Temperature generally determines microbial degradation rates. For microcystins, this is also generally true, but there is some experimental evidence that microbial communities may adapt to low temperatures and yield high degradation rates even at 5 °C: whereas laboratory experiments at less than 15 °C and 25 °C did not show different degradation rates, an experiment at 5 °C yielded the relevant microcystin breakthrough, though eventually degradation rates reached those obtained under higher temperatures (Chorus & Bartel, 2006). Hence, temperature monitoring in the well and in the surface water is useful in order to obtain an understanding of the prevailing temperature in the sediment flow path.

In summary, optimal conditions for biodegradation of microcystins are aerobic, with moderate to high temperatures (>10–15 °C) and an established microbial consortium (particularly in the clogging layer) capable of microcystin degradation. For degradation in the water phase, refer to Chapter 2.1.

9.2.3 Degradation of cylindrospermopsin during bank filtration

Removal of cylindrospermopsin (CYN) through subsurface passage or slow sand filtration schemes differs from that of microcystins in several ways:

1. A much higher fraction of CYN frequently occurs dissolved in water.
2. Dissolved CYN may persist in the waterbody for many weeks (Wörmer et al., 2008) after the producer bloom has subsided, particularly at low temperatures.
3. In temperate climates, CYN may occur about as frequently as microcystins, but concentrations rarely reach similarly high levels (maxima published from field samples are below 20 µg/L (Rücker et al., 1997; Bogialli et al., 2006; Rücker et al., 2007; Brient et al., 2009; see also section 2.2).

While points 1 and 2 increase the challenge for removal through sediment passage, point 3 may be a de-warning if high concentrations can be reliably excluded.

Under some conditions, CYN elimination in the subsurface or slow sand filter tends to be less effective than for microcystins: experiments by Klitzke et al. (2011b) showed that sorption of CYN to various sediments was very poor, and there was virtually no CYN retention on purely sandy sediments (Klitzke et al., 2010). Also, the presence of clay did not enhance CYN sorption (Klitzke et al., 2011b). The role of sediment organic carbon is still unclear, because at present, results on the role of organic carbon in CYN sorption are inconsistent, showing very high sorption on organic mud (Klitzke et al., 2011b) but no sorption on a "schmutzdecke" (Klitzke et al., 2011a). This uncertainty emphasises the importance of conditions conducive to biodegradation to ensure an efficient CYN removal – for CYN, these conditions are more crucial than for microcystins.

Redox conditions have been shown to be particularly crucial for CYN elimination: under anaerobic conditions, lag phases can last weeks, and even after the lag phase, degradation may remain incomplete even if residence times amount to many weeks (Klitzke & Fastner, 2012).

Preconditioning of sediments and flow rate also have a major impact: lag phases may last up to 3 weeks until a sufficiently large microbial community has developed and CYN breakdown commences (Klitzke et al., 2010). This implies that sediments need to have previous contact to CYN for at least 3 weeks before CYN is removed effectively. In practice, where CYN concentrations in the feed water build up over time, this may not prove to be a problem. However, additionally, contact times are sufficient for effective biodegradation only if the flow rate is sufficiently low (i.e., approximately 0.2 m/d); at higher flow rates (i.e., 0.7 m/d and 1.2 m/d), the shorter residence times in the sediment will not allow for CYN degradation, and hence, CYN breakthrough is likely (Klitzke et al., 2011a).

Low temperature (i.e., 10 °C or less) retarded CYN degradation by a factor of 10 in comparison with 20 °C (Klitzke & Fastner, 2012).

9.2.4 Degradation of other cyanotoxins during bank filtration

For the elimination of the other cyanotoxins by sediment passage, few experimental results have been published. For Anatoxin-a (ATX), rapid degradation already in the waterbody is well known (Rapala et al., 1994; see also section 2.3), and the half-life of ATX elimination in natural surface water studied in a batch system amounted to 4 weeks (Klitzke et al., 2011a).

In the sediment, ATX is eliminated through both sorption and degradation (Klitzke et al., 2011a): while on sandy sediments, ATX adsorbed very weakly, sorption was enhanced in sediments containing clay or organic carbon (Klitzke et al., 2011b). Column experiments with a filter velocity of approximately 1.4 m/d showed ATX retardation on sandy sediments due to sorption. However, sorption was reversible and only attenuated peak concentrations of ATX, but not the overall ATX breakthrough. Degradation took place only in the presence of DOC and hence resulted in ATX elimination of approximately 35% of the initial concentration (Klitzke et al., unpublished data).

So far, published results of systematic studies on ATX degradation in sediments are only available from laboratory batch studies (Klitzke et al., 2011a). With an initial concentration of 10–15 µg/L ATX (under oxic conditions and at room temperature of approximately 20 °C), it took 7 days for concentrations to drop below 1 µg/L following sediment contact. At 10 °C, elimination slowed down by approximately a factor of 2. Under anoxic conditions, ATX elimination in sediments was decelerated also by about a factor of 2. Degradation in sediments under all conditions investigated took place without a lag phase; that is, it commenced without any delay, suggesting that preconditioning is not required.

Burns et al. (2009) report an efficient sorption of saxitoxins (STX) on a sandy-silty sediment, parts of which had very high amounts of clay and silt (up to 89.6% silt) for an initial concentration of 5 µg/L. Besides, SXT sorbed on clay minerals with sorption increasing with increasing cation-exchange capacity. Romero et al. (2014) reported much lower sorption for SXT and neo-SXT (between 40% and 80%) on a sediment retrieved from a bank filtration site with only 4% fines (<0.063 mm). Saxitoxin removal in sand filters of both a water treatment plant and a wastewater treatment plant proved very inefficient (Kayal et al., 2008; Ho et al., 2012). In the wastewater treatment plant filter, SXT toxicity was even increased after filter passage (Ho et al., 2012). These findings suggest SXTs to be very persistent. The studies mentioned were conducted at filter velocities of 7.2 and 14.4 m/d, respectively.

As degradation processes are strongly influenced by residence times (Klitzke et al., 2011a), it remains unclear whether SXT breakdown would increase at lower flow rates (i.e., <1 m/d) as they may be encountered in riverbank filtration scenarios.

9.2.5 Planning, design and construction of sediment passage for cyanotoxin control

The general features of a managed aquifer recharge (MAR) site are determined by (i) subsurface characteristics, (ii) the design and operation of the drinking-water production wells (including distance from the bank/infiltration pond, filter depth and pumping rate) and (iii) hydrochemical and biological conditions. Similarly, performance of slow sand filters depends on the choice of sediment, size and depth of the filter as well as on the flow rate.

For bank filtration, the subsurface characteristics are given by the lithological parameters of the aquifer, and the only way to influence them is to choose the most suitable well locations along the banks of a river or lake. While coarse material (coarse sand and gravel) is often preferred for well construction in order to achieve highest productivity, finer material like middle- to fine-grained sand is more effective for the removal of particles and substances, with the challenge of combining a high straining effect with sufficient hydraulic conductivity. The same is true for other MAR techniques and slow sand filters, though cell removal may be achieved by filtration of the inlet water prior to infiltration so that grain size distribution will not be as important as during bank filtration. The basis for planning therefore will include a detailed hydrogeological site investigation prior to well construction.

Besides well position, screen depth also needs to be planned carefully, as the point of abstraction is crucial for residence times and these characteristics again are crucial for toxin removal rates – as well as for the fraction of groundwater relative to filtered surface water (leading to cyanotoxin dilution; see Figure 9.3). Basic information for the positioning of the wells and determining filter depth is derived from hydrogeological site investigations, including the total depth of the aquifer, the position of confining layers and a first assessment of the hydraulic conductivities (usually from grain size analysis). Hydraulic computer models (simple models obtainable as shareware, e.g., the "Bank Filtration Simulator"; Rustler et al., 2009) can then be used to simulate different well settings in order to find an optimum concerning productivity, residence time and share of ambient groundwater.

9.2.6 Critical aspects of operation, maintenance and monitoring

The residence time of surface water in the subsurface is crucial for an effective cyanotoxin degradation, and residence time depends on the flow rate in the subsurface as well as on the distance between the surface waterbody and the well. Well operation can vary with respect to the pumping rates and regimes – that is, continuously at a constant level or in an interval mode. It is therefore important to determine residence times for the range of potential operating conditions. Simple hydraulic models may provide a first assessment (Rustler et al., 2009). Tracer measurements, however, will reduce the uncertainty of this approach. Suitable tracers are characterised by their ability to be transported in the aquifer without sorption, degradation or any other reaction that may change the total load of this substance in the aqueous phase. A simple way of determining residence times is to monitor substances or parameters in the surface waterbody that show temporal variations (e.g., temperature), and to measure the offset of these variations over time in observation wells close to the waterbody (taking into account possible retardation coefficients).

If cyanotoxins break through the subsurface, their concentrations in the well water will depend on the proportion of bank filtrate relative to that of groundwater. Mixing ratios between surface water and groundwater can be assessed from the concentration of substances that show distinctive differences in concentration between groundwater and surface water (e.g., salinity or organic trace substances).

Flow rates as well as removal efficacy for many substances are affected by the gradual build-up of a clogging layer ("schmutzdecke"). Increasing build-up of a clogging layer will lead to reduced flow rates and eventually to the formation of anoxic zones. For this reason, the clogging layer is removed from slow sand filters at intervals which depend upon the infiltration rate loss. Some managed aquifer recharge (MAR) sites are also subject to regular cleaning of the uppermost layer, , that is, removing the clogging layer. In bank filtration settings, the degree of clogging may vary in time, due to variations in sedimentation and biological activity, but the existence of a clogging layer is likely in nearly all cases. A complete lack of a clogging layer is conceivable in rivers with high flow rates and erosion as predominant process, but in such situations cyanobacterial blooms are unlikely. Changes in clogging can be monitored by measuring the head loss (i.e., the difference in water level) between the surface water supply and the groundwater.

For many substances, clogging layers contribute significantly to both sorption and degradation processes: they provide fine particles with a high capacity for contaminant sorption and harbour a high share of the degrading microorganisms. This was also assumed for cyanotoxins; however, relevance of this mechanism was neither confirmed for the removal of microcystins nor for that of cylindrospermopsin: experimental results showed no effect of the presence or absence of a clogging layer on a slow sand filter on their removal (Grützmacher et al., 2007; Klitzke et al., 2011a) – possibly because the sorption of these toxins is poor or null and thus the time spent in the clogging layer was too short for biodegradation to be effective. Data on the role of the clogging layer for the removal of other cyanotoxins are lacking, but due to its higher sorption potential, a clogging layer may be relevant, for example, for the degradation of Anatoxin-a. Also, for coarse-grained material, the clogging layer may have an impact not through sorption, but by improving filter action as strainer to remove cells and thus cell-bound toxins.

A more crucial parameter for slow sand filters is the exchange of the entire sediment body, as biodegradation takes place throughout the filter body and is most effective if the filter is preconditioned, that is, colonised by microorganisms capable of degrading cyanotoxins.

Even if clogging layer removal is less relevant for the removal of the most frequently occurring cyanotoxins than generally assumed, documentation of removal is important to enable the assessment of filter flow rates. For slow sand filters, documentation is important for sediment exchange. Furthermore, timing of sediment exchange is preferably well before cyanobacterial blooms are expected in order to allow time for preconditioning of the filter.

9.2.7 Assessing the risk of cyanotoxin breakthrough where drinking-water is abstracted by magaged aquifer recharge and/or slow sand filtration

The following checklist outlines information needed to assess cyanotoxin removal through subsurface passage. It is neither complete nor designed as a template for direct use, but rather needs to be specifically adapted to local conditions. Support from hydrogeological expertise is highly valuable for this assessment.

CHECKLIST 9.2: COLLECTING INFORMATION ON THE RISK OF CYANOTOXIN BREAKTHROUGH WHERE DRINKING WATER IS ABSTRACTED BY MANAGED AQUIFER RECHARGE (MAR) AND/OR SLOW SAND FILTRATION

What are the hydrogeological characteristics in the area envisaged/used for infiltration?

- Determine the likely flow path from the surface waterbody to the well(s), including the screen depth of the production well(s).
- Determine the clay/silt content and fraction of organic matter of the soil or sediment through which water is likely to flow.
- Determine homogeneity of the soil or sediment.
- Determine likely redox and temperature conditions during subsurface passage.
- Estimate travel times of water between surface waterbody and drinking-water well (e.g., by using a conservative tracer or groundwater flow models).
- Estimate the share of ambient groundwater in the drinking-water well in relation to the share of bank filtrate (e.g., by using a conservative tracer or groundwater flow models).
- Assess whether previous contact of the sediments to cyanotoxins is likely.

How are managed aquifer recharge (MAR)/slow sand filtration operated?

- Are wells operated continuously or at intervals?
- How strongly does the clogging layer fluctuate (is it periodically removed)?
- For pond/trench/well infiltration/slow sand filtration: does an additional filtration step remove cells/particles prior to infiltration?

Which hazardous events are likely to affect raw water offtake?

- How heavy are cyanobacterial blooms? How long do they persist, and what seasonal patterns do they show? Which toxins do they contain? (See Chapter 2 and section 4.6.1)
- Are wellheads likely to be flooded?
- Are erosive events possible that could affect the clogging layer or filterbed structure (e.g., during snowmelt, repeated freezing and thawing cycles)?

What regulatory framework exists for abstracting drinking-water?

- Are there groundwater abstraction regulations that need to be considered when planning bank filtration and/or artificial recharge?

Documentation and visualisation of information on the abstraction scheme and regime:

- Compile a summarising report and consolidate information from your checklist.
- Map spatial distribution of existing or potential sites for wells abstracting bank filtrate, if available together with maps of hydrogeological conditions.

Outcome of system assessment:

- *Estimate the risk of cyanotoxin breakthrough in underground filtration*: What maximum levels do you expect to find, and for which time periods?
- *Estimate the uncertainty of this assessment*: Is the information base sufficient to make management decisions? If not, which information gaps should be closed with which priority?

9.2.8 Operational monitoring of sediment passage as control measure against cyanotoxins

As discussed in Chapter 6, the most effective way to ensure that control measures are operating as intended is to monitor easily observable indicators that show whether structures are intact and processes operating as they should – that is, operational monitoring. For controlling cyanotoxin concentrations by sediment passage, operational monitoring of managed aquifer recharge (MAR) and slow filters can use parameters indicating flow and redox conditions in order to ensure sufficient residence times in the subsurface. Surveillance will check the records of this monitoring as well as the adequacy of planning and design. Table 9.2 summarises selected examples of the measures proposed above to control cyanotoxin concentrations in raw water intake and suggests approaches to their monitoring and surveillance to check whether controls are operating as intended.

Table 9.2 Examples for control measures in sediment passage with options for monitoring their functioning

Process step	*Examples of control measures for sediment passage*	*Options for monitoring their functioning*
Planning	Choose site with optimum hydrogeological and technical prerequisites (fine- to medium-grained sand, land availability for pond/well construction)	Review plans and applications for permits in relation to hydrogeological information; inspect sites
	For bank filtration: optimise choice of locations and depths for production wells to ensure sufficient residence times in the subsurface	
	For other MAR techniques: assess soil characteristics, potential for rapid clogging during blooms and site of recharge in relation to production wells to ensure sufficient residence time in the subsurface	
	For slow filtration, to ensure sufficient residence time in the filter, consider area and depth needed in relation to water volume and filtration time	Review plans and applications for permits
Design, construction, maintenance	Ensure that wells are constructed according to best practice, avoiding short circuits	Assign experts, carry out maximum-capacity pumping test, TV inspection and borehole geophysical examination; inspect sites and records of maintenance
	Ascertain that minimum residence times are achieved	Use tracer investigations for validation, daily temperature measurements in surface water and bank filtrate (observation wells), water-level monitoring in surface and groundwater

(*Continued*)

Table 9.2 (Continued) Examples for control measures in sediment passage with options for monitoring their functioning

Process step	*Examples of control measures for sediment passage*	*Options for monitoring their functioning*
	Modify well locations/filter depth, if material proves coarser than expected	Grain size analysis of aquifer material prior to well lining to validate assumptions
	Remove clogging layer from infiltration pond/basin or river bank by dredging, in case infiltration rates decrease and anoxic conditions are established	Water-level monitoring in surface and groundwater, redox measurements in bank filtrate observation well
Operation	*For artificial recharge and slow sand filters*: avoid clogging by a regular removal of clogging layer (preferably before blooms are expected)	Monitor well production rates
	Avoid anoxic or anaerobic conditions by timely removal of clogging layer	Monitor oxygen content in bank filtrate, possibly also DOC in surface water, as indicators of oxygen consumption
	After sediment exchange, consider extending residence times by reducing pumping rates/hydraulic head	Inspect documentation of pumping rates/groundwater tables and records of sediment exchange
	For bank filtration, operational control options are limited to the pumping regime: If possible, during cyanobacterial blooms switch to production wells with a higher share of groundwater and reduce pumping rates at critical wells	Inspect records of well operation and pumping rates; measure tracer for the proportion of bank filtrate in production well regularly

In addition to the operational monitoring of the functioning of control measures, occasional cyanotoxin monitoring in the offtake is important to verify comprehensively that initial cyanotoxin concentrations do not exceed the concentration for which the system is designed.

9.2.9 Validation of control measures in sediment passage

Whether or not sediment passage is sufficiently effective in eliminating cyanotoxins can best be validated by following bloom events with samples from a drinking-water production well and analysing them for the cyanotoxins which occur in the waterbody. For this purpose, indicators such as cell counts or pigment analysis are not applicable, as their elimination rates will differ from those of the dissolved cyanotoxins. If observation wells are available or can be installed between the drinking-water production well and the waterbody, they provide an excellent opportunity to follow the concentration decline. Timing of sampling in relation to the travel time of the water in the subsurface is important in order not to miss the concentration peak as it moves through the subsurface. It is also useful to include tracer measurements to characterise travel times in the subsurface.

The handicap of validation by following bloom events is that this is difficult to plan for waterbodies in which blooms do not occur regularly, or if capacities for sampling and analysis are not available. In such cases, a first approach to validation is to characterise the sediment as well as residence times in the subsurface, redox conditions and temperature as discussed above in order to estimate the likelihood of effective cyanotoxin elimination. Thus, conditions likely to be safe may be identified and investigations can focus on the more critical situations.

REFERENCES

Bertone E, Burford MA, Hamilton DP (2018). Fluorescence probes for real-time remote cyanobacteria monitoring: a review of challenges and opportunities. Water Res. 141:152–162.

Bogialli S, Bruno M, Curini R, Di Corcia A, Fanali C, Lagana A (2006). Monitoring algal toxins in lake water by liquid chromatography tandem mass spectrometry. Environ Sci Technol. 40:2917–2923.

Brient L, Lengronne M, Bormans M, Fastner J (2009). First occurrence of cylindrospermopsin in freshwater in France. Environ Toxicol. 24:415–420.

Burns JM, Hall S, Ferry JL (2009). The adsorption of saxitoxin to clays and sediments in fresh and saline waters. Water Res. 43:1899–1904.

Chorus I, Bartel H (2006). Retention and elimination of cyanobacterial toxins (microcystins) through artificial recharge and bank filtration. Berlin: KWB:143 pp.

Corbel S, Bouaïcha N, Mougin C (2014). Dynamics of the toxic cyanobacterial microcystin-leucine-arginine peptide in agricultural soil. Environ Chem Lett. 12:535–541.

Dillon P (2005). Future management of aquifer recharge. Hydrogeol J. 13:313–316.

Ellis KV, Wood WE (1985). Slow sand filtration. CRC Critical Reviews in Environ Science and Technology. 15:315–354.

Fastner J, Chorus I, Willmitzer H, Rabe C (2001). Reducing intake of microcystins at the Deesbach Reservoir drinking-water abstraction system. In: Chorus I, editors: Cyanotoxins: occurrence, causes, consequences. Berlin: Springer:229–231.

Grützmacher G, Bartel H, Chorus I (2007). Cyanobakterientoxine bei der Uferfiltration. Berlin: Umweltbundesamt:345–353.

Grützmacher G, Böttcher G, Chorus I, Bartel H (2002). Removal of microcystins by slow sand filtration. Environ Toxicol. 17:386–394.

Grützmacher G, Wessel G, Klitzke S, Chorus I (2010). Microcystin elimination during sediment contact. Environ Sci Technol. 44:657–662.

Haig S, Collins G, Davies R, Dorea C, Quince C (2011). Biological aspects of slow sand filtration: past, present and future. Water Sci Technol Water Supply. 11:468–472.

Ho L, Tang T, Monis PT, Hoefel D (2012). Biodegradation of multiple cyanobacterial metabolites in drinking water supplies. Chemosphere. 87:1149–1154.

Kayal N, Newcombe G, Ho L (2008). Investigating the fate of saxitoxins in biologically active water treatment plant filters. Environ Toxicol. 23:751–755.

Klitzke S, Apelt S, Weiler C, Fastner J, Chorus I (2010). Retention and degradation of the cyanobacterial toxin cylindrospermopsin in sediments - The role of sediment preconditioning and DOM composition. Toxicon. 55:999–1007.

Klitzke S, Apelt S, Weiler C, Fastner J, Chorus I (2011a). The fate of cylindrospermopsin and anatoxin-a during sediment passage. In: Wiedner C, editor: Development of Toxic Nostocales (Cyanobacteria) in the Course of Declining Trophic State and Global Warming NOSTOTOX – Final project report. Berlin: KompetenzZentrum Wasser: 87–99.

Klitzke S, Beusch C, Fastner J (2011b). Sorption of the cyanobacterial toxins cylindrospermopsin and anatoxin-a to sediments. Water Res 45:1338–1346.

Klitzke S, Fastner J (2012). Cylindrospermopsin degradation in sediments–The role of temperature, redox conditions, and dissolved organic carbon. Water Res. 46:1549–1555.

Machado J, Campos A, Vasconcelos V, Freitas M (2017). Effects of microcystin-LR and cylindrospermopsin on plant-soil systems: a review of their relevance for agricultural plant quality and public health. Environ Res. 153:191–204.

Maeng SK, Ameda E, Sharma SK, Gruetzmacher G, Amy GL (2010). Organic micropollutant removal from wastewater effluent-impacted drinking water sources during bank filtration and artificial recharge. Water Res. 44:4003–4014.

Miller MJ, Critchley MM, Hutson J, Fallowfield HJ (2001). The adsorption of cyanobacterial hepatotoxins from water onto soil during batch experiments. Water Res. 35:1461–1468.

Miller MJ, Hutson J, Fallowfield HJ (2005). The adsorption of cyanobacterial hepatoxins as a function of soil properties. J Water Health. 3:339–347.

Pereira SP, de Cerqueira Martins F, Gomes LNL, do Vale Sales M, De Pádua VL (2012). Removal of cyanobacteria by slow sand filtration for drinking water. J Water, Sanit Hyg Dev. 2:133–145.

Rapala J, Lahti K, Sivonen K, Niemelä SI (1994). Biodegradability and adsorption on lake sediments of cyanobacterial hepatotoxins and anatoxin-a. Lett Appl Microbiol. 19:423–428.

Romero L, Mondardo R, Sens M, Grischek T (2014). Removal of cyanobacteria and cyanotoxins during lake bank filtration at Lagoa do Peri, Brazil. Clean Technol Environ Policy. 16:1133–1143.

Rücker J, Stüken A, Nixdorf B, Fastner J, Chorus I, Wiedner C (2007). Concentrations of particulate and dissolved cylindrospermopsin in 21 *Aphanizomenon*-dominated temperate lakes. Toxicon. 50:800–809.

Rücker J, Wiedner C, Zippel P (1997). Factors controlling the dominance of *Planktothrix agardhii* and *Limnothrix redekei* in eutrophic shallow lakes. Hydrobiologia. 342/343:107–15.

Rustler M, Boisserie-Lacroix C, Holzbecher E, Grützmacher G (2009). Combination of MAR and adjusted conventional treatment processes for an Integrated Water Resources Management. Deliverable 5.2.5, Bank Filtration Simulator. European Commission, Bruxelles. 50 pp. www.kompetenz-wasser.de/wp-content/uploads/2017/05/d5-2-5.pdf.

Sprenger C, Hartog N, Hernández M, Vilanova E, Grützmacher G, Scheibler F et al. (2017). Inventory of managed aquifer recharge sites in Europe: historical development, current situation and perspectives. Hydrogeol J. 25:1909–1922.

Tufenkji N, Ryan JN, Elimelech M (2002). The promise of bank filtration. Environ Sci Technol. 36:422A–428A.

Wörmer L, Cirés S, Carrasco D, Quesada A (2008). Cylindrospermopsin is not degraded by co-occurring natural bacterial communities during a 40-day study. Harmful Algae. 7:206–213.

Wu X, Xiao B, Li R, Wang C, Huang J, Wang Z (2011). Mechanisms and factors affecting sorption of microcystins onto natural sediments. Environ Sci Technol. 45:2641–2647.

Zamyadi A, Choo F, Newcombe G, Stuetz R, Henderson RK (2016). A review of monitoring technologies for real-time management of cyanobacteria: recent advances and future direction. Trends Anal Chem. 85:83–96.

Chapter 10

Controlling cyanotoxin occurrence

Drinking-water treatment

Gayle Newcombe, Lionel Ho, and José Capelo Neto

CONTENTS

INTRODUCTION

Drinking-water treatment is the last line of defence to ensure the provision of safe, clean water to consumers. The multibarrier approach – where control points within the overall treatment process are identified and optimised, and their efficiency is monitored and verified – is now globally accepted as best practice for this purpose. Water suppliers using raw water with cyanobacteria at levels causing a cyanotoxin risk need to identify the points in the plant where either removal or release of toxins can occur, optimise the controls and minimise the risks of toxin breakthrough. This chapter describes the current state of knowledge about the treatment measures that are available for the removal of cyanobacteria and the toxins they produce, the monitoring regimes that can be undertaken to ensure the optimum performance of those measures, as well as validation programmes that can be run to ensure optimum choice and design of measures.

10.1 TREATMENT OPTIONS FOR CYANOBACTERIA AND CELL-BOUND CYANOTOXINS

In many situations, most of the cyanotoxins will be cell-bound, while for cylindrospermopsin, a high fraction can occur in the dissolved state (see Chapter 2 and Box 5.1). Thus, any physical particle separation process that removes cyanobacterial cells without damage will offer an effective barrier to cyanotoxins, particularly microcystins. Section 10.1 describes the processes that can be applied to remove cells, while maintaining cell integrity. However, often pre-oxidation is applied for other treatment goals such as manganese removal or the improvement of coagulation, so it is also important for water suppliers to be aware of the potential risks of cell rupture and cyanotoxin release associated with the application of pre-oxidation.

10.1.1 Pre-oxidation

Chemical oxidation can have a range of effects on cyanobacteria cells, from minor cell wall damage to cell death and lysis (Pietsch et al., 2002).

Although improvement of the coagulation of algal cells through oxidation at the inlet of the treatment plant through a number of mechanisms has been reported (Petrusevski et al., 1996), this involves a high risk of damaging the cells and releasing metabolites into the dissolved state. For example, potassium permanganate, commonly used as a pre-oxidant to control manganese, can potentially damage cyanobacteria and release toxins without oxidising the released cyanotoxins (Dugan et al., 2018). Other common pre-oxidants include chlorine and ozone.

If pre-oxidation must be applied in the presence of cyanobacteria cells, the levels of oxidant should be sufficient to result in the residual required for the destruction of dissolved toxins (see section 10.2). If it is insufficient, this causes a risk of high levels of dissolved toxin and organic carbon adversely influencing subsequent removal processes and finished water quality.

Table 10.1 presents a summary of some of the literature on the oxidation of cyanobacteria and toxin release and destruction. Kinetic studies indicate clearly that the rate of cell membrane damage and toxin release is greater than the rate of toxin degradation. These results suggest that the oxidant doses required will vary depending on water quality parameters such as pH, dissolved organic carbon (DOC) concentration and characteristics, the abundance of cyanobacteria, the size of the cyanobacterial filaments or colonies and the amount of intracellular and extracellular organic materials (IOM and EOM) associated with the cyanobacteria. Natural cyanobacterial samples contain more EOM, and cells are more likely to occur in larger colonies or filaments than in the cultured samples used in many studies. Therefore, if pre-oxidation must be practiced in the presence of potentially toxic cyanobacteria, a regular laboratory testing is important to ensure that the oxidant demand is met and the released toxins are destroyed. If this is not possible, it is best to cease pre-oxidation for periods during which cyanobacteria are abundant in the raw water.

10.1.2 Physical separation processes

Ideally, the number of cyanobacterial cells in the raw water is minimised by appropriate measures in the waterbody (Chapter 8) and its catchment (Chapter 7). Pretreatment using bank filtration is also very effective (Chapter 9). However, even where these measures reduce the concentration of cyanobacteria entering the treatment plant, where they still occur, multiple barriers are important, and this requires optimising removal of cells and toxins through treatment.

Two main processes can be utilised for the physical removal of cyanobacteria from raw water: conventional processes (e.g., coagulation/clarification and sand filtration) and membrane filtration (e.g., microfiltration [MF] and ultrafiltration [UF]).

Table 10.1 Summary of studies on the effect of pre-oxidation of water containing cyanobacteria and the effect on cyanotoxins released by the process

Dominant cyanobacteria	*Toxin*	*Oxidant*	*Dose (mg/L)*	*CT (mg×min/L)*	*Toxin oxidation in %*	*Cell integrity*	*Reference*
Microcystis sp.	MC-LR	Cl_2	0.8–4	20–80	Dependent on dose and WQ	Rapid membrane rupture, toxin released faster than degradation	Ma et al. (2012)
Microcystis sp.	MC-LR	Cl_2	2	60	>90	Unclear due to a rapid oxidation	Ding et al. (2010)
Dol. circinale	SXT	Cl_2	3	7–50	95 (for CT>50)	Cell integrity compromised at CT<7	Zamyadi et al. (2010)
Dol. circinale	SXT	Cl_2	2.5	3–60	>87% (for CT>60)	<1% viability at CT<7	Zamyadi et al. (2012a)
R. raciborskii	CYN		2.5	3–50	83–99	<1% viability at CT<7	
Microcystis sp.	MC-LR			3–50	73–91	<1% viability at CT=31	
Microcystis sp.	MC-LR	Cl_2	4.5	130.3	100	76% compromised	Zamyadi et al. (2013b)
R. raciborskii	CYN	Cl_2		4.0	100	100% inactivated	Cheng et al. (2009)
Microcystis sp.	MC-LR	Cl_2	5	21.4	72	0% viable	Fan et al. (2014)
Microcystis sp. bloom	MC-LR	Cl_2		8	0	100% toxin release	He & Wert (2016)
Microcystis sp.	MC-LR	Cl_2	0.3–2.0		0–95; dependent on colony size and dose	0–98% cell viability dependent on colony size and Cl_2 dose	Fan et al. (2016)
Microcystis sp. bloom	MC-LR	Cl_2		<0.5	0	100% toxin release with no residual	Zhang et al. (2017)
Microcystis sp.	MC-LR	$KMnO_4$	1.5, 4.0		100	100% toxin release with no residual	

(*Continued*)

Table 10.1 (Continued) Summary of studies on the effect of pre-oxidation of water containing cyanobacteria and the effect on cyanotoxins released by the process

Dominant cyanobacteria	*Toxin*	*Oxidant*	*Dose (mg/L)*	*CT (mg×min/L)*	*Toxin oxidation in %*	*Cell integrity*	*Reference*
Microcystis sp.	MC-LR	Cl_2, O_3	0.6–5.0	40–410	100; at all doses for 50 000 cells/mL; for 200 000 cells/mL, dose dependent	Cell integrity compromised at all doses	Wert et al. (2014)
Microcystis sp.	MC-LR	NH_2Cl	0.6–5.0		0	Cell integrity compromised at all doses	
M. aeruginosa *Aph. flosaquae*	-	O_3	0.5	<0.2	-	Instant loss of cell integrity	Coral et al. (2013)
Microcystis sp.	MC-LR	O_3	1–2	25	100	>99% cell inactivation requires 55 mg×min/L	Ding et al. (2010)
Microcystis sp.	MC-LR	O_3	2–5		100	Decrease in the total cell number of 40–80%	Zamyadi et al. (2015)
Microcystis sp.	MC-LR	H_2O_2	51	*2 days	90	96% cell integrity compromised	Fan et al. (2014)
Microcystis sp.	MC-LR	$KMnO_4$	1–2	25	100	Unclear due to a rapid oxidation	Ding et al. (2010)
Dol. circinale	SXT	$KMnO_4$	0.5	*1 h	0	Negligible effect on cell integrity	Ho et al. (2009)
Microcystis sp.	MC-LR	$KMnO_4$	1–10	*3 h	20–74	2–98% cells compromised	Fan et al. (2014)
Microcystis sp.	MC-LR	ClO_2	2.5	>560	Oxidation below LoD	Cell integrity compromised at all doses	Wert et al. (2014)
Microcystis sp.	MC-LR	ClO_2	1.0	10	100	No intact cells left	Zhou et al. (2014)
Microcystis sp.	MC-LR	NH_2Cl	2.8, 3.5	10 000 14 000	0	Immediate release of intracellular toxins	Ho et al. (2010)

CT: chlorine exposure, concentration multiplied by time; $KMnO_4$: potassium permanganate; ClO_2: chlorine dioxide; LoD: limit of detection.

* no CT given, reaction time only.

10.1.2.1 Conventional processes

Background

Conventional treatment generally comprises coagulation and flocculation, followed by clarification and rapid media filtration. Coagulation and flocculation are processes that aggregate suspended particles through the addition of a chemical coagulant. Common coagulants used in water treatment include various aluminium and ferric salts, synthetic organic polymers or a combination of inorganic and organic coagulants. In the clarification step, the coagulated particles, or flocs, are separated from the water by processes such as sedimentation, dissolved air flotation (DAF) or upflow clarification processes. Two common alternatives to the full conventional process are direct filtration, where there is no clarification step, and contact filtration, where the flocculation and clarification steps are eliminated. While the coagulation process is ineffective for the removal of extracellular (dissolved) cyanotoxins, it is very effective in removing cell-bound cyanotoxins through the removal of the whole cyanobacterial cell (Drikas et al., 2001; Henderson et al., 2008; Newcombe et al., 2015).

Effect of the cyanobacteria's specific morphology and characteristics of the specific coagulant

The morphological characteristics of cells, in particular their size, shape and surface characteristics, may influence the efficiency of the coagulants used for the removal of cyanobacteria. One study showed that larger cells were more effectively removed, and within a size group, spherical cells were removed more effectively than elongated cells (Ma et al., 2007). Consequently, microscopy of cyanobacteria – even without identification to the species level – may be useful for optimising or predicting the effectiveness of the coagulation and flocculation process. Henderson et al. (2008) suggested that another indicator of the coagulant dose could be the surface area of the cells: smaller cells would require a higher dose than larger cells at an equivalent biovolume.

Some coagulants may be more effective than others for the removal of cyanobacteria, and the addition of polymers may, or may not, aid in the removal (Teixeira & Rosa, 2006a; Teixeira & Rosa, 2007; De Julio et al., 2010; Newcombe et al., 2015). In addition, it has been reported that proteins and other extracellular organic material (EOM) produced by some cyanobacteria may either inhibit (WHO, 2015) or enhance (Yap et al., 2012) the coagulation process. Cell removal efficiencies vary between species and even between strains of the same species, and depend on parameters such as the physiological stage of the cells, conditions of culturing (if grown in the laboratory), and characteristics of intracellular organic material (IOM) and EOM.

The inconsistent findings in the literature indicate that coagulation efficiencies strongly depend on cyanobacteria species and water quality conditions, and choices should be made on a case-by-case basis according to the raw water quality, available processes and the achievement of other water quality goals (e.g., dissolved organic carbon [DOC] removal).

Operational guidance for the coagulation step

Through a comprehensive study of the application of conventional treatment for the removal of cyanobacteria, Newcombe et al. (2015) developed some practical guidelines using conventional jar testing, which is usually implemented on a routine basis for the optimisation of treatment processes. This research involved several cyanobacterial taxa (*Microcystis* spp., *Raphidiopsis* (*Cylindrospermopsis*) *raciborskii*, *Dolichospermum circinale*, *Pseudanabaena* sp. and *Oscillatoria* sp.) at a range of cell numbers, a range of waters and three coagulants (aluminium sulphate – alum; ferric chloride and aluminium chlorohydrate – ACH). The authors demonstrated that cell removal of *Raphidiopsis raciborskii* (filamentous) was lower (<90%) than that of the other species; however, for all cyanobacteria, the authors found that optimisation of the coagulation process for the common water quality parameters (DOC, or total organic carbon – TOC and/or turbidity) resulted in the optimum removal of cyanobacteria. This finding is in agreement with further studies which demonstrated that the conditions for optimum turbidity removal corresponded with optimum cyanobacteria removal (De Julio et al., 2010; Şengül et al., 2016), although Newcombe et al. (2015) found that turbidity was a good indicator of the removal of cyanobacteria only in raw water with a turbidity of 10 NTU (nephelometric turbidity units) or above. While such literature is a useful starting point for planning the optimisation of coagulation, the most effective way forward for optimising coagulation will be to test efficacy under the respective local conditions during phases in which treatment is particularly challenged by blooms.

It should be noted that while the process of coagulation itself does not cause damage to cells, some cell damage and toxin release can occur if the pH of solution decreases to below 6 (Qian et al., 2014). Thus, in the presence of cyanobacteria, the pH of the coagulation step should be maintained above 6, even when a lower pH may be optimal for the removal of DOC or colour (e.g., when ferric salts are used as the coagulant).

Treatment steps following coagulation

After coagulation, the flocs must be removed by downstream processes. Mouchet and Bonnelye (1998) provided a summary of the types of clarifiers used by water suppliers. They determined that sludge blanket clarifiers (which keep the overflow rate such that it is less than the settling rate of the sludge, allowing the "blanket" of sludge to form) were more effective for cyanobacterial cell removal than static settlers, where the sludge is allowed to settle to the bottom of the clarifier and the clarified water is removed from the top through weirs. It is important to realise that this only applies if the sludge blanket clarifier is operated under optimal conditions to minimise clarified water turbidity. Dissolved air flotation (DAF), where small air bubbles are released from the bottom of the flotation tank and the coagulated/flocculated particles, or flocs, are captured by the bubbles and float to the surface, is particularly effective for the removal of cyanobacteria as many species contain gas vacuoles that provide buoyancy, leading to more efficient clarification by flotation than by sedimentation (Teixeira & Rosa, 2006a; Teixeira & Rosa, 2007; Aparecida Pera do Amaral et al., 2013). However, not all water sources impacted by cyanobacteria are suitable for DAF as, in general, only waters of high colour and low turbidity are amenable to flotation processes.

It is important to be aware that while optimisation of coagulation and clarification will maximise the removal of cyanobacteria under given conditions, 100% removal has seldom been reported in the literature. Even in the best-case scenario, where optimised removals may be in the range of 95–99% of cells, the presence of high cell numbers in the source water could result in significant cell concentrations remaining after clarification. For example, if only 50 000 cells/mL of toxic *Microcystis aeruginosa* entered an optimised treatment plant which achieves 98% removal through alum coagulation and sedimentation, the concentration of cells at the end of the clarification step could still be of the order of 1000 cells/mL, which (at a cell radius of 5 µm) would correspond to a biovolume of 0.5 mm^3/L. Although the barrier was optimised, this concentration of uncoagulated cells is above Alert Level 1 as described in section 5.1.2 for raw water risk. Due to the free-floating nature of cyanobacteria, this can lead to an accumulation within the treatment train, for example, in the clarifier, and a rapid increase in the cell number and toxin concentration, as has been described by Zamyadi et al. (2012b) and Zamyadi et al. (2013a).

Many plants carry out intermediate chlorination, prior to the filters, for manganese removal, or to reduce particle counts in filtered water. In the event of cell breakthrough, as described above, this practice should be either terminated or optimised to ensure the oxidation of cells and released toxins.

Filtration is usually employed immediately after the coagulation and clarification process. A variety of granular media are used in these filters, including sand, anthracite, coal and activated carbon. Although filtration is effective in the removal of cyanobacteria associated with flocs, individual cells and/or filaments are not always removed, resulting in breakthrough of cells into the filtered water. Different genera of cyanobacteria may also respond differently to granular filtration: Zamyadi et al. (2013b) reported poor coagulation and a significant breakthrough of

Aphanizomenon cells after filtration, while the removal was effective for *Microcystis*, *Anabaena* and *Pseudanabaena* in the same plant. Dugan and Williams (2006) evaluated the efficiency of downflow in-line filtration (coagulation followed by direct filtration) in the removal of cyanobacteria cells after abrupt increases in hydraulic loading rates. They observed a consistent impact of cell morphology on cyanobacteria cell breakthrough in all experimental trials, where effluent concentrations of *M. aeruginosa* (spherical shape) were consistently higher than for *Aphanizomenon flosaquae* (filamentous).

Sludge management after coagulation

While coagulation and clarification effectively separate the cyanobacteria from the treated water, up to 98% of the cell-bound cyanotoxins are consequently concentrated in the sludge, or float in the case of DAF; therefore, appropriate handling and disposal of the treatment plant residuals can become a challenge. Over a period of time, cell damage and lysis can occur in sludge produced by coagulation, releasing cyanotoxins (Drikas et al., 2001; Ho et al., 2012a; Zamyadi et al., 2018). This is an issue at long sludge detention times in treatment plant clarifiers where milligram concentrations of microcystins have been reported (Zamyadi et al., 2012b). When cell damage and toxin release from the accumulated sludge occurs during the clarification step, this may pose a significant risk if the treatment plant has no further barriers for dissolved toxin removal. To control this risk, it is therefore important to remove sludge frequently from within the treatment train during a toxic cyanobacterial challenge.

Also, more frequent backwashing of filters may be required, particularly in direct filtration plants, to prevent floc build-up and subsequent cyanotoxin release (Ho et al., 2012a), as cyanobacteria contained in flocs within the filter medium may lyse and release cell-bound toxins into the filter effluent. In particular, backwashing of filters prior to temporary filter shutdown could reduce the possibility of cyanotoxin release. The authors also demonstrated that cyanobacteria appear to be protected within the flocs and were not significantly damaged by rigorous backwashing procedures.

Once the sludge and backwash water are removed from the plant, care needs to be taken to manage the toxic waste appropriately. This issue has been the focus of a number of publications in the international literature due to the growing concern about conserving resources and reusing both treatment plant solids and supernatant water from the sludge treatment facilities (Ho et al., 2012b; Sun et al., 2013; Li et al., 2015; Dreyfus et al., 2016; Pestana et al., 2016).

The management of cyanobacteria-laden sludge is a complex challenge potentially multiple biological, chemical and physical processes taking place simultaneously in the sludge treatment facility. Some of the processes that may occur in sludge lagoons or sludge thickeners include (Pestana et al., 2016):

- reduced cell viability, with consequent lysis and metabolite release;
- cell multiplication in the sludge or supernatant;
- possible increase in metabolite production due to stress;
- biological degradation of metabolites;
- physical and chemical processes resulting in a decrease in metabolite concentrations.

All of these processes will be dependent on

- type of cyanobacteria and toxin;
- rate of biological and chemical degradation;
- rate of physical loss through adsorption;
- rate of production and release;
- water quality (nutrient levels, pH, DOC etc.);
- temperature.

Dreyfus et al. (2016) and Pestana et al. (2016) reported an unexpected additional risk associated with the storage and treatment of cyanobacteria-impacted sludge: these authors conducted a series of experiments, using cultured and environmental cyanobacteria, designed to simulate a sludge treatment lagoon. Within the closed systems containing cyanobacteria-laden sludge and supernatant, they reported an up to 2.8-fold increase of total metabolite concentration (MIB [2-methylisoborneol], geosmin and cyanotoxin concentration) over a period of 2–7 days. They attributed the increase to cell multiplication in the sludge or supernatant, increased metabolite production due to stress or a combination of both factors.

The findings of this research and previous literature show that, in a static (batch) system

- Cyanobacteria, once captured in the sludge, will generally begin to lyse within 0–2 days.
- Some cells will remain viable in the sludge, and the maximum release of toxins (indicative of total cell death and lysis) may take up to several weeks.
- The toxins released may represent up to 2.8 times the initial mass in the closed system.
- The time taken for the biodegradation of the toxins to half the observed maximum concentration may be a week or longer, depending on the toxin and the environmental conditions.

As a result, it is not possible to assess the risk posed by the reuse of sludge and sludge supernatant, and the assumption should be that the toxin concentration in the supernatant water may remain high for time spans of several weeks rather than days.

10.1.2.2 Membrane filtration

Most cyanobacterial cells and/or filaments or colonies are 2 µm in size or larger; therefore, membranes with a pore size smaller than this – such as microfiltration (MF) and ultrafiltration (UF) membranes – will remove the cells. However, a prior coagulation step is generally used in the application of membranes for water treatment, and the presence of cyanobacteria is likely to result in a rapid increase in transmembrane pressure (Dixon et al., 2012). The risk associated with any filtration process is damage to the cells and release of cell-bound toxin, which would not be removed by these membranes. In practice, some removal of dissolved toxins has been noted by MF and UF; however, this is most likely due to the adsorption of the cyanotoxins onto the membrane surface, which would decrease significantly with time as the adsorption sites are occupied by the toxin molecules (Chow et al., 1997; Dixon et al., 2012).

The extent of any damage to the cells will depend on operating parameters such as the flux through the membranes, pressure and the time period between backwashes. While some laboratory studies have shown that the cells are not damaged during filtration (Chow et al., 1997; Gijsbertsen-Abrahamse et al., 2006), full-scale data from a submerged UF membrane plant in South Australia suggest that the accumulation of floc in the membrane tank leads to metabolite release over time (unpublished data).

10.1.2.3 Assessing the risk of toxin release and breakthrough of cyanobacteria and cell-bound toxins

Checklist 10.1 outlines information needed to assess how effectively cyanobacterial cells can be removed intact without toxin release by optimising treatment; the higher the number of affirmative answers, the greater the likelihood of successful cyanobacteria removal. Adaptation of processes to specific local conditions is useful. Treatment plant operators will typically have the expertise and information needed for this assessment.

CHECKLIST 10.1: COLLECTING INFORMATION ON THE EFFICACY OF REMOVAL OF CYANOBACTERIA AND CELL-BOUND TOXINS

- Is data from cyanobacteria monitoring in the source water available at sufficiently regular time intervals to adapt treatment (e.g., fortnightly or weekly during seasons with likely occurrence)?

- Can pre-oxidation processes be suspended during a cyanobacterial bloom until the danger of release of dissolved toxins has passed?
- Are physical barriers optimised (such as coagulation/flocculation or membrane filtration) to achieve optimum particle and dissolved organic carbon (DOC) removal?
- Are treatment residuals removed rapidly from the system (e.g., is sludge removed from clarifiers, are filters backwashed frequently)?
- Are cyanobacterial concentrates isolated from the system (i.e., is sludge supernatant return and/or membrane backwash water return suspended during blooms)?
- Has the system been validated through measuring cyanotoxin concentrations after each of the steps in the treatment train during phases in which it is challenged with bloom material?

10.2 TREATMENT OPTIONS FOR DISSOLVED CYANOTOXINS

Despite the measures described above, a breakthrough of cyanobacterial toxins from the initial treatment steps cannot be avoided in all cases, and a treatment for the removal of dissolved cyanotoxins needs to be considered when planning operations.

Dissolved cyanotoxins can be removed using a range of treatment processes. As the effectiveness of each process depends on the raw water quality and the concentration and type of the cyanotoxins, a multibarrier approach is important for reliable removal.

Three main categories of treatment can be applied for the removal of dissolved cyanotoxins: physical, chemical and biological processes. Physical processes include adsorption and membrane filtration; chemical processes include oxidation by chlorine, ozone or other oxidants; and biological processes employ microorganisms fixed in biofilms, particularly on sand or other media used for filtration (rapid or slow), on granular activated carbon (GAC), or on sediment particles in slow sand filtration or bank filtration (see Chapter 8).

10.2.1 Physical processes

10.2.1.1 Adsorption

Powdered activated carbon

Powdered activated carbon (PAC) is a fine carbonaceous adsorbent with a high surface area (typically 800–1200 m^2/g) that can effectively remove a range of organic contaminants from water. As a treatment for dissolved cyanotoxins, it has the advantage that it can be dosed only when required, and at a range of concentrations. The most effective point for the addition of PAC is prior to coagulation to allow a contact time where the adsorbent is well dispersed and mixed before it is removed during the coagulation process. If this is not possible, PAC may be added with, or after, the coagulant and it will still achieve some removal.

There are many PACs commercially available; they vary in their properties according to the raw material from which they are produced (e.g., coal/anthracite, coconut shell or wood) and their mode and extent of activation. They will also vary in terms of effectiveness and cost. Other important things to note are that individual toxin variants will adsorb to different extents; for example, for MCs, the order of removal efficiency is MC-RR>MC-YR>MC-LR>MC-LA (Newcombe et al., 2003; He et al., 2017), and the efficiency of a particular carbon for a particular toxin will depend on the number and size of the adsorption sites, or pores, in the PAC. In addition, other DOC components compete for adsorption sites and will reduce the removal of cyanotoxins (Figure 10.1).

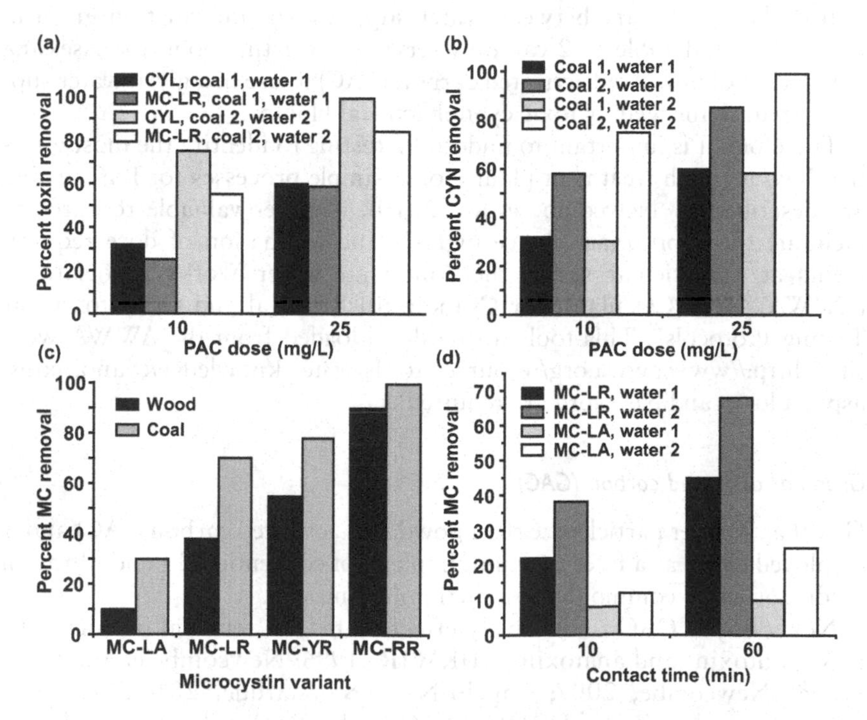

Figure 10.1 Examples of the effect of different factors in the application of PAC on cyanotoxin removal effectiveness. Coal – coal-based PAC; wood – wood-based PAC; water 1 – DOC= 10 mg/L; water 2 – DOC=5 mg/L. (a) Effect of toxin, PAC, DOC and dose; (b) Effect of DOC, PAC and dose; (c) Effect of PAC and MC variant; (d) Effect of MC variant, DOC and contact time. (Adapted from Newcombe & Nicholson, 2004; Cook & Newcombe, 2008; and Ho et al., 2008; Ho et al., 2011; with unpublished data.)

In summary, the major factors controlling the removal efficiency of cyanotoxins by PAC are as follows:

- type of cyanotoxin;
- type of PAC (raw material, manufacturing method, particle size);
- PAC dose;
- point of application;
- contact times;
- DOC concentration and characteristics.

Figure 10.1a–d demonstrates the significant differences in removals that are observed as a result of these factors – removal can range between 10% and close to 100% between cyanotoxins and the type of PAC that is used.

Table 10.2 presents a summary of some of the literature related to the application of PAC for the removal of cyanotoxins. Only very limited data are available for anatoxin-a (Vlad et al., 2014). As there is such a variety of factors influencing the effective application of PAC and the factors mentioned above will vary between water supplies, the information given in Figure 10.1 and Table 10.2 can only serve as a starting point for assessing which type of powdered activated carbon (PAC) to use in a given water supply system challenged by toxic cyanobacterial blooms.

Therefore, it is important to undertake testing to identify the most effective PAC for each treatment plant. Some simple processes for PAC testing are described by Newcombe et al. (2010). Another valuable resource to facilitate the appropriate choice of PAC and estimation of dose requirements at a particular site is the American Water Works Association's (AWWA) "PAC Calculator for Cyanotoxin Removal and Cyanotoxin Jar Testing Protocols". This tool can be downloaded from the AWWA website (http://www.awwa.org/resources-tools/water-knowledge/cyanotoxins.aspx; a login and password are required).

Granular activated carbon (GAC)

GAC has a larger particle size than powdered activated carbon (PAC) and is employed either as a filter medium, in place of conventional rapid filtration media, or, more commonly, as a final polishing step.

New (virgin) GAC is extremely effective for the removal of microcystins, saxitoxins and anatoxin-a (UKWIR, 1996; Newcombe et al., 2003; Ho & Newcombe, 2007; Capelo-Neto & Buarque, 2016); however, continuous adsorption of DOC (DOC preloading) reduces the adsorption capacity of a GAC filter for cyanotoxins and consequently reduces its operational lifetime. For example, virgin GAC removes cyanotoxins to below the detection limit in most cases, while after several months of operation, significant breakthrough usually occurs (Craig & Bailey,

Table 10.2 Effectiveness of removal of different cyanotoxins using various types of PAC

Type of PAC	*Dose (mg/L)*	*Contact time (min)*	*DOC (mg/L)*	*Percent removal of toxin*						*Reference*
				MC-LA	*MC-LR*	*MC-YR*	*MC-RR*	*SXTeq*	*CYN*	
Wood	20				95					Hart et al. (1998)
Coal	12				95					Bruchet et al. (1998)
Wood, coconut	25/50	30			98/60					Donati et al. (1994)
Wood	2	5 days	10	5	15	40	69			Newcombe et al. (2003)
Wood, coal	15	30	7	10/30	38/70	55/78	90/100			Newcombe & Nicholson (2004)
Wood	15	10/60	9.9	8/10	22/45					Cook & Newcombe (2008)
Wood	15	10/60	6.7	12/25	38/68					
Wood, coal	10	15	8.2					17/32		Ho et al. (2009)
Wood, coal	10	70	8.2					28/50		
Wood, coal	30	15	8.2					32/55		
Wood, coal	30	70	8.2					68/85		
Wood, coal	30	15	11.8					50/50		
Wood, coal	30	70	11.8					65/90		

(Continued)

Table 10.2 (Continued) Effectiveness of removal of different cyanotoxins using various types of PAC

Type of PAC	*Dose (mg/L)*	*Contact time (min)*	*DOC (mg/L)*	*Percent removal of toxin*						*Reference*
				MC-LA	*MC-LR*	*MC-YR*	*MC-RR*	*SXTeq*	*CYN*	
Coal	10/20/30	10	10.2						22/43/55	Ho et al. (2008)
Coal	10/20/30	30	10.2						34/60/74	
$Coal_a$, $coal_b$	10	30	4.3						32/58	Ho et al. (2011)
$Coal_a$, $coal_b$	25	30	4.3						60/95	
$Coal_a$, $coal_b$	10	30	5.0						30/75	
$Coal_a$, $coal_b$	25	30	5.0						58/99	
$Coal_a$	10/25	30	4.3	8/35	25/58	25/65	45/80			
$Coal_b$	10/25	30	5	42/60	55/84	65/92	76/95			
Wood	10/100	15			41/87					Drogui et al. (2012)

Subscripts a and b indicate two different PACs with the same raw material.

1995; Ho & Newcombe, 2007). Where no breakthrough is observed after 6 months' operation, the removal has been attributed to a combination of adsorption and biodegradation by biofilms established on the filter (UKWIR, 1996; Wang et al., 2007). There is an abundance of literature describing methods that may be used for the prediction of the lifetime of GAC filters for the removal of organic contaminants (e.g., Capelo-Neto & Buarque, 2016; Kennedy et al., 2017). In practice, it is very difficult to predict when the GAC filter may no longer provide a sufficient barrier for dissolved cyanotoxins. Therefore, it is recommended that if GAC is a major barrier within the plant, it is tested on a regular basis to demonstrate that it will be effective in the event of a toxin challenge. This can be accomplished by

1. full-scale investigative sampling through the plant during a cyanotoxin challenge (see section 9.7) and/or by
2. laboratory testing, accomplished through small-scale column trials with plant water spiked with the cyanotoxin(s) of interest (Sawade et al., 2012).

Laboratory testing also helps to determine whether any of the removal is due to biological degradation on the GAC. This can inform the operation of the filter; for example, if the majority of the removal is due to biological activity, the filter should be maintained as a biofilter (no disinfectant in the backwash water or influent to the filter) and the replacement of the GAC can be postponed.

The removal of cyanotoxins is also affected by the flow rate through the filter, in particular the empty bed contact time (EBCT, the length of time it takes for the volume of water equivalent to the filter volume to pass through). The longer the contact time, the more effective the removal, with an EBCT of 10–15 min considered to be optimal.

The major influences on the effectiveness of GAC for the removal of toxins are as follows:

- type of GAC;
- length of time since commissioning (dissolved organic carbon [DOC] loading time);
- EBCT;
- biological activity resulting in biodegradation.

Ozone can be used as a pretreatment step to granular activated carbon (GAC). The combined process is extremely effective as cyanotoxins are susceptible to ozonation (see following sections) and the GAC can remove any oxidation by-products that are formed.

Other adsorbents

While activated carbon is the most common adsorbent in use for the removal of cyanotoxins, the potential of novel adsorbents to remove cyanotoxins – in particular for the removal of microcystins – is the focus of a significant body of research. Table 10.3 summarises some of the findings for a range of studies on the adsorption of MC-LR onto these materials.

10.2.1.2 Membrane filtration

Membranes such as microfiltration (MF) and ultrafiltration (UF) have a pore range larger than the size of cyanotoxin molecules in solution so they are not an effective measure for the removal of dissolved cyanotoxins. Pore sizes of nanofiltration (NF) and reverse osmosis (RO) membranes do span the size of the cyanotoxin molecules; however, the rejection of the various toxins by these membranes is dependent on the molecular weight cut-off (MWCO) and the surface chemistry of the membrane as well as the relationship between these factors and the size and chemical characteristics (such as polarity, charge and hydrophilicity) of the toxins. That is, some NF or RO membranes will be effective, but others may be only partially effective for particular cyanotoxins (Gijsbertsen-Abrahamse et al., 2006; Teixeira & Rosa, 2006b; Dixon et al., 2012). In summary, it is expected that dissolved toxins would be rejected by RO membranes and NF membranes with a pore size distribution in the lower range in most cases. However, some membranes may allow smaller toxin molecules, like anatoxin-a, to permeate the membrane.

Table 10.3 Adsorption capacities of novel adsorbents for the adsorption of MC-LR

Adsorbent	*Capacity (mg/g)*	*Reference*
Activated carbon fibres	17.0	Pyo & Moon (2005)
Iron oxide nanoparticles	0.7	Lee & Walker (2011)
Magnetic macroporous silica	3.3×10^{-3}	Liu et al. (2010)
Fe_3O_4@CSNT	0.5	Chen et al. (2009)
Magnetic core mesoporous shell	20.0	Deng et al. (2008)
Microgel-Fe(III)	164.5	Dai et al. (2012)
HP20 resin	3.3	Zhao et al. (2013)
Peat	0.3	Sathishkumar et al. (2010)
Fe_3O_4@Al-B	161.3	Lian et al. (2014)
Graphene oxide	1.7	Pavagadhi et al. (2013)
Magnetophoretic polypyrrole nanoparticles	160	Hena et al. (2016)
KOH-activated semicoke	8430	Chen et al. (2015)
HNO_3-activated semicoke	4276	
PAM/SA-MMT [a]	32.7	Wang et al. (2015)

Source: Adapted from Lian et al. (2014).

[a] Polyacrylamide/sodium alginate montmorillonite.

10.2.2 Chemical processes

10.2.2.1 Chlorine

Chlorine has been demonstrated to be an effective oxidant for the destruction of microcystins, saxitoxins and cylindrospermopsin (e.g., Senogles et al., 2000; Acero et al., 2005; Merel et al., 2010), but not for anatoxin-a (Carlile, 1994; Rodriguez et al., 2007b).

The doses required for oxidation of the toxins to below the treatment goal, or the relevant guideline or regulatory value, depend on the conditions at the point of chlorination, which are as follows:

- DOC concentration and characteristics;
- the concentration of any other contaminant that may exert a chlorine demand such as ammonium, iron and manganese;
- reaction time and residual chlorine concentration;
- temperature;
- pH.

The most important criterion for the successful chlorination of toxins is the application of the dose required to overcome the chlorine demand and have sufficient residual chlorine to allow effective oxidation to occur. As this will vary depending on the chemical water characteristics, it is useful to use the concept of chlorine exposure, or CT, the chlorine concentration integrated over the reaction time, given in units of (mg×min)/L.

The pH has a significant effect on the reaction of chlorine with cyanotoxins as hypochlorous acid (HOCl) is a stronger oxidant than the hypochlorite ion (ClO^-), which is the major species of chlorine present at pH values above 7.5. In addition, pH can affect the degree of protonation of cyanotoxins, which may in turn affect their reactivity (Ho et al., 2006). The effect of pH on the chlorination of toxins is most likely a combination of both factors.

Table 10.4 summarises some of the literature relating to the chlorination of cyanotoxins under specific conditions. Anatoxin-a is not included as chlorination would not be recommended as a barrier for this cyanotoxin (Carlile, 1994; Rodríguez et al., 2007).

In summary, the susceptibility of individual microcystin congeners to chlorination was found to be (Ho et al., 2006):

MC-YR > MC-RR > MC-LR > MC-LA

and that of the most common cyanotoxins (Rodríguez et al., 2007):

CYN > MC-LR ≫ ATX.

It is important to note that the CT values and toxin oxidation values given in Table 10.4 are based on laboratory experiments only. It is recommended that caution be applied when considering chlorine as a major barrier, as the limited

Table 10.4 Contact time (CT) values for chlorination of cyanotoxins (mg min/L) (d)-Cl_2 dose mg/L

Toxin, water quality	*pH 6–6.9*	*pH 7–7.9*	*pH 8–8.9*	*Reference*
CT values for 95–100% oxidation				
MC-LR Reagent water 10 °C	47	68	187	Acero et al. (2005)
MC-LR Reagent water 20 °C	35	51	140	
MC-LR Reagent water 10 °C	46	220		Xagoraraki et al. (2006)
CYN DOC=0		30		Senogles et al. (2000)
CYN, DOC=3.0/4.1		2,2		Ho et al. (2008)
CYN, DOC=5		1.5 (d) [a]		Rodriguez et al. (2007c)
SXT_{eq} DOC=2.7		20	20	Ho et al. (2009)
CT values for 90-95% oxidation				
MC-LR Reagent water 10 °C	27	40	110	Acero et al. (2005)
MC-LR Reagent water 20 °C	21	30	82	
MC-LR, DOC=2.9		20		Ho et al. (2006)
MC-YR, DOC=2.9		<1		
MC-YR, DOC=5.0	1			
MC-RR, DOC=2.9		7		
MC-RR, DOC=5.0	3			
CYN, DOC=3.6			1(d) [b]	Rodríguez et al. (2007)
CT values for 75–90% oxidation				
MC-LR, coagulated water, 12 °C		65		Xagoraraki et al. (2006)
SXT_{eq} DOC=5.1		20	20	Ho et al. (2009)
CT values for 50–75% oxidation				
MC-LR, DOC=5.0	4			Ho et al. (2006)
MC-YR, DOC=5.0	<1			
MC-LA, DOC=2.9		15		
MC-LA, DOC=5.0	4			
MC-LR, DOC=3.6			2(d)	Rodríguez et al. (2007)
CYN, DOC=3.6			0.8(d) [b]	
MC-LR Reagent water 11 °C	11	51		Xagoraraki et al. (2006)

[a] Cl_2 demand met.
[b] Cl_2 demand not met.

literature describing full-scale chlorination of cyanotoxins suggests that these CT values may not be sufficient to achieve the desired results in the presence of a natural bloom (Zamyadi et al., 2012b; Mohamed et al., 2015; Mohamed, 2016). A potential issue with applying laboratory-based chlorination results to the full scale was outlined by Acero et al. (2005). These authors reported CT values for the oxidation of microcystins in batch experiments representing an ideal plug-flow reactor (PFR). Chlorination at the full scale does not take place under ideal flow conditions, and the authors suggested it would be better represented by a hybrid PFR and completely stirred tank reactor (CSTR) model. They described the two types of reactors as the most and least effective, respectively, and as a result, the laboratory-based experiments may underestimate the actual required CT by up to an order of magnitude (Acero et al., 2005).

Based on guidelines presented by the US EPA (US EPA, 2010), Stanford et al. (2016) also discussed the effect of nonideal conditions that may influence the application of oxidation data obtained from an ideal configuration and calculated the effect on percent oxidation of MC-LR by chlorine. They determined that the effective CT in the nonideal situation could be about one-third of that in an ideal reactor, and the removal of MC-LR could be approximately half the expected value. Therefore, it is recommended that CT values substantially higher than those suggested by laboratory data be applied at the full scale.

It is also important to note that the efficiency of chlorination is dependent on the chemical characteristics of the water at the chlorination point; for example, turbidity in filtered water>0.3 nephelometric turbidity units (NTU) could not only be an indicator of reduced filtration efficiency but also may reduce the effective chlorine CT for both toxin oxidation and disinfection (WHO, 2017).

10.2.2.2 Ozone

Ozone has been found to be a very effective oxidant for the destruction of dissolved cyanotoxins provided a residual is present (Rositano et al., 2001; Shawwa & Smith, 2001). Rodríguez et al. (2007) showed ozone to be effective for the elimination of a range of cyanotoxins and determined that the order of ease of oxidation followed the trend: MC-LR>CYN>ATX, while Rositano et al. (2001) reported a trend of MC-LR and MC-LA>ATX >STX.

As with chlorine, the doses required for ozonation of the toxins to below the treatment target, guideline value or regulation depend on the conditions at the point of ozonation, which are as follows:

- dissolved organic carbon (DOC) concentration and characteristics;
- reaction time and ozone concentration;
- temperature;
- pH.

In the case of ozone, other water quality parameters like alkalinity may also play a role as the carbonate ion can act as an inhibitor of the reaction by scavenging the hydroxyl radical, the major reactant for the oxidation of organic micropollutants by ozone (Ho et al., 2004).

Table 10.5 summarises some of the literature relating to the ozonation of cyanotoxins under specific conditions.

Table 10.5 Oxidation conditions and percent cyanotoxin reduction by the oxidants ozone, permanganate and monochloramine

Toxin	*Oxidant*	*Dose (mg/L)*	*Conditions (DOC and residual in mg/L)*	*Toxin removal (%)*	*Reference*
CYN	MnO_4^-	1.0	DOC 5.0, pH 7	0	Rodriguez et al. (2007b)
ATX		3.0	DOC 5.0, pH 7, 24h	8	
MC-LR	MnO_4^-	0.6/0.9	DOC 3.6, pH 8, experiment continued until oxidant was consumed	60/90	Rodríguez et al. (2007)
	O_3	0.25/0.3		60/90	
CYN	MnO_4^-	1.5		10	
	O_3	0.3/0.4		60/90	
ATX	MnO_4^-	0.3/0.4		60/90	
	O_3	0.5/0.75		60/90	
MC-LR, MC-RR, MC-YR	MnO_4^-	0.6	DOC 6.7, pH 7.3, 2h	43/50/52	Rodriguez et al. (2007a)
MC-LR, RR, YR	MnO_4^-	1.25	DOC 6.7, pH 7.3, 2h	95	
MC-LR	O_3	0.5	DOC 5.3, pH 7.8; no residual after 5′	100	Rositano et al. (2001)
		0.6	DOC 4.6, pH 7.5; no residual	100	
		0.7	DOC 5.7, pH 7.8; no residual	100	
		1.1	DOC 15.5, pH 7.1; no residual	100	
ATX		1.1	DOC 5.3, pH 7.8; residual 0.06	100	
		1.7	DOC 4.6, pH 7.5; residual 0.06	100	
		1.5	DOC 5.7, pH 7.8; residual 0.05	100	
		>2.2	DOC 15.5, pH 7.1; residual >0.03	100	

(*Continued*)

Table 10.5 (Continued) Oxidation conditions and percent cyanotoxin reduction by the oxidants ozone, permanganate and monochloramine

Toxin	*Oxidant*	*Dose (mg/L)*	*Conditions (DOC and residual in mg/L)*	*Toxin removal (%)*	*Reference*
ATX	O_3	0.1/1	DOC 1.6, pH 8, 30′	20/100	Onstad et al. (2007)
		0.1/1	DOC 13.1	0/30	
CYN		0.1/1	DOC 1.6, pH 8, 30′	35/100	
		0.1/1	DOC 13.1;	0/35	
MC-LR,		0.1/1	DOC 1.6, pH 8, 30′	95/100	
		0.1/1	DOC 13.1	0/60	
MC-LR, MC-LA	O_3	0.5	DOC 5.3, pH 7.4; residual 0.0	100	Brooke et al. (2006)
MC-LR	NH_2Cl	20	5 days	17	Nicholson et al. (1994)
MC-LR	NH_2Cl	2.8	Dam water, pH 8.5, 10 µg/L, CT 30 000	75	Ho et al. (2010)

10.2.2.3 Other oxidants

Chloramine and chlorine dioxide have been shown to be ineffective oxidants for cyanotoxins at CT values normally used in water treatment operations (Rodríguez et al., 2007; Ho et al., 2010).

Potassium permanganate has been reported to oxidise microcystins, anatoxin-a and cylindrospermopsin (Carlile, 1994; Rodriguez et al., 2007a; Rodríguez et al., 2007; Rodriguez et al., 2007b) showed slightly higher permanganate reactivity with MC-RR and MC-YR compared to MC-LR. This is in agreement with the order of oxidation of the microcystins by chlorine (MC-YR>MC-RR>MC-LR>MC-LA) reported by Ho et al. (2006). Although the dose required will be dependent on water chemistry, most studies found a dose of 1–2 mg/L to be very effective.

Table 10.5 summarises some of the literature relating to the use of the more common alternative oxidants for the oxidation of cyanotoxins under specific conditions.

Stanford et al. (2016) describe a tool designed to aid in the application of chlorine, monochlorine, ozone, chlorine dioxide and potassium permanganate for the oxidation of dissolved cyanotoxins (Hazen–Adams Cyanotoxin Tool for Oxidation Kinetics, CyanoTOX). This tool can be downloaded from the website of the American Water Works Association (AWWA, 2019). This tool is based on user-defined oxidant decay curves and desired final toxin concentrations, and nonideal plug flow is taken into account by the use of the baffling factor suggested by the US EPA for disinfection (US EPA, 2010).

In general, UV irradiation, as applied for disinfection of drinking-water, cannot be regarded as a practical method for an effective toxin removal. However, the combination of UV irradiation and catalysts such as hydrogen peroxide and titanium dioxide can be very effective for the destruction of dissolved toxins. These processes, and others that rely on the formation of hydroxyl radicals for the oxidation of chemical contaminants, are referred to as advanced oxidation processes (AOPs). A range of AOPs has been the focus of more recent research. In most cases, oxidation is very effective, but each process depends on the type and concentration of the catalyst, the chemical characteristics of the water and the type of toxin. The application of these processes is therefore very site and process specific. Table 10.6 presents some of the advanced oxidation techniques that have been studied for the destruction of cyanotoxins.

Although advanced oxidation techniques have been shown to be extremely effective at the laboratory scale, their use is very limited at the full scale; therefore, validation as an effective barrier to cyanotoxins is not possible at the time of the publication of this book. One example of AOP application at the full scale is the dosing of hydrogen peroxide or chlorine prior to UV disinfection. As UV irradiation is becoming more common in drinking-water treatment plants as an effective barrier against pathogens, a cost-effective option for some water utilities could be to provide an AOP barrier for organic micropollutants. As technology and cost-effectiveness improves, these processes may become more widespread.

Table 10.6 Advanced oxidation processes that have been studied for the destruction of cyanotoxins

Toxin	*Advanced Oxidation Process (AOP)*	*Reference*
MC-LR	UV/TiO_2	Feitz et al. (1999)
MC-LR	TiO_2/H_2O_2	Cornish et al. (2000)
CYN	UV/TiO_2	Senogles et al. (2001)
MC-LR	UV/TiO_2	Shephard et al. (2002)
MC-LR	UV/TiO_2, UV/TiO_2/ H_2O_2, UV/H_2O_2	Liu et al. (2002)
MC-LR	UV/TiO_2	Liu et al. (2003)
MC-LR	O_3/H_2O_2, O_3 /Fe(II), and Fenton oxidation	Al Momani et al. (2008)
MC-LR	UV/$S_2O_8^{2-}$	Antoniou et al. (2010)
MC-LR	UV/O_3	Liu et al. (2010)
CYN	UV/O_3	Song et al. (2012)
CYN	UV/H_2O_2	He et al. (2014)
MC-LR, CYN	Solar irradiation/TiO_2 nanoparticles	Pinho et al. (2015)
MCs	UV/H_2O_2, UV/$S_2O_8^{2-}$ and UV/HSO_5^-	He et al. (2015)
MCs	Simulated sunlight/H_2O_2	Huo et al. (2015)
MCs	Photoelectrooxidation – electrical current/UV	Garcia et al. (2015)
MCs	UV/microbubble O_3	Zhu et al. (2015)
MC-LR	Chlorine/UV	Zhang et al. (2016)
MC-LR, CYN	Visible-UV/carbon-doped TiO_2	Fotiou et al. (2016)
MCs, NOD	UV/TiO_2-coated glass spheres	Pestana et al. (2015)
CYN	Ozone/TiO_2	Wu et al. (2015)
CYN	Anatase–brookite heterojunction TiO_2/visible-UV	El-Sheikh et al. (2017)
MC-LR	TiO_2-coated carbon electrodes	Lobón et al. (2017)
MC-LR	Sulphur (S), nitrogen (N), and carbon (C)-codoped TiO_2 nanoparticles	Zhang et al. (2014)
MC-LR	Copper oxide-coated activated carbon	Karthikeyan et al. (2016)
MCs	Cold plasma and UV with TiO_2 coating	Jiang et al. (2017)

10.2.2.4 By-product formation

Chemical oxidation of organic compounds may form a range of by-products. In practice, when the oxidation of cyanotoxins takes place, a complex mixture of other organic compounds, such as DOC or natural organic matter (NOM), is also present (particularly in bloom situations) and will be oxidised simultaneously. As a consequence, many different types of by-products will be formed, some of which may be potentially harmful at high concentrations. In fact, many by-products of oxidation of NOM are currently unknown, so identifying the individual by-products of cyanotoxin oxidation separately from those produced from the oxidation of NOM is a difficult exercise. It is particularly problematic as other forms of DOC will be present at concentrations two or more orders of magnitude higher than the cyanotoxins. Table 10.7 presents some of the common disinfection by-products (DBPs) that have been identified after the oxidation of cyanotoxins and/or cyanobacteria.

Table 10.7 Overview of studies on the generation of disinfection by-products (DBPs) and changes in toxicity after oxidation of cyanobacterial cells and specific cyanotoxins

Toxin/cyanobacteria	*Oxidant*	*Presence of NOM (Y/N)*	*"Standard" DBPs*	*Reduced toxicity (Y/N)*	*Reference*
MC-LR, MC-LA, NOD	Cl_2	Y (AOM)	na	Y	Nicholson et al. (1994)
MC-LR, MC-LA	O_3	Y	na	Y	Brooke et al. (2006)
CYN	Cl_2	Y	THMs	n.a.	Rodriguez et al. (2007b)
MC-LR, MC-RR	MnO_4^-, Cl_2	N,Y	na	Y	Rodriguez et al. (2008)
AOM from *Dol. circinale*	Cl_2	Y	THMs, HAAs NDMA	n.a.	Zamyadi et al. (2010)
Diverse taxa	Cl_2	Y	THMs, HAAs NDMA	n.a.	Zamyadi et al. (2012a)
Microcystis sp.	Cl_2	N	THMs	n.a.	Zamyadi et al. (2013a)
CYN	O_3	N	na	Y	Yan et al. (2016)
Diverse taxa	O_3	Y (AOM)	THMs, HAAs	n.a.	Zamyadi et al. (2015)
AOM from *Aph. flosaquae, Ana. flosaquae, M. aeruginosa*	Cl_2	Y (AOM)	TCM, HAAs, DCAN, TCNM	n.a.	Goslan et al. (2017)
MC-LR	Cl_2/UV	N	na	Y	Zhang et al. (2016)
Microcystis sp.	Cl_2	Y (AOM)	TCM, TCNM, DCAN, 1,1,1,TCP, 1,1 DCP	n.a.	Liao et al. (2015)
MC-LR, MC-RR	Cl_2	Y	na	Y	Zong et al. (2015)

(Continued)

Table 10.7 (Continued) Overview of studies on the generation of disinfection by-products (DBPs) and changes in toxicity after oxidation of cyanobacterial cells and specific cyanotoxins

Toxin/cyanobacteria	*Oxidant*	*Presence of NOM (Y/N)*	*"Standard" DBPs*	*Reduced toxicity (Y/N)*	*Reference*
MC-LR	H_2O_2/UV	N	na	Y	Zong et al. (2015)
CYN	Cl_2	N	na	Y	Merel et al. (2010)

n.a.: not analysed; THM: trihalomethane; HAA: haloacetic acid; NDMA: nitrosodimethylamine; TCM: trichloromethane; DCAN: dichloroacetonitrile; TCNM: trichloronitromethane; DCP: dichloropropanone; AOM: intracellular (algal) organic material.

An important aspect of oxidation is whether or not it reduces the overall toxicity of a bloom. Table 10.7 also summarises some studies addressing this issue. Methods of toxicity testing have included mouse bioassay, protein phosphatase inhibition, human hepatoma cell line (HepG2), and mitochondrial and lysosomal activities measured on Caco-2 cells (see section 14.3.2). In all studies, the decrease in the concentration of the toxin due to oxidation has resulted in a decrease of toxicity, although these studies did not address genotoxicity or carcinogenicity which are also of a concern with DBPs.

Pre-oxidation during a cyanobacterial bloom may increase dissolved organic carbon (DOC) due to the release of intracellular organic material (IOMs), including cyanotoxins, which may result in increased concentrations of DBPs in the finished water. However, as discussed above, if pre-oxidation is avoided and cell removal through coagulation and/or filtration is optimised, the presence of cyanobacteria in the raw water should not result in a substantial increase in DBP formation.

10.2.3 Biological filtration

Biological filtration in drinking-water treatment occurs when a biofilm forms on granular filtration media particles such as sand, anthracite, filter coal or granular activated carbon (GAC). In the absence of a strong disinfectant residual in the inlet to the filters or the backwash water, all filter media surfaces will develop a biofilm within weeks to months, depending on the water quality. Reports in the literature describe very effective removal of cyanotoxin by biological filtration.

Microbial degradation during slow sand filtration has been reported to be very effective for the removal of microcystins (Grützmacher et al., 2002) and cylindrospermopsin (Smith et al., 2008), as has more rapid sand filtration (Wang et al., 2007; Somdee et al., 2014).

As discussed above in section 10.1, biological degradation of microcystins and anatoxin-a has also been reported in GAC filters (Carlile, 1994; Newcombe et al., 2003; Wang et al., 2007). GAC filters offer the advantage of two removal mechanisms, adsorption and biodegradation, and thus are an attractive treatment option for an effective cyanotoxin removal.

Biological filtration is not effective for the removal of the saxitoxins, and in one study, biological activity on an anthracite filter was shown to convert the less toxic variants to more toxic compounds, thus increasing the overall toxicity of the filtered water (Kayal et al., 2008).

Although biological filtration can be a very effective barrier, not all biological filters will remove cyanotoxins. For the removal of cyanotoxins to occur on biofilters, the following conditions are essential but not necessarily sufficient for an effective removal:

- Degrading microorganisms are present in the source water.
- They reach the filters.
- They adhere to the biofilm.
- They remain attached in the biofilm in sufficient numbers to accomplish an effective biological removal.

The type and abundance of bacteria, water chemistry, upstream treatment processes, filter media, filter contact time and hydraulic loading all have a major impact on biological filtration processes.

Perhaps the most challenging aspect of biological treatment processes is the delay for biodegradation to commence. This is often referred to as the lag period or lag phase and has been attributed to the degrading microorganisms "acclimating" or "acclimatising" to the conditions, or the numbers of degrading bacteria reaching a critical number after which degradation can be detected (see also Chapter 2). A more recent hypothesis is that bacteria may share genetic information associated with degradation, and the extent of the lag phase may depend on the copies of the genes responsible rather than the numbers of degrading bacteria (Ho et al., 2012a). Lag periods, ranging from days to more than a year, have been reported for some cyanotoxin biodegradation (Wang et al., 2007; Smith et al., 2008; Ho et al., 2012b; Somdee et al., 2014). The lag phase needs to be taken into account when planning control measures, as it may be a major hindrance for the application of biological filtration processes, particularly for the removal of contaminants that occur periodically like cyanotoxins.

In some cases, lag phases can be reduced or eliminated upon re-addition of the toxin in the filter influent, as has been shown for cylindrospermopsin (Smith et al., 2008) or MC-LR (Rapala et al., 1994; Christoffersen et al., 2002; Newcombe et al., 2003). As shown for slow sand filters in section 9.2, a filter that experiences regular toxin challenges may be more likely to display reliable removals with a reduced, or no, lag phase (Ho et al., 2012a).

If biological removal of cyanotoxins through filters within the treatment plant cannot be assured at all times, biological filtration may not be a reliable treatment barrier for the intermittent presence of cyanotoxins, and on-site validation is therefore critically important.

10.2.3.1 Assessing efficacy of treatment steps in eliminating dissolved cyanotoxins

The checklist below outlines the information needed to assess how effectively dissolved cyanotoxins can be removed by available treatment processes and how these can be optimised. The higher the number of affirmative answers, the greater the likelihood of successful cyanotoxin removal. It may be useful to adapt this checklist to specific local conditions. More than one of the treatment options addressed is likely to be available at many treatment plants, and the more the barriers that are present, the lower the risk of cyanotoxins reaching the consumer in critical concentrations. Treatment plant operators will typically have the expertise and information needed for this assessment:

CHECKLIST 10.2: COLLECTING INFORMATION ON THE EFFICACY DISSOLVED CYANOTOXIN REMOVAL

- Are powdered activated carbon (PAC) dosing facilities in place with
  - high-quality PAC, tested for the removal of cyanotoxins?
  - process control to achieve a contact time of 30 min, prior to chemical dosing? Or, if contact time is not available prior to coagulation, sufficiently higher PAC doses?
- Are granular activated filters in place with
  - good-quality GAC?
  - AC that has been tested regularly for an effective toxin removal and replaced when required?
  - empty bed contact time (EBCT) ≥ 10 min?
- Is ozone applied at a dose sufficient to maintain a residual concentration of at least 0.3 mg/L of ozone for 10 min?
- Is chlorine applied at a dose sufficient to allow a CT appropriate for the raw water quality?

10.3 SUMMARY OF TREATMENT MEASURES FOR THE REMOVAL OF CYANOBACTERIA AND ASSOCIATED CYANOTOXINS

In summary of the discussion above, the most common, cost-effective and reliable treatment processes for removing intra- and extracellular cyanotoxins are as follows:

- physical removal of cells, intact and without damage by coagulation or membrane filtration processes;
- adsorption of dissolved cyanotoxins onto activated carbon;
- oxidation, in particular using ozone and/or chlorine.

Table 10.8 Summary of treatment processes for the removal of cyanobacteria and individual cyanotoxins and their potential efficiency under optimum conditions

	Powdered activated carbon	*Coagulation, clarification, filtration*	*Membrane filtration*	*Ozone*	*Granular activated carbon*	*Biological degradation*	*Chlorine*	*Permanganate*
Cyanobacteria	na	+++	+++	-	na	na	-	-
Dissolved Cyanotoxins								
MC-LR	++	na	-	+++	++	+++	+++	+++
MC-LA	+	na	-	+++	+	+++	++	+++
MC-YR	+++	na	-	+++	+++	ie	+++	+++
MC-RR	+++	na	-	+++	+++	ie	+++	+++
STX_{eq}	++	na	-	++	++	-	++	ie
CYL	++	na	-	+++	ie	+++	+++	ie
ATX	ie	na	-	+++	ie	ie	-	++

Source: Adapted from Table A5.5 WHO (2017) and US EPA (2010).

For details on optimum conditions, see text. Note the importance of on-site validation for each individual process.

+++: >80% removal; ++: 50–80% removal; +: 20–50%; -: not recommended as a treatment barrier; na: not applicable; ie: insufficient evidence.

Table 10.8 presents a summary and an assessment of the main treatment measures that can be used for the removal of cyanobacteria and cyanotoxins in a water treatment plant.

10.4 AFTER THE WATER TREATMENT PLANT – RISKS ASSOCIATED WITH TREATED WATER STORAGE

After an effective treatment, it is important to ensure drinking-water remains safe and free of cyanobacterial regrowth. This can be accomplished by avoiding open channels and storages where cyanobacteria may proliferate, and by maintaining sufficient chlorine residual throughout the distribution system. Box 10.1 describes an incident of cyanobacterial growth in a small storage reservoir within a regional drinking-water distribution system.

BOX 10.1: CYANOBACTERIAL BLOOM, YORKE PENINSULA DRINKING-WATER SUPPLY (SOUTH AUSTRALIA)

In April 2000, a cyanobacterial bloom in a treated water storage within the distribution system on the Yorke Peninsula of South Australia led to drinking-water supplied to 15 000 people in 15 towns being declared unsafe for 8 days. In addition to permanent residents, the Yorke Peninsula is a popular vacation area for thousands of South Australian residents during holiday periods. The incident occurred over the Easter long weekend.

The incident began on 13–14 April when complaints from residents about musty tastes and odours led to the detection of the benthic cyanobacterium *Phormidium* aff. *formosum* in the Upper Paskeville Reservoir. The reservoir was an unroofed shallow 185 mL storage of filtered chloraminated drinking-water. The odours were caused by the nontoxic cyanobacterial metabolite 2-methyl isoborneol (MIB). The reservoir was taken out of service on 14 April and, although *Phormidium* was regarded as being nontoxic, precautionary testing of cell extracts using a mouse bioassay was initiated due to the unique nature of the detection. Positive bioassay results were reported on Tuesday 18 April. The State Health Department and the water utility (SA Water) immediately advised the public not to use the water for drinking and cooking. Free bottled water was supplied for all residents and visitors, and bulk water supplies were carted to major consumers of water, including local food manufacturers.

A mains flushing programme was commenced and further testing of cell extracts was initiated. The testing showed that the toxin was inactivated by boiling and chlorination, but not by chloramination. As a result, mains flushing with chloraminated water was replaced with chlorinated water, and the public was advised that the water could be used for drinking and cooking after being boiled. The public was given daily updates on the progress of flushing through joint media conferences convened by the health department and the water utility. Sections of the distribution system were gradually cleared from 21 April, and the whole system was declared safe on 25 April.

There was no evidence of any human health impacts caused by the incident, and a survey of the affected community and local businesses showed that actions undertaken by the health department and the water utility were supported and effective. Visitor numbers over the Easter long weekend were not reduced compared to previous years. Provision of alternative sources of drinking-water and the issuing of daily updates were seen as key factors in minimising concerns and impacts of the incident.

Subsequent investigations showed that the toxin was strongly associated with cellular material, was barely soluble and was not one of the established cyanotoxins (microcystin, cylindrospermopsin, anatoxin) or lipopolysaccharide (Baker et al., 2001). Oral dosing of mice did not produce evidence of toxicity. A roof was installed on the reservoir, and there has been no recurrence of the incident.

10.5 ASSESSING AND REDUCING THE RISK OF CYANOTOXIN BREAKTHROUGH IN DRINKING-WATER TREATMENT IN THE CONTEXT OF A WATER SAFETY PLAN

While optimising processes in the water treatment plant is an important measure for minimising the risk of cyanotoxins entering the drinking-water system, it is best integrated into the overall Water Safety Plan (WSP) for the supply, as introduced in Chapter 6. This includes an assessment of risks from cyanotoxins together with those from other hazards potentially challenging a water supply as well as identifying the critical points/processes within the supply chain that prevent occurrence, remove hazards through treatment and prevent regrowth in the distribution network. For cyanobacteria, this includes preventing toxin release from cells. A further essential part of the WSP concept is routine operational monitoring of the critical control measures and processes identified during the risk assessment to ensure their optimum operation, both in the presence and in the absence of a cyanobacteria challenge. Table 10.9 presents examples of some control measures that may be implemented in drinking-water treatment and some options for routine monitoring of their reliable operation.

A further important element of a WSP is validation of the efficacy of the control measures. Box 10.2 shows an example of how this was done for a specific water treatment plant. For cyanobacteria and their toxins, this is best achieved through investigative sampling when a bloom challenges the treatment. The most effective way to verify that the system of control measures is effective for cyanobacteria and cyanotoxin removal is systematic

Table 10.9 Examples of control measures for drinking-water treatment with options for monitoring their functioning

***Examples** of control measures for drinking-water treatment*	*Options for monitoring their functioning*
Terminate pre-oxidation measures during cyanobacteria bloom	On-line measurement of cyanobacterial cell density at intake (e.g., fluorometry) Inspection of operating records to monitor timely termination of pre-oxidation Regular visual inspection of waterbody at the raw water intake
Ensure a sufficient supply of the most effective PAC available for immediate use if required	Check PAC batches delivered for compliance to specification Check sufficient PAC available on site at the beginning of the cyanobacteria high-risk period
Determine approximate PAC dose based on toxin concentrations or maximum expected toxin concentrations estimated from 3 μg toxin per mm^3 biovolume or 1 μg toxin per μg chlorophyll-*a* (see Chapter 5)	Monitor intake cell numbers Record plant flow; inspect records of PAC dosing
Optimise coagulation for the removal of colour and turbidity	Record turbidity on-line and define corrective action if threshold level is exceeded
Maintain GAC contact time at ≥ 10 min Replace GAC when required to ensure cyanotoxin removal	Monitor GAC filter loading rates and verify that they result in a sufficient contact time for cyanotoxin removal Periodically test GAC for toxin removal [a]
Maintain ozone dose to produce a residual of ≥ 0.3 mg/L for a contact time ≥ 5 min	Record ozone concentration online at the outlet of this treatment step
Increase chlorine dose to produce a CT of ≥ 100 mg min/L	Record chlorine concentration on-line at the outlet of this treatment step

[a] Laboratory column testing of GAC can be used as an indication of the removals to be expected in the full scale. If this includes comparison with a sterilised sample, additional removal due to biological activity can be identified.

investigative sampling through the treatment plant during a bloom. The list below presents some examples of important measures to ensure the results are representative of the actual treatment process efficiencies. Note that this list is not comprehensive and needs to be adapted to the specific steps of the given treatment train:

- Develop a sampling procedure that identifies sampling points and describes sampling and sample handling practices.
- Have sampling packs (sampling procedure, sample bottles, filters, a template to record sample names and numbers, dates and times) ready

and several staff members trained so the response to a challenge can be immediate.

- Measure both total and dissolved toxins at the inlet to the plant to investigate the removal efficiency for each fraction.
- To quantify the efficacy of each step in the treatment train, realise that concentrations in the raw water can vary rapidly; therefore, prepare a list or table of the time each slug of water resides in a unit of the treatment process, and take each subsequent sample to quantify removals (or release) after the appropriate time lapse, equivalent to the detention time in the respective unit of the process, to ensure the results represent, as close as possible, the same slug of water.
- When powdered activated carbon (PAC) is used, take sample to determine the effectiveness for the removal of dissolved metabolites. As PAC is effective only while in suspension, samples should be taken prior to the sedimentation step.
- Samples taken after PAC dosing should be immediately filtered as the PAC may continue to adsorb metabolites over time and will thus give an inaccurate indication of plant performance.

BOX 10.2: CASE STUDY: A SYSTEM RISK ASSESSMENT FOR CYANOTOXIN CONTROL

When a water utility began to experience customer complaints due to earthy/musty tastes and odours caused by a cyanobacterial bloom in the raw water supply, water quality managers and plant operators realised there was also a potential risk of breakthrough of cyanotoxins into the drinking-water. While waiting for results of toxin analysis, they undertook an assessment of the barriers in place in the treatment plant for the removal of cyanobacterial cells and their metabolites as well as actions to minimise the risks of cyanotoxin breakthrough. The process took place in three steps:

- *Identify*: Identification of all of the points of potential control and risk;
- *Assess*: Assessment of the critical points of control and risk;
- *Optimise*: Optimisation of the control measures and minimisation of the risks.

After these steps had been completed, a verification process was undertaken to ensure the control measures were functioning as expected.

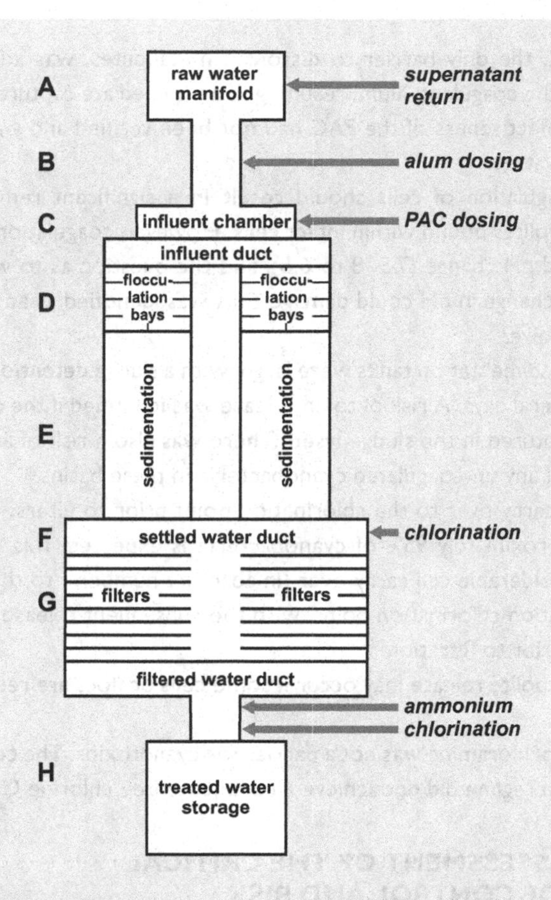

STEP1: IDENTIFICATION OF POINTS OF POTENTIAL CONTROL AND RISK

A schematic of the plant (see figure) was drafted to aid in the identification of points within the plant that might be either helping to control, or contributing to, the problem of toxin (or more general, metabolite) breakthrough into the distribution system.

The following points of potential control and risk were identified:

A – Cell breakup in pumps could cause the release of dissolved metabolites. Return of sludge supernatant could be contributing to the metabolite load within the plant.

B – Cell breakup in mixing chambers/flocculation bays could cause the release of dissolved metabolites.

C – PAC, the only barrier to dissolved metabolites, was added shortly after the coagulant, alum, resulting in an immediate capture in the floc. The effectiveness of the PAC had not been verified and was unknown at this stage.

D – Coagulation of cells should result in a significant removal of the metabolites bound within intact cells. However, coagulation resulted in a rapid pH change (7.5–8 to 6.5), and the question as to whether this rapid change in pH could damage cells was identified as an uncertainty to resolve.

E – The sedimentation tanks were large, with a sludge detention time of up to several days. A risk of toxin release was identified if the cyanobacteria captured in the sludge lysed. There was also a risk of an accumulation of any un-coagulated cyanobacteria in these basins.

F – Cell carry-over to the chlorination point prior to filters: if a removal of approximately 95% of cyanobacteria is expected, this could result in considerable cell carry-over (in absolute numbers) to the post-sedimentation chlorination point, with the subsequent release of metabolites prior to filtration.

G – Metabolite release may occur if some cells or flocs are retained in the filters.

H – Monochloramine was not a barrier for cyanotoxins. The current disinfection regime did not achieve an adequate free chlorine CT.

STEP 2: ASSESSMENT OF THE CRITICAL POINTS OF CONTROL AND RISK

Controls: This plant had three potential barriers to cyanotoxins:

- PAC application;
- coagulation;
- chlorination.

Risks: Using operator knowledge and previous monitoring, the three major risks were identified as follows:

- recycled sludge supernatant entering the plant inlet;
- cell lysis in the sedimentation tanks;
- accumulation of toxic cyanobacteria on the surface of the sedimentation tanks.

STEP 3: OPTIMISATION OF THE CONTROL MEASURES AND MINIMISATION OF THE RISKS

Controls:

- Coagulation at the plant was well managed, and the regular on-line monitoring of turbidity to optimise the coagulation process was considered to be sufficient for the optimisation of cell removal. pH was monitored closely to ensure it remained above 6.5.
- PAC was an expensive control method, and little was known regarding its effectiveness within the plant. As an interim measure, the dose was increased to the highest practicable within the plant until toxin analysis results were received.
- The disinfection process was modified temporarily to ensure a chlorine CT. The chlorine dose prior to the filters was increased, and the final chlorine dose was reduced. Although the CT remained below 20 mg×min/L due to engineering constraints, it was considered a more effective barrier than the previous process of chloramination.

Risks:

- Supernatant recycling was terminated until the risk could be quantified by toxin analysis.
- Sludge removal from the sedimentation basin was increased in frequency to ensure a sludge detention time of< 1 day.
- Visual monitoring of the surface of the sedimentation tanks was undertaken by the operator twice daily to allow the rapid identification of any accumulation of cyanobacteria on the surface. A portable pump that could be used to remove any cyanobacterial accumulation to waste was on stand-by if required.

Longer-term investigations and operational changes were undertaken to reduce future cyanotoxin risk using the findings of the system risk assessment.

LABORATORY STUDIES

Powdered activated carbon (PAC) testing was undertaken to determine

- the most effective PAC available on the market;
- expected metabolite removals under plant conditions;
- optimum dosing location and concentration of PAC.

Chlorination testing was undertaken to determine

- necessary CT values for the elimination of a range of toxins that could potentially challenge the water treatment plant;
- the appropriate configuration of the disinfection process to ensure a sufficient CT as well as an effective monochloramine production.

INFRASTRUCTURE MODIFICATION

- A PAC precoagulation contact tank was installed to ensure the optimum value from the adsorbent.
- A change in the disinfection regime was implemented to ensure a chlorine CT of at least 100 mg×min/L prior to ammonium addition.

PROCESS AND SYSTEM MODIFICATION

- A cyanotoxin response plan was developed by operators and water quality managers, and implemented at the plant.

Investigative sampling was undertaken on a regular basis to verify control measures were optimised.

IN-PLANT VERIFICATION OF THE EFFICIENCY OF THE CONTROL MEASURES

After the measures described above were put in place, the operators undertook a systematic investigative sampling through the plant to verify that each treatment step and control point was functioning to minimise the risk of cyanotoxin breakthrough into the distribution system. Duplicate samples were taken at each of the points A–H identified (see figure) for both total and dissolved metabolites to determine the removal of cells and cyanotoxins.

The results provided some useful insights into the efficiency of the control measures in place at the plant:

- Toxic *Microcystis* was present in the raw water. Microcystin-LR concentration was 3–5 µg/L, of which 75% was intracellular.
- PAC reduced the dissolved toxin by approximately 20%.
- Coagulation reduced the intracellular toxins to below detection.
- No increase in dissolved toxin was detected in the sedimentation basin.
- The available free chlorine CT reduced the dissolved toxin to the below detection limit.

No toxin was detected in the sludge treatment supernatant so recycling was reintroduced.

Note that taste and odour episodes caused by methylisoborneol (MIB) and/or geosmin do not necessarily indicate the presence of cyanotoxins; however, they may be more common than toxic blooms in the raw water source (see also section 2.9). Levels of MIB or geosmin can be measured through the plant using the procedure described above. These compounds will respond differently to the activated carbon and oxidation steps. However, for assessing the efficacy of some treatment steps, they can be used as a surrogate, for example, as an indicator of removals through coagulation, damage to cyanobacteria and release of metabolites.

A Water Safety Plan (WSP) supports day-to-day operations under normal circumstances, which may include "normal" amounts of cyanobacteria in the raw water. Heavy blooms may require additional control, and it is important to develop an emergency response plan that is integrated within the WSP framework for timely and effective responses, as discussed in Chapter 15. It is important that the staff of a treatment plant is familiar with both the WSP and the integrated emergency response plan. Audits are useful for this purpose and should include interviews with staff to check their familiarity with these plans and, for example, whether training exercises of responses to bloom events are periodically conducted.

10.6 ACHIEVEMENT OF CYANOTOXIN GUIDELINE VALUES

Clearly, the ultimate objective of the application of treatment measures for the control of cyanobacteria and cyanotoxins is the provision of safe drinking-water. For the cyanotoxins, this means achieving the provisional WHO guideline values of 1 µg/L for MC-LR and 0.7 µg/L for CYN – or for transient short episodes, at least the short-term guideline values for these toxins or the health-based reference value for ATX and the acute value for STX given in Table 5.1 (Note that while the provisional guideline values for MCs are given for MC-LR, the recommendation is to apply them to the sum of all MCs). As emphasised throughout this chapter, the removal achieved, and therefore the ability to achieve the guideline values, is strongly affected by site-specific conditions and therefore requires laboratory testing, monitoring of treatment processes and validation of treatment steps.

Once the effectiveness of treatment process is determined, it is possible to calculate the maximum tolerable concentrations (MTCs) of cyanobacteria and cyanotoxins in the raw water that can be controlled by the existing treatment measures to ensure the production of safe drinking-water. The calculation proposed by Schmidt et al. (2002) is:

$$\mathrm{MTC} = \frac{\mathrm{GV}}{1-\eta}$$

where GV is the guideline value and η is the achievable fraction removal for dissolved or cell-bound cyanotoxins.

For example, for CYN with GV=0.7 µg/L, for a plant with powdered activated carbon (PAC) achieving 70% CYN removal, the MTC of dissolved CYN in the raw water would be 0.7/0.3=2.3 µg/L (in face of the barriers in place in this given plant).

The application of the concept of MTC to cell-bound toxins requires a measure for the toxin content per cell, or cell quota. This can be determined locally by cell counts via microscopy (see section 13.3) and analysing cell-bound microcystin concentrations (see Chapter 14). In Chapter 2, Table 2.3 presents some literature values for MC content per cell ranging from 5 to 553 fg/cell with an average of 115 fg/cell (= 115×10^{-9} µg/cell). This range of variation is wide, and furthermore, published cell quota are largely limited to *Microcystis*. Operators of a treatment plant therefore best periodically determine the cell quota of the cyanobacteria currently present during a bloom. Using a cell quota of 115 fg/cell for intracellular MC, the guideline value is reached by a cell concentration (cell equivalent) of:

$$\text{Cell equivalent} = \frac{1\,\frac{\mu g}{L}}{115 \times 10^{-9}\,\frac{\mu g}{\text{cell}}} = 8695652\,\frac{\text{cells}}{L} = 8696\,\frac{\text{cells}}{\text{mL}}$$

If we estimate the cell removal by coagulation, η, at 90%, a conservative estimate of the MTC in cells/mL is given by

$$\text{MTC} = \frac{8696\,\frac{\text{cells}}{\text{mL}}}{0.10} = 86957\,\frac{\text{cells}}{\text{mL}}$$

A similar calculation for the minimum and maximum values for cell quotas given above amounts to MTCs of 20 000/0.1=200 000 cells/mL and 1808/0.1=18 080 cells/mL, respectively. Therefore, these calculations indicate that a treatment plant achieving 90% removal of cell-bound toxin through coagulation can achieve the guideline value of MC-LR through this one treatment step when challenged by concentrations between 18 080 cells/mL (minimum MTC) and 200 000 cells/mL (maximum MTC), provided the cell quota for microcystins is in the range given above.

In practice, both dissolved and cell-bound toxins will be present in raw water, and most treatment plants will have multiple barriers in place. A simple spreadsheet calculator as described by Zamyadi et al. (2018) supports these calculations. Cumulative removals of both dissolved and cell-bound toxins can then be taken into account when calculating the MTCs for individual treatment plants.

As these estimates indicate, guideline values should be achievable in an optimised treatment plant with multiple barriers in place where toxin removals are cumulative, under moderate conditions of cyanobacterial challenge. Calculations are best undertaken on a site-by-site basis as an important element of a cyanotoxin management plan (best developed as part of a Water Safety Plan [WSP]; see above). Furthermore, while estimates like those given by these calculations serve as point of departure, where mitigating cyanobacterial occurrence in the raw water is not successful or insufficient, the most effective way to ensure that guideline values are achieved is through periodic validation of the treatment process (most effectively when challenged by blooms) combined with monitoring of cyanotoxin concentrations in treated water during periods of cyanobacterial occurrence in the raw water.

REFERENCES

Acero JL, Rodriguez E, Meriluoto J (2005). Kinetics of reactions between chlorine and the cyanobacterial toxins microcystins. Water Res. 39:1628–1638.

Al Momani F, Smith DW, El-Din MG (2008). Degradation of cyanobacteria toxin by advanced oxidation processes. J Hazard Mat. 150:238–249.

Antoniou MG, de la Cruz AA, Dionysiou DD (2010). Intermediates and reaction pathways from the degradation of microcystin-LR with sulfate radicals. Environ Sci Technol. 44:7238–7244.

Aparecida Pera do Amaral P, Coral LA, Nagel-Hassemer ME, Belli TJ, Lapolli FR (2013) Association of dissolved air flotation (DAF) with microfiltration for cyanobacterial removal in water supply. Desal Water Treat. 51:1664–1671.

AWWA (2019). Resources & Tools. Available at: https://www.awwa.org/Resources-Tools/water-knowledge/cyanotoxins.

Baker PD, Steffensen DA, Humpage AR, Nicholson BC, Falconer IR, Lanthois B et al. (2001). Preliminary evidence of toxicity asociated with the benthic cyanobacetrium *Phormidium* in South Australia. Environ Toxicol. 16:506–511.

Brooke S, Newcombe G, Nicholson B, Klass G (2006). Decrease in toxicity of microcystins LA and LR in drinking water by ozonation. Toxicon. 48:1054–1059.

Bruchet A, Bernazeau F, Baudin I, Pieronne P (1998). Algal toxins in surface waters: analysis and treatment. Water Supply. 16:619–623.

Capelo-Neto J, Buarque NMS (2016). Simulation of saxitoxins adsorption in full-scale GAC filter using HSDM. Water Res. 88:558–565.

Carlile P (1994). Further studies to investigate microcystin-LR and anatoxin-a removal from water. Report FR 0458. Denver (CO): Water Research Foundation.

Chen H, Lu X, Deng C, Yan X (2009). Facile synthesis of uniform microspheres composed of a magnetite core and copper silicate nanotube shell for removal of microcystins in water. J Phys Chem C. 113:21068–21073.

Chen Y, Zhang X, Liu Q, Wang X, Xu L, Zhang Z (2015). Facile and economical synthesis of porous activated semi-cokes for highly efficient and fast removal of microcystin-LR. J Hazard Mat. 299:325–332.

Cheng XL, Shi HL, Adams CD, Timmons T, Ma YF (2009). Effects of oxidative and physical treatments on inactivation of *Cylindrospermopsis raciborskii* and removal of cylindrospermopsin. Wat Sci Technol. 60:689–697.

Chow C, Panglisch S, House J, Drikas M, Burch M, Gimbel R (1997). Study of membrane filtration for the removal of cyanobacterial cells. AQUA. 46:324–334.

Christoffersen K, Lyck S, Winding A (2002). Microbial activity and bacterial community structure during degradation of microcystins. Aquat Microb Ecol. 27:125–136.

Cook D, Newcombe G (2008). Comparison and modeling of the adsorption of two microcystin analogues onto powdered activated carbon. Environ Technol. 29:525–534.

Coral LA, Zamyadi A, Barbeau B, Bassetti FJ, Lapolli FR, Prevost M (2013). Oxidation of *Microcystis aeruginosa* and *Anabaena flos-aquae* by ozone: impacts on cell integrity and chlorination by-product formation. Water Res. 47:2983–2994.

Cornish BJPA, Lawton LA, Robertson PKJ (2000). Hydrogen peroxide enhanced photocatalytic oxidation of microcystin-LR using titanium dioxide. Appl Photocatal B: Environ. 25:59–67.

Craig K, Bailey D (1995). Cyanobacterial toxin microcystin-LR removal using activated carbon–Hunter Water Corporation Experience. Proceedings of the 16th AWWA Federal Convention. Sydney, Australia: AWWA:579–586.

Dai G, Quan C, Zhang X, Liu J, Song L, Gan N (2012). Fast removal of cyanobacterial toxin microcystin-LR by a low-cytotoxic microgel-Fe (III) complex. Water Res. 46:1482–1489.

De Julio M, Fioravante D, De Julio T, Oroski F, Graham N (2010). A methodology for optimising the removal of cyanobacteria cells from a brazilian eutrophic water. Braz J Chem Eng. 27:113–126.

Deng Y, Qi D, Deng C, Zhang X, Zhao D (2008). Superparamagnetic high-magnetization microspheres with an Fe_3O_4@ SiO_2 core and perpendicularly aligned mesoporous SiO_2 shell for removal of microcystins. J Am Chem Soc. 130:28–29.

Ding J, Shi H, Timmons T, Adams C (2010). Release and removal of microcystins from *Microcystis* during oxidative-, physical-, and UV-based disinfection. J Environ Eng. 136:2–11.

Dixon M, Ho L, Chow C, Newcombe G, Croue JP, Cigana J et al. (2012). Evaluation of integrated membranes for T&O and toxin control. Denver (CO): Water Research Foundation.

Donati C, Drikas M, Hayes R, Newcombe G (1994). Microcystin-LR adsorption by powdered activated carbon. Water Res. 28:1735–1742.

Dreyfus J, Monrolin Y, Pestana CJ, Reeve PJ, Sawade E, Newton K et al. (2016). Identification and assessment of water quality risks associated with sludge supernatant recycling in the presence of cyanobacteria. J Water Supply: Res Technol-AQUA. 65:441–452.

Drikas M, Chow CW, House J, Burch MD (2001). Using coagulation, flocculation and settling to remove toxic cyanobacteria. J - Am Water Works Assoc. 93:100–111.

Drogui P, Daghrir R, Simard M-C, Sauvageau C, Blais JF (2012). Removal of microcystin-LR from spiked water using either activated carbon or anthracite as filter material. Environ Technol. 33:381–391.

Dugan NR, Smith SJ, Sanan TT (2018). Impacts of potassium permanganate and powdered activated carbon on cyanotoxin release. J Am Water Works Assoc. 110:E31–E42.

Dugan NR, Williams DJ (2006). Cyanobacteria passage through drinking water filters during perturbation episodes as a function of cell morphology, coagulant and initial filter loading rate. Harmful Algae. 5:26–35.

El-Sheikh SM, Khedr TM, Zhang G, Vogiazi V, Ismail AA, O'Shea K et al. (2017). Tailored synthesis of anatase–brookite heterojunction photocatalysts for degradation of cylindrospermopsin under UV–Vis light. Chem Eng J. 310:428–436.

Fan J, Ho L, Hobson P, Daly R, Brookes J (2014). Application of various oxidants for cyanobacteria control and cyanotoxin removal in wastewater treatment. J Environ Eng. 140:04014022.

Fan J, Rao L, Chiu Y-T, Lin T-F (2016). Impact of chlorine on the cell integrity and toxin release and degradation of colonial *Microcystis*. Water Res. 102:394–404.

Feitz AJ, Waite TD, Jones GJ, Boyden BH, Orr PT (1999). Photocatalytic degradation of the blue green algal toxin microcystin-LR in a natural organic-aqueous matrix. Environ Sci Technol. 33:243–249.

Fotiou T, Triantis TM, Kaloudis T, O'Shea KE, Dionysiou DD, Hiskia A (2016). Assessment of the roles of reactive oxygen species in the UV and visible light photocatalytic degradation of cyanotoxins and water taste and odor compounds using C–TiO_2. Water Res. 90:52–61.

Garcia ACdA, Rodrigues MAS, Xavier JLN, Gazulla V, Meneguzzi A, Bernardes AM (2015). Degradation of cyanotoxins (microcystin) in drinking water using photoelectrooxidation. Braz J Biol. 75:45–49.

Gijsbertsen-Abrahamse A, Schmidt W, Chorus I, Heijman S (2006). Removal of cyanotoxins by ultrafiltration and nanofiltration. J Memb Sci. 276:252–259.

Goslan EH, Seigle C, Purcell D, Henderson R, Parsons SA, Jefferson B et al. (2017). Carbonaceous and nitrogenous disinfection by-product formation from algal organic matter. Chemosphere. 170:1–9.

Grützmacher G, Böttcher G, Chorus I, Bartel H (2002). Removal of microcystins by slow sand filtration. Environ Toxicol. 17:386–394.

Hart J, Fawell J, Croll B (1998). The fate of both intra- and extracellular toxins during drinking water treatment. Water Supply. 16:611–616.

He X, Armah A, Hiskia A, Kaloudis T, O'Shea K, Dionysiou DD (2015). Destruction of microcystins (cyanotoxins) by UV-254 nm-based direct photolysis and advanced oxidation processes (AOPs): influence of variable amino acids on the degradation kinetics and reaction mechanisms. Water Res. 74:227–238.

He X, Stanford BD, Adams C, Rosenfeldt EJ, Wert EC (2017). Varied influence of microcystin structural difference on ELISA cross-reactivity and chlorination efficiency of congener mixtures. Water Res. 126:515–523.

He X, Wert EC (2016). Colonial cell disaggregation and intracellular microcystin release following chlorination of naturally occurring *Microcystis*. Water Res. 101:10–16.

He X, Zhang G, de la Cruz AA, O'Shea KE, Dionysiou DD (2014). Degradation mechanism of cyanobacterial toxin cylindrospermopsin by hydroxyl radicals in homogeneous UV/H_2O_2 process. Environ Sci Technol. 48:4495–4504.

Hena S, Rozi R, Tabassum S, Huda A (2016). Simultaneous removal of potent cyanotoxins from water using magnetophoretic nanoparticle of polypyrrole: adsorption kinetic and isotherm study. Environ Sci Pollut Res. 23:14868–14880.

Henderson R, Parsons SA, Jefferson B (2008). The impact of algal properties and pre-oxidation on solid–liquid separation of algae. Water Res. 42:1827–1845.

Ho L, Croué J-P, Newcombe G (2004). The effect of water quality and NOM character on the ozonation of MIB and geosmin. Wat Sci Technol. 49:249–255.

Ho L, Kayal N, Trolio R, Newcombe G (2010). Determining the fate of *Microcystis aeruginosa* cells and microcystin toxins following chloramination. Wat Sci Technol. 62:442–450.

Ho L, Lambling P, Bustamante H, Duker P, Newcombe G (2011). Application of powdered activated carbon for the adsorption of cylindrospermopsin and microcystin toxins from drinking water supplies. Water Res. 45:2954–2964.

Ho L, Newcombe G (2007). Evaluating the adsorption of microcystin toxins using granular activated carbon (GAC). J Water Supply: Res Technol-AQUA. 56:281–291.

Ho L, Onstad G, von Gunten U, Rinck-Pfeiffer S, Craig K, Newcombe G (2006). Differences in the chlorine reactivity of four microcystin analogues. Water Res. 40:1200–1209.

Ho L, Sawade E, Newcombe G (2012a). Biological treatment options for cyanobacteria metabolite removal–A review. Water Res 46:1536–1548.

Ho L, Slyman N, Kaeding U, Newcombe G (2008). Optimizing PAC and chlorination practices for cylindrospermopsin removal. J - Am Water Works Assoc. 100:88.

Ho L, Tang T, Monis PT, Hoefel D (2012b). Biodegradation of multiple cyanobacterial metabolites in drinking water supplies. Chemosphere. 87:1149–1154.

Ho L, Tanis-Plant P, Kayal N, Slyman N, Newcombe G (2009). Optimising water treatment practices for the removal of *Anabaena circinalis* and its associated metabolites, geosmin and saxitoxins. J Water Health. 7:544–556.

Huo X, Chang D-W, Tseng J-H, Burch MD, Lin T-F (2015). Exposure of *Microcystis aeruginosa* to hydrogen peroxide under light: kinetic modeling of cell rupture and simultaneous microcystin degradation. Environ Sci Technol. 49:5502–5510.

Jiang X, Lee S, Mok C, Lee J (2017). Sustainable methods for decontamination of microcystin in water using cold plasma and UV with reusable TiO_2 nanoparticle coating. Int J Environ Res Public Health. 14:480.

Karthikeyan S, Dionysiou DD, Lee AF, Suvitha S, Maharaja P, Wilson K et al. (2016). Hydroxyl radical generation by cactus-like copper oxide nanoporous carbon catalysts for microcystin-LR environmental remediation. Catal Sci Technol. 6:530–544.

Kayal N, Newcombe G, Ho L (2008). Investigating the fate of saxitoxins in biologically active water treatment plant filters. Environ Toxicol. 23:751–755.

Kennedy AM, Reinert AM, Knappe DR, Summers RS (2017). Prediction of full-scale GAC adsorption of organic micropollutants. Environ Eng Sci. 34:496–507.

Lee J, Walker HW (2011). Adsorption of microcystin-LR onto iron oxide nanoparticles. Colloids Surf A: Physicochem Eng Asp. 373:94–100.

Li X, Pei H, Hu W, Meng P, Sun F, Ma G et al. (2015). The fate of *Microcystis aeruginosa* cells during the ferric chloride coagulation and flocs storage processes. Environ Technol. 36:920–928.

Lian L, Cao X, Wu Y, Sun D, Lou D (2014). A green synthesis of magnetic bentonite material and its application for removal of microcystin-LR in water. Appl Surf Sci. 289:245–251.

Liao X, Liu J, Yang M, Ma H, Yuan B, Huang C-H (2015). Evaluation of disinfection by-product formation potential (DBPFP) during chlorination of two algae species—blue-green *Microcystis aeruginosa* and diatom *Cyclotella meneghiniana*. Sci Tot Environ. 532:540–547.

Liu J, Cai Y, Deng Y, Sun Z, Gu D, Tu B et al. (2010). Magnetic 3-D ordered macroporous silica templated from binary colloidal crystals and its application for effective removal of microcystin. Micropor Mesopor Mat. 130:26–31.

Liu I, Lawton LA, Cornish BJPA, Robertson PKJ (2002). Mechanistic and toxicity tudies of the photocatalytic oxidation of microcystin-LR. J Photochem Photobiol. 148:349–354.

Liu I, Lawton LA, Robertson PK (2003). Mechanistic studies of the photocatalytic oxidation of microcystin-LR: an investigation of byproducts of the decomposition process. Environ Sci Technol. 37:3214–3219.

Lobón GS, Yepez A, Garcia LF, Morais RL, Vaz BG, Carvalho VV et al. (2017). Efficient electrochemical remediation of microcystin-LR in tap water using designer TiO_2 @ carbon electrodes. Sci Rep. 7:41326.

Ma J, Lei G, Fang J (2007). Effect of algae species population structure on their removal by coagulation and filtration processes–a case study. J Water Supply: Res Technol-AQUA. 56:41–54.

Ma M, Liu R, Liu H, Qu J (2012). Chlorination of *Microcystis aeruginosa* suspension: cell lysis, toxin release and degradation. J Hazard Mat. 217:279–285.

Merel S, Clement M, Mourot A, Fessard V, Thomas O (2010). Characterization of cylindrospermopsin chlorination. Sci Tot Environ. 408:3433–3442.

Mohamed ZA (2016). Breakthrough of *Oscillatoria limnetica* and microcystin toxins into drinking water treatment plants-examples from the Nile River, Egypt. Water SA. 42:161–165.

Mohamed ZA, Deyab MA, Abou-Dobara MI, El-Sayed AK, El-Raghi WM (2015). Occurrence of cyanobacteria and microcystin toxins in raw and treated waters of the Nile River, Egypt: implication for water treatment and human health. Environ Sci Pollut Res. 22:11716–11727.

Mouchet P, Bonnelye V (1998). Solving algae problems: French expertise and worldwide applications. J Water Supply: Res Technol-AQUA. 47:125–141.

Newcombe G, Cook D, Brooke S, Ho L, Slyman N (2003). Treatment options for microcystin toxins: similarities and differences between variants. Environ Technol. 24:299–308.

Newcombe G, Dreyfus J, Monrolin Y, Pestana C, Reeve P, Sawade E et al. (2015). Optimizing conventional treatment for the removal of cyanobacteria and toxins. Denver (CO): Water Research Foundation.

Newcombe G, House J, Ho L, Baker P, Burch M (2010). Management strategies for cyanobacteria (blue-green algae): A guide for water utilities. Adelaide: Water Quality Research Australia (WQRA).

Newcombe G, Nicholson B (2004). Water treatment options for dissolved cyanotoxins. J Water Supply Res. 53:227–239.

Nicholson BC, Rosinato J, Burch MD (1994). Destruction of cyanobacterial peptide hepatotoxins by chlorine and chloramine. Water Res. 28:1297–1303.

Onstad GD, Strauch S, Meriluoto J, Codd GA, von Gunten U (2007). Selective oxidation of key functional groups in cyanotoxins during drinking water ozonation. Environ Sci Technol. 41:4397–4404.

Pavagadhi S, Tang ALL, Sathishkumar M, Loh KP, Balasubramanian R (2013). Removal of microcystin-LR and microcystin-RR by graphene oxide: adsorption and kinetic experiments. Water Res. 47:4621–4629.

Pestana CJ, Edwards C, Prabhu R, Robertson PK, Lawton LA (2015). Photocatalytic degradation of eleven microcystin variants and nodularin by TiO_2 coated glass microspheres. J Hazard Mat. 300:347–353.

Pestana CJ, Reeve PJ, Sawade E, Voldoire CF, Newton K, Praptiwi R et al. (2016). Fate of cyanobacteria in drinking water treatment plant lagoon supernatant and sludge. Sci Tot Environ. 565:1192–1200.

Petrusevski B, van Breemen AN, Alaerts G (1996). Effect of permanganate pretreatment and coagulation with dual coagulants on algae removal in direct filtration. J Water Supply: Res Technol-AQUA. 45:316–326.

Pietsch J, Bornmann K, Schmidt W (2002). Relevance of intra-and extracellular cyanotoxins for drinking water treatment. CLEAN–Soil Air Water. 30:7–15.

Pinho LX, Azevedo J, Brito Â, Santos A, Tamagnini P, Vilar VJ et al. (2015). Effect of TiO_2 photocatalysis on the destruction of *Microcystis aeruginosa* cells and degradation of cyanotoxins microcystin-LR and cylindrospermopsin. Chem Eng J. 268:144–152.

Pyo D, Moon D (2005). Adsorption of microcystin LR by activated carbon fibers. B Korean Chem Soc. 26:2089.

Qian F, Dixon DR, Newcombe G, Ho L, Dreyfus J, Scales PJ (2014). The effect of pH on the release of metabolites by cyanobacteria in conventional water treatment processes. Harmful Algae. 39:253–258.

Rapala J, Lahti K, Sivonen K, Niemelä SI (1994). Biodegradability and adsorption on lake sediments of cyanobacterial hepatotoxins and anatoxin-a. Lett Appl Microbiol. 19:423–428.

Rodriguez E, Acero JL, Spoof L, Meriluoto J (2008). Oxidation of MC-LR and -RR with chlorine and potassium permanganate: toxicity of the reaction products. Water Res. 42:1744–1752

Rodriguez E, Majado ME, Meriluoto J, Acero JL (2007a). Oxidation of microcystins by permanganate: reaction kinetics and implications for water treatment. Water Res. 41:102–110.

Rodríguez E, Onstad GD, Kull TP, Metcalf JS, Acero JL, von Gunten U (2007). Oxidative elimination of cyanotoxins: comparison of ozone, chlorine, chlorine dioxide and permanganate. Water Res. 41:3381–3393.

Rodriguez E, Sordo A, Metcalf JS, Acero JL (2007b). Kinetics of the oxidation of cylindrospermopsin and anatoxin-a with chlorine, monochloramine and permanganate. Water Res. 41:2048–2056.

Rositano J, Newcombe G, Nicholson B, Sztajnbok P (2001). Ozonation of NOM and algal toxins in four treated waters. Water Res. 35:23–32.

Sathishkumar M, Pavagadhi S, Vijayaraghavan K, Balasubramanian R, Ong S (2010). Experimental studies on removal of microcystin-LR by peat. J Hazard Mat. 184:417–424.

Sawade E, Ho L, Hoefel D, Newcombe G (2012). Development and validation of the biological filtration potential test for the removal of cyanobacterial metabolites. AWA Ozwater Convention and Exhibition. Sydney: Water Research Australia.

Schmidt W, Willmitzer H, Bornmann K, Pietsch J (2002). Production of drinking water from raw water containing cyanobacteria – Pilot plant studies for assessing the risk of microcystin breakthrough. Environ Toxicol. 17:375–385.

Şengül AB, Tüfekçi N, Aktan S (2016). The use of alum as coagulant for removing cyanobacterial cells in drinking water. Desalin Water Treat. 57:25610–25616.

Senogles P, Shaw GR, Smith MJ, Norris RL, Chiswell RK, Mueller J et al. (2000). Degradation of the cyanobacterial toxin cylindrospermopsin, from *Cylindrospermopsis raciborskii*, by chlorination. Toxicon. 38:1203–1213.

Senogles P-J, Scott J, Shaw G, Stratton H (2001). Photocatalytic degradation of the cyanotoxin cylindrospermopsin, using titanium dioxide and UV irradiation. Water Res. 35:1245–1255.

Shawwa AR, Smith DW (2001). Kinetics of microcystin-LR oxidation by ozone. Ozone: Sci Eng. 23:161–170.

Shephard GS, Stockenstrom S, de Villiers D, Engelbrecht WJ, Wessels GFS (2002). Degradation of microcystin toxins in a falling film photocatalytic reactor with immobilized titanium dioxide catalyst. Water Res. 36:140–146.

Smith MJ, Shaw GR, Eaglesham GK, Ho L, Brookes JD (2008). Elucidating the factors influencing the biodegradation of cylindrospermopsin in drinking water sources. Environ Toxicol. 23:413–421.

Somdee T, Wibuloutai J, Somdee T, Somdee A (2014). Biodegradation of the cyanobacterial hepatotoxin [Dha^7] microcystin-LR within a biologically active sand filter. Water Supply. 14:672–680.

Song W, Yan S, Cooper WJ, Dionysiou DD, O'Shea KE (2012). Hydroxyl radical oxidation of cylindrospermopsin (cyanobacterial toxin) and its role in the photochemical transformation. Environ Sci Technol. 46:12608–12615.

Stanford BD, Adams C, Rosenfeldt EJ, Arevalo E, Reinert A (2016). CyanoTOX: tools for managing cyanotoxins in drinking water treatment with chemical oxidants. J Am Water Works Assoc. 108:41–46.

Sun F, Pei H-Y, Hu W-R, Li X-Q, Ma C-X, Pei R-T (2013). The cell damage of *Microcystis aeruginosa* in PACl coagulation and floc storage processes. Sep Purif Technol. 115:123–128.

Teixeira MR, Rosa MJ (2006a). Comparing dissolved air flotation and conventional sedimentation to remove cyanobacterial cells of *Microcystis aeruginosa*: part I: the key operating conditions. Sep Purif Technol. 52:84–94.

Teixeira MR, Rosa MJ (2006b). Neurotoxic and hepatotoxic cyanotoxins removal by nanofiltration. Water Res. 40:2837–2846.

Teixeira MR, Rosa MJ (2007). Comparing dissolved air flotation and conventional sedimentation to remove cyanobacterial cells of *Microcystis aeruginosa*: part II. The effect of water background organics. Sep Purif Technol 53:126–134.

UKWIR (1996). Pilot scale GAC tests to evaluate toxin removal. London: United Kingdom Water Industry Research. https://www.ukwir.org/reports/96-DW-07-1/66681/Pilot-Scale-GAC-Tests-to-Evaluate-Toxin-Removal.

US EPA (2010). Long term 2 enhanced surface water treatment rule toolbox. Guidance manual. Washington (DC): Environmental Protection Agency of the United States. https://www.epa.gov/dwreginfo/long-term-2-enhanced-surface-water-treatment-rule-documents.

Vlad S, Anderson WB, Peldszus S, Huck PM (2014). Removal of the cyanotoxin anatoxin-a by drinking water treatment processes: a review. J Water Health. 12:601–617.

Wang H, Ho L, Lewis DM, Brookes JD, Newcombe G (2007). Discriminating and assessing adsorption and biodegradation removal mechanisms during granular activated carbon filtration of microcystin toxins. Water Res. 41:4262–4270.

Wang Z, Wang C, Wang P, Qian J, Hou J, Ao Y (2015). Response surface modeling and optimization of microcystin-LR removal from aqueous phase by polyacrylamide/sodium alginate–montmorillonite superabsorbent nanocomposite. Desalin Water Treat. 56:1121–1139.

Wert EC, Korak JA, Trenholm RA, Rosario-Ortiz FL (2014). Effect of oxidant exposure on the release of intracellular microcystin, MIB, and geosmin from three cyanobacteria species. Water Res. 52:251–259.

WHO (2015). Management of cyanobacteria in drinking-water supplies: information for regulators and water suppliers. Geneva: World Health Organization. https://apps.who.int/iris/handle/10665/153970

WHO (2017). Technical Brief: Water quality and health-review of turbidity: information for regulators and water suppliers. Geneva: World Health Organization. http://https://apps.who.int/iris/handle/10665/254631

Wu C-C, Huang W-J, Ji B-H (2015). Degradation of cyanotoxin cylindrospermopsin by TiO2-assisted ozonation in water. J Environ Sci Health Part A. 50:1116–1126.

Xagoraraki I, Zulliger K, Harrington GW, Zeier B, Krick W, Karner DA (2006). CT values required for degradation of microcystin-LR by free chlorine. J Water Supply: Res Technol-AQUA. 55:233–245.

Yan S, Jia A, Merel S, Snyder SA, O'Shea KE, Dionysiou DD et al. (2016). Ozonation of cylindrospermopsin (cyanotoxin): degradation mechanisms and cytotoxicity assessments. Environ Sci Technol. 50:1437–1446.

Yap R, Whittaker M, Peirson W, Jefferson B, Stuetz R, Newcombe G et al. (2012). The impact of *Microcystis aeruginosa* strain on cell removal using bubbles modified with poly(diallyldimethyl ammonium chloride). 6th International Conference on Flotation for Water and Wastewater Systems. New York, USA.

Zamyadi A, Coral LA, Barbeau B, Dorner S, Lapolli FR, Prévost M (2015). Fate of toxic cyanobacterial genera from natural bloom events during ozonation. Water Res. 73:204–215.

Zamyadi A, Dorner S, Ndong M, Ellis D, Bolduc A, Bastien C et al. (2013a). Low-risk cyanobacterial bloom sources: Cell accumulation within full-scale treatment plants. J Am Water Works Assoc. 105.

Zamyadi A, Dorner S, Sauvé S, Ellis D, Bolduc A, Bastien C et al. (2013b). Species-dependence of cyanobacteria removal efficiency by different drinking water treatment processes. Water Res 47:2689–2700.

Zamyadi A, Henderson R, Newton K, Capelo-Neto J, Newcombe G (2018). Assessment of the water treatment process's empirical model predictions for the management of aesthetic and health risks associated with cyanobacteria. Water. 10:590.

Zamyadi A, Ho L, Newcombe G, Bustamante H, Prévost M (2012a). Fate of toxic cyanobacterial cells and disinfection by-products formation after chlorination. Water Res. 46:1524–1535.

Zamyadi A, Ho L, Newcombe G, Daly RI, Burch M, Baker P et al. (2010). Release and oxidation of cell-bound saxitoxins during chlorination of *Anabaena circinalis* cells. Environ Sci Technol. 44:9055–9061.

Zamyadi A, MacLeod SL, Fan Y, McQuaid N, Dorner S, Sauvé S et al. (2012b). Toxic cyanobacterial breakthrough and accumulation in a drinking water plant: a monitoring and treatment challenge. Water Res. 46:1511–1523.

Zhang G, Zhang YC, Nadagouda M, Han C, O'Shea K, El-Sheikh SM et al. (2014). Visible light-sensitized S, N and C co-doped polymorphic TiO_2 for photocatalytic destruction of microcystin-LR. Appl Catal B: Environ. 144:614–621.

Zhang H, Dan Y, Adams CD, Shi H, Ma Y, Eichholz T (2017). Effect of oxidant demand on the release and degradation of microcystin-LR from *Microcystis aeruginosa* during oxidation. Chemosphere. 181:562–568.

Zhang X, Li J, Yang J-Y, Wood KV, Rothwell AP, Li W et al. (2016). Chlorine/UV process for decomposition and detoxification of microcystin-LR. Environ Sci Technol. 50:7671–7678.

Zhao H, Qiu J, Fan H, Li A (2013). Mechanism and application of solid phase adsorption toxin tracking for monitoring microcystins. J Chromatogr A. 1300:159–164.

Zhou S, Shao Y, Gao N, Li L, Deng J, Zhu M et al. (2014). Effect of chlorine dioxide on cyanobacterial cell integrity, toxin degradation and disinfection by-product formation. Sci Tot Environ. 482:208–213.

Zhu G, Lu X, Yang Z (2015). Characteristics of UV-MicroO_3 reactor and its application to microcystins degradation during surface water treatment. J Chem. Article ID 240703.

Zong W, Sun F, Pei H, Hu W, Pei R (2015). Microcystin-associated disinfection by-products: The real and non-negligible risk to drinking water subject to chlorination. Chem Eng J. 279:498–506.

Chapter 11

Planning monitoring programmes for cyanobacteria and cyanotoxins

Martin Welker, Ingrid Chorus, Blake A. Schaeffer and Erin Urquhart

CONTENTS

INTRODUCTION AND GENERAL CONSIDERATIONS

As outlined in Chapters 3 and 4, cyanobacteria are likely to be present in any waterbody and hence so are cyanobacterial toxins. The critical issue for the protection of public health is whether concentrations are likely to exceed hazardous levels at points of human exposure.

For this overall objective, five different types of monitoring serve different specific purposes:

1. *Monitoring for risk assessment*: Monitoring of waterbodies for the purpose of assessing the risk of cyanobacteria to occur in amounts that may lead to hazardous concentrations does not only target cyanobacteria and cyanotoxins, but also target parameters describing the conditions leading to their proliferation and scum formation (e.g., nutrient concentrations, changes in waterbody stratification or water residence time). Time scales for monitoring in the context of risk assessment are typically once intensively, with periodic checking later on.
2. *Monitoring to trigger immediate responses*: For example, in the context of an Alert Levels Framework (ALF; see sections 5.1.2 and 5.2.3), it serves to recognise when levels triggering vigilance or alerts are exceeded and corresponding action needs to be taken. This is typically regular (e.g., monthly or weekly), focused on bloom seasons or triggered by exceedance of levels for vigilance or alerts.
3. *Monitoring for validation of the control measures in place:* It serves to assess whether they are adequate either to prevent cyanobacteria from proliferating to hazardous blooms or to prevent breakthrough of cells and dissolved toxins to the point of water use. For validation, monitoring is intensively done once, when establishing the control system or developing a Water Safety Plan (WSP; see Chapter 6) and is then periodically repeated when the system or WSP is revised. Validation is important for control measures from catchment to consumer, and aspects specific to the catchment, waterbody, site of use and treatment are discussed in the respective Chapters 7–10. While most monitoring for the validation of measures to control concentrations of cyanobacteria and/or cyanotoxins will address their adequacy, it may also include parameters they target, such as visual assessment of erosion in a catchment, nutrient concentrations or temperature profiles.
4. *Event-driven monitoring*: Monitoring may be triggered by events such as (unexpected) blooms as well as animal deaths or human illness suspected to have been caused by toxic cyanobacteria. The purpose of such event-driven monitoring is usually to identify the cause of the event. Specifically for drinking-water treatment, it may also be to validate efficiency of removal, which can best be done during the event of a heavy bloom. In face of the rapid variability of blooms, particularly of scum situations, sampling as soon as possible during or after the event is key to meaningful data collection: chances for capturing the agents that caused the effects dwindle as time progresses from hours to days. This is most likely to be possible and to provide robust results if sampling is prepared and preplanned, as discussed in

general in Chapter 15 and specifically for drinking-water treatment in Chapter 10.

5. *Monitoring for verification*: Confirming that the guideline values are met at the point of exposure involves regular sampling and analysis either of cyanotoxins or of parameters, indicating that cyanobacteria are unlikely to have been present. This is done regularly, possibly limited to the seasons in which they are known to occur.

Note that *operational monitoring* of control measures is fundamentally different from these five purposes: it serves to ensure that control measures are functioning as intended and that, should one fail, it is possible to respond quickly enough with corrective action to prevent human exposure. Operational monitoring is essential to ensure that systems provide safe water. It typically does not address individual hazards but rather uses a practical, easy-to-measure parameter that shows whether or not a control is functioning. Such parameters range from online recording of turbidity at the outlet of a filter in water treatment to weekly visual inspection of a fence to keep livestock out of a water course. Options for operational monitoring are proposed together with the respective control measures in Chapters 7–10 and are not discussed in the following.

While this chapter focuses on cyanobacteria, cyanotoxins and parameters describing growth conditions favourable for cyanobacteria, it also contains a section on satellite remote sensing. This technology has advanced significantly in the last few decades and is becoming more and more accessible. While remote sensing cannot replace traditional *in situ* cyanobacteria monitoring and subsequent laboratory analyses, it can be very a useful tool that complements field monitoring in supporting site selection and indicating the frequency of occurrence of blooms of cyanobacteria (or eukaryotic algae).

Water-use systems to be monitored for cyanobacteria and cyanotoxins vary widely, from small ponds to large lakes, from tropical to boreal regions, from small streams to big rivers. No monitoring scheme can be globally applicable to all types of waterbodies, and local conditions have to be taken into account, not least because available resources for a monitoring programme differ substantially (Strobl & Robillard, 2008; Srivastava et al., 2013). For this reason, this chapter does not propose detailed guidelines, but rather focuses on considerations for designing an appropriate monitoring programme for specific – and often unique – water-use systems.

Bertani et al. (2017) showed that monitoring strategies considerably affect the outcome of cyanobacterial monitoring. Even with expert planning, any monitoring programme inevitably renders data with inherent imprecision. For this reason, one important aspect for long-term monitoring is the continuity of sampling together with detailed documentation. This is crucial for the meaningful interpretation of the data later on – sometimes decades later (e.g., for assessing which changes in the catchment may have changed bloom occurrence). Documentation should include not only the sampling procedure and analytical methods, but also the considerations behind any deviation from commonly applied schemes. Further, meaningful data storage in an accessible database is the prerequisite for reliable evaluation of long-term trends.

A basic understanding of limnology is a prerequisite for planning an appropriate monitoring programme for cyanobacterial occurrence in waterbodies; therefore, persons trained in limnology should be consulted in the planning phase, preferably with knowledge of the local waterbodies. Likewise, planning monitoring of schemes for bank filtration or artificial groundwater recharge should involve soil scientists, and planning monitoring schemes for drinking-water treatment needs expertise in treatment technology.

11.1 DESIGNING A MONITORING PROGRAMME

Resources for monitoring can be focused on waterbodies at risk of blooms and, within these waterbodies, on time spans during which they are likely, if the purpose of monitoring is clearly defined. Where potentially toxic cyanobacteria are first recognised as potential risk, it is often possible to use data from past monitoring programmes for a first assessment. It may also be possible to integrate a targeted monitoring into other ongoing programmes.

11.1.1 Collecting and analysing existing knowledge

A first step is to explore which data are already available and whether they allow any estimate of the likelihood of potentially toxigenic cyanobacteria and hence cyanotoxins to occur in the waterbody of interest. In principle, all data on a particular waterbody can be relevant for this purpose, that is, from environmental monitoring programmes and public reports, scientific publications or satellite images (see section 11.4). This may include data on, for example:

- delineation of the catchment, land use and human activities therein (e.g., agricultural practices, waste-water treatment facilities capacity and functioning);
- waterbody morphology, in particular surface area, mean and maximum depth, stratification, water residence time or – for rivers – flow rate;
- types of water use such as drinking-water production, recreational, aquaculture and irrigation;
- location and depth of raw water intake sites (and of alternative locations);

- location of bathing sites and frequency of seasonal use;
- prevailing wind direction, especially when surface bloom-forming cyanobacteria are abundant (see Chapter 3);
- seasonal dynamics of phytoplankton occurrence and taxonomic composition;
- seasonality and timing of visible cyanobacterial blooms (surface blooms and scums);
- indication of suspected or proven water-related illnesses (human and animal);
- satellite images quantifying total phytoplankton (chlorophyll-*a*) and cyanobacterial (phycocyanin) biomass, as well as spatial distribution over time and space;
- nutrient concentrations, especially of total phosphorus (TP) and total nitrogen (TN), and their seasonal variation;
- potential major nutrient inputs and possible input fluctuations, for example, seasonality of surface run-off, and possible long-term changes.

Even if the information readily available is only fragmentary, this will support planning and may even allow a first assessment of the likelihood of blooms: specific phytoplankton communities are typical for particular types of waterbodies and seasons (see Chapters 3 and 4). Hence, occurrence of toxic cyanobacterial blooms and related health risks can be foreseen fairly well from basic limnological parameters, even where quantitative phytoplankton data may be lacking or are only rudimentary.

Cyanotoxins are commonly only one among several potential health hazards related to safe water use, and monitoring schemes addressing their occurrence will typically be part of more comprehensive programmes. Their relevance in relation to other hazards in such a programme is best prioritised in the context of overall risk assessment, as discussed in Chapter 6.

If no background data exist, a general limnological screening programme is recommended which may well serve as pilot for a monitoring programme to be subsequently implemented. This would consist of seasonal sampling for basic limnological parameters (e.g., TP, TN, Chl-a, temperature profiles, phytoplankton composition and whether cyanobacteria occur, basic cyanotoxin analysis, e.g., with ELISA) as well as site inspection for general observations (e.g., scum formation, fish and wildlife deaths, water-level fluctuations). The results serve as a starting point to fill some fundamental gaps in data and information, to estimate potential health hazards and potentially to design a full monitoring programme.

If the existing information or the outcomes of such a preliminary screening programme show that for a specific waterbody, a health hazard from cyanotoxins is unlikely (because cyanobacteria hardly occur) or of lower priority relative to other hazards (e.g., pathogens, pesticides or a spill of a hazardous chemical), cyanobacterial monitoring might be reduced to observation at low frequency (e.g., once annually during the season in which they are most likely to occur). This would serve to detect changes in the catchment or waterbody that might increase risk, for example, new upstream nutrient loads causing eutrophication or new impoundments changing the mixing regime (see Chapter 4).

11.1.2 Defining the objective of monitoring

The objective of monitoring, as outlined in points 1–4 in the introduction to this chapter, determines the information needed and time spans and intervals for which it will be necessary (discussed below in section 11.5). The objective also determines where and when samples will be taken in a specific waterbody or along the drinking-water production process as well as chemical and biological analyses to be performed (see Chapters 13 and 14) and the accuracy and sensitivity required. It is therefore very important to clarify the objective(s) early on when planning a programme, both to avoid dispensable efforts (e.g., detection of trace concentrations of cyanotoxins with expensive analytical methods in raw water) and to avoid missing essential information.

11.2 PLANNING FIELDWORK

Fieldwork, including site inspection, sample collection and, in most programmes, some on-site analyses, largely determines the quality of information obtained from the subsequent laboratory analyses. Fieldwork also causes a significant proportion of the total cost of a cyanobacterial or cyanotoxin monitoring programme. Well-planned and performed sampling is the prerequisite for meaningful results, and most shortcomings in the sampling design cannot be compensated later on: the most accurate and sensitive analytical procedures provide uncertain results if sampling was flawed. A well-designed and implemented fieldwork programme also improves cost-efficiency, that is, for the overall costs of personnel, transport and analytical procedures, by focusing on critical sites and critical periods, as there is little value in spending large amounts of effort on very small risks. This is especially important where the cyanotoxin risk is only one among other health risks from water.

While fieldwork to validate control measures in the catchment or waterbody typically requires a sampling strategy that observes the dynamic changes at a fixed sampling site over time, validating control measures along a drinking-water production process requires a sampling strategy that allows observing changes in a "slug" or "parcel" of water as it passes through the process (see Chapter 10 for details), with a focus on treatment steps that are expected to affect the cyanotoxin concentration.

For an efficient sampling programme, both sampling schemes need to be coordinated. Validation is most effective if sampling and analyses along the production line are conducted when the raw water at the site of abstraction contains a high concentration of cyanobacteria (possibly including extracellular toxins). Such a situation can be determined using an indicator parameter such as turbidity or fluorimetry (Chapter 5). It is important that sampling and analyses along the production line are launched immediately once a high amount of cyanobacteria is detected in the raw water, in order to have the best chance of determining elimination efficiencies of the individual steps of the treatment train.

The schemes in Figure 11.1 illustrate this, representing two lakes or reservoirs with differing phytoplankton communities (see Chapter 3): the lower scheme shows a lake with a perennial population of cyanobacteria, for example, *Planktothrix* sp., with a higher base frequency of sampling along the timeline. The upper scheme represents a lake with a strong seasonality of cyanobacteria and cyanotoxin occurrence, for example, the phytoplankton dynamics with a spring bloom of diatoms and dominance of *Microcystis* sp. in summer. Outside of the cyanobacterial season, the sampling intervals are extended to monthly, while they are reduced to weekly during the blooming season. If the drinking-water production line is to be validated, sampling along the production line is launched once blooms reach a threshold value at the raw water abstraction site, and the production process is then followed through the different steps of the treatment train to the finished drinking-water. If the outcome demonstrates that the treatment train

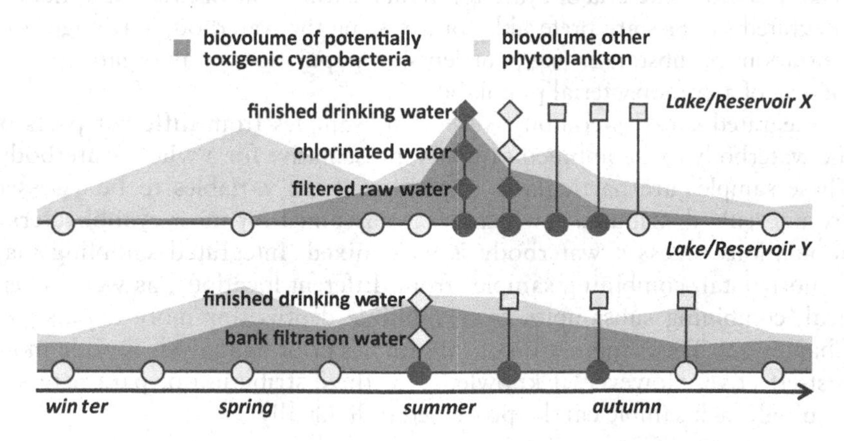

Figure 11.1 Schematic illustration of sampling strategies and frequencies in two waterbodies used for drinking- water production. Dots: sampling in the waterbody; diamonds: validation sampling along the treatment train ; open symbols: situation above Vigilance Level and below Alert Level 1; filled symbols: situation exceeding Alert Level 1; open squares: verification monitoring during Alert Level 1. Shaded area: occurrence of potentially toxigenic cyanobacteria in relation to other phytoplankton (light area). For details see text.

effectively controls cyanobacterial cells and toxins, sampling may then be limited to raw and finished water for the rest of the cyanobacterial season until concentrations are below the Vigilance Level of the Alert Levels Framework (ALF) (section 5.1.2). Monitoring may then be suspended until cyanobacteria reappear (possibly not until the following year) in the range of the Vigilance Level. Validation of the drinking-water production line will not be repeated every year if operational monitoring shows control measures to be working effectively and verification monitoring of finished drinking-water regularly shows concentrations well below the guideline values (see above and Chapter 6). However, as long as concentrations of cyanobacteria and/or cyanotoxins exceed the Vigilance Level or Alert Level 1 in the raw water, it is recommended to include cyanotoxins in a drinking-water utility's routine verification monitoring of finished drinking-water.

11.3 TYPES OF SAMPLES

For waterbodies, two principally different types of samples may be distinguished: a grab (or spot) sample and an integrated sample. A grab sample is a discrete volume of water taken at a selected location, depth and time. The simplest way to take a grab sample is to scoop water with a wide-mouthed vessel from or near the surface. Subsurface sampling is done with special sampling devices that are also used for integrated sampling. Whereas grab samples are suitable for analysing situations at specific sites (e.g., maximum density of cyanobacteria or cyanotoxins at a bathing site or raw water intake), integrated samples are preferable for assessing the waterbody's average concentrations of substances (e.g., nutrients) or populations of an organism (e.g., the size of a cyanobacterial population).

Integrated samples combine several subsamples from different parts of the waterbody to a combined sample representative for a whole waterbody. These samples are particularly important if the variables to be assessed are unevenly distributed – which is best assumed for most cyanobacterial populations unless a waterbody is well mixed. Integrated sampling may be horizontal, combining samples from different locations, as well as vertical, combining subsamples from multiple depths (for more details, see Chapter 12). The combination of subsamples prior to analysis is often more cost-effective. However, if knowledge of the distribution of parameters is required, each sample can be processed individually.

11.4 WHERE TO SAMPLE

Ideally, a sample from a waterbody is representative of the water compartment for which information is desired. The water compartment of interest can range from the whole waterbody volume to a mouthful of water swallowed by a child. Obviously, this needs to be considered when a monitoring programme is established.

Different waterbody compartments, sampled for different target information and requiring different sampling approaches, include:

- the entire waterbody;
- offtake sites of raw water for drinking-water production;
- sites of recreational activity.

For information to understand bloom development, samples representative of the entire waterbody are necessary. Respective data allow the estimation of, for example, carrying capacity for cyanobacterial biomass, average cyanobacterial abundance, cyanobacterial taxa present and average cyanotoxin concentrations. As outlined above, validating the effects of catchment management or waterbody restoration measures requires integrated sampling (see below and Chapter 12) at moderate frequency (monthly or bimonthly), but for several blooming seasons. Integrated sampling requires more effort than grab sampling because a boat, a water sampler (see Chapter 12), submersible oxygen and temperature probes, and other equipment are needed. Data from integrated samples cannot be used directly for assessing exposure risks at sensitive sites such as drinking-water intakes or bathing sites, where cyanotoxin concentrations can be orders of magnitude higher. The heterogeneous distribution of cyanobacteria in most waterbodies can lead to variability in abundance differing substantially even on narrow spatial scales, horizontally as well as vertically (see Chapters 4 and 9).

Sampling a drinking-water intake can be either in the waterbody at the point immediately before the water enters the drinking-water production system or directly from the raw water pipeline, that is, where it enters the waterworks. However, operators may wish to establish a wider understanding of the occurrence of the specific cyanobacterial population in the waterbody by taking samples representative of the water layer in which they primarily occur, for example, for *Planktothrix rubescens* in the metalimnion (see Chapter 3), or of a specific bay from which the utility abstracts raw water.

Sampling bathing sites includes shallow waters up to the shoreline and sometimes beyond when cyanobacterial scums have been washed ashore. Cyanobacterial abundance – and hence cyanotoxin concentrations – can fluctuate particularly at near-shore sites by orders of magnitude within days or even hours (see Chapter 4). It is therefore particularly important to clearly define the objective of sampling at respective sites; this determines the number of samples to be taken and the extent to which sampling can simulate the mouthful of water possibly ingested. This could, for example, be the determination of average concentrations for the bathing site up to a certain depth or the estimation of maximally expectable concentrations where scums accumulate. Sampling for compliance to the Vigilance Level, as proposed in the Alert Levels Framework (section 5.2.3), may, however, be more effective if it targets representing the entire water volume, as this reflects the overall size of the cyanobacterial population and thus the potential for scum accumulation.

In conclusion, improper sampling may lead to analytically accurate but nonetheless "false" results, which can trigger inappropriate actions or impede necessary steps because the data do not adequately reflect the health hazard.

Samples that cannot be preserved are preferably analysed as soon as possible (i.e., within hours) both for a timely and adequate reaction if results show Vigilance or Alert Levels to be exceeded and for avoiding changes (degradation) in the concentrations of parameters to be analysed.

Good documentation of monitoring is important, and it is valuable to include visual observation (photographs), comments on smells and reports from site users, etc. Where monitoring results of cyanotoxin concentrations or cyanobacterial biomass indicators exceed Alert Level 2 and this leads to restrictions in site use, this can have an immediate economic impact, leading to a high potential for conflicts of interest. Documentation of the rationale for such a decision can then become important. It is also a basis for clear communication between all stakeholders, which is essential for efficient health protection while keeping economic losses low (see Chapter 15).

11.5 FREQUENCY OF SITE INSPECTION AND SAMPLING

The frequency of site inspection and sampling also needs to be adapted to the objective of the programme. Table 11.1 summarises examples of sampling strategies for these monitoring objectives.

As indicated at the beginning of this chapter, monitoring for risk assessment is an intensive but short-lived exercise that can focus on a small number of sampling campaigns during situations in which blooms are expected to be most likely. It should be repeated at intervals of several years in the context of periodic revision of the management system or the Water Safety Plan, after unexpected blooms or any incident that suggests controls to be insufficient, or if changes in the catchment or other components of the system may have consequences for the adequacy of the control measures in place. A lower number of samples than in the initial campaign may well be sufficient for such repeats.

In contrast, where cyanobacteria are known to occur at potentially hazardous levels, monitoring needs to capture situations in which indicators for toxic cyanobacterial occurrence may exceed predefined thresholds, that is, the Vigilance Levels described in the Alert Levels Frameworks proposed in section 5.1.2 for drinking-water and in section 5.2.3 for recreational

Table 11.1 Summary of sampling strategies for examples of specific monitoring objectives

Objective	*Sampling sites*	*Sampling frequency*	*Sample type*	*Analytical targets*
Risk assessment: spatial distribution of cyanobacteria/ cyanotoxins	Multiple sites and multiple depths	Single or few sampling campaigns during bloom season	Multiple grab samples	Phytoplankton quantitatively Chlorophyll-*a* or fluorescence Cyanotoxins
Risk assessment: understanding the likelihood of cyanobacterial biomass to exceed Alert Levels	Central site or multiple sites and depths in the waterbody	Monthly or bimonthly Weekly during bloom season	Single grab samples or multiple integrated samples, depending on waterbody morphology	Nutrients Transparency Cyanobacteria (phytoplankton) taxa Cyanobacteria (phytoplankton) biovolumes Chlorophyll-*a* Cyanotoxins
Validation of nutrient load control: understanding the carrying capacity for cyanobacterial biomass	Major inflows Central site in waterbody	Monthly year-round (for first rough orientation once during spring mixing)	Single grab samples Integrated samples (vertically/ horizontally)	Nutrients
Event response: clarification of cyanotoxins as cause of animal poisoning	Multiple sites	Once (depending on outcome, continued till bloom is over)	Multiple grab samples along the lakeshore	Cyanobacteria (taxa, biovolume) Cyanotoxins

(*Continued*)

Table 11.1 (Continued) Summary of sampling strategies for examples of specific monitoring objectives

Objective	*Sampling sites*	*Sampling frequency*	*Sample type*	*Analytical targets*
Triggering immediate response: cyanobacteria/ cyanotoxins in raw water to inform treatment decisions	Raw water off-take site	Weekly or biweekly during seasons with cyanobacteria in the range of the Vigilance Level Increased frequency during bloom season	Depth-defined grab sample	Phytoplankton taxa Cyanobacteria (phytoplankton) biovolumes Chlorophyll-*a* Fluorescence Turbidity at intake Cyanotoxins
Verification of safe cyanotoxin levels in finished drinking-water	Finished drinking-water	Weekly or biweekly during seasons with cyanobacteria in the range of the Vigilance Level	Grab sample	Cyanotoxins
Verification of the safety of recreational activity from cyanotoxins	Bathing sites	Depending on season and use in response to visual inspection	Single or multiple surface grab samples	Transparency Cyanobacteria (phytoplankton) taxa Cyanobacteria (phytoplankton) biovolumes Chlorophyll-*a* Cyanotoxins

For details on analytical targets, see Chapters 12 (temperature, transparency), 13 (nutrients, transparency, chlorophyll, phytoplankton [i.e., cell counts and biovolume], fluorescence) and 14 (cyanotoxins). Sampling frequency is for rough orientation and may need to be adjusted to higher or lower frequencies.

water use. In waterbodies with a pronounced seasonality of cyanobacterial occurrence, this may require an increased frequency of site inspection and sampling during the development of cyanobacterial peak populations, the time of which can be fairly well estimated based on previous data and/or experiences in other, similar waterbodies in the same climatic region. During peak blooming, information on observations like scum formation at bathing sites is very important (see Chapter 4). If perennial persistence of cyanobacteria cannot be ruled out, drinking-water supply reservoirs may need to be monitored regularly throughout the year for compliance to the Vigilance Level. Persistent cyanobacterial populations can be expected in warm climates and in temperate zones in waterbodies populated by certain taxa like *Planktothrix* spp. Where monitoring may need to cover the entire year, frequency can be reduced in some season if growth rates are known to then be lower, for example, during the cold season.

Monitoring for compliance to the Vigilance Level is most effective if the time intervals are adapted to bloom occurrence on the basis of a good understanding of the waterbody: if cyanobacteria are known to appear at a certain time or in a certain season, this may initiate monitoring for compliance to Vigilance Levels. For situations exceeding the Vigilance Levels, both Alert Levels Frameworks (ALFs) give guidance for appropriate frequencies of further monitoring. However, once the necessary experience has been developed (typically on the basis of several years of data), it is useful to adapt the frequencies given in the ALFs in sections 5.1 and 5.2 to the bloom development in the given waterbody.

Time scales for monitoring for the purpose of validating the efficacy of control measures may vary widely, depending on the time span for different measures to take effect: for measures in catchment management, it may take years or even decades until reduced nutrient loads lead to reduced cyanobacterial biomass. This is because natural processes in ecosystems can limit or strongly delay responses to newly implemented control measures (Chapters 7 and 8). During this time span, monitoring may be sufficient at low frequency. In contrast, validation of technical measures may be much quicker, that is, in the range of a few days at two to three bloom occasions, for example, for validating whether the variation of raw water intake depth is optimal or the removal efficiency of drinking-water treatment train challenged by a heavy bloom is sufficiently effective (Chapters 7–10). As discussed above, monitoring for validation is conducted once intensively and repeated occasionally (usually after several years) in the context of the revision of the management system or Water Safety Plan. Repeats may also be triggered by changes in the system or incidents questioning whether the measures in place are sufficiently effective.

Monitoring for verification that guideline values for cyanotoxins are met at the points of human exposure is usually well established at regular intervals for microbial indicator organisms and selected chemicals (WHO, 2017).

While verification monitoring for toxic cyanobacteria should also occur regularly, sampling and analyses may be reduced to seasons in which they are known to occur (see above) or human exposure is likely (e.g., the bathing season). Particularly for the monitoring of bathing sites at which cyanobacteria are known or likely to occur, a high frequency of sampling (i.e., daily or weekly) may be necessary for relatively short periods during the peak bathing season.

11.6 PERSONNEL AND TRAINING

Properly trained field staff is the backbone of effective sampling and monitoring programmes. Training should include the handling of sampling and measuring devices as well as recognition of visible aspects of cyanobacterial blooms. For smooth and reliable sampling, continuity in staff is highly advantageous. Like with biomass estimation (see Chapter 13), the judgement of a person that knows a system can be very efficient for streamlining the monitoring programme. Experienced field staff should also be encouraged to take additional samples or to make records in case they have the impression that something unusual has occurred. This could be, for example, surface blooms of different colour, dead fish or blooms washed ashore.

Inclusion of the public as active participants in monitoring programmes is gaining acceptance and can contribute significantly to the quantity and quality of information obtained from a monitoring programme. Special interest groups (such as nongovernmental organisations and user associations) as well as concerned local populations in sensitive or affected areas can provide useful information. This is particularly valuable for monitoring sites of recreational activity (see Box 11.1).

BOX 11.1: CITIZEN SCIENTISTS AND COMMUNITY PARTICIPATION

The inclusion of volunteers in the collection of data can significantly support the assessment of the status of diverse environments (for a recent review, see Schröter et al., 2017). In lakes and rivers, it can significantly support the assessment of water quality conditions and cyanobacterial risks. Collected data can range from comparatively simple, such as Secchi depth readings to more complex data like taxonomic composition of phytoplankton communities.

The Secchi Dip-In programme (http://www.secchidipin.org/, sponsored by the North American Lake Management Society [NALMS]) hosts a website with instructions for Secchi depth readings and the option to upload data. The data can in turn be accessed for individual lakes. Originally the programme started in the USA but has been adopted in several other countries.

Three coordinated monitoring projects to locate and understand harmful cyanobacteria are conducted by Cyanobacteria Monitoring Collaborative (https://cyanos.org/). The most simple approach is followed by bloomWatch (https://cyanos.org/bloomwatch/) that consists of a free smartphone app and a platform to which digital pictures can be uploaded. The aim of the programme is to track the occurrence of cyanobacterial blooms in waterbodies that are not included in regular, institutional monitoring programmes. The project cyanoScope (https://cyanos.org/cyanoscope/) includes the microscopic observation of plankton samples. A nearly full limnological assessment is the subject of the project cyanoMonitoring (https://cyanos.org/cyanomonitoring/), which requires a more intensive training for interested citizens.

The Centre for Ecology & Hydrology (UK) provides a similar smartphone app ("Bloomin' Algae") enabling citizens to report algal (surface) blooms (https://www.ceh.ac.uk/algal-blooms/bloomin-algae). A local project primarily targeting public communication, CYANOBs in Potrero de Garay, Argentina, is described in Box 15.2.

Although for all programmes and initiatives the data for individual lakes are highly inconsistent – that is, for some lakes only a few datapoints are available while for others longer time series have been collected – encouraging citizens to collect data could help water managers to extend their knowledge on waterbodies in the region. In spite of reservations voiced about quality control, collected data can, for example, reveal pronounced trends over time (Lottig et al., 2014) or spatial patterns (Bigham-Stephens et al., 2015) in lake transparency.

Further information concerning planning and performing fieldwork can be found, for example, in Bartram and Ballance (1996), a volume published on behalf of United Nations Environment Programme and the World Health Organization or in United States Geological Survey Guidelines (Graham et al., 2008).

11.7 PREPARATIVE STEPS AND PILOT PHASE

A period of pilot testing before routine field visits begin helps to ensure that time requirements for inspection and sampling are understood and that activities are planned to make the best use of staff time and other necessary resources (e.g., vehicles). Realistic estimation of travelling time between laboratories and sampling sites is important to avoid exceeding tolerable sample storage times prior to analysis. Pilot testing should lead to the development of a detailed inventory and description of sampling sites. If changes in water quality with time are to be interpreted with confidence, samples must be taken consistently from the same locations and/or from other, precisely identified locations. Pilot testing also provides an opportunity for training personnel and familiarisation with the routine.

Coordination with the laboratory responsible for the analyses is an important aspect of preparation. In some cases, the laboratory will be responsible for the preparation of sample containers and chemical additives for sample preservation, and it may also be responsible for the provision and maintenance of equipment for on-site testing (see Chapter 12). Sampling tours also need to be coordinated with downstream analyses to ensure timely sample processing.

11.8 CONSIDERATIONS FOR DOWNSTREAM ANALYTICAL PROCEDURES

Balancing the costs of the procedures against the depth of information gained is important for an efficient hazard analysis and risk management. The methods for the detection and quantification of nutrients, cyanobacteria and their toxins reviewed in Chapters 13 and 14 range from simple manipulations that can be performed on site to complex techniques that require costly equipment and well-trained experts. The multitude of available methods for cyanotoxin analysis reflects the multitude of situations that demand respective methods. No analytical procedure is superior to others *per se* – rather, it is the context and purpose that renders an individual method appropriate – or inappropriate. Section 14.2 gives guidance on the scope of the methods for cyanotoxin analyses, and experts in the analytical laboratories will be familiar with most of the advantages and disadvantages discussed there. They should therefore be included when planning a sampling programme.

Laboratory capacity is a further important issue to be addressed in programme design and in pilot testing. It is essential that the workload generated by a sampling expedition is properly managed within the laboratory. Analysts need to know how many samples will be arriving, the approximate time of arrival and the analyses that are to be carried out. Excessive delays before sample processing and analysis may render the sample results invalid (and thereby useless) for the purposes for which they have been collected. Therefore, the timing of sample delivery to the laboratory and the workload management within the laboratory should be coordinated prior to fieldwork.

When planning monitoring programmes (or adjusting them, if necessary), it is important that all participating institutions, companies and laboratories agree on the type and number of samples that eventually are to be analysed. This should avoid that samples are taken that are not appropriate for downstream analyses. The following (nonexhaustive) list gives a number of key questions to consider:

- Have the analytical methods to be used been critically evaluated and agreed upon?
- Are the sample types and volumes appropriate for the desired laboratory analyses?
- Is the delay between taking samples and their arrival in the laboratory prone to cause analytical artefacts (e.g., lysis of cells, breakdown of toxins)?
- Could these effects be minimised by (more) appropriate sample handling (see also Chapter 12)?
- Will the samples arrive in the laboratory at a time that allows immediate and appropriate analysis or storage, respectively?
- Have the laboratories been well instructed on sample handling and which analyses are to be performed?

11.9 *AD HOC* SAMPLING FOLLOWING POISONING EVENTS

Unexpected events of poisoning particularly of domestic animals continue to be encountered, and comprehensive investigation of such cases can be important to prevent further exposure. It is also relevant to better understand exposure, toxic mechanisms, toxigenic taxa, yet unidentified toxic metabolites and more. For this aim, it is important that laboratories dealing with the analysis of cyanobacteria and their toxins be prepared for *ad hoc* sampling in case events such as animal deaths or human illness are suspected to be caused by cyanotoxins. The few published reports on such events (e.g., Gugger et al., 2005 and Wood et al., 2017) highlight the importance of having procedures in place to collect samples and information in time, as situations of severe cyanotoxin risks may be only short-lived and an extended delay prior to sampling may make the entire effort futile.

The following list suggests information and items that may help to react adequately to a request of event-triggered sampling:

- Where and when did the poisoning occur?
- Have cyanobacteria/algal mats, scums, pronounced turbidity or smells been observed at the site of poisoning?
- What symptoms have been observed in the casualty?
- Who proposed the diagnosis of cyanobacterial poisoning?
- Where is the casualty now? Is a detailed medical examination planned? Has the examinator been contacted?

A number of materials should be available ready to use for unexpected poisoning events, preferably in a dedicated "emergency kit":

- sample containers adequate for water, algal mats, macrophytes, phytoplankton microscopy etc;
- sample containers for animal tissue samples, stomach contents;
- contact details of physicians and veterinarians that may be involved or consulted;
- storage space in freezers and fridges, a cooling box for transport.

Chapter 15 gives further guidance for responding to events and emergencies.

11.10 SATELLITE REMOTE SENSING ANALYSES

Remote sensing can serve as a starting point to plan a monitoring programme, for example, by identifying the season of cyanobacterial blooms or locating sites of biomass accumulation. Once a monitoring programme is in place, remote sensing can serve to verify the validity of sampling points with respect to their representativeness of the monitored waterbody. Previous satellite data can provide historical assessments. As shown by the example in Box 11.2, remote sensing further provides a rather low-cost opportunity to intensify monitoring in time and space, particularly in areas for which a large number of sampling points and/or frequent visits for sampling would be necessary.

BOX 11.2: REMOTE SENSING OF CYANOBACTERIA: THE CASE STUDY OF SALTO GRANDE RESERVOIR

Andrea A. Drozd

Salto Grande reservoir, constructed in 1979, impounds the Uruguay River which divides Argentina and Uruguay. It is visited by thousands of tourists during summer in spite of recurrent and often heavy cyanobacterial blooms (O'Farrell et al., 2012; Bordet et al., 2017) with cyanobacterial cell density and microcystin concentrations frequently surpassing the recreational guideline levels given by WHO, and one case of severe liver damage was described after a young jet skier had spent many hours in the bloom (see section 5.2 and Giannuzzi et al., 2011). The reservoir's large area of 750 km^2 and its dendritic morphology with lateral arms renders sufficiently frequent and comprehensive sampling impossible. Since 2011, monitoring cyanobacteria blooms and water quality is being complemented with a remote sensing monitoring programme developed by the "Comisión Administradora del Río Uruguay" (CARU, 2016; 2017). It targets quantifying chlorophyll-*a* concentrations by satellite data at a scale of 1:50 000.

From 2012 to 2016, 10 field campaigns provided a basis for this by characterising spectral signatures of dam water under different conditions using samples from multiple sites for laboratory analyses (by Comisión Mixta de Salto Grande) of chlorophyll-*a* concentration, phytoplankton composition and abundance as well as turbidity, together with *in situ* hyperspectral signatures obtained by a hyperspectral radiometer (ASD Field Spec provided by the Comisión Nacional de Actividades Espaciales) in order to describe how suspended matter, chlorophyll-*a* concentration, phytoplankton composition and cyanobacteria cell density influence the spectral patterns. These spectral signatures were then used to build band algorithms for sensors of different satellites, that is, Landsat 7-8 and Sentinel 2, SPOT HRVIR, relating spectral pattern characteristics to concentrations of chlorophyll-*α* and cyanobacterial cell density (Drozd et al., 2020).

A first result was the spectral discrimination of phytoplankton communities. With Landsat and Sentinel 2, dominance of dinoflagellates showed a dark-blue colour (Figure 1, Panel A), and the absorption of the green spectral range increased with their cell density; in contrast, cyanobacteria showed an inverse response: the greater their biomass, the lower the absorption in the green spectral range, leading to a bright green colour when cyanobacteria dominated (Figure 1, Panel B). A second result was the relationship between chlorophyll-*a* concentration and a band index algorithm ($R^2 > 0.77$), allowing the monitoring of phytoplankton intensity and distribution as the basis for deciding on priority sites for field sampling, indicating the hotspots where blooms originated as well as beaches with potential health risks (Figure 1, Panel C).

As the next step, for situations with dominance of cyanobacteria, a cell density algorithm was developed, allowing estimation by satellite data. CARU has established Alert Levels for recreational use of waterbodies using observed colour patterns of the water which coincide with cyanobacteria cell density and toxin concentrations. This cyanobacterial algorithm is able to detect average densities of 200 cells/mL and hence to map CARU's Alert Levels (Figure 1, Panel D).

Since 2012, chlorophyll-*a* and cyanobacterial remote sensing proved a helpful tool for a synoptic understanding of spatiotemporal dynamics of blooms in Salto Grande and for providing an estimate of cell densities. In summertime, when rain is scarce, satellite data can be obtained at intervals of 3–5 days, enabling low-cost monitoring of phytoplankton communities, short-term reports and warnings as well as more effectively targeted sampling programmes. For more information, see www.caru.org.uy and www.saltogrande.org

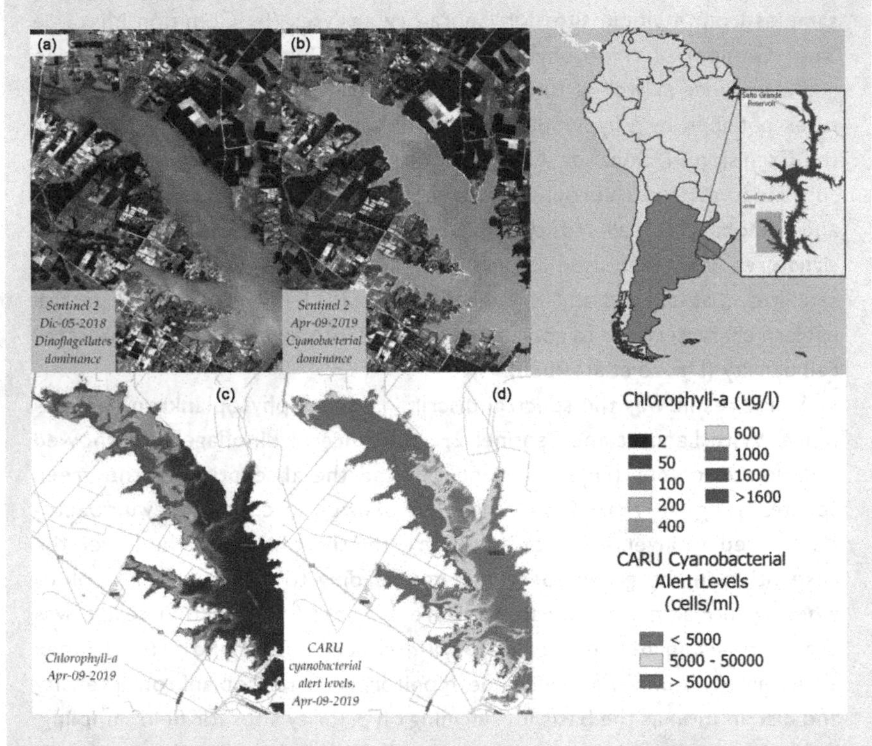

Figure 1 Salto Grande Reservoir, Gualeguaycito arm. Images obtained with Sentinel 2. Panel A: 05 December 2018 with dominance of dinoflagellates; Panel B: 9 April 2019 with cyanobacterial blooms. Panel C: Chlorophyll-*a* range estimation by Sentinel 2 on 9 April 2019. Panel D: Cyanobacterial cell density ranges estimated by Sentinel 2 on 9 April 2019, reflecting CARU cyanobacterial Alert Levels.

Available remote sensing systems are generally based on satellite images and may also be available from platforms such as drones and airplanes. The focus here will be satellites for monitoring water quality, instruments that orbit Earth in space. In contrast to drones and airplanes, data from many government operational satellite sensors are available free of charge.

Satellites can contain multiple sensors that provide a birds-eye view of the Earth's surface. Satellite sensors designed for water quality measures are typically passive sensors, which means they detect changes in sunlight reflected off the water surface. When light interacts with the water environment, it can either be absorbed or scattered in the water column (Figure 11.2). Dissolved and particulate matter in the water column absorb and scatter light differently across the wavelengths of the visible light spectrum. Changes in the visible light spectrum by materials present in the water column, like pigments in phytoplankton and cyanobacteria, can be quantified by these satellite sensors. Firstly, the sensor detects the spectral changes and then validated mathematical algorithms quantify concentrations of these water column materials. Algorithms are successful at quantifying both phytoplankton (chlorophyll-*a*) and cyanobacteria biomass concentrations (Figure 11.3). Satellite sensor technologies typically follow a transition pathway starting with research and development of theoretical

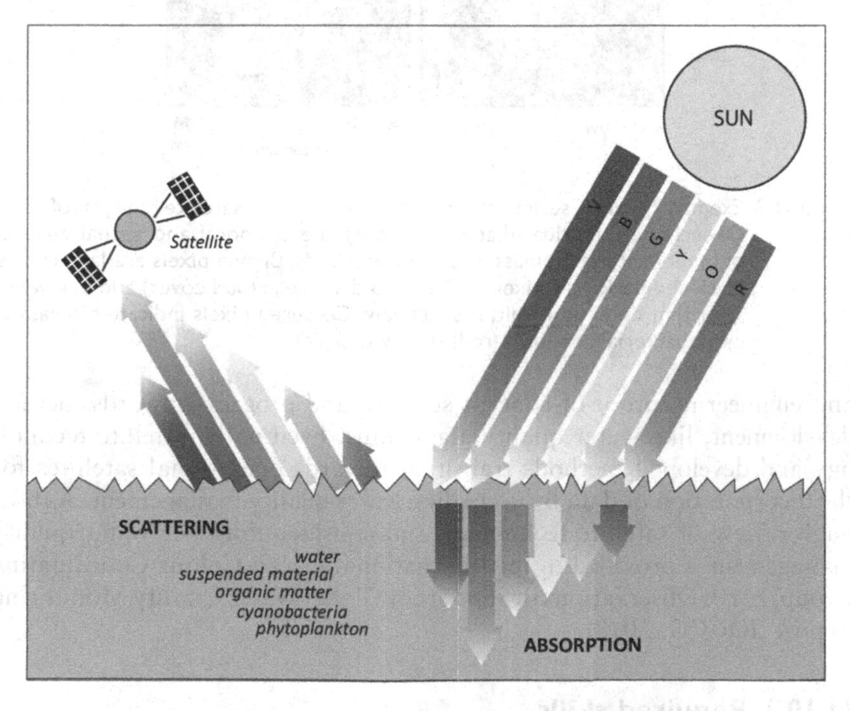

Figure 11.2 Conceptual diagram of how a typical satellite sensor detects water quality changes from sunlight reflected off the water surface. Light can either be absorbed or be scattered through interaction with water, phytoplankton, cyanobacteria, organic matter and suspended material across wavelengths of the visible spectrum. Changes in the visible light spectrum from scattering and absorption can be quantified with algorithms to derive measures of, for example, cyanobacteria biomass.

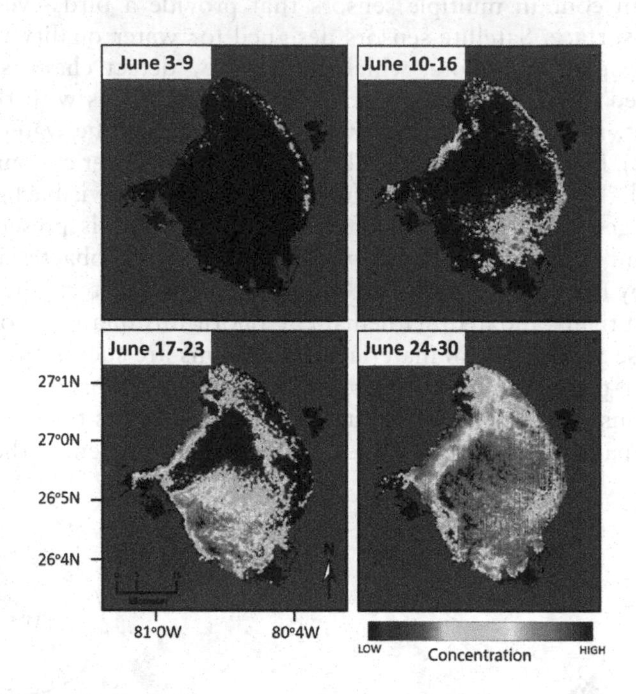

Figure 11.3 Exemplary time series of the Sentinel-3 OLCI satellite images of Lake Okeechobee, Florida, that can quantify the temporal and spatial changes of cyanobacteria biomass within a waterbody. Brown pixels are land masks; black and dark-grey pixels indicate no data (i.e., cloud cover) and below the algorithm detection limit, respectively. Coloured pixels indicate biomass of cyanobacteria from high (red) to low (purple).

and engineering proof-of-concept sensors, and progress towards method development, like water quality algorithms. Eventually, satellite technology and developed methods transition towards operational satellites for the incorporation of data by users, like water quality management. A thorough review of satellite technology and considerations for water quality management is provided in the International Ocean Colour Coordinating Group, Earth Observations in Support of Global Water Quality Monitoring Report (IOCCG, 2018).

11.10.1 Required skills

Typically, new users of satellite data will require computer hardware and expertise to enable adequate data processing and interpretation. Using satellite technology for water quality monitoring and assessment would likely require some new staff training, new software applications and at least a basic understanding of the fundamentals behind remote sensing technology (Schaeffer et al., 2013). Training should include understanding operational satellite platforms, data acquisition, data extraction, quality control and limitations of the applied methods such as algorithm accuracy, uncertainties, interferences and data quality. New software may include free and open source programs such as the National Aeronautics and Space Administration's (NASA) Sea-Viewing Wide Field-of-View Sensor (SeaWiFS) Data Analysis System (SeaDAS), European Space Agency's (ESA) Sentinel Application Platform (SNAP) or R computer language. Other software may include Geographic Information System (GIS) packages, web-based portals and various for purchase software options. Basic information from picture formats (JPEG, TIFF, etc.), without georeference, may also be informative.

As shown by the example in Box 11.2, it is important to validate the satellite-derived results with field measurements. It is also important to report accuracy or error estimates for the specific waterbodies.

11.10.2 Operational satellites

Current and future operational satellite sensors all have some limited ability to resolve the required geophysical variables but with significant trade-offs among spectral, spatial or temporal resolution (Mouw et al., 2015; Palmer et al., 2015). Here we only mention open-access operational satellites with the highest potential to inform management decisions for inland waters which exist at the time of the publication of this book. These operational satellites generally fall into two categories: (1) medium-resolution ocean colour sensors and (2) higher-resolution land imagers. The medium-resolution ocean colour sensors may include ESA's Ocean and Land Colour Instrument (OLCI) on the Sentinel-3 (3A launched 2016 and 3B launched in 2018) satellites. Historical data could be retrieved from the MEdium Resolution Imaging Spectrometer (MERIS) on the Envisat satellite from 2002 to 2012. The OLCI and MERIS sensors provide adequate spectral bands for inland water derivation of water quality parameters, with a typical revisit time of 2–3 days, but have spatial resolution (300 m pixel size) limits. Thus, they are useful for providing observations in larger lakes and reservoirs. The higher-resolution land imagers include the Multi-Spectral Instrument (MSI) on the Sentinel-2 (2A launched 2015, 2B launched in 2017) satellites and Landsat series satellites provide the best spatial resolution for inland waters but are at a disadvantage when it comes to spectral resolution, signal-to-noise ratio, and, to some extent, temporal coverage. Only by combining the observations from Landsat missions or Sentinel-2 missions would a temporal revisit of every 8 days and 5 days be possible, respectively. The satellite revisit time is defined as the time between measurements of the same location on the surface of the Earth.

11.10.3 Limitations

Satellite data may be used to infer surface bloom locations throughout the waterbody, albeit with some technological limits. For example, mixed land and water pixels, and bottom interference may confound the derivation of remote sensing results along the lakeshore. All satellite algorithms detect only near-surface concentrations, and the red and near-infrared part of the spectrum provide information from only the upper few metres of the water column. Atmospheric interference, cloud cover and ice formation limit the usability of satellite images at different rates, depending on the climate of the location. Overall waterbody size and optical complexity impact the application of satellites based on the native pixel resolution of the sensor and processing steps, respectively. There is no optical signal that can be detected by satellites to directly measure cyanotoxin concentrations.

11.10.4 Frequency of sampling

Satellite remote sensing presents a cost and time-effective approach complimentary to field-based cyanobacterial monitoring efforts for a more comprehensive assessment of inland waters. Remote sensing can provide water quality data with frequent revisit times for many lakes. These operational satellites provide daily, 2–3 day, weekly, monthly and seasonal assessments of water quality data. The near-real-time availability of water quality data from current satellites makes it possible to integrate such data into early warning systems to protect human health and ecosystems.

11.10.5 Applications for monitoring programmes

Historical satellite records may be used to contextualise background monitoring to identify the potential for cyanobacteria occurrence problems in waterbodies and the typical timing, location and extent of the bloom at local and regional scales. Near-real-time satellite records may be used for cyanobacteria monitoring to quantify abundance in recreational and drinking waterbodies.

11.10.6 Retrospective assessments

Satellite remote sensing may be used to quantify the spatial extent of the surface area covered by a cyanobacterial bloom (Urquhart et al., 2017). Relevant statistical tests and time-series analyses may be used to identify trends in satellite-derived extent of surface area covered by cyanobacteria. Trend analysis for surface area extent may be subdivided into categorical thresholds desired by the user, based on cyanobacteria concentration or chlorophyll-*a*, to help water managers effectively distribute resources to monitor and manage waters. Scalable assessments may permit the development of management objectives over different temporal periods and spatial scales. Improved multiscale assessment capability is desirable so that comparisons of condition may occur across local, regional and national scales to more adequately evaluate regional water quality, biological integrity and response to management actions.

The frequency of observed cyanobacteria (Figure 11.4) may be calculated as the fraction of total observations for which cyanobacteria biomass exceeded a specified threshold, for example, a Vigilance or Alert Level (sections 5.1.2 and 5.2.3; Clark et al., 2017). Values are summed for each pixel and divided by the total number of valid observations (i.e., those not flagged for clouds, land, mixed land water and lack of data). Finally, the magnitude of cyanobacteria biomass may be calculated based on the spatiotemporal mean of the biomass for a particular period of time such as a season or year. The spatial extent, temporal frequency and magnitude can all be used to rank waters in order of importance to prioritise management resources (Mishra et al, 2019).

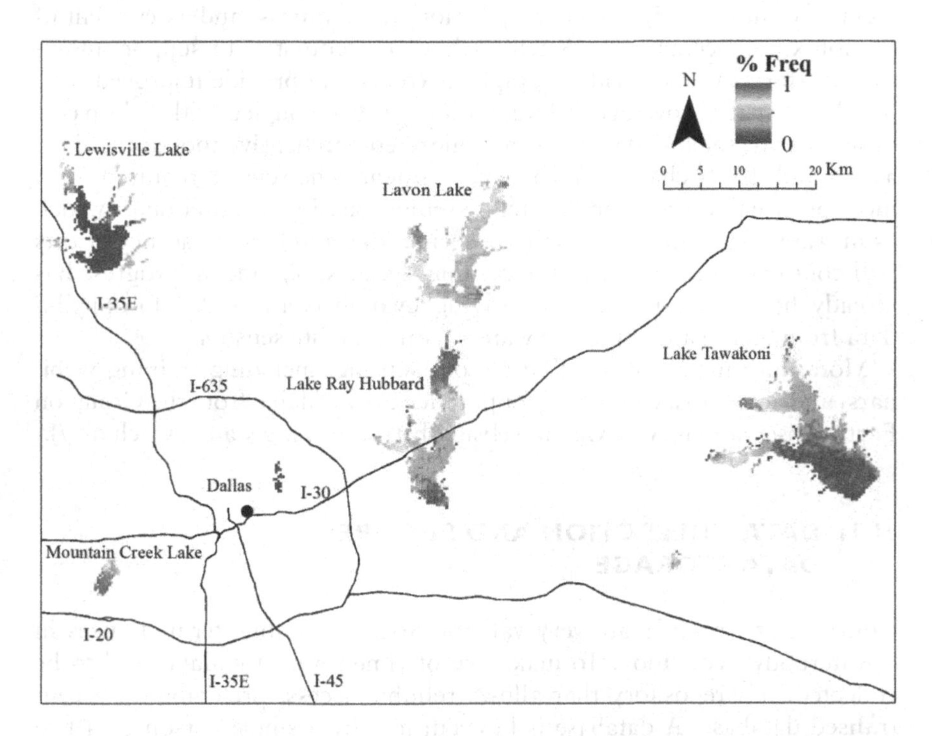

Figure 11.4 A demonstration of how ESA's Sentinel-3 OLCI satellite data can be used for a quantitative retrospective assessment across lakes near Dallas, Texas, USA. Here, the frequency of detected cyanobacteria biomass is calculated as the fraction of total observations throughout 2017. A value of 1 indicates cyanobacteria had a 100% frequency of occurrence in valid observations through 2017, and a value of 0 indicates cyanobacteria were not detected.

11.10.7 Near-real-time monitoring

Satellites provide a constant birds-eye view of the Earth's surface and can be used to identify changes in the environment across geopolitical boundaries by providing updated images the same day the data is acquired. This information may be used to identify events and locations during days of data acquisition, or at weekly, monthly and seasonal intervals. Management decisions such as updating recreational beach notices or modifying drinking-water treatment methods may benefit from access to near-real-time satellite-derived occurrence information.

11.10.8 Satellite support of monitoring programmes

The presence of cyanobacteria or chlorophyll-*a* biomass estimated from satellites may be used as a first-line indicator of potential ecological and human health risk that can be used to prioritise waterbodies requiring further evaluation for parameters such as visual inspection, laboratory assessment of cyanobacteria taxon composition and biomass, and assessment of cyanotoxin concentrations. Satellites have the potential to support monitoring efforts across broad geographic extents and provide improved temporal and spatial coverage at larger scales. When coupled with field-based observations, satellite data provide a more comprehensive tool to monitor, assess and detect changes in the environment. The science required for a more precise interpretation of satellite remote sensing of water quality such as in-water algorithms, atmospheric corrections and land adjacency effects will continue to mature over the coming decades. Significant progress has already been demonstrated in deriving cyanobacteria and chlorophyll-*a* data from inland and estuarine waters using satellite sensors.

More information on satellite remote sensing, including training webinars, and access to a community of practice are available from the Group on Earth Observations AquaWatch website (https://www.geoaquawatch.org/).

11.11 DATA COLLECTION AND SECURE DATA STORAGE

Consistent time series are very valuable to observe long-term changes in a waterbody's condition. To make use of time series, the data need to be collected in a repository that allows reliable access, preferably in a centralised database. A database is best curated by a single person or a few responsible persons/institutions, and any data entered in a database should be as uniform as possible. In particular, units for individual values have to be standardised – confusion may arise from different units when they are not clearly disclosed, that is, concentrations in µg/L or ng/L or ng/mL. This is especially important when data are compiled from several individual institutions and laboratories. Although this is in theory self-evident, the lack of uniformity of data is a constant source of hassle, particularly for supra-regional or supra-national data analysis and interpretation.

Monitoring programmes tend to require an adaption to changing circumstances as time progresses. This could be changes in sampling frequency, sampling point locations or analytical methods, for example. Any modification needs to be well documented to ensure traceability of the data. So do the analytical methods used. In addition to the laboratory quality assurance system implemented for sampling and analyses, a plausibility check of data prior to their final storage helps identify trivial errors that creep in, for example, due to misplacements of decimals. Again trivial, but often missed is that a timely check allows for questions back to those performing sampling and analyses in case data do not appear plausible, and this may well lead to further information explaining unusual data which is important to include in the documentation.

REFERENCES

Bartram J, Ballance R, editors (1996). Water quality monitoring: a practical guide to the design and implementation of freshwater quality studies and monitoring programmes. London: F & FN Spoon on behalf of United Nations Environment Programme and the World Health Organization:383 pp.

Bertani I, Steger CE, Obenour DR, Fahnenstiel GL, Bridgeman TB, Johengen TH et al. (2017). Tracking cyanobacteria blooms: Do different monitoring approaches tell the same story? Sci Tot Environ. 575:294–308.

Bigham-Stephens DL, Carlson RE, Horsburgh CA, Hoyer MV, Bachmann RW, Canfield DEJ (2015). Regional distribution of Secchi disk transparency in waters of the United States. Lake Reserv Manage. 31:55–63.

Bordet F, Fontanarrosa MS, O'farrell I (2017). Influence of light and mixing regime on bloom-forming phytoplankton in a subtropical reservoir. River Res Appl. 33:1315–1326.

CARU (2016). Estudio de la calidad del agua del Río Uruguay. Bienio 2013–2014. Paysandú: Comisión Administradora del Río Uruguay.

CARU (2017). Programa de vigilancias de playas del Río Uruguay. Paysandú: Comisión Administradora del Río Uruguay. http://www.caru.org.uy/web/2017/12/programa-de-vigilancia-de-playas-del-rio-uruguay/.

Clark JM, Schaeffer BA, Darling JA, Urquhart EA, Johnston JM, Ignatius AR et al. (2017). Satellite monitoring of cyanobacterial harmful algal bloom frequency in recreational waters and drinking water sources. Ecol Indicators. 80:84–95.

Drozd A, de Tezanos Pinto P, Fernandez V, Bazzalo M, Bordet F, Ibañez G (2020). Hyperspectral remote sensing monitoring of cyanobacteria blooms in a large South American reservoir: high-and medium-spatial resolution satellite algorithm simulation. Mar Freshwater Res. 71:593–605.

Giannuzzi L, Sedan D, Echenique R, Andrinolo D (2011). An acute case of intoxication with cyanobacteria and cyanotoxins in recreational water in Salto Grande Dam, Argentina. Mar Drugs. 9:2164–2175.

Graham J, Loftin A, Zeigler A, Meyer M (2008). Cyanobacteria in lakes and reservoirs—toxin and taste-and-odor sampling guidelines (version 1.0). Reston (VA): US Geological Survey.

Gugger MF, Lenoir S, Berger C, Ledreux A, Druart JC, Humbert JF et al. (2005). First report in a river in France of the benthic cyanobacterium *Phormidium favosum* producing anatoxin-a associated with dog neurotoxicosis. Toxicon. 45:919–928.

IOCCG (2018). Earth observations in support of global water quality monitoring. Dartmouth: International Ocean Colour Coordinating Group. www.ioccg.org.

Lottig NR, Wagner T, Henry EN, Cheruvelil KS, Webster KE, Downing JA et al. (2014). Long-term citizen-collected data reveal geographical patterns and temporal trends in lake water clarity. PloS One. 9:e95769.

Mishra S, Stumpf RP, Schaeffer BA, Werdell PJ, Loftin KA, Meredith A (2019) Measurement of cyanobacterial bloom magnitude using satellite remote sensing. Sci Rep 9:1–17

Mouw CB, Greb S, Aurin D, DiGiacomo PM, Lee Z, Twardowski M et al. (2015). Aquatic color radiometry remote sensing of coastal and inland waters: challenges and recommendations for future satellite missions. Rem Sens Environ. 160:15–30.

O'Farrell I, Bordet F, Chaparro G (2012). Bloom forming cyanobacterial complexes co-occurring in a subtropical large reservoir: validation of dominant eco-strategies. In: Salmaso N, Naselli-Flores L, Cerasino L et al., editors: Phytoplankton responses to human impacts at different scales. Cham: Springer Nature:175–190.

Palmer SC, Kutser T, Hunter PD (2015). Remote sensing of inland waters: challenges, progress and future directions. Remote Sens Environ. 157:1–8.

Schaeffer BA, Schaeffer KG, Keith D, Lunetta RS, Conmy R, Gould RW (2013). Barriers to adopting satellite remote sensing for water quality management. Int J Remote Sens. 34:7534–7544.

Schröter M, Kraemer R, Mantel M, Kabisch N, Hecker S, Richter A et al. (2017). Citizen science for assessing ecosystem services: status, challenges and opportunities. Ecosyst Serv. 28:80–94.

Srivastava A, Singh S, Ahn C-Y, Oh H-M, Asthana RK (2013). Monitoring approaches for a toxic cyanobacterial bloom. Environ Sci Technol. 47:8999–9013.

Strobl RO, Robillard PD (2008). Network design for water quality monitoring of surface freshwaters: a review. J Environ Manage. 87:639–648.

Urquhart EA, Schaeffer BA, Stumpf RP, Loftin KA, Werdell PJ (2017). A method for examining temporal changes in cyanobacterial harmful algal bloom spatial extent using satellite remote sensing. Harmful Algae. 67:144–152.

WHO (2017). Guidelines for drinking-water quality, fourth edition, incorporating the 1st addendum. Geneva: World Health Organization:631 pp. https://www.who.int/publications/i/item/9789241549950

Wood SA, Puddick J, Fleming RC, Heussner AH (2017). Detection of anatoxin-producing *Phormidium* in a New Zealand farm pond and an associated dog death. New Zeal J Botany. 55:36–46.

Chapter 12

Fieldwork

Site inspection and sampling

Martin Welker and Heather Raymond

CONTENTS

INTRODUCTION

This chapter is primarily intended for readers that are not running a routine monitoring programme but need to organise fieldwork and sampling from scratch. Local, national and regional guidelines may also exist and should be followed, when appropriate. This is especially important if the results of sampling are intended to inform any type of public advisory postings. If guidelines are not available for a specific region, it may be nonetheless helpful to consult the existing guidelines from neighbouring regions or regions with similar conditions. Under local conditions, however, it may be difficult to fully comply with the existing guidelines for various reasons, for example, lack of material or deviating seasonal patterns of phytoplankton dynamics. It is then preferable to organise fieldwork with the locally available means rather than to suspend fieldwork completely due to the lack of specific material requested in guidelines. This could be, for example, sample containers such as wide-mouthed amber glass bottles that are not available or unaffordable and have to be replaced by ubiquitous plastic bottles. This chapter is therefore more of a blueprint to develop locally adopted guidelines than a guideline itself.

The following is largely focused on the sampling of plankton and the measurement of hydrophysical parameters in the pelagic of waterbodies. The sampling of sediment and benthic cyanobacteria is briefly discussed.

A number of practical issues need to be considered when sampling for cyanobacteria and cyanotoxins and good preparation greatly facilitates on-site work. Sampling campaigns can be considerably impeded by weather conditions that make manoeuvers that appear very simple from behind a desk more challenging in the field. For this reason, sampling campaigns should be prepared in a way that reduces on-site handling steps to a minimum.

In addition, samplers should be prepared to address questions and concerns from the general public when sampling recreational sites (see Chapter 15).

12.1 PREPARATIVE STEPS

Before fieldwork is conducted, the monitoring programme should be consulted and for each task verified that it can be conducted as planned (see Chapter 11). Staff responsible for collecting samples needs to be trained on the entire process, including completing sampling protocols, handling of sampling devices and storage of samples during transport. Further, basic knowledge of cyanobacterial biology is favourable to decide on deviations from the sampling scheme or to collect additional samples when the actual conditions indicate this to be appropriate, for example, scum formation at an unexpected shore site due to unusual wind directions.

Preparative steps include:

- preparation of checklists for materials required for the on-site sampling (sampling devices, storage containers, vehicles, etc.);
- preparation of easy-to-fill-in protocol forms that can be completed under adverse conditions in the field (may include field data collection sheets, sample submission forms and chain of custody forms);
- consistent use of unique sampling location names for all sampling sites. This will greatly simplify data management and avoid confusion;
- establishing a sample labelling scheme that allows the unambiguous back-tracking of samples. This is especially important when samples

are diverted later, that is, to be sent to different laboratories/analysts. When multiple institutions and laboratories are involved, the consistent use of a labelling scheme must be asserted;
- verification that the sampling can be practically fulfilled as intended, with sufficient time buffer to compensate possibly occurring delays;
- planning and organisation of transport to and from sampling locations, including access permissions for restricted areas;
- planning of transport of samples to other laboratories when not analysed in-house, including measures to preserve samples appropriately;
- contact information for laboratories conducting analysis (in case questions arise during sampling, or sample transport will be delayed), sampling site owners or managers, and emergency contacts.

Good logistical preparation prior to fieldwork requires that equipment is checked to ensure that it is functioning properly, for example, regular testing and calibration of electrodes; testing and changing of batteries; and keeping operation, maintenance and calibration records, respectively. It is essential to prepare a sampling checklist that includes maps of sampling site locations, a list of required equipment, a detailed explanation of the methods for sample collection, lists of the types and numbers of samples to be taken at each site, as well as of the required volumes of samples. The labelling of sample containers (with water-proof markers) should follow a consistent system to make every sample traceable at any later time. As a minimum, sample container labels should include a unique and consistent sample site code (e.g., a code for the waterbody and a code for the sampling point) and the type of sample or the intended analysis, respectively. In combination with the collection date and time on the sample protocol, a unique sample identifier is created. It is imperative that unique and specific site names be established and consistently used by all sample collectors. Developing an electronic master site list linked to geographic information and other pertinent metadata (laboratory methods used, reporting limits, sample collector name, etc.) is encouraged. This will greatly simplify data storage, retrieval and future data analysis. Whenever possible, extra sample containers and labelling tools should be included in the material taken to collection sites. The extra containers can be used if additional samples are deemed necessary while on site (detection of scums or shifting bloom location) or if containers become broken or contaminated during transit or while on site.

12.2 DETERMINATION OF KEY HYDROPHYSICAL CONDITIONS

Among the hydrophysical conditions affecting cyanobacterial occurrence, the most important ones are turbidity, temperature profiles (stratification), pH, oxygen concentration and – for rivers or streams – flow rate (Chapter 3).

12.2.1 Turbidity

Turbidity is easily assessed with a Secchi disc. It is slowly submerged into the water at a line to the point where it is just still visible (or no longer visible) and this depth is termed "Secchi depth" (Figure 12.1). The depth down to which photosynthesis is possible in aquatic ecosystems, the euphotic depth, is 1.5–2.5 fold the Secchi depth (Preisendorfer, 1986), and in freshwater studies, the factor 2.3 is widely used (Chapter 4). More precise determinations of the euphotic depth are possible by photon flux measurements requiring a submersible quantum sensor (for photosynthetically active radiation; PAR). However, for the assessment of conditions favouring cyanobacterial proliferation, the much cheaper and simpler determination of Secchi depths is usually sufficient and allows reproducible measurements also by untrained persons after a brief introduction to the method (for an example, see Box 11.1).

12.2.1.1 Equipment

Secchi discs can be self-made, but convenient ones are available from companies that provide field-sampling equipment. They should be 25 cm in diameter, made of sufficiently heavy material to be readily submersible, may include holes to ensure easy horizontal sinking and be attached to a chain or cord of sufficient length with depth marks (Figure 12.1).

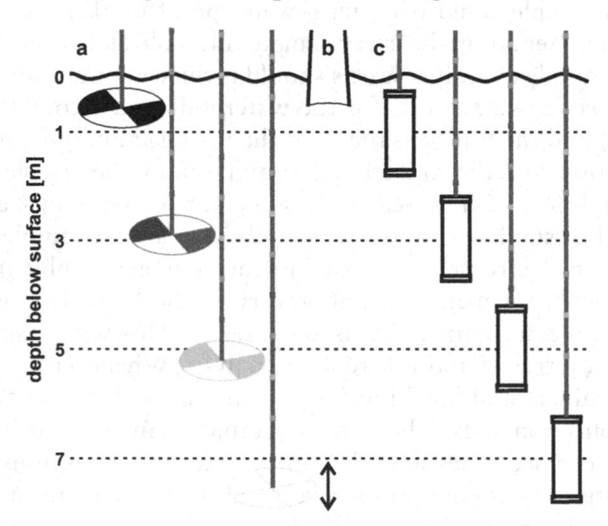

Figure 12.1 (a) Secchi depth measurement: the Secchi disc is lowered at the graduated rope to the depth where it is no longer visible. At this point, the disc is repeatedly lifted and lowered to determine accurately the depth at which the disc becomes visible, and this depth is read from the markings on the rope; the reading can be improved by using an underwater viewer to avoid reflection from the water surface (bathyscope, b). (c) Discontinuous depth-integrated sampling: with a water sampler, samples are taken at predefined (exemplary) depths and then combined.

12.2.1.2 Procedure

- Lower the disc into the water in the shade of a boat (or a pier) as reflections from the surface may distort the reading.
- Lower it to the depth at which it is just still visible; move it up and down several times to confirm that depth.
- If the water surface is very turbulent (e.g., through strong wind), it may help to create a quiet surface with a box without a bottom.
- Blooms may be very patchy, and immersing the disc will move them away from that spot. In such cases, wait a few seconds until they have redistributed.
- Do not wear sunglasses during the procedure as that may distort (i.e., reduce) the reading.
- Comparing readings between fieldworkers is an easy, but important exercise to reduce uncertainty, and it generally leads to remarkably similar results once the procedure has been discussed, understood and agreed.

For measuring transparency in shallow depth such as bathing sites, a Secchi disc with a smaller diameter can be mounted on a graduated rod instead of a rope. This allows rapid and precise measurements while wading in the water up to depth of about one meter.

For greater depths or under poor light conditions, the reading can be improved by using an underwater viewer or bathyscope (Figure 12.1) made of a wide box or tube with a transparent bottom on one side.

12.2.2 Temperature, oxygen and pH profiles

Whether a lake or reservoir is thermally stratified or totally or partially mixed can be determined from temperature, oxygen and pH depth profiles, usually measured at a central location. Modern fieldwork equipment includes multiprobes on long cables that can be lowered stepwise, taking readings at defined depths. A simplified approach is the measurement of temperatures in water samples taken at the defined depth, either directly after hauling the water sampler to the surface or with a thermometer mounted on the water sampler. For the latter approach, sufficient time needs to be allowed for an accurate reading and the haul to the surface has to be rapid enough to avoid errors through changes in the water's temperature when moved from deep layers to the surface.

More precise and continuous data are obtained by installing thermistor chains permanently in the water column. This may be of interest when raw water offtake sites are located at a depth close to the thermocline (see Chapter 8).

From such depth profiles, thermal and chemical stratification can be determined as described in Chapter 4 and Box 4.3.

12.2.3 Additional parameters measured on site

The availability of field-portable sensors enables quick data collection for a suite of informative water quality parameters. This includes multiparameter datasondes that are capable of simultaneously measuring chlorophyll-*a*, phycocyanin and turbidity along with the other parameters mentioned above. More sensitive multispectral sondes may also be able to discern between different types of phytoplankton and estimate their relative abundance through fluorescence measurements (see section 13.6). These tools can be used to help verify the presence of cyanobacteria while on site (through detecting phycocyanin/phycoerythrin) and can help direct sampling to locations of cyanobacteria maxima. For example, a datasonde profile can be collected throughout the water column or along a horizontal gradient, and samples can be collected at discrete depth or locations with elevated phycocyanin or chlorophyll-*a* concentrations.

12.2.4 Flow rate and discharge

In running waters such as rivers and streams, the determination of flow velocity and discharge is of interest for aspects such as estimates of nutrient input to a lake or reservoir, or turbulent mixing (see Chapter 7). Flow velocity is measured with a current meter. Current meters commonly can be mechanical with a propeller or based on Doppler acoustics. Since most running water show turbulent flow and pronounced gradients within the transversal section, a measurement of average flow velocity can be only achieved by measurements at multiple points in the profile. For some purposes, the temporal and seasonal variation at defined measurement points is more important than an exact determination of average flow velocity or discharge, and a measurement of flow velocity at a single, well-defined point in the middle of the stream or river may be sufficient for cyanobacterial monitoring and management purposes because in longer time series (frequent) data on relative changes in flow velocity are more meaningful than (a few) accurate measurements of absolute discharge. Correlating measured flow velocities with precipitation in the catchment may be helpful.

Discharge, the volume of water that flows through a transect per unit of time, usually in m^3/sec, is estimated from measurements at multiple points in the profile. This may require additional expertise or training. Discharge data may be available from regional water authorities.

12.3 ON-SITE INSPECTION AND DATA COLLECTION

A protocol for on-site inspection and data collection should be established, allowing a rapid and easy entry of data that are not logged electronically. Data to be registered include date and time, air and water temperature, wind and general weather conditions, observations such as surface blooms, smell, dead fish, reports from the local population or the like. For recurrent questions, multiple choice-type questionnaires are recommended as these allow a rapid entry, even under adverse conditions, and have the benefit of a consistent and comparable recording of key data, in particular when multiple institutions exchange data.

Sites used for drinking-water abstraction or recreation should be subject to inspection by trained staff, and preferably in conjunction with sampling expeditions. Careful inspection and reporting can assist in the interpretation of results from laboratory analysis. Moreover, the development of personal expertise in relation to specific waterbodies can provide the best form of early warning system, and hence, staff continuity has a high value.

When scums appear on the water surface, cyanobacteria may be present in densities hazardous to human health, and thus, appropriate responses should be initiated quickly (see Chapters 5 and 6), and samples for further analysis should be taken considering safety aspects (see section 12.10). Sampling of scums outside designated or habitual bathing sites is also of great value for determining and predicting hazards, for example, in case of a change in wind direction.

A well-prepared sampling protocol greatly facilitates on-site work. It should be easy to fill in under field conditions, that is, by using check-boxes or multiple-choice options. Information to be collected on site is as follows:

- *General information*: date, time, waterbody, sampling site, staff;
- *Weather conditions*: air temperature, precipitation, wind direction and speed;
- *Water conditions*: water temperature, water transparency, water colour, pH, conductivity, oxygen concentration;
- *Samples*: volume of specific samples, split samples;
- *General observations*: visibility of cyanobacterial (surface) blooms, odour, reports from local stakeholders;
- *Delivery of samples*: handover protocol to cooperating laboratories.

12.4 TAKING WATER SAMPLES

A variety of commercially available water sampling devices have been developed for specific purposes (Figure 12.2). Before purchasing a water sampler (or building one in-house), a limnologist should be consulted to select an adequate type. For practical reasons, the dimensions of a sampler should also be considered as the manual lifting of a filled sampler can be challenging, in particular when working from a small boat.

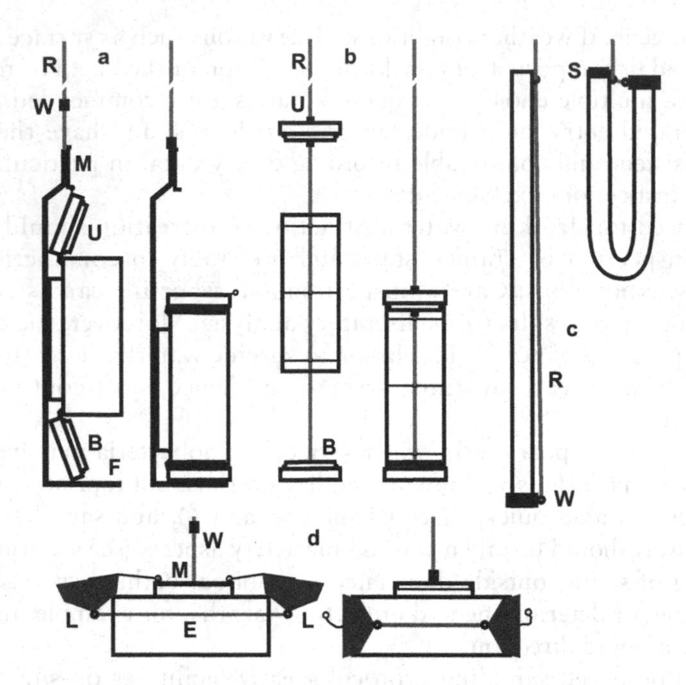

Figure 12.2 Water sampling devices. (a) Limnos-type sampler: the sampler is mounted on a frame (F); for sampling, the upper (U) and bottom (B) lids are held by a release mechanism (M). When the sampler is lowered to the desired depth, a weight (W) is let loose and slides down the rope (R) and hits the release mechanism, thereby unlocking the lids and closing the sampler. After bringing it to the surface, it is emptied. (b) (simplified) Ruttner- or Kemmerer-type water sampler: the bottom lid (B) is lowered to the desired depth, and the tube and the upper lid (U) are released and slide down along the rope to close the sampler. (c) Hosepipe sampler: the weighted end of the hosepipe (W) is lowered to the desired depth at a rope (R); the upper end is closed with a stopper (S) before the lower end is brought to the surface and the hosepipe is emptied. (d) Van Dorn-type sampler: the lids (L) are held open by a release mechanism (M) against the tension of an elastic strap (E); at the desired depth, the lids are released by the weight (W) and close the sampler.

The total volume of the water sample to be taken is determined by the sum of volumes of all subsamples for individual analyses (see below). The calculated total volume needed for all individual analyses is best exceeded about twice to ensure a sufficient sample volume even in case of accidental loss during sample processing.

Two principle types of samples are distinguished, grab samples and integrated samples. Grab samples, either from the surface or from a defined depth, provide information restricted to a specific spot in the waterbody, for example, for a site used for recreation or for drinking-water offtake, whereas for assessing conditions in the whole waterbody, integrated samples are more appropriate.

12.4.1 Grab samples

For surface samples, the easiest way is to submerse the sample container or another vessel. In the presence of surface blooms or scums, preferably multiple samples are taken to account for spatial heterogeneity that can be substantial within distances of a few metres. It is often helpful to collect grab samples at discrete depths, using either Van Dorn- or Kemmerer-type samplers. This is especially useful for determining source water conditions for drinking-water systems, which often draw water from deeper intake locations. For offtake systems with the flexibility to draw water from multiple intake depths, sampling each discrete intake depth can help inform water system operators of the region with the best water quality.

It is important to note exactly where and how the samples have been taken. When surface blooms or scums are present, highest concentrations are expectedly found in the uppermost centimetres, but once disturbed, for example, by wading in the water, the scum may be redispersed in the water column, thus lowering concentrations. It has therefore to be specified what individual samples should represent: maximum concentration or averaged for a water volume resembling the situation of a frequented bathing site.

12.4.2 Integrated samples

Thermal stratification usually results in inhomogeneous distributions of oxygen, nutrients and populations of algae, cyanobacteria and other organisms. For this reason, depth-integrated samples are generally more adequate than (surface) grab samples for the assessment of the size of a cyanobacterial population and nutrient concentrations. However, even when temperature profile is uniform throughout depth, stratification of organisms may develop on calm days. Depth gradients of oxygen concentration and pH are good indicators of this.

Depth-integrated samples are taken by either continuously sampling the entire water column from the surface to a defined depth or by taking several individual samples from defined depths and combining them into a larger volume. A vessel of sufficient volume needs to be available, such as a polyethylene barrel or canister.

Continuous depth-integrated samples are often adequate for shallow and waterbodies of moderate depths. A simple depth-integrating pipe or tube sampler for shallow water columns (up to 5 m depth) or for the surface layers of deeper waterbodies is shown in Figure 12.2. This sampler is made of a piece of flexible tubing of several centimetres in diameter and sufficient length, one end of which bears a weight and is open at both ends. Preferably, the tube is transparent to allow easy recognition of any contamination that may attach to the inner wall. The weighted end is lowered slowly into the water on an attached cord. When the tube has been lowered to the desired depth, it contains an integrated volume of the water column. Before hauling in the lower end with the attached line, the upper end is closed to avoid the loss of water once the lower end emerges from the surface. Hoses need to be thoroughly cleaned after use and stored preferably dry between sampling trips to avoid cross-contamination, for example, by microbial growth. In case multiple sites or waterbodies are sampled during a sampling trip, the hosepipe needs to be conditioned at each sampling site by repeatedly lowering it on one side of the sampling vessel or dock before the sample eventually is taken at the opposite side to avoid cross-contamination. Alternatively, continuous depth-integrated samples can be obtained using a submersible water pump attached to a hose that is operated at a steady pumping rate while the water inlet is drawn upwards between the desired depths at a uniform speed.

In deeper lakes or reservoirs with thermal stratification, depth-integrated samples can be obtained by taking multiple grab samples at defined depths, for example, at 1, 3, 5 and 7 m below the surface and combined to an integrated sample. If background information on the typical stratification characteristics of a given lake is available (e.g., from long-term monitoring), sample numbers can be reduced by selecting adequate depths to represent specific strata. If depth intervals are unequal and samples are to be integrated, the volume of each subsample must be chosen to represent the actual fraction of the vertical stratum it represents.

In the case of surface bloom-forming cyanobacteria, wind-driven inhomogeneity can be considerable with a variation in concentrations of cells and toxins by orders of magnitude across the lake's surface. Before a single sampling location is chosen as representative for a given waterbody – generally a central location is chosen – this should be confirmed by sampling at different locations and by visual inspection. When available, remote sensing data of the waterbody can give indications on heterogeneous horizontal distribution of phytoplankton or chlorophyll-*a*, respectively (see Chapter 11).

12.4.3 Sampling bulk material

For a number of purposes, the sampling of bulk cyanobacterial material is of interest, for example, in-depth chemical analysis of toxins and other metabolites, isolation of cyanobacterial strains or toxicological studies.

Sampling scums is carried out most easily with a wide-necked plastic or glass container submerged only to a depth corresponding to the thickness of the scum.

Cyanobacteria distributed in the water column can be concentrated with a plankton net. The plankton net is lowered to the desired depth and slowly hauled to the surface. The depth at which the plankton net is deployed depends on the taxa of algae and/or cyanobacteria present. Floating cells (e.g., *Microcystis, Dolichospermum, Aphanizomenon*) are harvested within the upper metres of the water column, while the sampling of well-mixed or stratified waterbodies with distinct depth distributions of cyanobacteria (e.g.,

Planktothrix) may include deeper water layers. The mesh size of the net needs to be appropriate for the taxa present, and for most cyanobacteria of interest, 20 µm will suffice. A plankton net sample is not fully representative for the sampled waterbody, especially not in quantitative terms because the efficiency with which the net can retain organisms depends on their size: it will be less effective for filaments with a small diameter (e.g., *Limnothrix* sp.) or picoplanktonic organisms (e.g., *Synechococcus* sp.), and this reduced efficiency cannot be quantified. Further, mucilaginous species (e.g., *Microcystis* sp.) may rapidly clog the mesh, thus reducing further passage of water.

12.5 SAMPLING IN THE DRINKING-WATER TREATMENT TRAIN

If cyanotoxins are detected at the raw water intake at concentrations of concern, treatment train samples are relevant for validating the efficiency of cyanotoxin removal at each treatment step (see also section 5.1, Chapters 10 and 11). Of critical importance for treatment optimisation is whether cyanotoxins are predominantly extracellular or intracellular and therefore at the raw water should be analysed for both intracellular and extracellular cyanotoxins. As cells may lyse and release toxins during treatment, analysing both fractions in every treatment train sample may be relevant.

Preferably, sampling is timed with the flow through the plant, so the effect of processes on the same parcel of water can be determined. This is most important for systems that experience large fluctuations in intake water quality. Most water plant operators understand flow rates and hydraulic residence time through their plant, and sampling times can be adjusted accordingly. If the entire treatment train cannot be sampled, at least the major processing steps that are anticipated to provide the bulk cyanotoxin removal should be sampled, for example, prior and after flocculation and filtration, and after oxidation, prior to distribution (see also Chapter 10).

Any sample collected after oxidant addition should be immediately quenched during sample collection. The quenching agent used will depend in part on the method selected to analyse the sample and must be chosen in contact with the laboratory. For example, sodium thiosulphate is a commonly used quenching agent when analysing a sample using an ELISA-based method, but ascorbic acid is more typically used when analysing a sample via an LC-based method (see Chapter 14).

Treatment train sampling may require some specialised sampling equipment. Swing samplers on telescoping poles are especially useful for sampling the top of deep sedimentation basins or filter beds if a dedicated sampling line is not available. A simple bucket attached to a rope can also work in many situations. Whichever sampling equipment is used, it should be cleaned and conditioned between sampling sites (at minimum, triple rinse).

Sampling programmes may include further water quality parameters, including those that serve as potential surrogates for, estimating cyanobacterial cell and cyanotoxin removal throughout the treatment plant: Operators have used portable multiparameter datasondes to collect real-time phycocyanin and chlorophyll-*a* measurements throughout the treatment train. Such real-time data can be useful for quickly estimating the presence of cyanobacterial cells and their removal throughout the plant – although not for dissolved toxins. If a datasonde is not available, grab samples can also be collected from the intake and throughout the treatment train, and analysed in a laboratory with a spectrophotometer. Turbidity reduction is associated with particle removal, including cyanobacterial cells, and is generally a valuable operational monitoring parameter for the efficacy of filtration methods. Critical turbidity limits are therefore frequently used in treatment plants.

12.6 SAMPLE CONTAINERS

It needs to be decided in advance whether it is more practical to subdivide a water sample into subsamples for each subsequent analysis (plankton, toxins, nutrients, etc.) prior to transportation, or whether a single larger sample is to be divided upon receipt in the laboratory (Figure 12.3). In both cases, for subdividing a larger sample, it needs to be ensured that the sample is well mixed. Especially, buoyant cyanobacteria (*Microcystis*, *Dolichospermum*, etc.) can float up within minutes and hence bias subsampling.

Bottles – or containers in general – used for the storage and transport of samples are ideally chosen by the laboratory that will conduct the analyses to avoid later problems due to inappropriate materials or insufficient volumes, respectively (see Chapter 14.1). Accordingly, the cleaning and preparation of the containers is most efficiently defined by the analysing laboratory because the staff can best estimate the risks of carryover effects due to inappropriately cleaned sample containers. This is particularly important for highly sensitive analytical procedures that can detect trace amounts (e.g., for soluble reactive phosphate).

Preferably, containers are prelabelled and well arranged in a suitable rack or box to allow rapid and easy handling under field conditions. To avoid cross-contamination, it is advisable to always use the same bottle for an individual site and individual parameter. For most samples, glass bottles are most appropriate due to the chemical inertness of glass. However, for safety reasons, plastic containers may be more adequate and can be used for most sampling purposes, for example, wide-mouthed polyethylene or polycarbonate bottles. Sample containers have to be checked for their appropriateness, including their volume, ease of cleaning and testing for possible adsorption of analytes (toxins, nutrients, etc.) to the material.

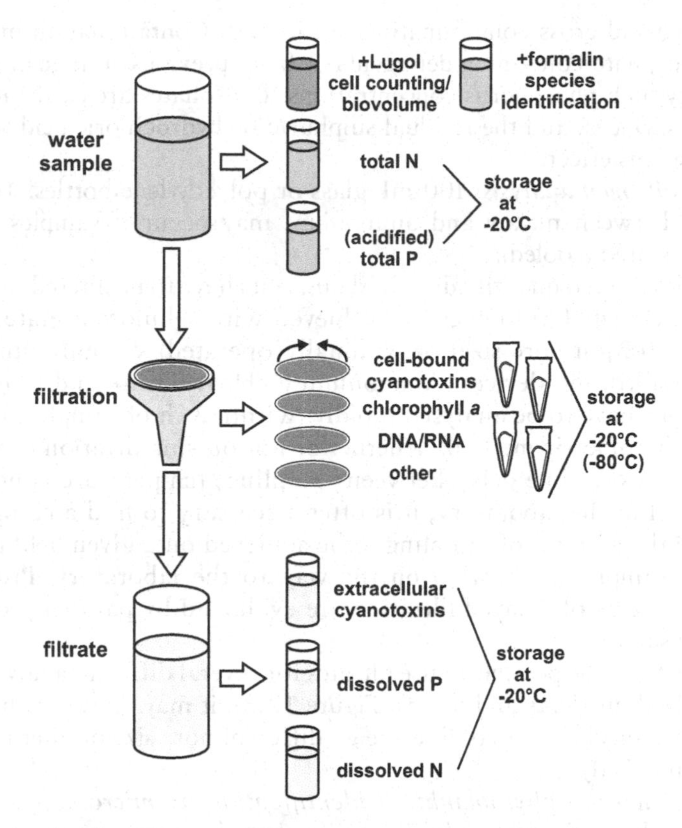

Figure 12.3 Scheme for the splitting of a water sample into multiple subsamples for particular analyses. The list of subsamples is not exhaustive; further parameters could be pigments, iron, dissolved organic carbon, etc. Filter types to collect cells must be chosen to be compatible with the particular downstream analyses. Subsamples can be kept at 4 °C; for later analyses, storage at −20 °C is adequate for most chemical analyses, while for molecular analyses (in particular, RNA), storage at −80 °C may be required.

All samples taken in the field should be stored cool and dark until returning to the laboratory. Sunlight and heat inevitably leads to changes of the samples and eventually to biased data. Insulating boxes such as camping boxes are widely used.

The following containers are recommended for the transport of samples taken for particular analyses. Before filling the individual containers with the samples for analysis, they need to be rinsed with the sample at least twice to minimise cross-contamination from previous samples.

Total phosphorus analysis (for various fractions, see Chapter 13): 100-mL glass bottles prewashed with and stored until usage containing a residual of sulphuric acid (4.5 M) or hydrochloric acid. Since the determination of dissolved phosphorus is done at low µg/L concentrations, care must be taken to avoid cross-contamination of samples. Contamination may arise from phosphate-containing detergents or from previous storage of samples with very high phosphorus concentrations. Phosphates are easily adsorbed to glass surfaces, and the residual sulphuric or hydrochloric acid serves to minimise this effect.

Total nitrogen analysis: 100-mL glass or polyethylene bottles. Transformations between nitrate and ammonium may occur if samples are not properly stored (cooled).

Samples used to quantify dissolved nutrients have to be filtered as soon as possible. On-site filtration can be achieved with cellulose acetate syringe filter (ca. 0.2 µm pore size) or (manually operated) vacuum pumps and adequate filtration devices. Ammonium (NH_4^+), nitrate and nitrite concentrations have to be analysed rapidly (within 24 h of sampling) using a spectrophotometric method. Whether or not on-site filtration is required depends on the time delay between sampling, temperature control and the arrival in the laboratory; it is often necessary to find a compromise between the amount of sampling to be achieved on a given field trip and possible sample degradation on the way to the laboratory. Preferably, possible effects of delayed filtration are evaluated by parallel processing of a few samples.

While it may be possible to use filtrates for several different analyses (e.g., of dissolved nutrients and toxins; Figure 12.3), it may, however, be necessary to use specific types of filters (e.g., different pore size or filter material) for specific analyses.

Cyanobacteria (phytoplankton) identification by microscopy: 100-mL wide-mouthed polyethylene bottles for fresh grab or net sample (see below). Samples can be stored with ethanol at a final concentration of 30% v/v or neutral-buffered formalin at a final concentration of 4% v/v.

Cyanobacteria (phytoplankton) quantification by microscopy: 100-mL clear glass bottles prefilled with 1 mL of Lugol's iodine solution (see below) or neutral-buffered formaldehyde solution (final concentration 4% v/v), respectively (in this case, of course, rinsing with samples is not done). Alternatively, the preservative is added immediately after filling the bottles with the sample. Bottles have to be stored in the dark to avoid photodegradation of iodine. Brown glass bottles are more protective but render later control of stored samples for sufficient iodine residue more difficult (see Chapter 13; also Catherine et al., 2017).

- *Preparation of Lugol's iodine solution*: Dissolve 20 g of potassium iodide (KI) in 200 mL of distilled water; add 10 g of sublimated iodine and 20 mL of glacial acetic acid. Test the solution by diluting 1 mL with 100 mL water. The diluted solution must be clear and have the colour of whisky. Samples fixed with Lugol's iodine and stored need to be checked regularly for decoloration (see Chapter 13).

Cyanotoxins: 1.0-L (minimum for some chemical analysis; depending on methods used in the laboratory) wide-mouthed glass or polyethylene bottles are preferred. For the detection of cyanotoxins by ELISA, smaller sample volumes are sufficient (100 mL). Cyanotoxins may bind to other types of plastic containers, which could reduce the measured concentrations. If only plastic containers are available, before filling the container, rinse it at least twice with the sample. This procedure will encourage binding during the rinsing steps and minimise potential for under-reporting cyanotoxin concentrations in the sample. Containers must be cleaned thoroughly with nonphosphate detergent and rinsed with distilled/deionised water to prevent contamination, especially from dried cells attached to internal surfaces, between sampling events. Generally, the denser a phytoplankton sample, the less volume is needed for a chemical analysis of cell-bound cyanotoxins. If dissolved cyanotoxin analysis was envisaged, a subsample can be filtered in the laboratory, possibly requiring a larger sample volume. To represent source water conditions, filtration should occur as soon as possible, preferably within 24 h of sample collection and prior to freezing the sample which could lyse cells and release cell-bound cyanotoxins. Filtration can also be done on site, but since this is a time-intensive step, it may not be feasible.

Chlorophyll-a analysis: 1.0-L (minimum) wide-mouthed bottles are preferred. Samples must be stored dark and cool to minimise chlorophyll degradation during transport.

Bulk cell material for toxin content, structural analysis of toxins or toxicity assays: Wide-mouthed bottles with volume according to the desired amount of sample material. For smaller volumes, containers for urine samples are particularly cheap and suitable. If samples are to be freeze-dried later on, the sample is preferably frozen in layers not thicker than 2 cm to reduce drying time. To produce frozen plaques, robust household plastic bags can be used when stored in watertight cooling boxes and immediately transferred to a freezer in the laboratory.

12.7 SEDIMENT SAMPLING

Sediment sampling may be helpful if there is a likelihood of high internal nutrient loads to the waterbody of interest (see Chapters 4 and 8). Waterbody management strategies that aim to limit internal nutrient loading often require baseline sediment nutrient data if they are to be successful. A limnologist should be involved in the selection of appropriate sample sites. In general, one to three sediment samples should be collected in small waterbodies, and more locations may be needed to collect representative data for larger waterbodies. Within larger waterbodies, both deep and shallow sites should be selected, representing inlets and some shallower bays. Ideally, sediment cores (at least 30 cm depth) should be collected and each 2 cm segment analysed for phosphorus (P) fractions, total aluminium, total iron and percent solids to help determine potential for sediment nutrient flux.

Sediment cores can also be analysed for cyanotoxins and may be able to provide a record of historic cyanotoxin occurrence within the waterbody (Waters, 2016; Zastepa et al., 2017).

Sediment corers, usually simple sampling devices, can be made in-house or purchased in a variety of materials. They are preferred compared to samples collected using a dredge sampler, because corers can maintain a representative vertical profile of the sediment stratigraphy, create less disturbance by shock waves and can collect more highly consolidated deposits. Sediment corers are slowly lowered to the substrate (gravity corers are released at the water surface and allowed to fall freely); they then penetrate the sediment under the sampler's own weight or are pushed or vibrated into the sediments. Commercial corers often contain core catcher inserts and one-way valves that allow the sample to enter the tube, but not exit and to hold it in place. Inserts should not be reused between sample locations unless properly cleaned. Inserts made of plastic should not be used when collecting samples for organic analysis. Upon retrieval, the corer can be disassembled (e.g., split spoons, some core tips unscrew) and the sample laid in a container or a prepared surface for further processing. Cores from simple tubes and most other corers often drop out or can be pushed out with a clean rod. Plastic or thin-walled metal corers (or core liners) can be cut, the ends capped and secured with tape, and the entire segment sent to the laboratory. This process and the split spoon sampler reduce contamination from one segment to another in vertically stratified samples (OEPA, 2018).

12.8 SAMPLING OF BENTHIC CYANOBACTERIA

Benthic cyanobacteria can be a source of cyanotoxins (as well as taste and odour compounds) and are typically more difficult to monitor than planktonic cyanobacteria. In shallow waters, such as wadable streams, a visual inspection to identify patches of possible benthic cyanobacterial growth is advised prior to the actual sampling. Wood et al. (2009) outline the estimation of cyanobacterial coverage of streambeds with the aid of an underwater viewer.

Limited established guidance is available on benthic sampling, but a variety of techniques have been used to varying degrees of success. Distribution of benthic cyanobacteria can be very heterogeneous, typically occurring in spatially limited patches with high density next to bare areas. Therefore, collecting samples from numerous sites and compositing may be appropriate for determining average conditions or assessing whether benthic cyanobacteria may be a concern. One method of collecting epilithic cyanobacteria (i.e., those growing attached to hard substrate) from streams or littoral zones is scraping a predefined area of representative rocks or substrate. Since variability in epilithic cyanobacteria can be high, multiple rock scrapes from a sampling site can be composited into a single representative sample (Bouma-Gregson et al., 2018). Samples can be collected by hand in wadable areas and by SCUBA divers in greater depths. Epiphytic cyanobacteria are collected together with the macrophytes to which they are attached (see Chapter 4).

Benthic cyanobacteria can also be sampled using a dredge sampler (for larger areas) or sediment corer (for fine-grained sediments). Discrete depth samplers can also be lowered to the bottom of a waterbody to collect samples near the benthic zone. Unfortunately, the dredge, corer and discrete depth sampling methods can displace benthic cyanobacteria during the sampling process and may result in underestimating benthic cyanobacteria occurrence. Since distribution of benthic populations is generally highly variable, these methods may also miss significant benthic mats that are not visible from the surface.

Analytical results of benthic cyanotoxins generally relate them to sediment area, for example, µg/cm^2, to (cyanobacterial) biomass, for example, µg/g fresh or dry weight. A transformation to volumetric units, this is, true concentrations, can only be tentative.

12.9 SAMPLES FOR MOLECULAR ANALYSES

DNA and RNA sample collection may require specific on-site sample preparation and handling protocols due to the potentially rapid degradation of DNA and, especially, RNA. Ideally, samples collected for molecular analyses should be filtered on-site and the filters placed on ice (or as cold as possible). In some cases, DNA sampling protocols may call for in-laboratory filtration, as long as samples are received and filtered by the laboratory within a narrow time frame (preferably within 24 h). RNA sample collection is typically more rigorous, requiring immediate filtration after sample collection, placement of samples onto dry ice to quickly flash freeze filtered material, and holding on dry ice until sample can be transferred to a laboratory or low-temperature freezer prior to extraction and analysis. Due to the extremely high sensitivity of molecular methods, care must also be taken to ensure all sampling equipment is thoroughly cleaned and sterilised. Preferably, sterile, disposable sampling supplies (syringes, cartridge filters, etc.) are used to avoid cross-contamination. In all cases, the validity of sampling protocols should be verified by the laboratory conducting the analyses (see Chapter 14).

12.10 SAFETY CONSIDERATIONS

Caution and attention are appropriate while working with cyanobacteria, particularly when they are highly concentrated in scums. It is wise to treat all blooms as potentially toxic. Contact with water should be minimised during sampling, and gloves and rubber boots should be worn because cyanobacteria (and organisms associated with them) might also have a high allergenic potential.

However, during sampling, cyanobacteria actually are often not the most important hazard and general safety considerations for water sampling need to be implemented. In some areas of the world, other water-based hazards (e.g., organisms causing schistosomiasis or bilharziosis) may also be present. In such circumstances, water contact should be minimised, and following contact, the skin should be immediately rinsed with clear water and dried.

Although glass is generally the most inert material, avoiding glassware for fieldwork enhances safety. For most samples, polyethylene bottles are appropriate.

Inhaling spray or getting spray in eyes from boats, wind or irrigation water from areas with cyanobacteria blooms has to be avoided. Under conditions that promote spray formation, eye protection and a mask are recommended, especially while sampling cyanobacteria scums.

Last but not the least, sampling preferably is always conducted in a team – a basic rule for fieldwork. This has practical reasons when handling water samples, sample bottles, conserving agents, field log sheets, etc. Furthermore, sampling a waterbody involves the risk of serious injury and drowning, even in shallow but turbid waters where dangerous objects may not be visible from the surface and especially when working from a boat. Wearing a life jacket on a boat is strongly recommended and may be mandatory by local safety regulations.

12.11 QUALITY ASSURANCE AND CONTROL

Quality control policies are required for many operators of water supplies, laboratories and public authorities conducting surveillance, and they are important. A subset of samples can be collected for quality control purposes. Duplicate samples can be used to determine laboratory method precision. Replicate samples can be used to determine representativeness of sampling. Field samples may also be split for interlaboratory comparisons. Field blanks consisting of distilled deionised water and preservative, where appropriate, should be submitted along with regular samples to establish practicable detection limits and to monitor for levels of contaminants to which field samples may be exposed. In addition, if sample bottles are being reused, after cleaning, a subset of reused sample bottles should be periodically filled with distilled or deionised water and analysed for the parameters of interest to verify the adequacy of the cleaning procedure. All field instruments used in the measurement of physical, chemical or biological parameters must be properly calibrated and maintained, with records kept of observations for each instrument. Laboratories should consider a regular participation in proficiency testing studies conducted by accredited providers or, more informally, in cooperation with other laboratories in the same region. Quality assurance sampling is especially important if the sample results will be used for regulatory purposes, to document human health impacts or where decisions based on the data could be disputed in court. The following sections describe different types of quality control samples and their intended purposes (OEPA, 2018).

Field duplicate samples (also known as field splits) are used to assess the variance of the total method of sampling and analytical procedures. Duplicate samples demonstrate the precision of the sampling system, from initial sample collection through analysis. A field duplicate is done by thoroughly mixing one sample, dividing it into two separate sets of containers and analysing as (blinded) independent samples.

Field replicate samples are used to measure sampling repeatability and natural variability within the sampled water. A field replicate is done by collecting two or more separate samples from the same site and time using the same sampling method (replicates A, B, ...) and analysed as independent blinded samples. The variability of replicates should be compared to duplicate variability. Replicate sampling is often used to estimate heterogeneity, for example, in sediments. Field duplicate and field replicate sampling may be combined to allow a full assessment of the validity of the entire sampling procedure.

Blank samples are used to evaluate the potential for contamination of a sample by contaminants from a source not associated with the water being tested. Blanks may be used to demonstrate that no contamination occurs from equipment, reagent water, preservatives, sample containers, ambient air, etc. Field blanks are used to evaluate the potential for contamination of a sample by site contaminants from a source not associated with the sample collected (i.e., air-borne dust, etc.). Equipment blanks are collected to verify that cleaning techniques are sufficient and that cross-contamination does not occur between sites, for example, by using the same water sampler. At least one equipment blank per equipment type per field season should be collected. One equipment blank container should be prepared for each type of preservative used. Container blanks are normally tested by the analysing laboratory (see Chapter 14).

12.12 PERMISSIONS AND DECLARATIONS

Sampling a waterbody may require permission, either because private property has to be accessed or because national regulations generally restrict the removal of organisms from the environment. The Nagoya Protocol (UN, 2011), an international agreement on the protection of economic interests possibly arising from natural biodiversity, has been implemented in the legislation of many countries.

REFERENCES

Bouma-Gregson K, Kudela RM, Power ME (2018). Widespread anatoxin-a detection in benthic cyanobacterial mats throughout a river network. PLoS One. 13:e0197669.

Catherine A, Maloufi S, Congestri R, Viaggiu E, Pilkaityte R (2017). Cyanobacterial samples: preservation, enumeration, and biovolume measurements. In: Meriluoto J, Spoof L, Codd GA et al., editors: Handbook of Cyanobacterial Monitoring and Cyanotoxin Analysis. Chichester: John Wiley & Sons:313–330.

OEPA (2018). Surface water field sampling manual. Columbus (OH): Ohio Environmental Protection Agency:39 pp. https://www.epa.ohio.gov/.

Preisendorfer RW (1986). Secchi disk science: visual optics of natural waters. Limnol Oceanogr. 31:909–926.

UN (2011). Nagoya protocol on access to genetic resources and the fair and equitable sharing of benefits arising from their utilization to the convention on biological diversity. Montreal: United Nations.

Waters M (2016). A 4700-year history of cyanobacteria toxin production in a shallow subtropical lake. Ecosystems. 19:426–436.

Wood SA, Hamilton DP, Paul WJ, Safi KA, Williamson WM (2009). New Zealand Guidelines for cyanobacteria in recreational fresh waters: Interim Guidelines. Wellington: Ministry for the Environment - Manatū Mō Te Taiao. https://www.mfe.govt.nz/publications/fresh-water-environmental-reporting/guidelines-cyanobacteria.

Zastepa A, Taranu Z, Kimpe L, Blais J, Gregory-Eaves I, Zurawell R et al. (2017). Reconstructing a long-term record of microcystins from the analysis of lake sediments. Sci Tot Environ. 579:893–901.

Chapter 13

Laboratory analyses of cyanobacteria and water chemistry

Judit Padisák, Ingrid Chorus, Martin Welker, Blahoslav Maršálek and Rainer Kurmayer

CONTENTS

INTRODUCTION

Identification and quantification of cyanobacteria in water resources is a basic component of cyanotoxin monitoring programmes to effectively allow early alerts of the type of toxins to expect as well as of bloom development. Further, data on concentrations of nutrients, that is, phosphorus and nitrogen, are valuable for assessing the potential for cyanobacteria to develop blooms (Chapter 4). Information on turbidity, waterbody mixing and flow rate supports this assessment. Methods for nutrient analysis have been extensively reviewed and internationally harmonised by the International Organization for Standardization (ISO). In contrast, approaches to identify and quantify cyanobacteria are very variable and can be undertaken at different levels of sophistication. Rapid and simple methods can be employed to analyse the composition of a sample at the taxonomic level of cyanobacterial genera rather than species. This is often sufficient for a preliminary assessment of potential hazard as well as for initial management decisions. Further investigation may be necessary in order to quantify cyanobacteria, for example, to determine whether they are present above a threshold biomass level. Quantitative counting methods can give useful estimates of cell numbers and biovolumes with a counting effort of less than one hour per sample and sometimes within minutes. Bulk methods such as biomass estimation by chlorophyll-*a* analysis or fluorimetric methods can be very time-effective with only moderate equipment demands. More detailed taxonomic resolution and biomass analysis is necessary to predict cyanobacterial bloom development. Prediction of toxin production carries uncertainties since the dominant species in multispecific cyanobacterial blooms are not necessarily the toxin producers. Distinction between these approaches is important because management must decide how available staff hours are most effectively allocated. In many cases, the priority is likely to be the evaluation of a larger number of samples at a lower level of precision and taxonomic detail.

The choice of methods further requires informed consideration of sources of variability and error at each stage of the monitoring process, particularly for sampling (see Chapter 12). Waterbodies with substantial temporal and spatial variation of cyanobacterial cell density may show variability of orders of magnitude in cyanobacterial biomass between samples taken within a few hours or within short distances. A highly accurate determination of biomass from singular weekly samples may not be adequate for the assessment of population size, while more useful information can be gained by investing the same effort into a less accurate evaluation of a larger number of samples, either multiple samples per date or samples taken at a higher frequency. Efficiency can further be optimised by regular interlaboratory calibrations of methods and their quality control as well as through testing the emerging new rapid molecular methods against the results of those obtained with accurate established methods.

This chapter describes methods for cyanobacterial identification and quantification at different levels of accuracy. Further, for determination of the key nutrients P and N, which control cyanobacterial biomass and species composition, it gives a brief overview of the ISO methods and guidance on how to assess key hydrophysical conditions.

13.1 HANDLING AND STORAGE OF WATER SAMPLES

Decisions on the type of analyses required should be made prior to sample collection (see Chapter 12). However, this is not always possible, particularly when a routine monitoring programme is not in place. Samples may therefore require immediate evaluation on arrival in the laboratory to determine if pretreatment is needed prior to appropriate sample storage.

Samples that have been taken for microscopic counting should ideally be preserved with Lugol's iodine solution at the time of collection (section 12.6). These samples will be relatively stable and no special storage conditions are required, other than protection from extreme temperatures and light, for example, in a nontransparent box at room temperature, although storage at 4–6 °C is preferred. However, samples should be examined and counted as soon as possible because some types of phytoplankton are sensitive to storage (Hawkins et al., 2005) and Lugol's iodine solution is chemically reduced by organic matter in the sample over extended storage periods (usually within months, but faster in very dense samples), visible by fading of the brownish colour of the Lugol's iodine solution. Therefore, a periodical visual check for loss of colour is recommended and add fresh Lugol's solution if decoloration becomes apparent.

Unpreserved samples for quantitative microscopic analysis require immediate attention in the laboratory either by the addition of preservative (e.g., Lugol's solution) or by following instructions of alternative quantification methods which do not use preserved cells. Where unpreserved samples cannot be analysed immediately, they should be stored in the dark at a temperature close to ambient field temperatures. Unpreserved samples are preferable for species identification because some characteristics cannot be recognised in preserved samples. For example, filaments of *Aphanizomenon flosaquae* aggregate to characteristic bundles, but preservatives tend to disintegrate bundles, and the isolated filaments are more difficult to distinguish from other species and genera. While samples for quantification must be preserved immediately, to samples for identification generally no preservative is added and these should be analysed within approximately 24 h because quantitative changes are less important.

Samples for chlorophyll-*a*, dissolved phosphorus, nitrate, ammonium and molecular analyses should be filtered as soon as possible. Storage for a few hours in the dark in glass bottles is usually acceptable if temperatures do not exceed 20 °C. Filtration at the sampling site is recommended, particularly in warm climates, or filtration should occur immediately upon arrival in the laboratory. Filtered samples for nutrient analysis may be stored at 4–6 °C for a few hours prior to analysis, or frozen at –20 °C for several days, or at –80 °C for several weeks. If extended storage (weeks to months) of loaded filters is intended, the filters are preferably freeze-dried and stored at –20 °C to minimise the degradation of chlorophyll-*a*, for example. Samples collected for RNA analysis or sequencing require immediate filtration and placement on dry ice until transferred to low-temperature freezer.

13.2 IDENTIFICATION OF CYANOBACTERIA

Microscopic examination of a bloom sample is very useful even when accurate counting is not being carried out. The information on cyanobacterial taxa in a sample can provide an instant alert that cyanotoxins may be present. This information can trigger the choice of the method for toxin analysis (see Chapter 14). Most cyanobacteria can be readily distinguished from other phytoplankton and particles under the microscope at a magnification of 100–400 times (see also Chapter 3).

However, organisms identified as belonging to a single species may be highly variable with respect to toxin content (see Chapter 4). Environmental populations of particular species generally consist of multiple genotypes that are not distinguishable based on morphological characteristics but with varying toxin contents. For the commonly occurring genera *Microcystis*, *Planktothrix*, *Aphanizomenon*, *Raphidiopsis* (*Cylindrospermopsis*) and *Dolichospermum* (*Anabaena*), identification on the genus level is often sufficient to allow a prediction on the presence of particular types of toxins (Chapters 2–4). Moreover, microscopic analysis often does not allow the differentiation of individual species for several reasons, such as uncertainties in the taxonomic scheme, absence of characteristic and stable morphological features or lack of experience. Preferably, identification results are then given at the genus level, for example, *Aphanizomenon* spp. (for "species pluralis": multiple species).

For establishing cyanobacterial identification in a laboratory, consultation with experts on cyanobacterial identification is helpful. Training courses for beginners should focus on the genera and species relevant in the region to be monitored. Experts can assist in initially deriving a list of these taxa and the criteria for their identification, and later by evaluating micrographs of the typical cyanobacteria, provided microscopes equipped with a camera are available. In the course of further monitoring, experienced experts should be consulted periodically for quality control and for updating such a list. Last but not the least, continuity of individual staff, that is, long-term responsibility for identification and enumeration, is highly valuable to facilitate identification and to allow the recognition of shifts in taxonomic composition.

Key morphological characteristics of cyanobacteria are described in Chapter 3, and Table 13.1 shows identification keys assisting in the determination of major cyanobacterial groups, genera and species.

13.3 QUANTIFICATION OF CYANOBACTERIA

Rapid methods for frequent monitoring of large numbers of waterbodies or sampling sites have been developed in some countries. These methods cannot be readily standardised and evaluated internationally, but can be adapted to regional or local conditions. Deciding on the appropriate classification of units to count depends upon variations such as whether prevalent taxa are filamentous, colony-forming or occur as single cells, and whether populations are very diverse or largely monospecific.

Cyanobacterial biomass can also be determined using indirect methods, the most common being the quantification of chlorophyll-*a*. The established

Table 13.1 Compilation of taxonomic keys for various taxa of cyanobacteria

Reference	*Covered taxa*
Anagnostidis & Komárek (1985)	Major groups
Komárek & Anagnostidis (1986) Komárek & Anagnostidis (2008)	"Chroococcales"
Anagnostidis & Komárek (1988) Komárek & Anagnostidis (2007)	"Oscillatoriales"
Komárek & Anagnostidis (1989)	"Nostocales"
Anagnostidis & Komárek (1990)	"Stigonematales"
Komárek (2013)	Heterocytous genera
Komárek (1991)	*Microcystis* in Japan
Komárek (1996)	Picocyanobacteria
Komárek (2003)	Planktonic "Oscillatoreales"
Komárek (2010)	Nostocaceae
Komárek & Cronberg (2001)	African "Oscillatoreales" and "Chroococcales"
Komárek & Zapomělová (2007)	*Anabaena/Dolichospermum*
Kaštovský et al. (2010)	Invasive cyanobacteria

methods (see section 13.5) are rapid and simple but also quantify chlorophyll-*a* from other phytoplankton; hence, it is best used when cyanobacteria are the main or dominant organisms present. Alternatively, submersible multiprobes measuring *in vivo* chlorophyll-*a* (Chl-*a*) fluorescence together with that of the pigment specific for cyanobacteria, that is, phycocyanin (PC), are increasingly used (e.g., Ziegmann et al., 2010; McQuaid et al., 2011; Zamyadi et al., 2012; Brentrup et al., 2016). Care should be taken to calibrate fluorimeters (by determining biovolume in selected samples; see section 13.3.2) since otherwise phycoerythrin-rich species might be missed (Selmeczy et al., 2016).

Approaches to monitoring cyanobacterial blooms are reviewed in Srivastava et al. (2013). The procedures and techniques described in the following can be considered as classical approaches. Techniques supported by digital image analysis and computation are emerging and may facilitate the determination of cyanobacterial or, more general, plankton biomass in future (Benfield et al., 2007; Saccà, 2016; Zohary et al., 2016). Automated methods are not yet widely applied for planktological studies and monitoring due to the complexity of the matter. A certain error in quantitative data on plankton biomass has to be accepted with any method (Saccà, 2017), although this error can be reduced by intercalibration exercises between laboratories and operators (Rott, 1981). Molecular methods can also be employed to quantitatively estimate total cyanobacterial biomass and toxigenic potential (see section 13.6).

13.3.1 Counting cyanobacterial cells

Microscopic counting of cyanobacterial cells, filaments or colonies has the advantage of directly assessing the abundance of potentially toxic taxa. Little equipment in addition to a microscope is required. The method may be rather time-consuming, ranging from a few minutes to several hours per sample, depending upon the accuracy required and the number of species to be differentiated. Further, counting time depends to a large degree on personal experience, and therefore, staff continuity is highly desired (Vuorio et al., 2007).

The following begins by outlining precise and widely accepted counting procedures which are more time-consuming and require a moderate level of expertise, but serve as a benchmark to assess the performance of simplified methods which can be developed to suit specific requirements of a given sampling programme. More details can be found, for example, in Olenina et al. (2006) and Karlson et al. (2010).

13.3.1.1 Sample concentration by sedimentation or centrifugation

Direct counting of preserved cells is typically carried out by Utermöhl's counting technique using a counting chamber and inverted microscope

(Utermöhl, 1958; CEN, 2006). This method is well suited for the assessment of a large variation in cell morphologies and is widely accepted as reliable. Counting chambers and sedimentation tubes are commercially available or can be built in-house (see Figure 13.1). The most commonly used chambers have a diameter of 2.5 cm and a height of 0.5 cm and can be fitted on the stage of an inverted microscope. If larger volumes of water need to be analysed, as is the case when cell density is low, a sedimentation tube can be used to increase the volume. The water volume used for an individual counting depends on density of cells, counting technique (fields or transects, see below) and microscopic magnification. If cell densities were high like in bloom samples, even a few millilitres could contain too many cells for accurate counting, and sample dilution is needed. Optimally, 10–30 items (cells, colonies, filaments) are present in an individual counting field. If less, search for cells in the view field consumes time, and if more, the investigator may get confused by the density and individual cells obscure each other, thus decreasing counting accuracy.

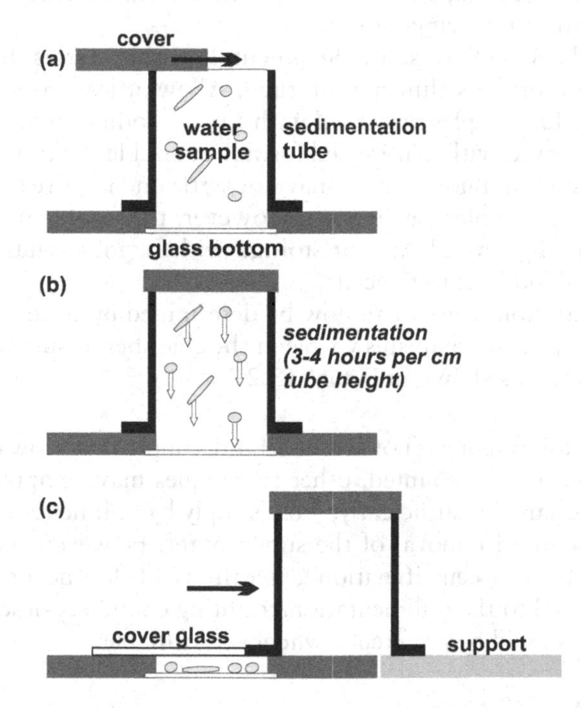

Figure 13.1 Preparation of samples in a sediment chamber for counting plankton with an inverted microscope according to Utermöhl (1958). (a) The sedimentation cylinder is placed on the microscope slide and filled completely with the water sample; a cover is slid on the cylinder. (b) The plankton in the sample fixed with Lugol's solution is allowed to sediment. (c) The sedimentation tube is pushed onto a support with a cover glass; the sample is ready for the counting procedure.

Equipment

- Inverted microscope with 100-, 400- and preferably 1000-fold magnification
- Counting chamber with sedimentation tubes of variable lengths, that is, variable volumes
- Cyanobacterial identification keys and images
- Samples preserved in Lugol's iodine solution (see section 11.3.1)

Procedure

1. Allow the sample to equilibrate to room temperature. If cold samples are placed directly in the counting chamber, gas bubbles develop disturbing sedimentation and interfere with visibility.
2. Gently invert the bottle containing the sample several times to ensure homogenous mixing.
3. Fill the sample into the sedimentation tube placed on the counting chamber.
4. Place the counting chamber on a stable horizontal surface at dark place with stable temperature.
5. Allow the sample to settle. Sedimentation time varies depending on the height of the sedimentation tube. Allow at least 3–4 h/cm height of liquid for samples preserved with Lugol's iodine solution. For samples preserved with neutralised formalin, double the time allowed for sedimentation. Buoyant cells may not settle and require the disruption of the gas vacuoles (see below). However, this problem is frequently overcome by several days of storage with Lugol's solution, through uptake of iodine into the cells.
6. Phytoplankton density can now be determined by counting either the total number of organisms visible in the chamber or subsections (transects, fields) as shown in Figure 13.2.

If an inverted microscope is not available and samples with low cyanobacterial density need to be counted, other techniques may be applied in order to concentrate samples sufficiently, most simply by sedimentation in a glass cylinder and careful removal of the supernatant. However, sedimentation in a glass cylinder or centrifugation generally yields less accurate counting results compared to the sedimentation/counting chambers described above and the latter should be preference whenever available.

Equipment

- Glass measuring cylinder, 100 mL
- Glass pipette with pipette bulb or filler
- Standard laboratory microscope with 10× and 40× objectives
- Sample preserved in Lugol's iodine solution (section 11.3.1)

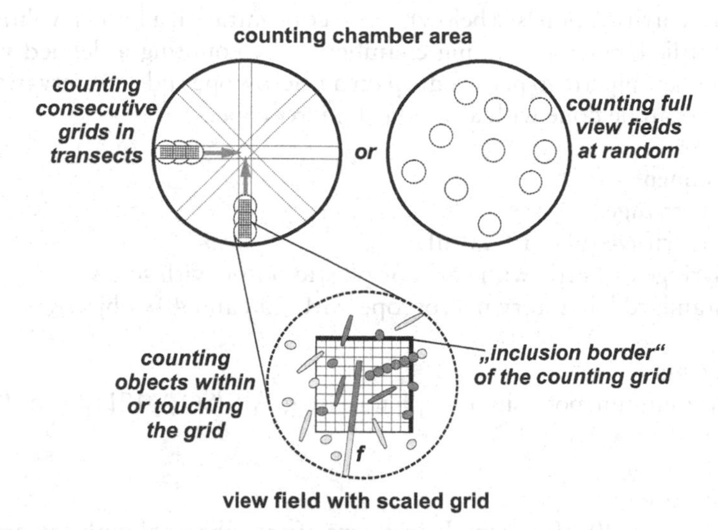

Figure 13.2 Quantitative enumeration of cyanobacterial and plankton cells with Utermöhl technique. Cell counting is done either in multiple consecutive counting grids in the view field following the chamber's transects or in full view fields that are randomly selected. In individual counting grids, two borders are defined as inclusion border. All cells touching these borders are included (dark grey), irrespective of the share of the cell lying within the grid. The opposite borders are defined as exclusion borders with all cells touching these lines to be excluded (light grey). For filaments in which individual cells cannot be distinguished (f), only the share of the filament lying within the grid is measured and counted (dark grey); individual filaments may extend over multiple counting grids or view fields. The grid is moved forward to the next position for the grid's width.

Procedure

1. Allow the sample to equilibrate to room temperature.
2. Gently invert the bottle containing the sample several times to ensure homogeneous mixing.
3. Fill 100 mL of the sample into the measuring cylinder.
4. Allow the sample to settle for an appropriate time (see above).
5. Using the glass pipette, carefully remove the supernatant, leaving only the last 5 mL undisturbed.
6. The sample has now been concentrated by a factor of 20 and can be counted using a counting chamber (e.g., Sedgewick-Rafter or haemocytometer).

Alternatively to sedimentation, centrifugation can offer a rapid and convenient method of concentrating a sample (Ballantine, 1953). Fixation with Lugol's iodine solution enhances the sedimentation. However, buoyant cells may still be difficult to pellet and may require the disruption of vacuoles prior to centrifugation (see below). Once concentrated, a known volume can be quantified using a counting chamber or by counting a defined volume using a micropipette to place a drop on a microscope slide. Observation and counting can be done with a standard microscope.

Equipment

- Centrifuge
- Centrifuge tube, 10–20 mL
- Syringe or bottle with cork, or plastic bottle with screw cap
- Standard laboratory microscope with 10× and 40× objectives

Reagents

- Aluminium potassium sulphate, 1.0 g $AlK(SO_4)_2 \cdot 12H_2O$ in 100 mL distilled water

Procedure

1. Place 10–20 mL of sample in a centrifuge tube, seal with cap and centrifuge at a minimum of 500 × g for 15 min.
2. If pelleting needs to be enhanced, add 0.05 mL of aluminium potassium sulphate solution per 10 mL of sample. Mix and centrifuge as described.
3. If pelleting of buoyant cells wasn't possible, try one of the following:
   i. Fill sample in a plastic syringe, ensure the end is tightly sealed, then apply pressure to the plunger.
   ii. Fill sample in a bottle with a tightly fitting cork, then bang the cork suddenly.
   iii. Fill sample in a well-sealed plastic bottle and drop it sharply onto a hard surface.
   Once subjected to this pressure shock, the gas vesicles should have been disrupted and cells should pellet when centrifuged.
4. After centrifugation, carefully remove the supernatant and resuspend the pellet in a small known volume (e.g., 0.5 mL).
5. Samples concentrated by centrifugation can be counted using a counting grid or haemocytometer.

13.3.1.2 Quantification of cyanobacteria using an inverted microscope

A prerequisite to the counting of cyanobacteria (phytoplankton in general) is the definition of the units to be counted, that is, individual cells, filaments or colonies. Globally, the accuracy of quantitative determination depends on the number of counted objects (Rott et al., 2007), and the relative error is approximately indirectly proportional to the square root of the number of counted objects (see below for more details).

The majority of bloom-forming, planktonic cyanobacteria form filaments (e.g., *Aphanizomenon* spp., *Dolichospermum* spp., *Planktothrix* spp.) or colonies (e.g., *Microcystis* spp., *Merismopedia* spp.) that consist of large numbers of cells which are often difficult to be counted individually. Both filaments and colonies can differ greatly in the number of cells per filament or colony, respectively. Hence, results given as number of colonies or filaments, respectively, per volume of sample do not provide a reliable estimate of the density of cyanobacterial cells or biomass in the sample (Alcántara et al., 2018).

Therefore, disintegration of colonies and subsequent counting of individual cells is preferable to counting colonies and estimating colony size (Box, 1981). Disintegration of colonies sometimes occurs spontaneously several days after fixation with Lugol's iodine solution. For more stable colonies, it can be achieved by heating at 80–90 °C for 15 min, paralleled by intensive mixing, or gentle ultrasonication may also help. These methods often separate cells very effectively, and even where colonies are not totally broken down into single cells, the colony size may be reduced sufficiently to allow individual cells to be counted. If these approaches failed, the volume of individual colonies could be determined as an estimate of cell density. If colonies were relatively uniform in size, the average number of cells per colony may be determined and used to transform colony numbers to cell numbers (Hötzel & Croome, 1999). The use of published values for numbers per colony is not recommended because the size of colonies varies greatly.

For filamentous species, a good estimate of cell numbers is calculated from the number of filaments multiplied by the average number of cells per filament. For the latter, the cells per filament are counted for the first 30 filaments encountered and then averaged. A caveat is that cell boundaries may be poorly visible in the microscope or that the average number of cells per filament is very variable, or both. In this case, it may be preferable to measure the length of a number of filaments to compute the average filament length to be used for calculations of biovolume or for estimates of cell number by dividing filament length through average cell length. The most precise approach is to measure the length of the part of the filament within the counting grid or boundaries of the transect (Figure 13.2). Other methods for the quantification of filamentous algae and (cyano)bacteria have been proposed by Burnham et al. (1973) and Ramberg (1988).

Most counting approaches aim at counting only parts of the entire sample (i.e., the entire area or parts of the sediment chamber) and then extrapolate to the volume of the entire sample (Catherine et al., 2017). The most common methods are as follows:

- total area counting, that is, the counting of all cells in the sediment chamber. For most samples, this is very time-consuming and it is usually only applied to the counting of large units like colonies at low magnification;

- counting of cells in transects from one edge of the chamber to the other, generally one vertical and one horizontal transect, passing through the centre of the chamber. Some inverted microscopes are equipped with special oculars so that the transect width can be adjusted as required. Alternatively, the horizontal or vertical sides of a simple counting grid can be used to indicate the margin of the transect. Extrapolation to the total sample requires measuring of the width of the transects and the diameter of the chamber or the counted total area, respectively, to calculate the area counted in proportion to the total area of the chamber bottom;
- counting of cells in randomly selected view fields. It is recommended that moving to consecutive view fields should be done without looking through the microscope to prevent a bias through subjective selection of fields. The view field area covered by a counting grid is usually considered as one field. However, if no counting grid is available, the total spherical view field can be considered as a single field. For an extrapolation to the total sample, the number of counted view fields, the area of a view field and the total area of the chamber are needed:

$$N_i = c_i \times \frac{A_C}{m \times A_F}$$

with N_i: number of cells of species i in the sample
c_i: counts for species i
A_C: total area of the chamber bottom [in mm^2]
A_F: area of a view field [in mm^2]
m: number of counted view fields.

In a final step, the total number of cells is divided through the initial volume of the sample to yield cell densities in the sample.

$$D_i = \frac{C_i}{V_s}$$

with D_i: density of species i in cells per litre
V_s: volume of sample [in L].

The density of different species in a sample can vary considerably for orders of magnitude, and there can also be difference in volume of individual species spanning orders of magnitude. It is therefore necessary to adopt the counting scheme for individual samples.

Total chamber area counting with low magnification (100×) is suitable for large cells or multicellular units, whereas transect or field counting with higher magnification (200×, 400×) is used for single cells or small units. Counting using transects or view fields assumes a homogeneous distribution of sedimented cells on the chamber's bottom. However, due to convection currents in the chamber, cells very rarely settle evenly on the bottom glass but often are found in higher densities in the centre and towards the borders of the chamber (Salmaso et al., 2017). Occasionally, density also varies between opposite borders of the chamber. Inhomogeneities can be minimised by stable ambient conditions during the sedimentation procedure, in particular by stable temperatures. Fortunately, transect counting of perpendicular diameters minimises the error and is hence the preferred method.

The accuracy of the counting increases with the ratio of counted to total area. But instead of counting the total area for each sample, a trade-off between time spent on counting and accuracy is to be made. The relation of accuracy to counting time is very effective when at least 100 units (cells, colonies, filaments) of a target taxon are counted (for simplification, see Box 13.1). In this case, the counting error is estimated at 20%, while for reducing the error to 10%, 400 units need to be counted (Lund et al., 1958). These percentages should be considered when deciding about the number of units of target species to be counted.

BOX 13.1: SIMPLIFICATION FOR BIOMASS ESTIMATES

With some experience and a flexible approach, the time needed for counting of cells and measuring cell dimensions can be considerably reduced without substantial loss of accuracy by applying the following procedures:

- If the difference in counts of dominant species in two perpendicular transects is less than 20%, no further transects need to be counted.
- If the relative standard deviation (standard deviation divided by the mean) of cell dimensions measured on 10 cells is less than 20%, no further cells need to be measured.
- If a set of samples from the same waterbody and only slightly differing sites (e.g., vertical or horizontal profiles, time series) is analysed, count all samples, but measure cell dimensions only in one. Visually estimate whether the size deviation of cells of the same species in other samples is not more than about 30% from the established average.

Objects often occur on the border of a view field, and it needs to be decided whether to count them or not. One simple solution is to ignore objects that touch left border while counting those that touch the right border or *vice versa*.

There are different recommendations regarding the number of units per species that must be counted to obtain sufficiently accurate data (Edler & Elbrächter, 2010; Karlson et al., 2010). As stated above, counting 100 units per taxon generally gives acceptable data. Yet, in natural plankton communities, several tens of taxa are normally present, of which only a small number makes up a large share of total cells. For these dominant taxa, fairly accurate counting can be rapidly achieved, while for subdominant or rare taxa, this may not be achieved or is very time-consuming. Unless subdominant or rare taxa are considered important for hazard assessment, counting of 400–800 units in total has been found to yield globally acceptable results with a total overall error of some 10%, an error of 10–20% for the dominant species and an error of 20–60% for subdominant species, respectively (Vuorio et al., 2007). For rare species (e.g., less than 10 counts out of 400), reliable cell density data can be derived only by increasing the number of counted view fields. In case of larger cells, the entire bottom area of the chamber may be counted at a lower magnification. If only cyanobacteria are to be counted, and only one or two taxa are present, counting with an error below 10% can be achieved within less than one hour by counting 400 individual units per taxon.

The use of mechanical or electronic counters for recording cell counts can shorten counting time considerably, especially if only a few taxa are counted. Computer keyboards can also be used together with suitable programmes for recording cell counts.

The use of an inverted microscope with counting chambers is generally the best approach for estimating cyanobacterial numbers. However, a standard microscope is sufficient for preconcentrated samples or for naturally dense samples from mass developments, provided the volume of the counted sample can be precisely defined, for example, by using a micropipette. Other counting chambers (e.g., Sedgewick-Rafter or haemocytometer) are available for use with a standard microscope. It can also be useful to monitor samples under high magnification with oil immersion (1000×) to check the sample for the presence of very small cells, which may be overlooked during normal counting.

13.3.1.3 Quantification of cyanobacteria using a standard microscope

An alternative counting method which has been found to be useful is syringe filtration. This method is considerably less time-consuming because it does not depend on lengthy sedimentation times and uses a standard laboratory microscope. However, the results generally are less accurate because the recognition of cells on the membrane is hampered considerably by the opaqueness of the membrane filter.

Equipment

- Syringe, 10 mL
- Membrane filters, 13 mm diameter with 0.45 m pore size
- Membrane filter holder adaptable to syringe (generally, a Luer connection)
- Glass microscope slides and cover glass
- Standard laboratory microscope with 100× and 400× magnification

Reagents
- Immersion oil

Procedure
1. Gently mix sample by inverting several times.
2. Take up a defined volume of sample into the syringe. The volume is to be adjusted to the estimated cell density in the sample. Ten millilitres is a good starting point for most samples.
3. Connect filter holder with filter to the syringe.
4. Pass the sample through the filter, but avoid heavy pressure. When the filter is clogged, repeat from step 1 with a smaller volume.
5. Once the complete sample volume has passed through the filter, remove the filter from the holder and place it on a glass microscope slide with the surface with the captured cells facing upwards.
6. Allow the filter to dry at room temperature, then carefully add one or two drops of immersion oil to the filter. The oil will make the filter appear transparent and permit observation of the plankton cells trapped on its surface.
7. Finally, cover the filter surface with a cover glass and examine under the microscope.
8. The density of cyanobacteria can be easily calculated from counts of cells on the filter (or part of it), the total area of the filter and the volume of sample filtered.

13.3.2 Estimation of cyanobacterial biomass by microscopy

For estimation of toxin concentrations, cell numbers may be only of limited value as cell size varies considerably between and within species. Hence, an estimate of biomass is a better parameter to assess potential toxicity. Two principle approaches are available: estimating biomass from cell counts and average cell volumes, or chemical analysis of pigment content.

13.3.1.4 Cyanobacterial counts and cell volumes

Biovolume can be obtained from cell counts by determining the average cell volume for each taxon or unit counted and then multiplying this value by the cell counts in the sample. While by assuming a specific weight of ca. 1 mg/mm^3 (= 1 g/cm^3; wet weight) for planktonic cells, the biovolume can be transformed to (fresh) biomass, giving results as biovolume avoids errors of this assumption and has become widely accepted. Average cell volumes are determined by assuming idealised geometric shapes for individual taxa like regular spheres for *Microcystis* cells and elongated cylinders for filaments of *Planktothrix*. Measuring the relevant geometric dimensions of 10–30 cells (depending upon variability) of each taxon allows computing of the corresponding average volume (Hillebrand et al., 1999; CEN, 2015). This is best done with the aid of a spreadsheet in which the computing steps have already been defined based on general formulae (Table 13.2). Cyanobacteria have fairly simple geometric shapes (in contrast to some

Table 13.2 Biovolume calculation for common shapes of cyanobacterial cells

Shape	*Taxon*	*Formula*	*Exemplary dimensions in µm*	*Biovolume in µm³*
Sphere	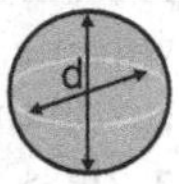 *Aphanocapsa*	$V=\pi/6\times d^3$	d=0.8	0.27
	Chroococcus		d=2	4.2
	Synechococcus		d=3	14
	Microcystis		d=4	34
	Microcystis		d=5	65
	Microcystis		d=6	113
Prolate Spheroid (rotational ellipsoid)	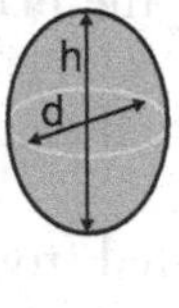 *Aphanothece* (cell)	$V=\pi/6\times d^2\times h$	d=1.2, h=2.5	1.9
	Radiocystis (cell)		d=3, h=4	19
	Dolichospermum (cell)		d=4, h=6	50
	Dolichospermum (cell)		d=5, h=7	92
	Dolichospermum (filament)	$V=\pi/6\times d^2\times h\times n$	d=3, h=4, n=80	942
	Dolichospermum (filament)		d=4, h=6, n=50	3770
Cylinder	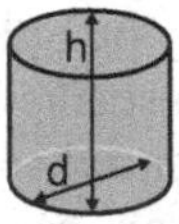 *Limnothrix* (cell)	$V=\pi/4\times d^2\times h$	d=2.5, h=10	49
	Planktothrix (cell)		d=5, h=5	98
	Planktothrix (cell)		d=8, h=5	251
	Moorea (cell)		d=20, h=3	942
	Planktothrix (filament)	$V=\pi/4\times d^2\times l$	d=5, l=300	5890
	Planktothrix (filament)		d=8, l=450	22 619
	Moorea (filament)		d=20, l=1500	471 238

For more complex shapes, see Hillebrand et al. (1999) and Napiórkowska-Krzebietke & Kobos (2016). The volumes presented here as examples should not be used for biovolume estimates in real samples. For samples to be analysed, the cell dimensions of encountered taxa have to be measured for biovolume calculations.

V: volume; *d*: cell diameter; *h*: cell height; *l*: filament length; *n*: number of cells in filament.

diatoms, Desmidiaceae or dinoflagellates, the shape of which needs to be approximate by combinations of simple shapes such as cylinders, cones and ellipsoids (Padisák & Adrian, 1999; Napiórkowska-Krzebietke & Kobos, 2016)). Table 13.2 gives exemplary shapes, dimensions and biovolumes of cyanobacteria. From the numbers in the table, it is evident that cell dimensions need to be determined as accurately as possible to minimise the error of biovolume estimates. Linear dimensions such as cell diameter and cell volume are related by a cubic function, and therefore, a measurement error of cell diameter of 25% (e.g., 5 versus 4 µm cell diameter) results in an error of 95% in biovolume. In consequence, while using mean cell volumes from literature, compiled from other waterbodies (e.g., as given in Kremer et al., 2014), provides more meaningful data than mere cell counts, the accurate measurement of cells in samples from the waterbody under study is more accurate.

Example 1: By measuring 20 *Microcystis* cells, an average diameter of 5 µm was established. Assuming spherical-shaped cells, the average cell volume is $\pi/6 \times 5^3$ $\mu m^3 = 65.4$ μm^3. Counting resulted in 100 000 cells per mL, and thus, the total biovolume is 65×10^5 $\mu m^3/mL = 6.5 \times 10^9$ $\mu m^3/L = 6.5$ mm^3/L.

Example 2: Measuring 30 *Planktothrix* filaments resulted in an average length of $L = 225$ µm and an average diameter of 6 µm. Assuming cylindrical filaments, the average filament volume is $\pi/4 \times 6^2 \times 225$ $\mu m^3 = 6362$ μm^3. Enumeration resulted in 1000 filaments per mL. Thus, the biovolume of *Planktothrix* was 6362×10^3 $\mu m^3/mL = 6.4 \times 10^9$ $\mu m^3/L = 6.4$ mm^3/L.

Thus, although the number of *Planktothrix* filaments was 100-fold less than that of *Microcystis* cells, biovolumes were similar. Both species often contain microcystins, and it is possible to compare the relative toxin content per biovolume or biomass, whereas there is little point in comparing toxin content in relation to the cell or filament counts, respectively.

13.4 ESTIMATION OF PHYTOPLANKTON BIOMASS USING CHLOROPHYLL-*A* ANALYSIS

The pigment chlorophyll-*a* generally contributes 0.5–1% of fresh weight of phytoplankton organisms (Kasprzak et al., 2008). Although the pigment content may vary depending on the physiological state of the organisms (see section 4.6.5), chlorophyll-*a* is a widely used and accepted measure of total phytoplankton biomass. It is an especially useful measure during cyanobacterial blooms, when the phytoplankton mainly consists of cyanobacteria, often of only one or a few taxa.

In modern laboratories, the analysis of chlorophyll and other pigments (carotenoids) is often done by HPLC (Bidigare et al., 2005). A number of methods have been described, and it is beyond the scope of this book to review these. In general, HPLC is the most accurate method for pigment analysis but much more expensive than the photometric approach described below. For most surveillance and monitoring practices, the latter is accurate and specific enough. Nevertheless, occasional parallel analysis of a single sample by HPLC and photometry is valuable to estimate the accuracy of the data.

Photometric analysis of chlorophyll-*a* requires relatively simple laboratory equipment, principally a filtration device, a centrifuge and a spectrophotometer. It is considerably less time-consuming than microscopic biomass determination but less specific. Standard protocols are available (e.g., ISO, 1992), but methods vary somewhat between laboratories. The main steps in most methods are essentially the same: solvent extraction of chlorophyll-*a*, determination of the concentration of the pigment by spectrophotometry and correction for pheophytin *a*, a degradation product of chlorophyll-*a*. The need for the latter, however, has been disputed (Stich & Brinker, 2005). Especially when chlorophyll-*a* concentrations are low, the correction for pheophytin *a* may introduce a bias and underestimate chlorophyll-*a* concentrations (or even lead to calculated negative concentrations). In case a correction for pheophytin *a* is not performed, the reported values should be declared as "chlorophyll-*a* not corrected for pheophytin *a*" or as "chlorophyll-*a* including pheophytin *a*".

A simple method following the ISO procedure involving an extraction step with 90% aqueous ethanol (Sartory & Grobbelaar, 1984), for the determination of chlorophyll-*a* in a field sample, is outlined here. Notably, extraction in 90% acetone instead of 90% ethanol according to Strickland & Parsons (1972) is applied in some studies.

Equipment

- Spectrophotometer suitable for readings up to 750 nm, or photometer with discrete wavelengths at 665 and 750 nm
- Glass cuvettes, typically of 1 cm path length, or 5 cm for expected very low concentrations
- Centrifuge
- 15-mL centrifuge tubes, graduated and with screw caps
- Water bath at 75 °C or other heating device for heating ethanol
- Glass fibre filters, ca. 50 mm diameter, fitting to the filtration apparatus
- Filtration apparatus and vacuum pump
- Tissue homogeniser or ultrasonication device
- Pipette or similar for the addition of acid

Reagents

- 90% aqueous ethanol
- 1 M hydrochloric acid

Procedure

Perform the following steps in low intensity of indirect light because light induces a rapid degradation of chlorophyll.

1. Filter a defined volume of water through a glass fibre filter as soon as possible and store the filter with the loaded face folded on itself in individual, labelled bags or tubes. If extraction cannot be performed immediately, filters should be stored –20 °C or better at –80 °C. For extended storage, freeze-drying of samples is strongly recommended to avoid degradation. Alternatively to freezing, samples can be stored in the extraction solvent (see below) for up to 4 days in the refrigerator.
2. Place the filter in a tissue homogeniser, add 2–3 mL of boiling ethanol (working with effective ventilation, preferably using a fume cupboard) and homogenise until the filter has been completely disintegrated. Samples can also be homogenised by ultrasonication or manual grinding using mortar and pestle. Pour the sample sludge into a centrifuge tube, rinse out the grinding tube with another 2 mL ethanol and add this to the centrifuge tube. Repeat this step. Make up to a total of 10 mL in the centrifuge tube with 90% ethanol. Seal the tube, label and store in darkness at approximately 20 °C for 24–48 h.
3. Centrifuge for 15 min at 3000–5000 g to clarify samples. Decant the clear supernatant into a clean vessel and record the volume.
4. Blank spectrophotometer with 90% ethanol over the wavelength range of 650–800 nm.
5. Transfer a volume of clear sample to the cuvette and record absorbance at 750 nm and 665 nm [readings A(750a) and A(665a)]. Absorbance (A) at 665 nm should range between 0.1 and 0.8. If higher, the sample should be diluted with 90% ethanol; if lower, a cuvette with a longer optical path should be used.
6. If correction for pheophytin was desired, add 30 μL of 1 M HCl per mL of sample volume in cuvette and agitate gently for 1 min. Record absorbance at 750 nm and 665 nm [readings A(750b) and A(665b)].

Calculation

1. Correct for turbidity: A(665a) – A(750a) = A(665a, corrected) and A(665b) – A(750b) = A(665b, corrected)
2. The concentrations of chlorophyll-*a* and pheophytin *a* are calculated:

$$\text{chlorophyll } a = \frac{29.62\left(A_{(665a,\ \text{corrected})} - A_{(665b,\ \text{corrected})}\right) \times V_c}{V_s \times l}\ \mu\text{g/L}$$

$$\text{pheophytin } a = \frac{20.73\left(A_{(665b,\ \text{corrected})}\right) \times V_e}{V_s \times 1}\ \mu\text{g/L}$$

with: V_e = volume of ethanol extract in mL
V_s = volume of water sample in L
l = path length of cuvette in cm.

Simplifications of the procedure may be applied. If no centrifuge for volumes of 10 mL is available, filtration may be used instead. In case neither tissue homogeniser nor ultrasonication device nor mortar and pestle are available, proceed without the homogenisation step. Underestimations of chlorophyll-*a* concentrations may occur, but for cyanobacteria, these are not likely to be substantial. Other solvents – *N, N*-dimethylformamide, dimethyl sulfoxide and acetone – have also been used for extraction (Speziale et al., 1984), but ethanol has the advantage of being less toxic and compatible with polymeric materials (Ritchie, 2006).

13.5 PHYTOPLANKTON AND CYANOBACTERIA QUANTIFICATION BY FLUORESCENCE ANALYSIS

As cyanobacterial biomass and community composition is highly inconstant in space and time, a quantification approach that is able to follow this variability is valuable. Standard method for phytoplankton quantification is based on the microscopic analyses of samples processed in the laboratory, complemented by chlorophyll-*a* analysis by spectrophotometer or spectrofluorimeter. Such results are based on the discrete sampling of individual localities at certain time and horizons. Equipment used in this type of monitoring is relatively cheap and has an acceptable sensitivity, but the analytical results are available only with a delay of hours or days and, depending on the frequency of sampling, by discrete sampling potential threats due to high cyanobacterial abundance can be missed such as short-lived surface blooms of cyanobacteria or the fast and rapid shifts of water quality due to quick hydrological or meteorological changes.

Advanced methods for phytoplankton quantification could be able to describe the variability, permanent changes and displacement of phytoplankton biomass and the spatial (vertical and horizontal) and temporal variability in a waterbody with the sensitivity and information frequency sufficient for the water management (raw water takeoff), or ecological understanding (developments and dynamic of phytoplankton assemblages).

Advanced methods for phytoplankton quantification include the following:

- remote sensing and satellite imagery based on radiometry (AVHRR – Advanced Very High-Resolution Radiometer, hyperspectral landscape imaging, etc. (Kahru & Brown, 1997));
- airborne- and satellite-based optical remote sensing including hyperspectral phytoplankton imaging, etc. (see section 11.10);

- optical *in situ* methods (*in situ* flow cytometry, analytical flow cytometry etc);
- *In situ* and online fluorescence quantification of dominant phytoplankton pigments including – automatic high-frequency monitoring (AHFM) systems.

While radiometry (AVHRR) was used for the water quality monitoring already 20 years ago (Kahru & Brown, 1997), the MODIS (Moderate Resolution Imaging Spectroradiometer) uses more and more detailed coefficients for correction of variabilities in the satellite phytoplankton quantification at the present time (Sayers et al., 2016). Airborne (airplane, drone etc.) remote sensing is based mostly on optical methods like hyperspectral imaging. This technology has recently been adopted for the advanced quantification of phytoplankton, including the correction for reflectance or humidity (Wang et al., 2016; Wolanin et al., 2016).

Besides monitoring of phytoplankton, remote sensing can be used to assess other properties of water ecosystems (see also section 11.10). For example, laser scanning can be used for bathymetry of shallow waters (Fernandez-Diaz et al., 2014) or for mapping sediments disposal (Montreuil et al., 2014). Hyperspectral data can be used for depth estimation in shallow waters (Ma et al., 2014), suspended inorganic particles (Giardino et al., 2015) or dissolved organic matter (Zhu et al., 2013). All these parameters are highly relevant for water quality monitoring; however, the objective of this chapter is to demonstrate possibilities of fluorescence and imaging spectroscopy to assess phytoplankton and cyanobacterial blooms by measuring concentrations of photosynthetical pigments.

The majority of real-time technologies employed for cyanobacterial management are based on fluorescence of pigments (Zamyadi et al., 2016). Each of the fluorescent pigments present in cyanobacterial or generally in algal cells, respectively, has a specific excitation and emission spectrum (see Table 13.3). chlorophyll-*a* is a photosynthetic pigment present in all species of phytoplankton, including eukaryotic (algae) and prokaryotic organisms (cyanobacteria), and thus, it is a good and commonly used indirect marker of the total phytoplankton biomass. Standard methods of its

Table 13.3 Excitation and emission maxima of dominant pigments and their general distribution among particular phytoplankton groups

Pigment	*Group*	*Excitation (nm)*	*Emission (nm)*
chlorophyll-*a*	Green algae,	440	685
Chlorophyll-*c*	Cryptophyceae	460	685
Carotenoids	Diatoms, Chrysophyceae	500–550	685
Phycoerythrin	Cryptophyceae, Cyanobacteria	560–585	590, 620, 685
Phycocyanin	Cyanobacteria	610–620	645, 685

quantification are based on the extraction of the pigment into an organic solvent and subsequent determination by spectrophotometry (Richards & Thompson, 1952), fluorimetry (Holm-Hansen et al., 1965) or chromatography (Jacobsen, 1978; Otsuki & Takamura, 1987). These methods have been routinely used for decades, but they are time-consuming and require a standard sampling, transport to the laboratory and immediate processing, as well as an experienced analyst. Furthermore, all steps of the process from water sampling to the final photometric determination of the chlorophyll-*a* content can be a source of variability. Other disadvantage is a comparatively large volume of sample needed and thus a limitation with respect to the number of samples taken and the possible changes during the sample transport and storage, namely, degradation.

One of the key characteristics of chlorophyll-*a* is its fluorescence. Photosystem II (PS II), which is mainly responsible for the chlorophyll fluorescence, consists of peripheral and core antenna. The first contains a species-dependent pigment absorbing quantum of light, the latter an evolutionary conserved molecule of chlorophyll-*a* (Beutler et al., 2002). Most of the energy transferred from the peripheral antenna to the core is used for photochemistry and thermal decay and several percent for fluorescence by emitting light at wavelength around 685 nm (red light). Measurement of this light serves as a tool for the *in vivo* determination of chlorophyll-*a*. Fluorescence of chlorophyll-*a* also enables its determination in the field studies, directly in the water column. Connecting the fluorimeter in continuous or stop flow mode to the pumping system, which brings water to the measuring cell of the fluorimeter, is one of the possibilities of the online monitoring of chlorophyll-*a* (Pinto et al., 2001; Odate et al., 2002; Goddard et al., 2005). The phytoplankton can be measured directly in the water column, and there is evidence from several studies that data are similar to those gained by standard microscopic analyses or chlorophyll-*a* quantification after solvent extraction and spectrophotometric analysis (Gregor & Maršálek, 2004; Gregor et al., 2005; Izydorczyk et al., 2005; Gregor et al., 2007). Estimation of cyanobacterial biomass or cell density is possible by measuring phycocyanin (PC) fluorescence (Figure 13.3).

Asai et al. (2001) presented a sensor with two fluorescence channels – the first one for detecting chlorophyll-*a* of eukaryotic algae (excitation 440 nm, emission 680 nm) and the second one for detecting the cyanobacteria-specific PC (excitation 620 nm, emission 645 nm). An *in situ* fluorimeter with three excitation bands and detection of emission from 546 to 733 nm was also designed (Desiderio et al., 1997). *In situ* fluorimeters include devices measuring each sample individually, with continuous circulation of water samples (flow-through) or with submersible probes. Submersible probes for detecting only chlorophyll-*a*, a combination of sensors for the detection of chlorophyll and phycocyanin (PC), or more phytoplankton classes are commercially available from manufacturers around the globe. They contain

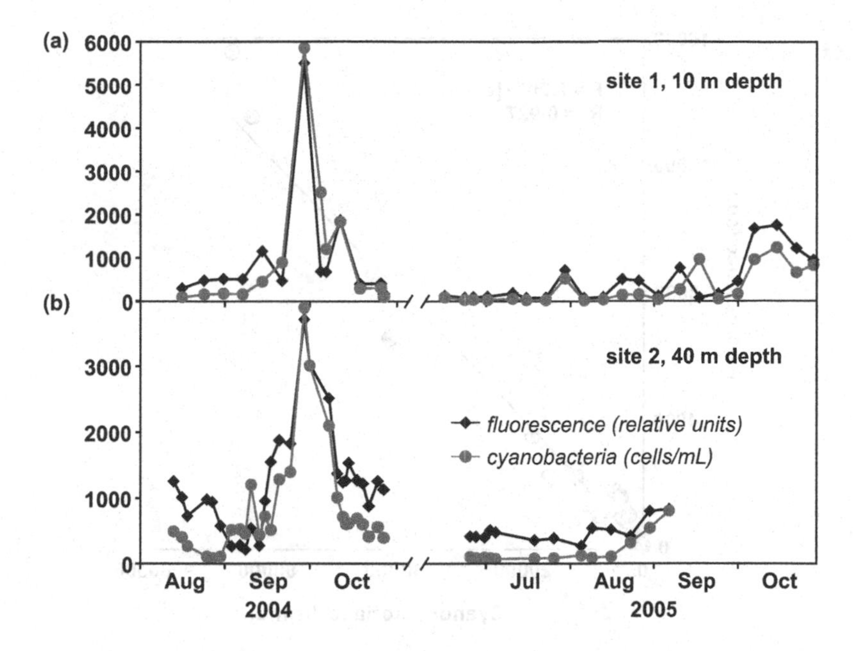

Figure 13.3 Seasonal dynamics of phycocyanin fluorescence (F590) and cyanobacteria cell counts at two individual sampling points in a drinking-water reservoir at 10 m depth (a) and 40 m depth (b).

diodes emitting light of defined wavelength bands for excitation of pigments and the response is measured as the fluorescence. These instruments are usually connected to a computer for operation and data storage. This design allows submersion of the probe to a defined depth, limited only by the cable length.

Submersible *in situ* fluorimeters are suited for online and real-time monitoring of spatiotemporal dynamics of cyanobacterial populations, for example, at raw water offtake sites for drinking-water production. They have acceptable sensitivity and are suitable for differentiation between algae and cyanobacteria (Zamyadi et al., 2016). When used for real-time management purposes, it is crucial that devices are well maintained, especially the regular cleaning of optical sensors is critical or the control of automatic cleaning system, respectively. Further, fluorescence measurements in a particular waterbody are preferably calibrated against other parameters of phytoplankton or cyanobacterial biomass, for example, cell counts (Figure 13.4).

Submersible fluorimeters are not suitable for species identification or assessment of the physiological status. The most important sources of variability of *in situ* measurements are interferences with weather (wind, sunshine), water turbidity, temperature, cyanobacterial morphology (colony, filaments, picocyanobacteria) and sensor types (Hodges et al., 2018). It is known that data

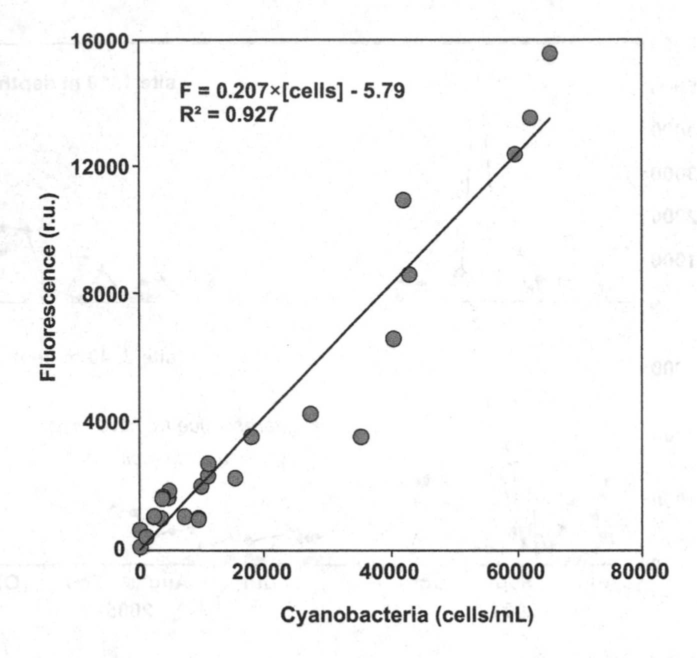

Figure 13.4 Comparison of quantitative assessment of cyanobacterial cell density by microscopic cell counts and chlorophyll fluorescence in samples from Brno reservoir ($n=43$). r.u.: relative units of fluorescence.

produced by these devices in dense cyanobacterial blooms may underestimate the real situation and should be corrected (Silva et al., 2016). Another parameter which was studied is optical interaction of algae and cyanobacteria in phytoplankton. The optimal correction method can be selected for each fluorimeter and cyanobacteria species pairs by validating against data from the investigation of green algae as an interference source (Choo et al., 2019).

Some general discrepancies can be found when comparing submersible fluoroprobes, spectrophotometric chlorophyll-*a* quantification and cell counts, even when the same or similar species are analysed. The probable explanation is a varying level of pigments among species and within species at different phases of growth. Other potential sources of variability include different sampling strategies (continuous and online measurement versus discrete sampling), sample treatment and mode of transport to the laboratory. Another source of the differences between results was observed where picocyanobacteria are present and less experienced and skilled laboratory staff overlooks their presence, but fluorescence probes takes them into account.

Submersible devices usually measure in a continuous mode; that is, they enable data to be obtained from the whole water column in a much shorter time compared to discrete sampling. This is especially useful when phytoplankton organisms occur in a narrow horizontal layer like *Planktothrix rubescens* in the metalimnion (see Chapters 4, 8 and 11) or to monitor the development of surface blooms at critical sites. In these cases, conventional methods based on discrete samples with the inherent lag time may miss ephemeral risks from cyanotoxins.

13.5.1 Recent advances and future direction in the phytoplankton and cyanobacteria quantification by fluorescence analysis

Cyanobacterial blooms that show a high and dynamic variability in time, space and abundance can be efficiently monitored by advanced fluorescence probe technology, which has become far more advanced in recent years with specific light-emitting diodes (LEDs) and optical filters. However, recent widespread application of *in situ* fluorimetric probes by both scientists and water managers has led to recognition of interferences, sources of variability and difficulties in comparison with the results. One common problem of *in situ* and online monitoring by fluorescence and optical sensors is biofouling. That is why the regular and preferably automatic cleaning of probes is extremely important.

Automatic high frequency monitoring (AHFM) systems are an important recent development which has increased not only the measurement frequency, but also the number of variables being monitored in waterbodies. Broad spectrum of sensors can be used on demand according to the monitoring purposes, like the ion-selective electrodes, UV absorbance, fluorescence and biochip probes (Marce et al., 2016). While full automation is already possible for probes based on optical properties (absorbance and fluorescence), this is still difficult for ion-selective electrodes and biochips. Main challenges are low limits of detection required for micropollutants and sensor maintenance requirements.

We can expect the combination of approaches for phytoplankton quantification in the near future. An approach useful in practice is to use submersible fluorescence probes for quantification of phytoplankton biomass based on AHFM systems in combination with hyperspectral or selective spectral cameras operated from airplanes or drones, which can describe the spatial variability of cyanobacterial biomass in waterbodies. As the data from AHFM systems are used by a number of institutions, calibration, validation and corrective coefficients for data comparison are particularly important (Bertani et al. 2017).

Further, new simple models of fluorometers for the quantification of algae and cyanobacteria using widely available devices like smart phones have been proposed (Friedrichs et al., 2017). Once calibration and variability in fluorescence signals are mastered, respective adapters can become effective tools.

New compact multiwavelength fluorimeters with modular design are highly versatile and flexible monitoring tools. Detection modules for green algae, cyanobacteria and dinophytes allow the continuous identification and quantification of the major relevant algal groups based on their spectral characteristics with a detection limit of 10 cells/L (Zieger et al., 2018). The sensitivity of most types of submersible fluorescence probes is equivalent to 1000–2000 cells/mL, which is acceptable for general monitoring. For sensors installed for *in situ* monitoring, biofouling is one of the most important sources of variability requiring a regular and thorough maintenance and verification. A disadvantage of several devices is that no correction on the turbidity is performed, which may be important in natural conditions.

There is new information highlighting the potential for multiparameter monitoring via fluorescence spectroscopy; fluorescence spectra can predict both microcystin-LR occurrence and disinfection by-products formation potential in the waterbody (Brophy et al., 2018).

An overview of new devices for *in situ* fluorescence phytoplankton quantification and discrimination, including the limitation and interference factors, is given by Bertone et al. (2018).

13.6 MONITORING TOXIGENIC CYANOBACTERIA BY MOLECULAR METHODS

Molecular methods have significantly increased our understanding on the distribution of genes involved in the production of toxins within the phylum cyanobacteria (see Sivonen & Börner, 2008 and Dittmann et al., 2013). This subchapter introduces the molecular detection of toxigenic cyanobacteria not only in surface waters such as lakes, rivers and drinking-water reservoirs but also in food supplements. Genetic methods are only able to indicate the potential of toxin synthesis and do not provide information about actual toxin production and concentrations. Nevertheless, applications in monitoring include early warning of the toxin-producing potential of a developing bloom and allow the identification of the toxin-producing taxa in mixed field populations of cyanobacteria. They also allow tracing the development of the genotype composition of a taxon, that is, whether the fraction of toxin-producing genotypes changes over time. Moreover, these methods allow high-throughput sample analysis.

This section provides an overview of the workflow for applying genetic methods. For more detailed information, the reader is referred to a handbook providing more details on the scientific basis for the use of molecular tools, protocols and the interpretation of respective results (Kurmayer et al., 2017). Section 13.6.8 reviews applications in practice.

A full sequence of a biosynthesis gene cluster of a cyanobacterial toxin was first reported for microcystin from *Microcystis* sp. (Tillett et al., 2000). Sequences from other taxa and encoding the synthesis of other toxins rapidly followed, and today, sequences of biosynthesis genes for all major types of cyanobacterial toxins are available (see Table 13.4; Figure 13.5)

Table 13.4 Overview on complete biosynthesis gene clusters reported for toxin types and various taxa of cyanobacteria

Toxin	*Organisms*	*Strain*	*Genes*	*Reference*
Microcystin	*Microcystis aeruginosa*	PCC 7806	*mcyA-J*	Tillett et al. (2000)
	Planktothrix agardhii	NIVA-CYA 126-8	*mcyA-J,T*	Christiansen et al. (2003)
	Planktothrix rubescens	NIVA-CYA 98		Rounge et al. (2009)
	Dolichospermum (Anabaena) sp.	90	*mcyA-J*	Rouhiainen et al. (2004)
	Nostoc sp.	152		Fewer et al. (2013)
	Fischerella sp.	PCC 9339		Shih et al. (2013)
Nodularin	*Nodularia spumigena*	NSOR10	*ndaA-G*	Moffitt & Neilan (2004)
Cylindrospermopsin	*Raphidiopsis (Cylindrospermopsis) raciborskii*	AWT205	*cyrA-O*	Mihali et al. (2008)
	Oscillatoria sp.	PCC 6506		Mazmouz et al. (2010)
	Aphanizomenon sp.	10E6		Stüken & Jakobsen (2010)
	Raphidiopsis curvata	CHAB1150, HB1		Jiang et al. (2014)
Saxitoxin	*Raphidiopsis (Cylindrospermopsis) raciborskii*	T3	*sxtA-X*	Kellmann et al. (2008)
	Microseira (Lyngbya) wollei	Carmichael / Alabama		Murray et al. (2011)
	Dolichospermum (Anabaena) circinale	AWQC131C		Murray et al. (2011)
	Aphanizomenon sp.	NH-5		Murray et al. (2011)
	Raphidiopsis brookii	D9		Stucken et al. (2010)
Anatoxin-a	*Oscillatoria* sp.	PCC 6506	*anaA-H*	Rantala-Ylinen et al. (2011)
	Dolichospermum (Anabaena) sp.	37		Rantala-Ylinen et al. (2011)
	Cylindrospermum sp.	PCC 7417		Calteau et al. (2014)
	Cuspidothrix issatschenkoi	CHAB D3, RM-6, LBRI48		Jiang et al. (2015)
Lyngbyatoxin	*Moorea producens (Lyngbya majuscula)*		*ltxA-D*	Edwards & Gerwick (2004)

The species name and strain identifier are given as reported in the original publication.

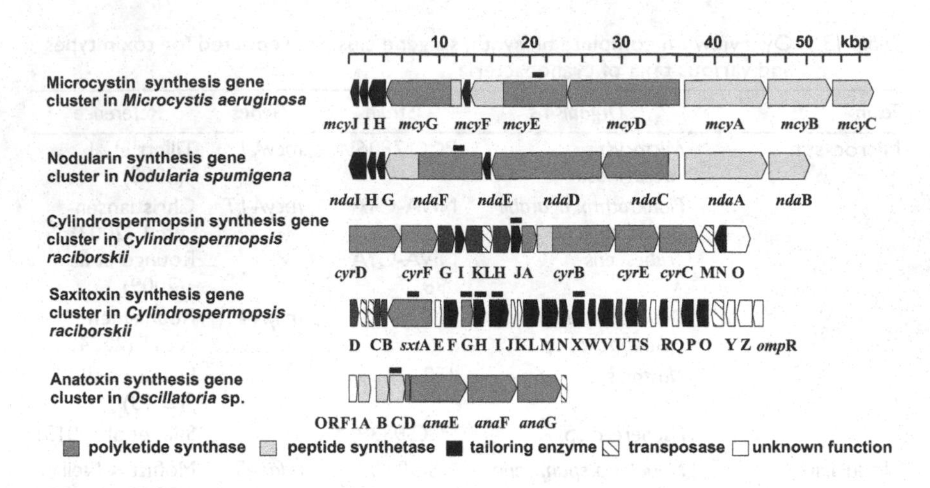

Figure 13.5 Overview of cyanotoxin synthesis gene clusters and PCR approach used for the detection of toxigenic genotypes. The PCR-amplified fragments and corresponding genes used to indicate potential toxin synthesis are indicated: *mcyE/ndaF* according to Rantala et al. (2004) and Jungblut and Neilan (2006); *cyrJ* (Mihali et al., 2008), *sxtA*, G, H, I, X (Casero et al., 2014), *anaC* (Rantala-Ylinen et al., 2011).

with the exception of anatoxin-a (S), the biosynthesis of which was still unknown by the time of publication of this book.

The elucidation of biosynthesis genes significantly increased the understanding on inheritance and evolution of cyanotoxin synthesis; that is, phylogenetic analyses lead to the conclusion that microcystin synthesis is an evolutionarily old feature that has been lost repeatedly during the evolution of cyanobacteria (Rantala et al., 2004). The genes involved in the synthesis of nodularin were probably derived from the genes encoding microcystin synthesis via a gene deletion event (Moffitt & Neilan, 2004; Rantala et al., 2004). Similar to microcystin synthesis, for saxitoxin synthesis genes, the comparison of gene synteny and phylogeny between taxa as well as the evidence of strong stabilising selection suggested that saxitoxin synthesis genes have been mostly inherited vertically (as opposed to horizontal gene transfer) and emerged at least 2 billion years ago (Murray et al., 2011). The saxitoxin-producing dinoflagellates are eukaryotic toxic algae containing a third membrane of endoplasmatic reticulum around the chloroplast organelles and evolved from cyanobacteria through secondary or tertiary endosymbiosis during the late Paleozoicum (Lee, 2018). Generally, it is understood that in dinoflagellates, genes forming the core genes for saxitoxin synthesis (i.e., *sxtA* and *sxtG*) have been acquired via horizontal gene transfer from bacteria and have been lost repeatedly in various lineages (Orr et al., 2013; Murray et al., 2015). For the cylindrospermopsin-producing taxa, phylogenetic congruence between taxonomic marker genes and cylindrospermopsin synthesis genes was reported, implying the dominant influence of vertical inheritance in the course of the evolution of the phylum of cyanobacteria (Jiang et al., 2014). Finally, for anatoxin-a synthesis, the comparison of anatoxin synthesis genes revealed that gene nucleotide variability was congruent with phylogenetic analysis across cyanobacterial taxa (Jiang et al., 2015; Brown et al., 2016). In summary, phylogenetic analyses rather support the role of vertical inheritance or the loss of cyanotoxin synthesis genes than the role of frequent lateral gene transfer events.

Although genetic methods are only able to indicate the potential of toxin synthesis, they are showing promising results supporting risk assessment. If genes for toxin production were absent in a cyanobacterial population, this population would not be able to produce a specific toxin. *Vice versa*, this dependence is not as certain, and cyanobacteria carrying the genes may or may not produce a particular toxin; only toxin analyses can show that they are indeed producing this toxin and to which extent. Thus, based on the presence of genes, the molecular tools indicate the occurrence of toxigenic genotypes in the environment or in food supplements, but the actual toxin content or concentration must be determined using chemical–analytical techniques (see section 14.1). Currently, all the molecular tools available are based on the principle of polymerase chain reaction (PCR). Due to the generally high sensitivity of PCR, it is possible to detect toxigenic genotypes in minute amounts, that is, long time before a toxic cyanobacterial bloom may occur or as minor component in food supplements. Consequently, waterbodies bearing a risk of toxic bloom formation could already be identified early on in the growing season possibly assisting in an economically more efficient application of cyanotoxin detection techniques (see Box 13.2). Furthermore, early identification of toxigenic genotype occurrence may lead to a more detailed recording of environmental factors potentially influencing the abundance of toxigenic genotypes. Finally, toxigenic genotypes can be detected from single cells: single colonies or filaments of cyanobacteria (*Dolichospermum (Anabaena)* sp., *Microcystis* sp., *Planktothrix* sp.) can be identified according to morphological criteria during counting via microscopy and then analysed by PCR for their potential of toxin production.

BOX 13.2: BENEFITS OF MOLECULAR TOOLS SUPPORTING THE MONITORING OF CYANOBACTERIA

EARLY WARNING

Molecular tools can support the identification of waterbodies at risk for toxic bloom formation early on in the growing season, thus supporting an economically efficient selection of situations for cyanotoxin analysis. Using quantitative PCR (qPCR)-based methods, it is possible to quantify toxigenic cyanobacteria occurring rarely in the plankton community or in food supplements that otherwise might be overlooked by microscopical methods.

UNDERSTANDING ENVIRONMENTAL DRIVERS

Early identification of toxigenic genotype occurrence may lead to a more detailed recording of environmental factors potentially influencing the abundance of toxigenic genotypes.

IDENTIFYING TOXIGENIC CYANOBACTERIA

Toxigenic cyanobacteria can be identified by sequencing of PCR-amplified indicative genes. Alternatively, single colonies of cyanobacteria are identified and taxonomically assigned by microscopical inspection according to morphological criteria and then analysed for toxigenic genotype occurrence.

In general, the application of molecular tools is possible either using biomass from isolated strains or isolated colonies/filaments grown in cultures or using biomass collected from field samples. The latter sample type will lead to rapid results on toxigenic cyanobacteria occurrence, however may contain a larger number of taxa carrying toxigenic genes which are not necessarily the dominant taxa in the respective sample (Rantala et al., 2006). Nevertheless, all sample types require a few mandatory steps, which include:

i. sampling of biomass or isolation of individual colonies/filaments under the microscope;
ii. extraction of the nucleic acid (DNA);
iii. amplification of gene fragments indicative of toxin synthesis based on conventional or quantitative PCR (qPCR);
iv. detection of PCR products using agarose gel electrophoresis (conventional PCR) or fluorescent dyes (qPCR or digital droplet PCR, ddPCR).

The sequencing of PCR products is optional and useful to confirm the results obtained or to identify the toxin-producing organism. More recent technology has enabled the so-called deep sequencing of PCR products which has become a widely applied technique to monitor the diversity of microorganisms and cyanobacteria in general (Pessi et al., 2016). The following gives an overview on the general workflow of applying molecular tools.

13.6.1 Sampling and nucleic acid extraction

In general, the basic sampling steps (e.g., stratified versus depth-integrated sampling, low vacuum filtration) are identical to the processing of samples for cyanotoxin detection and analysis (see Chapter 14). Food supplement samples should be handled as in food safety programmes for microbiological analysis (e.g., ISO, 2006).

However, molecular analyses require sampling precautions against cross-contamination or DNA degradation. For example, as all polymerase chain reaction (PCR)-based methods are highly sensitive (allowing to detect gene copies from single cells), the possibility of sample cross-contamination needs to be reduced as much as possible (e.g., by using new sample vessels or exhaustive rinsing of sample/filtration equipment between sampling dates or sites). Drying of biomass at high temperature (e.g., 100 °C) also should be avoided, as it leads to fragmentation of DNA. Similarly, nucleases released during cell lysis can lead to DNA fragmentation within a short time, and therefore, those enzymes need to be inactivated during the DNA extraction process. Typically, biomass for DNA extraction and subsequent PCR analysis is either used fresh or it has to been stored at –20 °C.

The conventional DNA extraction procedure uses a combination of osmotic shock and enzymatic treatment followed by chemical phase separation (e.g., Franche & Damerval, 1988). This DNA extraction procedure has been refined to obtain both qualitative and quantitative results, and today, robust protocols on cyanobacterial DNA extraction are available (e.g., Kurmayer et al., 2003). In some cases, extensive mucilage production as indicated by high viscosity of the DNA extract can inhibit the subsequent PCR amplification, and polysaccharides need to be selectively removed (Tillett & Neilan, 2000). In general, conventional DNA isolation procedure protocols are more time-consuming but cheaper than extraction with easy-to-use commercial DNA isolation kits which are widely available. Kit-based techniques typically include anion-exchange columns for DNA binding and purification. However, it is important to validate the efficiency of such techniques before using them for monitoring (Schober et al., 2007). DNA extraction from food supplements can be more difficult as for this purpose, cyanobacteria typically are processed by drying, and food supplements may contain additives that can affect DNA extraction efficiency, for example, pharmaceutical bulking agents with adsorbent properties (Costa et al., 2015). Thus, purification of DNA or alternatively the addition of substances reducing the effect of PCR inhibitors might be routinely required (Ramos et al., 2017a; b). Individual cyanobacterial colonies or filaments can be selected for PCR amplification of genes under a stereo microscope using a forceps or a micropipette (Kurmayer et al., 2002). Colonies or filaments are picked randomly from a subsample containing a few specimen only, washed by serial transfers in standard solution and stored in PCR buffer in the freezer (–20 °C). The DNA is extracted most efficiently by short sonication (Chen et al., 2016), and the obtained DNA quantities are sufficient for multiple individual PCR experiments.

13.6.2 Polymerase chain reaction (PCR) methodology

PCR is the technique that allows creating multiple copies of specific gene fragments through amplification by DNA polymerases. The most critical step for the reliable detection of toxigenic cyanobacteria is the selection of appropriate oligonucleotides (primers) which are used as molecular probes. Besides standard laboratory equipment, the instrumentation comprises a PCR cycling machine, a gel electrophoresis chamber and a gel documentation device. For detailed information on how to perform PCR, see the widely available laboratory manual revised by Sambrook and Russell (2001).

In contrast, quantitative PCR allows the determination of actual gene copy numbers (that can be used to approximate cell equivalents) and thus, by choosing the right targets for PCRs, the proportion of toxigenic genotypes present in a sample. In quantitative PCR (qPCR), amplification of the target gene is followed in real time via the detection of a fluorescent signal generated from DNA strand-intercalating dyes at each PCR cycle (e.g., SYBR Green). qPCR is based on the principle that the target DNA sequence is doubled in each cycle and that the dynamic increase of the recorded amplification reflects the amount of target sequence originally present. Theoretically, the more target sequence (or target genotypes) can be found in a specific sample, the earlier the amplification curve exceeds a predefined fluorescence threshold. The PCR cycle when this threshold is crossed is called a threshold cycle, C_t, or quantification cycle, Cq(-value). The use of fluorescent dyes makes qPCR most sensitive with calibration curves showing a wide dynamic range (up to seven orders of magnitude). Detailed information on the application of (q)PCR in the analysis of toxin genes has been compiled through the EU-initiative CyanoCOST (Rantala-Ylinen et al., 2017).

Digital droplet PCR (ddPCR) quantitates a target DNA sequence based on PCR of a partitioned DNA sample. The number of PCR-positive and PCR-negative partitions is used to determine the absolute number of target DNA molecules (Hindson et al., 2013). Reports on the use of ddPCR are only emerging, but first comparisons with qPCR revealed a comparable result (Te et al., 2015; Nshimyimana et al., 2019; Wood et al., 2019).

13.6.3 Detection of toxigenic cyanobacteria

Cyanotoxins are synthesised by large multifunctional enzyme complexes via the thiotemplate mechanism in a stepwise manner, known as nonribosomal peptide synthesis (NRPS). These NRPS enzyme complexes belong to the largest proteins within the prokaryotic cell (several thousand amino acids) and are often combined with polyketide synthases (PKS) (see Chapter 2; Dittmann et al., 2013). Large parts of gene sequences of particular biosynthesis gene clusters have been found to be variable and therefore not suitable for designing PCR primers with the desired specificity and sensitivity. For the microcystin (*mcy*) and nodularin (*nda*) biosynthesis gene cluster, conserved gene regions have been identified in the *mcyE*/*ndaF* gene encoding the enzymatic step condensing the Adda side chain with D-glutamate forming the conserved core of the peptides (Rantala et al., 2004; Jungblut & Neilan, 2006). With this region, PCR detection of microcystin and nodularin biosynthesis genes is possible in all cyanobacterial producers.

In contrast to microcystin synthesis, for the cylindrospermopsin synthesis (*cyr*) gene cluster, the *cyrJ* gene encoding a tailoring enzyme such as the sulfotransferase catalysing the sulphation of the C-12 atom of the cylindrospermopsin molecule was found only in cylindrospermopsin-producing strains (Mihali et al., 2008). However, the core genes encoding the synthesis of the cylindrospermopsin molecule itself, that is, *cyr*A/*aoa*A encoding the amidinotransferase or *cyr*B/*aoa*B and *cyr*C/*aoa*C encoding NRPS/PKS, were also detected in non-cylindrospermopsin-producing strains (Ballot et al., 2011; Hoff-Risseti et al., 2013) and thus are not considered reliable indicators. Thus, the *cyrJ* can be considered a gene marker to indicate potential cylindrospermopsin producers among the genera *Dolichospermum*, *Aphanizomenon*, *Raphidiopsis (Cylindrospermopsis)* and *Oscillatoria* (Mankiewicz-Boczek et al., 2012).

For saxitoxin biosynthesis, at the time of publication of this book, no gene loci are known that can unambiguously infer the *sxt* gene cluster from a diagnostic polymerase chain reaction (PCR) (Ballot et al., 2010; Casero et al., 2014). Indeed, all genes tested have also been detected in a variety of cyanobacterial strains not producing saxitoxins. A protocol of several genes serves to indicate saxitoxin synthesis potential: *sxtA* encoding a PKS, *sxtG* encoding an amidinotransferase, *sxtH* encoding hydroxylation of the C-12 atom, *sxtI* encoding carbamoylation and *sxtX* encoding hydroxylation of the N-1 atom, (Casero et al., 2014). The elucidation of the saxitoxin biosynthesis genes has been started from identifying a gene encoding an O-carbamoyltransferase (Kellmann et al., 2008), now called *sxtI*, that has been proposed as a reliable marker to indicate saxitoxin synthesis (Ballot et al., 2016) but requires further confirmation.

For the detection of an anatoxin-a synthesis gene, a conserved region of *ana*C encoding an NRPS module for proline activation (an initiating step of anatoxin-a synthesis) has been identified and primers able to detect *ana*C in both *Oscillatoria* sp. and in *Dolichospermum* sp. as well as in *Aphanizomenon* sp. have been designed (Rantala-Ylinen et al., 2011). The encoded NRPS AnaC has been heterologously expressed and a specificity for proline as a substrate under *in vitro* conditions has been reported (Mejean et al., 2009), which also makes it a reliable marker for PCR-based monitoring and early warning.

Notably PCR approaches can be combined with downstream applications such as direct sequencing of amplicons, or cloning and subsequent sequencing, or restriction fragment length polymorphism (RFLP) analysis. This type of post-PCR result analysis can inform about genera or species forming a particular toxigenic genotype. For example, Hisbergues et al. (2003) used a PCR-based detection technique for microcystin synthesis genes using *mcyA* followed by differentiation of genera by RFLP of the obtained PCR product. Similarly, Rantala-Ylinen et al. (2011) used a PCR-based detection technique for *anaC* synthesis gene and the differentiation of toxigenic genera by RFLP of the obtained PCR product.

13.6.4 Estimates of cyanobacterial biomass by molecular approaches

By using reference genes (see above), it is possible to estimate total cyanobacterial biomass or biomass of particular cyanobacterial genera in a water sample. In general, qPCR assays have been used either for 16S rDNA gene regions specific for cyanobacteria (Rinta-Kanto et al., 2005) or for gene loci specific for individual cyanobacterial genera known to produce cyanotoxins such as (i) the phycocyanin-intergenic spacer region (PC-IGS) for the genera *Microcystis* (Kurmayer & Kutzenberger, 2003) and *Planktothrix* (Ostermaier & Kurmayer, 2009; Kurmayer et al., 2011); (ii) the RNA polymerase gene loci *rpoC* for *Raphidiopsis* sp. (Fergusson & Saint, 2003; Rasmussen et al., 2008) and *Chrysosporum* (*Aphanizomenon*) *ovalisporum* (Campo et al., 2013); or (iii) the 16S rDNA for *Microcystis* sp. (Rinta-Kanto et al., 2005).

The molecular approach is an alternative to microscopic cell counting but has been primarily used to relate qPCR signals (Cq-values) indicative of cyanotoxin synthesis genes to the total cyanobacterial population, in order to quantify the proportion of a toxigenic subpopulation. For this purpose, it is required to compare Cq-values with microscopic cell counts. Standard curves need to be established (i) to quantify the target genotype using either cell numbers or gene copies and (ii) to determine the specificity and sensitivity of a specific quantitative PCR (qPCR) assay with isolates and background DNA (e.g., Kurmayer & Kutzenberger 2003). A more detailed protocol on calibration of qPCR results is available in Kurmayer et al. (2017).

qPCR assays have also been developed to quantify cyanobacterial taxa potentially producing cyanotoxins, such as the genus *Dolichospermum* (Doblin et al., 2007) and the species *Dolichospermum planktonicum* (Rueckert et al., 2007). For all the taxa mentioned above, quantifying cell numbers or biovolume using qPCR gave similar results to those obtained by cell counts via microscopy. For the genera *Microcystis*, *Planktothrix* and *Raphidiopsis*, close correlations have been reported from field samples between qPCR estimates and cell counts ranging across several orders of magnitude (i.e., 10^2–10^6 cells/mL). The limit of detection/quantification generally is in the range of a hundred to a few thousand cells per mL (Kurmayer & Kutzenberger, 2003; Rasmussen et al., 2008; Chiu et al., 2017). However, for cyanobacterial taxa classified as polyphyletic, for example, the genera *Aphanizomenon* or *Dolichospermum* (see Chapter 3), the qPCR approach is less feasible because it would require to run many qPCRs to quantify multiple individual taxa of a genetically diverse genus for each individual sample. Further insights in the molecular phylogeny of cyanobacteria will provide the basis for developing qPCR assays for their quantification.

13.6.5 Set up of PCR assays

In general, the presence of toxin genes is tested in one PCR, while another independent reaction is used to confirm the presence of cyanobacteria as well as to check for the quality of the DNA (or the presence of potential PCR inhibitors). In many studies, 16S ribosomal DNA (Taton et al., 2003) or gene loci encoding the synthesis of accessory pigments such as phycocyanin (*cpc*BA; Neilan, 1995) have been amplified, but not exclusively (Moreira et al., 2013). As a positive control for PCR, genomic DNA extracted from isolated strains should be used. A list of toxic strains containing the respective target genes, including information on their availability from international culture collections, is given in Kurmayer et al. (2017).

For the use of quantitative PCR (qPCR), the same principle has been applied to control for uncertainties in quantitative estimates of toxic genotypes, for example, due to a physiological variation of the gene or genome copy number (Kurmayer & Kutzenberger, 2003). Typically, primers specified to amplify the total population (generally at genus level) as well as those genotypes that carry the cyanotoxin synthesis genes have been designed. Several qPCR assays have been developed to quantify microcystin/nodularin genes in a given volume of water (for reviews, see Kurmayer & Christiansen (2009), Martins & Vasconcelos (2009), Kim et al. (2013), Pacheco et al. (2016), and Rantala-Ylinen et al. (2017)). In particular, the Taq nuclease assay (TNA) has been used to quantify microcystin genotypes in water samples (Kurmayer & Kutzenberger, 2003). This approach is based on the quantification of the total population of a specific cyanobacterium by a TNA targeted to the intergenic spacer region within the phycocyanin operon and another TNA targeted to the subpopulation carrying the *mcy* genes. A calibration curve based on defined cell concentrations or gene copy numbers is established by relating the known DNA concentrations to the threshold cycle of the diluted DNA extract. Similarly, qPCR assays have been developed for cylindrospermopsin synthesis genes (Rasmussen et al., 2008; Campo et al., 2013) and saxitoxin genes (Al-Tebrineh et al., 2010; Savela et al., 2015). For anatoxin synthesis genes, a qPCR approach was developed by Wang et al. (2015). For benthic cyanobacteria (*Microcoleus*

(Phormidium) autumnalis), Kelly et al. (2018) developed a specific qPCR assay targeting the *ana*C gene.

Another option is to use multiplex PCR and to run two, three or four reactions in the same test tube (Saker et al., 2007; Al-Tebrineh et al., 2010). However, it is emphasised that multiplex (q)PCR requires substantial optimisation (because of competitive effects of primers and targets) and its use is not widespread (see Pacheco et al. (2016) for a review). Multiplex qPCR has been used in monitoring of *Chrysosporum ovalisporum* in Australia integrating three methods: (i) microscopical identification and enumeration, (ii) multiplex qPCR for 16S rDNA, *mcy*E, *cyr*A and *sxt*A gene quantification (iii) and toxin analysis by LC-MS (Crawford et al., 2017). Such integrated approaches may contribute to more efficient handling of high sample numbers from large surveys in future, allowing faster and more precise monitoring (e.g., Lu et al., 2019) to support risk assessment.

13.6.6 Limits of PCR assays

Microcystin synthesis gene clusters may be inactive due to various mutations, and positive polymerase chain reaction (PCR) results thus overestimate the potential of microcystin production in water (Nishizawa et al., 1999). Although cyanotoxin production typically is constitutive if individual strains are analysed (see Chapter 4), inactivation of microcystin synthesis genes occurs regularly under natural conditions (e.g., by transposable elements or partial deletions; Chen et al. (2016)). Thus, it is not surprising that the qPCR-based quantification of *mcy/nda* genes is not always quantitatively related to the analysed microcystin/nodularin concentrations in field samples (reviewed by Pacheco et al., 2016). By analysing 38 studies published in peer-reviewed journals, the authors nevertheless concluded that qPCR can be proposed as a predictor for microcystin/nodularin concentrations.

Less data are available for the other cyanotoxins such as cylindrospermopsin, for which correlations between cylindrospermopsin concentrations and qPCR results have been reported from field samples (Pacheco et al., 2016; Lei et al., 2019). Data for saxitoxin are yet more scarce, with correlations between qPCR data (e.g., sxtA gene copies) and saxitoxin concentrations quite limited (e.g., Savela et al., 2016), in part because the gene loci used are not unambiguously indicative of saxitoxin synthesis and also because of the variety of saxitoxins (section 2.4). For anatoxin synthesis, Wood and Puddick (2017) report gene copy numbers estimated by qPCR with statistically significant positive relationships to anatoxin-a contents of benthic cyanobacteria. Since there are multiple reasons for poor correlations and these cannot always be differentiated (Ostermaier & Kurmayer, 2010), the current pragmatic approach in qPCR application as used by Pacheco et al. (2016) is to compare experience between research groups and waterbodies around the globe to improve molecular monitoring approaches.

An inherent limit of the accuracy of quantification in all qPCR-based techniques is the semilogarithmic nature of the calibration curves which potentially leads to overestimation or underestimation, respectively, by up to 70% (Schober et al., 2007) in estimating genotype numbers or proportions, because minor deviations on linear scale (Cq-values) are translated into larger deviations on logarithmic scale (DNA amount in the template). As a rule of thumb, deviations of <0.5 in Cq-value are considered due to experimental noise. This unspecific variation must be taken into account when translating the results from Cq-values into absolute numbers.

13.6.7 Sequencing of PCR products

Today DNA sequencing facilities offer rapid services at costs ranging at less than 2 € per sequence of approximately 800 base pairs as of 2020. Single PCR products obtained from isolated strains are sequenced directly using the conventional Sanger chain termination method allowing confirmation of the specificity of the obtained PCR products and identification of the source organism. PCR products obtained from field samples often contain mixtures of PCR fragments originating from various genotypes. Separating the individual genotypes requires a cloning approach before sequencing according to standard techniques (Sambrook & Russell, 2001). So-called cloning kits with cloning vectors (plasmids) are commercially available and allow for the amplification of individual genotypes. As a last step, vectors carrying the inserted sequence of individual genotypes are introduced into *Escherichia coli*, purified and sequenced.

13.6.8 Application of PCR-based methods in monitoring

In principle, PCR-based assays have the potential to guide a more efficient application of chemical–analytical tools. For example, toxigenicity (microcystin synthesis) has been detected in cyanobacterial food supplements and has been confirmed using ELISA techniques (Saker et al., 2005). The sequencing of the obtained PCR products revealed the occurrence of *Microcystis aeruginosa* in minor proportion, while the dominant organism *Aphanizomenon flosaquae* was found nontoxic. Similarly, Vichi et al. (2012) used an approach combining PCR-based tools with chemical–analytical detection to analyse cyanotoxins in food supplements from the Italian market and to identify the contaminating organisms. While *M. aeruginosa* was identified in *A. flosaquae* products, the contamination with *M. aeruginosa* was surprisingly, albeit less frequently, also confirmed from products derived from "*Spirulina*" cultivated at high pH and salt concentrations. A further application is the quality control of commonly used open pond mass cultures of eukaryotic microalgae food supplement production for contamination caused by cyanobacteria (Görs et al., 2010).

Analogously for environmental samples, PCR-based methods have been applied frequently to identify the various cyanotoxin (microcystin)-producing organisms. For example, in the temperate climatic zone, microcystin-producing genera such as *Microcystis*, *Planktothrix* and *Dolichospermum* frequently co-occur and diagnostic PCR has been used to differentiate and quantify the proportion of respective toxigenic genera (Rantala et al., 2006). Similarly, in tropical lakes in East Africa, PCR of *mcy* genes followed by sequencing showed that *Microcystis* was the dominant microcystin-producing genus, while co-occurring *Dolichospermum* sp. and *Planktothrix* sp. were not found to be toxigenic (Okello et al., 2010). Furthermore, the PCR-based analyses can give important clues on the stability or variability of the genetic structure of toxigenic subpopulations in aquatic habitats. For example, in lakes of the Alps, the changes occurring in toxigenicity of *Planktothrix* populations were observed to happen rather slowly over a period of three decades with nontoxic genotypes only showing a slow increase in proportion (Ostermaier et al., 2013). In the monitoring of Polish waterbodies, PCR methods have been routinely applied and qPCR results have been used to explain variable microcystin contents in *Microcystis* sp. biomass (Gągała et al., 2014). In conclusion, despite their limitations in absolute quantification, PCR-based methods might well increase the predictability of toxin concentrations by increasing the information on source organisms over time and space.

13.6.9 Identifying toxigenic cyanobacteria using high-throughput sequencing

The PCR-based tools described above cannot give comprehensive information on the taxonomic composition of cyanobacterial communities potentially including toxigenic species. In analogy to microscopy-based counting of cells (see section 13.3.1), the more recently developed deep amplicon (high-throughput) sequencing is able to sequence a very large number of PCR amplicons simultaneously and has been proposed as a tool for monitoring cyanobacteria in the environment (Eldridge et al., 2017). By obtaining at least several thousands of sequences from one amplified gene locus per sample (e.g., 16S rRNA), it is possible to monitor the presence of phytoplankton taxa and including bacteria, possibly including less abundant potentially toxigenic species. In general, the PCR products obtained using universal primers are barcoded via ligation of short nucleotides (MIDs, multiplex identifiers), clonally amplified (e.g., by the so-called bridge amplification of Illumina) and sequenced in parallel on plates. The large amount of sequence reads obtained requires bioinformatical processing following established standard algorithms and taxonomic reference databases available through various publically available international platforms, that is, the Ribosomal Database Project (Cole et al., 2013), or the "Greengenes" application, (DeSantis et al., 2006 (McDonald et al., 2012)) or the SILVA database (Glöckner et al., 2017). Further, several standard sequence-processing pipelines have been designed (e.g., (Schloss et al., 2009; Caporaso et al., 2010; Albanese et al., 2015; Bolyen et al., 2019). In general, the bioinformatics steps include (i) the quality trimming of sequences regarding the exact match of the MID code and the primer, the minimum length in base pairs, the frequency of ambiguous nucleotides in a sequence read, as well as chimera detection; (ii) the clustering of sequences by the genetic distance and assigning to operational taxonomic units (OTUs). Typically, for rDNA genes, a 3% genetic distance threshold is defined and OTUs will then be assigned taxonomically using reference databases as cited above; (iii) the calculation of rarefaction curves which are used to estimate additional sequencing effort as well as to standardise the comparison of diversity and richness estimates between samples; (iv) the calculation of diversity indices as well as richness estimators from the frequency of the OTUs and (v) the use of multivariate statistics to explain the variability in the data sets from recorded metadata (Deng et al., 2017).

Deep-sequencing application might be of relevance for monitoring of invasive species with toxigenic potential, for example, *Raphidiopsis raciborskii* or *Nodularia spumigena* (Sukenik et al., 2015). Currently, the reference taxonomic databases such as RDP have a relatively low resolution (Cole et al., 2013) and individual species of cyanobacteria are only rarely resolved. The relatively short read length (<400 bp) might be one cause of the low percentage of resolved OTUs, as environmental samples may contain a high share of OTUs which have not been characterised previously (Albanese et al., 2015). Further comparing resolved OTUs with the adjusted OTU composition in artificial communities can reveal a technical bias (Pessi et al., 2016). Comparing microscopical data with data obtained from deep sequencing also reveals discrepancies which show not only the limitation of microscopy (i.e., underestimating the abundance and diversity of picocyanobacteria such as *Synechococcus*), but also the limitation in deep sequencing, for example, because of low or uncertain resolution (Eiler et al., 2013; Xiao et al., 2014). In future, it will be important to standardise these emerging techniques (Hornung et al., 2019) to avoid systematic bias (Boers et al., 2016), for example, by using artificial (mock) communities (Pessi et al., 2016) as well as to create taxonomic reference databases from sequenced and morphologically described strains. Alternatively, as a way forward, the information obtained from both methodologies, microscopy and deep sequencing is combined and integrated into the community analysis of environmental samples.

13.7 DETERMINATION OF NUTRIENT CONCENTRATIONS

As discussed in detail in Chapter 4, the capacity for development of a cyanobacterial bloom depends on the available concentrations of nutrients, primarily of phosphorus and nitrogen. In freshwaters, often phosphorus concentrations limit the amount of biomass that can form in a given waterbody, but sometimes nitrogen is limiting. The chief sources of nitrogen are nitrate and ammonium, but to some extent, their lack can be compensated by some cyanobacteria through fixation of atmospheric nitrogen. Thus, even if phosphate is clearly the factor limiting carrying capacity, knowledge of nitrogen availability helps to predict whether nitrogen-fixing species are likely to dominate.

Cyanobacterial cells can store only some excess nitrogen, but can store phosphorus for up to four cell divisions, enabling a single cell to multiply into 16 cells without the need to take up further phosphorus. Information on dissolved phosphorus concentrations therefore only demonstrates that, if it can be detected, the phytoplankton population is not currently limited by phosphorus availability. In order to assess the capacity of the waterbody to support a cyanobacterial population, total phosphorus (TP) is a much better predictor. To assess whether nitrogen may be limiting, analysis of dissolved components (chiefly nitrate and ammonium) is sufficient.

In modern laboratories, various fractions of nitrogen and phosphorus are today quantified by automated technologies that allow high-throughput analyses generally based on photometry such as flow injection analysis (FIA) or continuous flow analysis (CFA). Respective, standard methods are available for nitrite and nitrate (ISO, 1996), ammonium (ISO, 2005), and total and ortho-phosphate (ISO, 2003). Since FIA or CFA is not available to many laboratories, simpler and largely manual methods for nutrient analysis are still in use. In the following, a brief description of such methods is given together with requirements to perform analyses.

Before any analysis is to be implemented in a laboratory, the national and international regulations concerning laboratory safety and environmental protection must be consulted and duly considered. Any laboratory analysis requires trained laboratory staff, especially when handling toxic or otherwise harmful chemicals is part of it.

Among the methods available, the procedure of Hansen and Koroleff (2007) for determining TP has proved to be most reliable and is the basis of an ISO protocol. For nitrate and ammonium, several methods are available, and the ISO method with the least demands on equipment is described below. For details on ISO methods, see the International Organization for Standardization's website (https://www.iso.org/).

13.7.1 Analysis of phosphorus

Phosphorus in various types of waters can be determined spectrometrically by the digestion of organic phosphorus compounds to inorganic soluble reactive phosphorus (SRP, largely comprising ortho-phosphate) and transforming this to an antimony–phosphomolybdate complex under acidic conditions, which is then reduced to a strongly coloured blue molybdenum complex. The internationally harmonised method as described in ISO (2004) is applicable to many types of waters (surface-, ground-, sea- and wastewater) in a concentration range of 0.005–0.8 mg/L. Differentiation by the following fractions is possible through filtration procedures:

- *SRP*: filtered sample, generally with a pore size of 0.45 µm.
- *Dissolved organic phosphorus* (*DOP*): digested filtered sample; dissolved organic P is converted into SRP.
- *TP*: digested unfiltered sample; all organic P is converted to SRP.
- *Particulate phosphorus* (*PP*): difference between TP and DOP.

For SRP sample preparation, it is important to note that filters can release phosphorus. To avoid a bias, the filters must be washed with the water sample (10–25 mL) and this filtrate be discarded.

Digestion or mineralisation of organophosphorus compounds to SRP for DOP and TP analysis is performed in tightly sealed screw-cap vessels with persulphate, under pressure and heat in an autoclave (or a household steamer), or simply by gentle boiling. The following gives an overview of the procedure, necessary equipment and chemicals; for details, see ISO (2004).

Equipment

- Photometer measuring absorbance in the visible and near-infrared spectrum above 700 nm; sensitivity is optimal at 880 nm; sensitivity is increased if optical cells of 50 mm optical pathlength are used
- Filter assembly and membrane filters, 45 mm diameter with 0.45 µm pore size
- For the digestion of samples (TP and DOP), an autoclave (or steamer) suitable for 115–120 °C
- For the digestion of samples, borosilicate vessels with heat-resistant caps that can be tightly sealed
- Bottles for samples as described in Chapter 14
- Precleaned glass bottles for filtered samples

13.7.2 Analysis of nitrate

Several methods for the determination of nitrate have been provided by the ISO, the simplest being a spectrometric measurement of the yellow compound formed by the reaction of sulphosalicylic acid with nitrate and subsequent treatment with alkali (ISO, 1988). The equipment required is a spectrometer operating at a wavelength of 415 nm and cuvettes with an optical path length of 40–50 mm, evaporating dishes, a water bath capable of accepting six or more dishes and a water bath capable of thermostatic regulation to 25 °C. This method is suitable for surface and potable water samples and has a detection limit of 0.003–0.013 mg/L (depending on optical equipment). Interference from a range of substances, particularly chloride, orthophosphate, magnesium and manganese (III), is possible. Interference problems can be avoided with other spectrometric methods (ISO, 1986b; c).

The equipment is similar to the one required for phosphorus analysis.

13.7.3 Analysis of ammonium

A manual spectrometric method is given in ISO (1984b), which analyses a blue compound formed by the reaction of ammonium with salicylate and hypochlorite ions in the presence of sodium nitrosopentacyanoferrate (III) at a limit of detection of 0.003–0.008 mg/L. An automated procedure is given in ISO (1986a). A distillation and titration method is given in ISO (1984a).

The equipment is similar to the one required for phosphorus analysis.

13.7.4 On-site analysis techniques for nutrients

A number of technologies are available for rapid on-site analysis of primarily dissolved nutrients such as SRP, nitrate and ammonium.

Most simple with respect to handling and required equipment are test strips that are submerged in (filtered) water and after a short incubation time, a colour change allows to estimate the concentration. Expectedly, the sensitivity is comparatively low as is the accuracy, but in some occasions, a rapid semiquantitative result may be more valuable than more accurate results that are available only after a considerable delay.

Selective electrodes are available for nitrate and ammonium (Cuartero & Bakker, 2017). Handling and data-logging is similar to that for pH or oxygen electrodes. This technique may be interesting in particular when a high variability of concentrations is suspected, for example, in individual inflows to a reservoir.

For on-site analyses, also fully functional photometers and ready-to-use reagents kits are available. Digestion of samples for analysis of total phosphorus is also possible on-site.

On-site analyses are generally less accurate and less sensitive compared to laboratory analyses but are, on the other hand, less expensive and faster with respect to time to result. This needs to be balanced for individual monitoring programmes. Preferably, any method eventually adopted is evaluated for accuracy and sensitivity by testing an individual sample with different methods.

REFERENCES

Al-Tebrineh J, Mihali TK, Pomati F, Neilan BA (2010). Detection of saxitoxin-producing cyanobacteria and *Anabaena circinalis* in environmental water blooms by quantitative PCR. Appl Environ Microbiol. 76:7836–7842.

Albanese D, Fontana P, De Filippo C, Cavalieri D, Donati C (2015). MICCA: a complete and accurate software for taxonomic profiling of metagenomic data. Sci Rep. 5:9743.

Alcántara I, Piccini C, Segura A, Deus S, González C, de la Escalera GM et al. (2018). Improved biovolume estimation of *Microcystis aeruginosa* colonies: A statistical approach. J Microbiol Methods. 151:20–27.

Anagnostidis K, Komárek J (1985). Modern approach to the classification system of Cyanophytes. 1-Introduction. Arch Hydrobiol Algol Stud. 38–39:291–302.

Anagnostidis K, Komárek J (1988). Modern approach to the classification system of Cyanophytes 3- Oscillatoriales. Arch Hydrobiol Algol Stud 80 (50–53):327–472.

Anagnostidis K, Komárek J (1990). Modern approach to the classification system of Cyanophytes. 5-Stigonematales. Arch Hydrobiol 86, Suppl:1–73.

Asai R, Horiguchi Y, Yoshida A, McNiven S, Tahira P, Ikebukuro K et al. (2001). Detection of phycobilin pigments and their seasonal change in Lake Kasumigaura using a sensitive in situ fluorometric sensor. Anal Lett. 34:2521–2533.

Ballantine D (1953). Comparison of the different methods of estimating nanoplankton. J Mar Biol Assoc UK. 32:129–147.

Ballot A, Cerasino L, Hostyeva V, Cirés S (2016). Variability in the *sxt* gene clusters of PSP toxin producing *Aphanizomenon gracile* strains from Norway, Spain, Germany and North America. PLoS One. 11:e0167552.

Ballot A, Fastner J, Wiedner C (2010). Paralytic shellfish poisoning toxin-producing cyanobacterium *Aphanizomenon gracile* in Northeast Germany. Appl Environ Microbiol. 76:1173–1180.

Ballot A, Ramm J, Rundberget T, Kaplan-Levy RN, Hadas O, Sukenik A et al. (2011). Occurrence of non-cylindrospermopsin-producing *Aphanizomenon ovalisporum* and *Anabaena bergii* in Lake Kinneret (Israel). J Plankton Res. 33:1736–1746.

Benfield MC, Grosjean P, Culverhouse PF, Irigoien X, Sieracki ME, Lopez-Urrutia A et al. (2007). RAPID: research on automated plankton identification. Oceanography. 20:172–187.

Bertani I, Steger CE, Obenour DR, Fahnenstiel GL, Bridgeman TB, Johengen TH et al. (2017). Tracking cyanobacteria blooms: Do different monitoring approaches tell the same story? Sci Tot Environ. 575:294–308.

Bertone E, Burford MA, Hamilton DP (2018). Fluorescence probes for real-time remote cyanobacteria monitoring: A review of challenges and opportunities. Water Res. 141:152–162.

Beutler M, Wiltshire KH, Meyer B, Moldaenke C, Luring C, Meyerhofer M et al. (2002). A fluorometric method for the differentiation of algal populations in vivo and in situ. Photosynth Res. 72:39–53.

Bidigare RR, Van Heukelem L, Trees CC (2005). Analysis of algal pigments by high-performance liquid chromatography. New York: Algal Culturing Techniques Academic Press:327–345.

Boers SA, Jansen R, Hays JP (2016). Suddenly everyone is a microbiota specialist. Clin Microbiol Infect 22:581–582.

Bolyen E, Rideout JR, Dillon MR, Bokulich NA, Abnet CC, Al-Ghalith GA et al. (2019). Reproducible, interactive, scalable and extensible microbiome data science using QIIME 2. Nat Biotechnol. 37:852–857.

Box J (1981). Enumeration of cell concentrations in suspensions of colonial freshwater microalgae, with particular reference to *Microcystis aeruginosa*. Brit Phycol J. 16:153–164.

Brentrup JA, Williamson CE, Colom-Montero W, Eckert W, de Eyto E, Grossart HP et al. (2016). The potential of high-frequency profiling to assess vertical and seasonal patterns of phytoplankton dynamics in lakes: an extension of the Plankton Ecology Group (PEG) model. Inland Waters. 6:565–580.

Brophy MJ, Trueman BF, Park Y, Betts RA, Gagnon GA (2018). Fluorescence spectra predict microcystin-LR and disinfection byproduct formation potential in lake water. Environ Sci Technol. 53:586–594.

Brown NM, Mueller RS, Shepardson JW, Landry ZC, Morré JT, Maier CS et al. (2016). Structural and functional analysis of the finished genome of the recently isolated toxic *Anabaena* sp. WA102. BMC Genomics. 17:457.

Burnham JC, Stetak T, Boulger J (1973). An improved method of cell enumeration for filamentous algae and bacteria. J Phycol. 9:346–349.

Calteau A, Fewer DP, Latifi A, Coursin T, Laurent T, Jokela J et al. (2014). Phylum-wide comparative genomics unravel the diversity of secondary metabolism in Cyanobacteria. BMC Genomics. 15:977.

Campo E, Lezcano M, Agha R, Cirés S, Quesada A, El-Shehawy R (2013). First TaqMan assay to identify and quantify the cylindrospermopsin-producing cyanobacterium *Aphanizomenon ovalisporum* in water. Adv Microbiol. 3:430–437.

Caporaso JG, Kuczynski J, Stombaugh J, Bittinger K, Bushman FD, Costello EK et al. (2010). QIIME allows analysis of high-throughput community sequencing data. Nat Meth. 7:335–336.

Casero MC, Ballot A, Agha R, Quesada A, Cirés S (2014). Characterization of saxitoxin production and release and phylogeny of sxt genes in paralytic shellfish poisoning toxin-producing *Aphanizomenon gracile*. Harmful Algae. 37:28–37.

Catherine A, Maloufi S, Congestri R, Viaggiu E, Pilkaityte R (2017). Cyanobacterial samples: preservation, enumeration, and biovolume measurements. In: Meriluoto J, Spoof L, Codd GA et al., editors: Handbook of cyanobacterial monitoring and cyanotoxin analysis. Chichester: John Wiley & Sons. 313–330.

CEN (2006). EN 15204: water quality – guidance standard on the enumeration of phytoplankton using inverted microscopy (Utermöhl technique). Brussels: European Committee for Standardization:46 pp.

CEN (2015). EN 16695: Water Quality – Guidance on the estimation of phytoplankton biovolume. Brussels, BE: European Committee for Standardization:40 pp.

Chen Q, Christiansen G, Deng L, Kurmayer R (2016). Emergence of nontoxic mutants as revealed by single filament analysis in bloom-forming cyanobacteria of the genus *Planktothrix*. BMC Microbiol 16:23.

Chiu Y-T, Chen Y-H, Wang T-S, Yen H-K, Lin T-F (2017). A qPCR-based tool to diagnose the presence of harmful cyanobacteria and cyanotoxins in drinking water sources. Int J Environ Res Public Health. 14:547.

Choo F, Zamyadi A, Stuetz R, Newcombe G, Newton K, Henderson R (2019). Enhanced real-time cyanobacterial fluorescence monitoring through chlorophyll-*a* interference compensation corrections. Water Res. 148:86–96.

Christiansen G, Fastner J, Erhard M, Börner T, Dittmann E (2003). Microcystin biosynthesis in *Planktothrix*: Genes, evolution, and manipulation. J Bacteriol. 185:564–572.

Cole JR, Wang Q, Fish JA, Chai B, McGarrell DM, Sun Y et al. (2013). Ribosomal Database Project: data and tools for high throughput rRNA analysis. Nucleic Acids Res. 42:D633–D642.

Costa J, Amaral JS, Fernandes TJ, Batista A, Oliveira MBP, Mafra I (2015). DNA extraction from plant food supplements: Influence of different pharmaceutical excipients. Mol Cell Probes. 29:473–478.

Crawford A, Holliday J, Merrick C, Brayan J, van Asten M, Bowling L (2017). Use of three monitoring approaches to manage a major *Chrysosporum ovalisporum* bloom in the Murray River, Australia, 2016. Environ Monit Assess. 189:202.

Cuartero M, Bakker E (2017). Environmental water analysis with membrane electrodes. Curr Opin Electrochem. 3:97–105.

Deng L, Sweetlove M, Blank S, Obbels D, Verleyen E, Vyverman W et al. (2017). Monitoring of toxigenic cyanobacteria using next-generation sequencing techniques. In: Kurmayer R, Sivonen K, Wilmotte A, Salmaso N, editors: Molecular tools for the detection and quantification of toxigenic Chichester: John Wiley & Sons:277–299.

DeSantis TZ, Hugenholtz P, Larsen N, Rojas M, Brodie EL, Keller K et al. (2006). Greengenes, a chimera-checked 16S rRNA gene database and workbench compatible with ARB. Appl Environ Microbiol. 72:5069–5072.

Desiderio RA, Moore C, Lantz C, Cowles TJ (1997). Multiple excitation fluorometer for in situ oceanographic applications. Appl Opt. 36:1289–1296.

Dittmann E, Fewer DP, Neilan BA (2013). Cyanobacterial toxins: biosynthetic routes and evolutionary roots. FEMS Microbiol Rev. 37:23–43.

Doblin MA, Coyne KJ, Rinta-Kanto JM, Wilhelm SW, Dobbs FC (2007). Dynamics and short-term survival of toxic cyanobacteria species in ballast water from NOBOB vessels transiting the Great Lakes – implications for HAB invasions. Harmful Algae. 6:519–530.

Edler L, Elbrächter M (2010). The Utermöhl method for quantitative phytoplankton analysis. In: Karlson B, Cusack C, Bresnan E, editors: Microscopic and molecular methods for quantitative phytoplankton analysis. Paris: Intergovernmental Oceanographic Commission of UNESCO:13–20.

Edwards DJ, Gerwick WH (2004). Lyngbyatoxin biosynthesis: Sequence of biosynthetic gene cluster and identification of a novel aromatic prenyltransferase. J Am Chem Soc. 126:11432–11433.

Eiler A, Drakare S, Bertilsson S, Pernthaler J, Peura S, Rofner C et al. (2013). Unveiling distribution patterns of freshwater phytoplankton by a next generation sequencing based approach. PLoS One. 8:e53516.

Eldridge SLC, Driscoll C, Dreher TW (2017). Using high-throughput DNA sequencing, genetic fingerprinting, and quantitative PCR as tools for monitoring bloom-forming and toxigenic cyanobacteria in Upper Klamath Lake, Oregon, 2013 and 2014. US Geological Survey.

Fergusson KM, Saint CP (2003). Multiplex PCR assay for *Cylindrospermopsis raciborskii* and cylindrospermopsin-producing cyanobacteria. Environ Toxicol. 18:120–125.

Fernandez-Diaz JC, Glennie CL, Carter WE, Shrestha RL, Sartori MP, Singhania A et al. (2014). Early results of simultaneous terrain and shallow water bathymetry mapping using a single-wavelength airborne LiDAR sensor. IEEE J. 7:623–635.

Fewer DP, Wahlsten M, Österholm J, Jokela J, Rouhiainen L, Kaasalainen U et al. (2013) The genetic basis for O-acetylation of the microcystin toxin in cyanobacteria. Chem Biol. 20:861–869.

Franche C, Damerval T (1988). Tests on *nif* probes and DNA hybridizations. Meth Enzymol. 167:803–808.

Friedrichs A, Busch J, van der Woerd H, Zielinski O (2017). SmartFluo: a method and affordable adapter to measure chlorophyll *a* fluorescence with smartphones. Sensors. 17:678.

Gągała I, Izydorczyk K, Jurczak T, Pawełczyk J, Dziadek J, Wojtal-Frankiewicz A et al. (2014). Role of environmental factors and toxic genotypes in the regulation of microcystins-producing cyanobacterial blooms. Microb Ecol. 67:465–479.

Giardino C, Bresciani M, Valentini E, Gasperini L, Bolpagni R, Brando VE (2015). Airborne hyperspectral data to assess suspended particulate matter and aquatic vegetation in a shallow and turbid lake. Rem Sens Environ. 157:48–57.

Glöckner FO, Yilmaz P, Quast C, Gerken J, Beccati A, Ciuprina A et al. (2017). 25 years of serving the community with ribosomal RNA gene reference databases and tools. J Biotechnol. 261:169–176.

Goddard VJ, Baker AC, Davy JE, Adams DG, De Ville MM, Thackeray SJ et al. (2005). Temporal distribution of viruses, bacteria and phytoplankton throughout the water column in a freshwater hypereutrophic lake. Aquat Microb Ecol. 39:211–223.

Görs M, Schumann R, Hepperle D, Karsten U (2010). Quality analysis of commercial *Chlorella* products used as dietary supplement in human nutrition. J Appl Phycol. 22:265–276.

Gregor J, Geris R, Maršálek B, Hetesa J, Marvan P (2005). In situ quantification of phytoplankton in reservoirs using a submersible spectrofluorometer. Hydrobiologia. 548:141–151.

Gregor J, Maršálek B (2004). Freshwater phytoplankton quantification by chlorophyll *a*: a comparative study of in vitro, in vivo and in situ methods. Water Res. 38:517–522.

Gregor J, Maršálek B, Sipkova H (2007). Detection and estimation of potentially toxic cyanobacteria in raw water at the drinking water treatment plant by *in vivo* fluorescence method. Water Res. 41:228–234.

Hansen HP, Koroleff F (2007). Determination of nutrients. In: Grasshoff K, Kremling K, Ehrhardt M et al., editors: Methods of seawater analysis, third edition. Weinheim: Wiley-VCH:159–228.

Hawkins PR, Holliday J, Kathuria A, Bowling L (2005). Change in cyanobacterial biovolume due to preservation by Lugol's Iodine. Harmful Algae. 4:1033–1043.

Hillebrand H, Dürselen C, Kirschtel D, Pollingher U, Zohary T (1999). Biovolume calculation for pelagic and benthic microalgae. J Phycol. 35:403–424.

Hindson CM, Chevillet JR, Briggs HA, Gallichotte EN, Ruf IK, Hindson BJ et al. (2013). Absolute quantification by droplet digital PCR versus analog real-time PCR. Nat Meth. 10:1003–1005.

Hisbergues M, Christiansen G, Rouhiainen L, Sivonen K, Börner T (2003). PCR-based identification of microcystin-producing genotypes of different cyanobacterial genera. Arch Microbiol. 180:402–410.

Hodges CM, Wood SA, Puddick J, McBride CG, Hamilton DP (2018). Sensor manufacturer, temperature, and cyanobacteria morphology affect phycocyanin fluorescence measurements. Environ Sci Pollut Res. 25:1079–1088.

Hoff-Risseti C, Dörr FA, Schaker PDC, Pinto E, Werner VR, Fiore MF (2013). Cylindrospermopsin and saxitoxin synthetase genes in *Cylindrospermopsis raciborskii* strains from Brazilian freshwater. PLoS One. 8:e74238.

Holm-Hansen O, Lorenzen CJ, Holmes RW, Strickland JD (1965). Fluorometric determination of chlorophyll. ICES J Mar Sci. 30:3–15.

Hornung BV, Zwittink RD, Kuijper EJ (2019). Issues and current standards of controls in microbiome research. FEMS Microbiol Ecol 95:fiz045.

Hötzel G, Croome R (1999). A phytoplankton methods manual for Australian freshwaters. Canberrra: Land and Water Resources Research and Development Corporation:58 pp.

ISO (1984a). ISO 5664. Determination of ammonium. Distillation and titration method. Geneva: International Organization for Standardization.

ISO (1984b). ISO 7150-1. Determination of ammonium Part 1: manual spectrometric method. Geneva: International Organization for Standardization.

ISO (1986a). ISO 7150-2. Determination of ammonium Part 2: automated spectrometric method. Geneva: International Organization for Standardization.

ISO (1986b). ISO 7890-1. Determination of nitrate Part 1: 2,6-Dimethylphenol spectrometric method. Geneva: International Organization for Standardization.

ISO (1986c). ISO 7890-2. Determination of nitrate Part 2: 4-Fluorophenol spectrometric method after distillation. Geneva: International Organization for Standardization.

ISO (1988). ISO 7890-3. Determination of nitrate Part 3: spectrometric method using sulfosalicylic acid. Geneva: International Organization for Standardization.

ISO (1992). ISO 10260. Measurement of biochemical parameters. Spectrometric determination of the chlorophyll-a concentrations. Geneva: International Organization for Standardization.

ISO (1996). ISO 13395. Determination of nitrite nitrogen and nitrate nitrogen and the sum of both by flow analysis (CFA and FIA) and spectrometric detection. Geneva: International Organization for Standardization.

ISO (2003). ISO 15681-2. Determination of orthophosphate and total phosphorus contents by flow analysis (FIA and CFA) Part 2: method by continuous flow analysis (CFA). Geneva: International Organization for Standardization.

ISO (2004). ISO 6878. Determination of phosphorus. Ammonium molybdate spectrometric method. Geneva: International Organization for Standardization.

ISO (2005). ISO 11732. Determination of ammonium nitrogen: method by flow analysis (CFA and FIA) and spectrometric detection. Geneva: International Organization for Standardization.

ISO (2006). ISO 20837. Microbiology of food and animal feeding stuffs - Polymerase chain reaction (PCR) for the detection of food-borne pathogens - Requirements for sample preparation for qualitative detection. ISO 20837. Geneva: International Organization for Standardization.

Izydorczyk K, Tarczynska M, Jurczak T, Mrowczynski J, Zalewski M (2005). Measurement of phycocyanin fluorescence as an online early warning system for cyanobacteria in reservoir intake water. Environ Toxicol 20:425–430.

Jacobsen T (1978). A quantitative method for the separation of chlorophylls a and b from phytoplankton pigments by high pressure liquid chromatography. Mar Sci Commun. 4:33–47.

Jiang Y, Song G, Pan Q, Yang Y, Li R (2015). Identification of genes for anatoxin-a biosynthesis in *Cuspidothrix issatschenkoi*. Harmful Algae. 46:43–48.

Jiang Y, Xiao P, Yu G, Shao J, Liu D, Azevedo SM et al. (2014). Sporadic distribution and distinctive variations of cylindrospermopsin genes in cyanobacterial strains and environmental samples from Chinese freshwater bodies. Appl Environ Microbiol. 80:5219–5230.

Jungblut A-D, Neilan BA (2006). Molecular identification and evolution of the cyclic peptide hepatotoxins, microcystin and nodularin, synthetase genes in three orders of cyanobacteria. Arch Microbiol. 185:107–114.

Kahru M, Brown CW (1997). Monitoring of algal blooms: New techniques for detecting large-scale environmental change. Berlin: Springer:190 pp.

Karlson B, Cusack C, Bresnan E (2010). Microscopic and molecular methods for quantitative phytoplankton analysis. Paris, FRA: Intergovernmental Oceanographic Commission of UNESCO:110 pp. http://hab.ioc-unesco.org/index.php?option=com_oe&task=viewDocumentRecord&docID=5440.

Kasprzak P, Padisák J, Koschel R, Krienitz L, Gervais F (2008). Chlorophyll a concentration across a trophic gradient of lakes: An estimator of phytoplankton biomass? Limnologica. 38:327–338.

Kaštovský J, Hauer T, Mareš J, Krautová M, Bešta T, Komárek J et al. (2010). A review of the alien and expansive species of freshwater cyanobacteria and algae in the Czech Republic. Biol Invasions. 12:3599–3625.

Kellmann R, Michali TK, Neilan BA (2008). Identification of a saxitoxin biosynthesis gene with a history of frequent horizontal gene transfers. J Mol Evol. 67:526–538.

Kelly LT, Wood SA, McAllister TG, Ryan KG (2018). Development and application of a quantitative PCR assay to assess genotype dynamics and anatoxin content in *Microcoleus autumnalis*-dominated mats. Toxins. 10:431.

Kim J, Lim J, Lee C (2013). Quantitative real-time PCR approaches for microbial community studies in wastewater treatment systems: applications and considerations. Biotechnol Adv. 31:1358–1373.

Komárek J (1991). A review of water-bloom forming *Microcystis* species, with regard to populations from Japan. Arch Hydrobiol Algol Stud. 64:115–127.

Komárek J (1996). Towards a combined approach for the taxonomy and species delimitation of picoplanktic cyanoprokaryotes. Arch Hydrobiol Algol Stud:377–401.

Komárek J (2003). Planktic oscillatorialean cyanoprokaryotes (short review according to combined phenotype and molecular aspects). Hydrobiologia. 502:367–382.

Komárek J (2010). Modern taxonomic revision of planktic nostocacean cyanobacteria: a short review of genera. Hydrobiologia. 639:231–243.

Komárek J (2013). Cyanoprokaryota Part 3: Heterocystous genera. In: Büdel B, Gärtner G, Krienitz L et al., editors: Süsswasserflora von Mitteleuropa. Heidelberg: Springer Spektrum:1131.

Komárek J, Anagnostidis K (1986). Modern approach to the classification system of Cyanophytes. 2-Chroococcales. Arch Hydrobiol Suppl. 73:157–226.

Komárek J, Anagnostidis K (1989). Modern approach to the classification system of Cyanophytes 4-Nostocales. Arch Hydrobiol Suppl. 82:247–345.

Komárek J, Anagnostidis K (2007). Cyanoprokaryota Part 2: Oscillatoriales. In: Büdel B, Gärtner G, Krienitz L et al., editors: Süßwasserflora von Mitteleuropa Vol 19. Heidelberg: Springer Spektrum:759.

Komárek J, Anagnostidis K (2008) Cyanoprokaryota Part 1: Chroococcales. In: Ettl H, Gärtner G, Heynig H et al., editors: Süwasserflora von Mitteleuropa. Heidelberg: Springer Spektrum:548.

Komárek J, Cronberg G (2001). Some chroococcalean and oscillatorialean Cyanoprokaryotes from southern African lakes, ponds and pools. Nova Hedwigia:129–160.

Komárek J, Zapomělová E (2007). Planktic morphospecies of the cyanobacterial genus *Anabaena*= subg. *Dolichospermum*–1. part: coiled types. Fottea. 7:1–31.

Kremer CT, Gillette JP, Rudstam LG, Brettum P, Ptacnik R (2014). A compendium of cell and natural unit biovolumes for> 1200 freshwater phytoplankton species. Ecology. 95:2984–2984.

Kurmayer R, Christiansen G (2009). The genetic basis of toxin production in cyanobacteria. Freshwater Rev. 2:31–50.

Kurmayer R, Christiansen G, Chorus I (2003). The abundance of microcystin-producing genotypes correlates positively with colony size in *Microcystis* sp. and determines its microcystin net production in Lake Wannsee. Appl Environ Microbiol. 69:787–795.

Kurmayer R, Dittmann E, Fastner J, Chorus I (2002). Diversity of microcystin genes within a population of the toxic cyanobacterium *Microcystis* spp. in Lake Wannsee (Berlin, Germany). Microb Ecol. 43:107–118.

Kurmayer R, Kutzenberger T (2003). Application of real-time PCR for quantification of microcystin genotypes in a population of the cyanobacterium *Microcystis* sp. Appl Environ Microbiol. 69:6723–6730.

Kurmayer R, Schober E, Tonk L, Visser PM, Christiansen G (2011). Spatial divergence in the proportions of genes encoding toxic peptide synthesis among populations of the cyanobacterium *Planktothrix* in European lakes. FEMS Microbiol Lett. 317:127–137.

Kurmayer R, Sivonen K, Wilmotte A, Salmaso N, editors (2017). Molecular tools for the detection and quantification of toxigenic cyanobacteria. Chichester: John Wiley & Sons:392 pp.

Lee RE (2018). Phycology, fifth edition. Cambridge, UK: Cambridge University Press:546 pp.

Lei L, Lei M, Lu Y, Peng L, Han B-P (2019). Development of real-time PCR for quantification of *Cylindrospermopsis raciborskii* cells and potential cylindrospermopsin-producing genotypes in subtropical reservoirs of southern China. J Appl Phycol. 31:3749–3758.

Lu K-Y, Chiu Y-T, Burch M, Senoro D, Lin T-F (2019). A molecular-based method to estimate the risk associated with cyanotoxins and odor compounds in drinking water sources. Water Res. 164:114938.

Lund J, Kipling C, Le Cren E (1958). The inverted microscope method of estimating algal numbers and the statistical basis of estimations by counting. Hydrobiologia. 11:143–170.

Ma S, Tao Z, Yang X, Yu Y, Zhou X, Li Z (2014). Bathymetry retrieval from hyperspectral remote sensing data in optical-shallow water. IEEE Trans Geosci Remote Sens. 52:1205–1212.

Mankiewicz-Boczek J, Kokociński M, Gagała I, Pawełczyk J, Jurczak T, Dziadek J (2012). Preliminary molecular identification of cylindrospermopsin-producing cyanobacteria in two Polish lakes (Central Europe). FEMS Microbiol Lett. 326:173–179.

Marce R, George G, Buscarinu P, Deidda M, Dunalska J, de Eyto E et al. (2016). Automatic high frequency monitoring for improved lake and reservoir management. Environ Sci Technol. 50:10780–10794.

Martins JC, Vasconcelos VM (2009). Microcystin dynamics in aquatic organisms. J Toxicol Environ Health Part B. 12:65–82.

Mazmouz R, Chapuis-Hugon F, Mann S, Pichon V, Méjean A, Ploux O (2010). Biosynthesis of cylindrospermopsin and 7-epicylindrospermopsin in *Oscillatoria* sp strain PCC 6506: identification of the *cyr* gene cluster and toxin analysis. Appl Environ Microbiol. 76:4943–4949.

McDonald D, Price MN, Goodrich J, Nawrocki EP, DeSantis TZ, Probst A *et al.* (2012) An improved Greengenes taxonomy with explicit ranks for ecological and evolutionary analyses of bacteria and archaea. ISME J. 6:610–618.

McQuaid N, Zamyadi A, Prévost M, Bird DF, Dorner S (2011). Use of in vivo phycocyanin fluorescence to monitor potential microcystin-producing cyanobacterial biovolume in a drinking water source. J Environ Monitor. 13:455–463.

Mejean A, Mann S, Vassiliadis G, Lombard B, Loew D, Ploux O (2009). In vitro reconstitution of the first steps of anatoxin-a biosynthesis in *Oscillatoria* PCC 6506: from free L-proline to acyl carrier protein bound dehydroproline. Biochemistry. 49:103–113.

Mihali TK, Kellmann R, Muenchhoff J, Barrow KD, Neilan BA (2008). Characterization of the gene cluster responsible for cylindrospermopsin biosynthesis. Appl Environ Microbiol. 74:716–722.

Moffitt MC, Neilan BA (2004). Characterization of the nodularin synthetase gene cluster and proposed theory of the evolution of cyanobacterial hepatotoxins. Appl Environ Microbiol. 70:6353–6362.

Montreuil A-L, Levoy F, Bretel P, Anthony EJ (2014). Morphological diversity and complex sediment recirculation on the ebb delta of a macrotidal inlet (Normandy, France): a multiple LiDAR dataset approach. Geomorphology. 219:114–125.

Moreira C, Vasconcelos V, Antunes A (2013). Phylogeny and biogeography of cyanobacteria and their produced toxins. Mar Drugs. 11:4350–4369.

Murray SA, Diwan R, Orr RJ, Kohli GS, John U (2015). Gene duplication, loss and selection in the evolution of saxitoxin biosynthesis in alveolates. Mol Phylogen Evol. 92:165–180.

Murray SA, Mihali TK, Neilan BA (2011). Extraordinary conservation, gene loss, and positive selection in the evolution of an ancient neurotoxin. Mol Biol Evol. 28:1173–1182.

Napiórkowska-Krzebietke A, Kobos J (2016). Assessment of the cell biovolume of phytoplankton widespread in coastal and inland water bodies. Water Res. 104:532–546.

Neilan BA (1995). Identification and phylogenetic analysis of toxigenic cyanobacteria by multiplex randomly amplified polymorphic DNA PCR. Appl Environ Microbiol. 61:2286–2291.

Nishizawa T, Asayama M, Fujii K, Harada K-I, Shirai M (1999). Genetic analysis of the peptide synthetase genes for a cyclic heptapeptide microcystin in *Microcystis* spp. J Biochem. 126:520–529.

Nshimyimana JP, Cruz MC, Wuertz S, Thompson JR (2019). Variably improved microbial source tracking with digital droplet PCR. Water Res. 159:192–202.

Odate T, Hirawake T, Kudoh S, Klein B, LeBlanc B, Fukuchi M (2002). Temporal and spatial patterns in the surface-water biomass of phytoplankton in the North Water. Deep Sea Res, Part II. 49:4947–4958.

Okello W, Portmann C, Erhard M, Gademann K, Kurmayer R (2010). Occurrence of microcystin-producing cyanobacteria in Ugandan freshwater habitats. Environ Toxicol. 25:367–380.

Olenina I, Hajdu S, Edler L, Andersson A, Wasmund N, Busch S et al. (2006). Biovolumes and size-classes of phytoplankton in the Baltic Sea. Baltic Marine Environment Protection Commission: Proc. No. 106, Helsinki. 144 pp.

Orr RJ, Stüken A, Murray SA, Jakobsen KS (2013). Evolutionary acquisition and loss of saxitoxin biosynthesis in dinoflagellates: the second "core" gene, sxtG. Appl Environ Microbiol 79:2128–2136.

Ostermaier V, Christiansen G, Schanz F, Kurmayer R (2013). Genetic variability of microcystin biosynthesis genes in *Planktothrix* as elucidated from samples preserved by heat desiccation during three decades. PLoS One 8:e80177.

Ostermaier V, Kurmayer R (2009). Distribution and abundance of nontoxic mutants of cyanobacteria in lakes of the Alps. Microb Ecol. 58:323–333.

Ostermaier V, Kurmayer R (2010). Application of real-time PCR to estimate toxin production by the cyanobacterium *Planktothrix* sp. Appl Environ Microbiol. 76:3495–3502.

Otsuki A, Takamura N (1987). Comparison of chlorophyll-a concentrations measured by fluorometric HPLC and spectrophotometry methods in highly eutrophic shallow Lake Kasumigaura: With 5 figures in the text. Verh - Int Ver Theor Angew Limnol. 23:944–951.

Pacheco ABF, Guedes IA, Azevedo SM (2016). Is qPCR a reliable indicator of cyanotoxin risk in freshwater? Toxins. 8:172.

Padisák J, Adrian R (1999). Biovolumen und Biomasse. In: von Tümpling W, Friedrich G, editors: Methoden der biologischen Wasseruntersuchung 2 Biologische Gewässeruntersuchung. Jena: Gustav Fischer:334–368.

Pessi IS, Maalouf PDC, Laughinghouse HD, Baurain D, Wilmotte A (2016). On the use of high-throughput sequencing for the study of cyanobacterial diversity in Antarctic aquatic mats. J Phycol 52:356–368.

Pinto A, von Sperling E, Moreira R (2001). Chlorophyll-a determination via continuous measurement of plankton fluorescence:: methodology development. Water Res. 35:3977–3981.

Ramberg L (1988). A simple method of quantifying filamentous algae in microscope. Swiss J Hydrol. 50:189–192.

Ramos V, Moreira C, Vasconcelos V (2017a). DNA extraction from food supplements SOP 5.6. In: Kurmayer R, Sivonen K, Wilmotte A et al., editors: Molecular tools for the detection and quantification of toxigenic cyanobacteria. Chichester: John Wiley & Sons:153–156.

Ramos V, Moreira C, Vasconcelos V (2017b). PCR detection of microcystin biosynthesis genes from food supplements SOP 6.9. In: Kurmayer R, Sivonen K, Wilmotte A, Salmaso N, editors: Molecular Tools for the Detection and Quantification of Toxigenic Cyanobacteria. Chichester: John Wiley & Sons:199–203.

Rantala-Ylinen A, Känä S, Wang H, Rouhiainen L, Wahlsten M, Rizzi E et al. (2011). Anatoxin-a synthetase gene cluster of the cyanobacterium *Anabaena* sp. strain 37 and molecular methods to detect potential producers. Appl Environ Microbiol. 77:7271–7278.

Rantala-Ylinen A, Savela H, Sivonen K, Kurmayer R (2017). Quantitative PCR. In: Kurmayer R, Sivonen K, Wilmotte a, Salmaso N, editors: Molecular tools for the detection and quantification of toxigenic cyanobacteria. Chichester: John Wiley & Sons:205–210.

Rantala A, Fewer D, Hisbergues M, Rouhiainen L, Vaitomaa J, Börner T et al. (2004). Phylogenetic evidence for the early evolution of microcystin synthesis. Proc Natl Acad Sci USA. 101:568–573.

Rantala A, Rajaniemi-Wacklin P, Lyra C, Lepistö L, Rintala J, Mankiewicz-Boczek J et al. (2006). Detection of microcystin-producing cyanobacteria in Finnish lakes with genus-specific microcystin synthetase gene E (*mcyE*) PCR and associations with environmental factors. Appl Environ Microbiol. 72:6101–6110.

Rasmussen JP, Giglio S, Monis PT, Campbell RJ, Saint CP (2008). Development and field testing of a real-time PCR assay for cylindrospermopsin-producing cyanobacteria. J Appl Microbiol. 104:1503–1515.

Richards FA, Thompson TG (1952). The estimation and characterization of planktonic populations by pigment analysis. II. A spectrophotometric method for the estimation of plankton pigments. J Mar Res. 11:156–172.

Rinta-Kanto J, Ouellette A, Boyer G, Twiss M, Bridgeman T, Wilhelm S (2005). Quantification of toxic *Microcystis* spp. during the 2003 and 2004 blooms in western Lake Erie using quantitative real-time PCR. Environ Sci Technol. 39:4198–4205.

Ritchie RJ (2006). Consistent sets of spectrophotometric chlorophyll equations for acetone, methanol and ethanol solvents. Photosynth Res. 89:27–41.

Rott E (1981). Some results from phytoplankton counting intercalibrations. Aquat Sci. 43:34–62.

Rott E, Salmaso N, Hoehn E (2007). Quality control of Utermöhl-based phytoplankton counting and biovolume estimates—an easy task or a Gordian knot? Hydrobiologia. 578:141–146.

Rouhiainen L, Vakkilainen T, Siemer BL, Buikema W, Haselkorn R, Sivonen K (2004). Genes coding for hepatotoxic heptapeptides (microcystins) in the cyanobacterium *Anabaena* strain 90. Appl Environ Microbiol. 70:686–692.

Rounge TB, Rohrlack T, Nederbragt AJ, Kristensen T, Jakobsen KS (2009). A genome-wide analysis of nonribosomal peptide synthetase gene clusters and their peptides in a *Planktothrix rubescens* strain. BMC Genomics. 10:396.

Rueckert A, Wood SA, Cary SC (2007). Development and field assessment of a quantitative PCR for the detection and enumeration of the noxious bloom-former *Anabaena planktonica*. Limnol Oceanogr Methods. 5:474–483.

Saccà A (2016). A simple yet accurate method for the estimation of the biovolume of planktonic microorganisms. PLoS One. 11:e0151955.

Saccà A (2017). Methods for the estimation of the biovolume of microorganisms: A critical review. Limnol Oceanogr Methods. 15:337–348.

Saker ML, Jungblut AD, Neilan BA, Rawn DFK, Vasconcelos VM (2005). Detection of microcystin synthetase genes in health food supplements containing the freshwater cyanobacterium *Aphanizomenon flos-aquae*. Toxicon. 46:555–562.

Saker ML, Welker M, Vasconcelos VM (2007). Multiplex PCR for the detection toxigenic cyanobacteria in dietary supplements produced for human consumption. Appl Microbiol Biotechnol. 73:1136–1142.

Salmaso N, Bernard C, Humbert JF, Akçaalan R, Albay M, Ballot A et al. (2017). Basic guide to detection and monitoring of potentially toxic cyanobacteria. In: Meriluoto J, Spoof L, Codd GA et al., editors: Handbook of cyanobacterial monitoring and cyanotoxin analysis. Chichester: John Wiley & Sons. 46–69.

Sambrook J, Russell DW, editors (2001). Molecular cloning: a laboratory manual, third edition. Cold Spring Harbor (NY): Cold Spring Harbor Laboratory Press:2100 pp.

Sartory D, Grobbelaar J (1984). Extraction of chlorophyll a from freshwater phytoplankton for spectrophotometric analysis. Hydrobiologia 114:177–187.

Savela H, Harju K, Spoof L, Lindehoff E, Meriluoto J, Vehniäinen M et al. (2016). Quantity of the dinoflagellate *sxtA*4 gene and cell density correlates with paralytic shellfish toxin production in *Alexandrium ostenfeldii* blooms. Harmful Algae. 52:1–10.

Savela H, Spoof L, Perälä N, Preede M, Lamminmäki U, Nybom S et al. (2015). Detection of cyanobacterial sxt genes and paralytic shellfish toxins in freshwater lakes and brackish waters on Åland Islands, Finland. Harmful Algae. 46:1–10.

Sayers M, Fahnenstiel GL, Shuchman RA, Whitley M (2016). Cyanobacteria blooms in three eutrophic basins of the Great Lakes: a comparative analysis using satellite remote sensing. Int J Remote Sens. 37:4148–4171.

Schloss PD, Westcott SL, Ryabin T, Hall JR, Hartmann M, Hollister EB et al. (2009). Introducing mothur: open-source, platform-independent, community-supported software for describing and comparing microbial communities. Appl Environ Microbiol. 75:7537–7541.

Schober E, Werndl M, Laakso K, Korschineck I, Sivonen K, Kurmayer R (2007). Interlaboratory comparison of Taq Nuclease Assays for the quantification of the toxic cyanobacteria *Microcystis* sp. J Microbiol Meth. 69:122–128.

Selmeczy GB, Tapolczai K, Casper P, Krienitz L, Padisák J (2016). Spatial-and niche segregation of DCM-forming cyanobacteria in Lake Stechlin (Germany). Hydrobiologia. 764:229–240.

Shih PM, Wu D, Latifi A, Axen SD, Fewer DP, Talla E et al. (2013). Improving the coverage of the cyanobacterial phylum using diversity-driven genome sequencing. Proc Natl Acad Sci USA. 110:1053–1058.

Silva T, Giani A, Figueredo C, Viana P, Khac VT, Lemaire BJ et al. (2016). Comparison of cyanobacteria monitoring methods in a tropical reservoir by in vivo and in situ spectrofluorometry. Ecol Eng. 97:79–87.

Sivonen K, Börner T (2008). Bioactive compounds produced by cyanobacteria. In: Herrero A, Flores E, editors: The cyanobacteria: molecular biology, genomics and evolution. Norfolk: Caister Academic Press:159–197.

Speziale BJ, Schreiner SP, Giammatteo PA, Schindler JE (1984). Comparison of N, N-dimethylformamide, dimethyl sulfoxide, and acetone for extraction of phytoplankton chlorophyll. Can J Fish Aquat Sci. 41:1519–1522.

Srivastava A, Singh S, Ahn C-Y, Oh H-M, Asthana RK (2013). Monitoring approaches for a toxic cyanobacterial bloom. Environ Sci Technol. 47:8999–9013.

Stich H, Brinker A (2005). Less is better: uncorrected versus pheopigment-corrected photometric chlorophyll-a estimation. Arch Hydrobiol. 162:111–120.

Strickland JD, Parsons TR (1972). A practical handbook of seawater analysis. Ottawa: Fisheries Research Board Canada.

Stucken K, John U, Cembella A, Murillo AA, Soto-Liebe K, Fuentes-Valdés JJ et al. (2010). The smallest known genomes of multicellular and toxic cyanobacteria: comparison, minimal gene sets for linked traits and the evolutionary implications. PLoS One. 5:e9235.

Stüken A, Jakobsen KS (2010). The cylindrospermopsin gene cluster of *Aphanizomenon* sp strain 10E6: organization and recombination. Microbiology. 156:2438–2451.

Sukenik A, Quesada A, Salmaso N (2015). Global expansion of toxic and non-toxic cyanobacteria: effect on ecosystem functioning. Biodivers Conserv. 24:889–908.

Taton A, Grubisic S, Brambilla E, De Wit R, Wilmotte A (2003). Cyanobacterial diversity in natural and artificial microbial mats of Lake Fryxell (McMurdo Dry Valleys, Antarctica): a morphological and molecular approach. Appl Environ Microbiol. 69:5157–5169.

Te SH, Chen EY, Gin KY-H (2015). Comparison of quantitative PCR and droplet digital PCR multiplex assays for two genera of bloom-forming cyanobacteria, *Cylindrospermopsis* and *Microcystis*. Appl Environ Microbiol. 81:5203–5211.

Tillett D, Dittmann E, Erhard M, von Döhren H, Börner T, Neilan BA (2000). Structural organization of microcystin biosynthesis in *Microcystis aeruginosa* PCC7806: an integrated peptide-polyketide synthetase system. Chem Biol. 7:753–764.

Tillett D, Neilan BA (2000). Xanthogenate nucleic acid isolation from cultured and environmental cyanobacteria. J Phycol. 36:251–258.

Utermöhl H (1958). Zur Vervollkommnung der quantitativen Phytoplankton-Methodik. Internationale Vereinigung für Theoretische und Angewandte Limnologie: Mitteilungen. 9:1–38.

Vichi S, Lavorini P, Funari E, Scardala S, Testai E (2012). Contamination by *Microcystis* and microcystins of blue-green algae food supplements (BGAS) on the Italian market and possible risk for the exposed population. Food Chem Toxicol. 50:4493–4499.

Vuorio K, Lepistö L, Holopainen A-L (2007). Intercalibrations of freshwater phytoplankton analyses. Boreal Environ Res. 12:561–569.

Wang GQ, Lee ZP, Mishra DR, Ma RH (2016). Retrieving absorption coefficients of multiple phytoplankton pigments from hyperspectral remote sensing reflectance measured over cyanobacteria bloom waters. Limnol Oceanogr Methods. 14:432–447.

Wang S, Zhu L, Li Q, Li G, Li L, Song L et al. (2015). Distribution and population dynamics of potential anatoxin-a-producing cyanobacteria in Lake Dianchi, China. Harmful Algae. 48:63–68.

Wolanin A, Soppa MA, Bracher A (2016). Investigation of spectral band requirements for improving retrievals of phytoplankton functional types. Remote Sensing. 8:871.

Wood SA, Pochon X, Laroche O, von Ammon U, Adamson J, Zaiko A (2019). A comparison of droplet digital polymerase chain reaction (PCR), quantitative PCR and metabarcoding for species-specific detection in environmental DNA. Mol Ecol Res. 19:1407–1419.

Wood SA, Puddick J (2017). The abundance of toxic genotypes is a key contributor to anatoxin variability in *Phormidium*-dominated benthic mats. Marine Drugs. 15:307.

Xiao X, Sogge H, Lagesen K, Tooming-Klunderud A, Jakobsen KS, Rohrlack T (2014). Use of high throughput sequencing and light microscopy show contrasting results in a study of phytoplankton occurrence in a freshwater environment. PLoS One. 9:e106510.

Zamyadi A, Choo F, Newcombe G, Stuetz R, Henderson RK (2016). A review of monitoring technologies for real-time management of cyanobacteria: Recent advances and future direction. Trends Anal Chem. 85:83–96.

Zamyadi A, McQuaid N, Prévost M, Dorner S (2012). Monitoring of potentially toxic cyanobacteria using an online multi-probe in drinking water sources. J Environ Monitor. 14:579–588.

Zhu W, Tian YQ, Yu Q, Becker BL (2013). Using Hyperion imagery to monitor the spatial and temporal distribution of colored dissolved organic matter in estuarine and coastal regions. Remote Sens Environ. 134:342–354.

Zieger SE, Mistlberger Gn, Troi L, Lang A, Confalonieri F, Klimant I (2018). Compact and low-cost fluorescence based flow-through analyzer for early-stage classification of potentially toxic algae and *in situ* semiquantification. Environ Sci Technol. 52:7399–7408.

Ziegmann M, Abert M, Müller M, Frimmel FH (2010). Use of fluorescence fingerprints for the estimation of bloom formation and toxin production of *Microcystis aeruginosa*. Water Res. 44:195–204.

Zohary T, Shneor M, Hambright KD (2016). PlanktoMetrix – a computerized system to support microscope counts and measurements of plankton. Inland Waters. 6:131–135.

Chapter 14

Laboratory analysis of cyanobacterial toxins and bioassays

Linda A. Lawton, James S. Metcalf, Bojana Žegura, Ralf Junek, Martin Welker, Andrea Törökné, and Luděk Bláha

CONTENTS

INTRODUCTION AND GENERAL CONSIDERATIONS

Cyanobacterial toxins or cyanotoxins are a diverse group of compounds with differing chemistries; hence, a single analytical method can rarely be used to evaluate all potential compounds. The general steps required to detect, identify, quantify and monitor the different classes of toxins do follow a common approach (Figure 14.1).

Before committing to a major cyanotoxin sampling campaign, it is important to evaluate the information and level of detail required to make appropriate management decisions (see also Chapters 11 and 12). Planning the work therefore is best done by a group including the different aspects involved, for example, field samplers, laboratory support, analysts, as well as

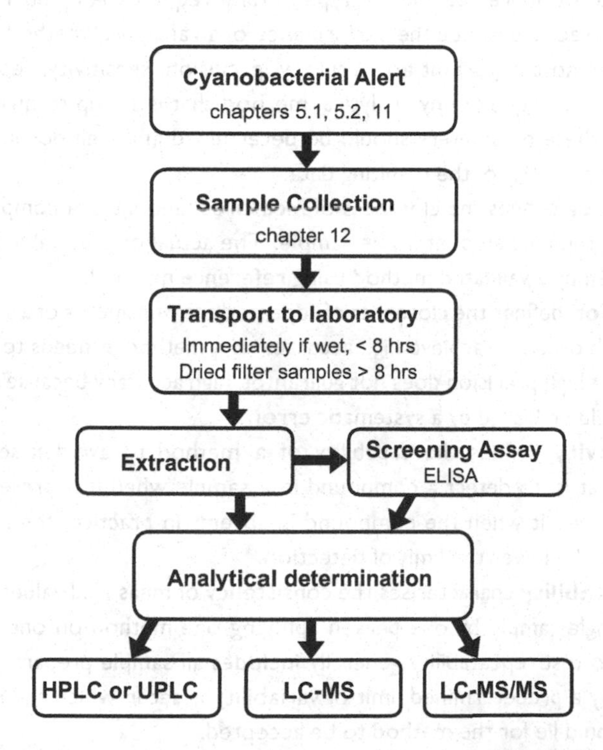

Figure 14.1 General scheme from cyanobacteria alert (see Chapter 5) to analytical determination of cyanotoxin concentrations.

managers leading the work, possibly in contact with the authority responsible for public health. This way a range of perspectives come together to design the most suitable and effective approach. An audit by others with expertise in the field (in-house or external) serves to ensure that all important areas have been considered and nothing essential is overlooked. Some of the key questions and topics which should be covered in planning and auditing are included in Checklist 14.1, and it is important to take time to make sure that all important requirements can be fulfilled.

The available methods for analysing cyanobacterial toxins are very diverse. The criteria for selecting the appropriate analytical method for a given monitoring programme include the consideration of reproducibility, sensitivity and selectivity (see Box 14.1) as well as factors such as the time to result, required training, capital investment and laboratory conditions needed, and consumable running costs per sample.

BOX 14.1: PARAMETERS OF ANALYTICAL PERFORMANCE

Before any analysis is done, it is important to understand the nature of the problem to be addressed and the type of data required. A number of terms are often used to describe the performance of an analytical method. Of these terms, the most important are accuracy, precision, sensitivity, repeatability and reproducibility. For any analytical method that is set up to quantify cyanotoxins, these parameters should be determined and well documented to support the validity of the resulting data.

Accuracy defines the closeness of measured amount of a compound in a sample to the true amount in the sample. The actual or true value has to be determined by a validated method using reference material.

Precision defines the closeness of repeated measurements of a single sample to each other. To achieve high accuracy of a method, it needs to be highly precise but high precision does not guarantee high accuracy because measurements could be biased by a systematic error.

Sensitivity defines the capability of a method to avoid false-negative results, that is, to detect a compound in a sample when it is present but to not give a result when the compound is absent. In practice, the higher the sensitivity, the lower the limit of detection.

Repeatability characterises the consistency of measured values obtained from a single sample by one person applying one method on one analytical system. To test repeatability generally includes all sample preparation steps, and usually a predetermined limit of variability is set in which the measured values should lie for the method to be accepted.

Reproducibility characterises the consistency of measured values obtained from a sample by different persons on different analytical systems. Reproducibility of a method that is applied in different laboratories is preferably validated with interlaboratory comparison tests, where a single sample of known toxin content is split and analysed by different laboratories.

Exact definitions of these and further terms related to analytical chemistry can be found at http://goldbook.iupac.org/index.html

Trained staff are needed, especially to operate complex analytical systems, primarily for the establishment of methods for analytes not yet analysed in the particular laboratory and their initial validation phase. For routine analyses, modern analytical systems generally offer some degree of automation that can be made use of once a method has been established.

As a rule of thumb, the most sensitive and specific methods tend to be the most demanding ones in terms of investment (capital, method optimisation and validation), required personnel training and running costs – but often with a substantial delay of result delivery of hours to days. On the other hand, methods that are less sensitive and selective may deliver results very fast. This can be essential in situations where an analytical result is needed to trigger immediate management actions to mitigate risks, for example, in Alert Level Frameworks (see Chapters 5.1 and 5.2).

CHECKLIST 14.1 CYANOTOXIN DETECTION

- Which toxin classes are to be analysed?
- From what type of samples (e.g., water, cells, tissue, sediments)?
- Does the detection have to be quantitative? If so, what detection limit is required?
- Do drinking-water samples need immediate quenching (e.g., sodium thiosulphate or ascorbic acid) to eliminate the continued action of oxidising agents like chlorine?
- What instrumentation and expertise is available (i) in-house or (ii) external?
- What capacity is available, that is, which number of samples can currently be analysed (including laboratory space, appropriate sample storage)?
- What training is needed – for sampling, for sample processing, for analysis and for data interpretation? How are the data going to be used and reported?

14.1 SAMPLE HANDLING, STORAGE AND SHIPPING

Following a sampling trip, the samples arriving in the laboratory need to be processed further for analysis or storage. Three aspects are important for sample handling and storage: safety, sample processing to ensure stability and traceability.

14.1.1 Safety

Laboratory staff handling samples potentially containing toxic cyanobacteria and cyanobacterial toxins are potentially exposed to health hazards (see also section 5.2), and appropriate protective measures need to be implemented. These measures will be based on two aspects: implementing general safety measures for hazardous material defined in national and international occupational health and safety guidelines, and an assessment of the risk of exposure to toxic cyanobacterial material potentially given in the specific procedures to be carried out. Any staff member handling potentially toxic cyanobacterial samples has to be accordingly trained and equipped with adequate protective equipment. Depending on the work to be done, this protective equipment will range from standard laboratory coats, gloves and safety glasses to – where there is a risk of inhalation exposure– breathing masks (Stewart et al., 2009). For water samples taken in monitoring programmes, the quantities of toxins are generally low, that is, in the low microgram per litre range, likely posing no risk of intoxication. Nonetheless, skin and eye contact, inhalation and ingestion have to be avoided by wearing an appropriate safety wear. Risks of exposure tend to be higher if larger quantities of bloom material are handled, for example, for toxin purification, with toxin quantities potentially in the low milligram range. The highest risk of exposure to toxic material is likely through the handling of dried bloom material, that is, exposure to dust, requiring the wearing of a breathing mask and/or the handling of sample material in an exhaust hood. Powdery freeze-dried bloom material, often statically charged, easily escapes containment, and this may cause a risk of exposure not only during laboratory work but also for cleaning staff.

14.1.2 Sample processing for storage

Samples that are not analysed immediately upon arrival at the laboratory need to be stored properly to avoid degradation of the cyanotoxins to be quantified (see Figure 12.3). Generally, cyanobacterial toxins are rather stable compounds, and storage at –20 °C largely prevents degradation. As guidance, the stability – or breakdown – as described in Chapter 2 may serve as a first orientation for conducting stability testing under the actual conditions, including all steps in transportation, storage and shipping. Since degradation mostly occurs through microbial activity, lower temperatures generally increase stability.

However, simply freezing entire samples may not be appropriate for practical and analytical reasons. For example, when extracellular and intracellular toxins are to be analysed separately, this requires prior separation of both fractions, which can no longer be done once samples have been frozen. Therefore, sample processing steps may be required that allow a reliable and efficient preservation of samples or portions of samples for later analyses.

If the toxin content in the particulate fraction is to be analysed, cyanobacterial cells and other particulate material ("seston") are best collected on a filter, thus concentrating the sample for space-efficient storage. One criterion for the choice of the type of filter (glass fibre, membranes of different types, cellulose) is the appropriate pore size to retain cyanobacteria. As most toxigenic taxa occur in colonies or filaments, pore sizes <2 µm are generally acceptable. Smaller pore sizes may slightly increase retention efficiency but at the cost of more rapid clogging of the filters – which is a factor that determines to a large extent the time needed to process samples. Further, it is important that the filter material be compatible with the downstream processing, in particular with the extraction procedure to avoid the release of any compounds that interfere with the downstream analysis during the extraction step in organic solvents; also, it must not be dissolved by these solvents. A protocol for the handling of such samples needs to be validated and tested with negative controls, that is, filters through which pure water has been passed.

For the filtration step, the water sample is well mixed immediately before measuring a volume to be filtered, for example, with a calibrated cylinder, because even within a very short time (i.e., minutes), buoyant cyanobacteria can float to the top (and other phytoplankton can settle), potentially leading to biased analytical results. Preferably, the total volume to be filtered is portioned into several smaller volumes to avoid increasingly longer filtration times on a gradually clogged filter. The volume to be filtered for an individual sample depends on the density of cells – that can hence be highly variable – and on the detection limit of the downstream analytical procedure. For the latter, the extraction and concentration steps need to be considered.

The filter should hold only residual moisture before it is frozen for storage or processed further for analysis, respectively, as discussed in the following section.

Filtrates are used for the analysis of dissolved (i.e., extracellular) cyanotoxins. To produce larger volumes of particle-free samples, it may be helpful to use a combination of two filters, for example, a glass fibre filter to retain larger particles and a second filter with smaller pores to remove small particles.

14.1.3 Sample storage and shipment

If filter samples are to be stored, they are folded loaded-face on loaded-face and placed in an appropriate container, for example, chemically inert reaction tubes or enveloped in aluminium foil and labelled correctly. For their long-term storage or shipment, it is advantageous to dry the loaded filters; otherwise, shipment on dry ice is recommended. Drying can be achieved by freeze drying, in a vacuum centrifuge or at moderate temperatures (<80 °C for microcystins and cylindrospermopsins) in a drying oven when lyophilisation is not available (Welker et al., 2005). For the latter, stability tests are recommended to check whether the drying process (that may last for several hours) causes any degradation. It is generally good practice to test the stability of the samples under the chosen storage conditions with a series of identical samples that are analysed at the different processing steps and after varying storage times. Performing the extraction procedure in the storage tube can allow safe handling, reduce processing time and minimise the risk of sample confusion.

For the storage of particle-free filtrate or extract, toxin adsorption to labware may be relevant. A few studies on this issue show that microcystin congeners differ in their tendency to adsorb to materials (Hyenstrand et al., 2001; Kamp et al., 2016; Altaner et al., 2017). These studies cover only some of the possible combinations of materials, solvents and toxins, and it is recommended to test the material used in an individual laboratory under the actual sample processing procedures to assess suitability. Once labware of a particular material and manufacturer has been found appropriate, it should not be changed without corresponding verification.

Shipping of samples requires consideration not only of stability but also of compliance to legal aspects, in particular declaration rules for transborder shipments (see Metcalf et al. (2006) for an overview).

14.1.4 Traceability

It is critically important to label all field samples arriving in the laboratory in a way that allows the results to be tracked back to the sampling site and sampling date at a later point in time. Similar considerations also apply for individual steps in sample processing, dilution of standards and quality control (QC) samples. For certified laboratories, the traceability of samples and materials is essential and generally follows guidelines such as ISO 9001 (ISO, 2015). Although a thorough quality management system according to ISO 9001 may not be necessary for non-certified laboratories, some principles can be implemented to ensure proper sample management.

As outlined in Chapter 11, a first important point is the labelling of individual samples. Sample names need to be unique, not only for a recent set of samples but also with respect to all samples that are expected in the future in a given laboratory and in cooperating laboratories. A naming system is therefore best defined prior to sampling campaigns and followed by the entire staff. Whichever system of sample naming is agreed upon, it is important to register all samples received in a laboratory in a central repository that is in turn backed up regularly. As barcode labelling and reading is becoming affordable, the introduction of these respective systems is now an option. Furthermore, labels need to be stable, that is, lastingly attached to sample containers under the actual storage conditions. Especially when samples are stored in the freezer, adhesive labels or markings need to be tested for resilience to repeated freeze–thaw cycles.

14.2 GENERIC METHODOLOGIES USED IN CYANOTOXIN ANALYSIS

14.2.1 Sample extraction for analysis

Most cyanotoxins are retained in the cell-bound fraction of samples taken from waterbodies (Chapter 2). Only when cyanobacterial populations experience cell lysis, for example, in surface scums exposed to high temperatures and light intensities, substantial amounts of toxins are released to the surrounding water. An exception is cylindrospermopsin that can be released from viable cell and found in large proportions in the cell-free fraction (section 2.2 Box 5.1 and Chapter 10). Further, extraction and sample clean-up is of critical importance for the analysis of cyanotoxins in foods to avoid both under- and overestimating concentrations (see section 5.3.4).

To make cell-bound cyanotoxins accessible to chemical analysis, they need to be extracted from the cells with an appropriate solvent. The solvent needs to be selected to efficiently extract toxins from cells in a few (maximum of three) extraction cycles and needs to be compatible with downstream analytical methods. Extraction procedures for individual classes of toxins will be discussed in the respective sections.

There is no single method available that is suitable for extracting all classes of cyanotoxins; hence, it may be necessary to collect several samples or subdivide samples prior to processing (see Chapter 12).

14.2.1.1 Solid-phase extraction (SPE)

While some immunoassay (ELISA) methods and LC-MS/MS (see below) may be sufficiently sensitive for monitoring in compliance with guideline values, trace analysis of cyanotoxins in water (i.e., concentrations below 1 µg/L) typically requires sample concentration using solid-phase extraction (SPE). This processing involves passing a known volume (typically 100–500 mL) of a (particle-free) water sample through a solid-phase cartridge to concentrate dissolved cyanotoxins. SPE requires a vacuum manifold system, PTFE (polytetrafluorethylen) connectors and tubing (for a minimum adsorption of analytes) and a vacuum pump. The equipment for SPE can be used for different classes of cyanotoxin; however, specific cartridges need to be used which retain the cyanotoxin of interest. A large selection of ready-to-use SPE cartridges is available with different sorbent materials and varying sorbent volumes. Sorbent materials primarily differ in the degree of hydrophobicity and have to be selected in correspondence with the cyanotoxin of interest, for example, C18 for microcystins or graphitised carbon for highly hydrophilic cylindrospermopsins.

Filtering water prior to SPE is essential to minimise clogging of the cartridges. Nonetheless, during times of blooms, the higher load of very small particles may significantly increase loading times. In this case, it is advised to note the volume of sample that has already passed and discontinue the sample loading. Once the water sample has passed through the cartridge, a washing step follows before the toxin adsorbed on the cartridge is eluted in a small volume (typically 3–5 times the sorbent bed volume) of solvent into a collection tube (Figure 14.2). This can be analysed directly or further concentrated by drying to enhance the detectability. It is tempting to load as large a volume of water sample as possible to allow the detection of low levels of cyanotoxin; however, this is not always appropriate. Larger volumes can take many hours to load onto the cartridge and matrix contaminants are also being concentrated; hence, larger sample volumes increase matrix interference. Loading samples onto SPE cartridges can be time-consuming so it is wise to determine the most appropriate volume of water. This can be done by processing some test samples of spiking water (preferably similar to the samples) and determining how long the process takes. Typically, a volume is chosen that takes no more than 3 h to pass through the SPE cartridge, allowing time for sample preparation, elution and processing before analysis.

It is useful to validate methods by spiking known amounts of the cyanotoxin of interest into water samples. Spiking should not be performed with high-purity water (e.g., Milli-Q) as this may lead to poor recoveries, and as this is in no way representative of the samples being processed. Using the typical raw or tap water to be tested is best (if tap water is chlorinated, using a chlorine-quenching agent, i.e., sodium thiosulphate or ascorbic acid) as this will provide a good indication of expected performance.

14.2.2 Enzyme-Linked Immunosorbent Assay (ELISA)

Immunoassays are based on the binding interaction between a highly specific antibody and the analytes of interest. The most common of these assays is the ELISA kit using antibodies raised to specific cyanotoxins. The toxins are detected by the modification in the colour reaction with the intensity of the colour being inversely related to the amount of toxin.

ELISAs can offer rapid results with a relatively low investment in capital equipment. As these assays do not identify specific cyanotoxin variants of a toxin class and give an indication of total toxin concentration – total

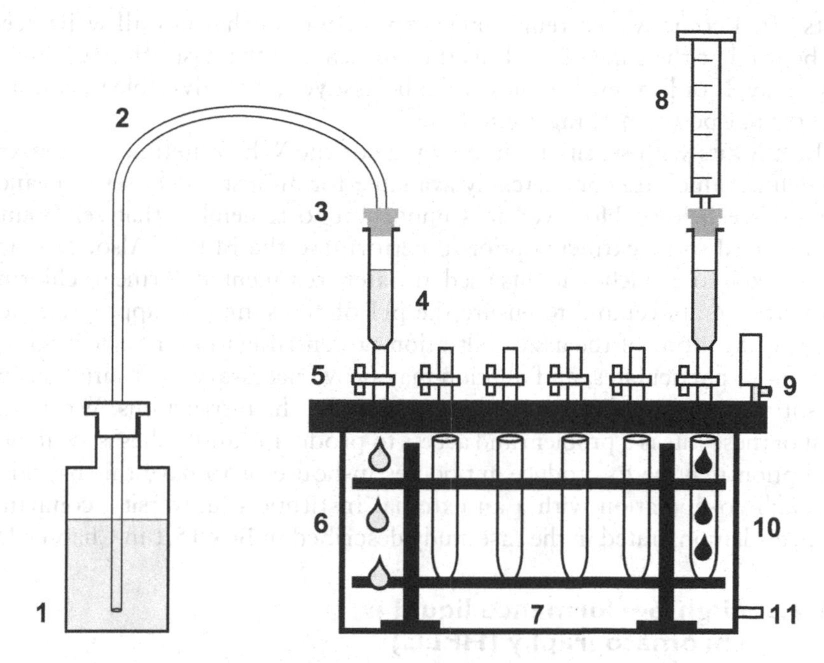

Figure 14.2 Setup for solid-phase extraction (SPE) for concentrating trace amounts of cyanotoxins dissolved in water. 1. Filtered water sample in stoppered glass bottle; 2. PTFE tubing carrying water under vacuum; 3. PTFE tubing; 4. SPE cartridge with sorbent; 5. PTFE stopcocks used to stop and start flow; 6. water flowing to waste while cyanotoxins are adsorbed on the cartridge; 7. vacuum manifold system with removable rack; 8: reservoir used to introduce solvents for conditioning and eluting the cartridge (syringe); 9. pressure gauge with needle valve; 10. concentrated sample eluted into a sample collection tube or vial; 11. vacuum line connection.

microcystins, for example – they are often used as a screening method. It is recommended to confirm toxin content and to routinely check for false negatives using instrumental methods (HPLC, LC/MS; Gaget et al., 2017). Where such methods are not available, periodic shipping of a few selected samples to a support laboratory elsewhere may be an option.

ELISA kits are very popular for a rapid, straightforward detection of most classes of cyanotoxins, although an individual kit is required for each class of cyanotoxin and even different kits may be necessary to cover the variants within one class. The kit-based formats provide a straightforward guidance on how to perform, calibrate and interpret the results. Multiple samples can be evaluated at one time, and results can typically be reported in less than a day. As with all biochemical test kits, care has to be taken with storage, since ambient or elevated temperature during prolonged transportation can reduce their reliability. The 96-well plate format allows samples to be read and quantified in a plate reader, facilitating the analysis of many samples and the calibration in a short space of time.

Kits often come with a removable strip format so that not all wells need to be used at the time of analysis thus increasing the cost-effectiveness of the assay. If only a few samples are to be assayed, it is advisable to confirm the format before making a purchase.

ELISA kits with sensitivity in the range of the WHO lifetime cyanotoxin guideline values are commercially available for almost all classes of cyanotoxins (see below). However, it is important to remember that cell-bound toxins need to be extracted prior to performing the ELISA. Also, care has to be taken to quench oxidants used in water treatment (chlorine or chlorine dioxide; see above) and to ensure the pH of the sample is appropriate for the specifications of the assay. Filtration or centrifugation may be required to remove particulates, and dilution may prove necessary to ensure that the quantification is in the approved range given in the instructions. Where the cost of these kits is a problem and access to producing antibodies is available, an option may be to produce antibodies in-house or to have this provided through co-operation with a an external institution (university, company, etc.), as demonstrated in the case study described in Box 15.1 in Chapter 15.

14.2.3 High-performance liquid chromatography (HPLC)

HPLC has become well established for the routine analysis of environmental pollutants. These systems consist of a solid-phase chromatography column through which analytes dissolved in liquid solvents are pumped and separated due to differences in the interaction of individual analytes with the solid phase. The flow then passes through a detector, for example, UV absorbance or fluorescence detectors, with the absorption proportional to the amount of analyte, with data collected on a computer. Most systems now include an autosampler to allow a set of samples to be loaded and automatically analysed. The number of samples that can be analysed in a given space of time depends primarily on the duration of a single sample run. For example, the run time for microcystins with a conventional HPLC is around 1 h per sample. Analyses with fewer target compounds, that is, less structural variants such as cylindrospermopsin or anatoxin-a, generally require shorter run times and hence allow a higher sample throughput. The separation of the analytes can be achieved by isocratic elution; this is when the solvent composition remains the same throughout the analysis. Isocratic elution is suitable for analyses that target only a few analytes and with a limited matrix interference, that is, with relatively low amounts of other, nontarget compounds. To allow for better separation of target analytes, gradient elution is commonly applied, where the proportions of the solvents change over the run time. This allows a wide range of analytes to be separated, such as multiple variants of microcystins. To ensure that analytes and contaminants are not carried over to the next sample, a washing step with 100% solvent is often included in the analytical run.

The most common detector used on HPLC systems is the photodiode array (PDA), which will provide an adsorption spectrum (200–600 nm) for the compounds being analysed. This is useful for the analysis of cyanotoxins as many of them have characteristic UV absorption spectra (Figure 14.4 and 14.5), thus providing an indication even of cyanotoxins in the sample for which no standard reference material is available (see Box 14.2).

BOX 14.2: REFERENCE MATERIAL FOR CALIBRATION

According to IUPAC, reference material is defined as "a substance or mixture of substances, the composition of which is known within specified limits [...] to be used for the calibration of an apparatus". For cyanobacterial toxins to be used as reference material for establishing a calibration curve for the quantification of these cyanotoxins, two criteria need to be fulfilled:

1. Purity defines the share of an individual compound of the total material. Purity is generally expressed in gravimetric percent that should be at least 95% in reference material.
2. Amount is generally defined in gravimetric units, and ideally with the specified limits, that is, a range of amount that should be as narrow as possible.

In this sense, not all cyanotoxins that are commercially available are reference materials. Hence, these compounds cannot be used directly to establish calibration curves. In particular, the nominal amount in a vial may deviate considerably from the true amount. In consequence, this means that a calibration curve established with such a "standard" would introduce a systematic error to all subsequent analytical quantifications.

The true amount hence needs to be determined. This can be done either by weighing with a sufficiently precise and sensitive balance or by spectrophotometric analysis (ISO, 2005). For the latter, wavelength-specific extinction coefficients need to be available, which is the case for some but not all cyanotoxin variants. Extinction coefficients are specific for solvents and temperature; that is, a compound dissolved in water cannot be quantified by using an extinction coefficient established for the compound dissolved in methanol.

For cyanobacterial toxins sold as certified reference material, the purity as well as the amount is well defined, and its can be used directly as standard for calibration.

The typical capital investment required for an HPLC is around $30,000 USD with relatively modest costs for maintenance, including replacement UV lamps and columns.

Training of staff and adoption of a standard protocol is easily achievable, while interpretation of samples and cyanotoxin identification (especially unknowns from their spectra alone) requires more time to develop confidence. This applies equally to UPLC (ultra-performance liquid chromatography) discussed below.

14.2.3.1 Ultra-performance liquid chromatography (UPLC)

UPLC offers a considerable advantage over conventional HPLC as it allows very rapid separation of analytes (run times of around 10 min) and a greatly reduced solvent usage, typically 0.3 mL/min compared to 1 mL/min for conventional HPLC systems. For example, these systems can achieve the separation of multiple microcystin variants in run times little over 10 min, thus providing high throughput of samples, substantial saving and results on the same day for samples with short extraction times (e.g., bloom material and filter discs with cells). For samples requiring longer extraction (SPE of water samples or tissue samples), it can yield results within 24 h.

UPLC systems are highly reliable with the response factor for microcystins in the UV detector changing little over time. The capital investment should typically be around $50 000 USD for the complete system at relatively low levels of maintenance, with the main component that needs replacement being the UV lamp (it is useful to have a spare in stock).

14.2.4 Liquid chromatography with mass spectrometry (LC-MS)

The addition of a mass detector to chromatography systems makes a very powerful tool for the analysis of cyanotoxins. Mass spectra can provide an indication of the elemental composition and structure of an analyte along with determining the quantity of analytes for which reference materials are available with high sensitivity (see Box 14.2). A range of differing systems is available, and very careful consideration is required to determine which fulfils analytical requirements and is within the budget available. Different ion sources are available with positive electrospray ionisation (ESI) most commonly used in the analysis of cyanotoxins. The type of mass analyser (Caixach et al., 2017) also varies and can have a significant impact on cost and the data obtained; hence, it is essential that background evaluation is carried out to ensure the system suits the needs defined during planning (see section 14.1.1). In general, LC-MS will provide data relating to chromatographic retention times, the parent ion masses and fragmentation patterns for each compound as they are eluted. More complex LC-MS/MS systems combine a series of more than one mass detector (e.g., a triple quadrupole mass detector). As analyte ions pass through mass analysers, the former allow the selection of an analyte based on parent ion mass, while the latter allow the selective detection of fragment ions. This makes LC-MS/MS a highly specific analytical technique.

Robust protocols are required for LC-MS/MS as the signal from the MS can be either enhanced or suppressed by matrix interference (salts, organics, etc.). Furthermore, the response (i.e., signal strength relative to analyte amount) can drift over a relatively short time, necessitating a regular and frequent calibration. A routine maintenance protocol is advisable with the interval for cone cleaning determined for different sample matrixes; for example, in studies for the analysis of mussel tissue, cone cleaning was required after 40 samples (Waack, 2017). This interval was determined by spiking an extracted sample and then carrying out repeated, identical sample injections and determining after how many samples the reliability of the detection and quantification diminished.

While LC-MS provides very powerful sensitivity and detection capabilities, the more advanced systems (LC-MS/MS) require a capital investment of around $500 000 USD and an annual running cost of $20–40 000. Furthermore, a high level of staff training is required to use, interpret and maintain these systems, but once established, they provide unrivalled analytical capabilities. It is advisable, where possible, to see one or several systems in operation and have an opportunity to analyse specific samples from the area to be monitored prior to committing to this significant capital investment.

14.2.5 Selecting an analytical system

The lack of suitable analytical equipment is typically a barrier to monitoring cyanotoxins, and a strong case is often made for capital investment. The influence of current scientific publications frequently draws attention to the significant capabilities of very advanced instrumentation. However, while these systems provide impressive capability, a robust evaluation of the analytical requirement, running costs and infrastructure should be made to inform purchasing decisions. Checklist 14.2 provides some key points to discuss both in-house and with those who have recently invested in cyanotoxin analysis before making decisions. In particular, it is advantageous to develop a regional network, sharing expertise and resources, for example, through a regional centre of competency. This may lead to a decision to use simpler techniques such as ELISA while validating the results periodically by having a small set of samples analysed with advanced techniques elsewhere. Where training is required, it is often more efficient to invite an expert to provide an in-house workshop as this ensures analyses are operational and staff develops confidence using the in-house system. (Note: small grants are often available for this, e.g., through international exchanges and workshop funding.) A further benefit of this approach is continued support from experienced international collaborators ensuring ongoing development of monitoring programmes. Support provided by the system's vendor generally is charged for. This should be considered in the budget for investment and running costs.

CHECKLIST 14.2 EVALUATING INSTRUMENTAL ANALYTICAL REQUIREMENTS AND SUITABILITY

- What information is needed? Which class of toxins will be the main focus? Is the main target monitoring compliance of cyanotoxin guideline levels?
- Check the cost of consumables, for example, vials, columns, SPE cartridges and solvents in relation to the number of samples expected over time. In many countries, solvents and even high-purity water can be prohibitively expensive (the benefits of ultra-performance liquid chromatography (UPLC) are low flow rates and short runtimes requiring little solvent).
- What are the costs of waste solvent disposal required by environmental legislation?
- Is the infrastructure appropriate? This includes a stable power supply as fluctuating or intermittent power can rapidly destroy equipment. If not, what are the costs of installing effective power surge protection such as uninterruptible power supply (UPS) systems?
- Can room temperature be kept within the range needed by the instruments and analyses (results can be affected by high or fluctuating temperature, so air conditioning is often required)?
- Do all purchases include installation and initial training, ensuring that there are available engineers in the area?
- What is the cost of a service contract, and is it essential?
- For planning to purchase a LC/MS (it requires a nitrogen generator and cannot run efficiently on regular laboratory gas cylinders), ensure that the contract includes either annual service – that is, the cost of service engineers visits – or, if you have technical expertise to carry out the service, the purchase of a service kit.
- Talk to users of different instrument manufacturers regarding their experience of service and support especially in your location.
- Consider the benefits of partnering with others rather than buying own equipment.

14.3 QUANTIFICATION OF MICROCYSTINS AND NODULARINS

Of all the cyanotoxins, most experience exists with the methodology for the extraction and detection of microcystins. Furthermore, due to its chemical similarity, many of the methods for microcystins will also readily detect nodularin (Lawton et al., 1994b); hence, it will be included in this section. In general, the term "microcystin" will be taken to refer to both these related classes of toxins unless the differentiation is required.

Methods range in complexity and sophistication, spanning the well-established "tried and tested" approaches through to preliminary research findings on novel detection strategies. While many of these novel methods offer exciting opportunities for the future, this chapter focuses on a few of the most relevant approaches for establishing routine methods suitable for the more widely available resources and common requirements.

14.3.1 Extraction methods for microcystins and nodularins

14.3.1.1 Cyanobacterial cells

All cell/bloom samples will require extraction as these toxins tend to be retained inside healthy cells. Many extraction protocols for microcystins have been described (e.g., various solvent combinations, cycles of freeze/thawing, sonication, freeze drying, including combined methods). Among these solvent combinations, aqueous methanol (typically 50–80%; (Barco et al., 2005)) has proven to be very effective for extracting microcystins in face of their wide range of polarities. This solvent can be used for extracting cell pellets once a sample has been centrifuged (and the supernatant discarded or assayed for extracellular microcystin) as well as for extracting cells concentrated on filters. Depending on the volume of cells, around 90% recovery of microcystins (Barco et al., 2005) can be achieved with the first extraction. Often this is sufficient, as this has to be balanced against the further time required for processing a second extraction, as this will typically yield less than 10% of the total microcystin; also if the two extractions are combined, this reduces the detection limit due to the additional volume of solvent used in the second extraction. Extraction time of around 1 h is sufficient for good recovery. With the increased availability of dispersive extractor systems (automated vortexers that shake samples vigorously at defined speeds and timed duration), however, extraction can be achieved in just a few minutes and with high reproducibility. Where samples are extracted in centrifuge tubes (typically 1.5-mL microfuge), these can be spun and the supernatant then directly analysed using instrumental methods.

When designing an extraction protocol, it is good to keep it as simple as possible as this will limit error and also potential workplace exposure to microcystins: for example, freeze drying is sometimes reported as a step during sample preparation if a specific dry weight of cells is to be determined, but this can produce powders that are difficult to contain and prone to static charge. Other methods also reported the use of a sonicator probe which may cause cross-contamination, but also produce aerosols.

The use of organic solvents (e.g., methanol) is not compatible with biochemical assays such as ELISA and enzyme inhibition tests. Some ELISA kit manufacturers provide a cell lysis kit, while other analysts have advocated aqueous extraction or dilution to limit the concentration of solvent: for example, a 1 in 10 dilution of a 50% aqueous methanol extract may be tolerated but should be checked with controls for the specific kit used. Since microcystins demonstrate high temperature stability, a brief exposure (5 min) of a small sample (e.g., 1 mL) to about 80 °C in a water bath followed by centrifugation (13 000×g; microfuge) can result in simple solvent-free extraction (Metcalf & Codd, 2000). Extracts can then be diluted in water or buffer as required.

Similarly, high organic solvent content in extracts to be analysed by chromatographic systems needs to be tested for compatibility, in particular when gradient elution is applied that generally starts with hydrophilic conditions.

14.3.1.2 Water samples

Some very sensitive methods (e.g., LC-MS/MS) may be able to detect microcystins at environmental concentrations. However, even then it may be desirable to carry out solid phase extraction (SPE) to limit matrix effects.

The most commonly used SPE material is end-capped C18 cartridges, which have demonstrated high recovery and reliability. Some users prefer newer resins (e.g., polymeric phases), which are good where MS is the detector of choice; however, the high recovery of polar compounds by these cartridges can interfere with the more polar microcystins (e.g., microcystin-RR) if detection is with photodiode array (PDA). Several published methods provide a good detail on establishing SPE extraction of microcystins (Lawton et al., 1994b; Triantis et al., 2017c).

Some researchers have developed online sample concentration for fully automated extraction and analysis of microcystins. This is typically an advanced option including LC/MS(MS) and a quite specialised approach; however, it may be desirable particularly for laboratories that need a high throughput, such as those of public authorities monitoring compliance to regulations or of drinking-water suppliers.

Recoveries are best if sample handling is limited, processing time is kept to a minimum and samples are analysed immediately or stored at –20 °C when this is not possible. There is some evidence that samples may change when stored for longer periods of time even at –20 °C, but further studies are required to clarify the extent of this problem. If samples are stored, they should be vortexed if a subsample is to be removed after storage.

14.3.1.3 Tissue samples

It is becoming increasingly important to evaluate microcystins in more complex matrixes such as animals that have become intoxicated, fish and aquaculture products that may be contaminated or even plant materials. Much work is still required to fully understand the efficiency of different extraction and toxin recovery protocols. This is particularly challenging for microcystins and nodularins as they are known to bind to proteins; furthermore, microcystins, in particular, can bind covalently to certain protein phosphatases in living cells. Further, the recovery of standards spiked to the material to be tested will only represent unbound toxin recovery efficiency.

While a range of processing strategies with varying degrees of complexity have been used, all of these strategies need to be tested and tailored to the specific requirements of the material to be studied. Simple blending of fresh tissue (mussels) followed by a single aqueous methanolic (80% methanol) extraction was found to give good recoveries in the range of 61–97% for 11 microcystins and nodularin (Turner et al., 2018). Very poor recovery was observed for hydrophobic microcystins when either the samples were acidified or water alone was used. The solvent extracts can be directly analysed by instrumental systems. In contrast, for biochemical tests (ELISA or protein phosphatase inhibition), samples will need to be dried to remove the solvent or sufficiently diluted with water or buffer.

Due to the difficulties in detecting bound microcystin, a method was developed which is designed to cleave part of the microcystin at the first double bond of the ADDA moiety liberating 2-methyl-3-methoxy-4-phenylbutyric acid (MMPB; see Figure 14.3). The assumption behind this approach is that one molecule of MMPB is liberated for each molecule of microcystin, hence predicting the total microcystin content. An oxidation step is used to liberate the MMPB fragment from the parent microcystin, which is assumed to be simpler than digesting the microcystin bound to protein. While this

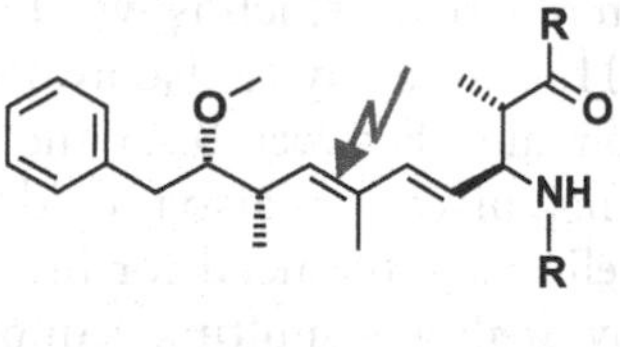

Figure 14.3 ADDA moiety of microcystins and nodularins with an indication of the site of MMPB cleavage. For the full structure, see section 2.1.

method has been used in a range of studies, it is very difficult to determine the degree of sample recovery as spiking will only represent free toxin. Most reported studies currently use MS detection of MMPB (*m/z* 208); however, this mass is not unique to this oxidation product (ChemSpider shows >6300 compound with this or very similar mass). Others have augmented the method to search for a product ion at 131, which again may not provide confident detection. However, Foss and Aubel (2015) have successfully used the MMPB method in comparison with the ADDA-ELISA, indicating good agreement.

In summary, the detection of microcystins and nodularins in tissue is important for assessing their possible role in animal poisoning or occurrence in food (fish, shellfish, vegetables, etc.), and while no current method will recover the total amount of microcystins, aqueous methanol extraction will give a good indication of whether microcystin is present.

14.3.2 Quantification of microcystins and nodularins by biochemical methods

14.3.2.1 Quantification by protein phosphatase inhibition assay

Microcystins and nodularins are known to be potent inhibitors of protein phosphatases PP1 and PP2A. This activity is central to their toxicity, and hence, detection of inhibition also indicates the potential biological activity of a sample. The assay can be performed relatively easily where the facilities are available for biochemical work. All the reagents and enzymes can be purchased for the colorimetric assay, which detects the enzymatic hydrolysis of the substrate (*p*-nitrophenyl phosphate) that liberates the coloured product *p*-nitrophenol (detected at 405 nm). The assay can be performed in a microtitre plate with the temperature, mixing and timing controlled by the plate reader and relatively straightforward protocol, typically resulting in good reproducibility.

Some challenges can arise if the sample inhibits the enzyme (depending on pH, solvents or other contaminants) or if it contains background colour; however, this rarely occurs as the enzyme assay is highly sensitive, allowing a significant dilution of, for example, a cell extract. Detection limits as low as 0.0039 µg/L have been reported, which is well below the WHO guideline value (Sassolas et al., 2011). This assay is also available as a commercial kit, which has a quantification range between 0.25 and 2.5 µg/L and an analysis time of only 30 min. The manufacturers also provide a tube format that could be used in the field and eliminate the need for investment in a plate reader, although users repeatedly analysing multiple samples will benefit from the multiwell plate format as this can be automatically analysed and is more practical for multiple samples and calibration points.

14.3.2.2 Immunoassays for microcystin and nodularin detection

The ADDA-based ELISA is particularly popular as it has been designed to detect the ADDA moiety, which is both very specific to these toxins and present in all variants, regardless of other chemical diversity. Sensitivity is reported as around 0.1 µg/L with the assay time of 2.5 h. There are a range of ELISA kits (usually with antibodies raised to microcystin-LR), and it is worthwhile checking which would be most appropriate for use in a specific situation. Considerations may be cost, format, location of suppliers and experience of other users in the vicinity. ELISA has been used for a range of samples demonstrating a good cross-reactivity for a number of variants and different sample matrices (Heussner et al., 2014); however, for tissue samples, sample processing needs a careful consideration as some studies have reported false positives and concentrations which are greatly in excess of the levels detected by LC-MS/MS (Brown et al., 2018).

14.3.3 Instrumental analytical methods for microcystins and nodularins

While a range of analytical approaches has been explored over the past 30 years, the central method of choice revolves around liquid chromatography (LC). Microcystins and nodularins can be separated very readily on C18 reverse-phase columns, although some closely eluting variants (e.g., microcystin-LR and desmethyl-microcystin-LR) require a careful column selection. Different methods of detection have also been evaluated, with a general consensus on UV detection and/or mass spectrometry, including MS/MS.

14.3.3.1 Analysis of microcystins and nodularins by HPLC-PDA

High-performance liquid chromatography with photodiode array (HPLC-PDA) provides an accessible, robust method for the detection and quantification of all microcystins by virtue of their distinct UV absorption spectra (Lawton et al., 1994b). Even in the absence of standards for every microcystin, confident quantification of total microcystins can be achieved. Most microcystins have very similar absorption spectra (although the overall characteristics between 200 and 300 nm can vary with concentration) with a maximum at 238 nm from the conjugated double bond in the ADDA moiety (Figure 14.4). The exception to this are microcystins that contain the variable amino acid tryptophan (e.g., microcystin-LW), which have an absorption maximum at 223 nm, and those that contain the variable amino acid tyrosine (e.g., microcystin-YR), which have an absorption maximum at 231 nm (Figure 14.4). A simple approach is to use microcystin-LR calibration for total microcystin quantification, assuming a similar molar

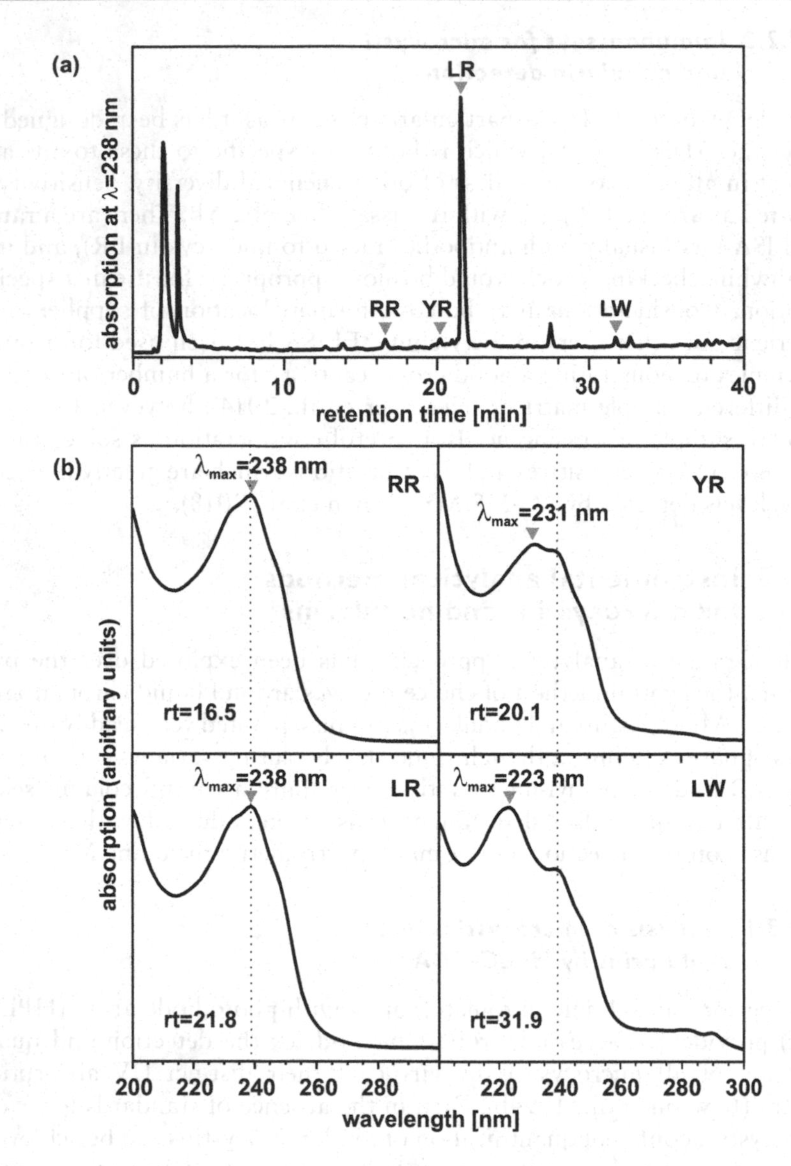

Figure 14.4 HPLC-PDA chromatogram of a bloom sample dominated by *Microcystis* sp. (a) and UV-absorption spectra of selected microcystin peaks (b). The individual variants are indicated by the standard two-letter code. rt: chromatographic retention time. The dotted line in the spectral plots indicates the wavelength $\lambda = 238$ nm; all microcystins show an absorption band at this wavelength that results in a shoulder in the spectra of variants containing tyrosine (Y) and tryptophan (W). A second absorption band at $\lambda = 246$ nm can be seen as shoulder in spectra of all microcystins. Both absorption bands are related to the conjugated double bond of the conserved ADDA moiety. For analytical details, see Welker et al. (2003).

absorption coefficient at 238 nm, that is, a similar response in PDA detection. An advantage of HPLC-PDA analysis is the fact that microcystin variants for which no reference material is available can be recognised based on a peak's absorption spectrum. Verification can be achieved by collecting the eluting peak and performing an offline analysis, for example, by MALDI-TOF MS (Welker et al., 2002b). Beyond this, it is preferable to set up the method with a range of microcystin variants of differing polarities, spanning from early to late retention, to provide confidence that a wide range of microcystins can be detected should they occur in samples.

The typical analytical set-up will be a gradient separation using high-purity water plus trifluoroacetic acid (TFA; 0.05%) and acetonitrile with TFA (0.05%). A gradient from 30% to 70% acetonitrile is usually required to separate all microcystins, and a rapid wash to 100% will eliminate carryover between runs.

14.3.3.2 Analysis of microcystins and nodularins by LC-MS(MS)

Both LC-MS and LC-MS/MS are powerful instruments for the analysis of microcystins. For developing a new method for microcystin analysis by LC-MS/MS, *the Handbook of Cyanobacterial Monitoring and Cyanotoxin Analysis* (Meriluoto et al., 2017) provides good guidance and a number of standard operating procedures (SOPs). Different approaches are possible, including the detection only of microcystins for which standards are available, using selected reaction monitoring (SRM), which is very sensitive and accurate (Turner et al. 2018). The drawback of this approach, however, is that a significant proportion of the microcystins in a sample could go unreported – those variants for which no calibration has been established based on standards. For example, USEPA Method 544 (Shoemaker et al., 2015) is limited to six microcystins, while other published methods have extended this to over 10. Birbeck et al. (2019) reported 40% of samples had more than 20% of their total microcystin variants not detected by the USEPA Method 544 and a number of these variants were the dominant microcystins. It is therefore advised to consider the taxonomic composition of the sample that allows a tentative prediction of the structural variants to be present. A way forward for routine monitoring is a detailed initial analysis to identify the spectrum of microcystins, and as long as bloom composition stays stable, such SRM can be a robust and sensitive approach, provided periodic checks are carried out to confirm that the overall microcystin profile is still covered by the standards used.

If microcystin variants are to be quantified for which no purified quantitative standards are available, estimates on the basis of their retention characteristics, mass and fragmentation pattern can serve for an initial assessment. However, as detector response intensities vary substantially between individual variants, a quantitation can be at best tentative; for example, MC-RR has a several-fold higher response than MC-LR, probably due to a higher ionisation efficiency (Krause et al., 1999). A valuable approach is to add a PDA detector to the system along with the MS detection: this will greatly enhance confident detection and in particular quantification of microcystins for which no specific calibration could be established. Since the PDA is robust and quantification varies little over time, it provides an excellent confirmation and quantification.

14.4 QUANTIFICATION OF CYLINDROSPERMOPSINS

Cylindrospermopsin (CYN), including its small number of variants, is highly polar due to its zwitterionic nature, and hence, is readily soluble in water. Unlike most of the other cyanotoxins, it is often found in significant concentrations outside the cell as well as within the cell (section 2.2). It appears to be relatively stable and to some extent resistant to a rapid biodegradation. Detection appears to be limited to either ELISA or chromatography (with photodiode array (PDA) and/or MS).

14.4.1 Extraction of cylindrospermopsins

14.4.1.1 Cyanobacterial cells

Extrtaction of cylindrospermopsins from dried cells can be simply achieved in water (Welker et al., 2002a). A known amount of freeze-dried cells can be weighted into a microcentrifuge tube and extracted with 1 mL of water added by vortexing intermittently for 1 h or placing in a dispersive extractor for 2 min at a full speed. The sample should then be centrifuged (13 000×g), and the supernatant can be directly analysed by either ELISA or chromatography. With fresh cells, a similar protocol can be followed by first centrifuging the fresh sample and retaining the supernatant for analysis. The cell pellet can then be extracted in 50% aqueous methanol, although care has to be taken either to remove the methanol by drying the sample or to dilute it, since the methanol will interfere with both ELISA and peak shape in chromatography (Metcalf et al., 2002a).

14.4.1.2 Water samples

Low concentrations of cylindrospermopsins in water will require sample concentration by solid phase extraction (SPE; Triantis et al., 2017a). Due to the polarity of the molecule, it is poorly retained by C18 and other solid media typically used for water analysis. Good recoveries can be achieved by the specialised cartridges such as graphitised carbon or polymeric resins. The typical protocol requires the filtration of a water sample (the filter should be extracted for cell-bound cylindrospermopsins) and then passing the sample through the preconditioned cartridge. Recovery can be significantly affected by the loading speed so this should be carefully controlled and optimised for the protocol being used. Cylindrospermopsins are eluted from the cartridge with methanol. It is useful to spike some samples of the water to be analysed as well as tap water (quenching any oxidant such as chlorine) to both become familiarised with the process and to define a standard operating procedure (SOP) within the laboratory.

14.4.1.3 Tissue and urine samples

Most accounts of studies investigating the localisation of cylindrospermopsins in experimental animals have indicated that unaltered cylindrospermopsins can be excreted in the urine (Norris et al., 2001); hence, methods to detect it in this matrix can be useful where human exposure is suspected. This has been successfully achieved through salt removal and SPE (carbograph or other hydrophobic analyte recovery solid-phase) clean-up and concentration (Foss & Aubel, 2013). Similarly, cylindrospermopsin has been recovered from serum samples although in these samples the focus is on protein precipitation with solvent (methanol) prior to SPE.

Good recovery of cylindrospermopsin from tissue (e.g., fish, mussels and vegetables) has been shown in a limited number of studies (Prieto et al., 2018) using aqueous solvents (typically methanol, aqueous methanol or acetic acid), although these methods may require further testing to determine the optimum protocol. Depending on the matrix and concentrations, direct analysis without SPE may give satisfactory results.

14.4.2 Quantification of cylindrospermopsins by ELISA

Cylindrospermopsin is a protein synthesis inhibitor, and as such, biological assays can be relatively slow and nonspecific. Therefore, the favoured assay is the cylindrospermopsin-specific ELISA kit. Several ELISA kits are commercially available for cylindrospermopsin with detection limits well below 1 µg/L, and these kits can be used directly on water samples. As the proportion of extracellular to cell-bound cylindrospermopsin can vary significantly, it is important to test for both the cell bound and free toxin. Use of kits will require relatively modest investment of a plate reader and the expense of the purchase of the kits. Full instructions for performing the assay, calibration and validation are provided with each kit. When establishing the use of ELISA, matching results with HPLC for selected samples and matrices will be valuable for determining the level of confidence as false positives have been shown with low-positive concentrations (Metcalf et al., 2017).

14.4.3 Instrumental analytical methods for cylindrospermopsins

As cylindrospermopsin is a very polar analyte, it is poorly retained on many C18 columns. Some columns have become available, which are better suited to the retention and separation of cylindrospermopsin and its variants (e.g., polar retention C18, graphitised carbon); therefore, it is advisable to select a column that is specifically designed for highly polar compounds (de la Cruz et al., 2013).

14.4.3.1 Analysis of cylindrospermopsins by HPLC-PDA

Cylindrospermopsins have a characteristic UV spectrum with an absorption maximum at 262 nm (Figure 14.5). This spectrum can be used to distinguish cylindrospermopsin from other peaks on the chromatogram in a way similar to that for microcystins. Chromatography is either carried out isocratically

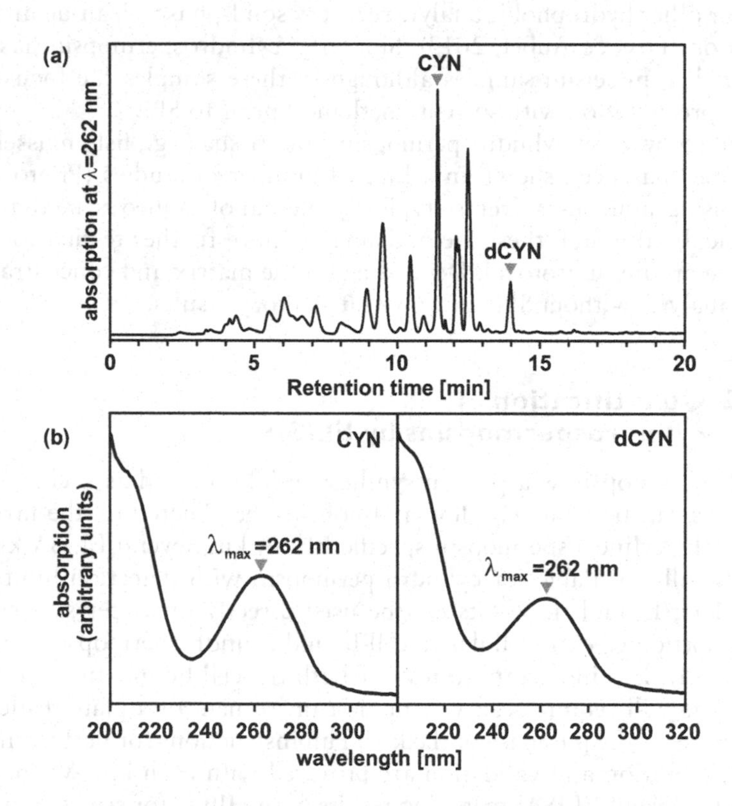

Figure 14.5 HPLC-PDA chromatogram of a bloom sample dominated by *Cylindrospermopsis* sp. (a) and absorption spectra of the two cylindrospermopsin variants (b), cylindrospermopsin (CYN) and 7-deoxy-cylindrospermosin (dCYN), showing an absorption maximum at $\lambda = 262$ nm. For analytical details, see Welker et al. (2002a).

(5% organic solvent, methanol or acetonitrile with 95% water) or using a slow gradient (e.g., from 0% to 10% methanol) with 100% solvent wash, which limits carryover of other contaminant peaks especially from crude extracts (e.g., cells, tissue). Extracts can be analysed directly, while water samples will require SPE-concentration. Where samples are in a high proportion of methanol, this can affect chromatography. It is therefore advisable to either dry and resuspend the extract in water or dilute in water (e.g., 1 in 10, if concentrations in relation to the detection limit are sufficiently high). Using a guard column can eliminate the negative impact of methanol on the chromatography.

Hydrophilic interaction liquid chromatography (HILIC) columns have also been evaluated for the analysis of cylindrospermopsin separation due to their suitability for highly polar compounds. These columns tend to be less robust and may not separate cylindrospermopsin as well as the high-polarity C18 phases; however, they are continually improving and increasingly becoming the column of choice for the polar cyanotoxins (cylindrospermopsin, anatoxin-a and saxitoxin), allowing the analysis of these cyanotoxins together in one run (Haddad et al., 2019).

14.4.3.2 Analysis of cylindrospermopsins by LC-MS(MS)

Cylindrospermopsin is readily detected by mass spectrometry using chromatography conditions similar to those for HPLC-PDA. Electrospray in positive ionisation mode yields the parent ion with *m/z* 416 and product ions with *m/z* 194, 176, 336 and 274. Selected reaction monitoring (SRM) can provide highly specific detection (Triantis et al., 2017a). Detection of cylindrospermopsin in drinking-water by this method with prior SPE concentration gave good recoveries at 0.01 µg/L (67%) and 0.1 µg/L (85%). Since there are only few other cylindrospermopsin variants, it is much less likely that MS detection will miss structural variants as in the case of microcystins. US EPA Method 545 based on LC-MS/MS has a minimum reporting level of 0.06 µg/L for finished drinking-water (US EPA, 2015); for ambient freshwaters, Shoemaker and Dietrich (2017) give a minimum reporting level of 0.23 µg/L. The analysis of cylindrospermopsin along with other more polar cyanotoxins (deoxycylindrospermopsin, anatoxin-a and saxitoxin) has been successfully achieved using HILIC-MS, demonstrating a robust detection of these toxins in cultured samples and bloom extracts (Dell'Aversano et al., 2004).

14.5 QUANTIFICATION OF ANATOXINS

Anatoxin-a and its analogues, homoanatoxin-a and dihydro-anatoxin-a, can readily be detected by both HPLC and LC-MS/MS, and while GC/MS can also be used, this is not commonly done. As with many other cyanotoxins, ELISA kits are available for rapid detection (less than 2h).

14.5.1 Extraction of anatoxins

Anatoxin-a can often be found in benthic cyanobacteria growing on surfaces of, for example rocks, riverbeds and submerged macrophytes. Anatoxins produced by these cyanobacteria has been implicated in animal fatalities. Addressing the occurrence of benthic cyanobacteria requires different sampling strategies than for pelagic cyanobacteria (Chapter 12).

14.5.1.1 Cyanobacterial cells

Anatoxin-a has been successfully extracted efficiently from cyanobacterial cells using acidified solvents, either just water or methanol or a mixture of both (e.g., 50% acidified aqueous methanol). This provides good recoveries and a relatively clean sample, although if microcystins are also going to be analysed from the same extract, it is advisable to omit the acid as it will adversely affect the recovery of more hydrophobic microcystins. Where samples are used for biological tests that would be sensitive to acid and/or organic solvents, extraction by multiple freeze/thawing cycles in water is preferable and has been successfully used.

14.5.1.2 Water samples

While anatoxins are largely cell-bound, it has also been observed to occur dissolved in water (Wood et al., 2018). Anatoxins are increasingly included in drinking-water monitoring during bloom episodes. ELISA kits can be used to determine toxin concentration without further extraction, merely after filtering the sample for cell removal. For instrumental analytical methods, particularly HPLC, concentration by SPE is required, and both C18 and graphitised carbon have been successfully used (Triantis et al., 2017b). Where mass spectrometry is going to be used for detection, a stable isotope-labelled phenylalanine-*d5* can be used as an internal standard to determine the recovery efficiency.

14.5.1.3 Tissue samples

Very few studies have determined the recovery of anatoxins from tissue samples, generally applying methods similar to those applied for analysis of anatoxins in cyanobacterial cells. Using acidified aqueous methanol can help provide a cleaner extract since both the solvent and acid will precipitate proteins. Further sample clean-up and concentration can be achieved by an additional SPE step.

14.5.2 Quantification of anatoxins by ELISA

ELISA kits are commercially available for the detection and quantification of anatoxin-a and homoanatoxin-a with quantification reported in the range of 0.15–5.0 µg/L. Typically, the assay takes around 2h and is suitable for extracts of cells and toxins dissolved in water. These kits have been used to determine anatoxin-a concentrations in field samples (John et al., 2019) and throughout the water treatment train (Almuhtaram et al., 2018). As always, when applying such kits to finished drinking-water, care should be taken to quench oxidants when sampling.

14.5.3 Instrumental analytical methods for anatoxins

The most commonly used analytical system for the detection, identification and quantification of anatoxin-a is HPLC-PDA or LC/MS(MS). The advantage of ultra-performance liquid chromatography (UPLC) columns and systems is that they allow short retention times, providing a rapid analysis. One of the main challenges for the analysis of anatoxin-a is the ubiquitous co-occurrence of phenylalanine, which has a very similar retention time and mass. It is important to ensure that the selected chromatography column and elution profile can separate the two compounds (see Box 5.3 for an example of this misinterpretation of analytical results).

14.5.3.1 Analysis of anatoxins by HPLC-PDA

Chromatography for anatoxin analysis is the same for HPLC-PDA and the LC-MS(MS), for example, a gradient mobile phase consisting of water/acetonitrile (both acidified with 0.1% formic acid) where the organic phase is increased from a low proportion of organic solvent (e.g., 2–5% to around 35% over 5min (for UPLC)) at a flow rate between 0.3 and 0.4 mL/min. Samples can be separated on a suitable UPLC C18 column typically maintained at 40°C (Colas et al., 2020). For photodiode array (PDA) detection, scanning between 200 and 300nm can be sufficient, with anatoxin-a showing a distinct absorption maximum at 227nm. Spiking of separate samples with both anatoxin-a and phenylalanine helps to ensure that both compounds are well separated and to establish a specific retention. A "similar retention time" is not sufficient to assign a peak to anatoxin-a as described in Box 5.3. Notably, dihydoanatoxin-a and dihydrohomoanatoxin-a do not show a distinct UV absorption spectrum due to the lack of the double-bond in the molecule, which makes it difficult to distinguish these variants from the background matrix by photometric detection without florescence derivatisation (James et al., 1998).

14.5.3.2 Analysis of anatoxins by LC-MS(MS)

The same chromatographic conditions are also appropriate for chromatographic separation prior to mass spectral detection via positive ESI. The parent ion $[M+H]^+$ 166.1 is identical for anatoxin-a and phenylalanine, although, as described above, with the suitable column the two compounds can be distinctly separated. This has been achieved on both C18 and hydrophilic interaction liquid chromatography (HILIC) columns, however, not necessarily on all brands. For anatoxin-a dissolved in water, spiking with stable isotope-labelled phenylalanine-*d5* allows estimates on recovery during SPE. The labelled phenylalanine elutes at the same retention time as phenylalanine occurring in the sample, but can be accurately quantified due to its altered mass (*m/z* 171). US EPA drinking and ambient water ATX method based on LC-MS/MS has minimum reporting levels below 0.1 µg/L (Shoemaker & Dietrich, 2017). Parent ions of homoanatoxin-a and dihydroanatoxin-a are $[M+H]^+$ 180.1 and $[M+H]^+$ 168.1, respectively (see section 2.3).

14.6 QUANTIFICATION OF SAXITOXINS

The identification and quantification of saxitoxins is challenging although there is a lot to be learned from the analysis of this toxin class in marine harmful algal blooms (HABs). In shellfish, monitoring with a mouse bioassay developed in the 1930s was still the benchmark until the validation of an analytical method in 2005 (Box 14.4). At least 57 saxitoxin variants have been reported (Wiese et al., 2010), but not all of these variants have been found in cyanobacteria (section 2.4). Furthermore, accurate information on the prevalence of different variants has been hampered by the complexity of analysis. It is known that saxitoxins can transform into different analogues (Wiese et al., 2010); hence, care has to be taken to ensure the stability of samples and standard solutions. In general, acidic solutions (e.g., HCl) are considered suitable (Alfonso et al., 1994).

14.6.1 Extraction of saxitoxins

Saxitoxins are highly polar, and extraction protocols tend to use acidic conditions. The extraction methods for saxitoxins extensively studied for marine shellfish are also suitable for freshwater saxitoxins. A protocol to extract STXs from a range of matrices is available from AOAC (AOAC, 2005b). Solid phase extraction (SPE) using graphitised carbon or HILIC resins, for example, can be used to concentrate STXs for achieving lower detection limits or for sample clean-up (Humpage et al., 2010; Testai et al., 2016).

14.6.1.1 Cyanobacterial cells

Saxitoxins typically occur cell bound, but up to 40% of the total amount has also been found extracellularly. As for cylindrospermopsin, it is therefore important to include both fractions – toxin dissolved in water and cell-bound toxin – in the determination of total toxin. Separation of fractions can be done either by filtration during sampling or soon afterwards in the laboratory by centrifugation or filtration. The cells preferably are extracted in acidified solvent, and the supernatant can be directly used for analysis.

14.6.1.2 Water samples

Saxitoxins in water need to be concentrated, which can be achieved by SPE (Imhof & Schmidt, 2017) with cartridges suitable for very polar compounds (e.g., porous graphitised carbon). It is recommended to evaluate the suitability of the selected cartridges and protocol by determining recovery using spiked (dechlorinated) tap water or filtrated lake water.

14.6.1.3 Tissue samples

Extraction of saxitoxins has been well described for marine shellfish samples. It involves homogenising tissue in a blender with the addition of acidified solvent, typically 1% acetic acid solution (AOAC, 2005b; 2011; Van De Riet et al., 2011). Prior to testing in a biological system, for example, ELISA, the pH will need to be adjusted to around 7.

14.6.2 Quantification of saxitoxins by biochemical methods

Several biochemical assays have been developed in the past but with the exception of ELISA, most have not been widely adopted, mainly because of their complexity and specialist expertise required to perform them.

14.6.2.1 Quantification of saxitoxins by ELISA

There are a number of (in 2019 at least six) manufacturers of different ELISA kits for saxitoxins. Most have been configured to detect saxitoxin, achieving good correlation with analytical results; however, there are challenges with cross-reactivity with other saxitoxin variants. This is of particular concern for neo-saxitoxin and gonyautoxin (GTX) variants since these variants represent as high a risk to health as saxitoxin (Papageorgiou et al., 2005). Some innovative methods are available which add an additional sample preparation step (e.g., incubation in the presence of *L*-cysteine) to transform most of the GTX variants to detectable saxitoxin or neo-saxitoxin (McCall et al., 2019). Some manufacturers produce an ELISA for saxitoxin and another kit for neo-saxitoxin. Recently, a multiplex ELISA has been demonstrated which detected nine saxitoxins in human plasma (Eangoor et al., 2019).

Careful selection of the most appropriate kit for screening purposes is important and should consider regional availability, cross-reactivity and the saxitoxin variants prevalent in the area of sampling (Harrison et al., 2016). As with all ELISA screening methods for cyanotoxins, it is wise to periodically confirm findings using an established instrumental analytical method. This may be achieved through collaboration with a centre of excellence rather than own investment in equipment for instrumental analytical methods and the corresponding expertise, as this can be considerable (see below).

14.6.3 Instrumental analytical methods for saxitoxins

The analysis of saxitoxins has been fraught with many difficulties as saxitoxins do not contain a chromophore (do not adsorb light) nor natural fluorescence; hence, typical HPLC detectors cannot be used to identify or quantify them. Further, they are very polar molecules that are not easily retained by reverse-phase chromatography (e.g., using C18 columns). Derivatisation to form a fluorescent analyte has proven valuable, with both precolumn and postcolumn (i.e., the saxitoxins are first chromatographically separated and then mixed with the derivatisation reagents before detection) methods developed and the precolumn method becoming an AOAC Official Method (AOAC, 2005b; 2011). While this method is validated for the analysis of paralytic shellfish poisoning in shellfish, it is also suitable for the analysis of saxitoxins from cyanobacteria in cells, water and tissues.

14.6.3.1 Prechromatographic oxidation and liquid chromatography with fluorescence detection

To overcome the difficulties of detecting saxitoxins, a preanalysis oxidation method was developed by Lawrence et al. (1995) allowing the saxitoxins to be analysed using a fluorescence detector (fluorescence excitation 340 nm and emission 395 nm). Around the same time, Oshima (1995) proposed a postcolumn derivatisation method. The so-called *Lawrence method* with precolumn sample oxidation has now been adopted for regulatory monitoring purposes and is being used in many laboratories as part of routine monitoring programmes for saxitoxins in shellfish. The attention this method has received has ensured significant performance testing, including interlaboratory studies (Turner et al., 2019). This analysis requires a significant commitment to setting up and maintaining the method, including purchasing a wide range of standards. Where intermittent confirmatory analysis of ELISA results is required, it is beneficial to approach a laboratory which is already well established in this field rather than committing to the onerous task of developing this method in-house.

14.6.3.2 Analysis of saxitoxins by LC-MS/MS

Ultra-performance liquid chromatography (UPLC) with MS/MS provides a very powerful analytical tool for the detection, identification and quantification of saxitoxins at very low limits of quantification in the sub-ng/mL range. Several chromatographic approaches can be used, one being the more traditional C18 columns with a high proportion of water in the gradient elution. Alternatively, by using hydrophilic interaction liquid chromatography (HILIC), which is well suited to highly polar analytes, a rapid analysis has become possible with fast UPLC columns (e.g., 11 min compared to a previous 40 min). Furthermore, the solvent-rich mobile phase used in HILIC can provide a significant advantage. One recent study demonstrated the separation, identification and quantification of 14 saxitoxins in less than 10 min (Turner et al., 2015).

Another promising approach to the detection of saxitoxins is the use of inline SPE coupled with C18 UPLC-MS/MS. This approach effectively combines sample extraction and concentration from a water sample which is then directly injected onto the analytical system. The full automation has many advantages in reducing staff time, handling errors and sample loss as well as giving a low limit of quantification (Imhof & Schmidt, 2017).

14.7 DETECTION AND QUANTIFICATION OF ANATOXIN-A(S)

Work on anatoxin-a(S) has been hampered by the limited availability of cultures which produce this cyanotoxin, which subsequently limits the availability of the purified toxin. This is further restricted by the difficulty in detecting anatoxin-a(S) in natural samples. As an organophosphate inhibitor, anatoxin-a(S) can potentially be detected with a biochemical screening assay using an acetylcholine esterase inhibition assay. However, since organophosphate pesticides may be present in environmental samples, confirmation is required. Furthermore, in the absence of any authentic purified anatoxin-a(S) and confirmatory analytical methods, few reports of anatoxin-a(S) have been confirmed.

With greater availability and use of HILIC columns along with MS/MS (Dörr et al., 2010), it may be expected that gradually anatoxin-a(S) will become more widely detected, isolated and investigated. When planning sampling campaigns, it may be useful to give consideration to the possible presence of anatoxin-a(S), preferably in collaboration with expert groups able to screen samples with advanced multitoxin methods.

14.8 METHODS FOR SYNCHRONOUS DETECTION OF MULTIPLE TYPES OF CYANOTOXIN

14.8.1 Multiplex antibody systems

Methods are likely to become available which allow the detection and quantification of multiple cyanotoxins in a single system (Eangoor et al., 2019). These systems are referred to as bioarray, microarray or multiplex systems.

One such system using fluorescence detection of antibody binding signals has been found to be very sensitive, allowing the detection of microcystins, saxitoxins and cylindrospermopsin along with two of the marine shellfish toxins, okadaic acid (OA) and domoic acid (McNamee et al., 2014). Assay time was around 15 min per sample, providing results for all five biotoxins at once and in the sub-µg/L range. While these systems are yet to become widely available, this type of screening is likely to become increasingly adopted in the near future with applications in screening of drinking-water quality and recreational waterbodies.

14.8.2 Multi-cyanotoxin analytical methods

An increasing number of instrumental methods which can analyse cyanotoxins in a single analytical chromatography run have been published (Dell'Aversano et al., 2004; Greer et al., 2016; Zervou et al., 2017). With the use of the optimal UPLC column and MS/MS detection, multitoxin methods are convenient where advanced analytical systems are routinely employed. Pekar et al. (2016) demonstrated the separation of anatoxin-a, cylindrospermopsin and microcystin variants, achieving the analysis of 22 cyanotoxins in both raw water and drinking-water Haddad et al. (2019) have added saxitoxin to the analysis, allowing the separation of four classes of cyanotoxin in a single analytical method.

14.9 FUTURE DEVELOPMENTS

With increasing legislation and pressure on water resources causing more demand to test for cyanobacterial toxins, rapid simple screening tests are increasingly likely to be required. These advances will most likely employ immunological, biosensors and related technologies to permit a rapid simple assessment of cyanobacterial and water samples after extracting cell-bound toxins. Immunological strip detection systems have been commercially developed (e.g., Kim et al., 2013; Weller, 2013), and such technologies are being adapted to furthering our understanding of cyanobacteria and their toxins in the environment.

Increasingly, cyanotoxin analytical methods will be required to analyse a wider complexity of novel materials that may require method development (e.g., in food; see section 5.3). In the case of cyanotoxins that do not covalently bind to proteins, simple extraction and clean-up methods with SPE should permit accurate analysis, although verification will still be required. Combined single-step SPE methods for sample preparation of, for example, cylindrospermopsins, anatoxin-a and saxitoxins should be developed to extract such cyanotoxins in an effort to reduce sample preparation for their subsequent measurement with, for example, mass spectrometry (Fayad et al., 2015).

Although more specialised analytical methods such as mass spectrometry will continue to be required to verify cyanotoxins, future needs will be to develop methods for not yet identified bioactive compounds produced by cyanobacteria and to develop robust multitoxin analytical methods.

Further advances may come with the use of biosensors, such as with recombinant PP1α showing an increased sensitivity for microcystin-LR (Catanante et al., 2015), 3D-graphene-based biosensors (Zhang et al., 2017) for the detection of microcystin-LR or DNA-based aptamer systems, either for the detection of *Microcystis* (Tong et al., 2015) or for the detection of microcystin (Li et al., 2016). Assay systems such as this show good promise and may be useful in future to provide quantitative and toxicological assessment of cyanobacterial toxins.

14.10 BIOASSAYS AND THEIR USE IN THE SURVEY OF TOXIC CYANOBACTERIA

Assessments of potentially toxic environmental samples, including blooms of cyanobacteria, most commonly rely on chemical monitoring of individual chemicals, that is, the targeted analytical or bioanalytical identification and quantification of known toxins as outlined above. With respect to quality assurance/quality control (QA/QC) criteria and straightforward interpretation of the results, chemical monitoring is an approach generally applied in all regulatory settings, including water quality and safety. The obvious limitation of chemical monitoring is the fact that many analytical methods detect only those toxins that they target, which could often be only a single or a few structural variants, while others remain undetected, thus potentially underestimating the sample's toxicity.

Several studies showed that field cyanobacterial samples may cause stronger toxicity in comparison with the effects of pure toxins when tested at equivalent concentrations, indicating the presence of other toxic components (see section 2.10). These may, in addition to diverse cyanobacterial metabolites, also be toxic metals or compounds of anthropogenic origin such as pesticides, polycyclic aromatic hydrocarbons and other emerging contaminants present in complex environmental samples. Based on chemical analysis alone, it is not possible to evaluate the overall toxicity of complex mixtures. To cope with these limitations, some environmental monitoring programmes have implemented toxicity testing with bioassay(s), for example, whole effluent toxicity testing in USA and Germany (Escher & Leusch, 2011) or the EU-supported effect-based monitoring programmes (Tousova et al., 2017).

This section firstly introduces some toxicology principles with respect to the interpretation of bioassay results; then summarises the existing experience; and critically discusses current state, limitations and recommendations on the applicability of bioassay for the monitoring of toxic cyanobacterial blooms with respect to possible impacts on human health. Other aspects such as testing for ecotoxicity with invertebrates or other aquatic biota are addressed only briefly because of their limitations for assessing human health end-points. This chapter does not cover other applications of bioassays such as the toxicological characterisation of individual toxins. The focus of this chapter is on bioassays employing cells, tissues or whole animals; other subcellular bioanalytical tools (ELISA or enzyme inhibitory assays such as the protein phosphatase assay) do not fit the "bioassay" criteria and are discussed in section 14.2.

14.10.1 Insights into interpretation of toxicity results

Before discussing examples and the practical applications of bioassays, those who plan to implement them in monitoring programmes need a good and common understanding of the terminology and how individual terms are used. Firstly, there is a central paradigm of toxicology, that is, "All things are poison and nothing is without poison. Solely, the dose determines that a thing is not a poison" (Paracelsus, 1493–1541). Whenever "toxicity" or "effect" is considered (e.g., animal death due to anatoxin-a neurotoxicity, microcystin-induced liver injury or decreased cell viability *in vitro* through cytotoxicity of cylindrospermopsin), observed effects are related to the defined test conditions. Most importantly, whether the effect manifests always depends on the dose, duration of exposure and biological system (organism). In this sense, toxicity of a compound is widely understood as causing adverse effects upon exposure as expected under normal conditions. For example, compounds like vitamins can cause adverse health effects when applied in high doses (hypervitaminoses) but vitamins are generally not considered a toxin, because, under normal conditions, an exposure leading to adverse effects is improbable (Hathcock et al., 1990; Vieth, 2007).

Current toxicology aims to establish links between the adverse health outcome (i.e., *in vivo* manifestation of the toxic effect) with the exposure to a toxicant through a chain of causal events formalised as an "adverse outcome pathway" (AOP; Patlewicz et al., 2015). Examples of relevant adverse health outcomes may be, for example, the death of an animal caused by a high dose of anatoxin-a, disruption of neurobehavioral abilities after chronic exposures to lower doses of anatoxin-a or eye irritation and skin rash after direct acute exposure to high doses of cyanobacterial biomass. "Toxicity" always starts at the molecular level; propagates through cells, tissues and organs; and eventually becomes apparent *in vivo* as systemic toxicity. The ultimate adverse outcome manifests only under specific preconditions (the toxin can reach the target, it is present in sufficiently high doses, etc.) and when the chain of events is not repaired by detoxification mechanisms (ADME – absorption, distribution, metabolism and excretion). An AOP relevant for cyanobacteria could be, for example, "inhibition of protein phosphatases (PPase) leading to hepatic hypertrophy and tumour promotion activity".

The interaction between the toxic chemical and its biological counterparts (molecular initiating event, MIE) can be either specific or nonspecific. Examples of specific interactions (key–lock principle) include binding of MC-LR to PPase, preferentially in liver cells, cylindrospermopsin interferences with the machinery of protein synthesis with no apparent preference to the cell type or binding of anatoxin-a to the nicotinic acetylcholine receptor on neuronal cells. Nonspecific interactions, when a chemical does not have a "specific target", are common and include, for example, the disruption of cell membrane function after the accumulation of chemicals also known as narcotic or basal toxicity, damage to proteins, membrane phospholipids or nucleic acids by reactive oxygen species or denaturation of proteins by acidic chemicals. One chemical may act through several modes of action that may lead to a single or multiple different adverse outcomes. For example, cylindrospermopsin may inhibit protein synthesis, react with DNA or induce oxidative stress leading to death or various chronic effects depending on the concentration, exposure duration, life stage, age or sex of the organism (Pichardo et al., 2017).

Toxicity of natural samples of cyanobacteria may thus be a complex response to – for example – unfavourable pH, presence of ions and metals, saccharides, peptides (including toxins, amino acids, nucleotides, phospholipids), components of other plankton organisms (Palíková et al., 2007a; Palíková et al., 2007b) or compounds of anthropogenic origin.

14.10.2 Bioassays in the assessment of toxic cyanobacteria

Bioassays have primarily been developed for the testing of chemical substances based on different regulatory frameworks. Most of these tests went through a validation process with standardisation by ISO or OECD assuring good characterisation of the studied chemical and testing conditions. Bioassays are mainly based on animal testing but there is an increasing demand to reduce animal experiments and use alternative methodologies such as *in silico* and *in vitro* methods often combined into so-called *integrated testing strategies (ITS)* or *integrated approaches to testing and assessment (IATA)*. Toxicity bioassays have been adapted to assess the toxicity of cyanobacterial samples. However, when testing complex samples like cyanobacterial crude extracts, the causative agent(s) inducing the toxicity cannot be easily identified.

Nevertheless, many studies explored the use of bioassays in toxicity screenings of natural cyanobacterial samples or explored their potential to serve as early warning tools. Positive bioassay responses could then trigger chemical analysis of cyanotoxins for more precise characterisation of the hazard. Testing with bioassays is expected to show whether the sample contains toxic substances and how toxic these substances may potentially be. Researchers can combine multiple bioassays to cover various end-points ranging from acute cytotoxicity and mortality to complex organ or systemic effects such as reproduction toxicity. Specific *in vitro* assays have been used to assess mechanisms of action, potential genotoxicity or endocrine disruptive effects. Unfortunately, complex research approaches can hardly be implemented for routine monitoring or screenings of potential health hazards. However, at least four cases can be listed in which toxicity testing remains relevant:

1. if illness of animals or humans is suspected to have been caused by cyanobacteria but symptoms cannot be attributed to known cyanotoxins found by chemical analyses;
2. for testing whether specific cyanobacterial strains show toxicity not attributable to known cyanotoxins;
3. to characterise the toxicity and/or mechanism of action of newly identified toxins or congeners of previously known cyanotoxins (Fischer et al., 2010);
4. to establish the data needed to derive guideline values for the concentrations of substances to which humans may be exposed, for example, cyanotoxins in drinking-water or in waterbodies used for recreation.

14.10.2.1 Nonmammalian bioassays

Ecotoxicity assays using bacteria, protozoans, invertebrates, plants or aquatic vertebrates such as fish or amphibians have been used in many studies for detecting cyanotoxins.

Bacterial bioassays have been used to screen complex cyanobacterial samples such as the Microtox bioluminescence assay using *Aliivibrio* (formerly *Vibrio*) *fischeri* or *Photobacterium* (*Vibrio*) *phosphoreum* (Lawton et al., 1994a; Vezie et al., 1996) but with poor correlations between the reduction of the measured end-point (emitted light) and the sample's content of cyanotoxins. Poor correlations were also revealed in the bioassay with *Serratia marcescens* despite promising original studies with pure saxitoxins and microcystins (Lawton et al., 1994a).

Cyanobacterial samples were also tested with protozoan assays such as *Tetrahymena thermophila* (commercially available as Protoxkit-F; Protoxkit-F, 1998), *T. pyriformis* and *T. thermophila* (Maršálek & Bláha, 2004) or *Spirostrum ambiguum* (Tarczyńska et al., 2001). Further, bioassays with aquatic or terrestrial plants were explored (Kós et al., 1995; Pflugmacher et al., 2001; Vasas et al., 2002).

Among the bioassays with aquatic invertebrate animals, cladocerans have been widely used due to their easy maintenance, small size, wide distribution and rapid growth rates. These bioassays include standardised 24- and 48-h immobilisation assays with *Daphnia magna* (OECD, 2004; 2012) or commercially available test kits (Daphtoxkit-F, 1995; Ceriodaphtoxkit-F, 1995). Complex cyanobacterial samples have been tested with species of *Daphnia* (DeMott et al., 1991; Okumura et al., 2007; Ferrão-Filho et al., 2009), *Ceriodaphnia* (Maršálek & Bláha, 2004; Okumura et al., 2007) or *Moina* (Ferrão-Filho et al., 2009). Further model organisms include *Artemia salina* (Kiviranta et al., 1991; Metcalf et al., 2002b; Beattie et al., 2003; Lindsay et al., 2006) or mosquito adults and larvae (Kiviranta et al., 1993). Also extensively used was the bioassay with larvae of fairy shrimp *Thamnocephalus platyurus* commercially available as Thamnotoxkit-F (MicroBioTests Inc., Mariakerke, Belgium). Box 14.3 provides more information and illustrates the difficulties and limitations in the interpretation of results for the *Thamnocepalus* bioassay and, correspondingly, for all invertebrate bioassays.

BOX 14.3: THE *THAMNOCEPHALUS PLATYURUS* BIOASSAY

The bioassay with larvae of fairy shrimp *Thamnocephalus platyurus* has been discussed in the past as a potential tool for routine screening of bloom toxicity. It has a number of advantages such as allowing for simple and practical use even in a format of a commercially available kit called *Thamnotoxkit*. It provides fast 24-h response with a possible reduction of exposure to 1 h (Törökné et al., 2007). The assay has been standardised (ISO, 2011), and it showed good performance in an interlaboratory test with cyanobacterial samples (Törökné et al., 2000a).

With respect to individual cyanobacterial metabolites, *T. platyurus* was generally reported to be highly sensitive. However, the reported IC50 values were surprisingly within a narrow – low micromolar – range for all the studied and structurally diverse cyanobacterial metabolites and toxins (including microcystins, cylindrospermopsin, microginin, aeruginosins, spumigins, cyanopeptolin, eucyclamides, oscillapeptin J) as well as fluoro-conjugated MCs or odour compounds such as sesquiterpenes (Blom et al., 2001; Blom & Jüttner, 2005; Portmann et al., 2008; Höckelmann et al., 2009; Gademann et al., 2010; Kohler et al., 2014; Grundler et al., 2015; Mazur-Marzec et al., 2015; Scherer et al., 2016; Bober & Bialczyk, 2017), while the assay was reported to be less sensitive to the neurotoxin anatoxin-a (Sieroslawska, 2013). With regard to screening of complex bloom samples, the literature provides a conflicting picture. One study (Tarczyńska et al., 2000) compared seven extracts and found a statistically significant relationship between the observed toxicity and the microcystin-LR content. On the other hand, several other studies showed high toxicity in *T. platyurus* irrespective of the content of major cyanotoxins (Maršálek et al., 2000; Törökné et al., 2000a; Törökné et al., 2000b; Keil et al., 2002; Nałęcz-Jawecki et al., 2002; Maršálek & Bláha, 2004; Törökné et al., 2007; Ács et al., 2013; Sieroslawska, 2013).

Despite apparently high sensitivity of the *T. platyurus* assay, its implementation into routine monitoring of bloom toxicity would not provide a major added value to toxin analyses with chemical or biochemical methods because it is generally not able to discriminate between toxic (in the sense of "containing cyanotoxins") and nontoxic cyanobacterial samples. Further, its responses were not correlated with toxicity observed with other organisms, including mouse *in vivo* assay (Tarczyńska et al., 2000; Tarczyńska et al., 2001).

In addition to invertebrates, fish and frog bioassays have also been explored. With respect to ethical concerns associated with the use of adult fish (namely, zebrafish *Danio rerio*, Japanese medaka *Oryzias latipes* or fathead minnow *Pimephales promelas*), assays with fish embryos have become popular (Berry et al., 2009). The zebrafish *Danio rerio* fish embryo toxicity (FET) assay has been standardised (OECD, 2013). The embryo fish tests were used in many studies of cyanobacteria (Oberemm et al., 1999; Wang et al., 2005; Lecoz et al., 2008), but some concerns related to the uptake of toxins through the chorion barrier or limited toxicokinetics in developing embryos have been raised. With respect to amphibians, frog embryo teratogenesis assay Xenopus (FETAX) (ASTM, 2017) using *Xenopus laevis* (African clawed frog) has also been explored but showed rather low correlation with the content of known cyanotoxins (Oberemm et al., 1999; Fischer & Dietrich, 2000; Burýšková et al., 2006).

14.10.2.2 Mouse bioassay

For many years, the mouse *in vivo* bioassay was used to determine toxicity of cyanobacterial blooms (Carmichael, 1992; Fastner et al., 2003; Sotero-Santos et al., 2006) or to detect phycotoxins in shellfish (Box 14.4). In testing of toxic cyanobacteria, male Swiss Albino mice were the most commonly used animals. Effects are assessed after intraperitoneal injection (i.p.) of 0.1–1.0 mL of a lysate of the cyanobacterial sample. Mice are observed for 24 h, sacrificed by an approved method and submitted to postmortem examination of tissue injury (Falconer, 1993). The observation period could be extended when the possibility of protracted symptom manifestation is expected, as it may be the case with cylindrospermopsin (see Chapter 2.2).

When more than one type of cyanotoxin is present, the more rapid-acting toxin may mask the symptoms of the others. Acute toxicity is expressed as the dose at which one half of the treated animals has died within the determined time period, that is, usually 24 h (LD_{50} in mg extract dry weight/kg mouse body weight). According to the Globally Harmonized System of Classification and Labelling of Chemicals (GHS; UNECE, 2017), five acute toxicity categories are recognised based on oral LD_{50} (mg/kg b). The most toxic is Category 1 with oral LD_{50} <5 mg/kg bw, while for Category 5 acute oral LD_{50} ranges between 2000 and 5000 mg/kg. LD_{50} higher than 5000 mg/kg b is considered as absence of acute toxicity.

BOX 14.4: MOUSE BIOASSAY IN TOXICITY TESTING OF MARINE BIVALVES

Biotoxins produced by marine (phyto)plankton which may accumulate in seafood remain a major public health issue in some parts of the world. Some regulatory approaches also refer to the use of mouse bioassays, but the bioassay is no longer used very often with regard to recent technology developments of chemical-specific analytical methods as well as ethical concerns.

For example, okadaic acid (so-called *OA toxins*, that is, OA and its analogues), the dinophysis toxins (DTX1, DTX2 and DTX3) can be found in tissues of molluscs such as oysters, mussels, scallops and clams, and cause diarrhetic shellfish poisoning (DSP). The inhibition of serine/threonine phosphoprotein phosphatases is their main mode of action, similar to, for example, cyanobacterial microcystins. To control for DSP, the mouse and the rat *in vivo* bioassays have been official reference methods in the EU (Commission Regulation (EC) No. 2074/20054) using the intraperitoneal (i.p.) injection of mussel tissue extract followed by 24-h monitoring of test animals. Despite the advantages of the bioassay (whole-organism toxicity response, no need for complex analytical equipment), an official opinion of the European Food Safety Authority (EFSA, 2009) highlighted several disadvantages, that is, a high variability and labour demand, needs of specialised animal facilities, false-positive results due to interferences with, for example, free fatty acids, not selective for solely the OA-group toxins, not quantitative, inappropriate i.p. administration route and ethical reasons. The EFSA therefore concluded that they are inappropriate for assessing compliance to the regulatory limit set for seafood by the EU. The same report also concluded that the phosphoprotein phosphatase assays and LC-MS/(MS)-based methods have the greatest potential to replace the mammalian assays, due to sufficient sensitivity and satisfactory validation performance.

A recommendation to replace the mouse bioassay for the assessment of broader groups of marine toxins (AZA, BTX, DA, OA, PTX, SXT, YTY) by alternative chemico-physical methods such as LC/MS has been prepared by German Federal Institute for Risk Assessment (BfR, 2005). The mouse bioassay is only envisaged as an additional analytical step when a positive result has been obtained and further clarification is needed in the interests of consumer protection (suitability of the test results for use in court, etc.). The LC/MS method has, for example, been recognised by the New Zealand Food Safety Authority (FSA) as an official method and successfully tested in an interlaboratory trial.

Nevertheless, for control of marine paralytic shellfish poison (PSP) biotoxins, the mouse bioassay remains a standardised method of the European Union Reference Laboratory for Marine Biotoxins (EURLMB) at Agencia Española de Consumo, Seguridad Alimentaria y Nutrición (AECOSAN, 2014), which is in line with the Association of Official Analytical Chemists Official Method (AOAC, 2005a). The mouse bioassay is mentioned in the context of PSP biotoxins in the food of animal origin within the frame of the European Regulation (EC) N° 853/2004, and related methodological regulations (Commission Regulation (EC) N° 2074/2005 amended by EC N° 1664/2006, EC N° 1244/2007 and EU N° 15/2011).

Among the other standardised methods for marine biotoxin detection, HPLC method – so-called *Lawrence method* (Lawrence et al., 2005) – is mentioned in the EU regulation and immunochemistry approaches remain to be discussed as an alternative for the future, after undergoing validation through interlaboratory exercises (Burrell et al., 2016; Dorantes-Aranda et al., 2018; Turner et al., 2019).

Using this mammalian model, observations of the target organs can help extrapolating to the effects in humans. However, the mouse bioassay is generally done through intraperitoneal injection, which may not be the most relevant route of exposure for such extrapolation to humans. Also, other aspects such as low sensitivity and selectivity, high rates of false positives, variability and ethical concerns created a demand for alternative tests. Nevertheless, in some countries, the mouse bioassays can still provide some guidance for managers, for example, to determine toxicity of marine bivalves considered for human consumption or when a bloom occurs in the raw water but chemical analyses do not reveal any known cyanotoxins.

For deriving WHO guideline values or regulatory standards issued by public authorities worldwide, if data on human populations are inadequate or insufficient, a preferred basis is chronic exposure studies with rodents, with toxin applied orally, that is, via food, drinking-water or gavage, and animal health observed during extended periods of time, particularly those based on standardised guidelines such as OECD Test No. 408 (Repeated Dose 90-Day Oral Toxicity Study in Rodents), Test No. 407 (Repeated Dose 28-day Oral Toxicity Study in Rodents). One-generation reproduction studies (OECD Tests No. 415 or No. 443) are particularly comprehensive and thus valuable, but rare because of high costs and demands. In practice, toxicological data from such chronic whole-animal studies may not be available, and risk assessors need to include other toxicological data, including those from acute oral exposure tests (e.g., OECD Tests No. 420, 423 or 425). These and further guidelines are freely available under https://www.oecd-ilibrary.org/environment (Book Series).

14.10.3 *In vitro* assays for determining toxicity and genotoxicity

In vitro bioassays using cell cultures have received wide attention for replacing animal tests. However, a single *in vitro* bioassay alone cannot cover all of the biological targets or processes found within an organism. Therefore, a hierarchic *in vitro* test strategy is necessary for characterising the type of toxicity induced by the unknown toxicants as proposed in different strategic documents and recommendations for water quality assessment (enHealth, 2012; Grummt et al., 2013). The following paragraphs provide examples as well as a summation of the advantages and disadvantages of *in vitro* bioassays, which could form part of such a hierarchic *in vitro* test strategy specifically for cyanotoxins.

The hepatotoxicity of microcystins triggered the use of hepatocytes (liver cells) from different fish or mammalian species (Aune & Berg, 1986). Freshly isolated hepatocytes may – for a certain period – retain necessary liver characteristics like active bile acid transport or phase I and II metabolising enzymes, and many studies showed high sensitivity to cyanotoxins in rat or mouse hepatocytes (Fladmark et al., 1998; Li et al., 2001; Boaru et al., 2006). A need for fresh isolation can be overcome by using cryopreserved hepatocytes, preferably of human origin or a specific cell line such as HepaRG (Bazin et al., 2010), which maintains most of the *in vivo* features. A basic prerequisite for microcystin uptake into the cells seems to be the presence of certain organic anion-transporting polypeptides (OATP) within the cell membrane, as a study with genetically modified OATP-competent HEK293 cells has shown (Fischer et al., 2010). Many other cell lines, such as HepG2, CaCo2, and V79, have been used to study cyanobacterial samples (Lawton et al., 1994a; Fastner et al., 2003; Žegura et al., 2003; Lankoff et al., 2006; Žegura et al., 2008; Fischer et al., 2010). Besides having a tumoral origin in most cases, these cell lines may lack certain metabolic enzymes important for activation and in particular for detoxification; therefore, they are prone to giving misleading results, thus limiting their informative value.

For assessing neurotoxicity, the Neuro-2A neuroblastoma cell test has been developed and used for monitoring of saxitoxins in freshwater cyanobacteria (Gallacher & Birkbeck, 1992; Humpage et al., 2007). Endocrine activity can be examined by oestrogen or androgen receptor-specific reporter gene assays (OECD, 2016a; b) and steroidogenesis assay in H295R cells (OECD, 2011). Within the OECD framework, these *in vitro* tests are part of the first tier and are considered as a screening tool, which is not sufficient to categorise a substance as an endocrine disruptor.

Genotoxicity and mutagenicity are important end-points for human and environmental hazard evaluation, and a number of assays, often adopted as ISO or OECD guidelines, are used in the assessment of toxic cyanobacteria. Among the bacterial assays, the Ames assay (OECD, 1997) showed the mutagenicity of various cyanobacterial extracts (Huang et al., 2007), while pure cyanotoxins were mostly negative in the assay (Žegura, 2016). Palus et al. (2007) showed genotoxicity of various extracts or cyanobacterial toxins in SOS chromo test with *Escherichia coli* PQ37 but negative results were reported with the SOS/umu-test (ISO, 2000) with *Salmonella* Typhimurium TA1535/pSK1002. However, the caveat of many publications is that cytotoxicity (resulting in DNA fragmentation) is not accounted for, which may lead to false positives or overestimation of the relevance of genotoxicity.

With regard to eukaryotic cell models, the mammalian cell gene mutation assay (OECD, 2016c) demonstrated that MC-LR preferentially induces clastogenic effects on DNA rather than point mutations (Zhan et al., 2004). Various cyanobacterial extracts induced micronuclei in the *in vitro* cytokinesis-block micronucleus assay (micronucleus test) (OECD, 2016d) in human lymphocytes (Palus et al., 2007). Cylindrospermopsin was shown to have clastogenic and aneugenic activities in human WIL2-NS lymphoblastoids (Humpage et al., 2000) and hepatic cells (Bazin et al., 2010; Štraser et al., 2011). The comet assay, also known as the single-cell gel electrophoresis (SCGE) assay – which detects DNA damage (in the form of strand breaks, or other lesions that are converted into strand breaks under alkaline conditions) and DNA repair activity, and gives an indication of the genotoxic insult – has gained broad attention in genetic toxicology of toxic cyanobacteria (Ding et al., 1999; Žegura et al., 2003; Humpage et al., 2005; Palus et al., 2007). Cyanobacterial extracts (Palus et al., 2007) and several pure cyanotoxins, including MC-LR, cylindrospermopsin and nodularin, have been shown to induce DNA strand breaks (see Žegura et al., 2011 for a review). MC-LR induces transiently present DNA strand breaks that can be repaired and most probably occur indirectly due to oxidative stress (Žegura, 2016), while CYN induces DNA strand breaks in metabolically active cells (Humpage et al., 2005; Hercog et al., 2017).

Since the 2010s, hazard identification shifted towards mechanistic assessment that enables predictions of adverse outcome pathways (AOPs) (Ankley et al., 2010; Schroeder et al., 2016). The "omic" biomarker approaches (Li et al., 2017) using high-throughput molecular biology and mass spectrometry tools allow us to identify biological targets and pathways affected by the toxic compounds, including cyanotoxins (Štraser et al., 2013; Hercog et al., 2017). The "omic" biomarkers complement the standard toxicity and genotoxicity assays but how to use these complex data in the risk assessment procedure remains to be clarified.

14.10.4 Summary

In summary, toxicity testing of complex samples such as cyanobacterial blooms, raw or tap water provides rather minor additional value to current risk assessment of cyanobacterial toxicity, particularly where known cyanotoxins are present and sensitive instrumental methods for the detection of multiple toxins, as discussed above, are available (Meriluoto et al., 2017; Zervou et al., 2017). Alternatively, immunoassays such as ELISA or enzyme-inhibitory assays may serve for semiquantitative and sufficiently selective screenings. Toxicity testing has its place in bloom situations in which targeted analyses do not reveal any known cyanotoxins and uncertainty about the safety of the water remains, as, for example, in the case of a South Australian water supply with an unidentified toxin from *Phormidium* (Baker et al., 2001). In such a situation, the results of bioassays with mammalian cells *in vitro* or animals *in vivo* (mouse test) are most likely to provide some immediate guidance for managers regarding the acute toxicity of water.

Bioassays are important for further exploring the effects of yet unknown or not sufficiently characterised substances produced by cyanobacteria. A wide range of bioassays is available at many different levels from molecular to cell cultures or whole organisms. However, one single test will rarely be sufficient to fully characterise the toxicity of a cyanobacterial bloom; this usually requires a set of assays. Bioassays can give rapid responses, but a thorough validation process is needed for testing their performances, particularly if they are to be applied in the investigation of complex samples such as blooms, raw or treated drinking-water. *In vitro* bioassays are useful for developing an understanding of the biochemical processes underlying toxicity, whereas *in vivo* studies, despite technical and ethical concerns, continue to have a key role in supporting risk assessment, including in guideline value derivation. For the identification of unknown toxic agents from cyanobacterial blooms, effect-based monitoring or effect-directed analyses (EDA) (Escher & Leusch, 2011; Tousova et al., 2017) efficiently combine both bioassays and chemical analyses.

REFERENCES

Ács A, Kovács AW, Csepregi JZ, Törő N, Kiss G, Győri J et al. (2013). The ecotoxicological evaluation of *Cylindrospermopsis raciborskii* from Lake Balaton (Hungary) employing a battery of bioassays and chemical screening. Toxicon. 70:98–106.

AECOSAN (2014). Standard Operating Procedure for PSP toxins by Mouse Bioassay. Vigo: Agencia Española de Consumo, Seguridad Alimentaria y Nutrición. http://www.aecosan.msssi.gob.es/.

Alfonso A, Louzao M, Vieytes M, Botana L (1994). Comparative study of the stability of saxitoxin and neosaxitoxin in acidic solutions and lyophilized samples. Toxicon. 32:1593–1598.

Almuhtaram H, Cui Y, Zamyadi A, Hofmann R (2018). Cyanotoxins and cyanobacteria cell accumulations in drinking water treatment plants with a low risk of bloom formation at the source. Toxins. 10:430.

Altaner S, Puddick J, Wood SA, Dietrich DR (2017). Adsorption of ten microcystin congeners to common laboratory-ware Is solvent and surface dependent. Toxins. 9:129.

Ankley GT, Bennett RS, Erickson RJ, Hoff DJ, Hornung MW, Johnson RD et al. (2010). Adverse outcome pathways: a conceptual framework to support ecotoxicology research and risk assessment. Environ Toxicol Chem. 29:730–741.

AOAC (2005a). Official method 959.08: paralytic shellfish poison - biological method. Gaithersburg (MD): Association of Official Analytical Chemists International. http://www.eoma.aoac.org/methods/.

AOAC (2005b). Official method 2005.06 paralytic shellfish poisoning toxins in shellfish. Gaithersburg (MD): Association of Official Analytical Chemists International. http://www.eoma.aoac.org/methods/.

AOAC (2011). Official method 2011.02 determination of paralytic shellfish poisoning toxins in mussels, clams, oysters and scallops. Gaithersburg (MD): Association of Official Analytical Chemists International. http://www.eoma.aoac.org/methods/.

ASTM (2017). E1439-12: standard guide for conducting the Frog Embryo Teratogenesis Assay-Xenopus (FETAX). ASTM Volume 1106, Environmental; Biological effects and environ fate; Industrial biotechnology. West Conshohocken (PA): American Society for Testing and Materials.

Aune T, Berg K (1986). Use of freshly prepared rat hepatocytes to study toxicity of blooms of the blue-green algae *Microcystis aeruginosa* and *Oscillatoria agardhii*. J Toxicol Environ Health Part A Current Issues. 19:325–336.

Baker PD, Steffensen DA, Humpage AR, Nicholson BC, Falconer IR, Lanthois B et al. (2001). Preliminary evidence of toxicity asociated with the benthic cyanobacetrium *Phormidium* in South Australia. Environ Toxicol. 16:506–511.

Barco M, Lawton LA, Rivera J, Caixach J (2005). Optimization of intracellular microcystin extraction for their subsequent analysis by high-performance liquid chromatography. J Chromatogr A. 1074:23–30.

Bazin E, Mourot A, Humpage AR, Fessard V (2010). Genotoxicity of a freshwater cyanotoxin, cylindrospermopsin, in two human cell lines: Caco-2 and HepaRG. Environ Mol Mutagen. 51:251–259.

Beattie KA, Ressler J, Wiegand C, Krause E, Codd GA, Steinberg CE et al. (2003). Comparative effects and metabolism of two microcystins and nodularin in the brine shrimp *Artemia salina*. Aquat Toxicol. 62:219–226.

Berry JP, Gibbs PDL, Schmale MC, Saker ML (2009). Toxicity of cylindrospermopsin, and other apparent metabolites from *Cylindrospermopsis raciborskii* and *Aphanizomenon ovalisporum*, to the zebrafish (*Danio rerio*) embryo. Toxicon. 53:289–299.

BfR (2005). Mouse bioassay not suitable as a reference method for the regular analysis of algae toxins in mussels. Berlin: Bundesanstalt für Risikobewertung:7 pp.

Birbeck J, Westrick J, O'Neill G, Spies B, Szlag D (2019). Comparative analysis of microcystin prevalence in Michigan lakes by online concentration LC/MS/MS and ELISA. Toxins. 11:13.

Blom JF, Jüttner F (2005). High crustacean toxicity of microcystin congeners does not correlate with high protein phosphatase inhibitory activity. Toxicon. 46:465–470.

Blom JF, Robinson JA, Jüttner F (2001). High grazer toxicity of [D-Asp(3) (E)-Dhb(7)] microcystin-RR of *Planktothrix rubescens* as compared to different microcystins. Toxicon. 39:1923–1932.

Boaru DA, Dragoş N, Schirmer K (2006). Microcystin-LR induced cellular effects in mammalian and fish primary hepatocyte cultures and cell lines: a comparative study. Toxicology. 218:134–148.

Bober B, Bialczyk J (2017). Determination of the toxicity of the freshwater cyanobacterium *Woronichinia naegeliana* (Unger) Elenkin. J Appl Phycol. 29:1355–1362.

Brown A, Foss A, Miller MA, Gibson Q (2018). Detection of cyanotoxins (microcystins/nodularins) in livers from estuarine and coastal bottlenose dolphins (*Tursiops truncatus*) from Northeast Florida. Harmful Algae. 76:22–34.

Burrell S, Crum S, Foley B, Turner AD (2016). Proficiency testing of laboratories for paralytic shellfish poisoning toxins in shellfish by QUASIMEME: a review. Trends Anal Chem. 75:10–23.

Burýšková B, Hilscherová K, Babica P, Vršková D, Maršálek B, Bláha L (2006). Toxicity of complex cyanobacterial samples and their fractions in *Xenopus laevis* embryos and the role of microcystins. Aquat Toxicol. 80:346–354.

Caixach J, Flores C, Spoof L, Meriluoto J, Schmidt W, Mazur-Marzec H et al. (2017). Liquid chromatography-mass spectrometry. In: Meriluoto J, Spoof L, Codd G et al., editors: Handbook of cyanobacterial monitoring and cyanotoxin analysis. Chichester: John Wiley & Sons:218–257.

Carmichael WW (1992). Cyanobacteria secondary metabolites - the cyanotoxins. J Appl Microbiol. 72:445–459.

Catanante G, Espin L, Marty J-L (2015). Sensitive biosensor based on recombinant PP1α for microcystin detection. Biosens Bioelectron. 67:700–707.

Colas S, Duval C, Marie B (2020). Toxicity, transfer and depuration of anatoxin-a (cyanobacterial neurotoxin) in medaka fish exposed by single-dose gavage. Aquatic Toxicol. 222:105422.

de la Cruz AA, Hiskia A, Kaloudis T, Chernoff N, Hill D, Antoniou MG et al. (2013). A review on cylindrospermopsin: the global occurrence, detection, toxicity and degradation of a potent cyanotoxin. Environ Sci Process Impacts. 15:1979–2003.

Dell'Aversano C, Eaglesham GK, Quilliam MA (2004). Analysis of cyanobacterial toxins by hydrophilic interaction liquid chromatography–mass spectrometry. J Chromatogr A. 1028:155–164.

DeMott WR, Zhang QX, Carmichael WW (1991). Effects of toxic cyanobacteria and purified toxins on the survival and feeding of a copepod and three species of *Daphnia*. Limnol Oceanogr. 36:1346–1357.

Ding WX, Shen HM, Zhu HG, Lee BL, Ong CN (1999). Genotoxicity of microcystic cyanobacteria extract of a water source in China. Mut Res Gen Toxicol Environ Mutagen. 442:69–77.

Dorantes-Aranda JJ, Tan JY, Hallegraeff GM, Campbell K, Ugalde SC, Harwood DT et al. (2018). Detection of paralytic shellfish toxins in mussels and oysters using the qualitative neogen lateral-flow immunoassay: an interlaboratory study. J AOAC Int. 101:468–479.

Dörr FA, Rodríguez V, Molica R, Henriksen P, Krock B, Pinto E (2010). Methods for detection of anatoxin-a (s) by liquid chromatography coupled to electrospray ionization-tandem mass spectrometry. Toxicon. 55:92–99.

Eangoor P, Indapurkar AS, Vakkalanka MD, Knaack JS (2019). Multiplexed ELISA screening assay for nine paralytic shellfish toxins in human plasma. Analyst. 144:4702–4707.

EFSA (2009). Scientific opinion: marine biotoxins in shellfish–saxitoxin group. EFSA J. 1019:1–76.

enHealth (2012). Environmental health risk assessment. Canberra: Environmental Health Standing Committee:244 pp. https://www.health.gov.au.

Escher B, Leusch F (2011). Bioanalytical tools in water quality assessment. London: IWA publishing. 272 pp.

Falconer IR (1993). Measurement of toxins from blue-green algae in water and foodstuffs. In: Falconer IR, editors: Algal toxins in seafood and drinking water. London: Academic Press.

Fastner J, Heinze R, Humpage AR, Mischke U, Eaglesham GK, Chorus I (2003). Cylindrospermopsin occurrence in two German lakes and preliminary assessment of toxicity and toxin production of *Cylindrospermopsis raciborskii* (Cyanobacteria) isolates. Toxicon. 42:313–321.

Fayad PB, Roy-Lachapelle A, Duy SV, Prévost M, Sauvé S (2015). On-line solid-phase extraction coupled to liquid chromatography tandem mass spectrometry for the analysis of cyanotoxins in algal blooms. Toxicon. 108:167–175.

Ferrão-Filho AdS, Soares MCS, de Freitas Magalhães V, Azevedo SM (2009). Biomonitoring of cyanotoxins in two tropical reservoirs by cladoceran toxicity bioassays. Ecotoxicol Environ Safety. 72:479–489.

Fischer A, Höger SJ, Stemmer K, Feurstein D, Knobeloch D, Nussler A et al. (2010). The role of organic anion transporting polypeptides (OATPs/SLCOs) in the toxicity of different microcystin congeners in vitro: a comparison of primary human hepatocytes and OATP-transfected HEK293 cells. Toxicol Appl Pharmacol. 245:9–20.

Fischer WJ, Dietrich DR (2000). Toxicity of the cyanobacterial cyclic heptapeptide toxins microcystin-LR and-RR in early life-stages of the African clawed frog (*Xenopus laevis*). Aquat Toxicol. 49:189–198.

Fladmark KE, Serres MH, Larsen NL, Yasumoto T, Aune T, Døskeland SO (1998). Sensitive detection of apoptogenic toxins in suspension cultures of rat and salmon hepatocytes. Toxicon. 36:1101–1114.

Foss AJ, Aubel MT (2013). The extraction and analysis of cylindrospermopsin from human serum and urine. Toxicon. 70:54–61.

Foss AJ, Aubel MT (2015). Using the MMPB technique to confirm microcystin concentrations in water measured by ELISA and HPLC (UV, MS, MS/MS). Toxicon. 104:91–101.

Gademann K, Portmann C, Blom JF, Zeder M, Jüttner F (2010). Multiple toxin production in the cyanobacterium *Microcystis*: isolation of the toxic protease inhibitor cyanopeptolin 1020. J Nat Prod. 73:980–984.

Gaget V, Lau M, Sendall B, Froscio S, Humpage AR (2017). Cyanotoxins: which detection technique for an optimum risk assessment? Water Res. 118:227–238.

Gallacher S, Birkbeck T (1992). A tissue culture assay for direct detection of sodium channel blocking toxins in bacterial culture supernates. FEMS Microbiol Lett. 92:101–107.

Greer B, McNamee SE, Boots B, Cimarelli L, Guillebault D, Helmi K et al. (2016). A validated UPLC–MS/MS method for the surveillance of ten aquatic biotoxins in European brackish and freshwater systems. Harmful Algae. 55:31–40.

Grummt T, Kuckelkorn J, Bahlmann A, Baumstark-Khan C, Brack W, Braunbeck T et al. (2013). Tox-Box: securing drops of life-an enhanced health-related approach for risk assessment of drinking water in Germany. Environ Sci Europe. 25:27.

Grundler V, Faltermann S, Fent K, Gademann K (2015). Preparation of fluorescent microcystin derivatives by direct arginine labelling and their biological evaluation. ChemBioChem. 16:1657–1662.

Haddad SP, Bobbitt JM, Taylor RB, Lovin LM, Conkle JL, Chambliss CK et al. (2019). Determination of microcystins, nodularin, anatoxin-a, cylindrospermopsin, and saxitoxin in water and fish tissue using isotope dilution liquid chromatography tandem mass spectrometry. J Chromatogr A. 1599:66–74.

Harrison K, Johnson S, Turner AD (2016). Application of rapid test kits for the determination of paralytic shellfish poisoning (PSP) toxins in bivalve molluscs from Great Britain. Toxicon. 119:352–361.

Hathcock JN, Hattan DG, Jenkins MY, McDonald JT, Sundaresan PR, Wilkening VL (1990). Evaluation of vitamin A toxicity. Am J Clin Nutr. 52:183–202.

Hercog K, Maisanaba S, Filipič M, Jos Á, Cameán AM, Žegura B (2017). Genotoxic potential of the binary mixture of cyanotoxins microcystin-LR and cylindrospermopsin. Chemosphere. 189:319–329.

Heussner AH, Winter I, Altaner S, Kamp L, Rubio F, Dietrich DR (2014). Comparison of two ELISA-based methods for the detection of microcystins in blood serum. Chem Biol Interact. 223:10–17.

Höckelmann C, Becher PG, von Reuß SH, Jüttner F (2009). Sesquiterpenes of the geosmin-producing cyanobacterium *Calothrix* PCC 7507 and their toxicity to invertebrates. Zeitschrift für Naturforschung C. 64:49–55.

Huang W-J, Lai C-H, Cheng Y-L (2007). Evaluation of extracellular products and mutagenicity in cyanobacteria cultures separated from a eutrophic reservoir. Sci Tot Environ. 377:214–223.

Humpage A, Magalhaes V, Froscio S (2010). Comparison of analytical tools and biological assays for detection of paralytic shellfish poisoning toxins. Anal Bioanal Chem. 397:1655–1671.

Humpage AR, Fenech M, Thomas P, Falconer IR (2000). Micronucleus induction and chromosome loss in transformed human white cells indicate clastogenic and aneugenic action of the cyanobacterial toxin, cylindrospermopsin. Mutat Res. 472:155–161.

Humpage AR, Fontaine F, Froscio S, Burcham P, Falconer IR (2005). Cylindrospermopsin genotoxicity and cytotoxicity: Role of cytochrome P-450 and oxidative stress. J Toxicol Environ Health Part A. 68:739–753.

Humpage AR, Ledreux A, Fanok S, Bernard C, Briand JF, Eaglesham G et al. (2007). Application of the neuroblastoma assay for paralytic shellfish poisons to neurotoxic freshwater cyanobacteria: interlaboratory calibration and comparison with other methods of analysis. Environ Toxicol Chem. 26:1512–1519.

Hyenstrand P, Metcalf JS, Beattie KA, Codd GA (2001). Effects of adsorption to plastics and solvent conditions in the analysis of the cyanobacterial toxin microcystin-LR by high performance liquid chromatography. Water Res. 35:3508–3511.

Imhof L, Schmidt W (2017). SOP20: extraction and chemical analysis of saxitoxin and analogues water. In: Meriluoto J, Spoof L, Codd GA et al., editors: Handbook of cyanobacterial monitoring and cyanotoxin analysis. Chichester: John Wiley & Sons.

ISO (2000). ISO 13829. Determination of the genotoxicity of water and waste water using the umu-test. Geneva: International Organization for Standardization.

ISO (2005). ISO 20179. Determination of microcystins: method using solid phase extraction (SPE) and high performance liquid chromatography (HPLC) with ultraviolet (UV) detection. Geneva: International Organization for Standardization.

ISO (2011). ISO 14380. Determination of the acute toxicity to *Thamnocephalus platyurus* (Crustacea, Anostraca). Geneva: International Organization for Standardization.

ISO (2015). ISO 9001. Quality management systems - Requirements. Geneva: International Organization for Standardization.

James KJ, Furey A, Sherlock IR, Stack MA, Twohig M, Caudwell FB et al. (1998) Sensitive determination of anatoxin-a, homoanatoxin-a and their degradation products by liquid chromatography with fluorimetric detection. J Chromatogr A 798:147–157.

John N, Baker L, Ansell BR, Newham S, Crosbie ND , Jex AR (2019). First report of anatoxin-a producing cyanobacteria in Australia illustrates need to regularly up-date monitoring strategies in a shifting global distribution. Sci Rep. 9:1–9.

Kamp L, Church JL, Carpino J, Faltin-Mara E, Rubio F (2016). The effects of water sample treatment, preparation, and storage prior to cyanotoxin analysis for cylindrospermopsin, microcystin and saxitoxin. Chem-Biol Interact. 246:45–51.

Keil C, Forchert A, Fastner J, Szewzyk U, Rotard W, Chorus I et al. (2002). Toxicity and microcystin content of extracts from a *Planktothrix* bloom and two laboratory strains. Water Res. 36:2133–2139.

Kim J, Lim J, Lee C (2013). Quantitative real-time PCR approaches for microbial community studies in wastewater treatment systems: applications and considerations. Biotechnol Adv. 31:1358–1373.

Kiviranta J, Abdel-Hameed A, Sivonen K, Niemelä S, Carlberg G (1993). Toxicity of cyanobacteria to mosquito larvae—screening of active compounds. Environ Toxicol. 8:63–71.

Kiviranta J, Sivonen K, Niemelä S, Huovinen K (1991). Detection of toxicity of cyanobacteria by *Artemia salina* bioassay. Environ Toxicol. 6:423–436.

Kohler E, Grundler V, Häussinger D, Kurmayer R, Gademann K, Pernthaler J et al. (2014). The toxicity and enzyme activity of a chlorine and sulfate containing aeruginosin isolated from a non-microcystin-producing *Planktothrix* strain. Harmful Algae. 39:154–160.

Kós P, Gorzo G, Suranyi G, Borbely G (1995). Simple and efficient method for isolation and measurement of cyanobacterial hepatotoxins by plant tests (*Sinapis alba* L.). Anal Biochem. 225:49–53.

Krause E, Wenschuh H, Jungblut PR (1999). The dominance of arginine-containing peptides in MALDI-derived tryptic mass fingerprints of proteins. Anal Chem. 71:4160–4165.

Lankoff A, Wojcik A, Fessard V, Meriluoto J (2006). Nodularin-induced genotoxicity following oxidative DNA damage and aneuploidy in HepG2 cells. Toxicol Lett. 164:239–248.

Lawrence J, Menard C, Cleroux C (1995). Evaluation of prechromatographic oxidation for liquid chromatographic determination of paralytic shellfish poisons in shellfish. J AOAC Int. 78:514–520.

Lawrence JF, Niedzwiadek B, Menard C (2005). Quantitative determination of paralytic shellfish poisoning toxins in shellfish using prechromatographic oxidation and liquid chromatography with fluorescence detection: collaborative study. J AOAC Int. 88:1714–1732.

Lawton L, Beattie K, Hawser S, Campbell D, Codd G (1994a). Evaluation of assay methods for the determination of cyanobacterial hepatotoxicity. In: Codd GA, Jefferies T, Keevil C et al., editors: Detection methods for cynobacterial toxins. Cambridge, UK: The Royal Society of Chemistry:111–116.

Lawton LA, Edwards C, Codd GA (1994b). Extraction and High-Performance Liquid Chromatography method for the determination of microcystins in raw and treated waters. Analyst. 119:1525–1530.

Lecoz N, Malécot M, Quiblier C, Puiseux-Dao S, Bernard C, Crespeau F et al. (2008). Effects of cyanobacterial crude extracts from Planktothrix agardhii on embryo-larval development of medaka fish, *Oryzias latipes*. Toxicon. 51:262–269.

Li H-H, Chen R, Hyduke DR, Williams A, Frötschl R, Ellinger-Ziegelbauer H et al. (2017). Development and validation of a high-throughput transcriptomic biomarker to address 21st century genetic toxicology needs. Proc Natl Acad Sci USA. 114:E10881–E10889.

Li X, Cheng R, Shi H, Tang B, Xiao H, Zhao G (2016). A simple highly sensitive and selective aptamer-based colorimetric sensor for environmental toxins microcystin-LR in water samples. J Hazard Mat. 304:474–480.

Li X, Liu Y, Song L (2001). Cytological alterations in isolated hepatocytes from common carp (*Cyprinus carpio* L.) exposed to microcystin-LR. Environ Toxicol. 16:517–522.

Lindsay J, Metcalf J, Codd G (2006). Protection against the toxicity of microcystin-LR and cylindrospermopsin in *Artemia salina* and *Daphnia* spp. by pre-treatment with cyanobacterial lipopolysaccharide (LPS). Toxicon. 48:995–1001.

Maršálek B, Bláha L (2004). Comparison of 17 biotests for detection of cyanobacterial toxicity. Environ Toxicol. 19:310–317.

Maršálek B, Bláha L, Hindák F (2000). Review of toxicity of cyanobacteria in Slovakia. Biologia. 55:645–652.

Mazur-Marzec H, Sutryk K, Hebel A, Hohlfeld N, Pietrasik A, Błaszczyk A (2015). *Nodularia spumigena* peptides—Accumulation and effect on aquatic invertebrates. Toxins. 7:4404–4420.

McCall JR, Holland WC, Keeler DM, Hardison DR, Litaker RW (2019). Improved accuracy of saxitoxin measurement using an optimized enzyme-linked immunosorbent assay. Toxins. 11:632.

McNamee SE, Elliott CT, Greer B, Lochhead M, Campbell K (2014). Development of a planar waveguide microarray for the monitoring and early detection of five harmful algal toxins in water and cultures. Environ Sci Technol. 48:13340–13349.

Meriluoto J, Spoof L, Codd GA, EU-COST (2017). Handbook of cyanobacterial monitoring and cyanotoxin analysis. Chichester: John Wiley & Sons:576 pp.

Metcalf JS, Beattie KA, Saker ML, Codd GA (2002a). Effects of organic solvents on the high performance liquid chromatographic analysis of the cyanobacterial toxin cylindrospermopsin and its recovery from environmental eutrophic waters by solid phase extraction. FEMS Microbiol Lett. 216:159–164.

Metcalf JS, Codd GA (2000). Microwave oven and boiling waterbath extraction of hepatotoxins from cyanobacterial cells. FEMS Microbiol Lett. 184:241–246.

Metcalf JS, Lindsay J, Beattie K, Birmingham S, Saker M, Törökné AK et al. (2002b). Toxicity of cylindrospermopsin to the brine shrimp *Artemia salina*: comparisons with protein synthesis inhibitors and microcystins. Toxicon. 40:1115–1120.

Metcalf JS, Meriluoto JAO, Codd GA (2006). Legal and security requirements for the air transportation of cyanotoxins and toxigenic cyanobacterial cells for legitimate research and analytical purposes. Toxicol Lett. 163:85–90.

Metcalf JS, Young FM, Codd GA (2017) Performance assessment of a cylindrospermopsin ELISA with purified compounds and cyanobacterial extracts. Environ Forensics 18:147–152.

Nałęcz-Jawecki G, Tarczyńska M, Sawicki J (2002). Evaluation of the toxicity of cyanobacterial blooms in drinking water reservoirs with microbiotests. Fresenius Environ Bull. 11:347–351.

Norris RL, Seawright AA, Shaw GR, Smith MJ, Chiswell RK, Moore MR (2001). Distribution of ^{14}C cylindrospermopsin in vivo in the mouse. Environ Toxicol. 16:498–505.

Oberemm A, Becker J, Codd GA, Steinberg C (1999). Effects of cyanobacterial toxins and aqueous crude extracts of cyanobacteria on the development of fish and amphibians. Environ Toxicol. 14:77–88.

OECD (1997). Test No. 471: bacterial reverse mutation test. Guidelines for the testing of chemicals. Paris: Organisation for Economic Cooperation and Development. https://read.oecd-ilibrary.org/environment/.

OECD (2004). Test No. 202: *Daphnia* acute immobilisation test and reproduction test. Guidelines for the testing of chemicals. Paris: Organisation for Economic Cooperation and Development. https://read.oecd-ilibrary.org/environment/.

OECD (2011). Test No. 456: H295R steroidogenesis assay. Guidelines for the testing of chemicals. Paris: Organisation for Economic Cooperation and Development. https://read.oecd-ilibrary.org/environment/.

OECD (2012). Test No. 211: *Daphnia* magna reproduction test. Guidelines for the testing of chemicals. Paris: Organisation for Economic Cooperation and Development. https://read.oecd-ilibrary.org/environment/.

OECD (2013). Test No. 236: Fish Embryo Acute Toxicity (FET) test. Guidelines for the testing of chemicals. Paris: Organisation for Economic Cooperation and Development. https://read.oecd-ilibrary.org/environment/.

OECD (2016a). Test No. 455: performance-based test guideline for stably transfected transactivation in vitro assays to detect estrogen receptor agonists and antagonists. Guidelines for the testing of chemicals. Paris: Organisation for Economic Cooperation and Development.

OECD (2016b). Test No. 458: stably transfected human androgen receptor transcriptional activation assay for detection of androgenic agonist and antagonist activity of chemicals. Guidelines for the testing of chemicals. Paris: Organisation for Economic Cooperation and Development. https://read.oecd-ilibrary.org/environment/.

OECD (2016c). Test No. 476: in vitro mammalian cell gene mutation test. Guidelines for the testing of chemicals. Paris: Organisation for Economic Cooperation and Development. https://read.oecd-ilibrary.org/environment/.

OECD (2016d). Test No. 487: in vitro mammalian cell micronucleus test. Guidelines for the testing of chemicals. Paris: Organisation for Economic Cooperation and Development. https://read.oecd-ilibrary.org/environment/.

Okumura DT, Sotero-Santos RB, Takenaka RA, Rocha O (2007). Evaluation of cyanobacteria toxicity in tropical reservoirs using crude extracts bioassay with cladocerans. Ecotoxicology. 16:263–270.

Oshima Y (1995). Postcolumn derivatization liquid chromatographic method for paralytic shellfish toxins. J AOAC Int. 78:528–532.

Palíková M, Krejčí R, Hilscherová K, Babica P, Navrátil S, Kopp R et al. (2007a). Effect of different cyanobacterial biomasses and their fractions with variable microcystin content on embryonal development of carp (*Cyprinus carpio* L.). Aquat Toxicol. 81:312–318.

Palíková M, Krejčí R, Hilscherová K, Burýšková B, Babica P, Navrátil S et al. (2007b). Effects of different oxygen saturation on activity of complex biomass and aqueous crude extract of cyanobacteria during embryonal development in carp (*Cyprinus carpio* L.). Acta Veterinaria Brno. 76:291–299.

Palus J, Dziubałtowska E, Stańczyk M, Lewińska D, Mankiewicz-Boczek J, Izydorczyk K et al. (2007). Biomonitoring of cyanobacterial blooms in Polish water reservoir and the cytotoxicity and genotoxicity of selected cyanobacterial extracts. Int J Occup Med Environ Health. 20:48–65.

Papageorgiou J, Nicholson BC, Linke TA, Kapralos C (2005). Analysis of cyanobacterial-derived saxitoxins using high-performance ion exchange chromatography with chemical oxidation/fluorescence detection. Environ Toxicol. 20:549–559.

Patlewicz G, Simon TW, Rowlands JC, Budinsky RA, Becker RA (2015). Proposing a scientific confidence framework to help support the application of adverse outcome pathways for regulatory purposes. Regul Toxicol Pharmacol. 71:463–477.

Pekar H, Westerberg E, Bruno O, Lääne A, Persson KM, Sundström LF et al. (2016). Fast, rugged and sensitive ultra high pressure liquid chromatography tandem mass spectrometry method for analysis of cyanotoxins in raw water and drinking water—first findings of anatoxins, cylindrospermopsins and microcystin variants in Swedish source waters and infiltration ponds. J Chromatogr A. 1429:265–276.

Pflugmacher S, Wiegand C, Beattie KA, Krause E, Steinberg CEW, Codd GA (2001). Uptake, effects, and metabolism of cyanobacterial toxins in the emergent reed plant *Phragmites australis* (cav.) trin. ex steud. Environ Toxicol Chem. 20:846–852.

Pichardo S, Cameán AM, Jos A (2017). In vitro toxicological assessment of cylindrospermopsin: a review. Toxins. 9:402.

Portmann C, Blom JF, Gademann K, Jüttner F (2008). Aerucyclamides A and B: isolation and synthesis of toxic ribosomal heterocyclic peptides from the cyanobacterium *Microcystis aeruginosa* PCC 7806. J Nat Prod. 71:1193–1196.

Prieto AI, Guzmán-Guillén R, Díez-Quijada L, Campos A, Vasconcelos V, Jos Á et al. (2018). Validation of a method for cylindrospermopsin determination in vegetables: application to real samples such as lettuce (*Lactuca sativa* L.). Toxins. 10:63.

Sassolas A, Catanante G, Fournier D, Marty JL (2011). Development of a colorimetric inhibition assay for microcystin-LR detection: comparison of the sensitivity of different protein phosphatases. Talanta. 85:2498–2503.

Scherer M, Bezold D, Gademann K (2016). Investigating the Toxicity of the Aeruginosin Chlorosulfopeptides by Chemical Synthesis. Ang Chem Int Ed. 55:9427–9431.

Schroeder AL, Ankley GT, Houck KA, Villeneuve DL (2016). Environmental surveillance and monitoring — The next frontiers for high-throughput toxicology. Environ Toxicol Chem. 35:513–525.

Shoemaker J, Dietrich W (2017). Single laboratory validated method for determination of cylindrospermopsin and anatoxin-a in ambient water by liquid chromatography/tandem mass spectrometry (LC/MS/MS). Washington (DC): United States Environmental Protection Agency; Office of Research and Development. https://cfpub.epa.gov/si/index.cfm.

Shoemaker J, Tettenhorst D, De la Cruz A (2015). Method 544: Determination of microcystins and nodularin in drinking water by solid phase extraction and liquid chromatography/tandem mass spectrometry (LC-MS/MS). Washington (DC): Environmental Protection Agency of the United States:70 pp. EPA/600/R-14/474.

Sieroslawska A (2013). Evaluation of the sensitivity of organisms used in commercially available toxkits to selected cyanotoxins. Polish J Environ Stud. 22.

Sotero-Santos RB, Silva CRDSE, Verani NF, Nonaka KO, Rocha O (2006). Toxicity of a cyanobacteria bloom in barra bonita reservoir (middle tiete river, Sao Paulo, Brazil). Ecotoxicol Environ Safety. 64:163–170.

Stewart I, Carmichael WW, Sadler R, McGregor GB, Reardon K, Eaglesham GK et al. (2009). Occupational and environmental hazard assessments for the isolation, purification and toxicity testing of cyanobacterial toxins. Environ Health. 8:52.

Štraser A, Filipič M, Žegura B (2011). Genotoxic effects of the cyanobacterial hepatotoxin cylindrospermopsin in the HepG2 cell line. Arch Toxicol. 85:1617–1626.

Štraser A, Filipič M, Žegura B (2013). Cylindrospermopsin induced transcriptional responses in human hepatoma HepG2 cells. Toxicol in Vitro. 27:1809–1819.

Tarczyńska M, Nałęcz-Jawecki G, Brzychcy M, Zalewski M, Sawicki J (2000). The toxicity of cyanobacterial blooms as determined by microbiotests and mouse assays. In: Persoone G, Janssen C, De Coen W, editors: New microbiotests for routine toxicity screening and biomonitoring. New York: Kluwer Academic:527–532.

Tarczyńska M, Nalecz-Jawecki G, Romanowska-Duda Z, Sawicki J, Beattie K, Codd G et al. (2001). Tests for the toxicity assessment of cyanobacterial bloom samples. Environ Toxicol. 16:383–390.

Testai E, Scardala S, Vichi S, Buratti FM, Funari E (2016). Risk to human health associated with the environmental occurrence of cyanobacterial neurotoxic alkaloids anatoxins and saxitoxins. Crit Rev Toxicol. 46:385–419.

Tong P, Shao Y, Chen J, He Y, Zhang L (2015). A sensitive electrochemical DNA biosensor for *Microcystis* spp. sequence detection based on an Ag@ Au NP composite film. Anal Meth. 7:2993–2999.

Törökné AK, László E, Chorus I, Fastner J, Heinze R, Padisák J et al. (2000a). Water quality monitoring by Thamnotoxkit F™ including cyanobacterial blooms. Wat Sci Technol. 42:381–385.

Törökné AK, László E, Chorus I, Sivonen K, Barbosa FA (2000b). Cyanobacterial toxins detected by Thamnotoxkit (a double blind experiment). Environ Toxicol. 15:549–553.

Törökné AK, Vasdinnyei R, Asztalos M (2007). A rapid microbiotest for the detection of cyanobacterial toxins. Environ Toxicol. 22:64–68.

Tousova Z, Oswald P, Slobodnik J, Blaha L, Muz M, Hu M et al. (2017). European demonstration program on the effect-based and chemical identification and monitoring of organic pollutants in European surface waters. Sci Tot Environ. 601:1849–1868.

Triantis TM, Kaloudis T, Hiskia A (2017a). SOP16: determination of cylindrospermopsin in filtered and drinking water by LC-MS/MS. In: Meriluoto J, Spoof L, Codd GA et al., editors: Handbook of cyanobacterial monitoring and cyanotoxin analysis. Chichester: John Wiley & Sons:400–404.

Triantis TM, Kaloudis T, Hiskia A (2017b). SOP17: solid-phase extraction of anatoxin-a from filtered and drinking water. In: Meriluoto J, Spoof L, Codd GA et al., editors: Handbook of cyanobacterial monitoring and cyanotoxin analysis. Chichester: John Wiley & Sons:405–407.

Triantis TM, Kaloudis T, Zervou S-K, Hiskia A (2017c). SOP7: solid-phase extraction of microcystins and nodularins from drinking water. In: Meriluoto J, Spoof L, Codd GA et al., editors: Handbook of cyanobacterial monitoring and cyanotoxin analysis. Chichester: John Wiley & Sons:354–357.

Turner AD, Hatfield RG, Maskrey BH, Algoet M, Lawrence J (2019). Evaluation of the new European Union reference method for paralytic shellfish toxins in shellfish: a review of twelve years regulatory monitoring using pre-column oxidation LC-FLD. TrAC Trends Anal Chem. 113:124–139.

Turner AD, McNabb PS, Harwood DT, Selwood AI, Boundy MJ (2015). Single-laboratory validation of a multitoxin ultra-performance LC-hydrophilic interaction LC-MS/MS method for quantitation of paralytic shellfish toxins in bivalve shellfish. J AOAC Int. 98:609–621.

Turner AD, Waack J, Lewis A, Edwards C, Lawton L (2018). Development and single-laboratory validation of a UHPLC-MS/MS method for quantitation of microcystins and nodularin in natural water, cyanobacteria, shellfish and algal supplement tablet powders. J Chromatogr B. 1074:111–123.

UNECE (2017). Globally harmonized system of classification and labelling of chemicals (GHS), seventh edition. Geneva: United Nations Economic Commission for Europe. United Nations Publications:527 pp.

US EPA (2015). Method 545: determination of cylindrospermopsin and anatoxin-a in drinking water by liquid chromatography electrospray ionization tandem mass spectrometry (LC/ESI-MS/MS). Washington (DC): United States Environmental Protection Agency:27.

Van De Riet J, Gibbs RS, Muggah PM, Rourke WA, MacNeil JD, Quilliam MA (2011). Liquid chromatography post-column oxidation (PCOX) method for the determination of paralytic shellfish toxins in mussels, clams, oysters, and scallops: Collaborative study. J AOAC Int. 94:1154–1176.

Vasas G, Gáspár A, Surányi G, Batta G, Gyémánt G, Márta M et al. (2002). Capillary electrophoretic assay and purification of cylindrospermopsin, a cyanobacterial toxin from *Aphanizomenon ovalisporum*, by plant test (Blue-Green Sinapis Test). Anal Biochem. 302:95–103.

Vezie C, Benoufella F, Sivonen K, Bertru G, Laplanche A (1996). Detection of toxicity of cyanobacterial strains using *Artemia salina* and MicrotoxR assays compared with mouse bioassay results. Phycologia. 35:198–202.

Vieth R (2007). Vitamin D toxicity, policy, and science. J Bone Miner Res. 22:V64–V68.
Waack J (2017). Uptake and depuration of cyanotoxins in the common blue mussel Mytilus edulis. Aberdeen: Institution:374 pp.
Wang P-J, Chien M-S, Wu F-J, Chou H-N, Lee S-J (2005). Inhibition of embryonic development by microcystin-LR in zebrafish, *Danio rerio*. Toxicon. 45:303–308.
Welker M, Bickel H, Fastner J (2002a). HPLC-PDA detection of cylindrospermopsin - opportunities and limits. Water Res. 36:4659–4663.
Welker M, Fastner J, Erhard M, von Döhren H (2002b). Application of MALDI-TOF MS in cyanotoxin research. Environ Toxicol. 17:367–374.
Welker M, Khan S, Haque MM, Islam S, Khan NH, Chorus I et al. (2005). Microcystins (cyanobacterial toxins) in surface waters of rural Bangladesh – pilot study. J Water Health. 3:325–337.
Welker M, von Döhren H, Täuscher H, Steinberg CEW, Erhard M (2003). Toxic *Microcystis* in shallow lake Müggelsee (Germany) - dynamics, distribution, diversity. Arch Hydrobiol. 157:227–248.
Weller MG (2013). Immunoassays and biosensors for the detection of cyanobacterial toxins in water. Sensors. 13:15085–15112.
Wiese M, D'agostino PM, Mihali TK, Moffitt MC, Neilan BA (2010). Neurotoxic alkaloids: saxitoxin and its analogs. Mar Drugs. 8:2185–2211.
Wood SA, Biessy L, Puddick J (2018). Anatoxins are consistently released into the water of streams with *Microcoleus autumnalis*-dominated (cyanobacteria) proliferations. Harmful Algae. 80:88–95.
Žegura B (2016). An overview of the mechanisms of microcystin-LR genotoxicity and potential carcinogenicity. Mini Rev Med Chem. 16:1042–1062.
Žegura B, Sedmak B, Filipič M (2003) Microcystin-LR induces oxidative DNA damage in human hepatoma cell line HepG2. Toxicon. 41:41–48.
Žegura B, Štraser A, Filipič M (2011). Genotoxicity and potential carcinogenicity of cyanobacterial toxins - a review. Mutat Res, Rev Mutat Res. 727:16–41.
Žegura B, Volčič M, Lah TT, Filipič M (2008). Different sensitivities of human colon adenocarcinoma (CaCo-2), astrocytoma (IPDDC-A2) and lymphoblastoid (NCNC) cell lines to microcystin-LR induced reactive oxygen species and DNA damage. Toxicon. 52:518–525.
Zervou S-K, Christophoridis C, Kaloudis T, Triantis TM, Hiskia A (2017). New SPE-LC-MS/MS method for simultaneous determination of multi-class cyanobacterial and algal toxins. J Hazard Mat. 323:56–66.
Zhan L, Sakamoto H, Sakuraba M, Wu DS, Zhang LS, Suzuki T et al. (2004). Genotoxicity of microcystin-LR in human lymphoblastoid TK6 cells. Mutat Res-Genet Toxicol Environ Mutagen. 557:1–6.
Zhang W, Han C, Jia B, Saint C, Nadagouda M, Falaras P et al. (2017). A 3D graphene-based biosensor as an early microcystin-LR screening tool in sources of drinking water supply. Electrochimica Acta. 236:319–327.

Chapter 15

Public health surveillance, public communication and participation

Lesley V. D'Anglada

CONTENTS

INTRODUCTION

The role of public health authorities is to protect, assess and ensure the health of people and communities. These agencies also play a role in promoting healthy environments, thus reducing the toll from illness due to exposure to pathogens or harmful substances such as cyanotoxins in drinking and recreational waters, in food or in water used for dialysis. Legal authority and regulations facilitate the control and management of blooms as well as public health responses and risk communication when they do occur.

The role of the responsible authority is likely to focus on surveillance, including independent verification of water quality and ideally, assessment that Water Safety Plans (WSPs) are being implemented effectively, rather than the day-to-day on-site management and monitoring. Operators of drinking-water supplies and managers of recreational sites or occupational water use are required for the day-to-day management (and assessment) of risks, including those from cyanotoxins. However, the role of authorities may be broader where regulations are sparse, water quality requirements such as limits for cyanotoxin concentrations have not been defined, institutional capacity is limited or the surveillance of water-use systems is challenging because of their high number, geographic spread or remoteness. Such situations may require an active role of public authorities in management, for example, in the development of WSPs (see Chapter 6). This chapter focuses on the role of public authorities in surveillance, the development and implementation of Incident Response Plans (IRPs) as well as in communicating risks to the public.

The WHO Framework for safe drinking-water outlines the key steps in providing safe drinking-water (Figure 15.1; see also the Guidelines for Drinking-Water Quality (WHO, 2017), Chapter 1), and these key steps can also be applied to safe design, operation and management of recreational or occupational water-use sites. Within this framework, public authorities have a role particularly at the "front end", that is, in setting targets, and at the "back end", that is, in surveillance. The authorities responsible for both may be different, operating on different levels: while setting targets often occurs on the national level by legislation, surveillance is typically local, requiring good knowledge of the local conditions and challenges.

Setting cyanotoxin water quality targets or action thresholds can be based on the guideline values summarised in Chapter 5 (see also Chapter 2 for their derivation), with the guideline values for short-term exposure through drinking-water being particularly relevant during bloom events. How the guideline values for lifetime exposure "translate" into targets for waterbody management is discussed in Chapters 6 and 7. In an event of a cyanotoxins incident, it is important to consider the risks from exposure to cyanotoxins

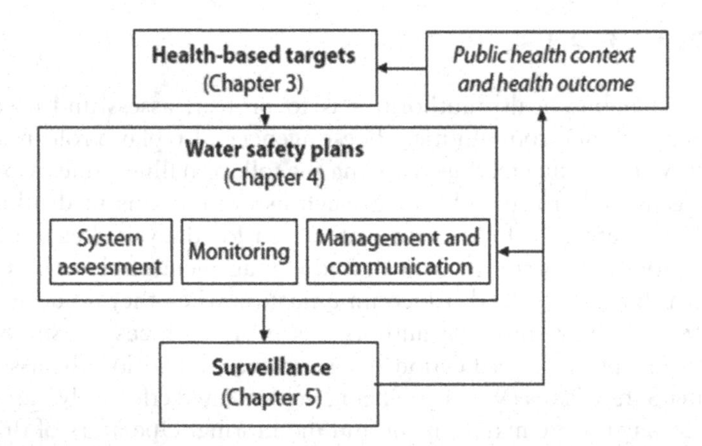

Figure 15.1 Framework for safe drinking-water (from the WHO Guidelines for Drinking-water Quality; WHO, 2017).

in relation to health risks from other microorganisms and chemicals (see section 5.1 and Chapter 6 for a discussion on target setting).

15.1 ASPECTS OF SURVEILLANCE

The other key role of public authorities, that is, surveillance, is often perceived to focus on assessing whether water quality meets the targets defined for a given parameter, such as cyanobacterial biomass or cyanotoxin concentrations. However, for drinking-water supply, surveillance is much more effective if it also includes a critical review of the facilities, their surroundings and operation, including operational parameters. This is best done through inspections of the site, review of records of operational parameters and conversations with operating staff. If management plans such as WSPs (see Chapter 6) and IRPs (see below) are in place, this greatly facilitates surveillance and provides a useful basis for discussions on potential improvement with operators and managers. This also applies to small supplies and situations with limited resources, where WSP development can be particularly useful (for more information, see WHO, 2012). As small supplies are typically less complex, system description, hazard analysis and risk assessment tend to be more straightforward and more readily accomplished even with a lower level of expertise, for example, by using a sanitary inspection as basis for the WSP. Outcomes may be highly valuable, allowing the water authority to prioritise its activities. For drinking-water supplies as well as for recreational sites or workplaces, surveillance should start with site inspections to assess the risk of cyanobacterial blooms, based on historical events and environmental conditions that lead to cyanobacterial bloom formation. Surveillance therefore requires an understanding of the systems – from catchment to the point of use and possible human exposure. The guidance given in Chapters 5–10 presents the necessary background, both for operators and for authorities performing surveillance, on assessing and managing risks of cyanobacterial blooms.

Through surveillance, public authorities gather a wide overview of conditions causing blooms and thus develop a locally and regionally specific understanding of the water systems. This enables them to effectively advise operators of drinking-water supplies and managers of recreational sites or workplaces on measures that have proven effective in similar cases. The operator of a drinking-water supply or manager of a recreational site is responsible for identifying hazards, assessing risks and identifying as well as implementing control measures, including organising collaboration with other public authorities and agencies. However, particularly in small-scale situations with limited resources, the role of public authorities can also involve triggering networking and exchange of experience between operators as well as organising collaboration.

Across the globe, different authorities may be responsible for responding to cyanotoxin occurrence, and responsibility may also be shared between environmental and health authorities. For managing cyanobacteria and cyanotoxins, contact and exchange are particularly important between health and environmental authorities, but in some cases also with those responsible for allocating water to specific uses and managing flow regimes (in some countries termed "water boards"). This is a basis for developing management strategies that address the problem at its source: that is, the causes for cyanobacterial proliferation and bloom formation.

To ensure appropriate responses to cyanobacterial bloom incidents, close coordination with all partners, including environmental authorities, is particularly critical so that those with responsibilities for specific incidence response actions are prepared to react quickly when contacted during the incident, to restore drinking-water service. IRPs help in providing the tools needed for an effective response and the protection of public health during a cyanobacterial bloom. Each cyanotoxin event is different, and correspondingly, the characteristics of the area, available resources, the interaction with outside partners and the response will be specific to the situation.

15.2 INCIDENT RESPONSE PLANS FOR CYANOBACTERIAL BLOOMS

The assessment of water-use systems according to Chapters 6–8 will show whether conditions are likely to support cyanobacterial dominance or blooms and whether they should be expected in surface waters used for drinking-water supplies, recreational or occupational use, particularly where there is a previous history of blooms. This can occur even when management measures to reduce their likelihood have recently been implemented because these measures usually require several years to start having an effect. In mildly eutrophic waterbodies, cyanobacterial dominance may occur only occasionally and as short-lived events, thus being perceived as an unusual incident. In more heavily eutrophic waterbodies, they may be a regular phenomenon throughout several months of the year, to which regular management actions such as drinking-water treatment or periodic warnings regarding recreational use have been adapted. Nonetheless, even in such settings, particularly dense blooms may constitute an "incident". Depending on the local conditions, Incident Response Plans (IRPs) will describe the actions and responses to be applied within a water-use system when events, such as a cyanobacterial bloom is not sufficiently controlled by normal operating procedures, occur. IRPs are typically developed by site operators or waterbody managers but approved by the public health authority. However, recreationally used waterbodies and beaches may not be formally managed or operated, and the responsibility for their monitoring may lie with the health authority which then also is responsible for coordinating the implementation of the IRP.

IRPs include incident criteria, roles and responsibilities, communication protocols, contact information of responsible authorities to involve in the response, mechanisms for monitoring and controlling the bloom and, where appropriate, the communication to the public: both about the risk and about actions to take to avoid exposure. It may be useful to prepare templates that site operators can adapt depending on the situation and the available resources (see Tables 15.1–15.3 for examples). Sections 15.3–15.5 outline the three important steps for the responsible authority to follow in response to cyanotoxin bloom incidents in drinking-water and water to which people are exposed during occupational or recreational use. These steps include monitoring, management to control the incident and risk communication. The Alert Levels Frameworks (ALFs) for drinking-water supplies (section 5.1) and for recreational waterbody use (section 5.2) give criteria for identifying an incident (i.e., Alert Level 2), provide a structure for responses to monitoring results that can be used directly when developing the IRP or adapted to local circumstances, as needed.

Table 15.1 Incident response and/or risk communication task force contacts

Principal Authority:					
Name	Title	Incident Role	Phone	E-mail	Address
			Office: Mobile:		
			Office: Mobile:		
Partner Authority:					
Name	Title	Incident Role	Phone	E-mail	Address
			Office: Mobile:		
			Office: Mobile:		
Partner Authority:					
Name	Title	Incident Role	Phone	E-mail	Address
(Cont.)					

Authority: e.g., Drinking-Water Supplier, Ministry of Health and its regional or departmental offices, Environmental Protection, Health Departments, Local Governments, Emergency Management, Environmental (or) Public Health Laboratories, etc.

Title: Drinking-water supplier, water treatment operator, recreational site manager, engineer, water quality officer, etc.

Table 15.2 Checklist of resources and capabilities for responses to a cyanobacterial incident

__ Roles and contact details for key personnel and other related partners are clearly stated;

__ Trigger levels for action to take during cyanobacterial biomass (in terms of biovolume or of chlorophyll-*a*) and cyanotoxins, including Alert Levels Framework (ALF), are established;

Monitoring

__ Appropriate personnel to perform
  __ Monitoring/Sampling __ Laboratory analysis
  is identified and contacts are documented;

__ Appropriate monitoring and sampling procedures have been established;

__ Appropriate public health laboratories to conduct sample analysis are identified;

__ Appropriate SOPs and QA/QC protocols have been established;

__ Monitoring and sampling records templates have been developed;

__ Required equipment and materials are available and their storage site is described in the incident response plan for
  __ Monitoring/sampling __ Laboratory analysis

Management and Control

__ Appropriate personnel to perform the control/mitigation and treatment techniques is identified and contacts are documented;

__ Clear description of the actions required in the event of a cyanotoxins incident in
  __ Drinking-water __ Recreational sites
  have been developed and are described in the IRP;

__ Appropriate mitigation/control measures for blooms in surface waters have been identified and are available;

__ Appropriate treatment techniques for the removal of cyanotoxins in drinking-water have been identified and are available;

__ Plans for alternative water supply including how to obtain, transport and distribute the alternate sources are available;

__ Templates to record the mitigation/control and treatment techniques have been developed;

Risk Communication

__ Appropriate personnel to perform risk communication is identified and prepared;

__ A risk communication plan with a list of contacts, communication steps and dissemination outlets is available;

__ Checklists, templates, Q and A, fact sheets and other reference materials including technical information (e.g., explanation of ALF) have been prepared and are up to date;

__ A post-incident comprehensive assessment is available.

Table 15.3 Post-cyanobacterial incident-response assessment checklist

Date of Assessment: _______ **Date and Location of Incident:** _________
Incident-Response Responsible Agency: ___________________
Responsible Point of Contact Information: ___________________
Assessment Committee or Task Force Members:
____
____

For each of the areas below, please check the factors that met the requirements of a successful response. In the Comments section, identify and describe those that require improvement.

Monitoring

__ Availability and skill level of personnel in charge of __ Monitoring/Sampling
__ Laboratory analysis

__ Appropriate monitoring and sampling procedures

__ Timely contact and services with public health laboratories to conduct sample analysis

__ Availability of clear and effective SOPs and QA/QC protocols

__ Availability of clear monitoring and sampling records and templates available

__ Availability and functionality of required equipment and materials

__ Monitoring/Sampling

__ Laboratory analysis

Comments:
____
____

Management and Control

__ Personnel in charge of control/mitigation and treatment techniques were available and skilled;

__ Description of the required steps to follow for incidents for:

__ Drinking-water __ Recreational sites

__ Mitigation/control measures for blooms in surface waters were available and effective;

__ Treatment techniques for the removal of cyanotoxins in drinking-water were available and effective;

__ The transportation and distribution of alternative water supply were effective;

__ Mitigation/control and treatment techniques records forms were appropriate;

Comments:
____
____

(*Continued*)

Table 15.3 (Continued) Post-cyanobacterial incident-response assessment checklist

Risk Communication

__ Personnel in charge of the Risk Communication were available and skilled;

__ The risk communication plan (list of contacts, communication steps and dissemination) was appropriate;

__ The checklists, templates, questions and answers, fact sheets and other reference materials including technical information (e.g., explanation of ALF) were appropriate and up to date;

__ Responses in the media met expectations;

__ New communication problems arose.

Comments:

____

____

Additional Discussion Questions:

1. What actions were successful that should be replicated in future incidents?
2. What actions did not work as planned? Why?
3. List any procedures, templates, checklists or communication materials that need revision.
4. Please list the remediation actions and who will be involved in doing them.
5. Who will inform the responsible agency and partner authority/agencies of the improvements and changes?
6. What is the time frame for making the revisions and informing others?
7. ____
8. ____

15.3 ROLES AND CAPACITIES OF THE RESPONSIBLE AUTHORITY IN INCIDENCE RESPONSES

Convening a multiagency and multidisciplinary committee, or task force, is essential for an effective and rapid surveillance and response to a cyanotoxin bloom. The IRP therefore should clearly define the responsible personnel, including roles, responsibilities and legal liabilities (Table 15.1). These contacts listed in the IRP are also responsible for coordinating with further partners who might be involved during the cyanotoxin incident. Stakeholders to consider to include in the IRP for further involvement, with clear roles and responsibilities, may include the ministry of health (or public health) and its regional or departmental offices, environmental protection authorities, health departments, local governments, emergency management agencies, medical and veterinary personnel, water suppliers, drinking-water consumers, recreational site operators and users, and the public. Other potential response partners include neighbouring environmental and/

or public health laboratories, other drinking-water utilities and the media. The roles outlined in the IRP should provide a description of the tasks for which each should be prepared and what is expected from the other agencies and supporting partners before, during and after the bloom. Their roles should be outlined clearly and regularly updated, together with contact information. The contact information should include the names, titles, addresses and all applicable phone numbers, as well as a secondary contact in case the primary contact cannot be reached.

The responsible authorities should also identify the resources, infrastructure and staff (Table 15.2) to effectively respond to the cyanotoxin incident. Available resources include necessary tools and equipment (e.g., sampling equipment) and laboratories that may be approached if needed. Drinking-water providers should determine their type of intakes and depths and establish if they are able to draw raw water from a different intake and/or depth, with approval from the drinking-water regulator, as appropriate. They also should be aware of treatment adjustments that are beneficial as well as those that may exacerbate problems: for example, inducing lysis when water with cyanobacterial cells is subjected to certain treatment steps (see Chapter 10). For recreational use of eutrophic waterbodies which often harbour some cyanobacteria but only sometimes develop blooms reaching Alert Level 2 (see section 5.2), it will be important to give renewed information to site users, emphasising the use restrictions that now apply under Alert Level 2 but were not yet in place under Alert Level 1. Where comprehensive monitoring of sites used for recreation is not possible due to their very large number or due to limited resources, an option may be to include volunteer citizens in observing and reporting blooms ("scum scouting"). Monitoring may also include tools already developed by partners such as satellite imagery, bulletins, systems for notifying other agencies and monitoring programmes (see below) that may be ongoing for other purposes, such as monitoring for parameters other than cyanobacteria.

Incident planning requires not only a list of the communication and public outreach mechanisms, such as websites, e-mail alerts and social media channels, but also convening a committee or task force with the staff responsible for coordinating public communication to ensure that conflicting information is avoided. The responsible authorities should have also confirmed that the staff involved in the response have the necessary skills to conduct monitoring and are capable of effective risk communication. It is further useful for such staff to understand the conditions leading to blooms (described in Chapters 4 and 8) in order to better anticipate bloom events and tailor intensified surveillance to such periods.

A specific aspect of incidence response is the occurrence of suspected cases of illness linked to cyanobacterial occurrence. As discussed in sections 5.1 and 5.2, the limitation with many of the published cases suspected to have been caused by cyanobacteria is the lack of data on cyanotoxin concentrations in the water to which people were exposed to. If samples were taken at all, this usually occurred days later. It is therefore useful for Incidence Response Plans to include contacts of medical services and requirements to report the incident to them for two reasons: to keep them informed about heavy blooms and possible human exposure, and to ensure that they inform those who can initiate immediate sampling if cases of illness due to cyanobacteria are suspected. Criteria for concluding a likely link between cyanotoxins and the illness include that the symptoms are typical for the respective cyanotoxin and that concentrations were in a range possibly causing them. In the United States of America, the Center for Disease Control developed a voluntary reporting system called *One Health Harmful Algal Bloom System* (*OHHABS*) to collect data on individual human and animal cases of illnesses from HAB-associated exposures, as well as environmental data, to support the understanding and prevention of HABs and HAB-associated illnesses. This reporting system is available to public health departments and their designated environmental health partners to help them better understand and identify the effects of cyanobacteria on humans, animals and the environment. Unspecific symptoms may be caused by other aetiological agents (including pathogenic microorganisms) that may or may not be associated with the bloom, or other unknown substances in cyanobacteria.

15.4 MONITORING

Where the responsible authority is involved in surveillance monitoring, it may already be positioned to include cyanobacteria or even cyanotoxins. However, many countries do not include cyanobacteria or cyanotoxins in regular surveillance, or have limited resources to conduct monitoring. In cases with limited resources, collaboration with expert support and/or creating partnerships at the regional or international level is useful for guidance on surveillance alternatives. As discussed in Chapter 11, an effective way forward may also be to create regional centres of excellence that can perform periodic cyanotoxin analyses on smaller numbers of selected samples that serve for orientation regarding the ratios of toxins to biomass or cell counts that then can be used locally for the bulk of samples. This is particularly useful where access to a chemical laboratory is missing or limited but microscopy is available for determining cell counts and biovolumes. Furthermore, seeking collaboration with research institutions can be very effective since they may have valuable expertise and analytical capacities. The example given in Box 15.1 illustrates how such collaboration can enable a low-cost approach to monitoring beaches. It is important that the Incident Response Plan (IRP) include contacts to laboratories that can analyse for cyanobacteria/toxins, and that agreements are in place for rapid reaction should a bloom occur that requires a rapid assessment.

BOX 15.1: THE EXPERIENCE OF URUGUAY WITH CYANOTOXIN RISK COMMUNICATION AND MANAGEMENT

Beatriz Brena

The Rio de la Plata, located between Argentina and Uruguay, is a broad, funnel-shaped estuary that drains the waters of two important rivers (Parana and Uruguay) into the Atlantic Ocean. With a basin of 3.2 million km^2, the second largest in South America after the Amazon, it has about 150 million inhabitants (35% of South America) and more than 75 big reservoirs for hydroelectric power generation. The main regional economic activities are agriculture and livestock production.

In the past 20 years, intense cyanobacterial blooms became frequent especially in the main reservoirs of the basin and reached Montevideo, the capital of Uruguay, a city with 1.3 million inhabitants located in the middle of the Rio de la Plata in the salinity and turbidity front of the estuary. The blooms had a great impact in the quality of life of people, since the estuary has a long coast of sandy beaches intensively used for recreation of both locals and tourists. Eutrophication could be associated with an intensification of agriculture; for example, the use of fertilisers increased more than threefold between 2000 and 2010.

The predominant cyanobacteria being *Microcystis*, mostly *M. aeruginosa*, considering its potential production of toxins, in the year 2000, the City Government implemented regular beach and coastal water monitoring. This included visual detection of blooms as well as analyses of chlorophyll-*a* and nutrients. The visual monitoring approach, performed at the beach, was based on a simple categorisation of samples in three groups: (i) "absence" of blooms when the operator does not detect any cyanobacterial colony by visual inspection and there are no signs of water discoloration; (ii) "presence of dispersed colonies" when colonies are observed from a close distance, for example, when entering into the water; and (iii) "scums", when the accumulation of colonies produces green colour like spilled paint, noticeable from several metres from the shore.

At the beginning, there was no public awareness of the risks associated with these blooms; they were mostly perceived as an aesthetic problem. The first scientific report of the presence of microcystins in the Rio de la Plata was in 2001, and the analysis was performed in Brazil, since no analytical capacity for cyanotoxins was installed in Uruguay. A collaboration with the University of the Republic of Uruguay was then established, starting in the summer of 2004, for the development and validation of an ELISA for microcystins, which was then included in the regular monitoring of beach water. The results demonstrated extremely high microcystin concentrations in the scums (up to 30 000 µg/L) and prompted the notification to the public to prevent recreational exposure.

The data accumulated over the first 6 years of monitoring showed that more than 95% of the scums were very toxic (mean 3300 µg/L), while microcystin concentrations in most of the samples in the "presence of dispersed colonies" were very low (<0.3 µg/L); however, even in this intermediate category 5.6% of the samples contained more than 20 µg/L. Noteworthy, when no blooms were detected (category "absence"), microcystin concentrations in the samples were below or equal to 0.3 µg/L.

These data serve to support the risk management approach, based on visual observation and a simple and fast method for microcystin determination. Since the antibodies and specific reagents of the ELISA kit were developed locally, the cost of analysis was very low and the approach is sustainable.

As Montevideo is in the salinity and turbidity front of the estuary, the intensity and frequency of the blooms is very variable and depends mainly on the inflow of water from the major rivers Paraná and Uruguay. An increase in the freshwater discharge due to rainfall in the upper basin is associated with the upcoming of blooms, originating upstream in the major reservoirs. In the summer of 2010, under the effect of "El Niño", scums in the coastal waters of Montevideo were very frequent (blooms occurred about 40% of the time along parts of the coast). Improving public communication and emphasising the need to prevent exposure was therefore indispensable. Recreational use of the beaches is a major activity in summers, but unfortunately, many people disregard the warning messages.

In consequence, a so-called sanitary flag (red flag with a green cross in the middle; see photo) was implemented. An intensive campaign in the public media explained its meaning and relevance. Furthermore, as the blooms can be highly dynamic, particularly in Montevideo where they can appear and disappear very rapidly, for example, within 1 h or even less, depending on the beach, lifeguards were trained to recognise the presence of blooms. Thus, a rapid on-site response was made possible at each beach at any moment during the course of a day, if necessary. Even so, in 2015, there was a report of a serious intoxication, most likely attributable to ingestion of bloom material (see section 5.2), of a 20-month-old girl who required a liver transplant. This means that there is still a lot to learn to prevent intoxication and to promote proper care of children and sensitive populations. At present, the National Environmental Direction generalised the use of the sanitary flag countrywide, and the information of beaches with a sanitary flag is published daily on the web so that the population can decide whether it is safe to go to the beach.

A further aspect of preparing for monitoring in the context of incident response is to clarify which threshold concentrations of cyanobacterial biomass (measured as biovolume, chlorophyll-*a* or other parameters chosen locally; see the Alert Level Frameworks (ALFs) in sections 5.1 and 5.2) or of cyanotoxins are to trigger which responses. This requires identifying whether regulations and/or guidance for cyanobacteria or cyanotoxins are in place, particularly for drinking-water, and if so, to include the respective statutory and regulatory requirements when developing the IRP. Where these are lacking, the drinking-water guidance values as well as the guidance values for recreational exposure, in Chapter 2 of this book, serve for orientation. Furthermore, the ALFs given for drinking-water in section 5.1 and for recreational or occupational exposure in section 5.2 highlight the sequence of events to follow for monitoring and management. The Alert Level values are intended for managers of water supplies, recreational or occupational sites and may be used both for normal day-to-day operations and for situations in which blooms escalate to be an incident that requires a quick response. When developing the IRP, the ALF templates given here are best adapted to the locally specific conditions, including availability of analytical capacity for determining concentrations of cyanobacterial biomass and/or cyanotoxins.

Monitoring may be tailored not only for verification of whether water quality targets or action thresholds for cyanobacteria or cyanotoxins are met, but also to assess whether the implemented control measures are achieving the desired objectives. Both for surveillance monitoring and for the IRP, it is important that operators as well as authorities develop and implement sampling procedures, sample analysis processes, and quality control and assurance plans (see Chapters 11–14). This involves coordination with the respective internal and/or external laboratories regarding sampling procedures, preservation, shipment and laboratory requirements. Documentation from the laboratory conducting the analyses should also be kept, including number of samples, a description of analytical methods, sampling, sample transport and analytical quality assurance procedures, and the results.

It is further worthwhile to consider partnering with others (e.g., citizens' monitoring programmes) for further support in monitoring. For example, swimmers and other users can contact local authorities if they see cyanobacterial scums, and householders can report unusual odours in their drinking-water supply. In the United States of America, community-based "Water-Watch" or "Stream-Watch" monitoring programmes undertaken by high school students and community groups have been initiated to monitor and report the presence of cyanobacterial blooms (see US EPA Citizen Science Projects for more information; https://www.epa.gov/citizen-science). In Argentina, a citizen science project "Why are our reservoirs green?" (abbreviated CIANOBs) involves school children in reporting blooms (see Box 15.3).

Monitoring data should be recorded and maintained. Monitoring procedures should be reviewed regularly and tailored to the current conditions of the surface water and/or treatment plant, including consideration of the available resources. This can be an effective component of the periodic review of a Water Safety Plan (WSP).

If monitoring results indicate the presence of cyanotoxins in the surface water, further monitoring may be needed. Monitoring frequency as well as communication procedures will vary depending on the ALF and other factors such as the cost of monitoring and available resources.

15.5 MANAGEMENT AND CONTROL MEASURES

If cyanotoxins or substantial amounts of cyanobacteria are detected, the responsible authority should work together with operators of drinking-water supplies and managers of recreational sites, as appropriate, and with health and cyanobacteria specialist or experts as well as public health laboratories to determine whether immediate or short-term responses are needed and which longer-term measures are appropriate. Many larger-scale operations, that is, drinking-water supplies, recreational sites or workplaces, are responsible for day-to-day operations, including monitoring and incidence response planning as well as the notification of exceedances and the proposed response to the public surveillance authority. In such operations, the role of the public authority is to assess and approve the IRP proposed by the operators as well as to facilitate and support its implementation. However, in smaller-scale operations, a more active, coordinating role of the health authority may be necessary to fulfil these responsibilities.

Consultation with key technical (e.g., scientific and engineering) experts will help in both assessing the risk based both on the information about the waterbody (Chapter 9) and on the available laboratory data as well as in determining the locally appropriate control measures. These measures may encompass interventions taking immediate effect, particularly in drinking-water treatment (Chapter 10), shifting water use to sites less prone to scum accumulation (Chapter 8) or restricting recreational use (section 5.2 and the ALF in Figure 5.4). However, a bloom incident should also trigger planning measures for prevention of blooms in future, using the momentum of expertise and experience with the waterbody that has already come together for the immediate response. Measures addressing the cause of blooms typically take longer to take effect, for example, controlling nutrient loads from the catchment (Chapter 7) or managing hydrophysical or food-chain conditions in the waterbody (Chapter 8). Some control measures (e.g., shifting recreational sites, applying algaecides or artificial mixing) may be subject to specific requirements or regulations, thus requiring consultation with the respective regulatory body that may need to issue a permit for conducting the measure.

Where immediate or short-term actions cannot be taken, or when short-term water quality targets or action thresholds for toxins are exceeded (or bloom biomass indicates this potential), a temporary switch to an alternative drinking-water supply may be appropriate. Where resources are lacking for upgrading drinking-water treatment during blooms and bloom incidents occur regularly, the incident response plan (IRP) should include the identification, if feasible, of potential alternative water supplies. This may include plans for transporting clean, treated water from other areas or deploying portable water treatment systems, if available. Special precautions (e.g., portable water treatment systems or transported safe water supplies) may be advisable for "at-risk" groups especially susceptible to cyanotoxins, such as bottle-fed infants, small children and patients with previous acute liver and kidney damage (preferably identified in the course of developing a WSP; see Chapter 6). Boil water advisories are not recommended as boiling water will not remove the cyanotoxins. Other options are providing water in tanks or bottles to the affected population. This also requires specific planning when preparing an IRP, including pathways for providing information on distribution locations for bottled water to the affected communities.

When selecting a treatment or other control measure, the responsible operators or authority should consider any regulatory restrictions (such as mentioned above, e.g., for algicide application), specific characteristics of the waterbody, human resources, effectiveness, adverse impacts, short-term versus long-term results and costs versus benefits.

Once the situation is under control, sampling and monitoring is best continued as long as the bloom occurs in the waterbody in order to confirm that the measures taken are effective. The results of laboratory analyses provide a sound basis upon which the responsible authority can determine whether the cyanotoxins are now effectively under control and the water system can be returned to normal operations.

15.6 RISK COMMUNICATION

The Incidence Response Plan should include the communication steps to follow when cyanobacteria and related toxin incidents occur, including the personnel responsible for initiating the communication, the order in which the notification should occur and the different communication methods to be used. Furthermore, information to the public needs to be given in formats that the respective public can read and understand. This may include tailoring to specific populations speaking different languages dialects, as well as knowledge and literacy levels. It may be useful to engage with knowledgeable regional partners to develop and or customise appropriate communication messages and materials. It may also be important to consider information formats for people with hearing and/or vision impairments as well as for persons with specific medical needs (such as people who are on dialysis) and for specific stages of life that may make people particularly sensitive to cyanotoxins such as pregnant and nursing mothers and those taking care of babies and young children.

Risk communication materials (see the *Additional Tools and Resources for the Development of an Incident Response Plan in* Box 15.2) with core messages can be customised for different countries and groups during different phases of the risk communication steps to ensure that drinking-water consumers, those using recreational sites or people potentially exposed at their workplace will obtain the information they need to protect themselves from cyanobacteria and cyanotoxins.

BOX 15.2: ADDITIONAL TOOLS AND RESOURCES FOR THE DEVELOPMENT OF AN INCIDENT RESPONSE PLAN

Tool or Resource Link (last accessed on 3 February 2020)
Monitoring and Responding to Cyanobacteria and Cyanotoxins, USEPA https://www.epa.gov/ground-water-and-drinking-water/cyanotoxin-management-plan-template-and-example-plans-0
Drinking Water Cyanotoxin Risk Communication Toolbox, USEPA https://www.epa.gov/ground-water-and-drinking-water/drinking-water-cyanotoxin-risk-communication-toolbox-templates
Recommendations for Public Water Systems to Manage Cyanotoxins in Drinking Water, USEPA https://www.epa.gov/ground-water-and-drinking-water/recommendations-public-water-systems-manage-cyanotoxins-drinking
Drinking Water Advisory Communication Toolbox, CDC https://www.cdc.gov/healthywater/emergency/dwa-comm-toolbox/
Guidelines for Safe Recreational Waters Volume 1 – Coastal and Fresh Waters, WHO https://apps.who.int/iris/handle/10665/42591
International Guidance Manual for the Management of Toxic Cyanobacteria: A Guide for Water Utilities, Australia https://www.waterra.com.au/cyanobacteria-manual/PDF/GWRCGuidanceManualLevel1.pdf

An important basis for promoting information that is clear and consistent is to convene a multiagency and multidisciplinary committee or task force across all responsible parties (as mentioned in section 15.2), that is, including drinking-water suppliers or recreational sites managers, communities and public health authorities as well as environmental and water quality regulators, before, during and after a cyanotoxin incident.

15.6.1 Communication preparedness before blooms occur

The responsible authority for managing the incidence response may vary depending on whether drinking-water, recreational water use or water use at workplaces is primarily affected. Particularly where multiple routes of exposure may be relevant, it is, however, important to clarify which authority will take the lead under which circumstances. Each authority potentially involved should determine the designated personnel to be part of the committee or task force responsible for developing the communication materials and for issuing the appropriate information. This includes determining the responsible person to lead the committee or task force in case the respective authority is to take the lead. The person in charge of the communications may be the one regularly in charge of the authority's public communication.

The committee or task force should develop a list of contacts within each authority potentially involved as well as the incident-related partners. The partners relevant for the bloom incident may include consumers, media, visitor centres, recreational parks and veterinarians – that is, representatives of those affected as well as of those to involve in the incident response. It is also important to include experts on cyanobacteria and cyanotoxins for two reasons: one is to gain their support for understanding potential health impacts and exposure routes as well as for determining the most effective control measures and appropriate actions. The other, which is sometimes challenging, is to integrate them in joint communication to the public in order to avoid disparities between the messages given.

If toxins occur at public health concerns levels (i.e., Alert Level 2 in the Alert Level Frameworks given in sections 5.1 and 5.2), the committee or task force will immediately need to ensure that information reaches critical partners such as dialysis and health care centres, childcare and critical care facilities, hospital and clinics, nursing homes, schools, food and beverages businesses and, if waterbodies affected are used for recreation, managers of sites such as lake shore recreational areas, visitor centres and recreational parks. This includes both the operators of such facilities and the consumers or people potentially affected, as listed above. Because of the diversity of those affected, different methods and pathways of communication will be effective for the respective audience. It is therefore effective to develop a contact list and/or decision tree similar to Table 15.1 with the responsible personnel that will be in charge of the communications, and this should include the contact information and the communication steps.

Generic communication materials are best developed prior to any bloom incident to guide drinking-water suppliers and managers of recreational sites to communicate to the public as appropriate (e.g., use of alternative water supply, do not drink advisory, recreational site closure) and kept updated in such a way that they can be readily adapted to any specific situation. The communication materials to consider developing include templates, news releases, beach postings, frequent questions and answers, fact sheets and other background materials. Pathways for distribution should also be identified, considering multiple outlets or media of communication to reach the greatest number of people in a timely manner. This could include media releases and briefings, e-mail and text message alerts, broadcasting, mass distribution through social media via Facebook, Instagram, Twitter, texts, others, posting on beaches and on websites, listservs by e-mail, phone messages, fliers, community meetings and any other locally effective way of communication.

15.6.2 Communication during a cyanotoxin incident

If a cyanobacterial bloom is observed and cyanobacterial toxins are suspected to be present in surface water, the committee or task force should be called together for an emergency meeting to first initiate communication with a smaller group of those directly involved (e.g., drinking-water operators and/or managers of recreational sites). Coordination by the responsible public authority and the site operator is important to confirm that the resources needed for the response are available and that a quick, accurate, effective and harmonised response will take place once the exposure risk is confirmed.

If monitoring results then show cyanobacteria and/or toxins to be above water quality targets or action thresholds (e.g., WHO guideline values (see Table 5.1) or Alert Levels (see Figure 5.2 and 5.4 – or any values derived nationally)), even after control measures are applied, the committee or task force should verify that communication materials already prepared (see above) are appropriate for the specific situation or adjust them as needed, making sure that the message is consistent across all partners involved in the response. The committee or task force will determine the appropriate content, format and frequency of the risk communication. Public notification regarding restrictions on water use may be required to minimise the potential for exposure. "Do not drink" advisories are recommended only if they are necessary to reduce a relevant public health risk, and this decision is usually the responsibility of the drinking-water system or public health authority. Likewise, "do not swim" advisories should be balanced against the health benefits of outdoor water-related activities.

An example of a communication material is given in Figure 5.7 in section 5.2. Further information materials with more detail should give specific information about the current event, including information about the extent of occurrence in relation to water quality targets or action thresholds, if available, on the species of cyanobacteria and toxin concentrations detected, how humans and animals are affected, when the incident started, who is the responsible agency in-charge of the response, date and specific location of the incident, name and phone number of a contact person, steps taken to respond and mitigate/control the incident, as well as precautionary measures such as avoiding contact with contaminated water, rinsing with clean water after coming in contact, keeping pets or livestock away from the bloom and any other relevant information.

15.6.3 Communication after a cyanotoxin incident

Once the bloom is over or under control, the committee or task force should notify the public and other related partners that the incident is resolved and that the water is safe from cyanobacteria and their toxins. The committee or task force could use the same communication outlets contacting the same partners that were notified of the cyanotoxins incident. Communication will be trusted best if it includes information about the final decision, control measures applied, monitoring results, future follow-up steps, longer-term prevention approaches and related outreach materials.

15.7 FOLLOW-UP ASSESSMENT OF INCIDENT MANAGEMENT

A post-incident comprehensive assessment (Table 15.3) to identify the adequacy of the cyanotoxins incident response and assess the effectiveness of the risk communication activities during and after the incident will improve the basis for next time. A debrief with all the involved agencies (e.g., drinking-water supplies and managers of recreational sites) after the incident helps to identify problems and flaws during the incident and to determine areas that need improvement, as well as those actions that contributed to a successful response and that should be repeated in future cyanotoxins contamination events.

It is also useful for the committee or task force to assess the effectiveness of the risk communication during the incident, for example – in the case of toxins in drinking-water – by conducting a customer survey. This can include questions on how well the type of information provided met information needs and how they learned about the incident. The results of the debrief and customer survey should be used to update or modify the incidence response plan (IRP), if appropriate.

15.8 PUBLIC PARTICIPATION

Involving observations of the population using a waterbody in the context of site inspection and when developing a monitoring programme or a Water Safety Plan (WSP) can be highly useful for obtaining information that otherwise might be missed. This may help with focusing attention to high-risk bloom situations as well as practices causing nutrient loads to a waterbody (see section 6.2.2). However, there are numerous situations (including settings with a high level of surveillance) in which the capacity of the responsible authority is not sufficient to ensure that the water does not contain hazardous concentrations of cyanotoxins. This applies particularly for surveillance of waterbodies for recreational use. Monitoring and surveillance at sufficiently tight intervals both in time and in space to ensure capturing high-risk situations may not be feasible, for example, in areas with a large number of waterbodies used for recreation under the responsibility of one public authority. Also, implementing management plans and seeing results may take time. For example, it may take years for an ecosystem to respond to the point where blooms are effectively prevented, and scum situations are not always captured by monitoring. Particularly in such situations, involving site users – that is, the general public – in schemes of monitoring and reporting can be important to avoid exposure. This requires giving users the information they need to develop an understanding of risk situations and to be able to make decisions on water use for their own health (see Figure 5.7 in section 5.2.5).

BOX 15.3: PUBLIC COMMUNICATION THROUGH PUBLIC PARTICIPATION IN DATA COLLECTION – THE CIANO PROJECT IN POTRERO DE GARAY

Raquel del Valle Bazán

Los Molinos reservoir in the province of Córdoba, Argentina, is used for many purposes, that is, generation of hydropower, drinking-water production and recreational activities, including extreme sports such as kitesurfing, but also hiking, sports fishing, horseback riding and outdoor festivals. It is a favourite destination both for tourists and for the local population. The largest town on the reservoir shore, Potrero de Garay, has an estimated population of 5000 inhabitants, but during summer, tourism triples this. The town uses drinking-water directly from the reservoir, without prior treatment.

Since approximately 2010, the reservoir has increasingly been afflicted by blooms of *Microcystis* and *Dolichospermum*, and citizens of Potrero de Garay are concerned about the quality of the water they use. This gave rise to the citizen science project "*Why are our reservoirs green?*" abbreviated CIANO project – a collaborate effort between the University of Cordoba and the local lifeguard organisation (GERS) and the Alfonsina Storni School. CIANO targets a combination of education and collecting data by involving citizens in the observation of meteorological conditions, water temperature, Secchi disc, type of turbidity (algal or nonalgal), water colour, water odour, appearance of cyanobacterial blooms and the geographic coordinates. Citizens, in particular students (11 and 12years old) and staff of GERS, are introduced to the project in workshops and are then invited to report their observations in a WhatsApp group (composed of members of the three participating institutions) following a simple form introduced at the workshop (Figure 1). The project management processes the data and shares a report on the results of the cyanobacterial monitoring as well cyanobacteria alert levels (Table 1). This information is communicated both to government authorities and to private organisations involved in water treatment for the city of Córdoba.

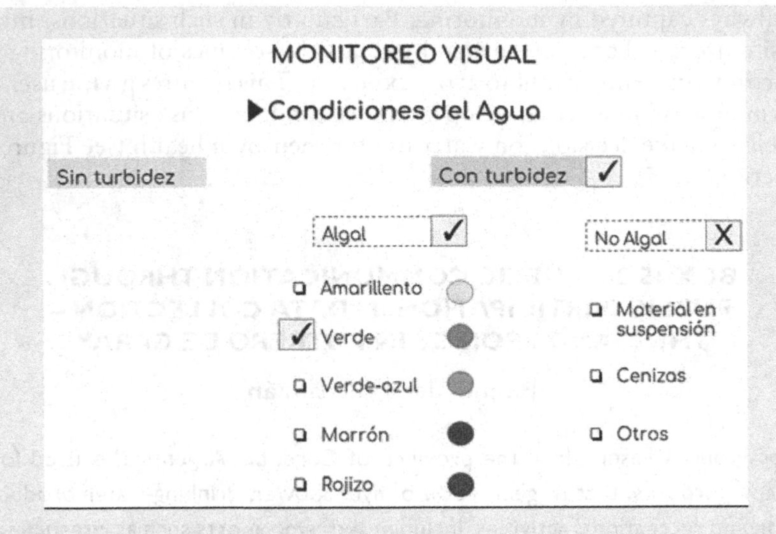

Figure 1 Reporting form for the "condition of the water" by colour and turbidity (note that "cenizas" means "ashes from forest fires" and "otros" means "others" – i.e., macrophytes).

Table 1 Example of data collected through the project app and reported to the public and authorities

Date	*Site*	*Secchi depth (m)*	*Water temp. (°C)*	*Air temp. (°C)*	*Wind speed (km/h)*	*Cyanobacterial blooms*
22 October 2019	Centro	2.1	17.0	19.7	25.6	**Absence**
23 October 2019	Los Espinillos	1.5	21.0	18.8	5	**Absence**
25 October 2019	Centro	4.0	18.0	20.2	9.7	**Absence**
26 October 2019	Garganta	6.5	18.0	s/d	s/d	**Absence**
31 October 2019	Presa	5.0	18.0	20.5	7.2	**Absence**
3 November 2019	Los Espinillos	2.3	19.0	23.0	11.0	**Absence**
4 November 2019	Los Espinillos	3.0	19.0	21.6	19.4	**Absence**

One of the tools introduced at the workshops is the "cyanosemaphore" poster, provided by the *Ministerio de Salud de la Nación* (Health Ministry) for dissemination and prevention of exposure in coastal areas of reservoirs, lakes and rivers. It indicates whether or not specific recreational activities are possible or should be avoided (Figure 2). It is also disseminated in hospitals and Primary Health Care Centres.

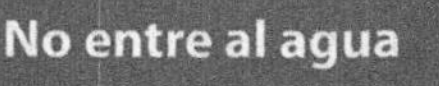

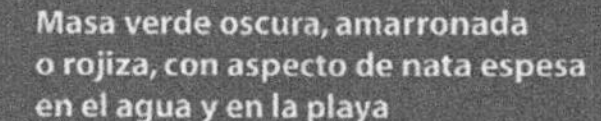
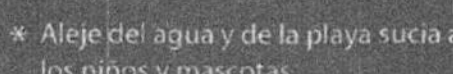

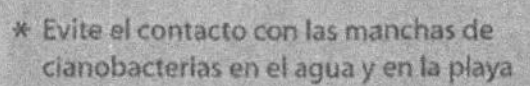

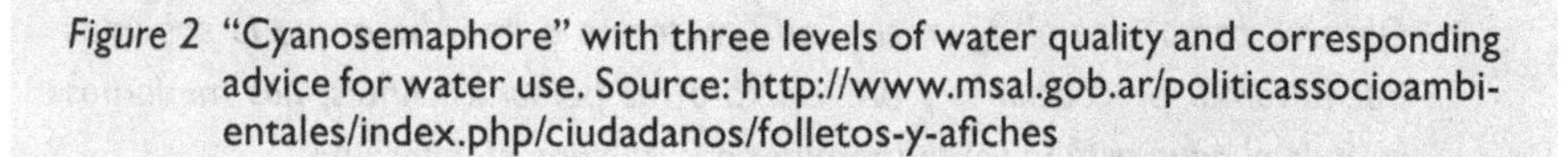

Ciano Semáforo
Prevención de riesgos por contacto con cianobacterias (algas verde-azules) en agua ambiente
No tenga miedo, tenga cuidado
Nivel de riesgo
Aspecto del agua en la playa
Precauciones en el uso
ALTO
No entre al agua
Masa verde oscura, amarronada o rojiza, con aspecto de nata espesa en el agua y en la playa
Alta presencia de cianobacterias potencialmente tóxicas en estado de floración
* No consuma el agua
* Aleje del agua y de la playa sucia a los niños y mascotas
* El agua no es apta hasta que desaparezca la floración de cianobacterias
MEDIO
Busque sectores de agua limpia
Masa verde brillante en la superficie del agua y en la orilla, similar a "mancha de pintura"
Mediana densidad de cianobacterías potencialmente tóxicas en el agua. Puede aparecer depositada sobre la arena de la playa
* Evite el contacto con las manchas de cianobacterias en el agua y en la playa
* Si lo tuvo, lávese con agua limpia lo antes posible
* No consuma el agua
* Cuide a los niños y a las mascotas
BAJO
Puede bañarse en el embalse, lago, río
Superficie del agua: apariencia de "yerba dispersa"
Baja densidad de algas y cianobacterias
* Lávese con agua limpia después
* No consuma el agua
* Cuide a los niños y a las mascotas
Si siente náuseas, diarrea y cualquier otro síntoma consulte a su médico
o llame las 24 hs. al 0-800-333-0160: Centro Nacional de Intoxicaciones - Hospital "Prof. A. Posadas"
Si su mascota tiene vómitos, diarrea o convulsiones, consulte a su veterinario
Dirección Nacional de Determinantes de la Salud e Investigación
Ministerio de Salud
Presidencia de la Nación

Figure 2 "Cyanosemaphore" with three levels of water quality and corresponding advice for water use. Source: http://www.msal.gob.ar/politicassocioambientales/index.php/ciudadanos/folletos-y-afiches

The students act as multipliers of knowledge in their homes and in the community, particularly for communicating bloom alarms. They exhibited information at their stand at the annual school fair where they handed out a brochure they had developed to explain what cyanobacteria are, how they affect health, what precautions to take, as well as the meaning of the different alert levels of the "cyanosemaphore" (Figure 3).

Figure 3 (a) Stand at the annual school fair, (b) brochure by sixth grade students, (c and d) explanation and visualisation of the different alert levels of the "cyanosemaphore".

While the input for water quality monitoring is yet to be evaluated, the first year of the project already achieved a positive impact on awareness of cyanobacterial blooms and their implications for health. Responses and comments of sixth grade students include the following:

> *Observo como está el agua, si puedo meterme o no. Ahora que lo sabemos puedo tomar precauciones y prevenir a otras personas. Antes, nos metíamos cuando el agua estaba verde (porque no lo sabíamos), ahora no.*
>
> I observe how the water is to see how it is, if I can get in or not. Now that we know I can take precautions and prevent others. Before we got in when water was green (because we didn't know better), but not anymore

Miro el agua, también puedo comunicar a los demás que tengan cuidado para que no se enfermen, puedo ver el estado del agua y me fijo primero si me puedo meter o no. Además, ahora sé que no puedo tomar directamente el agua (aunque no tenga muchas cianobacterias).

I look at the water, and I can also tell others to be careful not to get sick, because I can see the condition of the water, and I first look whether I can get in or not. Also, now I know that I can't drink water directly (even if it does not have many cyanobacteria).

Dependiendo del color del agua, puedo entrar al agua o no. Si está con una coloración verde oscuro no puedo entrar, pero si está claro si puedo hacerlo!

Depending on the color of the water I can get into the water or not. If it is dark green I cannot get in, but it is clear I can.

The example in Box 15.3 shows how a research project uses a citizen science approach involving school children in observations and informing others, while at the same time contributing to the collection of data that will describe the bloom situation in the reservoir. Such public participation can also serve to generate political initiatives and interest in waterbody management towards preventing blooms.

A further aspect of public participation is generating broader support for protection of the waterbody and/or its catchment. The example in Box 15.4 shows how broad involvement of the public served to improve the vegetation cover of a riparian buffer zone and fencing around the reservoir, thus keeping animals out of the water and targeting improved retention of nutrients. Involving citizens develops a sense of ownership and responsibility and thus supports the implementation of use restrictions that might otherwise meet resistance.

BOX 15.4: RESTORING RIPARIAN AREAS OF PASO SEVERINO RESERVOIR, URUGUAY, WITH CITIZEN'S PARTICIPATION

Rafael Bernardi, Eduardo Andrés, Elisa Dalgalarrondo, Cesar García, Natalia Jara

In 2013, the Government of Uruguay issued an action plan to address water quality issues in the Santa Lucía Basin, which provides water to 60% of the country's population (see Box 7.3 in Chapter 7). One of the measures it stipulates is the establishment of riparian buffer zones with no agricultural activity around the main waterbodies of the basin, and the National Direction of Environment (DINAMA) has accordingly initiated and led a programme to manage and restore the buffer zone of Paso Severino, the main reservoir in the basin, in partnership with several institutions: the national water company (OSE), local governments, the ministry of agriculture (MGAP), and the Botanical Garden of Montevideo, among others.

The reservoir has a capacity of 70 M m^3 and a perimeter of ~110 km. The riparian land up to the flood level is owned by OSE, but has been traditionally used by local producers that extended the agricultural and livestock use up to the reservoir shore, affecting the water quality. The first measure was to fence the reservoir, restricting access to the shore. This resulted in an initial conflict with local producers which has gradually been solved, although some contentious issues persist. However, the government offers assistance to affected producers, including financial support for installing drinking-water supply for livestock as alternative to direct access to the reservoir.

A key component of the measure was to build a strong participation programme, partnering with the volunteer programme of the ministry of social development (MIDES) and with local schools and institutions, staff associations and local actors. Trees were first provided by national and local governmental nurseries and later from nurseries established by the community with the support of the Small Grant Programme. In total, more than 200 volunteers have participated, many of them attending several days, together with children of several schools of the region and staff from partnering institutions, totalling approximately 1000 working days over a 4-year period.

The programme planted approximately 5000 trees and constructed infrastructure for their protection. It has placed posters with information about the measures implemented and to discourage unauthorised use of the area, and a lookout platform was built together with the community of the nearby town "25 de Mayo".

Native saplings and juvenile trees were planted in the perimeter of the reservoir, with a choice of species and their spatial distribution according to ecological conditions of the sites. Natural regeneration is being monitored. Initial estimates show a survival of over 90% of individuals planted, although monitoring efforts are still underway, the design of which was developed together with the local community. Also, a partnership was established with the University of the Republic of Uruguay to assess the effects of different natural covers on the retention of nutrient loads to the reservoir.

The success of the programme up to the end of 2019 has been twofold: one is the active participation and sense of ownership by the local community and volunteers from Montevideo who can now visit their water source and contribute to its restoration. The other is that this active management has prevented lands from being reclaimed for their previous use. Maintaining these two objectives is challenging, but key to ensure the long-term management of this water source.

FURTHER READING

Bartram J, Corrales L, Davison A, Deere D, Drury D, Gordon B et al. (2009). Water safety plan manual: step-by-step risk management for drinking-water suppliers. Geneva: World Health Organization. Available online at: https://apps.who.int/iris/handle/10665/75141

CDC (2016). Drinking water advisory communication toolbox. Centers for Disease Control and Prevention, Atlanta, GA. Available online at: http://www.cdc.gov/healthywater/emergency/dwa-comm-toolbox/

Chorus, I. (2012). Current approaches to Cyanotoxin risk assessment, risk management and regulations in different countries. Dessau-Roßlau: Federal Environment Agency. Available online at: https://www.umweltbundesamt.de/sites/default/files/medien/374/publikationen/4390.pdf.

Gianuzzi, L. (2009). Cianobacterias y Cianotoxinas. Identificación, Toxicología, Monitoreo y Evaluación de Riesgo. Moglia S.R.L., Corrientes, Argentina: Instituto Correntino del Agua y Administración de Obras Sanitarias de Corrientes. 237p. Available at: http://www.icaa.gov.ar/2010/gerencias/gest_ambiental/manualcianobacterias.htm

Health Canada (2012a). Guidelines for Canadian recreational water quality, third edition. Ottawa: Water, Air and Climate Change Bureau, Healthy Environments and Consumer Safety Branch, Health Canada. (Catalogue No H129-15/2012E). Available at: https://www.canada.ca/en/health-canada/services/environmental-workplace-health/water-quality/recreational-water-water-quality.html

Health Canada (2012b). Guidelines for Canadian drinking water quality: guidance technical document -cyanobacterial toxins. Ottawa: Water, Air and Climate Change Bureau, Healthy Environments and Consumer Safety Branch, Health Canada. Available online at: https://www.canada.ca/en/health-canada/services/publications/healthy-living/guidelines-canadian-drinking-water-quality-guideline-technical-document-cyanobacterial-toxins-document.html

Ministry for the Environment and Ministry of Health (2009). New Zealand guidelines for cyanobacteria in recreational fresh waters – interim guidelines. Prepared for the Ministry for the Environment and the Ministry of Health by SA Wood, DP Hamilton, WJ Paul, KA Safi and WM Williamson. Wellington: Ministry for the Environment. Available online at: https://www.mfe.govt.nz/sites/default/files/nz-guidelines-cyanobacteria-recreational-fresh-waters.pdf

Newcombe, G. (ed) (2009). International guidance manual for the management of toxic cyanobacteria: a guide for water utilities. Global Water Research

Coalition, Unley, SA. Available online at: https://www.waterra.com.au/cyanobacteria-manual/PDF/GWRCGuidanceManualLevel1.pdf.

Newcombe, G., House, J., Ho, L., Baker, P., and Burch, M. (2010). Managing strategies for cyanobacteria (blue-green algae): a guide for water utilities. Water Quality Research Australia: Research Report 74. Available online at: http://www.waterra.com.au/publications/document-search/?download=106.

New Zealand Ministry of Health (2017). Guidelines for drinking-water quality management for New Zealand, third edition. Wellington: Ministry of Health. Available online at: https://www.health.govt.nz/publication/guidelines-drinking-water-quality-management-new-zealand; Datasheets for cyanotoxins available online at: https://www.health.govt.nz/system/files/documents/publications/vol_3_ds_2.4_cyanotoxins_0.doc

Scottish Government Health and Social Care Directorates Blue-Green Algae Working Group (2012). Cyanobacteria (blue-green algae) in inland and inshore waters: assessment and minimization of risks to public health. Available online at: https://www.gov.scot/publications/cyanobacteria-blue-green-algae-inland-inshore-waters-assessment-minimisation-risks-public-health/

UNESCO (2009). Cianobacterias Planctónicas del Uruguay. Manual para la identificación y medidas de gestión. Sylvia Bonilla, (Ed). Documento Técnico PHI. LAC, N° 16. 94p. Available online at: https://www.academia.edu/32964769/Cianobacterias_Planct%C3%B3nicas_del_Uruguay._Manual_para_la_identificaci%C3%B3n_y_medidas_de_gesti%C3%B3n

USEPA (2003). Drinking Water Utility Response Protocol Toolbox (DWRPTB). Office of Water. Available online at: https://www.epa.gov/waterutilityresponse/drinking-water-and-wastewater-utility-response-protocol-toolbox#DWRPTB

USEPA (2008). Water security initiative: interim guidance on developing consequence management plans for drinking water utilities. Office of Water. EPA 817-R-08-001. Available online at: https://www.epa.gov/sites/production/files/2015-06/documents/wsi_interim_guidance_on_developing_consequence_management_plans_for_drinking_water_utilities.pdf

USEPA (2016) Drinking water cyanotoxin risk communication toolbox - templates. Available online at: https://www.epa.gov/ground-water-and-drinking-water/drinking-water-cyanotoxin-risk-communication-toolbox-templates.

USEPA (2017). Sampling guidance for unknown contaminants in drinking water. Office of Water. EPA-817-R-08-003. Available online at: https://www.epa.gov/sites/production/files/2017-02/documents/sampling_guidance_for_unknown_contaminants_in_drinking_water_02152017_final.pdf.

WHO (2003). Guidelines for safe recreational water environments: Coastal and fresh waters (Vol. 1). Geneva: World Health Organization. Available online at: https://apps.who.int/iris/handle/10665/42591

WHO (2012). Water safety planning for small community water supplies: step-by-step risk management guidance for drinking-water supplies in small communities. Geneva: World Health Organization. Available online at: https://apps.who.int/iris/handle/10665/75145

WHO (2017). Guidelines for drinking-water quality, fourth edition, incorporating the 1st addendum. Geneva: World Health Organization. Available online at: https://www.who.int/publications/i/item/9789241549950

Index

Note: **Bold** page numbers refer to tables and *italic* page numbers refer to figures.

Printed in the United States
by Baker & Taylor Publisher Services